TEACHER'S EDITION

GEOMETRY

for Enjoyment and Challenge

NEW EDITION

GEOMETRY

for Enjoyment and Challenge

NEW EDITION

Richard Rhoad
New Trier High School
Winnetka, Illinois

George Milauskas
Illinois Mathematics and
Science Academy
Aurora, Illinois

Robert Whipple
New Trier High School
Winnetka, Illinois

ML McDougal, Littell & Company

Evanston, Illinois
New York   Dallas   Sacramento   Columbia, SC

Authors

Richard Rhoad

Teacher, New Trier High School
  Winnetka, Illinois
Recipient, Presidential Award for Excellence in Teaching
  Mathematics, Illinois, 1985
Recipient T. E. Rine Award, Illinois, 1981
Former Governor, Mu Alpha Theta
Head Coach, State Championship Math Team, 1981–1988
Chairman, ICTM State Contest Committee, 1978–1984
Previously taught at Homewood-Flossmoor High School
  Homewood, Illinois
  Avon Lake High School

George Milauskas

Teacher, Illinois Mathematics and
  Science Academy
  Aurora, Illinois
Author, Algebra 1, An Integrated Approach
  Algebra 2 and Trigonometry
Editorial Advisory Board member and writer,
  National Council of Teachers of Mathematics
  Yearbook on Geometry, 1987
Previously taught at Barrington High School
  Barrington, Illinois
  New Trier High School
  Winnetka, Illinois

Robert Whipple

Teacher, New Trier High School
  Winnetka, Illinois
Geometry for Teachers 1989
  Northwestern University
  Evanston, Illinois

2000 Impression
Copyright © 1991 by McDougal, Littell & Company
Box 1667, Evanston, Illinois 60204

ISBN 0-86609-966-2

03 02 01 / 15 14 13 12 11 10 9 8 7

CONTENTS

STUDENT TEXT

A Letter to Teachers

This textbook is the result of many years of classroom work with students—writing lessons and problems, teaching them, changing them, and then teaching them again. After the first edition of *Geometry for Enjoyment and Challenge* was published, we expanded our "classroom," talking to and coming in contact with teachers who use the book nationally. When we sat down together last year, we had a wealth of experience, suggestions, and thoughts about what made our book special and what would make it even better.

The result is this new edition of *Geometry for Enjoyment and Challenge*. We believe it is even more enjoyable and challenging than previous editions. We have had the freedom to completely revise our book, rearranging ideas for better learning, incorporating new topics that we feel are appropriate, but retaining the book's considerable strengths. The result is a book we are very proud of.

This is still a book written to be read by students. It maintains and augments algebra skills by using algebra to teach geometry. It introduces logical argument and proof early and simply. It presents a range of proofs—paragraph proofs, flow diagrams, coordinate proofs—and does it so that the student becomes comfortable with all styles. By the end of the year, *all* students will be able to analyze a problem in order to supply a diagram, set up a proof, and complete it successfully.

The book is clearly organized to help students become more independent in their learning.

Objectives

are clearly stated so that the student knows exactly what he or she will begin to master.

Part One: Introduction

presents the lesson clearly in a style that students find easy to read and understand.

Part Two: Sample Problems

helps students develop the ability to analyze geometric properties and prepares them for independent homework assignments.

Part Three: Problem Sets

contains a wide variety of interesting, varied problems that students can master and enjoy. Three levels of difficulty provide flexibility, ensure mastery, and foster enjoyment. In the Teacher's Edition, short answers to problems appear immediately following each problem. Outline solutions to some B and all C level problems can be found in the margin of the problem page.

We hope you and your students are successful and enjoy this book. We would be happy to hear about your experience.

Good luck!

CALCULATORS AND COMPUTERS IN GEOMETRY

Calculators and computers can empower students mathematically, permitting them to focus on ideas rather than on calculations. For example, a drawing tool such as any of the programs in *The Geometric Supposer* series lets students discover geometric concepts. One possibility is that by looking at the altitude and the median of many triangles and measuring the intersected segments and angles, students can discover that an altitude forms right angles and a median divides a side into equal segments.

Scientific calculators free a student from troublesome computation in number-based problems so that the emphasis remains on the concept. In general, students can learn to make available hardware work for them and enhance their learning. It is also important that students learn when it is appropriate to use a calculator or computer and when a paper and pencil is most appropriate.

In the Teacher's Edition of *Geometry for Enjoyment and Challenge,*
The Geometric Supposer
provides suggestions for integrating the use of programs in the Geometric Supposer series (or any other drawing and measuring software that may be available) into work with the section. These Supposer activities are available as reproducible student worksheets in the *Teacher's Resource Book.*

The suggestions for using *The Geometric Supposer* series are a place to begin. Experience has shown that once a class begins working with this kind of tool, there are many different activities and directions that follow.

While most of the terminology used in *The Geometric Supposer* series is actually mathematical terminology, some special terms that are used refer to the programs. Following is a brief summary of some *Geometric Supposer* terminology.

Extension: a segment that extends an existing side of a figure you have drawn, either by a specified length or to intersect another line, segment, or circle.

Label: to label a point on a figure you have drawn, or a point anywhere on the drawing screen.

Moveable point: a dot that you can move to any position in the drawing portion of the screen using the arrow keys. When you press RETURN, a label will appear in that location.

Random point: A point and label that will be selected randomly by the program and displayed in the drawing portion when you choose this option.

Repeat: the option to automatically repeat a construction. When you have selected the option to repeat, the construction in progress will be repeated step by step as you press the space bar.

Your own (triangle, quadrilateral, or circle): a figure you can define by providing information requested by the program, such as the lengths of the sides of a triangle or the center and radius of a circle.

For information about *The Geometric Supposer* series, contact:

Sunburst Communications
101 Castleton Street
Pleasantville, New York 10570-3498
(800) 628-8897

LASSROOM MANAGEMENT

As the three of us worked together to write and rewrite this textbook, we talked a great deal about what works best for us in our own classrooms. You are the best judge of what works for you and your class. What we would like to share with you is some of our most successful ideas. These are included in the Teacher's Edition.

An overview of each chapter on the opening page provides some background you may be able to use as you plan for each chapter.

Class Planning

Time Schedule

Resource References

opens each section to help you plan most efficiently and prepare for each class, especially during the first year you use the book.

Class Opener

is intended as a problem that students work on as soon as they enter the classroom. It can be on an overhead projector, on the board, or copied for each student. A Class Opener is offered for each of two levels—A and B—in the *Teacher's Resource Book*. The Level A Opener is in the Teacher's Edition. Students work independently or in small groups on these problems. In our classrooms, we also expect students to have their homework out and available for us to check when the bell rings. As we circulate, we record the homework and note where students are having difficulties. In this way, within the first few minutes of class, students are at work, we know the results of last night's assignment, and we know where the learning blocks are.

Class Openers are available in reproducible form in the TEACHER'S RESOURCE BOOK.

Lesson Notes

expand on the student text.

Vocabulary

that is introduced in each section is listed at the bottom of the first lesson page. You may choose to review the words with your students before starting the lesson or after they have gone through the material.

Assignment Guide

is based on our experience as we have taught the lessons. Since each class is different, you may find it helpful to modify the assignments as you proceed. The Assignment Guides listed here are available as blackline masters in the *Teacher's Resource Book*.

Problem-Set Notes and Additional Answers

■ In problem **18**, there are four 30° angles, three 60° angles, two 90° angles, and one 120° angle. Of these ten angles, 45 pairs can be made.

18a 6 pairs of 30° angles; $\frac{6}{45} = \frac{2}{15}$

b A 30° angle can be matched with a 60° angle in $4 \cdot 3 = 12$ ways; $\frac{12}{45} = \frac{4}{15}$.

Problem-Set Notes and Additional Answers

are comments and/or suggestions about specific problems; and answers that do not fit next to the respective problems. Other answers can be found in the back of your book, on the tabbed pages.

Many of us have found success with techniques presented by David R. Johnson in his brief booklets *Every Minute Counts* and *Making Every Minute Count Even More* (Palo Alto, Calif.: Dale Seymour, 1982).

We have tried to include material to make you and your class most successful as you work through our book, especially the first year. We hope you and your students have a rewarding time.

RITING IN THE MATHEMATICS CLASSROOM

GEOMETRY for Enjoyment and Challenge provides student writing opportunities in each lesson. Designed to help students internalize mathematical concepts, these brief assignments draw their topics from the lesson material and actively involve the student in the concepts that underlie the problems in the lesson.

Much of language theory is based on the premise that communication is basic to thinking. Vygotsky in *Thought and Language* (1962) writes that putting thoughts into words is the only way we can give form to the myriad images that assault our minds. Britton in *Language and Learning* (1970) theorizes that speaking and writing are "commentary," or a way of making sense out of our random perceptions, and that we must write or speak about an experience to understand it. And Donald Murray in *Write to Learn* (1987) states that writing is the most disciplined way in which we organize our thoughts. As mathematics education continues to emphasize problem solving, we seek ways to translate the symbolic and theoretical language of mathematics into real-world situations. The writing exercises in *GEOMETRY for Enjoyment and Challenge* give students a solid foundation for using everyday language to record their thoughts about mathematics. With this experience a student should be better able to work through everyday situations as well as solve multiple-step word problems.

Communicating Mathematics

Have students write a paragraph that explains how to determine whether a point is on a circle, in the interior of a circle, or in the exterior of a circle.

In addition to facilitating a relationship between mathematics and everyday life, writing empowers the student by giving him or her a sense of ownership about the material. When the writer restates an idea or concept, that idea belongs to the writer and becomes a part of his or her experience. Thus internalized, concepts become more useful and less likely to be forgotten.

Communicating Mathematics

Have students write an explanation of the term *mean proportional*.

As a teacher, you may find completed student writing assignments helpful in making informal assessments. Reading a piece of writing can be an excellent way of targeting weakness or misunderstanding at the concept level. A student's writing may also shed light on why certain kinds of errors are made and what might be done to correct them.

Metacognition, or the ability to recognize what you do and do not know, is an important thinking skill. Writing can be a valuable tool for arriving at this type of self-knowledge. Confused writing reveals confused thinking on the part of the writer, and difficulty in writing about a concept probably signals incomplete understanding. Many teachers encourage students to use their writing for discovering where comprehension is weak or thinking is muddy. You may also suggest that students keep their GEOMETRY writings in a spiral notebook, which can be used for reference and to remind them of their progress throughout the year.

COOPERATIVE LEARNING

Cooperative learning is not new to schools. Recently, cooperative groups have again emerged as an effective way to organize classes for maximum learning. While it is more common to think of elementary school classrooms organized for cooperative learning, recent research shows that high school students benefit from the interaction of a small group. For those of us who have seen the rewards, small-group activities make sense.

Some students are more willing to participate in a small group than in a large one. It has been our experience that when an activity is open-ended, more students participate in problem solving. The smaller group is often less intimidating to the uncertain student. Through small-group participation, many students gain the confidence to contribute in the large group.

Structuring a class for group work can be achieved in a variety of ways:

▼ by designating time to work in class with a suggestion such as "You and Tom have different answers. Why don't you talk about it?"

▼ by suggesting that students form groups of two, three, or four informally

▼ by assigning groups of two, three, or four with a designated meeting place and time limit

The method you choose will depend upon your class, the physical structure of your room (movable vs. fixed seats), your own experience with small-group activities, and a variety of other factors.

We have included cooperative learning activities with each problem set.

Cooperative Learning

will be self contained:

or will be an extension of a problem in the problem set:

You may find that cooperative group work does not come naturally to your students at first. But with practice, cooperative activities can provide opportunities for student involvement, problem solving, and higher-order thinking that may be lacking in a strictly traditional approach. For further reading and information on cooperative learning, the following list of resources may be helpful.

Cooperative Learning

Give students a sheet with three parallel lines cut by a transversal. The transversal should be cut in a common ratio (e.g., 2 to 5). Have students first individual-

Cooperative Learning

Have students work in small groups to solve problem **26.** Suggest that students draw diagrams. Then rework the problem with one more ray added. Adding one more ray reduces the maximum number of pairs of complementary angles and increases the minimum number and the maximum number of pairs of supplementary angles.

Reading

Aronson, E. *The Jigsaw Classroom.* Beverly Hills, Calif.: Sage, 1978.

Dewey, John. *Experience and Education.* New York: Collier, 1977.

Glasser, William, MD. *Control Theory in the Classroom.* New York: Harper and Row, 1986.

Newman, F.M., and J. A. Thompson. *Effects of Cooperative Learning on Achievement in Secondary Schools.* Madison, Wis.: National Center on Effective Secondary Schools.

Piaget, Jean. *Language and Thought of the Child.* New York: Meridian, 1926.

Sharan, S., and Y. Sharan. *Small-Group Teaching.* Englewood Cliffs, N.J. Educational Technology Publications, 1976.

Slavin, R. E. "Cooperative Learning." *Review of Educational Research* 50 (2): 315-42.

Tobias, Sheila. *Overcoming Math Anxiety.* New York: W. W. Norton, 1978.

Vygotsky, L.S. *Thought and Language.* Cambridge, Mass.: MIT Press, 1962.

White, Merry. *The Japanese Educational Challenge.* New York: Macmillan, 1987.

Organizations

CENTER FOR RESEARCH ON ELEMENTARY AND MIDDLE SCHOOLS
The Johns Hopkins University
3505 North Charles Street
Baltimore, MD 21218

THE COOPERATIVE LEARNING CENTER
David and Roger Johnson
202 Pattee Hall
University of Minnesota
Minneapolis, MN 55455

INTERNATIONAL ASSOCIATION FOR THE STUDY OF COOPERATION IN EDUCATION
136 Liberty Street
Santa Cruz, CA 95060

Cooperative Learning Activities are available in reproducible form in the TEACHER'S RESOURCE BOOK.

PROOF IN GEOMETRY

In the Curriculum and Evaluation Standards for School Mathematics, the National Council of Teachers of Mathematics recommends a reevaluation of the role of proof in the geometry course. In GEOMETRY *for Enjoyment and Challenge*, we introduce proof early along with a wide variety of other types of exercises such as making conjectures, formulating generalizations and drawing conclusions. Beginning with exceedingly simple proofs of one or two steps, proof is developed gradually, giving students the opportunity to become familiar and comfortable with the concept of proof and with different forms of proof, including paragraph and indirect proof in addition to the formal two-column proof.

In this way, proof in GEOMETRY *for Enjoyment and Challenge* is not an end in itself but one of many important avenues by which students come to understand geometric relationships and, even more, mathematical relationships that can be represented geometrically.

# USING MANIPULATIVES

In the past, the use of hands-on material has been associated with teaching young children. But there is a growing recognition that older students can also increase success in understanding the abstract concepts of geometry. Thus increasing attention is being paid to the use of manipulative materials in the teaching and learning of algebra and geometry.

In *GEOMETRY for Enjoyment and Challenge*, we have tried to accommodate a variety of learning styles. It has been our experience that while some students are comfortable with the abstraction of a diagram and set of givens, some benefit from more experiences focusing on visual representation. Others may need still more concrete experiences.

For this reason, in the *Teacher's Resource Book* you will find a set of 17 manipulative activities correlated to appropriate lessons throughout the book. These activities involve completing constructions, exploring symmetry relationships using a reflective surface, and a variety of activities involving objects that exist in the world. These activities are referenced in the Teacher's Edition under the heading

Using Manipulatives

We hope you and your students find this approach helpful.

A GUIDE TO ADDITIONAL RESOURCES

	TEACHER'S RESOURCE BOOK				ADDITIONAL RESOURCES		
Section	Practice	Class Openers	Manipulatives	Cooperative Learning	Transparencies	Computer Practice	Tests/Quizzes
1.1		1.1A, B		1.1			
1.2		1.2A, B		1.2			Quiz 1 Series 1, 2
1.3	1	1.3A, B		1.3	1, 2		Quiz 1 Series 3
1.4	2	1.4A, B		1.4			Quiz 2 Series 2
1.5		1.5A, B	1	1.5		1	Quiz 3 Series 2
1.6		1.6A, B		1.6			
1.7		1.7A, B		1.7			
1.8		1.8A, B		1.8	3		
1.9		1.9A, B		1.9			Quiz 2, 3 Series 1 Quiz 2 Series 3
1		Review 1A					Test 1 Series 1, 2, 3
2.1		2.1A, B	2	2.1			
2.2	3	2.2A, B		2.2			Quiz 1 Series 1, 3
2.3		2.3A, B		2.3			
2.4		2.4A, B		2.4		2	
2.5		2.5A, B		2.5			Quiz 1 Series 2 Quiz 2 Series 1, 3
2.6		2.6A, B		2.6			
2.7		2.7A, B		2.7			Quiz 2 Series 2 Quiz 3 Series 1, 3
2.8	4	2.8A, B		2.8	4		Quiz 3 Series 2
2							Test 2 Series 1, 2, 3

TEACHER'S RESOURCE BOOK					ADDITIONAL RESOURCES		
Section	Practice	Class Openers	Manipulatives	Cooperative Learning	Transparencies	Computer Practice	Tests/ Quizzes
3.1		3.1A, B	3, 4	3.1	5		Quiz 1 Series 3
3.2		3.2A, B		3.2			Quiz 1 Series 1, 2
3.3		3.3A, B		3.3		3	
3.4	5	3.4A, B	5	3.4			
3.5		3.5A, B		3.5			Quiz 2 Series 1, 2, 3
3.6		3.6A, B	6	3.6		4	Quiz 3 Series 2
3.7	6	3.7A, B		3.7			Quiz 3 Series 1
3.8		3.8A, B		3.8	6		
3		Review 3A					Test 3 Series 1, 2, 3
4.1	7	4.1A, B		4.1	7		
4.2		4.2A, B		4.2			Quiz 1 Series 2
4.3	8	4.3A, B		4.3			Quiz 2 Series 2 Quiz 1 Series 1
4.4		4.4A, B		4.4		5	Quiz 1 Series 3 Quiz 2 Series 1
4.5		4.5A, B		4.5			Quiz 3 Series 1, 2
4.6		4.6A, B		4.6	8		
4							Test 4 Series 1, 2, 3

A complete Assignment Guide will be found on the first page of each problem set. A complete Assignment Guide in reproducible form for all problem sets will be found in the TEACHER'S RESOURCE BOOK.

ADDITIONAL RESOURCES, *continued*

	TEACHER'S RESOURCE BOOK				ADDITIONAL RESOURCES		
Section	Practice	Class Openers	Manipulatives	Cooperative Learning	Transparencies	Computer Practice	Tests/ Quizzes
5.1		5.1A, B		5.1			
5.2		5.2A, B	7	5.2			
5.3	9	5.3A, B		5.3	9		Quiz 1, 2 Series 1 Quiz 1 Series 2
5.4		5.4A, B	8	5.4			
5.5	10	5.5A, B	9	5.5		6	Quiz 2, 3 Series 2 Quiz 3 Series 1 Quiz 1, 2 Series 3
5.6		5.6A, B		5.6			
5.7		5.7A, B, C	10	5.7			
5							Test 5 Series 1, 2, 3
6.1		6.1A, B		6.1	10		
6.2	11	6.2A, B		6.2	11		Quiz 1, 2 Series 1, 2
6.3	12	6.3A, B		6.3			Quiz 1 Series 3
6		Review 6A					Test 6 Series 1, 2, 3

	TEACHER'S RESOURCE BOOK				ADDITIONAL RESOURCES		
Section	Practice	Class Openers	Manipulatives	Cooperative Learning	Transparencies	Computer Practice	Tests/Quizzes
7.1		7.1A, B	11	7.1		7	
7.2	13	7.2A, B		7.2			Quiz 1 Series 1, 2
7.3	14	7.3A, B	12, 13	7.3			Quiz 2 Series 2 Quiz 1, 2 Series 3
7.4		7.4A, B		7.4			Quiz 2, 3, 4 Series 1
7		Review 7B					Test 7 Series 1, 2, 3 Midyear Test Series 1
8.1		8.1A, B		8.1			
8.2		8.2A, B		8.2	12		
8.3	15	8.3A, B		8.3	13		
8.4		8.4A, B	14	8.4			Quiz 1, 2, 3 Series 1 Quiz 1-4 Series 2
8.5	16	8.5A, B		8.5		8, 9, 10	Quiz 4 Series 1 Quiz 1 Series 3
8		Review 8B					Test 8 Series 1, 2, 3 Midyear Test Series 2

A complete Assignment Guide will be found on the first page of each problem set. A complete Assignment Guide in reproducible form for all problem sets will be found in the TEACHER'S RESOURCE BOOK.

ADDITIONAL RESOURCES, *continued*

	TEACHER'S RESOURCE BOOK				ADDITIONAL RESOURCES		
Section	Practice	Class Openers	Manipulatives	Cooperative Learning	Transparencies	Computer Practice	Tests/ Quizzes
9.1		9.1A, B		9.1			
9.2		9.2A, B	15	9.2			
9.3	17	9.3A, B		9.3	14, 15		
9.4		9.4A, B		9.4	16		Quiz 1-3 Series 2, 3 Quiz 1-5 Series 1
9.5		9.5A, B		9.5			Quiz 4 Series 2 Quiz 4 Series 3 Quiz 6 Series 1 Quiz 7, 8 Series 1
9.6		9.6A, B		9.6	17		Quiz 5 Series 2 Quiz 7-8 Series 1
9.7	18	9.7A, B		9.7			Quiz 5 Series 3 Quiz 6-9 Series 2 Quiz 9-10 Series 1
9.8		9.8A, B		9.8			Quiz 6-7 Series 3 Quiz 10 Series 2
9.9		9.9A, B		9.9		11	
9.10		9.10A, B		9.10			
9		Review 9A					Test 9 Series 1, 2, 3 Midyear Test Series 3
10.1		10.1A, B		10.1			
10.2		10.2A, B		10.2			
10.3	19	10.3A, B		10.3			
10.4	20	10.4A, B	16	10.4			Quiz 1, 2 Series 3 Quiz 1 Series 1, 2
10.5		10.5A, B	17	10.5			Quiz 2-3 Series 2 Quiz 2-5 Series 1 Quiz 3-4 Series 3
10.6		10.6A, B		10.6	18		Quiz 4 Series 2
10.7		10.7A, B	18	10.7		12	Quiz 5-6 Series 2 Quiz 6 Series 1 Quiz 5 Series 3
10.8		10.8A, B		10.8			Quiz 6-7 Series 3 Quiz 7 Series 2

	TEACHER'S RESOURCE BOOK				ADDITIONAL RESOURCES		
Section	Practice	Class Openers	Manipulatives	Cooperative Learning	Transparencies	Computer Practice	Tests/ Quizzes
10.9		10.9A, B	19	10.9			Quiz 8 Series 2 Quiz 7-10 Series 1
10		Review 10B					Test 10 Series 1, 2, 3
11.1		11.1A, B		11.1			
11.2	21	11.2A, B		11.2			
11.3		11.3A, B		11.3			Quiz 1 Series 1, 2
11.4		11.4A, B		11.4			Quiz 2-3 Series 2 Quiz 2-4 Series 1
11.5		11.5A, B		11.5			
11.6	22	11.6A, B		11.6			Quiz 4-6 Series 2 Quiz 5 Series 1 Quiz 1 Series 3
11.7		11.7A, B	20	11.7		13	Quiz 7 Series 2 Quiz 6-9 Series 1 Quiz 2 Series 3
11.8		11.8A, B		11.8			Quiz 10 Series 1 Quiz 3 Series 3
11							Test 11 Series 1, 2, 3
12.1		12.1A, B		12.1	19		
12.2		12.2A, B		12.2	20		
12.3		12.3A, B		12.3	21		
12.4	23	12.4A, B		12.4			
12.5	24	12.5A, B		12.5			
12.6		12.6A, B		12.6			Quiz 1-3 Series 2 Quiz 1, 2 Series 1, 3
12							Test 12 Series 1, 2, 3

A complete Assignment Guide will be found on the first page of each problem set. A complete Assignment Guide in reproducible form for all problem sets will be found in the TEACHER'S RESOURCE BOOK.

ADDITIONAL RESOURCES, *continued*

TEACHER'S RESOURCE BOOK					ADDITIONAL RESOURCES		
Section	Practice	Class Openers	Manipulatives	Cooperative Learning	Transparencies	Computer Practice	Tests/ Quizzes
13.1		13.1A, B		13.1			
13.2	25	13.2A, B		13.2			Quiz 1 Series 1, 2, 3
13.3		13.3A, B		13.3			
13.4		13.4A, B		13.4			Quiz 2 Series 1, 2, 3
13.5		13.5A, B		13.5	22		
13.6	26	13.6A, B		13.6			Quiz 3 Series 1, 2, 3
13.7		13.7A, B		13.7			
13		Review 13A					Test 13 Series 1, 2, 3 Final Test Series 1
14.1		14.1A, B	21	14.1			Quiz 1 Series 1
14.2		14.2A, B		14.2			
14.3		14.3A, B	22	14.3	23	14	Quiz 1 Series 3 Quiz 1, 2 Series 2
14.4		14.4A, B		14.4			
14.5	27	14.5A, B		14.5			Quiz 2 Series 3
14.6	28	14.6A, B		14.6			Quiz 3 Series 3
14							Test 14 Series 1, 2, 3 Final Test Series 2

ADDITIONAL RESOURCES, *continued*

	TEACHER'S RESOURCE BOOK				ADDITIONAL RESOURCES		
Section	Practice	Class Openers	Manipulatives	Cooperative Learning	Transparencies	Computer Practice	Tests/ Quizzes
15.1		15.1A, B		15.1			
15.2	29	15.2A, B		15.2	24	15	Quiz 1 Series 3
15.3	30	15.3A, B		15.3			Quiz 2 Series 3
15							Test 15 Series 1, 2, 3 Final Test Series 3
16.1							
16.2							
16.3							
16.4							
16.5							
16.6							
16.7							
16							

GEOMETRY
for Enjoyment and Challenge

NEW EDITION

Y ou are invited to cross the bridge
into the exciting world of geometry,
for enjoyment and challenge.

GEOMETRY

for Enjoyment and Challenge

NEW EDITION

Richard Rhoad
New Trier High School
Winnetka, Illinois

George Milauskas
Illinois Mathematics and
Science Academy
Aurora, Illinois

Robert Whipple
New Trier High School
Winnetka, Illinois

ML McDougal, Littell & Company

Evanston, Illinois
New York   Dallas   Sacramento   Columbia, SC

Credits

Cover

Cover collage: Carol Tornatore, photograph by Tom Petrillo

1: © Peter Mauss/Esto, Mamaronock, New York, all rights reserved; 2,5,6,13,15: © Chip Clark, Washington, D.C.; 3,7,8,12: Adrienne McGrath, North Barrington, Illinois; 4,10: (detail) Reprinted by Permission of E.P. Dutton, New York; 9: (detail) Reprinted with Permission of *Better Homes and Gardens* ®, Des Moines, Iowa, American Patchwork and Quilting; 11: JPL, NASA's Regional Planetary Image Facilities, Pasadena, California; 14: Ric Ergenbright, Bend, Oregon.

Fine Art and Photography

Frontispiece: *Suspension Bridge, Kaibab Trail, Grand Canyon, Arizona*, Paul A. Otto, Lake St. Louis, Missouri

2: *Open Book*, 1930, Paul Klée, Guggenheim Museum, New York; 35: Courtesy of Wendell Griffen; 48: © 1985 Sidney Harris, New Haven, Connecticut; 60: Richard Sullivan, Los Angeles; 81: J. Nettis/H. Armstrong Roberts, Chicago; 94: Stember Photography, Lawrenceville, New Jersey; 110: *Sheba*, 1980, Dorothea Rockburne, National Museum of Women in the Arts, Washington, D.C., Holladay Collection; 130: Norman McGrath, New York; 137: Courtesy of Dorothy Washburne; 168: Algimantas Kezys, Stickney, Illinois; 175: *Ad Parnassum*, 1932, Paul Klée, Kunstmuseum, Bern, Switzerland; 191: (all) E.D. Getzoff/Scripps Clinic and Research Foundation, LaJolla, California; 194: Carmine Fantasia, Chicago; 197: Historical Pictures Service, Chicago; 198: Algimantas Kezys, Stickney, Illinois; *Credits* continued on p. 769

CONTENTS

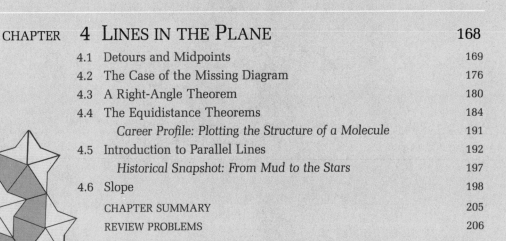

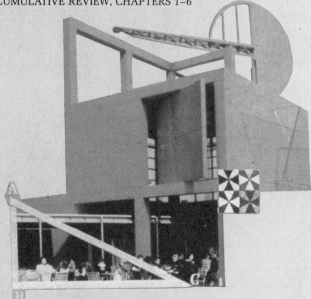

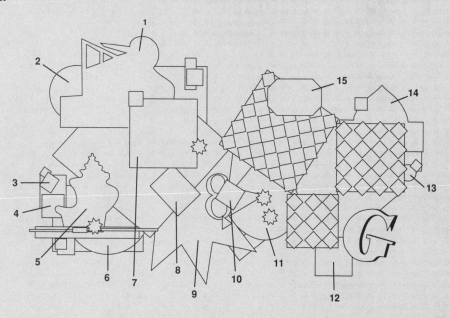

1. Café at Parc de la Villette, Paris
2, 13. Cuban tree snails
3, 8, 12. Pythagorean Theorem proofs
4, 10. Pieced quilt, Broken Dishes variation, silk,
        made by Susan Simpson, New Jersey, c. 1870
5. Miraculous Thatcheria, a marine snail from Taiwan
6. Round blue topaz
7. Amusement-park ride
9. Amish quilt, Star of Bethlehem pattern
11. Saturn's B- and C-rings
14. Hazart Ali Shrine, Mazari-i-Sharif, Afghanistan
15. Malaya, a variety of garnet, from East Africa

LETTER TO STUDENTS

Why Study Geometry?

Reason 1: Geometry is useful. Engineers, architects, painters, carpenters, plumbers, teachers, electricians, machinists, and homebuilders are only a few of the people who use geometry in their daily lives. Geometric principles are important in the construction of buildings and roads, the design and use of machinery and scientific instruments, the operation of airplanes, and the planning of new inventions plus many other activities.

Reason 2: Geometry is challenging. Many people enjoy the challenge of solving riddles and other types of puzzles. The study of geometry offers similar intriguing challenges—challenges that are particularly appealing because they involve visible figures as well as words and ideas.

Here's a first challenge for you. How many squares are in the figure shown? (The answer is upside down at the bottom of the next page.)

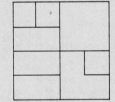

Reason 3: Geometry is logical. As we become educated, we learn to rely more on reason and proof and less on superstition, prejudice, and guesswork. One of the main purposes of this book is to help you appreciate the power of logic as a tool for understanding the world around you. For this reason, the first six chapters focus on the concept of proof. Although proofs may seem difficult to you for a few weeks, with reasonable effort on your part the feeling of difficulty will soon pass. You will be amazed at your skill in forging a chain of reasoning and will appreciate as never before the uses of logic in mathematics and in your daily life.

Reason 4: Geometry gives visual meaning to arithmetic and algebra. Here is a
problem that does so:

If angle 2 is five times as large as angle 1, what is the size of each of the angles?

A little thought might lead us to write the equation

$$x + 5x = 180$$

which we can use to solve the problem. (Where do you think the x, the 5x, and
the number 180 came from?)

The important ideas of geometry must be developed gradually. Almost
every day, your geometry homework will include a few problems involving
areas, perimeters, and the measures of angles and segments. You will begin to
learn about the significant concepts of probability, rotation, and reflection in
Chapter 1. In Chapter 2, you will review and begin to extend what you have
learned about coordinate graphs in your algebra studies. As you progress, you
will become more and more familiar with these topics.

You will find that some of the problems in this book can be solved in your
head, some require paper and pencil, and some are most easily solved with the
aid of a scientific calculator.

The geometry course on which you are about to begin is one that we hope
you will find fun, exciting, and powerful. We, the authors, wish you well on the
year's journey.

Richard Rhoad   *George Milauskas*   *Robert Whipple*

There are nine squares in the figure.

Chapter 1 Schedule

Basic
Problem Sets: 10 days
Review: 1 day
Test: 1 day

Average
Problem Sets: 9 days
Review: 1 day
Test: 1 day

Advanced
Problem Sets: 8 days
Review: 1 day
Test: 1 day

*T*HE TEXT IS written to be read, daily, by students. Acuity in reading, like mastery of the material, should come to students over time. The text has built-in repetition so students gradually develop, and continually maintain, these skills.

While few students will master all the material in the chapter before beginning Chapter **2**, all students should study the sample problems carefully, read the text, and become acquainted with the correct form of the early proofs.

For many students, constructions contribute to an intuitive understanding of the properties of geometric figures. Constructions are presented in Chapter **14** but can be integrated into appropriate chapters. Page references and assignments are included on Chapter Summary pages.

CHAPTER

1 INTRODUCTION TO GEOMETRY

T his painting, *Open Book* by Paul Klee, incorporates geometric shapes and relationships.

1.1 GETTING STARTED

Objectives

After studying this section, you will be able to
- Recognize points
- Recognize lines
- Recognize line segments
- Recognize rays
- Recognize angles
- Recognize triangles

Part One: Introduction

Points

In the diagram at the right, five **points** are represented by five dots. The names of the points are A, B, C, D, and E. (We use capital letters to name points.)

Lines

The diagram below represents three **lines.** Lines are made up of points and are straight. The arrows on the ends of the figures show that the lines extend infinitely far in both directions.

All lines are straight and extend infinitely far in both directions.

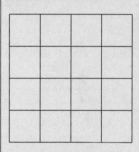

- The line on the left is called line m.
- Since we can name a line in terms of any two points on it, the line in the middle can be called by a variety of names.

$$\overleftrightarrow{BD} \quad \overleftrightarrow{BC} \quad \overleftrightarrow{CD} \quad \overleftrightarrow{CB} \quad \overleftrightarrow{DB} \quad \overleftrightarrow{DC}$$

- The line on the right can be called by any of three names.

$$\text{line } \ell \quad \overleftrightarrow{EF} \quad \overleftrightarrow{FE}$$

In algebra you learned that a **number line** is formed when a numerical value is assigned to each point on a line.

Section 1.1   Getting Started   **3**

Class Planning

Time Schedule
All levels: 1 day

Resource References
Teacher's Resource Book
   Class Opener 1.1A, 1.1B

Class Opener

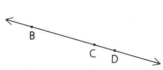

1. How many squares are there in the 4 × 4 square in the diagram?   30
2. How many squares are there in a 5 × 5 square?   55
3. How many squares are there in a 6 × 6 square?   91
4. How many squares are there in an n × n square?
   $1^2 + 2^2 + 3^2 + \ldots + n^2$

Lesson Notes

- Students can read Parts One and Two of the lesson on their own. Be sure they understand that the text in Part One helps answer the sample problems in Part Two.

Vocabulary

angle	line	ray	union
endpoint	line segment	segment	vertex
intersection	number line	side	
	point	triangle	

Communicating Mathematics

The questions below referring to the diagram are good discussion questions.

1 Name line PR by two letters in six different ways.
$\overleftrightarrow{PQ}, \overleftrightarrow{PR}, \overleftrightarrow{QR}, \overleftrightarrow{RQ}, \overleftrightarrow{RP}, \overleftrightarrow{QP}$

2 Name segment PR by two letters in two different ways.
$\overline{PR}, \overline{RP}$

3 Name ray PR by two letters in two different ways.
$\overrightarrow{PQ}, \overrightarrow{PR}$

The coordinate of A is -2. The coordinate of B is $1\frac{1}{2}$.

Line Segments

The following diagram represents several **line segments,** or simply **segments.** Like lines, segments are made up of points and are straight. A segment, however, has a definite beginning and end.

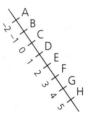

A segment is named in terms of its two **endpoints**

- The segment on the left can be called either $\overline{RS}$ or $\overline{SR}$.
- In the middle figure there are two segments. The vertical (up-and-down) segment can be called either $\overline{PX}$ or $\overline{XP}$. The horizontal (crosswise) segment can also be named in two ways. Can you name these two ways?
- How might we name the segment whose endpoints have coordinates 3 and 0 in the figure on the right?

Rays

In the diagram below, three **rays** are represented. Rays, like lines and segments, are made up of points and are straight. A ray differs from a line or a segment in that it begins at an endpoint and then extends infinitely far in *only one* direction.

When we name a ray, we must name the endpoint first so that it is clear where the ray begins.

- The ray on the left is called $\overrightarrow{AB}$.
- The ray in the middle can be called $\overrightarrow{CD}$ or $\overrightarrow{CE}$. (As long as the endpoint is given first, any other point on the ray can be used in its name.)
- The ray on the right can be named in only one way. Do you know what its name is?

Angles

Two rays that have the same endpoint form an **angle**.

Definition   An **angle** is made up of two rays with a common endpoint. This point is called the **vertex** of the angle. The rays are called **sides** of the angle.

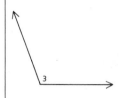 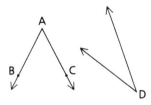 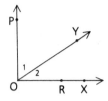

- In the diagram above, the angle on the left is called ∠3. The 3 placed inside the angle near the vertex names it.
- The second angle in the diagram can be called by any of three names.

$$\angle BAC \qquad \angle CAB \qquad \angle A$$

   (Notice that when we use three letters, the vertex must be named in the middle.)
- The third angle is called ∠D.
- In the last figure above, there are three angles. Can you tell which angle is ∠O? Because names might refer to more than one angle in a diagram, we *never name an angle in a way that could result in confusion.*

   ∠1 can also be called ∠POY or ∠YOP.

   ∠2 can also be called ∠YOR, ∠YOX, ∠ROY, or ∠XOY.

   The other angle in this figure can be named ∠POR. See if you can find three other names for this angle.

Triangles

We shall call the following figure **triangle** ABC (△ABC).

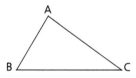

A triangle has three segments as its sides. You may wonder whether we can talk about an ∠B in the triangle, since there are no arrows in the diagram. The answer is yes. We shall often talk about rays, lines, and angles in a diagram of a triangle. So a triangle not only

Lesson Notes, continued

- The first definition given in the text is for *angle*. Many defined terms, such as *perpendicular*, *median*, and *midpoint*, will be used in the "Reasons" column of proofs and should be memorized. Students must be able to recognize and use other terms, such as *angle*, *triangle*, and *congruent*, even though they may not be used as reasons in a proof.
- Students should already be familiar with much of the terminology, symbolism, and notation presented in this section. You may, however, need to stress the differences between the symbols for line, segment, and ray. You should also emphasize that although geometric figures can often be named in different ways, sometimes care must be taken to name an endpoint first or a vertex in the middle.

Lesson Notes, continued

■ If necessary, review the meaning of *union* and *intersection*
for sets of points.

has three sides but has three angles as well. Can you name the angle
at the top of the triangle shown on the preceding page in three
ways? ∠A, ∠BAC, ∠CAB

The triangle is the **union** (∪) of three segments.

$$\triangle ABC = \overline{AB} \cup \overline{BC} \cup \overline{AC}$$

The **intersection** (∩) of any two sides is a **vertex** of the triangle.

$$\overline{AB} \cap \overline{BC} = B$$

Part Two: Sample Problems

Problem 1

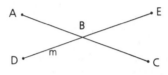

a *How many lines are shown? (Imagine that there are arrows in the
diagram.)*

b *Name these lines.*

c *Where do* $\overleftrightarrow{AC}$ *and* $\overleftrightarrow{DE}$ *intersect?*

d *Where does* $\overrightarrow{AC}$ *intersect* $\overleftrightarrow{BC}$? ($\overrightarrow{AC} \cap \overleftrightarrow{BC} = \underline{\ ?\ }$)

e *What is the union of* $\overrightarrow{BA}$ *and* $\overrightarrow{BD}$? ($\overrightarrow{BA} \cup \overrightarrow{BD} = \underline{\ ?\ }$)

Answers

a 2

b Line m, $\overleftrightarrow{DB}$, $\overleftrightarrow{DE}$, $\overleftrightarrow{BD}$, $\overleftrightarrow{BE}$, $\overleftrightarrow{EB}$, or $\overleftrightarrow{ED}$;
$\overleftrightarrow{AB}$, $\overleftrightarrow{AC}$, $\overleftrightarrow{BA}$, $\overleftrightarrow{BC}$, $\overleftrightarrow{CA}$, or $\overleftrightarrow{CB}$

c B

d $\overrightarrow{AC}$ (Remember sets? If P and Q are two sets of points, then
P ∩ Q = {all points in P *and* in Q}.)

e ∠ABD (P ∪ Q = {all points in P *or* in Q *or* in both}.)

Problem 2

A   B   C

a *Name the ray that has endpoint A and goes in the direction of C.*

b *Name the segment joining A and B.*

Answers

a $\overrightarrow{AB}$ or $\overrightarrow{AC}$

b $\overline{AB}$ or $\overline{BA}$

Problem 3

Draw a diagram in which the intersection of $\overrightarrow{AB}$ *with* $\overrightarrow{CA}$ *is* $\overline{AC}$
($\overrightarrow{AB} \cap \overrightarrow{CA} = \overline{AC}$).

Solution

A   C B

Problem 4   *Draw a diagram in which △ABC ∩ $\overrightarrow{DE}$ = F.*

Solution

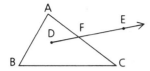

There are other correct answers, and a lot of wrong ones.

Part Three: Problem Sets

Problem Set A

In the back of the book, you will find answers to many of the problems. It will help you learn to check your answer in the back after you solve a problem. Then rethink your work if necessary.

1 What are three possible names for the line shown? $\overleftrightarrow{AB}$, $\overleftrightarrow{BA}$, line ℓ

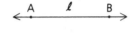

2 What are four possible names for the angle shown? ∠CED, ∠DEC, ∠E, ∠7

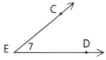

3 Can the ray shown be called $\overrightarrow{XY}$? No

4 Name the sides of △RST. $\overline{RS}$, $\overline{ST}$, $\overline{RT}$

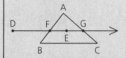

5 a $\overline{AB} \cap \overline{BC}$ = __?__ B
  b $\overrightarrow{EC} \cup \overrightarrow{EA}$ = __?__ $\overleftrightarrow{AC}$ or ∠CEA
  c $\overleftrightarrow{AC} \cap \overleftrightarrow{DB}$ = __?__ E
  d $\overline{DC} \cap \overline{AB}$ = __?__ ∅
  e $\overrightarrow{AC} \cap \overrightarrow{EC}$ = __?__ $\overrightarrow{EC}$
  f $\overrightarrow{BA} \cup \overrightarrow{BC}$ = __?__ ∠ABC
  g $\overline{EC} \cup \overline{CB} \cup \overline{BE}$ = __?__ △BEC

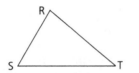

6 a Name ∠OPR in all other possible ways. ∠RPO, ∠RPS, ∠SPR
  b What is the vertex of ∠TOS? O
  c How many angles have vertex R? 3
  d Name ∠TSP in all other possible ways.
  e How many triangles are there in the figure? 8

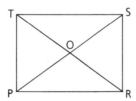

Cooperative Learning

Refer to Sample Problem **4**. Have students work in small groups and draw as many different diagrams as possible to demonstrate the correct answer. Extend the problem by asking students to show diagrams for △ABC ∩ $\overrightarrow{DE}$ = {F, G}.
Answers may vary. One possible diagram is given below.

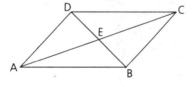

Then have students investigate whether they can draw a diagram for △ABC ∩ $\overrightarrow{DE}$ = {F, G, H}. Not possible

Assignment Guide

Basic
1–12
Average
1–12, 14
Advanced
5–14

Problem-Set Notes and Additional Answers

- Each problem set provides an overabundance of problems; those not assigned can be used as "extra credit" problems or as a periodic review in conjunction with the Cumulative Review every third chapter.
- The student text contains answers to selected problems. Students should be encouraged to use these answers for self-checking as they complete their homework assignments.

6d ∠PST, ∠TSO, ∠OST

7

9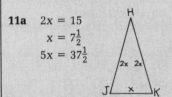

11a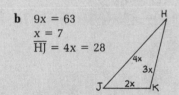
$2x = 15$
$x = 7\frac{1}{2}$
$5x = 37\frac{1}{2}$

b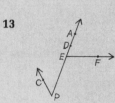
$9x = 63$
$x = 7$
$\overline{HJ} = 4x = 28$

12
A   C   B   D

13

14 8 △s: △EXA, △EXC,
    △AXC, △TXC, △TAC,
    △XET, △EAT, △CTE
  a △TAC, △TCX, △CTE;
     $\frac{3}{8} = 37\frac{1}{2}\%$
  b △AXC, △TAC; $\frac{2}{8} = 25\%$

T 8

Problem Set A, continued

7 Figure 1 shows the reflection of the letter
F over a line. Copy Figure 2 and draw
the reflections of the letters P, A, and J
over the given line.

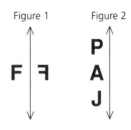

8 a A line is made up of ___?___. Points
   b An angle is the union of two ___?___ with a common ___?___.
      Rays; endpoint
9 Draw a number line and label points F, G, H, and J with the
   coordinates $-4\frac{2}{3}$, 2, 5, and 3.5 respectively. One of these points is
   the *midpoint* (the halfway point) between two others. Which is it? J

10 Given a rectangle with sides 2.5 cm and
   8.6 cm long, find

   a The rectangle's area 21.5 sq cm

   b The rectangle's perimeter (the distance
      around it) 22.2 cm

Problem Set B

11 a In △HJK, $\overline{HJ}$ is twice as long as $\overline{JK}$ and
   exactly as long as $\overline{HK}$. If the length of
   $\overline{HJ}$ is 15, find the perimeter of (the
   distance around) △HJK. $37\frac{1}{2}$

   b If the length of $\overline{HJ}$ were 4x, the length
   of $\overline{HK}$ were 3x, the length of $\overline{JK}$ were
   2x, and the perimeter of △HJK were
   63, what would the length of $\overline{HJ}$ be? 28

12 Draw a diagram in which $\overline{AB} \cap \overline{CD} = \overline{CB}$.

Problem Set C

13 Draw a diagram in which the intersection of ∠AEF and ∠DPC is
   $\overrightarrow{ED}$.

14 a What percentage of the triangles in the
   diagram have $\overline{CT}$ as a side? 37.5%

   b What percentage have $\overline{AC}$ as a side? 25%

MEASUREMENT OF SEGMENTS AND ANGLES

Objectives

After studying this section, you will be able to
- Measure segments
- Measure angles
- Classify angles by size
- Name the parts of a degree
- Recognize congruent angles and segments

Part One: Introduction

Measuring Segments

We measure segments by using such instruments as rulers or metersticks. We may use any convenient length as a unit of measure. Some of the units that are currently in common use are inches, feet, yards, millimeters, centimeters, and meters. To indicate the measure of $\overline{AB}$, we write AB.

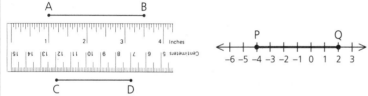

On the ruler shown, find the length of $\overline{AB}$ in inches and the length of $\overline{CD}$ in centimeters. On the number line, find PQ. $2\frac{1}{2}$ in.; 5 cm; 6

Measuring Angles

Angles are commonly measured by means of a **protractor**. (The diagram at the top of the next page shows how a protractor can be used to measure a 117° angle.) We shall measure angles (∠s) in **degrees** (°). In later courses, you may use other units, such as radians or grads.

The **measure**, or size, of an angle is the amount of turning you would do if you were at the vertex, looking along one side, and then turned to look along the other side. (A surveyor's transit works in much the same way.)

Class Planning

Time Schedule
All levels: 1 day

Resource References
Teacher's Resource Book
   Class Opener 1.2A, 1.2B
Evaluation
   Tests and Quizzes
      Quiz 1, Series 1, 2

Class Opener

Name all angles in the diagram.

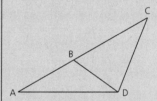

∠A, ∠ADB, ∠ADC, ∠BDC, ∠C, ∠CBD, ∠ABD, ∠ABC

Lesson Notes

- All students should be able to give accurate descriptions of the six terms defined in this section. Since the definitions will not be explicitly given as reasons in proofs, students need not be required to memorize the definitions verbatim.

Vocabulary

acute angle	degree	obtuse angle	second
congruent angles	measure	protractor	straight angle
congruent segments	minute	right angle	tick marks

Lesson Notes, *continued*

■ The text uses informal terms, such as "quarter turn," to discuss angle measurement. Students who have little experience with protractors may benefit from additional class time spent in measuring given angles and drawing angles of specified measure.

Cooperative Learning

Have each student use a ruler to draw a large triangle and measure the angles of the triangle. Then have students work in small groups and compare their results. Ask them to make a conjecture about the sum of the measures of the angles of a triangle.

The sum of the measures of the angles of a triangle is 180.

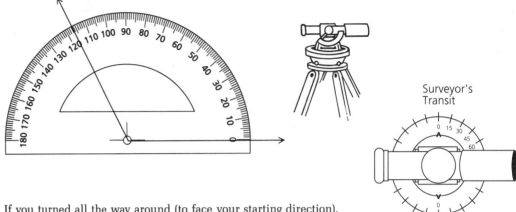

Surveyor's Transit

If you turned all the way around (to face your starting direction), you would turn 360°. You can use this fact to estimate the size of an angle.

$\overrightarrow{AP}$ appears to have been turned one fourth of the way around from $\overrightarrow{AR}$, so you might guess that $\angle A$ is approximately a 90° angle.

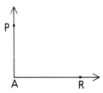

Angle 1 required less than a quarter turn. A good guess would be that it is a 60° angle.

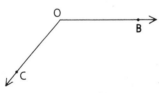

Angle BOC required more than a quarter turn, so its size could be estimated at 130°.

Some math courses deal with negative angles, zero angles, and angles greater than 180°. In this course, you will usually be working with angles greater than 0° and less than or equal to 180°.

$$0 < \text{angle measure} \leq 180$$

10 Chapter 1   Introduction to Geometry

Classifying Angles by Size

As shown below, we classify angles into four categories according to their measures.

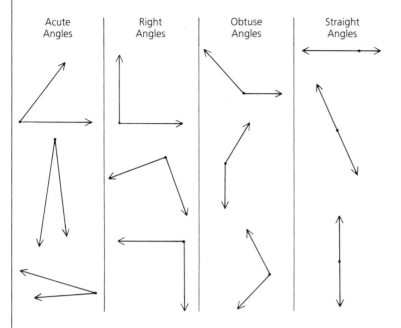

| Acute Angles | Right Angles | Obtuse Angles | Straight Angles |

Definitions   An *acute angle* is an angle whose measure is greater than 0 and less than 90.

A *right angle* is an angle whose measure is 90.

An *obtuse angle* is an angle whose measure is greater than 90 and less than 180.

A *straight angle* is an angle whose measure is 180. (As you can see, a straight angle forms a straight line.)

Lesson Notes, continued

■ From these definitions, the measure of an angle must be greater than 0 and less than or equal to 180. Emphasize that these restrictions are arbitrary and that angles of greater measure and of negative measure are used in other courses.

Parts of a Degree

As you know, each hour of the day is divided into 60 minutes, and each minute is divided into 60 seconds. Similarly, each degree (°) of an angle is divided into 60 *minutes* ('), and each minute of an angle is divided into 60 *seconds* (").

60' = 1°   (60 minutes equals 1 degree.)
60" = 1'   (60 seconds equals 1 minute.)

Thus, $87\frac{1}{2}° = 87°30'$

$60.4° = 60°24'$

$90° = 89°60'$ (since 60' = 1°)

$180° = 179°59'60''$ (since 60" = 1' and 60' = 1°)

- The degree-minute-second conversions could be omitted. Students might encounter such problems in navigation and time measurement. The problems also expose students to a base-60 (sexagesimal) system. Examples **1–3** present arithmetic concepts in a geometric setting. Throughout the text, geometric examples are used to present concepts and skills from arithmetic, algebra, and probability. The text follows these conventions.

  1. Angles are congruent, while their measures are equal. For example, $\angle A \cong \angle B$, while $m\angle A = m\angle B$.
  2. Angles can be described in degrees, but measures of angles are dimensionless. For example, $\angle A = 45°$, but $m\angle A = 45$. Similar distinctions will be made later between arcs and their measures.
  3. Distances are numbers and may be equal; segments may be congruent. For example, RS = ST, while $\overline{RS} \cong \overline{ST}$.

- These distinctions are implicit in the text; you may want to discuss them explicitly with advanced students.
- Tick marks are used throughout the text to indicate congruent segments, angles, and arcs.

Communicating Mathematics

Ask students to prepare a lesson that explains how to use tick marks. Choose a few students to present their lessons to the class.

Study the following examples closely.

Example 1 Change $41\frac{2}{5}°$ to degrees and minutes.

Since there are 60′ in 1°, $\frac{2}{5}°$ is $\frac{2}{5}(60)$ minutes, or 24′.

Hence, $41\frac{2}{5}° = 41°24′$.

Example 2 Given: $\angle ABC$ is a right angle.
  $\angle ABD = 67°21′37″$

Find: $\angle DBC$

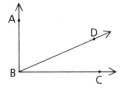

Subtract $67°21′37″$ from $90°$ as follows.

$$89°59′60″ \quad (90° = 89°59′60″)$$
$$-67°21′37″$$
$$\overline{22°38′23″}$$

Example 3 Change $60°45′$ to degrees.

We must change 45′ to a fractional part of a degree. Since 60′ = 1°, 45 is divided by 60, and the fraction is reduced.

$$\frac{45}{60} = \frac{3}{4} \qquad \text{So } 60°45′ = 60\frac{3}{4}°.$$

Congruent Angles and Segments

In the diagram below, $\angle$s A, B, and C are ***congruent***. We write $\angle A \cong \angle B \cong \angle C$.

Definition ***Congruent ($\cong$) angles*** are angles that have the same measure.

In a similar way, segments can be congruent.

Definition ***Congruent ($\cong$) segments*** are segments that have the same length.

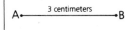  

In the diagram above, segments $\overline{AB}$, $\overline{CD}$, and $\overline{EF}$ are congruent. We write $\overline{AB} \cong \overline{CD} \cong \overline{EF}$.

Often, we use identical ***tick marks*** to indicate congruent angles and segments. In the following diagram, the identical tick marks

indicate that there are four pairs of congruent parts. Can you name
them? $\overline{GH} \cong \overline{SR}$; $\angle K \cong \angle T$; $\overline{WZ} \cong \overline{XY}$; $\angle W \cong \angle X$

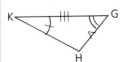

Part Two: Sample Problems

Problem 1 Classify each of the angles below as acute, right, or obtuse. Then
estimate the number of degrees in the angle.

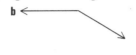

Answers **a** Acute; 40°   **b** Obtuse; 150°   **c** Right; 90°

Problem 2 In the diagram below, $\angle DEG = 80°$, $\angle DEF = 50°$, $\angle HJM = 120°$, and
$\angle HJK = 90°$. Draw a conclusion about $\angle FEG$ and $\angle KJM$.

Solution $\angle FEG = 30°$ and $\angle KJM = 30°$, so $\angle FEG \cong \angle KJM$.

Problem 3 Given: $\angle ABC$ is a right angle.
$\angle 1 = (3x + 4)°$,
$\angle 2 = (x + 6)°$
Find: $m\angle 1$ (the measure of $\angle 1$)

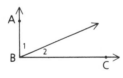

Solution Since $\angle ABC$ is a right $\angle$, $m\angle 1 + m\angle 2 = 90$.

$(3x + 4) + (x + 6) = 90$
$4x + 10 = 90$
$4x = 80$
$x = 20$

Since $m\angle 1 = 3x + 4$, $m\angle 1 = 3(20) + 4$, or 64.

Problem 4 $\angle B$ is acute.
a What are the restrictions on $m\angle B$?
b What are the restrictions on x?

Lesson Notes, continued

■ Sample Problem **3** introduces
the notation "$m\angle 1$."

Lesson Notes, continued

- There are several problems like Sample Problem **5** in the text. Some are quite difficult, but most are appropriate for all students. The solutions can be approximated with a clock and a protractor, or they can be arrived at algebraically.

Assignment Guide

Basic
1–3, 5–8, 9b, c, 10, 11, 13, 16, 18

Average
1, 2, 4–8, 9b, d, 10, 11, 13–18

Advanced
8, 9b, d, 10, 11, 15–20

Problem-Set Notes and Additional Answers

- The sample problems are similar to the problems in the problem sets. Students should study the sample problems *before* starting their problem-set assignments.

Solution   **a** Since ∠B is acute, m∠B > 0 and m∠B < 90 (0 < m∠B < 90).

b 2x + 14 > 0 and 2x + 14 < 90
  2x > −14   and   2x < 76
   x > −7   and   x < 38

Thus, −7 < x < 38.

Problem 5 *Find the angle formed by the hands of a clock at each time.*

a 4:00   **b** 5:15

Solution **a** Since 360° is divided into 12 intervals on a clock, each interval is 30°. From 12 to 4 there are 4 intervals, so the angle is 4(30°), or 120°.

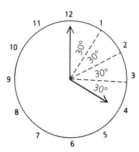

b Remember that the hour hand is on 5 only when the minute hand is on 12. At 5:15 the hour hand is one fourth of the way from 5 to 6. Since $\frac{1}{4}(30°) = 7\frac{1}{2}°$, the hands form an angle of $60 + 7\frac{1}{2}$, or $67\frac{1}{2}$ degrees.

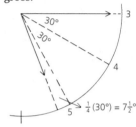

Part Three: Problem Sets

Problem Set A

1 Change each of the following to degrees and minutes.
  a $61\frac{2}{3}°$ 61°40′   **b** 71.7° 71°42′

2 Change each of the following to degrees.
  a 132°30′ $132\frac{1}{2}°$ **b** 19°45′ $19\frac{3}{4}°$

3 Which two of the angles below *appear* to be congruent? ∠1 and ∠2

4 a $\overrightarrow{QV} \cap \overleftrightarrow{TS} = $ _?_ T
  b $\overline{WP} \cap \overline{VR} = $ _?_ $\overline{VW}$
  c $\overrightarrow{WP} \cup \overrightarrow{VR} = $ _?_ $\overleftrightarrow{PR}$
  d $\overrightarrow{SQ} \cup \overrightarrow{SR} = $ _?_ ∠QSR
  e How many angles have vertex Q? 6

5 a Evaluate 49°32′55″ + 37°27′15″. 87°10″

  b Evaluate 123°15′ − 40°26′. 82°49′

6 There is a right angle at each corner of PRST. (Later in the course you will learn that PRST is a rectangle.)

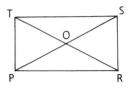

  a If ∠TPO = 60°, how large is ∠RPO? 30°

  b If ∠PTO = 70°, how large is ∠STO? 20°

  c If ∠TOP = 50°, how large is ∠POR? 130°

  d Classify ∠TOS as acute, right, or obtuse. Obtuse

7 a Which angle *appears* to have the same measure as ∠1? ∠5

  b Which angle *appears* larger, ∠2 or ∠3?   **b** Same size

  c Does ∠3 *appear* to be congruent to ∠4 or to ∠5? ∠4

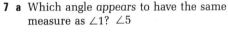

8 If ∠CBD ≅ ∠DBE, find m∠A. 55

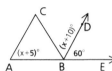

9 Find the measure of the angle formed by the hands of a clock at each time.

  a 3:00 90   **b** 4:30 45   **c** 7:20 100   **d** 1:45 $142\frac{1}{2}$

10 a Find PQ.

  b If R's coordinate is 7, why is $\overline{PQ} \neq \overline{QR}$?

  c What must the coordinate of R be in order for Q to be the midpoint of $\overline{PR}$?
  a 5; **b** PQ ≠ QR; **c** 8

11 Given: ∠CAR is a right angle.
        m∠CAT = 37°66′10″

Find: m∠RAT
51°53′50″

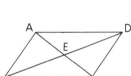

Problem Set B

12 a How many triangles (△) are in the diagram? 8

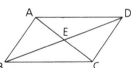

  b How many angles (∠s) in the figure *appear* to be right? 2

  c How many angles in the figure *appear* to be acute? 10

  d How many angles in the figure *appear* to be obtuse? 4

  e Name the straight angles in the figure. ∠AEC, ∠BED

Problem-Set Notes and Additional Answers, continued

■ The text frequently previews material that students will study later in the course. In problem **8** and throughout the text, students are required to use algebraic skills to solve geometric problems.

12a △ABC, △ADE, △ABD, △BCD, △ABE, △BCE, △ACD, △CDE

  b ∠BAE, ∠DCE

  c ∠EBA, ∠DCE, ∠EBC, ∠AEB, ∠ECB, ∠EDA, ∠EDC, ∠EAD, ∠ABC, ∠ADC

  d ∠BCD, ∠BEC, ∠AED, ∠BAD

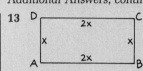

Problem Set B, *continued*

13 The perimeter of (the distance around) ABCD is 66, and $\overline{DC}$ is twice as long as $\overline{CB}$. How long is $\overline{AB}$? 22

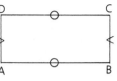

14 Given: $\overline{XS} \cong \overline{YT}$, $\overline{YS} \cong \overline{XT}$,
XT = 2r + 5,
XS = 3m + 7,
YS = $3\frac{1}{2}r + 2$,
YT = 4.2m + 5
Solve for r and m. 2; $\frac{5}{3}$

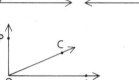

15 Given: ∠1 ≅ ∠2,
m∠1 = x + 14,
m∠2 = y − 3
Solve for y in terms of x. y = x + 17

16 If ∠POA is a right angle and if ∠POC is three times as large as ∠COA, find m∠POC. $67\frac{1}{2}$

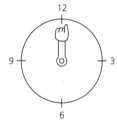

17 ∠P is acute.
a What are the restrictions on m∠P? 0 < m∠P < 90
b What are the restrictions on x? 20 < x < 50

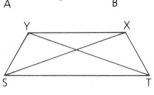
(3x − 60)°

18 The hand is at 12 on the clock.
a If the hand were rotated 90° clockwise, at what number would it point? 3
b If the hand were rotated 150° clockwise and then 30° counterclockwise, at what number would it point? 4

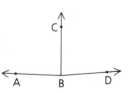

Problem Set C

19 ∠ABC and ∠CBD have the same measure. If ∠ABC = $\left(\frac{3x}{2} + 2\right)°$ and ∠CBD = $\left(2x - 29\frac{1}{4}\right)°$, is ∠ABD a straight angle? No

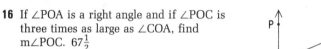

20 Change $15\frac{2}{9}°$ to degrees, minutes, and seconds. 15°13′20″

16 Chapter 1 Introduction to Geometry

21 Given: ∠TRS is a straight angle.
∠TRX is a right angle.
m∠TRS = 2x + 5y,
m∠XRS = 3x + 3y

Solve for x and y. −10; 40

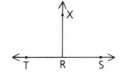

*Problem-Set Notes and
Additional Answers, continued*

21 $\begin{cases} 3x + 3y = 90 \\ 2x + 5y = 180 \end{cases}$

22 $\begin{cases} 3x + 2y = 90 \\ 3x + y = 60 \end{cases}$
$x = 10; y = 30$

22 Maxie and Minnie were taking a stroll in the Arizona desert when a spaceship from Mars landed. A Martian walked up to them and pointed to Figure 1. "XLr8r, XLr8r, XLr8r plus YBcaws, YBcaws," she said. Pointing to Figure 2, she said, "YBcaws plus XLr8r, XLr8r, XLr8r." What might *XLr8r* mean? 10°

Figure 1 90°   Figure 2 60°

23 Change 72°22′30″ to degrees. $72\frac{3}{8}°$

MATHEMATICAL EXCURSION

GEOMETRY IN NATURE
Orange sections and spiraling leaves

If you cut a cross section of an orange, you will see that it is divided into sections that together form a 360° angle. The mathematician Johannes Kepler (1571–1630) thought that all fruits and flowers that grew on trees had five sections or petals. You can see that this isn't true, but the sections of an orange do appear to be the same size and shape.

Flower petals, and leaves on stems grow in a spiral pattern and form angles of consistent sizes. *Phyllotaxis* is the distribution of leaves around the stem of a plant. The measure of the angle formed by any two leaves in succession on a stem is equal to the measure of the angle between any two other leaves in succession.

The most common angles seem to be 144° and 135°. A 144° angle is characteristic

144°

for rose leaves. Suppose you draw a series of 144° angles with a protractor, using one of the sides of the last angle you drew for each new angle and proceeding in a clockwise direction. You will see that the angles eventually divide a circle into five equal parts.

Botanists say that these angles exist because each bud grows where it will have the most room between the bud before it and the one that will come after it.

COLLINEARITY, BETWEENNESS, AND ASSUMPTIONS

Class Planning

Time Schedule
All levels: 1 day

Resource References
Teacher's Resource Book
    Class Opener 1.3A, 1.3B
    Additional Practice Worksheet 1
Transparencies 1, 2
Evaluation
    Tests and Quizzes
        Quiz 1, Series 3

Class Opener

Find m∠DAT.

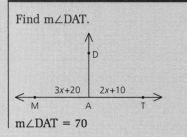

m∠DAT = 70

Lesson Notes

■ The text illustrates *betweenness* without a formal definition.

Objectives
After studying this section, you will be able to
■ Recognize collinear and noncollinear points
■ Recognize when a point can be said to be between two others
■ Recognize that each side of a triangle is shorter than the sum of the other two sides
■ Correctly interpret geometric diagrams

Part One: Introduction

Collinearity
It is often useful to know that a group of points lie on the same line.

Definition   Points that lie on the same line are called ***collinear***. Points that do not lie on the same line are called ***noncollinear***.

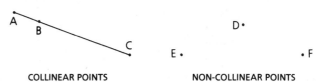

COLLINEAR POINTS   NON-COLLINEAR POINTS

In the diagram at the right, R, S, and T are collinear points. P, O, and X are also collinear. M, O, X, and Y are noncollinear.

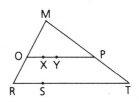

Betweenness of Points
In order for us to say that a point is between two other points, all three of the points must be collinear.

T is between A and R   We do *not* say O is between X and Y

Vocabulary
collinear
noncollinear

Triangle Inequality

For any three points, there are only two possibilities:

1 They are collinear. (One point is between the other two. Two of the distances add up to the third.)

2 They are noncollinear. (The three points determine a triangle.)

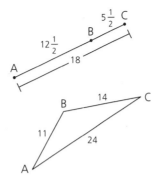

   Notice that in the triangle, $14 + 11 > 24$. This is an example of an important characteristic of triangles: *The sum of the lengths of any two sides of a triangle is always greater than the length of the third.*

Assumptions from Diagrams

You may wonder what you should and should not assume when you look at a diagram. The chart below gives the general rules you should follow as you work with this book. (There are, however, occasional exceptions, as in Section 1.2, problem 19.)

How to Interpret a Diagram	
You Should Assume	**You Should Not Assume**
Straight lines and angles	Right angles
Collinearity of points	Congruent segments
Betweenness of points	Congruent angles
Relative positions of points	Relative sizes of segments and angles

The following example will help you understand what assumptions can be made.

Example

Given: *Diagram as shown*

Question: *What should we assume?*

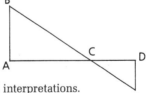

The following are some of the many valid interpretations.

Do Assume
$\overleftrightarrow{ACD}$ and $\overleftrightarrow{BCE}$ are straight lines.
∠BCE is a straight angle.
C, D, and E are noncollinear.
C is between B and E.
E is to the right of A.

Do Not Assume
∠BAC is a right ∠.
$\overline{CD} \cong \overline{DE}$
∠B ≅ ∠E
∠CDE is an obtuse angle.
$\overline{BC}$ is longer than $\overline{CE}$.

Reread and study the chart and the example carefully, for it is important that you know what to assume from a diagram.

Lesson Notes, continued

■ The chart about assumptions from diagrams is vital. Students do not necessarily need to be able to recite the two lists, but they must be able to classify each entry in the chart.

Cooperative Learning

Have students work in small groups. Give each group a diagram similar to the one below. Then have students make a list of things they can assume from the diagram and a list of things they cannot assume.

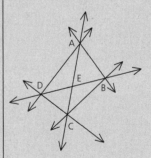

Answers will vary. Possible answers: can assume: AB, BC, CD, DA, AC, and DB are lines; points D, A, and B are noncollinear; point E is between points D and B. Cannot assume: m∠DEC = m∠BEC = 90

Part Two: Sample Problems

Problem 1   For each diagram, tell whether X is between P and R. (*Answer Yes or No.*)

a P   X   R

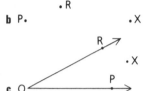

b P.   •R   •X

c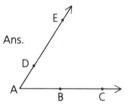

O — R — X — P

Answers   **a** Yes   **b** No   **c** No

Problem 2   Draw a diagram in which A, B, and C are collinear, A, D, and E are collinear, and B, C, and D are noncollinear.

Solution   The diagram at the right shows one of the possible solutions.

Ans.

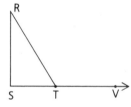

Problem 3   **a** Should we assume that S, T, and V are collinear in the diagram?

b Should we assume that ∠S = 90°?

Answers   **a** Yes

b No

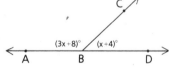

Part Three: Problem Sets

Problem Set A

1 Find m∠ABC (the measure of ∠ABC). 134

$(3x+8)°$ $(x+4)°$

A   B   D

C

2 Draw a diagram showing four points, no three of which are collinear.

Communicating Mathematics

In Sample Problem **2**, have students draw a different diagram that fulfills the conditions of the problem and a diagram that does not fulfill the conditions. Have them explain why their second diagram does not.

Assignment Guide

Basic
1–10, 12, 13
Average
1–14
Advanced
3–16

Problem-Set Notes and Additional Answers

2   Answers will vary. Possible answer shown.

•C
•A   •B   •D

3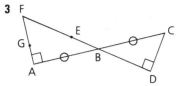

a Name all points collinear with E and F. B and D
b Are G, E, and D collinear? Are F and C collinear? No; yes
c Which two segments do the tick marks indicate are congruent? $\overline{AB}$ and $\overline{BC}$
d Is $\angle A \cong \angle D$? Yes
e Is $\angle F \cong \angle ABF$? Not necessarily
f Where do $\overleftrightarrow{AC}$ and $\overleftrightarrow{FE}$ intersect? B
g $\overline{AG} \cap \overline{GF} = \underline{\ ?\ }$ G
h $\overline{AG} \cup \overline{GF} = \underline{\ ?\ }$ $\overline{AF}$
i B lies on a ray whose endpoint is E. Name this ray in all possible ways. $\overrightarrow{EB}, \overrightarrow{ED}$
j Name all points between F and D. E and B

4

a Should we assume that angles E, F, G, and H are right angles?
Explain your answer. No; right angles cannot be assumed unless they are marked.
b Should we assume that points E, F, and G are noncollinear?
Explain your answer. Yes; collinearity of points can be assumed from a diagram.

5 Draw a number line and shade all points that are at or between −5 and 2. Find the length of this shaded segment. 7

6 $\angle ABC$ is a right angle. The ratio of the measures of $\angle ABD$ and $\angle DBC$ is 3 to 2. Find $m\angle ABD$. (Hint: Let $m\angle ABD = 3x$ and $m\angle DBC = 2x$.) 54

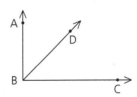

7 Explain how the sum of two acute angles could be
a Acute
e.g., 33° and 40°
b Obtuse
e.g., 60° and 70°
c Right
e.g., 45° and 45°

8 a Change $124\frac{3}{5}°$ to degrees and minutes. 124°36′
b Change 84°50′ to degrees. $84\frac{5}{6}°$

Problem-Set Notes and Additional Answers, continued
5
−6 −5 −4 −3 −2 −1 0 1 2 3

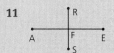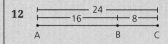
Problem Set A, *continued*

9 ∠ABD = (3x)°
   ∠DBC = x°
   Find: m∠ABD 135

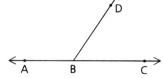

Problem Set B

10 A, K, O, and Y are collinear points. K is between O and A, the
length of $\overline{AO}$ added to the length of $\overline{AY}$ is equal to the length of
$\overline{OY}$ (OA + AY = OY), and A is to the right of O. Draw a diagram
that correctly represents this information.

11 Draw a diagram in which F is between A and E, F is also
between R and S, and A, E, R, and S are noncollinear.

12 If AB = 16, BC = 8, and AC = 24, which point is between the
other two? B

13 a AC must be smaller than what
      number? 15

   b AC must be larger than what number? 3

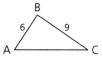

14 Q is between P and R on a number line. P = −8, and R = 4.
   a What do we know about the coordinate of Q? −8 < Q < 4
   b What do we know about the length PQ + QR? PQ + QR = PR = 12

Problem Set C

15 Given: m∠1 = 2x + 40,
         m∠2 = 2y + 40,
         m∠3 = x + 2y
   Find: m∠1, m∠2, and m∠3
         80; 100; 80

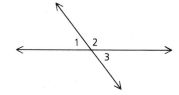

16 When Brock Clock was asked what time it was, he said, "Well,
the minute hand is pointing directly at one of the twelve numbers on the clock, the hour hand is pointing toward a spot whose
nearest number is at least five greater than the number the
minute hand is pointing toward, the angle formed by the hands
is acute, the sun is shining in the east, and it is not five minutes
past the hour." Wow! What time was it? 11:10 A.M.

17 To the nearest second, what is the first time after 12:00 that the
hour hand and the minute hand of a clock are together? ≈1:05:27

1.4 BEGINNING PROOFS

Objective

After studying this section, you will be able to
- Write simple two-column proofs

Part One: Introduction

Much of the enjoyment and challenge of geometry is found in "proving things." In this section, we shall give examples of ***two-column proofs.*** The two-column proof is the major type of proof you will use as you study this book.

We shall also introduce our first ***theorems***.

Definition A ***theorem*** is a mathematical statement that can be proved.

This section also illustrates a procedure that we shall use numerous times in this textbook:

Theorem Procedure

1 We present a theorem or theorems.
2 We prove the theorem(s).

   Note Although all theorems presented can be proved, we shall omit the proofs of certain theorems.

3 We use the theorems to help prove sample problems.
4 You are then given the challenge of using the theorems to prove homework problems. Theorems will save you much time if you learn them and then use them.

We now present our first two theorems.

Class Planning

Time Schedule
All levels: 1 day

Resource References
Teacher's Resource Book
   Class Opener 1.4A, 1.4B
   Additional Practice Worksheet 2
Evaluation
   Tests and Quizzes
   Quiz 2, Series 2

Class Opener

Is ∠STR a right angle?

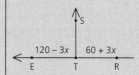

Since x can be any real number, ∠STR could be a right angle, but it does not have to be. It is a right angle if and only if x = 10.

Lesson Notes

- Few students will have seen two-column proofs prior to this section. Many students develop only a beginning understanding of proof in Chapters **1** and **2.** Often their understanding of and skill in writing proofs will improve substantially in Chapter **3,** where many proofs related to congruent triangles are presented.

Vocabulary
theorem
two-column proof

Lesson Notes, continued

While students can be encouraged to paraphrase theorems, such as "Right angles are congruent" (Theorem 1), the "If . . . , then . . ." form is often more helpful in proofs.

■ The parenthetical remarks in reason 5 of the proofs of Theorems 1 and 2 indicate that statements 2 and 4 have been combined to conclude statement 5. This use of parentheses is explained to students in a note in Section **3.2,** page 119. As students learn to apply their first theorems to solve problems, they sometimes unnecessarily repeat all the steps in proving a theorem. Instead, they should learn to use the theorem itself as the justification for a step in solving a problem or proving a theorem.

Theorem 1   *If two angles are right angles, then they are congruent.*

Given: $\angle A$ is a right $\angle$.
   $\angle B$ is a right $\angle$.
Prove: $\angle A \cong \angle B$

Proof:

Statements	Reasons
1 $\angle A$ is a right angle.	1 Given
2 $m\angle A = 90$	2 If an angle is a right angle, then its measure is 90.
3 $\angle B$ is a right angle.	3 Given
4 $m\angle B = 90$	4 Same as 2
5 $\angle A \cong \angle B$	5 If two angles have the same measure, then they are congruent. (See steps 2 and 4.)

Theorem 2   *If two angles are straight angles, then they are congruent.*

Given: $\angle ABC$ is a straight angle.
   $\angle DEF$ is a straight angle.
Prove: $\angle ABC \cong \angle DEF$

Proof:

Statements	Reasons
1 $\angle ABC$ is a straight angle.	1 Given
2 $m\angle ABC = 180$	2 If an angle is a straight angle, then its measure is 180.
3 $\angle DEF$ is a straight angle.	3 Given
4 $m\angle DEF = 180$	4 Same as 2
5 $\angle ABC \cong \angle DEF$	5 If two angles have the same measure, then they are congruent. (See steps 2 and 4.)

Now that we have presented and proved two theorems, we are ready to use them to help prove some sample problems.

We will use the theorems themselves as reasons in our proofs. You should also use the theorems as reasons in your homework problems.

Remember, the purpose of a theorem is to shorten your work. Therefore, when doing homework problems, do not use the proofs of theorems as a guide. Use the sample problems as a guide.

Part Two: Sample Problems

Problem 1   Given: ∠A is a right angle.
            ∠C is a right angle.
        Conclusion: ∠A ≅ ∠C

Proof

Statements	Reasons
1 ∠A is a right angle.	1 Given
2 ∠C is a right angle.	2 Given
3 ∠A ≅ ∠C	3 If two angles are right angles, then they are congruent.

You probably recognize that reason 3 is Theorem 1. Although it may seem easier merely to write "Theorem 1," *do not do so!* Eventually, such a shortcut would make it harder for you to learn the concepts of geometry.

Problem 2   Given: Diagram as shown
        Conclusion: ∠EFG ≅ ∠HFJ

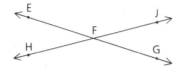

Proof

Statements	Reasons
1 Diagram as shown	1 Given
2 ∠EFG is a straight angle.	2 Assumed from diagram
3 ∠HFJ is a straight angle.	3 Assumed from diagram
4 ∠EFG ≅ ∠HFJ	4 If two angles are straight angles, then they are congruent.

Problem 3   Given: ∠RST = 50°,
            ∠TSV = 40°;
            ∠X is a right angle.
        Prove: ∠RSV ≅ ∠X

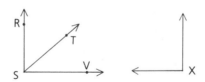

Proof

Statements	Reasons
1 ∠RST = 50°	1 Given
2 ∠TSV = 40°	2 Given
3 ∠RSV = 90°	3 Addition (50° + 40° = 90°)
4 ∠RSV is a right angle.	4 If an angle is a 90° angle, it is a right angle.
5 ∠X is a right angle.	5 Given
6 ∠RSV ≅ ∠X	6 If two angles are right angles, then they are congruent.

Lesson Notes, continued
- Remind students that sample problems should be studied carefully, since their solutions will serve as guides when students complete homework assignments.
- For Sample Problem **1**, point out that the statement labeled "Conclusion" is the statement to be proved.

Cooperative Learning

Give small groups of students a logic problem similar to the following one. Then have the groups try to write a two-column proof for each conclusion.

The Archimedean Club of Higher Mathematics is planning its annual banquet. The club gives awards to members of 5, 10, and 15 years. The club's officers, Mrs. Martinez, Mr. Washington, and Ms. Chung, are due to receive awards. From the following information, which award will each receive and which post does each hold?

1 Ms. Chung has been a member for 10 years.
2 The president has been a member for fewer years than Ms. Chung.
3 Mrs. Martinez has been a member for more years than the secretary.

Mrs. Martinez is the vice president and has been a member for 15 years; Mr. Washington is the president and has been a member for 5 years; Ms. Chung is the secretary and has been a member for 10 years.

Part Three: Problem Sets

Problem Set A

In problems 1 and 2, copy the figure and the incomplete proof. Then complete the proof by filling in the missing reasons.

1 Given: ∠1 is a right ∠.
   ∠2 is a right ∠.

   Prove: ∠1 ≅ ∠2

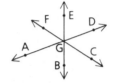

Statements	Reasons
1 ∠1 is a right angle.	1 _____
2 ∠2 is a right angle.	2 _____
3 ∠1 ≅ ∠2	3 _____

2 Given: Diagram as shown

   Prove: ∠AGD ≅ ∠EGB

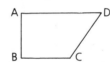

Statements	Reasons
1 Diagram as shown	1 _____
2 ∠AGD is a straight angle.	2 _____
3 ∠EGB is a straight angle.	3 _____
4 ∠AGD ≅ ∠EGB	4 _____

In problems 3–7, use the two-column form of proof.

3 Given: ∠A is a right angle.
   ∠B is a right angle.

   Prove: ∠A ≅ ∠B

4 Given: ∠CDE = 110°,
   ∠FGH = 110°

   Conclusion: ∠CDE ≅ ∠FGH

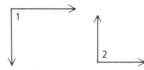

5 Given: JK = 2.5 cm, NO = 2.5 cm
   Conclusion: $\overline{JK} ≅ \overline{NO}$

6 Given: Diagram as shown
   Prove: ∠APR ≅ ∠SPB

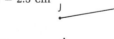

7 Given: $\angle 1 = 20°$,
　　　$\angle 2 = 40°$,
　　　$\angle 3 = 30°$

　Prove: $\angle XYZ$ is a right angle.

8 Draw the figure ABCDEF.

　a Draw its reflection over $\overleftrightarrow{AF}$.

　b Draw its reflection over $\overleftrightarrow{AB}$.

　c Draw a 90° clockwise rotation of the figure about B.

9 Find the angle formed by the hands of a clock at 11:40.　110°

10 The square has a perimeter of 42.

　a Solve for x.　7.5

　b If the perimeter were greater than 42, what would we know about the value of x?
　　$x > 7.5$

$x + 3$

Problem Set B

11 Given: $\angle ABD = 10°$,
　　　$\angle ABC = 100°$,
　　　$\angle EFY = 70°20'$,
　　　$\angle XFY = 19°40'$

　Prove: $\angle DBC \cong \angle XFE$

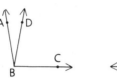

12 Point P has a coordinate of 7 on a number line. If you "slide" P 15 units in the negative direction, what are the coordinates of the resulting point P'?　-8

13 a Draw a number line, labeling points $A = (-1)$ and $B = (5)$. Then label point A', the reflection of A over B.

　b Does AB = BA'?　Yes

　c What do we know about point B?　It is the midpt. of $\overline{AA'}$.

Problem Set C

14 The measure of an obtuse angle is $5y + 45$. What are the restrictions on y?　$9 < y < 27$

15 Given: $\angle 1 = (x + 7)°$,
　　　$\angle 2 = (2x - 3)°$,
　　　$\angle ABC = (x^2)°$,
　　　$\angle D = (5x - 4)°$

　Show that $\angle ABC \cong \angle D$.

Problem-Set Notes and Additional Answers, continued

■ See *Solution Manual* for answers to problems **7** and **11**.

8a

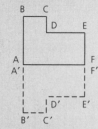

b

c

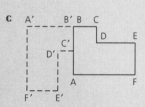

■ Notice that problem **12** introduces the idea of a slide. Remind students that a slide translates a given number of units in a given direction. Transformations are covered formally in Chapter **13**.

13a

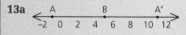

15　$x^2 = x + 7 + 2x - 3$
　　　$x^2 - 3x - 4 = 0$
　　　$(x - 4)(x + 1) = 0$
　　　$x = 4$ or $x = -1$
　　　$m\angle D = 5(4) - 4 = 16$
　　　$m\angle ABC = x^2 = 16$
　　　$\therefore \angle ABC \cong \angle D$

T 27

1.5 DIVISION OF SEGMENTS AND ANGLES

Objectives

After studying this section, you will be able to
■ Identify midpoints and bisectors of segments
■ Identify trisection points and trisectors of segments
■ Identify angle bisectors
■ Identify angle trisectors

Part One: Introduction

Midpoints and Bisectors of Segments

We shall often work with segments that are divided in half.

Definition   A point (or segment, ray, or line) that divides a segment into two congruent segments *bisects* the segment. The bisection point is called the *midpoint* of the segment.

X is not a midpoint   Y is not a midpoint

Only segments have midpoints. It does not make sense to say that a ray or a line has a midpoint. Do you understand why?

How many midpoints does $\overline{PQ}$ have?
How many bisectors could $\overline{PQ}$ have?

Study the following examples.

Example 1   If $\overline{XY}$ bisects $\overline{AC}$ at B, *what conclusions can we draw?*
Conclusions:
   B is the midpoint of $\overline{AC}$.
   $\overline{AB} \cong \overline{BC}$

Example 2   If D is the midpoint of $\overline{FE}$, *what conclusions can we draw?*
Conclusions:
   $\overline{FD} \cong \overline{DE}$
   Point D bisects $\overline{FE}$.
   $\overrightarrow{DG}$ bisects $\overline{FE}$.

Example 3 *If $\overline{OK} \cong \overline{KP}$, what conclusions can we draw?*

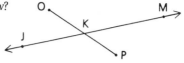

Conclusions:
  K is the midpoint of $\overline{OP}$.
  $\overleftrightarrow{JM}$ is a bisector of $\overline{OP}$.
  Point K bisects $\overline{OP}$.

Trisection Points and Trisecting a Segment

A segment divided into *three* congruent parts is said to be **trisected**.

Definition Two points (or segments, rays, or lines) that divide a segment into three congruent segments **trisect** the segment. The two points at which the segment is divided are called the **trisection points** of the segment.

  Again, only segments have trisection points; rays and lines do not have trisection points.

Example 1 *If $\overline{AR} \cong \overline{RS} \cong \overline{SC}$, what conclusions can we draw?*

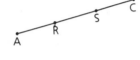

Conclusions:
  R and S are trisection points of $\overline{AC}$.
  $\overline{AC}$ is trisected by R and S.

Example 2 *If E and F are trisection points of $\overline{DG}$, what conclusions can we draw?*

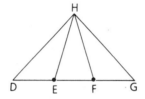

Conclusions:
  $\overline{DE} \cong \overline{EF} \cong \overline{FG}$
  $\overline{HE}$ and $\overline{HF}$ are trisectors of $\overline{DG}$.

Angle Bisectors

An angle, like a segment, can be bisected.

Definition A ray that divides an angle into two congruent angles **bisects** the angle. The dividing ray is called the **bisector** of the angle.

If $\angle ABD \cong \angle DBC$, then $\overrightarrow{BD}$ (not $\overrightarrow{DB}$) is the bisector of $\angle ABC$.

If $\angle NOP \cong \angle POR$ and $\overrightarrow{OQ}$ bisects $\angle POR$, then $\overrightarrow{OP}$ (not $\overrightarrow{PO}$) is the bisector of $\angle NOR$, and $\angle 1 \cong \angle 2$.

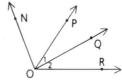

Section 1.5   Division of Segments and Angles   **29**

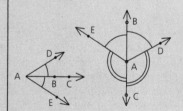

Communicating Mathematics

Ask students to give two different "If . . . , then . . ." statements for the definitions of *segment trisectors*, *angle bisectors*, and *angle trisectors*.

Angle Trisectors

Two rays can divide an angle into three equal parts.

Definition   Two rays that divide an angle into three congruent angles ***trisect*** the angle. The two dividing rays are called ***trisectors*** of the angle.

If ∠ABC ≅ ∠CBD ≅ ∠DBE, then $\overrightarrow{BC}$ and $\overrightarrow{BD}$ trisect ∠ABE.

If $\overrightarrow{SV}$ and $\overrightarrow{SX}$ are trisectors of ∠TSY, then ∠TSV ≅ ∠VSX ≅ ∠XSY.

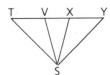

Part Two: Sample Problems

Problem 1   *The tick marks indicate that* $\overline{RS} ≅ \overline{ST}$. *Is S the midpoint of* $\overline{RT}$?

Answer   No, the points are not collinear.

Problem 2   *If* $\overrightarrow{BD}$ *bisects* ∠ABC, *does* $\overrightarrow{DB}$ *bisect* ∠ADC?

Answer   No. We need more information.

Problem 3   *If B and C trisect* $\overline{AD}$, *do* $\overrightarrow{EB}$ *and* $\overrightarrow{EC}$ *trisect* ∠AED?

Answer   No! It is true that $\overline{AB} ≅ \overline{BC} ≅ \overline{CD}$, but the fact that the *segment* has been trisected does not mean that the *angle* has been trisected.

Problem 4   *Given:* $\overrightarrow{PS}$ *bisects* ∠RPO.
Prove: ∠RPS ≅ ∠OPS

Proof

Statements	Reasons
1 $\overrightarrow{PS}$ bisects ∠RPO.	1 Given
2 ∠RPS ≅ ∠OPS	2 If a ray bisects an angle, it divides the angle into two congruent angles.

Lesson Notes, continued

- Students should realize that much of geometry coincides with their commonsense interpretations in the physical world. Sample Problem **3**, however, is an example of a case where common sense would lead students astray. This is a good reason for the important role that proof plays in learning geometry.
- Be sure students understand that the "If . . . , then . . ." clauses in reason 2 of Sample Problems **4–6** must be in the correct order; the "then" clause must agree with statement 2.

30   Chapter 1   Introduction to Geometry

Problem 5 Given: $\overleftrightarrow{CM}$ bisects $\overline{AB}$ (In Chapter 3 we shall call $\overline{CM}$ a median of the triangle.)

Conclusion: $\overline{AM} \cong \overline{MB}$

Proof

Statements	Reasons
1 $\overleftrightarrow{CM}$ bisects $\overline{AB}$.	1 Given
2 $\overline{AM} \cong \overline{MB}$	2 If a line bisects a segment, it divides the segment into two congruent segments.

Problem 6 Given: $\overline{DH} \cong \overline{HF}$

Prove: H is the midpoint of $\overline{DF}$.

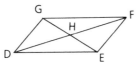

Proof

Statements	Reasons
1 $\overline{DH} \cong \overline{HF}$	1 Given
2 H is the midpoint of $\overline{DF}$.	2 If a point divides a segment into two congruent segments, it is the midpoint of the segment.

Problem 7 $\overline{EH}$ is divided by F and G in the ratio $5:3:2$ from left to right. If EH = 30, find FG and name the midpoint of $\overline{EH}$.

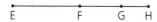

Solution According to the ratio, we can let EF = 5x, FG = 3x, and GH = 2x. First we draw a diagram and place the algebra on it as part of the solution.

$$5x + 3x + 2x = 30$$
$$10x = 30$$
$$x = 3$$

Thus, FG = 3(3), or 9. Since EF = 15 and FH = 15, F is the midpoint of $\overline{EH}$.

Problem 8 Given: $\overrightarrow{KO}$ bisects $\angle JKM$.

$\angle JKM = 41°37'$

Find: m$\angle OKM$

Solution $\frac{1}{2}(41°37') = 20\frac{1}{2}°18\frac{1}{2}'$

$= 20°48\frac{1}{2}'$ (since $\frac{1}{2}° = 30'$)

$= 20°48'30''$ (since $\frac{1}{2}' = 30''$)

Lesson Notes, continued

- The solution to Sample Problem **7** illustrates a standard procedure for ratio problems that is followed throughout the text. Problem **18** of the problem set requires the use of this procedure.

The Geometric Supposer

Have students use *The Geometric preSupposer: Points and Lines* or *The Geometric Supposer: Circles* to explore the concepts of the measurement of segments and angles, of collinearity, and of segment and angle division. Students should construct angles by drawing an initial segment, labeling a movable point (see Geometric Supposer terminology on page Tix) not on that segment, and drawing a segment connecting that point to an endpoint of the segment.

The program allows students to label a segment's midpoint or trisection points (under Subdivide Segment) and to draw angle bisectors. Students can draw a segment bisector by first labeling the midpoint, then drawing a line through an external point and the midpoint.

The following is a game that you can play with students. Have students label two points outside a segment on a line that they think is the segment's bisector. The score can be a function either of how close the students come to actually bisecting the segment or of how many tries it takes to bisect the segment. You can play as a class or students can compete against each other. You could also do this as a cooperative activity by dividing the class into teams.

Part Three: Problem Sets

Problem Set A

1 Name the congruent segments.

a O is the midpoint of $\overline{CD}$. $\overline{CO} \cong \overline{DO}$

b $\overline{SW}$ bisects $\overline{XV}$. $\overline{WX} \cong \overline{WV}$

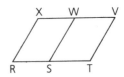

2 Name the congruent angles.

a $\overrightarrow{RO}$ bisects $\angle NRP$. $\angle NRO \cong \angle PRO$

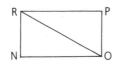

b $\overrightarrow{XT}$ and $\overrightarrow{XV}$ trisect $\angle SXW$. $\angle SXT \cong$ $\angle TXV \cong$ $\angle WXV$

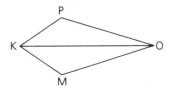

3 Name the angle bisector.

a $\overrightarrow{JG}$

b $m\angle POK = m\angle MOK$ $\overrightarrow{OK}$

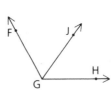

4 Find $\angle XTZ$ if $\overrightarrow{TZ}$ bisects $\angle XTY$ and $\angle XTY$ equals

a 60° 30°

b 48°50′ 24°25′

c $36\frac{1}{2}°$ $18\frac{1}{4}°$

d 85°74′ 43°7′

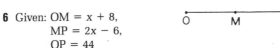

5 B and C trisect $\overline{AD}$.

a Find the coordinates of B and C. 2; 9
b Find AC. 14

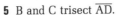

6 Given: OM = x + 8,
   MP = 2x − 6,
   OP = 44

Is M the midpoint of $\overline{OP}$? Yes

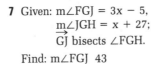

7 Given: $m\angle FGJ = 3x − 5$,
   $m\angle JGH = x + 27$;
   $\overrightarrow{GJ}$ bisects $\angle FGH$.

Find: $m\angle FGJ$ 43

8 B and C are trisection points of $\overline{AD}$, and AD = 12.

a Find AB. 4

b Find AC. 8

c If AB = x + 3, solve for x. 1

d If AB = x + 3 and AE = 3x + 6, find AE. 9

e What segment is C the midpoint of? $\overline{BD}$

f Do $\overrightarrow{EB}$ and $\overrightarrow{EC}$ trisect ∠AED? No

9 Given: ∠ABC = 90°,
∠1 = (2x + 10)°,
∠2 = (x + 20)°,
∠3 = (3x)°

Has ∠ABC been trisected? Yes

In problems 10 and 11, reason 2 in each proof is stated incorrectly. Supply the correct final reason for each problem.

10 Given: ∠DEG ≅ ∠FEG
Prove: $\overrightarrow{EG}$ bisects ∠DEF.

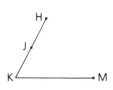

Statements	Reasons
1 ∠DEG ≅ ∠FEG	1 Given
2 $\overrightarrow{EG}$ bisects ∠DEF.	2 If a ray divides an angle into two angles, the ray bisects the angle. (*What is the correct reason?*)

11 Given: $\overline{KJ} \cong \overline{HJ}$
Prove: J is the midpoint of $\overline{HK}$.

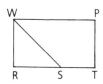

Statements	Reasons
1 $\overline{KJ} \cong \overline{HJ}$	1 Given
2 J is the midpoint of $\overline{HK}$.	2 If a point is the midpoint of a segment, it divides the segment into two congruent segments. (*What is the correct reason?*)

In problems 12–17, write a proof in two-column form.

12 Given: $\overrightarrow{WS}$ bisects ∠RWP.
Prove: ∠RWS ≅ ∠PWS

Section 1.5   Division of Segments and Angles   **33**

Problem-Set Notes and Additional Answers

10 If a ray divides an ∠ into 2 ≅ ∠s, the ray bisects the ∠.

11 If a point divides a segment into 2 ≅ segments, the point is the midpoint of the segment.

■ See *Solution Manual* for answer to problem **12**.

*Problem-Set Notes and
Additional Answers, continued*

■ See *Solution Manual* for answers to problems **13–17.**

Problem Set A, *continued*

13 Given: $\overline{XY} \cong \overline{YZ}$
Prove: Y is the midpoint of $\overline{XZ}$.

14 Given: $\angle AEB \cong \angle BEC \cong \angle CED$
Conclusion: $\overrightarrow{EB}$ and $\overrightarrow{EC}$ trisect $\angle AED$.

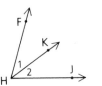

15 Given: $\angle 1 \cong \angle 2$
Conclusion: $\overrightarrow{HK}$ bisects $\angle FHJ$.

16 Given: $\angle TXW$ is a right angle.
   $\angle TYV$ is a right angle.

Prove: $\angle TXW \cong \angle TYV$

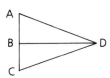

17 Given: B is the midpoint of $\overline{AC}$.
Prove: $\overline{AB} \cong \overline{BC}$

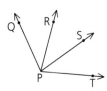

Problem Set B

18 $\overrightarrow{OG}$ and $\overrightarrow{OH}$ divide straight angle FOJ
into three angles whose measures are in
the ratio 4:3:2. Find m$\angle$FOG. 80

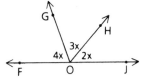

■ In problem **18,** students may need a reminder to refer to Sample Problem **7,** page 31, for the "standard procedure" for ratio problems.

19 Given: $\overleftrightarrow{TP}$ bisects $\overline{VS}$ and $\overline{MR}$.
   $\overline{VM} \cong \overline{SR}$,
   MP = 9, VT = 6,
   perimeter of MRSV = 62

Find: VM 16

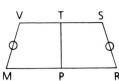

20 $\overrightarrow{PR}$ and $\overrightarrow{PS}$ trisect $\angle QPT$.

a If m$\angle RPS = 23°50'$,
   find m$\angle QPT$. 71°30'

b If m$\angle QPT = 120°48'30''$,
   find m$\angle QPS$. 80°32'20''

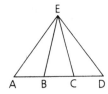

21 a Find the value of x. 2

   b Is Q the midpoint of $\overline{PR}$? No

Problem Set C

22 Given: $\overrightarrow{OP}$ and $\overrightarrow{OR}$ trisect $\angle NOS$.

        $m\angle NOP = 3x - 4y$,

        $m\angle POR = x - y$,

        $m\angle ROS = y - 10$

   Find: $m\angle ROS$ 10

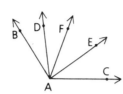

23 $\angle BAC = 120°$, and points D, E, and F are in the interior of $\angle BAC$ as shown. $\overrightarrow{AD}$ bisects $\angle BAF$. $\overrightarrow{AE}$ bisects $\angle CAF$. Find $m\angle DAE$. 60

24 The measures of two angles are in the ratio 5:3. The measure of the larger angle is 30 greater than half the difference of the angles. Find the measure of each angle. 25 and 15 or 37.5 and 22.5

Problem-Set Notes and Additional Answers, continued

22 $\begin{cases} 3x - 4y = x - y \\ x - y = y - 10 \end{cases}$

      $x = 30; y = 20$

23

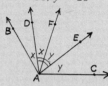

   $x + x + y + y = 120$

   $2x + 2y = 120$

   $x + y = 60$

24 Let the angles measure 5x and 3x.

   Case 1:

   $5x = \frac{1}{2}(5x - 3x) + 30$

   $5x = x + 30$

   $x = 7.5$

   $\therefore m\angle s = 37.5$ and 22.5

   Case 2:

   $5x = \frac{1}{2}(3x - 5x) + 30$

   $5x = -x + 30$

   $x = 5$

   $\therefore m\angle s = 25$ and 15

PARAGRAPH PROOFS

Class Planning

Time Schedule
All levels: 1 day

Resource References
Teacher's Resource Book
   Class Opener 1.6A, 1.6B

Class Opener

If $\overrightarrow{AM}$ bisects $\angle FAD$, find the measure of $\angle RQS$.

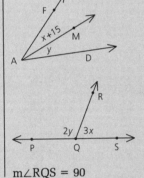

$m\angle RQS = 90$

Lesson Notes

- Paragraph proofs are introduced in this section. While paragraph proofs are used sparingly in this text, especially in the first half, most proofs in later mathematics courses use that form.

Objective

After studying this section, you will be able to
- Write paragraph proofs

Part One: Introduction

Although most of the proofs you will encounter this year will be in two-column form, you also need to be familiar with **paragraph proofs**. They are important because the proofs in journals, more-advanced mathematics courses, and other areas of study are usually in paragraph form.
    The sample problems that follow demonstrate how to write paragraph proofs, as well as how to show that a particular conclusion cannot be proved true or can be proved false.

Part Two: Sample Problems

Problem 1   Given: $\angle O = 67\frac{1}{2}°$,
                           $\angle P = 67°30'$
                  Prove: $\angle O \cong \angle P$

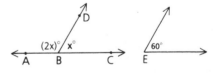

Proof   Since there are 60 minutes in 1 degree, $67°30'$ equals $67\frac{1}{2}°$.
                 Since $\angle O$ and $\angle P$ have the same measure, they are congruent.

Problem 2   *Given: Diagram shown*
                  Prove: $\angle DBC \cong \angle E$

Proof   According to the diagram, $\angle ABC$ is a straight angle. Therefore,

$$2x + x = 180$$
$$3x = 180$$
$$x = 60$$

Since $\angle DBC = 60°$ and $\angle E = 60°$, the angles are congruent.

36   Chapter 1   Introduction to Geometry

Vocabulary
counterexample
paragraph proof

Problem 3   Given: ∠1 *is acute.*
                 ∠2 *is acute.*
                 Conclusion: ∠1 ≅ ∠2

Proof   This conclusion *cannot be proved.* For example, if m∠1 = 20 and
                 m∠2 = 30, they are both acute but ∠1 is not congruent to ∠2. (An
                 example, like this, of a case in which a conclusion is false is called a
                 counterexample.)

Problem 4   Given: ∠D = 90°;
                 ∠E *is obtuse.*
                 Prove: ∠D ≅ ∠E

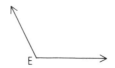

Proof   This conclusion can be proved to be *false.* Since ∠E is obtuse, its
                 measure is greater than 90. Since ∠D and ∠E have different mea-
                 sures, they are not congruent (∠D ≇ ∠E).

Part Three: Problem Sets

Problem Set A

In problems 1–6, write paragraph proofs.

1 Given: $\angle V = 119\frac{2}{3}°$,
       $\angle S = 119°40'$
       Conclusion: ∠V ≅ ∠S

2 Given: Diagram shown
       Prove: ∠FEH ≅ ∠JKM

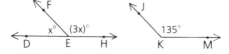

3 Given: Diagram shown, ∠OPT = 90°
       Prove: The measure of ∠VAY is twice that
       of ∠RPT.

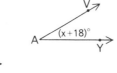

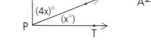

4 Given: AB = x + 4,
       BC = 2x,
       AC = 16
       Conclusion: $\overline{AB} \cong \overline{BC}$

Lesson Notes, continued

■ You might want to stress that a
  good paragraph proof, like a
  good English paper, should
  have an introduction, a body,
  and a conclusion. The introduction should include the
  given conditions and any conclusions that can be immediately drawn from the given
  conditions. In the body, students should solve any equations and complete steps that
  lead to the conclusion. The
  conclusion should summarize
  the paragraph proof.
  Throughout the text, there are
  conclusions, like the one in
  Sample Problem **3**, that can
  be proved false by counterexample, and conclusions, like
  the one in Sample Problem **4**,
  that can be proved to be gener-
  ally false.

Cooperative Learning

Refer to the logic problem in the
Cooperative Learning activity in
Section **1.4** (or a similar problem). Have students work in
small groups to write a paragraph proof leading to the conclusion of the problem.

Assignment Guide

Basic

1–5

Average

1–4, 7–9

Advanced

3, 6–11

Problem-Set Notes
and Additional Answers

■ See *Solution Manual* for answers to problems **1–4**.

Problem-Set Notes and Additional Answers, continued

- See Solution Manual for answers to problems **7–9**.

10

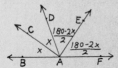

From the given, the diagram can be labeled as shown above. Then $m\angle CAE = x + \frac{180 - 2x}{2} = 90$. Thus, $\angle CAE$ is a right angle.

11a $m\angle J + m\angle H + m\angle JKH = 180$
Then $m\angle J + m\angle H = 180 - m\angle JKH$; but $\angle 1 = 180 - m\angle JKH$. Thus, $m\angle J + m\angle H = m\angle 1$.

b From part **a**, we know that $m\angle 1 = m\angle J + m\angle H$. Since each measure is positive, $m\angle 1 > m\angle J$.

12

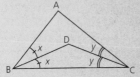

From the given,
$\begin{cases} m\angle A + 2x + 2y = 180 \\ m\angle D + x + y = 180 \end{cases}$
$\therefore m\angle D + x + y = m\angle A + 2x + 2y$
$m\angle D = m\angle A + x + y$
But $x + y = 180 - m\angle D$.
Then $m\angle D = m\angle A + 180 - m\angle D$.
$2m\angle D = 180 + m\angle A$
$\therefore m\angle D = 90 + \frac{1}{2} m\angle A$

Communicating Mathematics

Ask several students to read to the entire class their paragraph proofs for various problems. Ask for comments on the proofs.

Problem Set A, *continued*

5 Given: $\angle D$ is obtuse.
$\angle C$ is greater than 90° ($\angle C > 90°$).
Conclusion: $\angle D \cong \angle C$ Cannot be proved

6 Given: $\angle 1$ is obtuse.
$\angle 2$ is acute.
Prove: $\angle 1 \cong \angle 2$ Can be proved false

Problem Set B

In problems 7–9, write paragraph proofs.

7 Prove that if $\angle 1 \cong \angle 2$, they are both right angles.

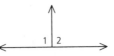

8 Prove the following statement: "If an obtuse angle is bisected, each of the two resulting angles is acute."

9 Given: $\overrightarrow{CE}$ bisects $\angle BCD$.
$\angle A$ is a right angle.
$m\angle BCE = 45$
Prove: $\angle A \cong \angle BCD$

Problem Set C

In problems 10 and 11, write paragraph proofs.

10 Given: Diagram shown;
$\overrightarrow{AC}$ bisects $\angle BAD$.
$\overrightarrow{AE}$ bisects $\angle DAF$.
Prove: $\angle CAE$ is a right angle.

11 Given: $m\angle J + m\angle H + m\angle JKH = 180$
Prove: **a** $m\angle 1 = m\angle J + m\angle H$
b $m\angle 1 > m\angle J$

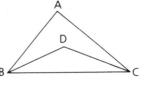

Problem Set D

12 Given: $m\angle A + m\angle ABC + m\angle ACB = 180$,
$m\angle D + m\angle DBC + m\angle DCB = 180$;
$\overrightarrow{BD}$ bisects $\angle ABC$. $\overrightarrow{CD}$ bisects $\angle ACB$.
Prove: $m\angle D = 90 + \frac{1}{2}(m\angle A)$
(Write a paragraph proof.)

1.7 DEDUCTIVE STRUCTURE

Objectives

After studying this section, you will be able to
- Recognize that geometry is based on a deductive structure
- Identify undefined terms, postulates, and definitions
- Understand the characteristics of theorems and the ways in which they can be used in proofs

Part One: Introduction

The Structure of Geometry

You have just spent a few days writing two-column proofs and paragraph proofs. Since you have learned how to prove a few statements, you may be interested in knowing something about the theory of proofs.

Geometry is based on a ***deductive structure***—a system of thought in which conclusions are justified by means of previously assumed or proved statements. Every deductive structure contains the following four elements.
- Undefined terms
- Assumptions known as ***postulates***
- Definitions
- Theorems and other conclusions

Undefined Terms, Postulates, and Definitions

Undefined terms, postulates, and definitions form the foundation on which the rest of a deductive structure is based. Examples of the undefined terms you have already encountered are *point* and *line*. Although we have not defined these terms, we have described points and lines, so that everyone should have a fairly clear idea of what they are.

As yet, we have not formally presented any postulates. We have, however, used some algebraic postulates in solving problems.

Definition   A ***postulate*** is an unproved assumption.

The postulates presented in this book will be preceded by the heading **Postulate**.

Class Planning

Time Schedule
Basic: 1 day
Average: 1 day
Advanced: $\frac{1}{2}$ day

Resource References
Teacher's Resource Book
  Class Opener 1.7A, 1.7B

Class Opener

Discuss the validity of the proof given below.

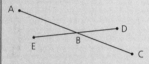

Given: $\overline{AB} \cong \overline{BC}$,
       $\overline{EB} \cong \overline{BD}$
Prove: B is the midpoint of ED.

Statements	Reasons
1 $\overline{EB} \cong \overline{BD}$	1 Given
2 B is the midpoint of ED.	2 A midpoint divides a segment into two ≅ segments.

The proof fails because of the use of the converse of the required statement in reason two. The reason should be, If a point divides a segment into two ≅ segments, then it is a midpoint.

Lesson Notes

- Advanced courses can cover Sections **1.7** and **1.8** in one day.
- This section is intended as an introduction to deductive structure; mastery is not expected at this time. In Chapters **2** and **3**, as students learn how to construct simple proofs, the elements of the deductive structure of geometry will take on concrete meaning.

You have already seen a number of definitions, such as the definitions of *acute angle*, *right angle*, *obtuse angle*, and *straight angle*.

Definition   A **definition** states the meaning of a term or idea.

In this book, important definitions are identified by the heading **Definition**

One very important characteristic of definitions is that they are *reversible*. For example, the definition of *midpoint* (of a segment) can be expressed in either of two ways:

1 If a point is the midpoint of a segment, then the point divides the segment into two congruent segments.
2 If a point divides a segment into two congruent segments, then the point is the midpoint of the segment.

In some problems, form 1 of the definition of *midpoint* must be used. In other problems the definition must be *reversed*, as in form 2 above.

Notice that this definition is stated in the form

$$\text{"If } p, \text{ then } q\text{"}$$

where p and q are declarative statements. Such a sentence is called a **conditional statement** or an **implication.** The "if" part of the sentence is called the **hypothesis**. The "then" part of the sentence is called the **conclusion**. "If p, then q" can be symbolized $p \Rightarrow q$ (also read "p implies q").

The **converse** of $p \Rightarrow q$ is $q \Rightarrow p$. To write the converse of a conditional ("If . . . , then . . . ") statement, you reverse parts p and q. The converse of "If p, then q" is "If q, then p," so forms 1 and 2 of the definition of *midpoint* are converses of each other.

Theorems

As you have seen, a theorem is a mathematical statement that can be proved. Almost all the theorems presented in this book will be numbered for ease of reference. Each theorem will be preceded by a heading such as the following:

Theorem 78

You will prove some theorems and other relationships as homework problems. As you work, remember that you must prove conclusions by using conclusions previously assumed or proved. Thus, you cannot use Theorem 1 in order to prove Theorem 1.

Theorems and postulates are not always reversible. For example, "If two angles are right angles, then they are congruent" is true. The converse statement, "If two angles are congruent, then they are right angles," is false.

Remember,
- *Definitions* are always reversible
- *Theorems* and *postulates* are not always reversible

If you are to be successful in writing proofs, you must memorize postulates, definitions, and theorems. There is no easier way.

A complete mastery of the deductive structure of geometry is not possible in a short time. However, we do wish to point out the most common error that students make—using the converse of a statement at the wrong time.

It is important to pay attention to the direction of the flow of logic in order to avoid this error. The theorem "If two angles are right angles, then they are congruent" means that whenever we encounter right angles, we can conclude that they are congruent. There is a flow from *right angles* to *congruent*.

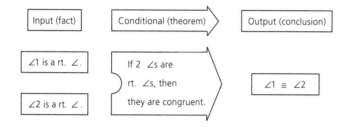

In this case, the flow works in only one direction—the converse of the statement, "If two angles are congruent, then they are right angles" is not true. *Remember, only definitions are always reversible. Theorems and postulates are not always reversible.*

The major purpose of this section and the next is to acquaint you with some terminology. As you study Chapter 2 and Chapter 3, you will grow to appreciate and understand these sections even more. The homework problems in these sections are rather different from those you have been solving, and we think you will enjoy them.

Part Two: Sample Problem

Problem   *State the converse of each of the following statements and tell whether the converse is true or false.*

a *If an angle contains 90 degrees, it is a right angle.*
b *If Mary received a B on her history test, then she passed the test.*

Answers   **a** *If an angle is a right angle, it contains 90 degrees. (True)*
b *If Mary passed her history test, she received a B on the test. (False)*

Lesson Notes, *continued*

- Reversibility is an important feature of a definition. For example, the sentence "If a quadrilateral is a square, then all four sides are equal" is true; but its converse, "If all four sides of a quadrilateral are equal, then the figure is a square" is not true. Thus, the first sentence could not be a definition of a square.

Communicating Mathematics

Have students express the definitions of *acute, right, obtuse,* and *straight angles,* given in Section **1.2,** as two conditional statements that are converses of each other.

Cooperative Learning

Have students work in small groups and write sets of conditional sentences for which both the statement and its converse are true and sets of sentences for which the statement is true but the converse is not.

Answers will vary. Possible answers: If a student received a B on the test, then his or her score was between 80 and 90 (both statement and converse are true); if an animal is a dog, then it has four legs (statement is true; converse is not).

Assignment Guide

Basic

1–5, 8–12, 14

Average

1–5, 8–12, 14

Advanced

($\frac{1}{2}$ day) Section **1.7** 5, 14

($\frac{1}{2}$ day) Section **1.8** 3–5, 7–10

Problem-Set Notes and Additional Answers

5a **i** If B, then A.
ii Wet ⇒ rain
iii If an angle is acute, then it is a 45° angle.
iv If a point divides a segment into two congruent segments, it is the midpoint of the segment.

b **i** May be either true or false
ii False
iii False
iv True

6 Possible

7 Not true if it is a fair coin; past outcomes do not influence current outcomes.

8 Correct

9 Not correct, since we do not know that ∠C is acute

10 Not correct, since we were reasoning from the converse

Part Three: Problem Sets

Problem Set A

1 What four elements are found in any deductive structure?
Undefined terms, postulates, definitions, theorems

2 Which of the following kinds of statements are always reversible? a

a Definitions   **b** Theorems   **c** Postulates

3 Answer each question Yes or No.

a Do we prove theorems? Yes   **b** Do we prove definitions? No

4 Tell whether each of the following statements is a theorem or a definition.

a If two angles are right angles, then they are congruent. Theorem

b If a ray bisects an angle, it divides the angle into two congruent angles. Definition

5 **a** Write the converse of each of the following statements.

   i If A, then B.
   ii Rain ⇒ wet
   iii If an angle is a 45° angle, then it is acute.
   iv If a point is the midpoint of a segment, it divides the segment into two congruent segments.

b Discuss the truth of each of the converses in part **a**.

In problems 6 and 7, comment on the reasoning used.

6 The school colors are orange and black, so I'll wear my orange skirt to the game and everyone will notice me.

7 I've flipped this silver dollar five times and the toss has come up heads each time. Thus, the odds are greater than 50–50 that the toss will come up tails next time.

Problem Set B

In problems 8–12, study each of the arguments and state whether or not the conclusion is deducible. If it is not, comment on the error in the reasoning.

8 If a student at Niles High has room 303 as his or her homeroom, the student is a freshman. Joe Jacobs is a student at Niles High and has room 303 as his homeroom. Therefore, Joe Jacobs is a freshman.

9 If the three angles of a triangle are acute, then the triangle is acute. In triangle ABC, angle A and angle B are acute. Therefore, triangle ABC is acute.

10 All school buses stop at railroad crossings. A vehicle stopped at the Santa Fe railroad crossing. Therefore, that vehicle is a school bus.

11 All cloudy days are depressing. Therefore, since I was depressed on Thursday, Thursday was cloudy.

12 If two angles of a triangle are congruent, then the sides opposite them are congruent. In △ABC, ∠A ≅ ∠B. Therefore, in △ABC, $\overline{BC} \cong \overline{AC}$.

Problem Set C

13 Study the following five statements.

    1 Spoof is the set of all purrs.
    2 Spoof contains at least two distinct purrs.
    3 Every lilt is a set of purrs and contains at least two distinct purrs.
    4 If *A* and *B* are any two distinct purrs, there is one and only one lilt that contains them.
    5 No lilt contains all the purrs.

  a Show that each of the following statements is true.
    i There is at least one lilt.
   ii There are at least three purrs.
  iii There are at least three lilts.

  b If the lilt "girt" contains the purr "pil" and the purr "til" and if the lilt "mirt" contains the purr "pil" and the purr "til" then the lilt "girt" is the same as the lilt "mirt" except in one case. What is this case?

14 The Bronx Zoo has a green lizard, a red crocodile, and a purple monkey. They are the only animals of their kind in existence. One violently windy Saturday, their name tags blew off, and their keeper's journal was torn to shreds. Inasmuch as they were to appear on television at 7:30 Sunday morning, the night watchman had to replace their name tags. He managed to piece together the following information from the mangled journal.

    1 Wendy cannot get along with the lizard.
    2 Katie playfully took a bite out of the monkey's ear one month ago.
    3 Wendy never casts a red reflection in the mirror.
    4 Jody has the personality of a crocodile, but she isn't one.

Match the animals with their names.   Wendy—purple monkey; Katie—red crocodile; Jody—green lizard

Problem-Set Notes and Additional Answers, continued

11 Not correct, since we were reasoning from the converse.
12 Correct

■ See *Solution Manual* for answer to problem **13a**.

13b If the two purrs "pil" and "til" are the same (not distinct), then more than one lilt could contain them.

14

	Wendy	Katie	Jody
green lizard	O	O	X
red crocodile	O	X	O
purple monkey	X	O	O

STATEMENTS OF LOGIC

Class Planning

Time Schedule
Basic: 1 day
Average: 1 day
Advanced: $\frac{1}{2}$ day

Resource References
Teacher's Resource Book
  Class Opener 1.8A, 1.8B
Transparency 3

Class Opener

Given: $\overrightarrow{BD}$ bisects $\angle ABC$.

Prove: $\angle ABC$ is a right angle.
Since x is any real number that makes sense in the expression $-2 < x \le 58$, $\angle ABC$ can be a right angle, but this conclusion is not warranted.

Lesson Notes

- Point out that p and q are variables for statements or clauses. You might also point out that the inverse can be thought of as the negation of the original conditional statement and that the contrapositive is the converse of the inverse.

Objective

After studying this section, you will be able to
- Recognize conditional statements
- Recognize the negation of a statement
- Recognize the converse, the inverse, and the contrapositive of a statement
- Use the chain rule to draw conclusions

Part One: Introduction

Review of Conditional Statements

In this section, we will review and extend the discussion of conditional statements in Section 1.7. Recall that a conditional statement is a sentence that is in the form "If . . ., then . . ." Many declarative sentences can be rewritten in conditional form.

Declarative Sentence:
- Two straight angles are congruent.

Conditional Form:
- If two angles are straight angles, then they are congruent.

Remember that
- The clause following the word *if* is called the hypothesis
- The clause following the word *then* is called the conclusion
- The conditional statement "If *p*, then *q*" can be written in symbols as $p \Rightarrow q$

Negation

The ***negation*** of any statement *p* is the statement "not *p*." (Thus, the negation of "It is raining" is "It is not raining.") The symbol for "not *p*" is ~*p*. Notice also that the negation of "It is not raining" is "It is raining"—in general, not (not *p*) = *p*, or ~~*p* = *p*.

Converse, Inverse, and Contrapositive

Every conditional statement "If *p*, then *q*" has three other statements associated with it. (You have already been introduced to the first of these—the converse).

1 A converse (If *q*, then *p*.)
2 An ***inverse*** (If ~*p*, then ~*q*.)
3 A ***contrapositive*** (If ~*q*, then ~*p*.)

Vocabulary

chain of reasoning
chain rule
contrapositive

inverse
negation
Venn diagram

Example   *Find the converse, the inverse, and the contrapositive of the statement
"If you live in Atlanta, then you live in Georgia."*

The statement is in the form "If *p*, then *q*," with *p* being "You live
in Atlanta" and *q* being "You live in Georgia."

Converse: "If you live in Georgia, then you live in Atlanta."
     (If *q*, then *p*.)

Inverse: "If you don't live in Atlanta, then you don't live in Georgia."
     (If ~*p*, then ~*q*.)

Contrapositive: "If you don't live in Georgia, then you don't live in
     Atlanta." (If ~*q*, then ~*p*.)

You may have noticed that some of the statements in the preceding
example are not necessarily true, although the original statement is
true. A useful tool for determining whether or not a conditional
statement is true or false is a **Venn diagram**. Assume that the
following statement is true: "If Jenny lives in Atlanta, then Jenny
must live in Georgia."

   All the people who live in Georgia are
represented by points on the large circle and
in its interior (G).
   All the people who live in Atlanta are
represented by points on the small circle
and in its interior (A).
   Notice that every person in set A, including Jenny (J), is also in set G.

   The Venn diagram for this conditional statement may be used to
test whether its converse, inverse, and contrapositive are true or
false.

Converse: "If Jenny lives in Georgia, then she must live in Atlanta."

   This statement is not necessarily true, as
shown by the diagram. Notice that point J
may lie in G but not in A. This means that
Jenny could live in Georgia and yet not live
in Atlanta.

   In general, the converse of a conditional statement is not necessarily true. Try a similar argument with the same Venn diagram to
convince yourself that the inverse of a conditional statement is also
not necessarily true.

Contrapositive: "If Jenny does not live in Georgia, then she does not
     live in Atlanta."

   This time point J lies outside of G, so it
cannot lie in A. Any point that is not in G is
also not in A. Therefore, the contrapositive
is true.

Cooperative Learning

Have students compare diagrams
and answers for problem **4** on
page 47. Then have groups of
students draw a Venn diagram
for the following conditional
sentence.
   If Joe is a member of the
   (school name) soccer team,
   then he is a student at
   (school name).
Sample diagram:

Then have students write the
converse, the inverse, and the
contrapositive of the original
conditional statement. Finally,
have students use their diagrams
to determine the truth value of
the three new statements.

Converse: If Joe is a member of
the student body, then he is a
member of the soccer team;
false.

Inverse: If Joe is not a member of
the soccer team, then he is not a
member of the student body;
false.

Contrapositive: If Joe is not a
member of the student body,
then he is not a member of the
soccer team; true.

Lesson Notes, continued

- Theorem **3** is extremely important in mathematical proof. Be sure students understand its significance. Explain that the symbol ⟺ is read as "is equivalent to."
- Emphasize that the chain rule requires that all steps leading to the final conclusion must be accepted as true. In terms of a geometric proof, this means that all steps prior to the last one represent given information, definitions, assumptions, or theorems.

- Work through the sample problems with the students. Before making a homework assignment, clear up any misunderstanding they might have about the chain rule or a chain of reasoning.

Communicating Mathematics

Have each student do the following.

1 Write a conditional statement.
2 Write the converse, the inverse, and the contrapositive of the conditional statement.
3 Use the symbols ⟹ and ~ and the variables p and q to symbolize the conditional and its converse, inverse, and contrapositive.
4 Determine which of the four conditional statements are true.
   Answers will vary.

This analysis suggests the following important theorem:

Theorem 3 *If a conditional statement is true, then the contrapositive of the statement is also true.*
(If p, then q ⟺ If ~q, then ~p.)

In other words, a statement and its contrapositive are *logically equivalent.*

Chains of Reasoning

Each proof that you do involves a series of steps in a logical sequence. In many cases, the sequence will take the following form.

$$\text{If } p \Rightarrow q \text{ and } q \Rightarrow r, \text{ then } p \Rightarrow r.$$

This is called the ***chain rule***, and a series of conditional statements so connected is known as a ***chain of reasoning***.

Example *If we accept the two statements "If you study hard, then you will earn a good grade" (p ⟹ q) and "If you earn a good grade, then your family will be happy" (q ⟹ r), what can we conclude?*

We can conclude that p ⟹ r—that is, if you study hard, then your family will be happy.

Part Two: Sample Problems

Problem 1 Write the converse, the inverse, and the contrapositive of the following true statement: "If two angles are right angles, then they are congruent."

Solution Converse: "If two angles are congruent, then they are right angles." (The converse is false; for example, each angle may have a measure of 60.)

Inverse: "If two angles are not right angles, then they are not congruent." (The inverse is also false.)

Contrapositive: "If two angles are not congruent, then they are not right angles." (The contrapositive is true—the statements are logically equivalent.)

Problem 2 Draw a conclusion from the following statements:
   If gremlins grow grapes, then elves eat earthworms.
   If trolls don't tell tales, then wizards weave willows.
   If trolls tell tales, then elves don't eat earthworms.

Solution First, we rewrite the statements in symbolic form.

(1) $g \Rightarrow e$

(2) $\sim t \Rightarrow w$

(3) $t \Rightarrow \sim e$

To complete the chain of reasoning, we can rearrange the statements and use contrapositives as needed to match symbols. Thus,

(1) $g \Rightarrow e$

(3) $e \Rightarrow \sim t$ ($t \Rightarrow \sim e$ is equivalent to $e \Rightarrow \sim t$.)

(2) $\sim t \Rightarrow w$

$\therefore g \Rightarrow w$ (The symbol $\therefore$ means "therefore.")

Hence, if gremlins grow grapes, then wizards weave willows.

Part Three: Problem Sets

Problem Set A

1 Write each sentence in conditional ("If . . ., then . . .") form.

a Eighteen-year-olds may vote in federal elections.

b Opposite angles of a parallelogram are congruent.

2 Write the converse, the inverse, and the contrapositive of each statement. Determine the truth of each of the new statements.

a If each side of a triangle has a length of 10, then the triangle's perimeter is 30.

b If an angle is acute, then it has a measure greater than 0 and less than 90.

3 If a conditional statement and its converse are both true, the statement is said to be *biconditional*. Which of these statements is biconditional? a

a If two angles are congruent, then they have the same measure.

b If two angles are straight angles, then they are congruent.

4 Draw a Venn diagram for the true conditional statement "If a person lives in Chicago, then the person lives in Illinois." Assuming that each of the following "Given . . ." statements is true, determine the truth of the conclusion.

a Given: Penny lives in Chicago.
Conclusion: Penny lives in Illinois. True

b Given: Benny lives in Illinois.
Conclusion: Benny lives in Chicago. False

c Given: Kenny does not live in Chicago.
Conclusion: Kenny must live in Illinois. False

d Given: Denny does not live in Illinois.
Conclusion: Denny lives in Chicago. False

Assignment Guide

Basic

1–5, 8

Average

1–5, 8, 9

Advanced

($\frac{1}{2}$ day) Section **1.7** 5, 14

($\frac{1}{2}$ day) Section **1.8** 3–5, 7–10

Problem-Set Notes and Additional Answers

1a   If a person is 18 years old, then he or she may vote in a federal election.

b   If two angles are opposite angles of a parallelogram, then they are congruent.

2a   Converse: If a triangle has a perimeter of 30, then each side has a length of 10; false.
Inverse: If each side length of a triangle is not 10, then the perimeter is not 30; false.
Contrapositive: If the perimeter of a triangle is not 30, then each side length is not 10; true.

b   Converse: If an angle has a measure greater than 0 and less than 90, then the angle is acute; true.
Inverse: If an angle is not acute, then it does not have a measure greater than 0 and less than 90; true.
Contrapositive: If an angle does not have a measure greater than 0 and less than 90, then it is not acute; true.

4

Problem-Set Notes and
Additional Answers, continued

6 Converse: If M, A, and B are collinear, then M is the midpoint of $\overline{AB}$; false.
Inverse: If M is not the midpoint of $\overline{AB}$, then M, A, and B are noncollinear; false.
Contrapositive: If M, A, and B are noncollinear, then M is not the midpoint of $\overline{AB}$; true.

7 If a polygon is a square, then it is a quadrilateral with four congruent sides.
Converse: If a quadrilateral has four congruent sides, then it is a square.
Inverse: If a polygon is not a square, then it is not a quadrilateral with four congruent sides.
Contrapositive: If a quadrilateral does not have four congruent sides, then it is not a square.

8a Converse: If a ray divides an angle into two congruent angles, then it bisects the angle.
Inverse: If a ray does not bisect an angle, then it does not divide the angle into two congruent angles.
Contrapositive: If a ray does not divide an angle into two congruent angles, then it does not bisect the angle.

b Converse: If two angles of a triangle are congruent, then the sides opposite those angles are congruent.
Inverse: If two sides of a triangle are not congruent, then the angles opposite those sides are not congruent.
Contrapositive: If two angles of a triangle are not congruent, then the sides opposite those angles are not congruent.

Problem Set A, *continued*

5 Write a concluding statement for each of the following chains of reasoning.

a $a \Rightarrow b \quad d \Rightarrow f$
$d \Rightarrow {\sim}c$
${\sim}c \Rightarrow a$
$b \Rightarrow f$

b $p \Rightarrow {\sim}q \quad s \Rightarrow {\sim}p$
$r \Rightarrow q$
$s \Rightarrow r$

c If weasels walk wisely, then cougars call their cubs.
If goats go to graze, then horses head for home.
If cougars call their cubs, then goats go to graze.
If bobcats begin to browse, then weasels walk wisely.
If bobcats begin to browse, then horses head for home.

Problem Set B

6 Write the converse, the inverse, and the contrapositive of "If M is the midpoint of $\overline{AB}$, then M, A, and B are collinear." Are these statements true or false?

7 Rewrite the following sentence in conditional form and find its converse, inverse, and contrapositive: "A square is a quadrilateral with four congruent sides."

8 Write the converse, the inverse, and the contrapositive of each statement.

a If a ray bisects an angle, it divides the angle into two congruent angles.

b If two sides of a triangle are congruent, then the angles opposite those sides are congruent.

9 What conclusion can be drawn from the following?
${\sim}c \Rightarrow {\sim}f \quad g \Rightarrow b \quad p \Rightarrow f \quad c \Rightarrow {\sim}b \quad p \Rightarrow {\sim}g$

Problem Set C

10 What conclusion can be drawn from the following?
If the line is long, then Quincy will go home.
If it is morning, then Quincy will not go home.
If the line is long, then it is morning.
At least one of the given statements is false.

PROBABILITY

IF YOU HAVE 5 DOGS, 3 WILL BE ASLEEP

1.9 PROBABILITY

Objective

After studying this section, you will be able to
- Solve probability problems

Part One: Introduction

A knowledge of **probability** is obviously important to an insurance company, to a card player or a backgammon expert, and to an operator of a gambling casino. Moreover, setting up and solving probability problems requires the precision and the organized, ordered thinking needed by secretaries, accountants, doctors, filing clerks, computer programmers, and geometry students.

Although probability is not one of the major topics in this book, you will occasionally encounter probability problems in the problem sets. You can analyze such problems by following a simple two-step procedure.

Two Basic Steps for Probability Problems
1 Determine all possibilities in a logical manner. Count them.
2 Determine the number of these possibilities that are "favorable." We shall call these *winners*.

You can then calculate the probability by means of the following formula.

$$\text{Probability} = \frac{\text{number of winners}}{\text{number of possibilities}}$$

Part Two: Sample Problems

Problem 1 *If one of the four points is picked at random, what is the probability that the point lies on the angle?*

Section 1.9   Probability   **49**

Class Planning

Time Schedule
All levels: 1 day

Resource References
Teacher's Resource Book
  Class Opener 1.9A, 1.9B
Evaluation
  Tests and Quizzes
    Quizzes 2–3, Series 1
    Quiz 2, Series 3

Class Opener

List all triangles in the diagram.

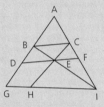

△ABC, △ABI, △ADF, △AGI,
△BCE, △BCI, △BDE, △BGI,
△CEF, △CHI, △CEI, △EFI, △EHI

Lesson Notes

- Probability problems occur throughout the text. Solving such problems helps students develop organized, orderly work habits and can help them develop problem-solving skills. All students should be encouraged to solve some of these problems.

Vocabulary
probability

■ The solutions for these introductory probability problems
make use of counting techniques. You might want to develop algebraic techniques
with some of your students.

Communicating Mathematics

Have students write a paragraph
that explains how the probability
of an event can be 0 or 1.

Solution   We follow the two basic steps by listing all the possibilities and
circling the winners.

$$\frac{\text{Winners}}{\text{Possibilities}} = \frac{4}{4} = 1$$

Problem 2   *If two of the four points are selected
at random, what is the probability
that both lie on $\overrightarrow{CA}$?*

Solution   We follow the two basic steps by listing all the possibilities and
circling the winners. (Notice how we have attempted to list the
possibilities in an orderly manner.)

AB   BC   CD
AC   BD
AD

$$\frac{\text{Winners}}{\text{Possibilities}} = \frac{3}{6} = \frac{1}{2}$$

Problem 3   *If three of the four points are selected
in a random order, what is the proba-
bility that the ordered letters will correctly name the angle shown?*

Solution   We follow the two basic steps by listing all the possibilities and
circling the winners. (This problem is harder than the first two
examples because the *order* of the points is important. Notice how
we have listed the possibilities in an orderly manner.)

ABC	BAC	CAB	DAB
ABD	BAD	CAD	DAC
ACB	BCA	CBA	DBA
ACD	BCD	CBD	DBC
ADB	BDA	CDA	DCA
ADC	BDC	CDB	DCB

$$\frac{\text{Winners}}{\text{Possibilities}} = \frac{4}{24} = \frac{1}{6}$$

Problem 4   *A point Q is randomly chosen on $\overline{AB}$.
What is the probability that it is with-
in 5 units of C?*

Solution   Even though there are infinitely many points on the segment, we can
find the probability by comparing the length of the "winning" region
with the total length of $\overline{AB}$.

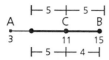

The "winning" region is 9 (not 10) units long. $\overline{AB}$ is 12 units long.

Probability $= \dfrac{9}{12} = \dfrac{3}{4}$

Part Three: Problem Sets

Problem Set A

In problems 1–4, refer to the following diagram.

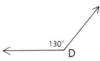

1 If one of the five angles is selected at random, what is the probability that the angle is acute? $\frac{3}{5}$

2 If one of the five angles is selected at random, what is the probability that the angle is right? $\frac{1}{5}$

3 If one of the five angles is selected at random, what is the probability that the angle is obtuse? $\frac{1}{5}$

4 If one of the five angles is selected at random, what is the probability that the angle is straight? 0

5 If a point is randomly chosen on $\overline{PR}$, what is the probability that it is within 2 units of R? $\frac{1}{3}$

P   R
4   10

Problem Set B

In problems 6–9, use the five angles shown at the beginning of Problem Set A.

6 If two of the five angles are selected at random, what is the probability that both are acute? $\frac{3}{10}$

7 If two of the five angles are selected at random, what is the probability that one of them is obtuse? $\frac{2}{5}$

8 If two of the five angles are selected at random, what is the probability that one is right and the other is obtuse? $\frac{1}{10}$

Assignment Guide

Basic
1–7, 10, 11
Average
1–11, 15
Advanced
1–11, 14, 15

Cooperative Learning

Have each student in a group of four students draw and label two angles. Then have each group pool its angles and write problems similar to those in Problem Set **A,** where the probabilities for the students' sets of angles will be 0, $\frac{1}{4}$, $\frac{1}{2}$, $\frac{3}{4}$, and 1.

Answers will vary, depending on student drawings.

12 ∠A, ∠B; ∠B, ∠C; ⟨∠C, ∠D⟩
∠A, ∠C; ∠B, ∠D; ∠C, ∠E;
∠A, ∠D; ∠B, ∠E; ∠A, ∠E;
∠D, ∠E

13
  • • •
  1 3 4
  •
  2

If point 1 is A, then there are
two choices for point 3, and
one choice for point 4. Thus,
there are six different ways
to label the diagram with A
as point 1. Similarly, there
are six different ways if A is
point 3 or point 4. Thus,
there is a total of 18 possible
ways to label the diagram.

14a Any two points are
collinear.

  b ABC   ACD   ADE
     ABD   ACE
     ABE
     BCD   BDE
     BCE
     ⟨CDE⟩

  c No four points in the
diagram are collinear.

15 Area ABCD = 9(5) = 45
    Area SQUA = 3(3) = 9
  a Probability = $\frac{9}{45}$ = $\frac{1}{5}$
  b Probability = $\frac{36}{45}$ = $\frac{4}{5}$,
    or $1 - \frac{1}{5} = \frac{4}{5}$

Problem Set B, *continued*

9 An angle is selected at random from the five angles and then
replaced. A second selection is then made at random. (Thus, the
same angle might be selected twice.) What is the probability that
an acute angle is selected both times? $\frac{9}{25}$

10 If a point B is chosen on $\overline{AC}$, what is the
probability that $-5 \leq B \leq 7$? $\frac{2}{5}$

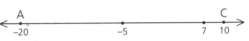

11 The second hand of a clock sweeps continuously around the face
of the clock. What is the probability that at any random moment
the second hand is between 7 and 12? $\frac{5}{12}$

Problem Set C

12 If two of the five angles shown in Problem Set A are selected at
random, what is the probability that neither angle is acute? $\frac{1}{10}$

13 If the four points shown are to be labeled
with the letters A, B, C, and D in such a
way that A and two of the other points
are collinear, in how many different
ways can the diagram be labeled? 18

14 Consider points A, B, C, D, and E as shown.   E   D   C
                         A   B

  a If two of these points are selected at random, what is the
probability that they are collinear? 1

  b If three of these points are selected at random, what is the
probability that they are collinear? $\frac{1}{10}$

  c If four of these points are selected at random, what is the
probability that they are collinear? 0

15 If a point is chosen at random in rectangle ABCD, what is the probability that

  a It is in square SQUA? $\frac{1}{5}$

  b It is not in square SQUA? $\frac{4}{5}$

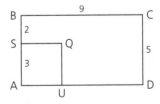

CHAPTER SUMMARY

CONCEPTS AND PROCEDURES

After studying this chapter, you should be able to
- Recognize points, lines, segments, rays, angles, and triangles (1.1)
- Measure segments and angles (1.2)
- Classify angles and name the parts of a degree (1.2)
- Recognize congruent angles and segments (1.2)
- Recognize collinear and noncollinear points (1.3)
- Recognize when a point is between two other points (1.3)
- Apply the triangle-inequality principle (1.3)
- Correctly interpret geometric diagrams (1.3)
- Write simple two-column proofs (1.4)
- Identify bisectors and trisectors of segments and angles (1.5)
- Write paragraph proofs (1.6)
- Recognize that geometry is based on a deductive structure (1.7)
- Identify undefined terms, postulates, and definitions (1.7)
- Understand the characteristics and application of theorems (1.7)
- Recognize conditional statements and the negation, the converse, the inverse, and the contrapositive of a statement (1.8)
- Use the chain rule to draw conclusions (1.8)
- Solve probability problems (1.9)

VOCABULARY

acute angle (1.2)	implication (1.7)	protractor (1.2)
angle (1.1)	intersection (1.1)	ray (1.1)
bisect, bisector (1.5)	inverse (1.8)	right angle (1.2)
chain rule (1.8)	line (1.1)	second (1.2)
collinear (1.3)	line segment (1.1)	segment (1.1)
conclusion (1.7)	measure (1.2)	straight angle (1.2)
conditional statement (1.7)	midpoint (1.5)	theorem (1.4)
congruent angles (1.2)	minute (1.2)	tick mark (1.2)
congruent segments (1.2)	negation (1.8)	triangle (1.1)
contrapositive (1.8)	noncollinear (1.3)	trisect, trisectors (1.5)
converse (1.7)	number line (1.1)	trisection points (1.5)
counterexample (1.6)	obtuse angle (1.2)	two-column proof (1.4)
deductive structure (1.7)	paragraph proof (1.6)	union (1.1)
definition (1.7)	point (1.1)	Venn diagram (1.8)
endpoint (1.1)	postulate (1.7)	vertex (1.1)
hypothesis (1.7)	probability (1.9)	

1 REVIEW PROBLEMS

Chapter 1 Review

Class Planning

Time Schedule
All levels: 1 day

Resource References
Evaluation
   Tests and Quizzes
      Test 1, Series 1, 2, 3

Assignment Guide

Basic
1, 3–11, 13, 14, 16–18, 21b, 33

Average
1h–l, 4, 13, 14, 16–18, 21b, 22,
23, 25, 28, 29, 32, 35

Advanced
13, 14, 18, 19, 21c, 24, 25, 28–
33, 35, 39, 40

To integrate constructions, discuss Section **14.4** on pages 667–
668. Constructions 1, 2, and 3
can be included at this point.

■ You may wish to have students
study all review problems rather than just those included in
the Assignment Guide.

Problem Set A

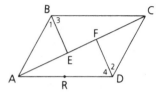

1 a Name in all possible ways, the line containing A, R, and D. $\overleftrightarrow{AR}$, $\overleftrightarrow{AD}$, $\overleftrightarrow{RA}$, $\overleftrightarrow{RD}$, $\overleftrightarrow{DA}$, $\overleftrightarrow{DR}$

b Name the sides of ∠ABC. $\overrightarrow{BA}$, $\overrightarrow{BC}$

c What side do ∠2 and ∠4 have in common? $\overrightarrow{DF}$

d Name the horizontal ray with end-point C. $\overrightarrow{CB}$

e Estimate the sizes of ∠BAD, ∠2, and ∠ABC. 60°; 52°; 120°

f Are angles FCD and DCE different angles? No

g Which angle in the figure is ∠B? No angle can be called ∠B since 3 angles have B as a vertex.

h $\overrightarrow{EC} \cup \overrightarrow{FA} = \underline{\quad?\quad}$ $\overleftrightarrow{AC}$
i $\overrightarrow{EC} \cap \overrightarrow{FA} = \underline{\quad?\quad}$ $\overline{EF}$
j $\overrightarrow{BA} \cup \overrightarrow{BE} = \underline{\quad?\quad}$ ∠1
k $\overleftrightarrow{AC} \cap \overleftrightarrow{DR} = \underline{\quad?\quad}$ A
l ∠AFD $\cap$ $\overline{CE}$ $= \underline{\quad?\quad}$ $\overline{FE}$

2 Tell whether each of the following angles *appears* to be acute, right, obtuse, or straight. Which angle's classification can be assumed from the diagram? ∠DEF is straight.

a ∠H Right

b ∠G Obtuse

c ∠GFE Acute

d ∠DEF Straight

e ∠HDF Right

3 a 43°15′17″ + 25°49′18″ = $\underline{\quad?\quad}$ 69°4′35″

b 90° − 39°17″ = $\underline{\quad?\quad}$ 50°59′43″

4 a Change $46\frac{7}{8}°$ to degrees, minutes, and seconds. 46°52′30″

b Change 132°6′ to degrees. $132\frac{1}{10}°$

5 a According to the diagram, which two segments are congruent? $\overline{BC} \cong \overline{RT}$

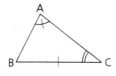

b According to the diagram, which two angles are congruent? $\angle A \cong \angle S$

6 a If $\angle EFG$ is obtuse and $\angle HJK$ is right, is $\angle 1 \cong \angle 2$? No

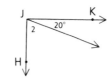

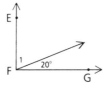

b If $\angle EFG \cong \angle HJK$, is $\angle 1 \cong \angle 2$? Yes

7 If $\angle A \cong \angle B$, find m$\angle A$. 20

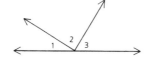

8 The measures of $\angle 1$, $\angle 2$, and $\angle 3$ are in the ratio 1:3:2. Find the measure of each angle. 30; 90; 60

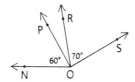

9 Is it possible for both $\angle NOR$ and $\angle POS$ to be right angles? No

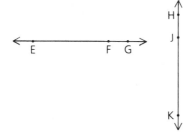

In problems 10 and 11, copy each figure and incomplete proof. Then complete the proof by filling in the missing reasons.

10 Given: Diagram as shown

Prove: $\angle EFG \cong \angle HJK$

Statements	Reasons
1 Diagram as shown	1 _____
2 $\angle EFG$ is a straight angle	2 _____
3 HJK is a straight angle	3 _____
4 $\angle EFG \cong \angle HJK$	4 _____

Review Problems **55**

Problem-Set Notes and Additional Answers

■ See *Solution Manual* for answer to problem **10**.

Review Problem Set A, *continued*

*Problem-Set Notes and
Additional Answers, continued*

■ See *Solution Manual* for answers to problems **11–16**.

11 Given: ∠ABC = 130°,
∠ABD = 60°

Prove: ∠DBC is acute.

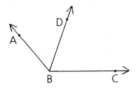

Statements	Reasons
1 ∠ABC = 130°	1 _____
2 ∠ABD = 60°	2 _____
3 ∠DBC = 70°	3 _____
4 ∠DBC is acute	4 _____

In problems 12–15, write each proof in two-column form.

12 Given: ∠X is a right angle.
∠Y is a right angle.

Prove: ∠X ≅ ∠Y

13 Given: $\overline{AB} \cong \overline{BC}$

Prove: B is the midpoint of $\overline{AC}$.

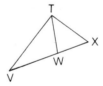

14 Given: $\overrightarrow{DF}$ and $\overrightarrow{DG}$ trisect ∠EDH.

Conclusion: ∠EDF ≅ ∠FDG ≅ ∠GDH

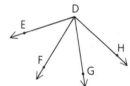

15 Given: $\overrightarrow{TW}$ bisects ∠VTX.

Prove: ∠VTW ≅ ∠XTW

16 Given: ∠1 = 61.6°,
∠2 = $61\frac{3}{5}°$

Prove: ∠1 ≅ ∠2 (Write a paragraph
proof.)

17 a Find coordinate of C (the midpoint
of $\overline{BD}$). −3

b If AD = 15, find the coordinate of A. −13

18 Copy the diagram and draw △A'B'C', the reflection of △ABC, over ℓ_2. What is the length of A'B'? 5

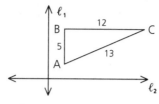

19 a If one of the five labeled points is selected at random, what is the probability that it is a midpoint? $\frac{2}{5}$

  b If two of the five points are randomly chosen, what is the probability that both are midpoints? $\frac{1}{10}$

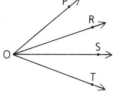

20 Given: $\overrightarrow{OR}$ and $\overrightarrow{OS}$ trisect ∠TOP.
    ∠TOP = 40.2°

Find: m∠POR 13°24'

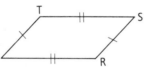

21 Find the angle formed by the hands of a clock at each time.

  a 1:00 30°   **b** 11:20 140°   **c** 4:45 $127\frac{1}{2}°$

22 Write the converse, the inverse, and the contrapositive of the statement "If the time is 2:00, then the angle formed by the hands of a clock is acute." Are these statements true or false?

Problem Set B

23 The perimeter of PRST is 10 more than 5(RS). If PR = 26, find RS. 14

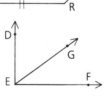

24 Given: ∠DEG = $(x + 3y)°$,
      ∠GEF = $(2x + y)°$;
      ∠DEF is a right angle.

  a Solve for y in terms of x. $y = -\frac{3}{4}x + 22\frac{1}{2}$

  b If ∠DEG ≅ ∠GEF, find the values of x and y. 18; 9

25 Given: WY = 25;
      The ratio of WX to XY is 3:2.

Find: WX 15

26 The measure of ∠A is 6 greater than twice the measure of ∠B. If the angles' sum is 42°, find the measure of ∠A. 30

Review Problems | 57

Problem-Set Notes and Additional Answers, continued

18

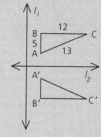

22 Converse: If the angle formed by the hands of a clock is acute, then the time is 2:00; false.
Inverse: If the time is not 2:00, then the angle formed by the hands is not acute; false.
Contrapositive: If the angle formed by the hands of the clock is not acute, then the time is not 2:00; true.

Problem-Set Notes and
Additional Answers, continued

■ See *Solution Manual* for answers to problems **27** and **28**.

Review Problem Set B, *continued*

27 Given: ∠ABC is a right angle.
      ∠DBC = 20°,
      ∠FEG = 40°,
      ∠GEH = 30°
   Prove: ∠ABD ≅ ∠FEH
      (Write a two-column proof.)

28 Given: ∠OMK = 50°,
      ∠OKM = $(2x)°$,
      ∠OKJ = $(5x + 5)°$
   Conclusion: ∠OKJ ≅ ∠OMN
      (Write a paragraph proof.)

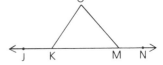

29 Find m∠1. 48°50′44″

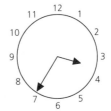

30 The diagram shows Kara's watch. If Kara
   cannot go home until 4:15, how many
   degrees must the hour hand travel before
   she can go home? 20°

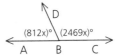

31 Find the measure of ∠ABD to
   a The nearest tenth of a degree ≈44.5°
   b The nearest minute ≈44°33′

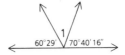

32 If a point is chosen at random on $\overline{PR}$,
   what is the probability that it is within 6
   units of Q? $\frac{1}{3}$

33 The characteristics of a triangle require
   that PR be between what two values?
   7 < PR < 31

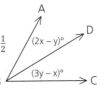

34 Given: $\overrightarrow{BD}$ bisects ∠ABC.
      m∠ABC = 25
   Solve for x and y. x = 10; y = $7\frac{1}{2}$

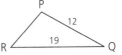

Chapter 1 Introduction to Geometry

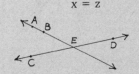

38

x + y = 90
z + y = 90
∴ x + y = z + y
x = z

39

35 ∠Q is obtuse.

  a What are the limitations on m∠Q?
    (Write two inequalities.) $90 < m\angle Q < 180$

  b What are the restrictions on x? $59 < x < 104$

36 Given that ∠R is a right angle, solve for x.
    $x = 30$ or $x = -3$

37 The perimeter of a rectangle is 20. If the rectangle's length is less than 4, what is the range of possible values of its width? $6 < w < 10$

Problem Set C

38 Given: ∠ABC is a right angle.
          ∠DBE is a right angle.

  Prove: ∠ABD ≅ ∠CBE
      (Write a paragraph proof.)

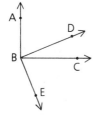

39 Draw a diagram in which $\overleftrightarrow{AB}$ and $\overleftrightarrow{CD}$ intersect at E but in which ∠AEC does not appear to be congruent to ∠DEB.

40 Jennie's teacher told her to select two problems from a list of two C-level problems, five B-level problems, and one A-level problem. If she selected at random, what is the probability that she selected two B-level problems? $\frac{5}{14}$

41 At 3:00 the hands of a clock form an angle of 90°. To the nearest second, at what time will the hands of the clock next form a 90° angle? $\approx 3{:}32{:}44$

Problem Set D

42 If six points are represented on a sheet of paper in such a way that any four of them are noncollinear,

  a What is the maximum number of lines determined? 15

  b What is the minimum number of lines determined? 7

43 To the nearest second, what is the first time after 2:00 that the hands of a clock will form an angle $2\frac{1}{2}$ times as great as the angle formed at 2:00? $\approx 2{:}38{:}11$

Problem-Set Notes and Additional Answers, continued

40 There are
$7 + 6 + 5 + 4 + 3 + 2 + 1 = 28$
possibilities and
$4 + 3 + 2 + 1 = 10$
winners; so, probability =
$\frac{10}{28} = \frac{5}{14}$.
Note By a probability diagram, $\frac{5}{8} \cdot \frac{4}{7} = \frac{5}{14}$.

41 Some thought will show that the hands of the clock will form 22 right angles during a 12-hour period. This means one right angle every $\frac{6}{11}$ hours.
$\frac{6}{11} \cdot 60 = 32\frac{8}{11}$ min
$\frac{8}{11} \cdot 60 \approx 44$ sec
Thus, the next right angle will be about 32 min, 44 sec past 3:00.

42a

  b

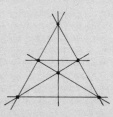

43 The angle at 2:00 is 60°. The required angle is 150°.
$\frac{150}{360} = \frac{5}{12}$
Since the hands of a clock are together every $\frac{12}{11}$ hr, the required time will occur at $\frac{5}{12} \cdot \frac{12}{11} = \frac{5}{11}$ hr after they last came together, which was at $2\frac{2}{11}$ hr. Thus, the required time is $2\frac{7}{11}$ hr, which is approximately 38 min, 11 sec past 2:00.

T 59

Chapter 2 Schedule

Basic
Problem Sets:	12 days
Review:	1 day
Test:	1 day

Average
Problem Sets:	10 days
Review:	1 day
Test:	1 day

Advanced
Problem Sets:	7 days
Review:	1 day
Test:	1 day

STUDENTS SHOULD DEVELOP their ability to use the correct form and proper wording for proofs in this chapter, but mastery need not be expected until students work on the congruent-triangle proofs in Chapter **3.** Be sure to point out to students that each reason in a two-column proof should be a single sentence or less.

It would be helpful to discuss with students exactly what they can use for reasons. At this early stage, you may want to encourage students to keep a list of possible reasons. Be sure to emphasize, however, that students should understand the meaning of what they are doing and not just memorize or look at their lists.

Throughout the text, alternative forms of many statements are used (e.g., the Division Property: Theorem 15, p. 90, and reason 4, p. 91). This is done to encourage students' facility in using geometric terms and to eliminate nonessential steps.

CHAPTER

2 BASIC CONCEPTS AND PROOFS

People encounter the geometric concepts of perpendicularity, complementary angles, and supplementary angles on a leisurely stroll.

PERPENDICULARITY

Objectives

After studying this chapter, you will be able to
- Recognize the need for clarity and concision in proofs
- Understand the concept of perpendicularity

Part One: Introduction

A Look Back and a Look Ahead

If you feel somewhat confused at this time, you need not feel discouraged. Some confusion is inevitable at the start of geometry. Be patient! Read the lessons carefully, study the sample problems closely, and the confusion will begin to go away. Also, see your teacher for help as you need it.

In Chapter 1, you concentrated on two-column proofs but were also exposed to paragraph proofs. When writing either type, remember that understanding what you are trying to say is the most important element.

From now on, when you write a two-column proof, try to state each reason in a single sentence or less. To help you, the problems in Problem Set A of this section and the next will include a hint when a proof requires more than two steps.

This chapter contains more definitions and theorems for you to memorize and use. Toward the end of the chapter, the proofs will begin to get a little longer. As the proofs become more challenging, you will find more satisfaction in completing them.

Perpendicular Lines, Rays, and Segments

Perpendicularity, right angles, and 90° measurements all go together.

Definition   Lines, rays, or segments that intersect at right angles are ***perpendicular*** ($\perp$).

Below are some examples of perpendicularity.

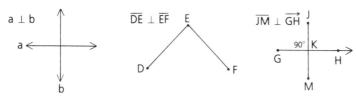

Section 2.1   Perpendicularity   **61**

Vocabulary

coordinates   perpendicular
oblique lines   x-axis
origin   y-axis

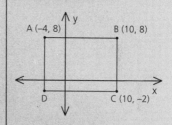
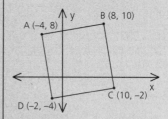

In the figure at the right, the mark inside the angle (⌐) indicates that ∠B is a right angle. It is also true that $\overline{AB} \perp \overline{BC}$ and ∠B = 90°.

Do not assume perpendicularity from a diagram! In △DEF it appears that $\overline{DE} \perp \overline{EF}$, but we may not assume so.

In your algebra studies, you learned that two perpendicular number lines form a two-dimensional coordinate system, or coordinate plane. (The horizontal line is called the **x-axis**; the vertical line, the **y-axis**.) Each point on the plane can be represented by an ordered pair in the form (x, y). The values of x and y in the pair, called the point's **coordinates**, represent the point's distances from the y-axis and the x-axis respectively. In the diagram, point A is represented by (−1, 3).

The intersection of the axes is called the **origin**. Its coordinates are (0, 0).

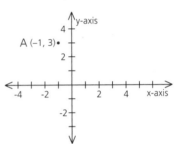

Part Two: Sample Problems

Problem 1

Given: $\overline{AB} \perp \overline{BC}$,
$\overline{DC} \perp \overline{BC}$

Conclusion: ∠B ≅ ∠C

Proof

Statements	Reasons
1 $\overline{AB} \perp \overline{BC}$	1 Given
2 ∠B is a right angle.	2 If two segments are ⊥, they form a right angle.
3 $\overline{DC} \perp \overline{BC}$	3 Given
4 ∠C is a right angle.	4 Same as 2
5 ∠B ≅ ∠C	5 If angles are right angles, they are ≅.

The braces joining steps 1 and 2 emphasize the logical flow of reasoning from one step to the other. There is a similar logical flow from step 3 to step 4.

Problem 2

Given: $\overleftrightarrow{EH} \perp \overleftrightarrow{HG}$
Name all the angles you can prove to be right angles.

Answer

Only ∠EHG (Why not ∠EFH and ∠HFG?)

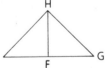

Lesson Notes, continued

- Sample Problem **1** illustrates a three-step logical argument, from perpendicular segments to right angles to congruent angles.

- Sample Problem **2** emphasizes the importance of not assuming perpendicularity from a diagram.

Problem 3   Given: $\vec{KJ} \perp \vec{KM}$;
$\angle JKO$ is four times as large as $\angle MKO$.
Find: $m\angle JKO$

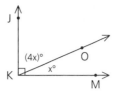

Solution   Since $\vec{KJ} \perp \vec{KM}$, $m\angle JKO + m\angle MKO = 90$.

$$4x + x = 90$$
$$5x = 90$$
$$x = 18$$

Substituting 18 for x, we find that $m\angle JKO = 72$.

Problem 4   Given: $a \perp b$,
$c \not\perp d$ (c is not $\perp$ to d.)
Conclusion: $\angle 1 \cong \angle 2$

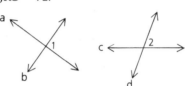

Solution   This conclusion is false. Since $a \perp b$, $\angle 1 = 90°$. Since $c \not\perp d$, $\angle 2 \neq 90°$. Since $\angle 1$ and $\angle 2$ have different measures, $\angle 1 \not\cong \angle 2$.

Problem 5   Given: $\overline{EC} \parallel$ to x-axis
$\overline{RT} \parallel$ to x-axis
Find: Area of rectangle RECT

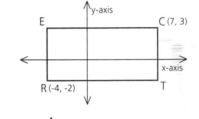

Solution   The remaining coordinates are $T = (7, -2)$ and $E = (-4, 3)$. So $RT = 11$ and $TC = 5$ as shown. We shall concentrate on area in Chapter 12, but from previous courses you should know how to find a rectangle's area.

$$\text{Area of rectangle} = \text{base} \times \text{height}$$
$$A = bh$$
$$= 11(5)$$
$$= 55$$

The area of RECT is 55 square units.

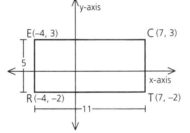

Part Three: Problem Sets

Problem Set A

1 Name all the angles in the figures to the right that appear to be right angles.

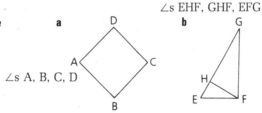

∠s A, B, C, D

∠s EHF, GHF, EFG

Communicating Mathematics

Ask students to write a paragraph explaining why perpendicularity cannot be assumed from a diagram.

Assignment Guide

Basic	
1–11	
Average	
2–12	
Advanced	
4–15	

Problem-Set Notes and Additional Answers

■ Be sure to emphasize to students that each reason in a two-column proof should be a single sentence or less.

Section 2.1   Perpendicularity   **63**

Problem Set A, *continued*

2 In each of the following, name the angles that can be *proved* to be right angles.

a Given: $\overline{JM} \perp \overline{JK}$

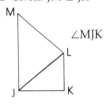

∠MJK

b Given: $\overrightarrow{RO} \perp \overrightarrow{PN}$

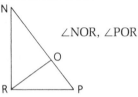

∠NOR, ∠POR

c Given: $\overline{OT} \perp \overline{SW}$

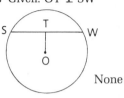

None

3 In each of the following, find the measure of ∠1.

a $\overline{AB} \perp \overline{BC}$,
∠2 = 68°17′34″
21°42′26″

b $\overleftrightarrow{DE} \perp \overleftrightarrow{EF}$;
$\overrightarrow{EG}$ bisects ∠DEF.
45

4 a ∠1 is five times as large as ∠2. Find m∠2. 30

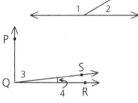

b ∠3 is 72 times as large as ∠4, and $\overleftrightarrow{PQ} \perp \overleftrightarrow{QR}$. Find m∠4 to the nearest tenth. (Hint: Use a calculator to do the arithmetic.) ≈1.2

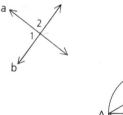

5 On a graph, point A is at (0, 4). Point A is then rotated 90° clockwise about the origin to point A′. What are the coordinates of A′? (4, 0)

6 Given: a ⊥ b

Prove: ∠1 ≅ ∠2 (Hint: This proof takes more than two steps. Remember, each reason should be a single sentence or less.)

7 Given: ∠ACB = 90°,
$\overline{AD} \perp \overline{BD}$

Prove: ∠C ≅ ∠D (Hint: This proof takes more than three steps.)

8 Given: ∠MOR = (3x + 7)°,
∠ROP = (4x − 1)°,
$\overline{MO} \perp \overline{OP}$
Which angle is larger, ∠MOR or ∠ROP?
∠ROP

---

64   Chapter 2   Basic Concepts and Proofs

---

9 You are the engineer for the development of a new subdivision in your town. When you design your street intersections, is it better to make the intersections perpendicular or oblique? Explain why.
Note When two lines intersect and are not perpendicular, they are called **oblique lines.**

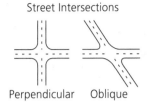

Street Intersections

Perpendicular   Oblique

10 PQRS is a rectangle.
  a Find the coordinates of point P. $(-3, 2)$
  b Find the area of the rectangle. 117

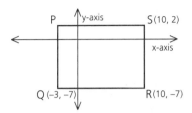

P   y-axis   S(10, 2)

x-axis

Q (−3, −7)   R(10, −7)

Problem Set B

11 $\overleftrightarrow{AB} \perp \overleftrightarrow{BC}$ and angles 1, 2, and 3 are in the ratio $1:2:3$. Find the measure of each angle. 15; 30; 45

12 Line DE is perpendicular to line EF. The resulting angle is trisected, then one of the new angles is bisected, and then one of the resulting angles is trisected. How large is the smallest angle? 5°

13 Given: $\angle HGJ = 37°20'$,
        $\angle KGJ = 52°40'$,
        $\overline{KJ} \perp \overline{HJ}$

Conclusion: $\angle HGK \cong \angle HJK$ (Use a paragraph proof.)

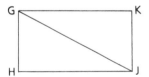

Problem Set C

14 Given: $\overline{AB} \perp \overline{BC}$,
      $\angle ABO = (2x + y)°$,
      $\angle OBC = (6x + 8)°$,
      $\angle AOB = (23y + 90)°$,
      $\angle BOC = (4x + 4)°$

Find: $m\angle ABO$ 22

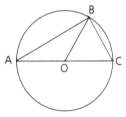

15 If a ray, $\overrightarrow{BD}$, is chosen at random between the sides of $\angle ABC$, where $m\angle ABC = 100$, what is the probability that
  a $\angle ABD$ is acute? $\frac{9}{10}$
  b $\angle DBC$ is acute? $\frac{9}{10}$
  c Both $\angle ABD$ and $\angle DBC$ are acute? $\frac{4}{5}$

Problem-Set Notes and Additional Answers, continued

  9   Answers will vary. Some reasons: better organization and use of space; easier for motorists to see other cars coming; easier for motorists to turn the corner.

  13 $\angle HGJ + \angle KGJ = 90°$, so $\angle HGK$ is a right angle. $\overline{KJ} \perp \overline{HJ}$, so $\angle HJK$ is a right angle. Then $\angle HGK \cong \angle HJK$, since all right angles are $\cong$.

  14 $\begin{cases} 2x + y + 6x + 8 = 90 \\ 23y + 90 + 4x + 4 = 180 \end{cases}$
       $x = 10; y = 2$
       $m\angle ABO = 2x + y = 22$

  15a

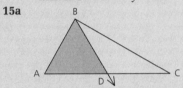

      In the limiting case as shown, $m\angle ABD = 90$.
      Prob. $= \frac{\text{acute region}}{\text{entire region}} =$
      $\frac{90}{100} = \frac{9}{10}$

  b   Same reasoning as part **a.**

  c

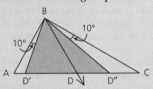

      From the figure, D can range from D′ to D″. $m\angle ABD$ must be at least 10 so that $m\angle CBD$ will not be more than 90. Use similar reasoning for $m\angle CBD$. Then the area of the shaded region is $100 - 10 - 10 = 80$.
      Prob. $= \frac{\text{shaded region}}{\text{entire region}} =$
      $\frac{80}{100} = \frac{4}{5}$

Class Planning

Time Schedule
Basic: 2 days
Average: 1 day
Advanced: 1 day

Resource References
Teacher's Resource Book
    Class Opener 2.2A, 2.2B
    Additional Practice Worksheet 3
Evaluation
    Tests and Quizzes
        Quiz 1, Series 1, 3

Class Opener

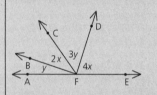

Given: ∠BFC = 2x, ∠CFD = 3y,
    $\overline{BF} \perp \overline{DF}$, ∠BFA = y,
        ∠DFE = 4x
Find: ∠DFE.   ∠DFE = 4 · 18°
        = 72°

2.2

COMPLEMENTARY AND SUPPLEMENTARY ANGLES

Objective

After studying this section, you will be able to
- Recognize complementary and supplementary angles

Part One: Introduction

We frequently see pairs of angles whose measures add up to a right angle or a straight angle. In this section we will study such pairs of angles—those with sums of 90° and 180°.

Definition   **Complementary angles** are two angles whose sum is 90° (a right angle). Each of the two angles is called the **complement** of the other.

The following are examples of pairs of complementary angles.

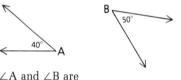

∠A and ∠B are
complementary.

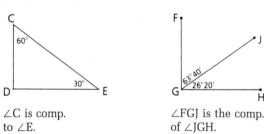

∠C is comp.   ∠FGJ is the comp.
to ∠E.   of ∠JGH.

In the first diagram, ∠A is the complement of ∠B, and ∠B is the complement of ∠A. In the second diagram, two angles of a triangle, ∠C and ∠E, are complementary. In the third diagram, you can see how two complementary angles can share a side to form a right angle.

66   Chapter 2   Basic Concepts and Proofs

Vocabulary
complement
complementary angles
supplement
supplementary angles

Definition   ***Supplementary angles*** are two angles whose sum is 180° (a straight angle). Each of the two angles is called the ***supplement*** of the other.

The following are examples of pairs of supplementary angles.

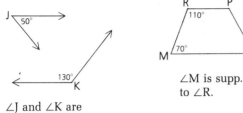

∠J and ∠K are supplementary.

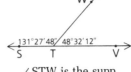

∠M is supp. to ∠R.

∠STW is the supp. of ∠WTV.

In the first diagram, ∠J is the supplement of ∠K, and vice versa. In 'the middle diagram, which angle is the supplement of ∠M?

Sometimes, two supplementary angles will form a straight angle by sharing a side. See if you can verify that ∠STW + ∠WTV = 180°.

Part Two: Sample Problems

Problem 1   Given: ∠TVK is a right ∠.
Prove: ∠1 is comp. to ∠2.

Proof

Statements	Reasons
1 ∠TVK is a right ∠.	1 Given
2 ∠1 is comp. to ∠2.	2 If the sum of two ∠s is a right ∠, they are comp.

Problem 2   Given: Diagram as shown
Conclusion: ∠1 is supp. to ∠2.

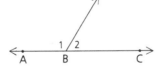

Proof

Statements	Reasons
1 Diagram as shown	1 Given
2 ∠ABC is a straight angle.	2 Assumed from diagram
3 ∠1 is supp. to ∠2.	3 If the sum of two ∠s is a straight ∠, they are supp.

Lesson Notes

■ In Sample Problem **2**, point out that a straight angle can be assumed from the diagram.

Communicating Mathematics

Have students write a paragraph explaining the expressions x, $90 - x$, and $180 - x$ used in Sample Problem **4**. Then have them write and solve a similar problem.

Problem 3   *The measure of one of two complementary angles is three greater than twice the measure of the other. Find the measure of each.*

Solution   Draw the angles and place your algebra on the figure.

Let x = the measure of the smaller angle and $2x + 3$ = the measure of the larger angle.

$$x + 2x + 3 = 90 \quad \text{(The sum of two comp. } \angle s \text{ is } 90°.)$$
$$3x + 3 = 90$$
$$3x = 87$$
$$x = 29$$

The measure of one angle is 29. The measure of the other is $2(29) + 3$, or 61.

Problem 4   *The measure of the supplement of an angle is 60 less than 3 times the measure of the complement of the angle. Find the measure of the complement.*

Solution   Draw the three angles and place your algebra on the figure.

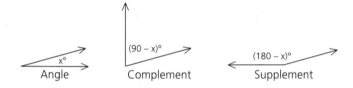

Let x = the measure of the angle.
So $90 - x$ = the measure of the complement.
          (Do you know why?)
So $180 - x$ = the measure of the supplement.
          (Do you know why?)

$$180 - x = 3(90 - x) - 60$$
$$180 - x = 270 - 3x - 60$$
$$180 - x = 210 - 3x$$
$$2x = 30$$
$$x = 15$$

The measure of the complement is $90 - 15$, or 75.

Note   This is a key sample problem. The expressions used at the start of the solution (x, $90 - x$, and $180 - x$) are used in many problems throughout the book.

Part Three: Problem Sets

Problem Set A

1 Which two angles are complementary? ∠A and ∠C

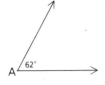 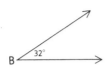 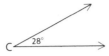

2 What is the supplement of a 70° angle? 110°

3 ∠1 is complementary to ∠3. If ∠3 = y°, how large is ∠1? (90 − y)°

4 Find the complement of a 61°21′13″ angle. 28°38′47″

5 One of two complementary angles is twice the other. Find the measures of the angles. 30 and 60

6 Copy the figure and the proof below. Then complete the proof by filling in the missing statements.
Given: ∠1 is comp. to ∠2.
Prove: $\overleftrightarrow{AB} \perp \overleftrightarrow{BC}$

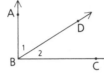

Statements	Reasons
1 _____	1 Given
2 _____	2 If a ray divides an ∠ into two comp. ∠s, then the original ∠ is a right ∠.
3 _____	3 If two lines intersect to form a right ∠, the two lines are ⊥.

7 Given: $\overleftrightarrow{CD} \perp \overleftrightarrow{DE}$

Prove: ∠CDF is comp. to ∠FDE. (Hint: This proof takes more than two steps.)

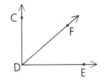

8 Given: Diagram as shown

Prove: ∠GHK is supp. to ∠KHJ. (Hint: This proof takes more than two steps.)

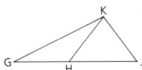

9 Given: ∠MRO is comp. to ∠PRO.
Prove: ∠MRP is a right angle.

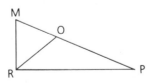

Section 2.2   Complementary and Supplementary Angles   **69**

Assignment Guide

Basic	
Day 1	1–6, 10–13
Day 2	7–9, 14–16, 21

Average	
1–7, 11–16, 21, 24	

Advanced	
7–9, 12–17, 20, 22, 24	

Problem-Set Notes and Additional Answers

■ See *Solution Manual* for answers to problems **6–9.**

Problem Set A, *continued*

10 Find the measure of ∠XVS. 60

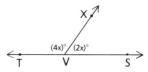

11 One of two supplementary angles is 70° greater than the second. Find the measure of the larger angle. 125

Problem Set B

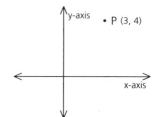

12 a Point P is reflected over the y-axis to point A. Find the coordinates of A. (−3, 4)

b Point P is reflected over the origin to point B. Find the coordinates of B. (−3, −4)

c If C is the midpoint of $\overline{PA}$, find the coordinates of C. (0, 4)

13 Complete each of the following conditional statements and justify your completion with an explanation.

a If two angles are supplementary and congruent, then ___?___.

b If two angles are complementary and congruent, then ___?___.

14 Find the measures of ∠AOC and ∠COB in the graph. 45; 45

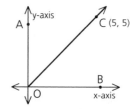

15 Find, to the nearest hundredth, the area of the rectangle. ≈94.84

16 Two supplementary angles are in the ratio 11:7. Find the measure of each. 110 and 70

17 Write a paragraph proof to show that ∠ABF is complementary to ∠EBD.

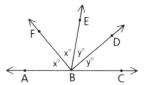

---

18 The larger of two supplementary angles exceeds 7 times the smaller by 4°. Find the measure of the larger angle. 158

19 One of two complementary angles added to one-half the other yields 72°. Find half the measure of the larger. 27

20 Given: $\overline{XY} \perp \overline{YW}$,
$\overline{AB} \perp \overline{BC}$

Find: m∠DBC
Impossible

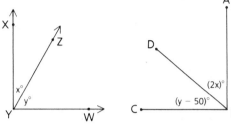

21 The supplement of an angle is four times the complement of the angle. Find the measure of the complement. 30

22 Five times the complement of an angle less twice the angle's supplement is 40°. Find the measure of the supplement. $163\frac{1}{3}$

23 The measure of the supplement of an angle is 30° less than five times the measure of the complement. Find two-fifths the measure of the complement. 12

24 Arnex has a 30°, a 60°, a 150°, a 45°, and a 135° angle in his pocket. He takes out two of the five angles. Find the probability that

a The two angles are supplementary $\frac{1}{5}$

b The two angles are complementary $\frac{1}{10}$

Problem Set C

25 The supplement of an angle is 60° less than twice the supplement of the complement of the angle. Find the measure of the complement. 70

26 Debbie has drawn distinct rays $\overrightarrow{BA}$, $\overrightarrow{BC}$, $\overrightarrow{BD}$, $\overrightarrow{BE}$, and $\overrightarrow{BF}$ on a piece of paper, with ∠ABC being a straight angle.

a What is the minimum number of pairs of complementary angles that she could have drawn? 0

b What is the maximum number of pairs of complementary angles that she could have drawn? 6

c What is the minimum number of pairs of supplementary angles that she could have drawn? 3

d What is the maximum number of pairs of supplementary angles that she could have drawn? 16

Problem-Set Notes and
Additional Answers, continued

20 $\begin{cases} x + y = 90 \\ 2x + y - 50 = 90 \end{cases}$
Solving, x = 50, which is impossible.

25 $180 - x = 2[180 - (90 - x)] - 60$
$x = 20$
comp. = 70

26a No pairs of comp. ∠s

b 6 pairs of comp. ∠s: ∠ABF, ∠FBE; ∠ABF, ∠EBD; ∠ABF, ∠DBC; ∠FBE, ∠EBD; ∠FBE, ∠DBC; ∠EBD, ∠DBC

c 3 pairs of supp. ∠s: ∠ABF, ∠FBC; ∠ABE, ∠EBC; ∠ABD, ∠DBC

d 16 pairs of supp. ∠s: ∠CBD, ∠CBE; ∠CBD, ∠DBA; ∠CBD, ∠EBF; ∠CBD, ∠CBF; ∠DBE, ∠CBE; ∠DBE, ∠DBA; ∠DBE, ∠EBF; ∠DBE, ∠CBF; ∠EBA, ∠CBE; ∠EBA, ∠DBA; ∠EBA, ∠EBF; ∠EBA, ∠CBF; ∠ABF, ∠CBE; ∠ABF, ∠DBA; ∠ABF, ∠EBF; ∠ABF, ∠CBF

DRAWING CONCLUSIONS

Class Planning

Time Schedule
Basic: 1 day
Average: 1 day
Advanced: $\frac{1}{2}$ day

Resource References
Teacher's Resource Book
Class Opener 2.3A, 2.3B

Class Opener

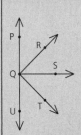

In the diagram, $\overleftrightarrow{PQ} \perp \overleftrightarrow{QS}$ and $\overrightarrow{QR} \perp \overrightarrow{QT}$. If two of the non-straight angles are selected at random, find the probability that the angles are

1 Supplementary $\frac{7}{36}$
2 Complementary $\frac{1}{9}$
3 Congruent $\frac{5}{36}$

Lesson Notes

- Students in advanced courses can read Sections **2.3** and **2.4** in one day.
- This material is intended to help students who leave blank those problems they cannot immediately solve. This procedure helps students analyze each given condition.

Objective

After studying this section, you will be able to
- Follow a five-step procedure to draw logical conclusions

Part One: Introduction

There wouldn't be much progress in this world if all we did was justify conclusions that someone else had already drawn. Neither will you make much progress as a student of geometry if all you can do is justify conclusions the textbook has already stated. Although the following procedure may not work every time, it will be helpful to you in drawing conclusions.

Procedure for Drawing Conclusions

1 Memorize theorems, definitions, and postulates.
2 Look for key words and symbols in the given information.
3 Think of all the theorems, definitions, and postulates that involve those keys.
4 Decide which theorem, definition, or postulate allows you to draw a conclusion.
5 Draw a conclusion, and give a reason to justify the conclusion. Be certain that you have not used the *reverse* of the correct reason.

Example   Given: $\overrightarrow{AB}$ bisects $\angle CAD$.
Conclusion: ___?___

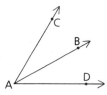

Thinking Process:
  The key word is *bisects*.
  The key symbols are $\rightarrow$ and $\angle$.
  The definition of *bisector* (of an angle) contains those keys.
  An appropriate conclusion is that $\angle CAB \cong \angle DAB$.

Statements	Reasons
1 $\overrightarrow{AB}$ bisects ∠CAD.	1 Given
2 ∠CAB ≅ ∠DAB	2 If a ray bisects an ∠, then it divides the ∠ into two ≅ angles.

Note The "If . . ." part of the reason matches the given information, and the "then . . ." part matches the conclusion being justified. *Be sure not to reverse that order.*

Part Two: Sample Problems

For each of these problems, we will write a two-column proof, supplying a correct conclusion and reason.

Problem 1 Given: ∠A is a right angle.
  ∠B is a right angle.
  Conclusion: ___?___

Proof

Statements	Reasons
1 ∠A is a right angle.	1 Given
2 ∠B is a right angle.	2 Given
3 ∠A ≅ ∠B	3 If two ∠s are right ∠s, then they are ≅.

Problem 2 Given: E is the midpoint of $\overline{SG}$.
  Conclusion: ___?___

Proof

Statements	Reasons
1 E is the midpoint of $\overline{SG}$.	1 Given
2 $\overline{SE} ≅ \overline{EG}$	2 If a point is the midpoint of a segment, the point divides the segment into two ≅ segments.

Problem 3 Given: ∠PRS is a right angle.
  Conclusion: ___?___

Proof

Statements	Reasons
1 ∠PRS is a right ∠.	1 Given
2 $\overleftrightarrow{PR} ⊥ \overleftrightarrow{RS}$	2 If two lines intersect to form a right ∠, they are ⊥.

In sample problem 3, we could have drawn a different conclusion. Do you know what that other conclusion is?

Lesson Notes, continued
- In Sample Problem **3,** another possible conclusion is that m∠PRS = 90.

Communicating Mathematics

Have students complete the following "If . . . , then . . ." statement: If the two parts of a statement are reversed, then _____. Then have students write a paragraph defending their statement.

Assignment Guide

Basic

1–7

Average

1–4, 8, 10, 11

Advanced

($\frac{1}{2}$ day) Section **2.3**   optional

($\frac{1}{2}$ day) Section **2.4**   8, 10, 11, 13–15, 17, 20, 21

Problem-Set Notes and Additional Answers

■ Students in advanced courses can read and solve some problems orally in class.

■ See *Solution Manual* for answers to problems **1–7.**

Part Three: Problem Sets

Problem Set A

In problems 1–7, write a two-column proof, supplying your own correct conclusion and reason.

1 Given: $\overleftrightarrow{AB} \perp \overleftrightarrow{BC}$

Conclusion: ___?___

2 Given: ∠DEF is comp. to ∠HEF.

Conclusion: ___?___

3 Given: ∠WXZ ≅ ∠YXZ

Conclusion: ___?___

4 Given: $\overrightarrow{QS}$ and $\overrightarrow{QT}$ trisect ∠PQR.

Conclusion: ___?___

5 Given: E is the midpoint of $\overline{AC}$.

Conclusion: ___?___

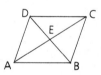

6 Given: A and R trisect $\overline{CD}$.

Conclusion: ___?___

7 Given: Diagram as shown

Conclusion: ___?___

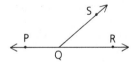

Problem Set B

In problems 8–12, draw at least two conclusions for each "given" statement, and give reasons to support them in two-column-proof form.

8 Given: $\overleftrightarrow{WZ}$ bisects $\overline{VY}$.
Conclusions: ___?___

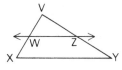

9 Given: $\overline{PA} \perp \overline{AR}$
Conclusions: ___?___

10 Given: $\overleftrightarrow{CG}$ bisects $\overline{BD}$.
Conclusions: ___?___

11 Given: $\angle AEN \cong \angle GEN \cong \angle GEL$
Conclusions: ___?___

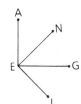

12 Given: $m\angle PQS = 90$
Conclusions: ___?___

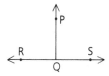

Problem Set C

13 Given: Two intersecting lines as shown
Conclusions: ___?___ (Find as many as you can.)

14 The right angle of a right triangle is bisected. Draw a diagram and set up the given information. Then discuss all possible conclusions.

Problem-Set Notes and Additional Answers, continued

8 $\overline{VZ} \cong \overline{ZY}$, or Z is the midpoint of $\overline{VY}$.

9 $\angle PAR$ is a right angle, or $\angle PAR = 90°$.

10 $\overline{BF} \cong \overline{FD}$, or F is the midpoint of $\overline{BD}$.

11 $\overline{EN}$ and $\overline{EG}$ trisect $\angle AEL$, or $\overline{EN}$ bisects $\angle AEG$, $\overline{EG}$ bisects $\angle NEL$.

12 $\overrightarrow{PQ} \perp \overleftrightarrow{RS}$, or $\angle PQS$ is a right angle.

13 $\angle VZX$ is a straight angle; $\angle WZY$ is a straight angle; $\angle WZV$ is supp. to $\angle VZY$; $\angle YZX$ is supp. to $\angle VZY$; $\angle WZV$ is supp. to $\angle WZX$; $\angle YZX$ is supp. to $\angle WZX$.

14

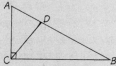

Given: right $\triangle ACB$, right $\angle ACB$, $\overline{CD}$ bisects $\angle ACB$
Conclusions: $\overline{AC} \perp \overline{CB}$; $\angle ACD \cong \angle BCD$; $m\angle ACB = 90$; $\angle ACD$ is comp. to $\angle BCD$; $m\angle ACD = m\angle BCD = 45$; $\angle ADC$ is supp. to $\angle BDC$.

Cooperative Learning

Problems **13** and **14** are good small-group discussion questions. These problems can be extended by having groups make diagrams and having other groups draw conclusions.

Class Planning

Time Schedule
Basic: 2 days
Average: 2 days
Advanced: $\frac{1}{2}$ day

Resource References
Teacher's Resource Book
   Class Opener 2.4A, 2.4B
   Supposer Worksheet 2

Class Opener

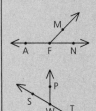

Given: ∠MFN ≅ ∠SWP
Conclusion: Answers may
vary. One possible conclusion is
that ∠AFM ≅ ∠PWT.

Lesson Notes

■ Students are exposed to much
new material in Chapters **1**
and **2**, and most students experience initial difficulties in
writing proofs. Students will
have many opportunities in
Chapter **3** to master this material on supplementary and
complementary angles; there
they use the material in congruent-triangle proofs.

Objective
After studying this section, you will be able to
■ Prove angles congruent by means of four new theorems

Part One: Introduction

In the diagram below, ∠1 is supplementary to ∠A, and ∠2 is also
supplementary to ∠A.

How large is ∠1? Now calculate ∠2. How does ∠1 compare with
∠2? Your results will illustrate (but not prove) the following
theorem.

Theorem 4   *If angles are supplementary to the same angle, then*
they are congruent.

Given: ∠3 is supp. to ∠4.
       ∠5 is supp. to ∠4.
Prove: ∠3 ≅ ∠5

Proof: ∠3 is supp. to ∠4, so m∠3 + m∠4 = 180.
       Therefore, m∠3 = 180 − m∠4.
       ∠5 is supp. to ∠4, so m∠5 + m∠4 = 180.
       Therefore, m∠5 = 180 − m∠4.
       Since ∠3 and ∠5 have the same measure, ∠3 ≅ ∠5.

A companion to Theorem 4 follows.

Theorem 5   *If angles are supplementary to congruent angles,*
then they are congruent.

Given: ∠F is supp. to ∠G.
     ∠H is supp. to ∠J.
     ∠G ≅ ∠J
Conclusion: ∠F ≅ ∠H

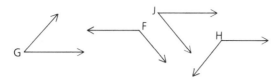

The proof of Theorem 5 is similar to that of Theorem 4.
Two similar theorems apply to complementary angles.

Theorem 6   *If angles are complementary to the same angle,*
then they are congruent.

Theorem 7   *If angles are complementary to congruent angles,*
then they are congruent.

When studying the definitions of such terms as *right angle*, *bisect*,
midpoint, and *perpendicular*, you will master the concepts more
quickly if you try to understand the ideas involved without memorizing the definitions word for word. The theorems in this section,
however, are different. Unless you *memorize* Theorems 4–7, you will
have difficulty remembering the concepts they contain.
    Therefore, before you begin your homework,
      1 Memorize Theorems 4–7
      2 Read the sample problems carefully, so that you understand
        which of the theorems is used in each type of problem

Part Two: Sample Problems

Problem 1   Given: ∠1 is supp. to ∠2.
           ∠3 is supp. to ∠4.
           ∠1 ≅ ∠4
        Conclusion: ∠2 ≅ ∠3

Proof

Statements	Reasons
1 ∠1 is supp. to ∠2.	1 Given
2 ∠3 is supp. to ∠4.	2 Given
3 ∠1 ≅ ∠4	3 Given
4 ∠2 ≅ ∠3	4 If angles are supplementary to ≅ angles, they are ≅. (Short form: Supplements of ≅ ∠s are ≅.)

- Good overall results can be obtained by *not* drilling for mastery in Chapter **2.** Students'
proof-writing skills will greatly
improve in Chapter **3,** when
the material becomes more
meaningful.
- Students should realize that
the keywords *complementary*
and *supplementary* signal the
four theorems in this section.

Communicating Mathematics

Have students write a lesson
plan for teaching Theorems 4–7.
Choose one or two students to
present their lesson to the class.

- Have students list the three steps of the proof that were condensed into step 4 of Sample Problem **4.**

The Geometric Supposer

Using either *The Geometric pre-Supposer* or *The Geometric Supposer: Circles,* have students do the following.

Draw a line. Label two movable points on the line. Label a second pair of movable points, not on that line, such that a line through the second pair of points will intersect the first line at an angle other than 90°. Draw a line through those points. Label the intersection.

Ask students to identify three pairs of supplementary angles. Measure to verify that the angles in each pair are supplementary.

(Measure the first angle, then choose "+," and complete the expression by inputting the name of the second angle.)

To preview the concept of vertical angles (developed in Section **2.8**), ask students to make conjectures about relationships among the four angles formed by the intersecting lines. Other than supplementary angles, what pairs of angles are there? How do they seem to be related?

Modify the activity to explore relationships among angles that are complements of the same angle.

Problem 2   *Given: ∠A is comp. to ∠C.*
                      ∠DBC is comp. to ∠C.
                Conclusion: __?__

Proof

Statements	Reasons
1 ∠A is comp. to ∠C.	1 Given
2 ∠DBC is comp. to ∠C.	2 Given
3 ∠A ≅ ∠DBC	3 If angles are complementary to the same angle, they are ≅. (Short form: Complements of the same ∠ are ≅.)

Problem 3   *Given: Diagram as shown*
                Prove: ∠HFE ≅ ∠GFJ

Proof

Statements	Reasons
1 Diagram as shown.	1 Given
2 ∠EFG is a straight ∠.	2 Assumed from diagram
3 ∠HFE is supp. to ∠HFG.	3 If two ∠s form a straight ∠, they are supplementary.
4 ∠HFJ is a straight ∠.	4 Same as 2
5 ∠GFJ is supp. to ∠HFG.	5 Same as 3
6 ∠HFE ≅ ∠GFJ	6 If angles are supplementary to the same angle, they are ≅. (Short form: Supplements of the same ∠ are ≅.)

Problem 4   *Given: $\overline{KM} \perp \overline{MO}$,*
                      $\overline{PO} \perp \overline{MO}$,
                      ∠KMR ≅ ∠POR
                Prove: ∠ROM ≅ ∠RMO

Proof

Statements	Reasons
1 $\overline{KM} \perp \overline{MO}$	1 Given
2 ∠KMO is a right ∠.	2 If segments are ⊥, they form right ∠s.
3 ∠RMO is comp. to ∠KMR.	3 If two ∠s form a right ∠, they are complementary.
4 In a similar manner, ∠ROM is comp. to ∠POR.	4 Reasons 1–3
5 ∠KMR ≅ ∠POR	5 Given
6 ∠ROM ≅ ∠RMO	6 If angles are complementary to ≅ angles, they are ≅. (Short form: Complements of ≅ ∠s are ≅.)

78   Chapter 2   Basic Concepts and Proofs

Part Three: Problem Sets

Problem Set A

Before starting the assignment, memorize Theorems 4–7. The key to the use of these theorems is to look for the double use of the word *complementary* or *supplementary* in a problem.

1 Given: ∠2 is comp. to ∠3.
∠4 = 131°

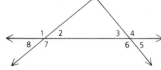

Find the measure of each of the following angles.

a ∠3 49   **c** ∠5 49   **e** ∠1 139   **g** ∠7 139
b ∠6 131   **d** ∠2 41   **f** ∠8 41

2 Given: ∠1 is supp. to ∠3.
∠2 is supp. to ∠3.
Prove: ∠1 ≅ ∠2

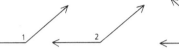

3 Given: ∠4 is comp. to ∠6.
∠5 is comp. to ∠6.
Prove: ∠4 ≅ ∠5

4 One of two supplementary angles is four times the other. Find the larger angle. 144°

5 One of two complementary angles is 20° larger than the other. Find the measure of each. 35 and 55

6 Given: ∠4 is supp. to ∠6.
∠5 is supp. to ∠7.
∠4 ≅ ∠5
Conclusion: ___?___

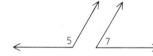

7 Given: ∠FKJ is a right ∠.
∠HJK is a right ∠.
∠GKJ ≅ ∠GJK
Conclusion: ∠FKG ≅ ∠HJG

8 Given: Diagram as shown,
∠6 ≅ ∠7
Prove: ∠5 ≅ ∠8

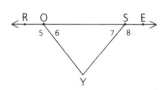

Section 2.4   Congruent Supplements and Complements   **79**

2.4

Assignment Guide

Basic

| Day 1 | 1–6 |
| Day 2 | 8–14 |

Average

| Day 1 | 1–7 |
| Day 2 | 9, 11, 13–16, 18, 21 |

Advanced

($\frac{1}{2}$ day) Section **2.3**   optional
($\frac{1}{2}$ day) Section **2.4**   8, 10, 11, 13–15, 17, 20, 21

Problem-Set Notes and Additional Answers

- See *Solution Manual* for answers to problems **2**, **3**, and **6–8**.

4 $x + 4x = 180$
  $x = 36$
  supp. $= 4 \cdot 36° = 144°$

5 $x + (x + 20) = 90$
  $x = 35°$
  $x + 20 = 55°$

Problem-Set Notes and
Additional Answers, continued

■ See *Solution Manual* for answers to problems **9–11, 13**,
and **16**.

Problem Set A, *continued*

9 Given: $\overrightarrow{SV}$ bisects $\angle RST$.

Conclusion: $\angle RSV \cong \angle TSV$

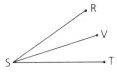

Problem Set B

10 Given: $\overleftrightarrow{OA} \perp \overleftrightarrow{OC}$,
$\overleftrightarrow{OB} \perp \overleftrightarrow{OD}$

Prove: $\angle 1 \cong \angle 3$

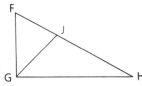

11 Given: $\angle F$ is comp. to $\angle FGJ$.
$\angle H$ is comp. to $\angle HGJ$.
$\overrightarrow{GJ}$ bisects $\angle FGH$.

Conclusion: $\angle F \cong \angle H$

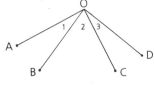

12 The measure of the supp. of an $\angle$ exceeds 3 times the measure
of the comp. of the $\angle$ by 10. Find the measure of the comp. 40

13 Draw the reflection of right angle ABC
over line $\overleftrightarrow{AB}$.

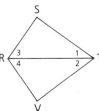

14 RECT is a rectangle.

a Find the coordinates of R. $(-10, 0)$

b What do we know about $\angle RTE$ and
$\angle CTE$? They are comp.

c Find the area of $\triangle ERT$. 35

15 Given: $\overline{PQ} \perp \overline{QR}$
Find: $m\angle PQS$ 37 or 61

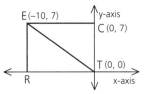

16 Given: $\angle 1$ is comp. to $\angle 4$.
$\angle 2$ is comp. to $\angle 3$.
$\overrightarrow{RT}$ bisects $\angle SRV$.

Prove: $\overrightarrow{TR}$ bisects $\angle STV$.

Cooperative Learning

Have students work in small
groups to write the proof for
problem **16**. Then have them
take a second look at the diagram and the information to determine what other conclusions
can be drawn from the information.
Answers may vary. Students may
conclude that $\angle RST \cong \angle RVT$, or
that $\triangle RST$ is a reflection of
$\triangle RVT$.

T 80

17 If three times the supp. of an $\angle$ is subtracted from seven times the comp. of the $\angle$, the answer is the same as that obtained by trisecting a right $\angle$. Find the supplement. 165°

18 Given: $\angle WXZ \cong \angle VXY$
Conclusion: $\angle 1 \cong \angle 3$

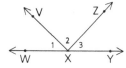

19 Given: $\angle PQR$ supp. $\angle QRS$, $\angle QRS$ supp. $\angle TWX$,
$\angle PQR = (5x - 48)°$, $\angle TWX = (2x + 30)°$
Find: m$\angle QRS$ 98

Problem Set C

20 Given: $\angle 1 = (x^2 + 3y)°$,
$\angle 2 = (20y + 3)°$,
$\angle 3 = (3y + 4x)°$

Find: m$\angle 1$
$23\frac{2}{23}$ or 37
21 The ratio of an angle to its supplement is 3:7. Find the ratio of the angle to its complement. 3:2

MATHEMATICAL EXCURSION

GEOMETRY IN COMPUTERS
Three-Dimensional views on a flat screen

Designers, architects, and draftspeople are putting away their T squares and doing more of their work with computers. A wide variety of software for computer-aided drafting and design (CADD) has made it possible to do accurate work on a computer screen. Using a computer makes exploring solutions to design problems, as well as making corrections and revising, more efficient. A computer also performs calculations and offers a system for filing alternative versions of a plan.

One of the most exciting features of CADD software is that it allows you to create a three-dimensional design and then rotate it on the screen, still in three dimensions. This enables an architect or designer to see, with the press of a key or the click of a mouse, how his or her design would look from any direction or angle.

Using a CADD program, you can see the measure of an angle displayed as you draw the angle. You can instruct the program to automatically bisect an angle you have drawn.

Simpler geometric drawing programs such as The Geometric Supposer offer some of the drawing and measuring capabilities of the CADD programs, including the opportunity for experimenting with geometric concepts such as angle sizes and relationships.

Problem-Set Notes and Additional Answers, continued

17 $7(90 - x) - 3(180 - x) = \frac{1}{3}(90)$
$x = 15$
supp. = 165°

■ See *Solution Manual* for answer to problem **18**.

20 $\begin{cases} x^2 + 3y + 20y + 3 = 180 \\ 20y + 3 + 3y + 4x = 180 \end{cases}$
$\begin{cases} x^2 + 23y = 177 \\ 4x + 23y = 177 \end{cases}$
$x = 0$ or 4; $y = \frac{177}{23}$ or $\frac{161}{23}$
m$\angle 1 = 23\frac{2}{23}$ or 37

21 $3x + 7x = 180$
$x = 18$
acute $\angle = 3x = 54°$
comp. = 36°
ratio $= \frac{54}{36} = \frac{3}{2}$

ADDITION AND SUBTRACTION PROPERTIES

Class Planning

Time Schedule
Basic: 2 days
Average: $1\frac{1}{2}$ days
Advanced: 1 day

Resource References
Teacher's Resource Book
    Class Opener 2.5A, 2.5B
Evaluation
    Tests and Quizzes
        Quiz 2, Series 1
        Quiz 1, Series 2
        Quiz 2, Series 3

Class Opener

If the complement of an angle is 40° less than the supplement of the angle, find the measure of the angle.
Impossible; the complement of an angle must be 90° less than the supplement of the angle.

Lesson Notes

■ The theorems in Sections 2.5–2.7 are based on algebraic properties of equality. Some students may appreciate that these algebraic properties can be considered as postulates.

Objectives

After studying this section, you will be able to
■ Apply the addition properties of segments and angles
■ Apply the subtraction properties of segments and angles

Part One: Introduction

Addition Properties

In the diagram below, AB = CD. Do you think that AC = BD? Suppose that BC were 3 cm. Would AC = BD? If AB = CD, does the length of BC have any effect on whether AC = BD?

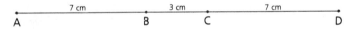

Your answers should be that AC = BD in each case and the length of BC does not effect that equality. This is a geometric application of the algebraic Addition Property of Equality (AB + BC = CD + BC).

Theorem 8   *If a segment is added to two congruent segments, the sums are congruent. (Addition Property)*

Given: $\overline{PQ} \cong \overline{RS}$
Conclusion: $\overline{PR} \cong \overline{QS}$

Proof: $\overline{PQ} \cong \overline{RS}$, so by definition of congruent segments, PQ = RS. Now, the Addition Property of Equality says that we may add QR to both sides, so PQ + QR = RS + QR. Substituting, we get PR = QS. Therefore, $\overline{PR} \cong \overline{QS}$ by the definition of *congruent segments* (reversed).

82 | Chapter 2   Basic Concepts and Proofs

Vocabulary
Addition Property
Subtraction Property

Does a similar relationship hold for angles? Is ∠EFH necessarily congruent to ∠JFG?

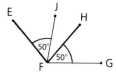

The next theorem confirms that the answer is yes. Its proof is like that of Theorem 8.

Theorem 9 *If an angle is added to two congruent angles, the sums are congruent. (Addition Property)*

In the figures below, identical tick marks indicate congruent parts.

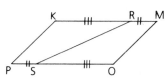 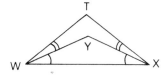

Do you think that $\overline{KM}$ is necessarily congruent to $\overline{PO}$? In the right-hand diagram, is ∠TWX necessarily congruent to ∠TXW? The answer to these questions is yes.

These congruencies are established by the following two theorems. Their proofs are similar to that of Theorem 8.

Theorem 10 *If congruent segments are added to congruent segments, the sums are congruent. (Addition Property)*

Theorem 11 *If congruent angles are added to congruent angles, the sums are congruent. (Addition Property)*

Subtraction Properties

We now have four addition properties. Because subtraction is equivalent to addition of an opposite, we can expect four corresponding subtraction properties.

If AC = BD, is AB = CD?
Let AC = 12 and BC = 3.
How long is $\overline{BD}$?
Is AB = CD?

If ∠EFH ≅ ∠GFJ, is ∠EFG ≅ ∠HFJ?
Let m∠EFH = 50 and m∠GFH = 10.
How large is ∠GFJ?
Is ∠EFG ≅ ∠HFJ?

Communicating Mathematics

Use the questions at the bottom of this page and at the top of page 84 to generate small-group discussions. Ask students to write and defend their conclusions.

If KO = KP and NO = RP,
is KN = KR?
Try this on your own and
see what you think.

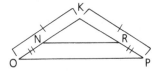

If ∠STE ≅ ∠WET and ∠STW ≅ ∠WES,
is ∠WTE ≅ ∠SET?
Try this on your own.

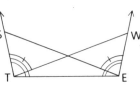

Your results should agree with the next two theorems.

Theorem 12 *If a segment (or angle) is subtracted from congruent segments (or angles), the differences are congruent. (Subtraction Property)*

Theorem 13 *If congruent segments (or angles) are subtracted from congruent segments (or angles), the differences are congruent. (Subtraction Property)*

Using the Addition and Subtraction Properties in Proofs

1 An addition property is used when the segments or angles in the conclusion are *greater than* those in the given information.
2 A subtraction property is used when the segments or angles in the conclusion are *smaller than* those in the given information.

Part Two: Sample Problems

Problem 1   Given: $\overline{AB} \cong \overline{FE}$,
$\overline{BC} \cong \overline{ED}$

Prove: $\overline{AC} \cong \overline{FD}$

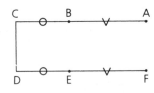

Proof

	Statements		Reasons
1	$\overline{AB} \cong \overline{FE}$	1	Given
2	$\overline{BC} \cong \overline{ED}$	2	Given
3	$\overline{AC} \cong \overline{FD}$	3	If ≅ segments are added to ≅ segments, the sums are ≅. (Addition Property)

84 | Chapter 2   Basic Concepts and Proofs

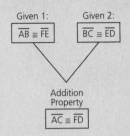

Problem 2   Given: $\overline{GJ} \cong \overline{HK}$
Conclusion: $\overline{GH} \cong \overline{JK}$

Proof

Statements	Reasons
1 $\overline{GJ} \cong \overline{HK}$	1 Given
2 $\overline{GH} \cong \overline{JK}$	2 If a segment ($\overline{HJ}$) is subtracted from $\cong$ segments, the differences are $\cong$. (Subtraction Property)

Problem 3   Given: $\angle NOP \cong \angle NPO$,
   $\angle ROP \cong \angle RPO$

Prove: $\angle NOR \cong \angle NPR$

Proof

Statements	Reasons
1 $\angle NOP \cong \angle NPO$	1 Given
2 $\angle ROP \cong \angle RPO$	2 Given
3 $\angle NOR \cong \angle NPR$	3 If $\cong$ angles are subtracted from $\cong$ angles, the differences are $\cong$. (Subtraction Property)

Problem 4   Given: $\overline{AB} \cong \overline{CD}$
Conclusion: ___?___

Proof

Statements	Reasons
1 $\overline{AB} \cong \overline{CD}$	1 Given
2 $\overline{AC} \cong \overline{BD}$	2 If a segment ($\overline{BC}$) is added to $\cong$ segments, the sums are $\cong$. (Addition Property)

Problem 5   Given: $\angle HEF$ is supp. to $\angle EHG$.
   $\angle GFE$ is supp. to $\angle FGH$.
   $\angle EHF \cong \angle FGE$,
   $\angle GHF \cong \angle HGE$

Conclusion: $\angle HEF \cong \angle GFE$

Proof

Statements	Reasons
1 $\angle HEF$ is supp. to $\angle EHG$.	1 Given
2 $\angle GFE$ is supp. to $\angle FGH$.	2 Given
3 $\angle EHF \cong \angle FGE$	3 Given
4 $\angle GHF \cong \angle HGE$	4 Given
5 $\angle EHG \cong \angle FGH$	5 If $\cong$ angles are added to $\cong$ angles, the sums are $\cong$. (Addition Property)
6 $\angle HEF \cong \angle GFE$	6 Supplements of $\cong$ $\angle$s are $\cong$.

Lesson Notes, continued

■ Some teachers introduce the Reflexive Property now (see Chapter **3**) and do Sample Problem **2** as follows.

Statements	Reasons
1 $\overline{GJ} \cong \overline{HK}$	1 Given
2 $\overline{HJ} \cong \overline{HJ}$	2 Reflexive Prop.
3 $\overline{GH} \cong \overline{JK}$	3 If $\cong$ segments are subtracted from $\cong$ segments, the results are $\cong$. (Subtraction Prop.)

■ If you introduce the Reflexive Property now, here is an alternative proof for Sample Problem **4**.

Statements	Reasons
1 $\overline{AB} \cong \overline{CD}$	1 Given
2 $\overline{BC} \cong \overline{BC}$	2 Reflexive Prop.
3 $\overline{AC} \cong \overline{BD}$	3 If $\cong$ segments are added to $\cong$ segments, the results are $\cong$. (Addition Prop.)

Assignment Guide

Basic

Day 1	1–7
Day 2	8–11, 15, 16

Average

Day 1	4–10
Day 2	($\frac{1}{2}$ day) Section **2.5** 11, 12, 15, 16
	($\frac{1}{2}$ day) Section **2.6** 1, 4–10

Advanced

4, 5, 11, 15–18

Problem-Set Notes and Additional Answers

- Students may find it helpful to refer to the paragraph at the beginning of the problem set.
- See *Solution Manual* for answers to problems **3–5.**

Part Three: Problem Sets

Problem Set A

Throughout this problem set, think of addition when you are asked to prove that segments or angles are larger than the given segments or angles. Think of subtraction when you are asked to prove that segments or angles are smaller than the given segments or angles.

1 Name the angles or segments that are congruent by the Addition Property.

a

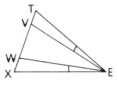

$\overline{AD} \cong \overline{AC}$

b

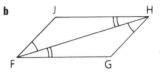

$\angle JFG \cong \angle JHG$

2 Name the angles or segments that are congruent by the Subtraction Property.

a

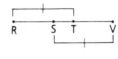

$\overline{RS} \cong \overline{TV}$

b

$\angle YXZ \cong \angle CBD$

3 Given: $\overline{PQ} \cong \overline{SR}$,
$\overline{QN} \cong \overline{RN}$
Conclusion: $\overline{PN} \cong \overline{SN}$

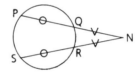

4 Given: $\angle TEV \cong \angle XEW$
Prove: $\angle TEW \cong \angle XEV$

5 Given: $\overline{AC} \cong \overline{DF}$,
$\overline{BC} \cong \overline{EF}$
Prove: $\overline{AB} \cong \overline{DE}$

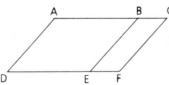

6 Given: $\overline{GH} \cong \overline{JK}$, GH = x + 10, HJ = 8, JK = 2x − 4

Find: GJ 32

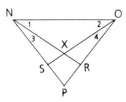

7 Given: $\angle PNO \cong \angle PON$, $\angle 1 \cong \angle 2$

Conclusion: ___?___

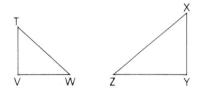

Problem-Set Notes and Additional Answers, continued

■ See *Solution Manual* for answers to problems **7, 8,** and **10–12.**

8 Given: $\angle T$ is comp. to $\angle W$. $\angle X$ is comp. to $\angle Z$. $\angle Z \cong \angle W$

Prove: ___?___

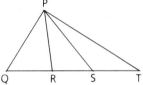

9 Given: $\overline{QR} \cong \overline{ST}$, QS = 5x + 17, RT = 10 − 2x, RS = 3

Find: QS and QT   12; 21

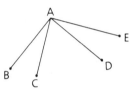

■ In problem **9,** students must first conclude that $\overline{QS} \cong \overline{RT}$. Then 5x + 17 = 10 − 2x. Solving, x = −1. QS = 5x + 17 = 12; QT = QS + RT − RS = 12 + 12 − 3 = 21

10 Given: $\angle BAD$ is a right $\angle$. $\overline{CA} \perp \overline{AE}$

Prove: $\angle BAC \cong \angle EAD$

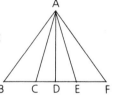

Problem Set B

11 Given: $\angle BAD \cong \angle FAD$; $\overrightarrow{AD}$ bisects $\angle CAE$.

Conclusion: $\angle BAC \cong \angle FAE$

12 Given: J and K are trisection points of $\overline{HM}$. $\overline{GH} \cong \overline{MO}$

Conclusion: $\overline{GJ} \cong \overline{KO}$

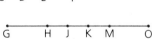

Problem-Set Notes and Additional Answers, continued

- See Solution Manual for answer to problem **13**.

14 In problem **14**, three systems of equations can be set up.

$$\begin{cases} 3x + y = 3y - 3 \\ 3x + y + x + 4y + 2 = 90 \end{cases}$$

$$\begin{cases} 3x + y = 3y - 3 \\ x + 4y + 2 + 3y - 3 = 90 \end{cases}$$

$$\begin{cases} 3x + y + x + 4y + 2 = 90 \\ x + 4y + 2 + 3y - 3 = 90 \end{cases}$$

Solving, $x = 7$; $y = 12$; $m\angle B = 57$.

17a $3x + 7 = 5x - 35$
$x = 21$, so $2x - 22 = 20$ and $\angle ABF \cong \angle CBF$. $\therefore \overrightarrow{BF}$ bisects $\angle CBA$.

b $\angle ABC$ is a straight $\angle$; $\overrightarrow{BF} \perp \overline{AC}$

- In problem **18**, there are four 30° angles, three 60° angles, two 90° angles, and one 120° angle. Of these ten angles, 45 pairs can be made.

18a 6 pairs of 30° angles; $\frac{6}{45} = \frac{2}{15}$

b A 30° angle can be matched with a 60° angle in $4 \cdot 3 = 12$ ways; $\frac{12}{45} = \frac{4}{15}$

19 The hour hand is $\frac{1}{12}$ of 30° $= 2\frac{1}{2}°$ before the 6. $\therefore m\angle = 5 \cdot 30 + 2\frac{1}{2} = 152\frac{1}{2}$

Problem Set B, continued

13 Given: $\angle NPR$ is a right $\angle$.
$\overline{WE} \perp \overline{ET}$,
$\angle SPR \cong \angle XET$
Prove: $\angle NPS \cong \angle WEX$

14 Given: $\angle A$ is comp. to $\angle B$.
$\angle C$ is comp. to $\angle B$.
$\angle A = (3x + y)°$,
$\angle B = (x + 4y + 2)°$,
$\angle C = (3y - 3)°$
Find: $m\angle B$ **57**

15 Draw a right angle ABC. Then draw a dotted line such that the reflection of $\overrightarrow{BA}$ over the dotted line is $\overrightarrow{BC}$. How would you describe this dotted line? **Bisector of $\angle ABC$**

16 On a graph, carefully locate points $A = (1, 4)$ and $B = (11, 10)$. Now locate the point with coordinates $\left(\frac{1 + 11}{2}, \frac{10 + 4}{2}\right)$. Does this point appear to be on $\overline{AB}$? Where? **Yes; midpoint**

Problem Set C

17 $\overrightarrow{BF}$ bisects $\angle DBE$.

a Does $\overrightarrow{BF}$ bisect $\angle CBA$? **Yes**

b What did you discover about $\angle ABC$ and $\overrightarrow{BF}$? $\angle ABC = 180°$; $\overrightarrow{BF} \perp \overleftrightarrow{AC}$

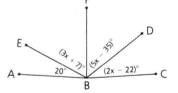

18 If two angles are chosen at random from the ten angles in the diagram, what is the probability that

a The sum of their measures is less than 90? $\frac{2}{15}$

b They are complementary? $\frac{4}{15}$

19 Find the measure of the angle formed by the hands of a clock at 5:55 A.M. $152\frac{1}{2}$

88   Chapter 2   Basic Concepts and Proofs

2.6 MULTIPLICATION AND DIVISION PROPERTIES

Class Planning

Time Schedule
Basic: 2 days
Average: $1\frac{1}{2}$ days
Advanced: 1 day

Resource References
Teacher's Resource Book
    Class Opener 2.6A, 2.6B

Objective

After studying this section, you will be able to
- Apply the multiplication and division properties of segments and angles

Part One: Introduction

In the figure below, B, C, F, and G are trisection points.

If AB = EF = 3, what can we say about $\overline{AD}$ and $\overline{EH}$?

If $\overline{AB} \cong \overline{EF}$, is $\overline{AD}$ congruent to $\overline{EH}$?

In the figure at the right, $\overrightarrow{KO}$ and $\overrightarrow{PS}$ are angle bisectors.

If m∠JKO = m∠NPS = 25, what can we say about ∠JKM and ∠NPR?

If ∠JKO ≅ ∠NPS, is ∠JKM congruent to ∠NPR?

The examples above illustrate a property whose proof is similar to the proof of Theorem 8.

Theorem 14 *If segments (or angles) are congruent, their like multiples are congruent. (**Multiplication Property**)*

Also, because division is equivalent to multiplication by the reciprocal of the divisor, it is easy to prove the next theorem.

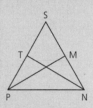
Vocabulary
Division Property
Multiplication Property

■ Students should realize that the keyword *midpoint* can signal either the Multiplication Property or the Division Property.

Theorem 15   *If segments (or angles) are congruent, their like divisions are congruent. (**Division Property**)*

Using the Multiplication and Division Properties in Proofs

1 Look for a double use of the word *midpoint* or *trisect* or *bisects* in the given information.
2 The Multiplication Property is used when the segments or angles in the conclusion are *greater than* those in the given information.
3 The Division Property is used when the segments or angles in the conclusion are *smaller than* those in the given information.

Part Two: Sample Problems

Problem 1   Given: $\overline{MP} \cong \overline{NS}$;
 O is the midpoint of $\overline{MP}$.
 R is the midpoint of $\overline{NS}$.
Prove: $\overline{MO} \cong \overline{NR}$

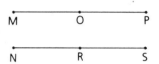

Proof

Statements	Reasons
1 $\overline{MP} \cong \overline{NS}$	1 Given
2 O is the midpoint of $\overline{MP}$.	2 Given
3 R is the midpoint of $\overline{NS}$.	3 Given
4 $\overline{MO} \cong \overline{NR}$	4 If segments are ≅, their like divisions (halves) are ≅. (Division Property)

Problem 2   Given: $\angle TRY \cong \angle ABE$;
 $\overrightarrow{RW}$ and $\overrightarrow{RX}$ trisect $\angle TRY$.
 $\overrightarrow{BC}$ and $\overrightarrow{BD}$ trisect $\angle ABE$.
Conclusion: $\angle TRW \cong \angle CBD$

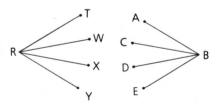

Proof

Statements	Reasons
1 $\angle TRY \cong \angle ABE$	1 Given
2 $\overrightarrow{RW}$ and $\overrightarrow{RX}$ trisect $\angle TRY$.	2 Given
3 $\overrightarrow{BC}$ and $\overrightarrow{BD}$ trisect $\angle ABE$.	3 Given
4 $\angle TRW \cong \angle CBD$	4 If angles are ≅, their like divisions (thirds) are ≅. (Division Property)

Problem 3 Given: $\overline{MK} \cong \overline{FG}$;
$\overline{KG}$ bisects $\overline{MJ}$ and $\overline{FH}$.
Prove: $\overline{MJ} \cong \overline{FH}$

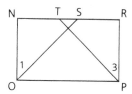

Proof

Statements	Reasons
1 $\overline{MK} \cong \overline{FG}$	1 Given
2 $\overline{KG}$ bisects $\overline{MJ}$ and $\overline{FH}$.	2 Given
3 $\overline{MJ} \cong \overline{FH}$	3 If segments are $\cong$, their like multiples (doubles) are $\cong$. (Multiplication Property)

Problem 4 Given: $\angle NOP \cong \angle RPO$;
$\overrightarrow{PT}$ bisects $\angle RPO$.
$\overrightarrow{OS}$ bisects $\angle NOP$.
$\angle NSO$ is comp. to $\angle 1$.
$\angle RTP$ is comp. to $\angle 3$.
Prove: $\angle NSO \cong \angle RTP$

Proof

Statements	Reasons
1 $\angle NOP \cong \angle RPO$	1 Given
2 $\overrightarrow{PT}$ bisects $\angle RPO$.	2 Given
3 $\overrightarrow{OS}$ bisects $\angle NOP$.	3 Given
4 $\angle 1 \cong \angle 3$	4 Halves of $\cong$ angles are $\cong$. (An alternative form of the Division Property)
5 $\angle NSO$ is comp. to $\angle 1$.	5 Given
6 $\angle RTP$ is comp. to $\angle 3$.	6 Given
7 $\angle NSO \cong \angle RTP$	7 Complements of $\cong$ $\angle$s are $\cong$.

Part Three: Problem Sets

Problem Set A

Before starting the proofs in this problem set, reread the chart on page 90.

1 Given: $\angle KMR \cong \angle VTW$;
$\overrightarrow{MR}$ and $\overrightarrow{MP}$ trisect $\angle KMO$.
$\overrightarrow{TX}$ and $\overrightarrow{TW}$ trisect $\angle STV$.
Prove: $\angle KMO \cong \angle STV$

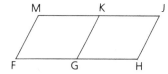

Communicating Mathematics

Have students write their own explanations of the Multiplication Property and the Division Property and defend their explanations.

Assignment Guide

Basic

Day 1	1–5
Day 2	6–10, 12

Average

Day 1	($\frac{1}{2}$ day) Section **2.5** 11, 12, 15, 16
	($\frac{1}{2}$ day) Section **2.6** 1, 4–10
Day 2	11, 12, 15

Advanced

1, 4, 9, 11, 12, 14, 15

Problem-Set Notes and Additional Answers

- See *Solution Manual* for answer to problem **1**.

Problem-Set Notes and
Additional Answers, continued

2a Students must conclude that
∠HGK ≅ ∠ONR. Thus,
50 = 2x + 10. Solving,
x = 20.

b Students must conclude that
$\overline{ST} ≅ \overline{YZ}$. Thus, 12 = x − 4.
Solving, x = 16.

■ See *Solution Manual* for answer
to problems **3–6.**

Cooperative Learning

Have small groups of students
write ten algebraic equations
that require the use of the Multiplication or Division Property
of Equality to solve. Have groups
trade and solve problem sets.
Each group should be sure that
all members can apply the properties to solve each of the problems.
Answers may vary. Possible
equations for each property
might be Multiplication
Property: $\frac{x}{3.2} = 9$; Division
Property: 6x = 12.

Problem Set A, *continued*

2 Use the given information to find the value of x.

a ∠HGJ ≅ ∠ONP;
$\overrightarrow{GJ}$ and $\overrightarrow{NP}$ are ∠ bisectors.
∠HGK = 50°,
∠ONR = (2x + 10)° 20

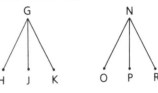

b $\overline{SW} ≅ \overline{SZ}$;
$\overleftrightarrow{TX}$ and $\overleftrightarrow{VY}$ trisect $\overline{SW}$ and $\overline{SZ}$.
ST = 12,
YZ = x − 4 16

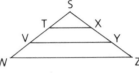

3 Given: $\overline{DF} ≅ \overline{GJ}$;
E is the midpoint of $\overline{DF}$.
H is the midpoint of $\overline{GJ}$.
Prove: $\overline{DE} ≅ \overline{GH}$

4 Given: ∠AFE ≅ ∠DEF;
$\overrightarrow{FC}$ bisects ∠AFE.
$\overrightarrow{EB}$ bisects ∠DEF.
Conclusion: ∠1 ≅ ∠2

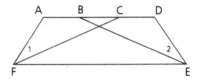

5 Given: $\overline{JK} ≅ \overline{MK}$;
$\overleftrightarrow{OP}$ bisects $\overline{JK}$ and $\overline{MK}$.
Prove: $\overline{JO} ≅ \overline{PK}$

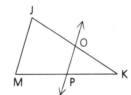

6 Given: ∠TNR ≅ ∠TRN,
∠NRS ≅ ∠RNS
Conclusion: ___?___

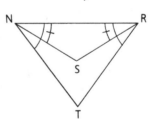

7 a If $\overline{PQ} ≅ \overline{PR}$ in △PQR, what can we
conclude? x = 6

b If AC = AB + 3 in △ABC, what can
we conclude? y = 8

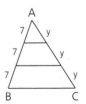
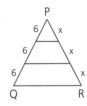

8 Given: M is the midpoint of $\overline{GH}$.
Conclusion: $\overline{GM} \cong \overline{MH}$

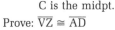
G M H

9 Given: $(x_1, y_1) = (5, 1)$,
$(x_2, y_2) = (9, 3)$
Find: $\left(\dfrac{x_1 + x_2}{2}, \dfrac{y_1 + y_2}{2}\right)$ (7, 2)

*Problem-Set Notes and
Additional Answers, continued*

■ See *Solution Manual* for answers to problems **8** and **10–13**.

10 Copy the diagram and the proof. Then complete the proof by filling in the missing reasons.

Given: $\overline{VW} \cong \overline{AB}$, $\overline{WX} \cong \overline{BC}$;
X is the midpt. of $\overline{VZ}$.
C is the midpt. of $\overline{AD}$.
Prove: $\overline{VZ} \cong \overline{AD}$

V W X Z

A B C D

Statements	Reasons
1 $\overline{VW} \cong \overline{AB}$	1 _____
2 $\overline{WX} \cong \overline{BC}$	2 _____
3 $\overline{VX} \cong \overline{AC}$	3 _____
4 X is the midpt. of $\overline{VZ}$.	4 _____
5 C is the midpt. of $\overline{AD}$.	5 _____
6 $\overline{VZ} \cong \overline{AD}$	6 _____

Problem Set B

11 Given: $\overline{SZ} \cong \overline{ST}$,
$\overline{XY} \cong \overline{VW}$;
Y is the midpt. of $\overline{ZX}$.
V is the midpt. of $\overline{TW}$.
Prove: $\overline{SX} \cong \overline{SW}$

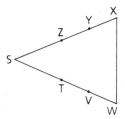

12 Given: $\overrightarrow{PR}$ bisects $\angle QPS$.
$\overrightarrow{KO}$ bisects $\angle JKM$.
$\angle 1$ is supp. to $\angle JKM$.
$\angle 1$ is supp. to $\angle QPS$.
Conclusion: $\angle 2 \cong \angle 3$

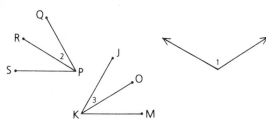

13 Given: $\angle 1 \cong \angle 2$;
$\overrightarrow{BG}$ bisects $\angle ABF$.
$\overrightarrow{CE}$ bisects $\angle FCD$.
Prove: $\angle 3 \cong \angle 4$

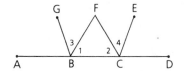

Problem-Set Notes and
Additional Answers, continued

14 Let x = the measure of the angle.
Then 4(180 − x) + 8(90 − x) = 3(180). Solving, x = 75.
Then the comp. of this angle is 15°. The supp. of this angle is 165°.

16 ∠1 and ∠APN and ∠2 and ∠ANP are supplementary. ∠APN ≅ ∠ANP because ∠s supplementary to ≅ ∠s are ≅. But by addition, ∠APN = ∠3 + ∠4 and ∠ANP = ∠5 + ∠6. ∴ ∠3 + ∠4 = ∠5 + ∠6. Since ∠3 ≅ ∠4 and ∠5 ≅ ∠6 by definition of ∠ bisector, ∠3 ≅ ∠5 by the Subtraction Property. Then, by the Addition Property, ∠1 + ∠3 = ∠2 + ∠5. ∴ ∠XPE ≅ ∠ENY

Problem Set B, *continued*

14 If four times the supplement of an angle is added to eight times the angle's complement, the sum is equivalent to three straight angles. Find the measure of the angle that is supplementary to the complement. 165

Problem Set C

15 Point T is located on the graph so that $\overleftrightarrow{RT}$ is perpendicular to the x-axis and 3 < RT < 5. Find the restrictions on the coordinates of T.
x = −5; 4 < y < 6 or −4 < y < −2

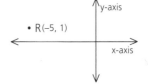

16 Given: ∠1 ≅ ∠2,
$\overrightarrow{PE}$ bis. ∠APN,
$\overrightarrow{NE}$ bis. ∠ANP
Prove: ∠XPE ≅ ∠ENY

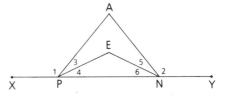

CAREER PROFILE

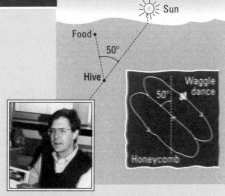

BEE GEOMETRY
James L. Gould shows that bees do indeed know about angles

In the early 1900's, zoologist Karl von Frisch showed that bees convey information geometrically by "waggle dancing." The duration of a dance conveys the distance from the hive of a new food source. The *angle* of the axis of symmetry of the dance relative to the honeycomb conveys the angle of the food measured from the sun line.

Behaviorists didn't accept Frisch's conclusions. Recently, however, James L. Gould, a professor of biology at Princeton University, has conducted research that seems to confirm Frisch's results.

Opponents of the theory argued that new recruits simply observed the direction from which dancers returned to the hive and then flew off in that direction," explains Gould. "I've found a

way to make dancers lie. Recruits still followed the dance directions."

Geometry comes naturally to bees. "They're wired for it," explains Gould. "It's like a computer program in their brains."

Gould has a bachelor's degree in molecular biology from the California Institute of Technology and a doctorate from Rockefeller University. Today he is a professor of biology at Princeton University.

2.7 TRANSITIVE AND SUBSTITUTION PROPERTIES

Objectives

After studying this section, you will be able to
- Apply the transitive properties of angles and segments
- Apply the Substitution Property

Part One: Introduction

Transitive Properties

Suppose that $\angle A \cong \angle B$ and $\angle A \cong \angle C$. Is $\angle B \cong \angle C$?

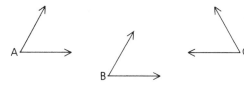

The transitive property of algebra can be used to prove this general rule.

Theorem 16	*If angles (or segments) are congruent to the same angle (or segment), they are congruent to each other. (**Transitive Property**)*

Theorem 16 can be used twice to prove the next theorem.

Theorem 17	*If angles (or segments) are congruent to congruent angles (or segments), they are congruent to each other. (**Transitive Property**)*

Substitution Property

In your algebra studies and in some of the problems you have worked this year, you have solved for a variable such as x and then **substituted** the value you found for that variable.

Class Planning

Time Schedule
All levels: 1 day

Resource References
Teacher's Resource Book
   Class Opener 2.7A, 2.7B
Evaluation
   Tests and Quizzes
      Quiz 2, Series 2
      Quiz 3, Series 1, 3

Class Opener

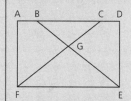

1   Given: $\overline{FC}$ bisects $\angle AFE$.
         $\overline{EB}$ bisects $\angle DEF$.
    Prove: $\angle AFE \cong \angle DEF$
    The given information is
    insufficient to prove
    $\angle AFE \cong \angle DEF$.
2   What additional information
    would be sufficient to prove
    the angles are congruent?
    $\angle CFE \cong \angle BEF$ or
    $\angle AFC \cong \angle DEB$

Lesson Notes

- The Transitive Property is presented independently of the Substitution Property, rather than as a special case of substitution, because transitivity is later extended to order relations, parallel lines, and similar polygons.

Vocabulary
Substitution Property
Transitive Property

Example   If $\angle A \cong \angle B$, find m$\angle A$.

$$2x - 4 = x + 10$$
$$x = 14$$

We can now substitute 14 for x in m$\angle A = x + 10$ to find that m$\angle A = 14 + 10 = 24$.

The Substitution Property can also be applied when no variables are involved.

If $\angle 1$ is comp. to $\angle 2$ and $\angle 2 \cong \angle 3$, then $\angle 1$ is comp. to $\angle 3$ by **substitution.**

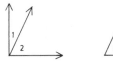

Part Two: Sample Problems

Problem 1   Given: $\overline{FG} \cong \overline{KJ}$,
                   $\overline{GH} \cong \overline{KJ}$
          Prove: $\overleftrightarrow{KG}$ bisects $\overline{FH}$.

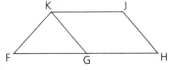

Proof

Statements	Reasons
1 $\overline{FG} \cong \overline{KJ}$	1 Given
2 $\overline{GH} \cong \overline{KJ}$	2 Given
3 $\overline{FG} \cong \overline{GH}$	3 If segments are $\cong$ to the same segment, they are $\cong$. (Transitive Property)
4 $\overleftrightarrow{KG}$ bisects $\overline{FH}$.	4 If a line divides a segment into two $\cong$ segments, it bisects the segment.

Problem 2   Given: $\angle 1 + \angle 2 = 90°$,
                   $\angle 1 \cong \angle 3$
          Prove: $\angle 3 + \angle 2 = 90°$

Proof

Statements	Reasons
1 $\angle 1 + \angle 2 = 90°$	1 Given
2 $\angle 1 \cong \angle 3$	2 Given
3 $\angle 3 + \angle 2 = 90°$	3 Substitution (step 2 in step 1)

Problem 3   If $\angle P \cong \angle R$ and $\angle Q \cong \angle R$, express $m\angle Q$ in terms of x and a.

Solution

$$2y + a = x + y + a$$
$$2y = x + y$$
$$y = x$$

$$m\angle P = x + y + a = x + x + a$$

$$m\angle Q = 2x + a$$

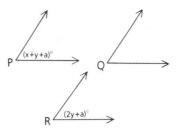

Part Three: Problem Sets

Problem Set A

1 Given: $\angle X \cong \angle Y$,
     $\angle X \cong \angle Z$
   Conclusion: $\angle Y \cong \angle Z$

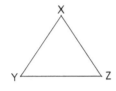

2 Given: $\angle 1 \cong \angle 2$,
     $\angle 2 \cong \angle 3$
   Conclusion: $\angle 1 \cong \angle 3$

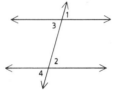

3 Given: $\angle 1 \cong \angle 3$,
     $\angle 2 \cong \angle 3$,
     $\angle 2 \cong \angle 4$
   Prove: $\angle 1 \cong \angle 4$

4 Given: $BC + BE = AD$,
     $BE = EF$
   Prove: $BC + EF = AD$

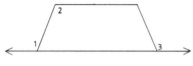

5 Given: O is the midpt. of $\overline{NP}$.
     R is the midpt. of $\overline{SP}$.
     $\overline{NP} \cong \overline{SP}$
   Conclusion: $\overline{SR} \cong \overline{NO}$

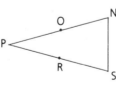

6 Given: $\overline{GJ} \cong \overline{HK}$
   Conclusion: $\overline{GH} \cong \overline{JK}$

Assignment Guide

Basic
1–6
Average
2–6, 15
Advanced
3–5, 10, 12–14

Problem-Set Notes and Additional Answers

■ See *Solution Manual* for answers to problems **1–6.**

- See Solution Manual for answers to problems **7** and **10–12.**

8 $(90 - x) = 24 + 2x$
$x = 22$; comp. $= 68°$

9 $x + 15 = 2x - 5$
$x = 20$
$x + 15 = 35$
$\angle STW = 70°$

Problem Set A, *continued*

7 Given: $\angle OMP \cong \angle RPM$;
$\overrightarrow{MP}$ bisects $\angle OMR$.
$\overrightarrow{PM}$ bisects $\angle OPR$.
Prove: $\angle OMR \cong \angle OPR$

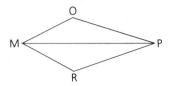

8 The complement of an angle is 24° greater than twice the angle. Find the measure of the complement. **68**

9 $\angle W \cong \angle STV$;
$\overrightarrow{TV}$ bisects $\angle STW$.
$\angle W = (2x - 5)°$,
$\angle VTW = (x + 15)°$
Find: $m\angle STW$ **70**

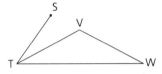

Problem Set B

10 Given: $\overline{VW} \cong \overline{RS}$,
$\overline{XY} \cong \overline{RS}$
Prove: $\overline{VX} \cong \overline{WY}$

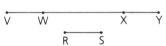

11 Given: $\angle 1 \cong \angle 2$
Conclusion: $\angle 1$ is supp. to $\angle 3$.

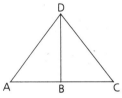

12 Given: $\angle A$ is comp. to $\angle ADB$.
$\angle C$ is comp. to $\angle CDB$.
$\overrightarrow{DB}$ bisects $\angle ADC$.
Conclusion: $\angle A \cong \angle C$

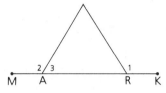

13 Find the measures of each of the following angles in terms of x and y.
a $\angle HFK$ $180 - x - y$
b $\angle EFK$ $180 - y$
c $\angle HFG$ $180 - x$

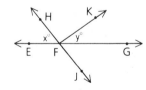

14 $\frac{1}{2}(180 - x) + (90 - x) = 120$
$x = 40$; comp. $= 50°$

14 When one-half the supplement of an angle is added to the complement of the angle, the sum is 120°. Find the measure of the complement. **50**

15 Given: ∠A is a right ∠.
∠B is a right ∠.
∠B ≅ ∠D
Prove: ∠A ≅ ∠D

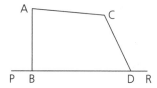

Problem Set C

16 Given: $\overline{AB} \perp \overline{PR}$,
$\overline{AB} \cong \overline{CD}$

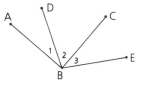

Fool Proof said that since $\overline{AB} \perp \overline{PR}$ and $\overline{AB} \cong \overline{CD}$, he could prove that $\overline{CD} \perp \overline{PR}$ by substitution. What is wrong with Fool's proof?

17 Given: $\overline{AB} \perp \overline{BC}$,
∠1 ≇ ∠3

Prove: ∠DBE is a right angle.
Can be proved false

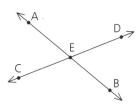

Problem Set D

18 $\overleftrightarrow{AB}$ and $\overleftrightarrow{CD}$ intersect at E, and the ratio of m∠AEC to m∠AED is 2:3. Write an argument to show that it is impossible for m∠DEB to be 80.

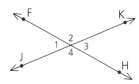

19 If two of the four nonstraight angles formed by the intersection of $\overleftrightarrow{FH}$ and $\overleftrightarrow{JK}$ are selected at random, what is the probability that the two angles are congruent?
$\frac{1}{3}$

20 Find all possible values of x if x is the measure of an angle that satisfies the following set of conditions:

The angle must have a complement, and three fourths of the supplement of the angle must have a complement. $\{x \mid 60 < x < 90\}$

*Problem-Set Notes and
Additional Answers, continued*

- See *Solution Manual* for answers to problems **15** and **18**.

16 The Substitution Property does not apply to statements of perpendicularity.

17 While this argument uses indirect reasoning, students may accept it.
∠1 + ∠2 = 90°,
so ∠1 = 90° − ∠2.
It is given that ∠3 ≠ ∠1,
so ∠3 ≠ 90° − ∠2 by substitution and ∠3 + ∠2 ≠ 90° by "addition."

20 To have a comp., an angle must be between 0° and 90°. Thus,
0 < x < 90
$0 < \frac{3}{4}(180 - x) < 90$
The second equation gives
0 < 180 − x < 120
−180 < −x < −60
180 > x > 60
Combining equations yields
60 < x < 90.

Class Planning

Time Schedule
All levels: 1 day

Resource References
Teacher's Resource Book
   Class Opener 2.8A, 2.8B
   Additional Practice Worksheet 4
Transparency 4
Evaluation
   Tests and Quizzes
     Quiz 3, Series 2

Class Opener

Given: Diagram as shown

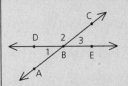

Prove: ∠1 ≅ ∠3

Statements	Reasons
1 ∠ABC is a straight ∠.	1 Assumed from diagram
2 ∠1 is supp. to ∠2.	2 If 2 adjacent ∠s form a straight ∠, they are supp.
3 ∠DBE is a straight ∠.	3 Same as 1
4 ∠2 is supp. to ∠3.	4 Same as 2
5 ∠1 ≅ ∠3	5 Supplements of the same ∠ are ≅.

Lesson Notes

- In proofs, students can assume the existence of vertical angles from a diagram. Students may want to copy the chart on page 19 and add this assumption.

2.8 VERTICAL ANGLES

Objectives
After studying this section, you should be able to
- Recognize opposite rays
- Recognize vertical angles

Part One: Introduction

Opposite Rays
$\overrightarrow{AB}$ and $\overrightarrow{AC}$ are **opposite rays**.

$\overrightarrow{ED}$ and $\overrightarrow{EF}$ are also opposite rays,
as are $\overrightarrow{EG}$ and $\overrightarrow{EH}$.

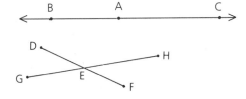

Definition	Two collinear rays that have a common endpoint and extend in different directions are called **opposite rays**.

Some pairs of rays that are *not* opposite rays are shown below.
$\overrightarrow{JK}$ and $\overrightarrow{MO}$ are not parts of the same line.
$\overrightarrow{PT}$ and $\overrightarrow{RS}$ are not opposite, since
they do not have a common endpoint.

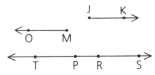

Vertical Angles
Whenever two lines intersect, two pairs of **vertical angles** are
formed.

Definition	Two angles are **vertical angles** if the rays forming the sides of one and the rays forming the sides of the other are opposite rays.

100 | Chapter 2   Basic Concepts and Proofs

Vocabulary
opposite rays
vertical angles

∠1 and ∠2 are vertical angles.
∠3 and ∠4 are vertical angles.

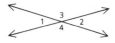

Are ∠3 and ∠2 vertical angles? How do vertical angles compare in size?

Theorem 18   *Vertical angles are congruent.*

Given: Diagram as shown
Prove: ∠5 ≅ ∠7

We proved Theorem 18 in Section 2.4, sample problem 3.

Part Two: Sample Problems

Problem 1   Given: ∠2 ≅ ∠3
Prove: ∠1 ≅ ∠3

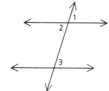

Proof

Statements	Reasons
1 ∠2 ≅ ∠3	1 Given
2 ∠1 ≅ ∠2	2 Vertical angles are congruent.
3 ∠1 ≅ ∠3	3 If ∠s are ≅ to the same ∠, they are ≅. (Transitive Property)

Problem 2   Given: ∠O is comp. to ∠2.
∠J is comp. to ∠1.
Conclusion: ∠O ≅ ∠J

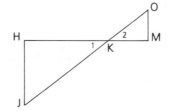

Proof

Statements	Reasons
1 ∠O is comp. to ∠2.	1 Given
2 ∠J is comp. to ∠1.	2 Given
3 ∠1 ≅ ∠2	3 Vertical angles are congruent.
4 ∠O ≅ ∠J	4 Complements of ≅ ∠s are ≅.

Communicating Mathematics

Have students summarize the different angles they have studied. Encourage them to be creative. Suggest that they construct posters that can be displayed in your classroom.

Problem 3 Given: m∠4 = 2x + 5,
m∠5 = x + 30

Find: m∠4

Solution 2x + 5 = x + 30
x = 25

Therefore, m∠4 = 2(25) + 5, or 55.

Part Three: Problem Sets

Problem Set A

1 a Name three pairs of opposite rays in the diagram. $\overrightarrow{FE}$, $\overrightarrow{FC}$; $\overrightarrow{FD}$, $\overrightarrow{FA}$; $\overrightarrow{BA}$, $\overrightarrow{BC}$
b Name two pairs of vertical angles.
∠EFA and ∠CFD;
∠EFD and ∠CFA

2 Given: ∠1 = 60°32′

Find: **a** ∠2 119°28′
b ∠3 60°32′
c ∠4 119°28′

3 Given: ∠5 = (2x + 7)°,
∠6 = (x + 25)°

Find: m∠5 43

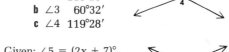

4 Given: ∠A ≅ ∠ACB
Prove: ∠A ≅ ∠DCE

5 Given: ∠1 ≅ ∠4
Conclusion: ∠2 ≅ ∠3

6 Given: $\overline{FH}$ ≅ $\overline{GJ}$
Prove: $\overline{FG}$ ≅ $\overline{HJ}$

7 Is this possible? No
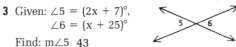
(3x−3)°
(4x)°

8 Given: $\angle 4 \cong \angle 6$

Prove: $\angle 5 \cong \angle 6$

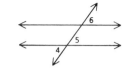

Problem Set B

9 Given: $\angle 1 \cong \angle 3$

Prove: $\angle 2$ is supp. to $\angle 3$.

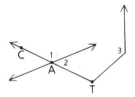

10 Given: $\angle V \cong \angle YRX$,

$\angle Y \cong \angle TRV$

Prove: $\angle V \cong \angle Y$

11 Given: $\overleftrightarrow{GD}$ bisects $\angle CBE$.

Conclusion: $\angle 1 \cong \angle 2$

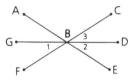

12 Angles 4, 5, and 6 are in the ratio 2:5:3.
Find the measure of each angle. 36; 90; 54

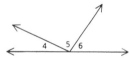

13 If a pair of vertical angles are supp., what
can we conclude about the angles? They are right $\angle$s.

14 Graph the five points A = (3, −4), B = (0, 5), C = (0, −5),
D = (−3, 4), and O = (0, 0). Which of the following are opposite
rays? **a** and **b**

a $\overrightarrow{OC}$, $\overrightarrow{OB}$ **b** $\overrightarrow{OA}$, $\overrightarrow{OD}$ **c** $\overrightarrow{BC}$, $\overrightarrow{CB}$ **d** $\overrightarrow{OB}$, $\overrightarrow{OD}$

Problem Set C

15 Find m$\angle 1$.
$132\frac{3}{4}$ or 140

*Problem-Set Notes and
Additional Answers, continued*

■ See *Solution Manual* for answers to problems **8–11**.

12 $2x + 5x + 3x = 180$

$x = 18$

$2x = 36$
$5x = 90$
$3x = 54$

15 $x^2 - 6x = \frac{1}{2}x + 42$,

so $x = \frac{21}{2}$ or $x = -4$.

$x^2 - 6x = 47\frac{1}{4}$ or $x^2 - 6x = 40$

$m\angle 1 = 180 - 47\frac{1}{4} = 132\frac{3}{4}$ or

$m\angle 1 = 140$

CHAPTER SUMMARY

CONCEPTS AND PROCEDURES

After studying this section, you should be able to
- Recognize the need for clarity and concision in proofs (2.1)
- Understand the concept of perpendicularity (2.1)
- Recognize complementary and supplementary angles (2.2)
- Follow a five-step procedure to draw logical conclusions (2.3)
- Prove angles congruent by means of four new theorems (2.4)
- Apply the addition properties of segments and angles (2.5)
- Apply the subtraction properties of segments and angles (2.5)
- Apply the multiplication and division properties of segments and angles (2.6)
- Apply the transitive properties of angles and segments (2.7)
- Apply the Substitution Property (2.7)
- Recognize opposite rays (2.8)
- Recognize vertical angles (2.8)

VOCABULARY

complement (2.2)
complementary angles (2.2)
coordinates (2.1)
oblique lines (2.1)
opposite rays (2.8)
origin (2.1)
perpendicular (2.1)

substitute (2.7)
substitution (2.7)
supplement (2.2)
supplementary angles (2.2)
vertical angles (2.8)
x-axis (2.1)
y-axis (2.1)

REVIEW PROBLEMS

Problem Set A

1 Given: $\overline{JK} \perp \overline{KM}$

Prove: $\angle JKO$ is comp. to $\angle OKM$.

2 Given: $\overline{PV} \cong \overline{PR}$,
$\overline{VT} \cong \overline{RS}$

Conclusion: $\overline{PT} \cong \overline{PS}$

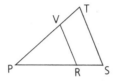

3 Given: $\angle WXT \cong \angle YXZ$

Prove: $\angle WXZ \cong \angle TXY$

4 Given: $\overline{FG} \cong \overline{JH}$;
N is the midpt. of $\overline{FG}$.
O is the midpt. of $\overline{JH}$.

Prove: $\overline{NG} \cong \overline{OH}$

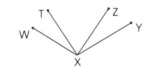

5 Given: $\overline{RV}$ and $\overline{SW}$ trisect $\overline{PT}$ and $\overline{PX}$.
$\overline{ST} \cong \overline{WX}$

Conclusion: $\overline{PT} \cong \overline{PX}$

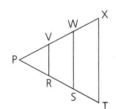

6 Given: Diagram as shown,
$\angle 1 \cong \angle 4$

Prove: $\angle 2 \cong \angle 3$

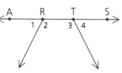

7 Point E divides $\overline{DF}$ into segments in a
ratio (from left to right) of 5:2.
If DF = 21 cm, find EF. 6 cm

CHAPTER 2 REVIEW

Class Planning

Time Schedule
All levels: 1 day

Resource References
Evaluation
  Tests and Quizzes
    Test 2, Series 1, 2, 3

Assignment Guide

Basic
1–8, 11–18, 20, 21

Average
4, 5, 7, 11–18, 20, 21, 32, 33, 35

Advanced
17, 18, 23–27, 30, 33–36

Problem-Set Notes
and Additional Answers

■ See *Solution Manual* for answers to problems **1–6.**

7 5x + 2x = 21
       x = 3

*Problem-Set Notes and
Additional Answers, continued*

■ See *Solution Manual* for answers to problems **8–11**.

12 $(90 - x) = 6 + 2x$

13 $(180 - x) = 5(90 - x)$

Review Problem Set A, *continued*

8 Given: ∠A is supp. to ∠D.
     ∠A ≅ ∠C
   Prove: ∠C is supp. to ∠D.

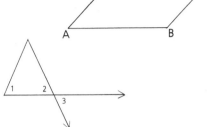

9 Given: ∠1 ≅ ∠3
   Conclusion: ∠1 ≅ ∠2

10 Given: ∠EGF ≅ ∠EFG,
      ∠EGH ≅ ∠EFJ
    Conclusion: ∠HGF ≅ ∠JFG

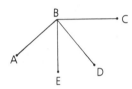

11 Given: ∠ABD is a right ∠.
      ∠CBE is a right ∠.
    Conclusion: ∠ABE ≅ ∠CBD

12 One of two complementary angles has a measure that is six more than twice the other's. Find the measure of the larger angle. 62

13 The meaure of the supplement of an angle is five times that of the angle's complement. Find the measure of the complement. $22\frac{1}{2}$

14 Two nonperpendicular intersecting lines are called __?__. Oblique

15 Point A is the midpoint of $\overline{DE}$, and DA = 12. Points I and N are trisection points of $\overline{DE}$. Find AN. 4

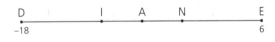

16 Find the supplement and the complement of each angle.

  a 83°   **b** 42°15′38″   **c** 97°
     97°; 7°   137°44′22″; 47°44′22″   83°; none

17 If $\overline{AB}$ is reflected over the x-axis, what will the coordinates of the endpoints of the reflection be? (2, −5), (6, −11)

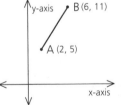

18 A point, R, was rotated about the origin, first 180° clockwise and then 90° counterclockwise. It ended at R' = (5, 0). Find the coordinates of point R. (0, 5)

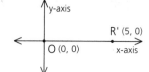

19 ∠PQR is a right angle. If $\overrightarrow{QS}$ is drawn at random between the sides of ∠PQR, what is the probability that

a ∠PQS and ∠SQR are complementary? 1
b ∠PQS is between 0° and 45°? $\frac{1}{2}$

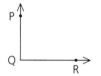

20 ABCD is a rectangle.

a Find the coordinates of B and D. (0, 8); (6, −5)
b If a point within ABCD is picked at random, what is the probability that it is in the shaded region? $\frac{5}{13}$

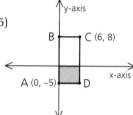

21 Given: ∠PQR ≅ ∠PRQ;
$\overrightarrow{QS}$ bisects ∠PQR.
$\overrightarrow{RS}$ bisects ∠PRQ.
∠PQR = 87°26′
Find: ∠PRS 43°43′

Problem Set B

22 Given: ∠1 is comp. to ∠3.
∠4 is comp. to ∠2.
Conclusion: ∠1 ≅ ∠4

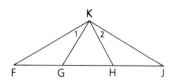

23 Given: O is the midpoint of $\overline{NP}$.
$\overline{RN} ≅ \overline{PO}$
Conclusion: $\overline{RN} ≅ \overline{NO}$

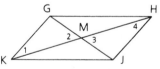

24 Given: ∠F ≅ ∠1,
∠J ≅ ∠2,
$\overline{FK} ⊥ \overline{KH}$,
$\overline{GK} ⊥ \overline{KJ}$
Prove: ∠F ≅ ∠J

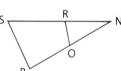

Problem-Set Notes and Additional Answers, continued

■ See *Solution Manual* for answers to problems **22–24**.

Problem-Set Notes and
Additional Answers, continued

■ See *Solution Manual* for answers to problems **25–27**.

28 Solve any two of the three
equations,
$2x + 2y + x = 180$,
$2y + x + y = 180$,
and $y = 2x$.

29 $2x = x + 4 = y + 5$,
so $x = 4$, $y = 3$.
∴ Perimeter = 50

Review Problem Set B, *continued*

25 Given: $\overrightarrow{VY}$ bisects $\angle TVZ$.
$\overrightarrow{ZY}$ bisects $\angle TZV$.
$\angle TVZ \cong \angle TZV$

Conclusion: $\angle 3 \cong \angle 1$

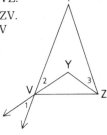

26 Given: $\overrightarrow{BC}$ bisects $\angle DBE$.
Prove: $\angle ABD \cong \angle ABE$

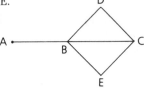

27 Given: $\angle NOP \cong \angle SRP$;
$\angle NOP$ is comp. to $\angle POR$.
$\angle SRP$ is comp. to $\angle PRO$.
Prove: $\angle POR \cong \angle PRO$

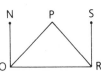

28 Solve for x and y.
$x = 25\frac{5}{7}$; $y = 51\frac{3}{7}$

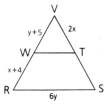

29 Given: $\overline{VS} \cong \overline{VR}$;
$\overline{WT}$ bisects $\overline{VS}$ and $\overline{VR}$.
Find: The perimeter of $\triangle VRS$ 50

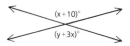

30 Solve for y in terms of x
$y = -2x + 10$

31 By how much does x exceed y? 110

32 The measure of the supplement of an angle exceeds twice the measure of the complement of the angle by 20. Find the measure of half of the complement. 35

33 $\overrightarrow{BD}$ bisects ∠ABC.

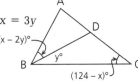

a Write an equation that relates x and y. x − 2y = y or x = 3y
b If ∠DBC ≅ ∠C, write another equation relating x and y. y = 124 − x
c Use substitution with parts **a** and **b** to find m∠C. 31

34 Given: $\overrightarrow{OP}$ bisects ∠MOE.
  m∠MOP = 10 − 3x,
  m∠POE = x² − 6x
Find: m∠MOE 32

35 a Find the area of △BDE. 28
b How does the area of △ABC compare with the area of △EDC? Same

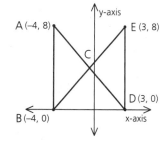

Problem Set C

36 With respect to the origin, point A = (1, 2) is rotated 100° clockwise, then 80° counterclockwise, then 210° clockwise, and finally 50° counterclockwise to point B.

a Find the coordinates of point B. (−1, −2)
b After which of the four rotations was the point in Quadrant I? After the second

37 Tippy Van Winkle is awakened from a deep sleep by the cuckoo of a clock that sounds every half hour. Before Tippy can look at the clock, his brother Bippy enters the room and offers to bet $10 that the hands of the clock form an acute angle. Assuming that the hands have not moved since the cuckoo sounded, how much should Tippy put up against Bippy's $10 so that it is an even bet? $14

38 Given: ∠ABD is supp. to ∠EDB.
  $\overrightarrow{BC}$ bisects ∠ABD.
  $\overrightarrow{DC}$ bisects ∠BDE.
Prove: ∠CBD is comp. to ∠BDC.
(Use a paragraph proof.)

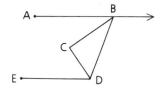

Review Problems | **109**

Problem-Set Notes and Additional Answers, continued

37

12:00	6:00
12:30	(6:30)
(1:00)	7:00
1:30	(7:30)
(2:00)	8:00
2:30	(8:30)
3:00	9:00
(3:30)	9:30
4:00	(10:00)
(4:30)	10:30
5:00	(11:00)
(5:30)	11:30

Probability of acute = $\frac{10}{24} = \frac{5}{12}$.
Thus, $\frac{5}{12}x = \frac{7}{12}($10)$; x = $14

38

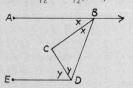

2x + 2y = 180,
so x + y = 90

3 | CONGRUENT TRIANGLES

Chapter 3 Schedule

Basic
Problem Sets:	13 days
Review:	2 days
Test:	1 day

Average
Problem Sets:	12 days
Review:	2 days
Test:	1 day

Advanced
Problem Sets:	10 days
Review:	2 days
Test:	1 day

*T*HE CHAPTER SCHEDULE indicates one day for the Chapter Review (p. 162) and one day for the Cumulative Review (p. 165). Other cumulative reviews follow Chapters **6, 9, 12,** and **15**; these problems can also be used as additional practice or extra credit.

Students learned the rudiments of writing proofs in the two previous chapters. In this chapter, they should learn how to analyze a problem in Euclidean geometry and logically organize its solution.

The short forms of theorems and postulates introduced in this chapter can reduce the amount of time needed to write out a proof. Students who use the short forms should occasionally be required to describe the abbreviations, to retain the meanings of the original statements.

Congruent triangles create a geometric design in this painting by Dorothea Rockburne.

WHAT ARE CONGRUENT FIGURES?

Objectives

After studying this section, you will be able to
- Understand the concept of congruent figures
- Accurately identify the corresponding parts of figures

Part One: Introduction

Congruent Figures

Although you learned a bit about the art of proof in Chapters 1 and 2, you may still be uneasy about proofs. You will, however, find your confidence growing as you work with triangles in this chapter. What you discover about congruent triangles will help you understand the characteristics of the other geometric figures you will meet in your studies.

In general, two geometric figures are congruent if one of them could be placed on top of the other and fit exactly, point for point, side for side, and angle for angle. *Congruent figures have the same size and shape.*

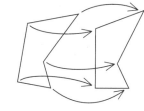

Every triangle has six parts—three angles and three sides. When we say that $\triangle ABC \cong \triangle FED$, we mean that $\angle A \cong \angle F$, $\angle B \cong \angle E$, and $\angle C \cong \angle D$ and that $\overline{AB} \cong \overline{FE}$, $\overline{BC} \cong \overline{ED}$, and $\overline{CA} \cong \overline{DF}$.

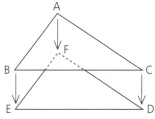

Definition **Congruent triangles** $\Leftrightarrow$ all pairs of corresponding parts are congruent.

Remember, an arrow symbol ($\Rightarrow$) means "implies" ("If . . ., then . . ."). If the arrow is double ($\Leftrightarrow$), the statement is reversible.

Vocabulary

congruent polygons
congruent triangles
reflection

Reflexive Property
rotate
slide

Class Planning

Time Schedule
All levels: 1 day

Resource References
Teacher's Resource Book
   Class Opener 3.1A, 3.1B
   Using Manipulatives 3, 4
Transparency 5
Evaluation
   Tests and Quizzes
     Quiz 1, Series 3

Class Opener

In problems **1–3**, find the number of triangles in each diagram.

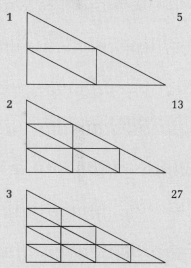

1   5

2   13

3   27

Lesson Notes

- As students read this introductory section in class, it can be emphasized that the properties listed on page 112 apply to both segments and angles.

Lesson Notes, continued

- A brief preview of the concept of similarity as "same shape" may help students see that the congruence symbol is a combination of same shape (~) and same size (=).
- Students should realize that, when they write a statement such as △ABC ≅ △DEF, they are not only stating that two triangles are congruent, but also stating which parts are congruent.
- While some students might think a statement based on the Reflexive Property is too obvious to bother writing in a proof, the statement must be included to complete the proof. Students will see this need when they read the note on page 119 about identifying statements in which congruent parts are found.

Cooperative Learning

This is a traditional quilt pattern made from many congruent pieces of fabric.

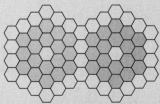

Have students work in small groups and find four congruent figures within the patterns. Have them label the vertices of two of the figures and determine the correspondence based on one of the following.   Answers will vary. Possible answers shown.

1   A reflection over a shared line

Would the statement △ABC ≅ △DEF be correct? The answer is no! Corresponding letters must match in the correspondence.

Correct: △ ABC ≅ △ FED   Incorrect: △ ABC ≅ △ DEF

To say that △ABC ≅ △DEF is incorrect because △ABC cannot be placed on △DEF so that A falls on D, B on E, and C on F. Would it be correct to say that △EDF ≅ △BCA?

In later chapters we will use the following definition.

Definition   ***Congruent polygons*** ⟺ all pairs of corresponding parts are congruent.

Writing proofs involving congruent triangles will be unnecessarily tedious unless we shorten some of the reasons. From now on, therefore, we will refer to many theorems and postulates in proofs only by the names or abbreviations we have assigned. You may wish to review the following properties, presented in Chapter 2:

Addition Property   Multiplication Property   Transitive Property
Division Property   Subtraction Property   Substitution Property

More About Correspondences

Notice that △KET is a ***reflection*** of △KIT over $\overline{KT}$.

∠I reflects onto ∠E.
∠1 reflects onto ∠2.
∠3 reflects onto ∠4.
$\overline{KI}$ reflects onto $\overline{KE}$.
$\overline{IT}$ reflects onto $\overline{ET}$.

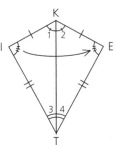

Notice also that $\overline{KT}$ is the sixth corresponding part. $\overline{KT}$ reflects onto itself. In fact, it is actually a side shared by the two triangles. We often need to include a shared side in a proof. Whenever a side or an angle is shared by two figures, we can say that the side or angle is congruent to itself. This property is called the ***Reflexive Property.***

Postulate   ***Any segment or angle is congruent to itself.***
(Reflexive Property)

A corresponds to H;   D corresponds to D;
B corresponds to G;   E corresponds to J;
C corresponds to C;   F corresponds to I.

∠PQR, in △PQR, is congruent to ∠SQT, in △SQT, by the Reflexive Property.

    Notice that ∠SQT and ∠PQR are actually different names for the same angle. We used different names so that you could see that the angle belonged to two different triangles.

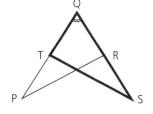

The two figures shown are congruent.

        PQRST ≅ VWXYZ

The correspondence is evident if we **slide** PQRST onto VWXYZ.

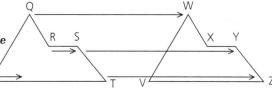

The triangles at the right are congruent. To determine the correspondence of the triangles, we can **rotate** △FGH onto △LKH about H.

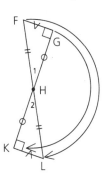

Angle 1 at H rotates onto angle 2 at H.

Thus, all six pairs of corresponding parts are congruent.

Part Two: Sample Problems

In the following two problems, try to justify each conclusion with one of the properties presented in Chapter 2 and in this section.

Problem 1   *Given:* M *and* N *are midpoints.*
$\overline{DC} \cong \overline{AB}$, $\overline{AB} \cong \overline{DB}$,
∠1 ≅ ∠4, ∠2 ≅ ∠3

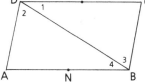

    Conclusions: **a** ∠ADC ≅ ∠ABC
                b $\overline{CM} \cong \overline{AN}$
                c $\overline{BD} \cong \overline{DB}$
                d $\overline{DC} \cong \overline{DB}$

Answers   **a** Addition Property
           b Division Property
           c Reflexive Property
           d Transitive Property

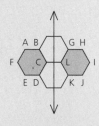

Problem 2   Given: $\overrightarrow{FP}$ and $\overrightarrow{GP}$ are angle bisectors.
∠5 is an acute angle.
∠5 ≅ ∠7, $\overline{PF}$ ≅ $\overline{PG}$, $\overline{QG}$ ≅ $\overline{FR}$

Conclusions: **a** ∠QFG ≅ ∠RGF
b $\overline{QP}$ ≅ $\overline{PR}$
c ∠7 is an acute angle.
d ∠FER ≅ ∠GEQ

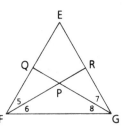

Answers   **a** Multiplication Property
b Subtraction Property
c Substitution
d Reflexive Property

Part Three: Problem Set

In problems 1–3, indicate which triangles are congruent. Be sure to have the correspondence of letters correct.

1 △ERC ≅ _?_ △TCR
Why is $\overline{RC}$ ≅ $\overline{RC}$?
Reflexive Property

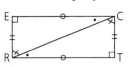

2 E is the midpt. of $\overline{TP}$.
△SPE ≅ _?_ △RTE

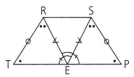

3 △BOW ≅ _?_ △YTW
Why is ∠1 ≅ ∠2? Vert. ∠s are ≅.

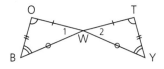

4 a Copy △PQR. Draw its reflection over
the x-axis and give the coordinates of
the vertices. P′ = (−5, 0); Q′ = (−2, −6); R′ = (7, 0)

b Copy △PQR. Draw its reflection over
the y-axis and give the coordinates of
the vertices. P′ = (5, 0); Q′ = (2, 6); R′ = (−7, 0)

c Copy △PQR. Slide it 3 units to the left
and give the coordinates of the vertices. P′ = (−8, 0); Q′ = (−5, 6); R′ = (4, 0)

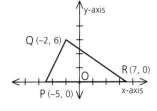

5 a Draw the rotation of △PQR 180° clockwise about O. Label its
vertices with their coordinates. P′ = (5, 0); Q′ = (2, −6); R′ = (−7, 0)

b Draw the slide of △PQR along ray $\overrightarrow{PR}$ so that P is at O, and
label its vertices with their coordinates. P′ = (0, 0); Q′ = (3, 6); R′ = (12, 0)

c Draw the reflection of △PQR over the y-axis and label its
vertices with their coordinates.
P′(5, 0); Q′(2, 6); R′(−7, 0)

114   Chapter 3   Congruent Triangles

THREE WAYS TO PROVE TRIANGLES CONGRUENT

Objectives

After studying this section, you will be able to
- Identify included angles and included sides
- Apply the SSS postulate
- Apply the SAS postulate
- Apply the ASA postulate

Part One: Introduction

Included Angles and Included Sides

In the figure at the right, ∠H is **included** by the sides $\overline{GH}$ and $\overline{HJ}$. Side $\overline{GH}$ is included by ∠H and ∠G. Can you name the sides that include ∠G? Can you name the angles that include side $\overline{HJ}$?

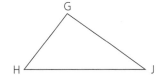

The SSS Postulate

Proving triangles congruent could be a very tedious task if we had to verify the congruence of every one of the six pairs of corresponding parts. Fortunately, triangles have some special properties that will enable us to prove two triangles congruent by comparing only three specially chosen pairs of corresponding parts. One of these sets of pairs consists of the corresponding sides.

This is the triangle that Jill built.

These are the three sticks that make up the triangle that Jill built.

Jill knows that there is only one triangle that can be constructed from three given sticks. In other words, if Jack has three sticks that are the same size as Jill's sticks, the only triangle he can build is one congruent to the triangle that Jill built.

Class Planning

Time Schedule
Basic: 3 days
Average: 3 days
Advanced: 2 days

Resource References
Teacher's Resource Book
    Class Opener 3.2A, 3.2B
Evaluation
    Tests and Quizzes
        Quiz 1, Series 1–2

Class Opener

1 Draw a triangle with the following specifications.
  a One side 5 cm
  b One side 8 cm
  c A 40° angle
2 Compare the triangle you drew to the triangles drawn by some of your classmates. Why are some triangles congruent and others not? Answers may vary. Triangles drawn with the 40° angle included between the sides with lengths 5 cm and 8 cm are congruent by the SAS postulate.

Lesson Notes

- Students usually need several days not only to properly understand the concepts of congruency and correspondence and the SSS, SAS, and ASA statements, but also to learn how they interrelate.

Vocabulary
included angles
included sides

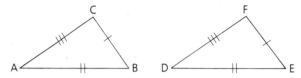

The tick marks on △ABC and △DEF show sufficient conditions for us to know that △ABC ≅ △DEF. This special property of triangles can be expressed as a postulate, which we will refer to as the SSS postulate. Each S stands for a pair of congruent corresponding sides, such as $\overline{AC}$ and $\overline{DF}$.

Postulate   *If there exists a correspondence between the vertices of two triangles such that three sides of one triangle are congruent to the corresponding sides of the other triangle, the two triangles are congruent. (SSS)*

The SSS relationship can be proved by methods that are not part of this course; we shall assume it and use the abbreviation SSS in proofs.

In the figure, is △GHJ congruent to △GKJ by SSS? The tick marks give us *two* pairs of congruent sides, but that is not enough. However, since $\overline{GJ}$ is a common side of both triangles, $\overline{GJ} \cong \overline{GJ}$ by the Reflexive Property. So we actually do have SSS!

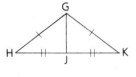

The following diagram illustrates the flow of logic that proves that △GHJ and △GKJ are congruent.

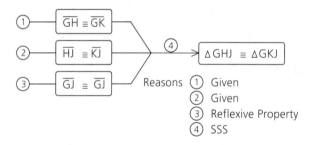

The SAS Postulate

It can also be shown that only two pairs of congruent corresponding sides are needed to establish the congruence of two triangles if the angles included by the sides are known to be congruent.

Postulate   *If there exists a correspondence between the vertices of two triangles such that two sides and the included angle of one triangle are congruent to the corresponding parts of the other triangle, the two triangles are congruent. (SAS)*

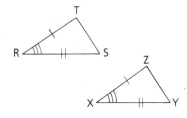

The fact that the *A* is between the *S*'s in *SAS* should help you remember that the congruent angles in the triangles must be the angles *included by* the pairs of congruent sides. Although this relationship, like SSS, can be proved, we shall assume it and use the abbreviation SAS in proofs.

The ASA Postulate

The following postulate will give us a third way of proving triangles congruent.

Postulate   *If there exists a correspondence between the vertices of two triangles such that two angles and the included side of one triangle are congruent to the corresponding parts of the other triangle, the two triangles are congruent. (ASA)*

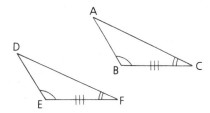

Again, ASA can be proved, although we shall assume it. The arrangement of the letters in *ASA* matches the arrangement of marked parts in the triangles; the congruent sides must be the ones included by the pairs of congruent angles.

If you are curious, you may be wondering whether SSS, SAS, and ASA are the only shortcuts for proving that triangles are congruent. Not quite. These three postulates, however, are enough to get us started on proofs that triangles are congruent.

Study the sample problems carefully before you attempt the problem sets. Notice that we call SSS, SAS, and ASA methods of proof. Any definition, postulate, or theorem can be called a method if it is a key reason in proofs.

Lesson Notes, continued

■ SAS and ASA are presented as postulates, although each can be derived from SSS, in order to get quickly into proving triangles congruent. Advanced courses may want to go into these derivations after the students are skilled in devising proofs.

Communicating Mathematics

Have students write a paragraph that explains the difference between the ASA postulate and the SAS postulate.

Part Two: Sample Problems

In problems 1–3 and 5, you are given the congruent angles and sides shown by the tick marks. Name the additional congruent sides or angles needed to prove that the triangles are congruent by each specified method.

Problem 1   **a** SSS
               b SAS

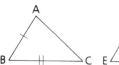

Answers   **a** $\overline{AC} \cong \overline{DF}$
            b $\angle B \cong \angle E$

Problem 2   **a** SAS
               b ASA

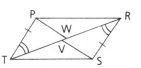

Answers   **a** $\overline{GJ} \cong \overline{OM}$
            b $\angle H \cong \angle K$

Problem 3   Prove: $\triangle PWT \cong \triangle SVR$
               a SAS
               b ASA

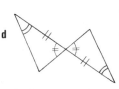

Answers   **a** $\overline{TW} \cong \overline{RV}$
            b $\angle TPW \cong \angle RSV$

Problem 4   Using the tick marks for each pair of triangles, name the method (SSS, SAS, or ASA), if any, that can be used to prove the triangles congruent.

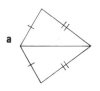
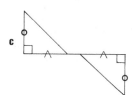
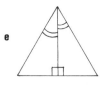

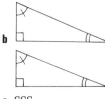

Answers   **a** SSS   **c** SAS   **e** ASA
            b None   **d** ASA   **f** None

Problem 5 Prove: △AEC ≅ △DEB

a SSS

b SAS

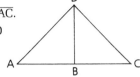

Answers **a** $\overline{AC} \cong \overline{BD}$

b ∠AEC ≅ ∠DEB

Problem 6 Given: $\overline{AD} \cong \overline{CD}$;

B is the midpoint of $\overline{AC}$.

Conclusion: △ABD ≅ △CBD

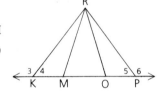

Proof

Statements	Reasons
1 $\overline{AD} \cong \overline{CD}$	1 Given
2 B is the midpt. of $\overline{AC}$.	2 Given
3 $\overline{AB} \cong \overline{CB}$	3 If a point is the midpoint of a segment, it divides the segment into two ≅ segments.
4 $\overline{BD} \cong \overline{BD}$	4 Reflexive Property
5 △ABD ≅ △CBD	5 SSS (1, 3, 4)

Note After SSS, SAS, or ASA we shall identify the numbers of the statements in which the pairs of congruent parts were found.

Problem 7 Given: ∠3 ≅ ∠6,

$\overline{KR} \cong \overline{PR}$,

∠KRO ≅ ∠PRM

Prove: △KRM ≅ △PRO

Proof

Statements	Reasons
1 ∠3 ≅ ∠6	1 Given
2 ∠3 is supp. to ∠4.	2 If two ∠s form a straight ∠ (assumed from diagram), they are supp.
3 ∠5 is supp. to ∠6.	3 Same as 2
4 ∠4 ≅ ∠5	4 Angles supp. to ≅ ∠s are ≅.
5 $\overline{KR} \cong \overline{PR}$	5 Given
6 ∠KRO ≅ ∠PRM	6 Given
7 ∠KRM ≅ ∠PRO	7 Subtraction Property
8 △KRM ≅ △PRO	8 ASA (4, 5, 7)

Note The assumption of straight angles and the fact that two angles that form a straight angle are supplementary may now be combined in one step (as in step 2 above).

Lesson Notes, continued

- The parenthesis in step 5 of Sample Problem **6** is a check that the proof is complete. The idea can be extended to any "If . . . , then . . ." statement, to identify which step in the proof satisfies the "if" clause.

- Refer to Sample Problem **7**. Students may want to add this assumption about straight angles to their copy of the chart on page 19.

Assignment Guide

Basic

Day 1	1–4, 7
Day 2	5, 6, 8–11
Day 3	12–17

Average

Day 1	3, 4, 6–10
Day 2	11, 12, 14, 15, 18
Day 3	20–26

Advanced

| Day 1 | 3–5, 7, 10–12 |
| Day 2 | 17, 19–24, 26, 28 |

Problem-Set Notes
and Additional Answers

■ See Solution Manual for answer
to problem **3**.

Part Three: Problem Sets

Problem Set A

1 Study the congruent sides and angles shown by the tick marks, then identify the additional information needed to support the specified method of proving that the indicated triangles are congruent.

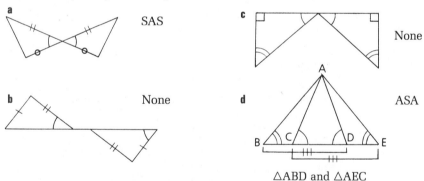

	Triangles	Method	Needed Information
a	△HGJ and △OKM	SAS	__?__ $\overline{GH} \cong \overline{KO}$
		ASA	__?__ $\angle J \cong \angle M$
b	△PSV and △TRV	SAS	__?__ $\overline{PS} \cong \overline{TR}$
		ASA	__?__ $\angle PVS \cong \angle TVR$
c	△WBZ and △YAX	SSS	__?__ $\overline{BZ} \cong \overline{AX}$
		SAS	__?__ $\angle BWZ \cong \angle AYX$

2 Using the tick marks for each pair of △, name the method (SSS, SAS, or ASA), if any, that will prove the △ to be ≅.

a SAS

b None

c None

d ASA

△ABD and △AEC

3 Given: $\overline{AB} \cong \overline{CB}$,
  $\angle ABD \cong \angle CBD$
Prove: △ABD ≅ △CBD

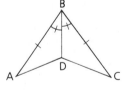

120 | Chapter 3 Congruent Triangles

4 Given: $\angle 1 \cong \angle 2$,
$\overline{EF} \cong \overline{HF}$

Prove: $\triangle EFJ \cong \triangle HFG$

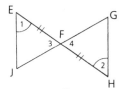

5 Given: $\overline{RO} \perp \overline{MP}$,
$\overline{MO} \cong \overline{OP}$

Prove: $\triangle MRO \cong \triangle PRO$

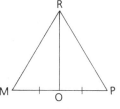

6 Given: $\overrightarrow{SV}$ bisects $\angle TSB$.
$\overrightarrow{VS}$ bisects $\angle TVB$.

Prove: $\triangle TSV \cong \triangle BSV$

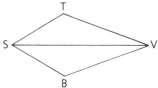

7 Given: $\overline{TV} \cong \overline{XW}$,
$\overline{VA} \cong \overline{WA}$,
$\overline{TA} \cong \overline{XA}$

Prove: $\triangle TVA \cong \triangle XWA$

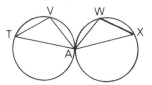

8 Given: $\overline{BC} \cong \overline{FE}$,
$\overline{DC} \cong \overline{DE}$,
$\angle 5 \cong \angle 6$

Prove: $\triangle BDG \cong \triangle FDG$

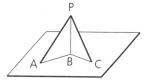

9 Two triangles are standing up on a table-top as shown. $\overline{PA} \cong \overline{PC}$ and $\overline{BA} \cong \overline{BC}$.

Prove: $\triangle PBA \cong \triangle PBC$

10 The perimeter of ABCD is 85. Find the value of x. Is $\triangle ABC$ congruent to $\triangle ADC$?
$x = 5\frac{1}{2}$; yes, by SSS

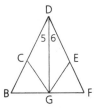

*Problem-Set Notes and
Additional Answers, continued*

■ See *Solution Manual* for answers to problems **4–9.**

Problem Set A, *continued*

11 Given: ∠N is comp. to ∠NPO.
∠S is comp. to ∠SPR.
∠NPO ≅ ∠SPR,
$\overline{NP} \cong \overline{SP}$

Conclusion: △NOP ≅ △SRP

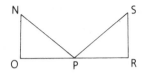

12 Given: O is the midpt. of $\overline{AY}$.
O is the midpt. of $\overline{ZX}$.

Conclusion: △ZOA ≅ △XOY

13 Given: $\overline{EO} \cong \overline{KM}$,
$\overline{FO} \cong \overline{JM}$,
$\overline{EG} \cong \overline{KH}$;
F is the midpt. of $\overline{EG}$.
J is the midpt. of $\overline{KH}$.

Conclusion: △EFO ≅ △KJM

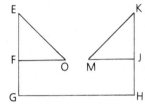

14 Given: ∠1 ≅ ∠4,
$\overline{PR} \cong \overline{TS}$,
$\overline{NP} \cong \overline{NT}$

Prove: △NPR ≅ △NTS

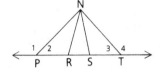

15 Given: $\overline{GH} \cong \overline{KJ}$,
$\overline{HM} \cong \overline{JO}$,
$\overline{GO} \cong \overline{KM}$

Prove: △GOJ ≅ △KMH

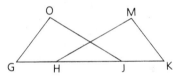

16 Given: ∠R ≅ ∠N,
$\overline{RP} \cong \overline{NT}$,
$\overline{RT} \cong \overline{NP}$,
$\overline{TS} \cong \overline{OP}$

Conclusion: △NOT ≅ △RSP

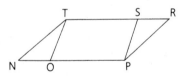

Problem Set B

17 Given: ∠1 ≅ ∠6,
$\overline{BC} \cong \overline{EC}$

Conclusion: △ABC ≅ △DEC

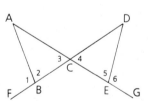

18 Given: $\overline{FH} \cong \overline{FK}$,
  ∠H ≅ ∠K;
  G is the midpt. of $\overline{FH}$.
  M is the midpt. of $\overline{FK}$.
  J is the midpt. of $\overline{HK}$.

Conclusion: △GHJ ≅ △MKJ

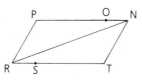

19 Given: $\overline{PR} \cong \overline{NT}$,
  $\overline{NO} \cong \overline{SR}$;

  O is $\frac{1}{3}$ of the way from N to P.

  S is $\frac{1}{3}$ of the way from R to T.

Prove: △NRT ≅ △RNP

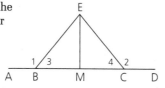

20 Study the problem below, then copy the
flow diagram and fill in the reason for
each statement.

Given: ∠1 ≅ ∠2;
  M is the midpt. of $\overline{BC}$.
  $\overline{BE} \cong \overline{CE}$

Prove: △EMB ≅ △EMC

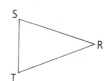

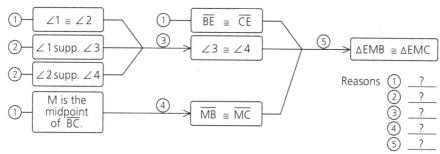

Reasons ① ____?____
         ② ____?____
         ③ ____?____
         ④ ____?____
         ⑤ ____?____

21 In problem 20, what given information is not needed to prove the
triangles congruent? ∠1 ≅ ∠2

22 Given: $\overline{RS} \cong \overline{RT}$
Conclusion: △RST ≅ △RTS

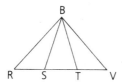

23 Given: S and T trisect $\overline{RV}$
  ∠R ≅ ∠V
  ∠BST ≅ ∠BTS
Conclusion: △BRS ≅ △BVT

■ See *Solution Manual* for answers to problems **18–20, 22,**
and **23.**

Problem-Set Notes and
Additional Answers, continued

Section 3.2 Three Ways to Prove Triangles Congruent **123**

24 Since $\overrightarrow{PY}$ bisects $\angle VPZ$,
$\angle VPY \cong \angle ZPY$;
$\therefore 2x + 7 = 3x - 9$
$x = 16$
$PZ = \frac{1}{2}x + 5 = 13$;
$PV = x - 3 = 13$,
so $\overline{PZ} \cong \overline{PV}$. Since $\overline{PY} \cong \overline{PY}$,
the $\triangle$ are $\cong$ by SAS.

25 $\angle DAC \cong \angle BAC$ by
substitution, $\overline{AC} \cong \overline{AC}$ by
the Reflexive Property, so
$\triangle CAD \cong \triangle CAB$ by SAS.

26 $\overline{AB} \cong \overline{AE}$; $\angle B \cong \angle E$, since
both are right $\angle$s.
$m\angle BAC = m\angle BAE + m\angle EAC$
$\qquad = m\angle CAD + m\angle EAC$
$\qquad = m\angle EAD$
Thus, $\triangle ABC \cong \triangle AED$ by
ASA.

27 $\overline{JK} \cong \overline{GM}$ by division;
$\angle FJK \cong \angle HGM$ by addition;
$\angle FKJ \cong \angle HMJ$ by $\angle$s supp.
to $\cong \angle$s are $\cong$. Thus,
$\triangle FJK \cong \triangle HGM$ by ASA.

28

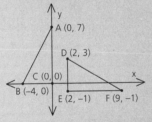

$BC = 4$, $DE = 4$, so
$\overline{BC} \cong \overline{DE}$. $\angle BCA$ and $\angle DEF$
are right $\angle$s, so they are $\cong$.
$CA = 7$, $EF = 7$, so
$\overline{CA} \cong \overline{EF}$. $\triangle ABC \cong \triangle FDE$
by SAS.

Cooperative Learning

Use this suggestion to extend
problem **28.** Have students work
in small groups to try to prove
that $\triangle ABC \cong \triangle FDE$ by SSS.
(Students can use the distance
formula or the Pythagorean Theorem to find the length of the
third sides of the triangles.)

Problem Set B, *continued*

24 Given: $\overrightarrow{PY}$ bisects $\angle VPZ$.
$\angle VPY = (2x + 7)°$,
$\angle ZPY = (3x - 9)°$,
$PZ = \frac{1}{2}x + 5$,
$PV = x - 3$
Prove: $\triangle VPY \cong \triangle ZPY$
(Use a paragraph proof.)

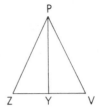

25 Given: $\angle 3 \cong \angle 1$, $\angle 4 \cong \angle 2$,
$\angle DAC \cong \angle 3$, $\angle BAC \cong \angle 1$,
$\overline{AD} \cong \overline{AB}$
Prove: $\triangle CAD \cong \triangle CAB$

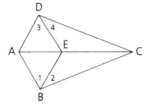

Problem Set C

26 Given: $\overline{AB} \cong \overline{AE}$;
$\overrightarrow{AE}$ and $\overrightarrow{AC}$ trisect $\angle BAD$.
$\overline{AB} \perp \overline{BC}$,
$\overline{AE} \perp \overline{DE}$
Conclusion: $\triangle ABC \cong \triangle AED$

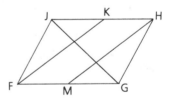

27 Given: $\overline{JH} \cong \overline{FG}$;
K and M are midpoints.
$\angle HKF \cong \angle FMH$,
$\angle KJG \cong \angle MGJ$,
$\angle JGH \cong \angle FJG$
Conclusion: $\triangle FJK \cong \triangle HGM$

28 Consider two triangles, $\triangle ABC$ and $\triangle FDE$, with vertices
$A = (0, 7)$, $B = (-4, 0)$, $C = (0, 0)$, $D = (2, 3)$, $E = (2, -1)$, and
$F = (9, -1)$. Draw a diagram and explain why $\triangle ABC \cong \triangle FDE$.

3.3 CPCTC AND CIRCLES

Objectives

After studying this section, you will be able to
- Apply the principle of CPCTC
- Recognize some basic properties of circles

Part One: Introduction

CPCTC

Suppose that in the figure △ABC ≅ △DEF. Is it therefore true that ∠B ≅ ∠E? If you refer to Section 3.1, you will find that we have already answered yes to this question in the definition of *congruent triangles*.

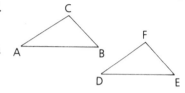

In the portions of the book that follow, we shall often draw such a conclusion *after* knowing that some triangles are congruent. We shall use CPCTC as the reason. *CPCTC* is short for "*Corresponding Parts of Congruent Triangles are Congruent.*" By corresponding parts, we shall mean only the matching angles and sides of the respective triangles.

Introduction to Circles

Point O is the center of the circle shown at the right. By definition, every point of the circle is the same distance from the center. The center, however, is not an element of the circle; the circle consists only of the "rim." A circle is named by its center: this circle is called circle O (or ⊙O).

Points A, B, and C lie on circle P (⊙P).
$\overline{PA}$ is called a *radius*.
$\overline{PA}$, $\overline{PB}$, and $\overline{PC}$ are called *radii*.

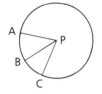

Class Planning

Time Schedule
Basic: 2 days
Average: 2 days
Advanced: 1 day

Resource References
Teacher's Resource Book
   Class Opener 3.3A, 3.3B
   Supposer Worksheet 3

Class Opener

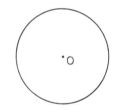

Given: $\overline{SM} \cong \overline{PM}$,
   ∠SMW ≅ ∠PMW
Prove: $\overline{SW} \cong \overline{WP}$

Statements	Reasons
1 $\overline{SM} \cong \overline{PM}$	1 Given
2 ∠SMW ≅ ∠PMW	2 Given
3 $\overline{MW} \cong \overline{MW}$	3 Reflexive Prop.
4 △SMW ≅ △PMW	4 SAS
5 $\overline{SW} \cong \overline{WF}$	5 CPCTC

Some students may not be able to complete the proof. However, they should understand that if two triangles are congruent, corresponding sides are also congruent. This is a nice introduction to CPCTC.

Lesson Notes

- Students should realize that they never use CPCTC in proofs until after proving that triangles are congruent.
- The definition of *circle* is presented in Chapter **10** (p. 439). This section provides a brief introduction to the circle.

From previous math courses you may remember formulas for the area and the circumference of a circle:

$$A = \pi r^2$$
$$C = 2\pi r$$

By pressing the $\boxed{\pi}$ key on a scientific calculator, you can find that $\pi \approx 3.141592654$.

Theorem 19 *All radii of a circle are congruent.*

Lesson Notes, *continued*

■ The proof of Theorem 19 follows from the definitions of circle and radius (p. 439).

The Geometric Supposer

Note It is most effective to do this activity as a class. *The Geometric Supposer: Triangles* offers two menu options to draw the altitudes and medians of a triangle. Have students draw any nonequilateral, nonisosceles triangle. Have students draw a median and an altitude. On the basis of this example, what seems to define a median and an altitude? Have students test their conjectures by measuring.

Part Two: Sample Problems

Problem 1   Given: ⊙P
Conclusion: $\overline{AB} \cong \overline{CD}$

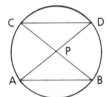

Proof

Statements	Reasons
1 ⊙P	1 Given
2 $\overline{PA} \cong \overline{PB} \cong \overline{PC} \cong \overline{PD}$	2 All radii of a circle are ≅.
3 ∠CPD ≅ ∠APB	3 Vertical angles are ≅.
4 △CPD ≅ △APB	4 SAS (2, 3, 2)
5 $\overline{AB} \cong \overline{CD}$	5 CPCTC (Corresponding parts of congruent triangles are congruent.)

Problem 2   Given: ⊙O;
∠T is comp. to ∠MOT.
∠S is comp. to ∠POS.
Prove: $\overline{MO} \cong \overline{PO}$

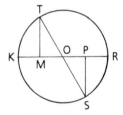

Proof

Statements	Reasons
1 ⊙O	1 Given
2 $\overline{OT} \cong \overline{OS}$	2 All radii of a circle are ≅.
3 ∠T is comp. to ∠MOT.	3 Given
4 ∠S is comp. to ∠POS.	4 Given
5 ∠MOT ≅ ∠POS	5 Vertical angles are ≅.
6 ∠T ≅ ∠S	6 Complements of ≅ ∠s are ≅.
7 △MOT ≅ △POS (Watch the correspondence.)	7 ASA (5, 2, 6)
8 $\overline{MO} \cong \overline{PO}$	8 CPCTC

Part Three: Problem Sets

Problem Set A

1 Given: $\overline{AB} \cong \overline{DE}$,
$\qquad$ $\overline{BC} \cong \overline{EF}$,
$\qquad$ $\overline{AC} \cong \overline{DF}$
$\quad$ Prove: $\angle A \cong \angle D$

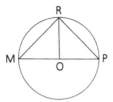

2 Given: $\angle HGJ \cong \angle KJG$,
$\qquad$ $\angle KGJ \cong \angle HJG$
$\quad$ Conclusion: $\overline{HG} \cong \overline{KJ}$

3 Given: $\odot O$,
$\qquad$ $\overline{RO} \perp \overline{MP}$
$\quad$ Prove: $\overline{MR} \cong \overline{PR}$

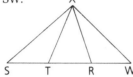

4 Given: T and R trisect $\overline{SW}$.
$\qquad$ $\overline{XS} \cong \overline{XW}$,
$\qquad$ $\angle S \cong \angle W$
$\quad$ Prove: $\overline{XT} \cong \overline{XR}$

5 Given: $\angle B \cong \angle Y$;
$\qquad$ C is the midpt. of $\overline{BY}$.
$\quad$ Conclusion: $\overline{AB} \cong \overline{YZ}$

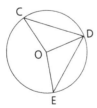

6 Given: $\odot O$,
$\qquad$ $\overline{CD} \cong \overline{DE}$
$\quad$ Prove: $\angle COD \cong \angle DOE$

7 Find, to the nearest tenth, the area and the circumference of a circle whose radius is 12.5 cm. $A \approx 490.9$ sq cm; $C \approx 78.5$ cm

8 $\triangle ABC \cong \triangle DEF$,
$\quad$ $\angle A = 90°$, $\angle B = 50°$, $\angle C = 40°$,
$\quad$ $m\angle E = 12x + 30$, $m\angle F = \frac{y}{2} - 10$,
$\quad$ $m\angle D = \sqrt{z}$
$\quad$ Solve for x, y, and z. $\frac{5}{3}$; 100; 8100

Problem-Set Notes and Additional Answers, continued

■ See *Solution Manual* for answers to problems **9, 10, 13,** and **14.**

11 Since the length of the radius is 5, the length of the circle's diameter is 10. Thus, the length of the side of the square is also 10.

Cooperative Learning

In problem **11,** consider another square, inscribed inside the circle. What is the area of the space between the circle and the smaller square? $25\pi - 50$
What is the ratio of the area of the larger square to the smaller square? 2:1
Does this ratio hold for any two such squares? Yes

12 $\overline{GO} \cong \overline{JK}, \angle G \cong \angle K,$
$\overline{GJ} \cong \overline{OK},$ and points H and M are midpts. ∴ $\overline{GH} \cong \overline{KM}$ by the Division Property.
$\triangle GHO \cong \triangle KMJ$ by SAS.
Then $\angle GOH \cong \angle MJK$ by CPCTC. So $x + 24 = 4x - 105; x = 43.$
Then $m\angle GOH = x + 24 = 67.$
$\angle GHO \cong \angle JMK$ by CPCTC.
So $2y - 7 = 3y - 23;$
$y = 16.$
Then $m\angle GHO = 2y - 7 = 25.$
$GH = KM = \frac{1}{2}(OK) = \frac{1}{2}(27)$
$= 13\frac{1}{2}$

Communicating Mathematics

Have students write a paragraph that explains what CPCTC represents. They should include examples in the explanation.

Problem Set A, *continued*

9 Given: $\overleftrightarrow{FH}$ bisects $\angle GFJ$
and $\angle GHJ.$
Conclusion: $\overline{FG} \cong \overline{FJ}$

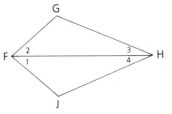

10 Given: $\angle M \cong \angle R,$
$\angle RPS \cong \angle MOK,$
$\overline{MP} \cong \overline{RO}$
Conclusion: $\overline{KM} \cong \overline{RS}$

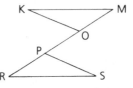

11 Explain why the area of the shaded region is $100 - 25\pi.$

Problem Set B

12 Given: H is the midpt. of $\overline{GJ}.$
M is the midpt. of $\overline{OK}.$
$\overline{GO} \cong \overline{JK},$
$\overline{GJ} \cong \overline{OK},$
$\angle G \cong \angle K,$
$OK = 27,$
$m\angle GOH = x + 24, m\angle GHO = 2y - 7,$
$m\angle JMK = 3y - 23, m\angle MJK = 4x - 105$
Find: $m\angle GOH, m\angle GHO,$ and GH 67; 25; $13\frac{1}{2}$

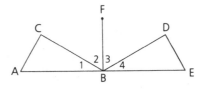

13 Given: $\angle A \cong \angle E,$
$\overline{AB} \cong \overline{BE},$
$\overline{FB} \perp \overline{AE},$
$\angle 2 \cong \angle 3$
Prove: $\overline{CB} \cong \overline{DB}$

14 Given: $\angle 5 \cong \angle 6,$
$\angle JHG \cong \angle O,$
$\overline{GH} \cong \overline{MO}$
Conclusion: $\angle J \cong \angle P$

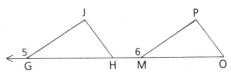

15 Given: ∠RST ≅ ∠RVT,
    ∠RVS ≅ ∠TSV
Conclusion: $\overline{RS} \cong \overline{VT}$

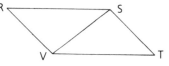

16 Given: ∠7 ≅ ∠8,
    $\overline{ZY} \cong \overline{WX}$
Prove: ∠W ≅ ∠Y

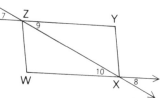

17 Given: ∠AEC ≅ ∠DEB,
    $\overline{BE} \cong \overline{CE}$,
    ∠ABE ≅ ∠DCE
Prove: $\overline{AB} \cong \overline{CD}$

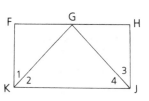

18 Given: $\overline{KG} \cong \overline{GJ}$,
    ∠2 ≅ ∠4,
    ∠1 is comp. to ∠2.
    ∠3 is comp. to ∠4.
    ∠FGJ ≅ ∠HGK
Conclusion: $\overline{FG} \cong \overline{HG}$

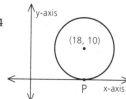

19 a Find the coordinates of point P. (18, 0)
   b Find the area of the circle. 100π, or ≈ 314

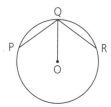

20 Given: ⊙O,
    $\overrightarrow{PQ} \cong \overrightarrow{QR}$
Prove: $\overrightarrow{QO}$ bisects ∠PQR.

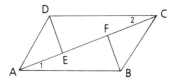

Problem Set C

21 Given: $\overline{AE} \cong \overline{FC}$,
    $\overline{FB} \cong \overline{DE}$,
    ∠CFB ≅ ∠AED
Prove: ∠1 ≅ ∠2

Problem-Set Notes and
Additional Answers, continued

■ See *Solution Manual* for answers to problems **15–18** and
20.

21 $\overline{AF} \cong \overline{EC}$ by addition.
∠DEF ≅ ∠BFE since ∠s
supp. to ≅ ∠s are ≅.
∴ △DEC ≅ △BFA by SAS.

Section 3.3   CPCTC and Circles   **129**

Problem Set C, *continued*

22 Prove that if $\overline{GJ}$ and $\overline{KH}$ bisect each other, then ∠MHO is larger than ∠K. (Write a paragraph proof.)

23 A radio antenna is kept perpendicular to the ground by three wires. They are staked at three points on a circle whose center is at the base of the antenna. Justify that the wires are equal in length.

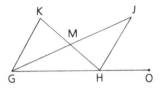

MATHEMATICAL EXCURSION

STRUCTURAL CONGRUENT TRIANGLES
Humanizing skyscrapers

The structural engineer William Le Messurier (born 1926) is a pioneer who has used geometric shapes to make taller, lighter, and more spacious skyscrapers that are structurally sound. One of his techniques is to use congruent triangles. In the 915-foot-tall Citicorp Center in Manhattan, he used triangles to more efficiently distribute the downward pressure exerted by each of the building's vertical sections. Each triangle absorbs the stress—the straining forces resulting from weight and gravity—from its section of the building and transfers it to a vertical column down the center of that side of the building. While we might take for granted the congruence of the triangles, it is important to the design of this building. If the triangles were not congruent, then the building's stress would be distributed unevenly. That would make it difficult to predict what would happen to the building as gravity acted upon it over time, or in extreme conditions, such as high winds.

The building's design and structural efficien-cy make possible the sunken, skylit plaza that sits underneath it, an inviting place to visit. Thus, congruent triangles contribute not only to safety but also to making our cities more pleasant and livable.

3.4 BEYOND CPCTC

Objectives

After studying this section, you will be able to
- Identify medians of triangles
- Identify altitudes of triangles
- Understand why auxiliary lines are used in some proofs
- Write proofs involving steps beyond CPCTC

Part One: Introduction

Medians of Triangles

Three *medians* are shown:

$\overline{AD}$ is a median of △ABC.
$\overline{EH}$ is a median of △EFG.
$\overline{FJ}$ is a median of △EFG.

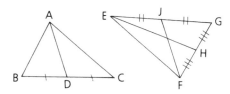

Every triangle has three medians.

Definition A *median* of a triangle is a line segment drawn from any vertex of the triangle to the midpoint of the opposite side. (A median divides into two congruent segments, or bisects the side to which it is drawn.)

Altitudes of Triangles

In the first diagram below, $\overline{AD}$ and $\overline{BE}$ are altitudes of △ABC.

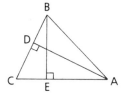 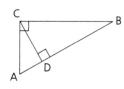 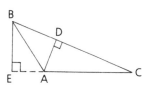

In the middle diagram, $\overline{AC}$ and $\overline{BC}$ and $\overline{CD}$ are altitudes of △ABC. Notice that in this case, two of the altitudes are sides of the triangle.

In the diagram on the right, $\overline{AD}$ and $\overline{BE}$ are altitudes of △ABC. Notice that altitude $\overline{BE}$ falls outside the triangle. Where does the third altitude lie?

Every triangle has three altitudes.

Section 3.4   Beyond CPCTC   **131**

Vocabulary
altitude
auxiliary lines
median

Class Planning

Time Schedule
All levels: 1 day

Resource References
Teacher's Resource Book
   Class Opener 3.4A, 3.4B
   Using Manipulatives 5
   Additional Practice Worksheet 5

Class Opener

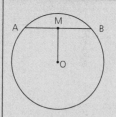

Given: $\overline{AM} \cong \overline{BM}$
Prove: ∠AMO ≅ ∠BMO
After some thought, students will want to know whether they can draw in $\overline{AO}$ and $\overline{BO}$. This is the appropriate time to talk about auxiliary lines.

Statements	Reasons
1 ⊙O	1 Given
2 $\overline{AM} \cong \overline{BM}$	2 Given
3 $\overline{MO} \cong \overline{MO}$	3 Reflexive Prop.
4 Draw $\overline{AO}$ and $\overline{BO}$.	4 Two points determine a segment.
5 $\overline{AO} \cong \overline{BO}$	5 All radii of a circle are ≅.
6 △AMO ≅ △BMO	6 SSS (2, 3, 5)
7 ∠AMO ≅ ∠BMO	7 CPCTC

Definition An *altitude* of a triangle is a line segment drawn from any vertex of the triangle to the opposite side, extended if necessary, and perpendicular to that side. (An altitude of a triangle forms right [90°] angles with one of the sides.)

Could an altitude of a triangle be a median as well?

Auxiliary Lines

Consider the following problem.

Given: $\overline{AB} \cong \overline{AC}$,
$\overline{BD} \cong \overline{CD}$

Conclusion: $\angle ABD \cong \angle ACD$

This proof would be easy if a line segment were drawn from A to D. We could then proceed to prove that $\triangle ABD \cong \triangle ACD$ (by SSS) and that $\angle ABD \cong \angle ACD$ (by CPCTC).

You will find that many proofs involve lines, rays, or segments that do not appear in the original figure. These additions to diagrams are called *auxiliary lines.* Most auxiliary lines connect two points already in the diagram, although you will see other types of auxiliary lines later in the course. Whenever we use an auxiliary line in a proof, we must be able to show that such a line can be drawn.

It is a postulate that *one and only one line, ray, or segment can be drawn through any two distinct points.*

Postulate *Two points determine a line (or ray or segment).*

The word *determine* indicates that there *is* a line through the given points and there is *no more than one* such line.

Steps Beyond CPCTC

Consider the following problem.

Given: $\overline{AD} \cong \overline{CD}$,
$\angle ADB \cong \angle CDB$

Prove: $\overline{DB}$ is the median to $\overline{AC}$.

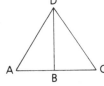

In this problem, we can prove that $\triangle ABD \cong \triangle CBD$ by SAS. Do you see how? Therefore, $\overline{AB} \cong \overline{CB}$ by CPCTC. Now we shall go one step beyond CPCTC. Since $\overline{AB} \cong \overline{CB}$, we may call $\overline{DB}$ a median of $\triangle ACD$, and the proof is complete.

Many proofs involve steps beyond CPCTC. By using CPCTC first, we can identify altitudes, angle bisectors, midpoints, and so forth. You will see some examples in the sample problems to follow.

A fascinating type of proof involves showing that one pair of triangles are congruent and then using CPCTC to show that another pair of triangles are congruent. Such proofs, called *detour proofs*, are explained in detail in Chapter 4.

Part Two: Sample Problems

Problem 1 Given: $\overline{AC} \cong \overline{BC}$,
$\overline{AD} \cong \overline{BD}$
Prove: $\overrightarrow{CD}$ bisects $\angle ACB$.

Proof *Flow of logic:*

$$\overline{AC} \cong \overline{BC}$$
$$\overline{AD} \cong \overline{BD} \rightarrow \triangle ACD \cong \triangle BCD \rightarrow \angle ACD \cong \angle BCD \rightarrow \overrightarrow{CD} \text{ bis. } \angle ACB$$
$$\overline{CD} \cong \overline{CD}$$

Statements	Reasons
1 $\overline{AC} \cong \overline{BC}$	1 Given
2 $\overline{AD} \cong \overline{BD}$	2 Given
3 $\overline{CD} \cong \overline{CD}$	3 Reflexive Property
4 $\triangle ACD \cong \triangle BCD$	4 SSS (1, 2, 3)
5 $\angle ACD \cong \angle BCD$	5 CPCTC
6 $\overrightarrow{CD}$ bisects $\angle ACB$.	6 If a ray divides an $\angle$ into two $\cong \angle$s, the ray bisects the $\angle$.

Problem 2 Given: $\overline{CD}$ and $\overline{BE}$ are altitudes of $\triangle ABC$.
$\overline{AD} \cong \overline{AE}$
Prove: $\overline{DB} \cong \overline{EC}$

Proof

Statements	Reasons
1 $\overline{CD}$ and $\overline{BE}$ are altitudes of $\triangle ABC$.	1 Given
2 $\angle ADC$ is a right $\angle$.	2 An altitude of a $\triangle$ forms right $\angle$s with the side to which it is drawn.
3 $\angle AEB$ is a right $\angle$.	3 Same as 2
4 $\angle ADC \cong \angle AEB$	4 If $\angle$s are right $\angle$s, they are $\cong$.
5 $\angle A \cong \angle A$	5 Reflexive Property
6 $\overline{AD} \cong \overline{AE}$	6 Given
7 $\triangle ADC \cong \triangle AEB$	7 ASA (4, 6, 5)
8 $\overline{AB} \cong \overline{AC}$	8 CPCTC
9 $\overline{DB} \cong \overline{EC}$	9 Subtraction Property (6 from 8)

Cooperative Learning

Have students work in small groups. Have each student draw one large triangle. For half the triangles drawn, the group should draw the three medians of each triangle. They should draw the three altitudes for each of the remaining triangles. Have groups discuss the following question: Could an altitude of a triangle be a median as well? Have students draw a diagram to support their answers.
Triangles will vary. An altitude can be a median as well. For example, in an isosceles triangle, the altitude drawn from the vertex angle to its base is also a median of the triangle.

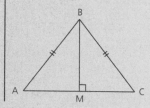

Lesson Notes, continued

■ In reason 5 of Sample Problem **2,** notice that the Reflexive Property has been used for angles. Students sometimes apply the Reflexive Property only to segments.

Problem 3   Given: G is the midpt. of $\overline{FH}$.
$\overline{EF} \cong \overline{EH}$

Prove: ∠1 ≅ ∠2

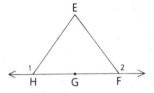

Proof

Statements	Reasons
1 G is the midpt. of $\overline{FH}$.	1 Given
2 $\overline{FG} \cong \overline{HG}$	2 If a point is the midpt. of a segment, it divides the segment into two ≅ segments.
3 $\overline{EF} \cong \overline{EH}$	3 Given
4 Draw $\overline{EG}$.	4 Two points determine a segment.
5 $\overline{EG} \cong \overline{EG}$	5 Reflexive Property
6 △EFG ≅ △EHG	6 SSS (2, 3, 5)
7 ∠EFG ≅ ∠EHG	7 CPCTC
8 ∠2 is supp. to ∠EFG.	8 If two ∠s form a straight ∠, they are supplementary.
9 ∠1 is supp. to ∠EHG.	9 Same as 8
10 ∠1 ≅ ∠2	10 Supplements of ≅ ∠s are ≅.

Problem 4   Given: ∠T ≅ ∠Y,
∠SVZ ≅ ∠SXZ,
$\overline{TV} \cong \overline{YX}$

Conclusion: $\overline{SZ}$ is the median to $\overline{TY}$.

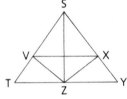

Proof

Statements	Reasons
1 ∠T ≅ ∠Y	1 Given
2 ∠SVZ ≅ ∠SXZ	2 Given
3 ∠SVZ is supp. to ∠TVZ.	3 If two ∠s form a straight ∠, they are supplementary.
4 ∠SXZ is supp. to ∠YXZ.	4 Same as 3
5 ∠TVZ ≅ ∠YXZ	5 Supplements of ≅ ∠s are ≅.
6 $\overline{TV} \cong \overline{YX}$	6 Given
7 △TVZ ≅ △YXZ	7 ASA (1, 6, 5)
8 $\overline{TZ} \cong \overline{YZ}$	8 CPCTC
9 $\overline{SZ}$ is the median to $\overline{TY}$.	9 If a segment from a vertex of a △ divides the opposite side into two ≅ segments, it is a median.

Lesson Notes, continued

■ In Sample Problem **3**, students should note that the form of reason 4 (the postulate on p. 132) is chosen because statement 4 establishes a segment.

Part Three: Problem Sets

Problem Set A

1 For the following figures, identify $\overline{AD}$ as a median, an altitude, neither, or both according to what can be proved.

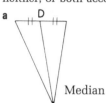
a

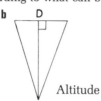
b

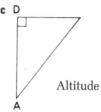
c

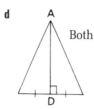
d Both

Median   Altitude   Altitude

2 Given: $\overline{HJ} \cong \overline{KJ}$,
$\angle MJH \cong \angle MJK$
Prove: $\overrightarrow{MJ}$ bisects $\angle HMK$.

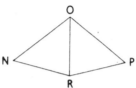

3 Given: $\overline{NR} \cong \overline{PR}$;
$\overrightarrow{RO}$ bisects $\angle NRP$.
Prove: $\overrightarrow{OR}$ bisects $\angle NOP$. (Draw a logical flow diagram for this problem and then give the proof.)

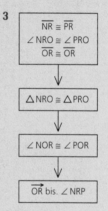

4 Given: $\angle CFD \cong \angle EFD$;
$\overline{FD}$ is an altitude.
Prove: $\overline{FD}$ is a median.

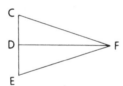

5 Given: $\odot O$,
$\overline{GJ} \cong \overline{HJ}$
Prove: $\angle G \cong \angle H$

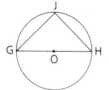

6 Given: $\overline{TW}$ is a median.
ST = x + 40,
SW = 2x + 30,
WV = 5x − 6
Find: SW, WV, and ST
54; 54; 52

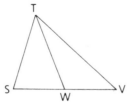

Section 3.4   Beyond CPCTC   **135**

Assignment Guide

Basic
2, 4–7
Average
3–7, 9
Advanced
3, 4, 11, 13–15

Problem-Set Notes and Additional Answers

■ See *Solution Manual* for answers to problems **2**, **4**, and **5**.

3

$\overline{NR} \cong \overline{PR}$
$\angle NRO \cong \angle PRO$
$\overline{OR} \cong \overline{OR}$

↓

$\triangle NRO \cong \triangle PRO$

↓

$\angle NOR \cong \angle POR$

↓

$\overrightarrow{OR}$ bis. $\angle NRP$

6 $\overline{SW} \cong \overline{WV}$, since $\overline{TW}$ is a median.
2x + 30 = 5x − 6
x = 12
ST = x + 40 = 52
SW = 2x + 30 = 54
WV = 5x − 6 = 54

Problem-Set Notes and
Additional Answers, continued

■ *See Solution Manual for an-*
swers to problems 7–11.

Problem Set A, *continued*

7 Given: $\overline{KP}$ is a median.
  $\overline{MK} \cong \overline{RK}$

  Conclusion: $\angle 3 \cong \angle 4$

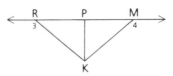

Problem Set B

8 Given: $\angle AEB \cong \angle DEC$,
  $\overline{AE} \cong \overline{DE}$,
  $\angle A \cong \angle D$
  Conclusion: $\overline{AC} \cong \overline{BD}$

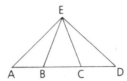

9 Given: $\odot O$,
  $\angle NOG \cong \angle POG$
  Conclusion: $\overrightarrow{RO}$ bisects $\angle NRP$.

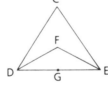

10 Given: $\overline{AZ} \cong \overline{ZB}$;
  Z is the midpt. of $\overline{XY}$.
  $\angle AZX \cong \angle BZY$,
  $\overline{XW} \cong \overline{YW}$
  Prove: $\overline{AW} \cong \overline{BW}$

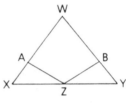

11 Given: $\overrightarrow{DF}$ bisects $\angle CDE$.
  $\overrightarrow{EF}$ bisects $\angle CED$.
  G is the midpt. of $\overline{DE}$.
  $\overline{DF} \cong \overline{EF}$
  Prove: $\angle CDE \cong \angle CED$

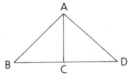

Problem Set C

12 Given: $\overline{AC}$ is the altitude to $\overline{BD}$.
  $\overline{AC}$ is a median.
  $\angle BAC$ is comp. to $\angle D$.
  Conclusion: $\angle DAC$ is comp. to $\angle B$.

13 In the graph of $\triangle ABC$, A = (−2, 6) and B = (8, 6). The altitude
  from C is 5. Where is point C located?
  At any point (x, y) where y = 11 or y = 1

12 Prove $\triangle BAC \cong \triangle DAC$ by
  SAS. Then $\angle BAC \cong \angle DAC$
  and $\angle D \cong \angle B$. We get the
  conclusion by substitution.
13 Since $\overline{AB}$ is a horizontal
  segment, the altitude from C
  will be a vertical segment
  and C can be any point
  (x, y) such that y = 6 ± 5;
  so y = 11 or y = 1.

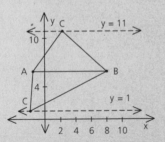

14 Given: ⊙O and ⊙P;
Perimeter of △AOP = 80.
OC + DP = 16;
$\overline{CD}$ is 2 units longer than $\overline{OC}$.

Find: OB + BP 48

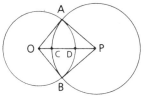

15 Given: $\overline{AB} \cong \overline{AC}$,
$\overline{BD} \cong \overline{CE}$

Prove: ∠1 ≅ ∠2

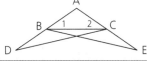

---

*Problem-Set Notes and
Additional Answers, continued*

14

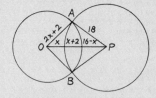

(2x + 2) + (18) + (x) +
(x + 2) + (16 − x) = 80
x = 14
OB = 30
BP = 18
OB + BP = 48

15 This problem is the historically famous "Bridge of Donkeys" proof of Euclid's *Propositions,* Book 1, which proves the base angles of an isosceles triangle are congruent.
Euclid's proof: AD = AE so
△ADC ≅ △AEB. Now
△BCD ≅ △CBE by SSS.
Then ∠1 ≅ ∠2 by ∠s supp. to ≅ ∠s are ≅. Modern-day proof:
△ABC ≅ △ACB, so
∠1 ≅ ∠2 by CPCTC.

CAREER PROFILE

SYMMETRY UNLOCKS CULTURE
Dorothy Washburn uses patterns to decode the past

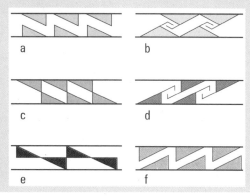

Archaeologists have traditionally classified decorative basket, cloth, and pottery patterns through reference to the design elements of the patterns. Archaeologist Dorothy Washburn decided to take a different approach. She explains her theory: "Structure is important in every culture. Instead of focusing on design elements I decided to look at the underlying *structure* of

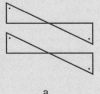

a

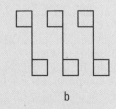

b

the patterns. It appeared that one fundamental rule guiding artists in their choices of patterns was pattern structure, so I proposed using symmetry as a basis for pattern classification."

Using Washburn's system, the two designs above would have identical classifications, since each has 180° (bifold) rotational symmetry. "We've uncovered a remarkable consistency in the choice of symmetries within a given cultural group. In my study of the Anasazi people of the American Southwest, I found that at most sites at least 50 percent of their decorative patterns were structured just by bifold rotational symme-

try." Most cultural groups use a small number of symmetries, sometimes for hundreds of years. If the group undergoes some major upheaval, the artists might then adopt a new series of symmetries.

Washburn majored in American history at Oberlin college, but one day she overheard another student discussing an upcoming archaeological dig. "Can I come along?" she inquired. Her future was altered. She joined a summer dig in Wyoming, then entered graduate school at Columbia University, where she earned her doctorate in anthropology. Today she is a research associate in anthropology at the University of Rochester.

Which strip patterns display bifold rotational symmetry?

a	b
c	d
e	f

OVERLAPPING TRIANGLES

Class Planning

Time Schedule
Basic: 2 days
Average: 1 day
Advanced: 1 day

Resource References
Teacher's Resource Book
　Class Opener 3.5A, 3.5B
Evaluation
　Tests and Quizzes
　　Quiz 2, Series 1–3

Class Opener

Given: $\overline{PW} \cong \overline{TM}$,
　　　$\overline{PM} \cong \overline{TW}$
Prove: $\angle P \cong \angle T$

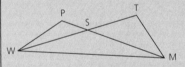

Statements	Reasons
1 $\overline{PW} \cong \overline{TM}$	1 Given
2 $\overline{PM} \cong \overline{TW}$	2 Given
3 $\overline{WM} \cong \overline{MW}$	3 Reflexive Prop.
4 $\triangle PWM \cong$ $\triangle TMW$	4 SSS (1, 2, 3)
5 $\angle P \cong \angle T$	5 CPCTC

This proof requires no new theorem. Students need to recognize the overlapping triangles.

Lesson Notes

■ Basic courses may benefit from a combination of colored triangles (as in the text) and "separating" the triangles.

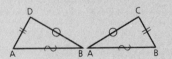

Objective
After studying this section, you will be able to
■ Use overlapping triangles in proofs

Part One: Introduction

Consider the following problem.

Given: $\overline{DB} \cong \overline{AC}$,
　　　$\overline{AD} \cong \overline{BC}$
Conclusion: $\angle D \cong \angle C$

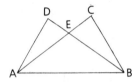

At first glance you would probably think of showing that $\triangle ADE \cong \triangle BCE$, thus proving that $\angle D \cong \angle C$ by CPCTC.

Soon you would realize that there is not enough information to prove that $\triangle ADE \cong \triangle BCE$. There must be another way.

In this case the problem can be solved by finding two other triangles to which $\angle D$ and $\angle C$ belong. We can prove that the overlapping triangles ABD and BAC are congruent by SSS, and thus that $\angle D \cong \angle C$ by CPCTC.

At first, you may have trouble recognizing which triangles to use in a proof. You may want to outline triangles in color, as in the sample problems. Just be willing to draw figures several times to find the triangles that serve best.

Almost all the problems in this section involve overlapping triangles. Elsewhere, the triangles of interest may or may not overlap.

Part Two: Sample Problems

Problem 1 Given: $\overline{AC} \cong \overline{AB}$,
$\overline{AE} \cong \overline{AD}$

Conclusion: $\overline{CE} \cong \overline{BD}$

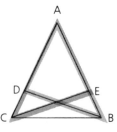

Proof

Statements	Reasons
1 $\overline{AC} \cong \overline{AB}$	1 Given
2 $\overline{AE} \cong \overline{AD}$	2 Given
3 $\angle A \cong \angle A$	3 Reflexive Property
4 $\triangle ADB \cong \triangle AEC$	4 SAS (1, 3, 2)
5 $\overline{CE} \cong \overline{BD}$	5 CPCTC

Problem 2 Given: $\overline{FH} \cong \overline{MJ}$;
G is the midpt. of $\overline{FH}$.
K is the midpt. of $\overline{MJ}$.
$\angle GHJ \cong \angle KJH$

Prove: $\overline{GJ} \cong \overline{HK}$

Proof

Statements	Reasons
1 $\overline{FH} \cong \overline{MJ}$	1 Given
2 G is the midpt. of $\overline{FH}$.	2 Given
3 K is the midpt. of $\overline{MJ}$.	3 Given
4 $\overline{GH} \cong \overline{KJ}$	4 Division Property
5 $\angle GHJ \cong \angle KJH$	5 Given
6 $\overline{HJ} \cong \overline{HJ}$	6 Reflexive Property
7 $\triangle GHJ \cong \triangle KJH$	7 SAS (4, 5, 6)
8 $\overline{GJ} \cong \overline{HK}$	8 CPCTC

Part Three: Problem Sets

Problem Set A

1 Given: $\overline{AB} \cong \overline{DC}$,
$\overline{AC} \cong \overline{DB}$

Prove: $\triangle ABC \cong \triangle DCB$

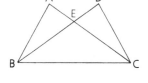

2 Given: $\angle FGH$ is a right $\angle$.
$\angle JHG$ is a right $\angle$.
$\overline{FG} \cong \overline{JH}$

Prove: $\triangle FGH \cong \triangle JHG$

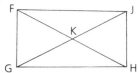

Cooperative Learning

Refer to the three diagrams below. Have students work in small groups to outline the overlapping triangles with different colored pencils or markers. Can more than one pair of overlapping triangles be found in any one diagram?   Yes

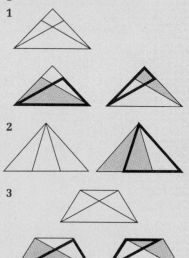

Assignment Guide

Basic	
Day 1	1–4
Day 2	5, 7, 8, 11

Average
4, 5, 7, 9–11

Advanced
5, 9–12, 14

Problem-Set Notes and Additional Answers

■ See *Solution Manual* for answers to problems **1** and **2**.

Problem-Set Notes and Additional Answers, continued

■ See *Solution Manual* for answers to problems **3–8.**

Communicating Mathematics

Have students draw two or three figures that contain overlapping triangles. Have them trade diagrams with each other and identify the overlapping triangles.

Problem Set A, *continued*

3 Given: $\overline{PM} \cong \overline{RM}$,
        $\angle SPM \cong \angle ORM$
    Prove: $\triangle PSM \cong \triangle ROM$

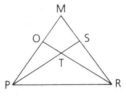

4 Given: $\angle 1 \cong \angle 3$,
        $\angle 2 \cong \angle 4$
    Conclusion: $\overline{BC} \cong \overline{ED}$

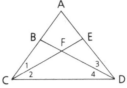

5 Given: $\overline{JH} \cong \overline{KH}$,
        $\overline{HG} \cong \overline{HM}$,
        $\angle 5 \cong \angle 6$
    Conclusion: $\triangle JHG \cong \triangle KHM$

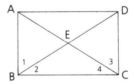

Problem Set B

6 Given: $\angle 1$ is comp. to $\angle 2$.
        $\angle 3$ is comp. to $\angle 4$.
        $\angle 1 \cong \angle 3$
    Conclusion: $\overline{AB} \cong \overline{CD}$

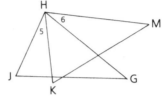

7 Given: Figure NOPRS is *equilateral*
        (all sides are congruent).
        $\angle OPR \cong \angle PRS$,
        $\overline{PT} \cong \overline{TR}$
    Prove: $\overline{OT} \cong \overline{ST}$

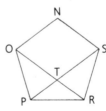

8 Given: $\angle 9 \cong \angle 10$,
        $\angle GFH \cong \angle HJG$
    Conclusion: $\overline{FG} \cong \overline{JH}$

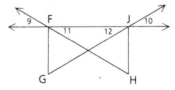

9 Given: $\overline{YW}$ bisects $\overline{AX}$.
$\angle A \cong \angle X$,
$\angle 5 \cong \angle 6$
Conclusion: $\overline{ZW} \cong \overline{YW}$

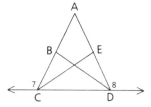

10 Given: B is the midpt. of $\overline{AC}$.
E is the mdpt. of $\overline{AD}$.
$\angle 7 \cong \angle 8$,
$\angle ECD \cong \angle BDC$
Prove: $\overline{AC} \cong \overline{AD}$

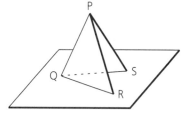

11 Given: Two triangles, joined along $\overline{PQ}$
and standing on a desktop,
$\overline{PS} \cong \overline{PR}$, $\angle QPS \cong \angle QPR$
Prove: $\overline{QR} \cong \overline{QS}$

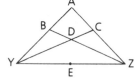

Problem Set C

12 Given: $\overline{HO} \cong \overline{MO}$,
$\overline{JO} \cong \overline{KO}$;
$\overline{HJ}$ is an altitude of $\triangle HJK$.
$\overline{MK}$ is an altitude of $\triangle MKJ$.
Prove: $\angle 1 \cong \angle 2$

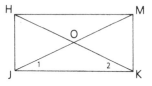

13 Given: $\overline{NR} \cong \overline{NV}$;
P and Q are midpoints.
$\angle R \cong \angle V$,
$\overline{PX} \cong \overline{QX}$
Prove: $\triangle XST$ is *isosceles* (at least two sides are $\cong$).

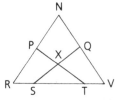

14 Given: $\overline{YD} \cong \overline{ZD}$,
$\overline{BD} \cong \overline{CD}$;
E is the midpt. of $\overline{YZ}$.
Conclusion: $\angle BYZ \cong \angle CZY$

Problem-Set Notes and Additional Answers, continued

■ See *Solution Manual* for answers to problems **9–11**.

12 Prove $\triangle HOJ \cong \triangle MOK$ by SAS to get $\overline{HJ} \cong \overline{MK}$. Then use addition and Reflex. Prop. to prove $\triangle HJK \cong \triangle MKJ$ by SSS.

13 Use division and draw $\overline{NX}$ to prove $\triangle NPX \cong \triangle NQX$. $\therefore \angle NPX \cong \angle NQX$, so $\angle RPT \cong \angle VQS$, since they are $\angle$s supp. to $\cong \angle$s. Since $\overline{PR} \cong \overline{QV}$ by division, $\triangle PRT \cong \triangle QVS$ by ASA, so $\overline{QS} \cong \overline{PT}$. Then $\overline{XS} \cong \overline{XT}$ by subtraction.

14 Draw $\overline{DE}$. Prove $\triangle BYD \cong \triangle CZD$ and $\triangle DYE \cong \triangle DZE$. Then $\angle BYD \cong \angle CZD$ and $\angle DYE \cong \angle DZE$ (use addition). Or (without using E) prove $\triangle BYD \cong \triangle CZD$, then $\overline{BZ} \cong \overline{CY}$, $\angle YBD \cong \angle ZCD$, and $\overline{BZ} \cong \overline{CY}$ (by addition). $\therefore \triangle BYZ \cong \triangle CZY$, and the conclusion follows by CPCTC.

Objective
After studying this section, you will be able to
■ Name the various types of triangles and their parts

Part One: Introduction

A number of names are used to distinguish triangles having special
characteristics.

Definition   A *scalene triangle* is a triangle in which
            no two sides are congruent.

Scalene Triangle

Definition   An *isosceles triangle* is a triangle in which at least
            two sides are congruent.

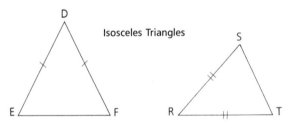

Isosceles Triangles

In △DEF above, $\overline{DE} \cong \overline{DF}$. $\overline{DE}$ and $\overline{DF}$ are called *legs* of the isosceles
triangle, $\overline{EF}$ is called the *base*, ∠E and ∠F are called *base angles*,
and ∠D is called the *vertex angle*. Can you name these parts in
△RST?

Definition   An *equilateral triangle* is a triangle
            in which all sides are congruent.

Equilateral Triangle

142   Chapter 3   Congruent Triangles

Vocabulary

acute triangle	equilateral triangle	obtuse triangle
base	hypotenuse	right triangle
base angles	isosceles triangle	scalene triangle
equiangular triangle	legs	vertex angle

The word *equilateral* can be applied to any figure in which all sides are congruent.

Definition   An *equiangular triangle* is a triangle in which all angles are congruent.

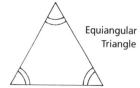
Equiangular Triangle

The word *equiangular* can be applied to any figure in which all angles are congruent.

Looking at the diagrams, you may wonder if there is any real difference between an equilateral triangle and an equiangular triangle. You will find out in Section 3.7, where you will also learn whether any differences exist between equilateral and equiangular figures of other numbers of sides.

Definition   An *acute triangle* is a triangle in which all angles are acute.

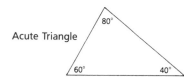
Acute Triangle

Definition   A *right triangle* is a triangle in which one of the angles is a right angle. (The side opposite the right angle is called the **hypotenuse.** The sides that form the right angle are called **legs.**)

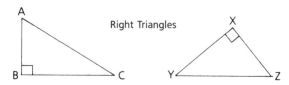
Right Triangles

In △ABC above, $\overline{AB}$ and $\overline{BC}$ are the legs, and $\overline{AC}$ is the hypotenuse. Can you name these parts in △XYZ?

Definition   An *obtuse triangle* is a triangle in which one of the angles is an obtuse angle.

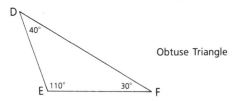
Obtuse Triangle

Section 3.6   Types of Triangles   **143**

The Geometric Supposer

Do this activity in conjunction with teaching the lesson.

With their books open to Section **3.6,** have students draw the various types of triangles described in the lesson. Have them measure sides and angles, as appropriate, to test the triangle definitions given. For example, for a right triangle, have them identify the right angle in the triangle on the screen. Students probably will say that they can't draw a scalene triangle. Ask them to draw an acute triangle. Then ask whether or not it is scalene. Also point out that students can draw a scalene triangle by choosing "Your own" from the triangle menu. They should choose the side-side-side option, since the definition of a scalene triangle is based on the lengths of its sides.

Assignment Guide

Basic
_____
1, 3–6

Average
_____
1, 4, 5, 9, 10, 13, 15

Advanced
_____
7–10, 13, 15, 16

Part Two: Sample Problems

Problem 1   Given: ∠CBD = 70°
Prove: △ABC is obtuse.

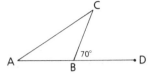

Proof   ∠CBD = 70° and ∠ABD is a straight angle, so ∠ABC = 110°. Since △ABC contains an obtuse angle, it is an obtuse triangle.

Problem 2   Given: EG = FH,
EF > EG
Prove: △EFG is scalene.

Proof   Since EG = FH and $\overline{FH}$ is clearly longer than $\overline{FG}$, $\overline{EG}$ is also longer than $\overline{FG}$. It is given that EF > EG, so $\overline{EF}$ is also longer than $\overline{FG}$. Since no two sides of △EFG are congruent, the triangle is scalene.

Problem 3   Given: ∠1 ≅ ∠3,
∠2 ≅ ∠4,
$\overline{JP} ≅ \overline{PO}$
Prove: △KPM is isosceles.

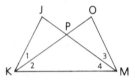

Proof

Statements	Reasons
1 ∠1 ≅ ∠3	1 Given
2 ∠2 ≅ ∠4	2 Given
3 ∠JKM ≅ ∠OMK	3 Addition Property
4 $\overline{KM} ≅ \overline{KM}$	4 Reflexive Property
5 △JKM ≅ △OMK	5 ASA (2, 4, 3)
6 $\overline{JM} ≅ \overline{KO}$	6 CPCTC
7 $\overline{JP} ≅ \overline{PO}$	7 Given
8 $\overline{KP} ≅ \overline{MP}$	8 Subtraction Property
9 △KPM is isosceles.	9 If at least two sides of a △ are congruent, the △ is isosceles.

Part Three: Problem Sets

Problem Set A

1 If the perimeter of △EFG is 32, is △EFG scalene, isosceles, or equilateral? Scalene

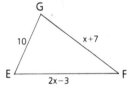

144   Chapter 3   Congruent Triangles

2 Classify each of the triangles as scalene, isosceles, or equilateral.

a

10   8
9

Scalene

c

7   7
7

Equilateral

e
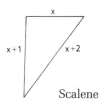
x
x+1   x+2

Scalene

b

Isosceles

d
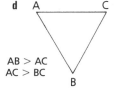
A   C

AB > AC
AC > BC
B

Scalene

f
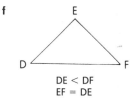
E
D   F
DE < DF
EF = DE

Isosceles

3 Classify each of the triangles as acute, right, or obtuse.

a Right

d Acute

60°
40°   100°

b   Obtuse

50°
100°   30°

e $\overline{GH} \perp \overline{HJ}$
Right

H
G   J

c   Right

70°
89°60'   20°

f $\frac{1}{2}(m\angle K) = 30$,
$\frac{1}{3}(m\angle M) = 20$,
$\frac{1}{4}(m\angle O) = 15$
Acute
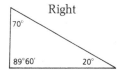
K
M   O

4 Using the figure as marked, write a paragraph proof showing that △ABC is acute.

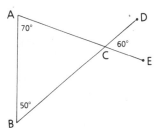
A
70°   D
60°
C   E
50°
B

**Problem-Set Notes
and Additional Answers**

4 Since vertical ∠s are ≅,
∠ACB = 60°. Since ∠A = 70°
and ∠B = 50°, all ∠s
measure less than 90°. By
definition △ABC is acute.

Section 3.6   Types of Triangles   **145**

T 145

- See *Solution Manual* for answers to problems **5** and **7–9.**

Problem Set A, *continued*

5 Given: ⊙O

Prove: △COD is isosceles.

6 If △HJK is equilateral, what are the values of x and y? 7; 63

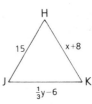

Problem Set B

7 Given: $\overline{AD}$ and $\overline{CD}$ are legs of isosceles △ACD.
    B is the midpt. of $\overline{AC}$.

Prove: ∠A ≅ ∠C

8 Given: $\overline{BI} \cong \overline{RD}$, $\overline{RI} \cong \overline{BD}$;
    ∠3 is comp. to ∠2.

Prove: △RIB is a right △.

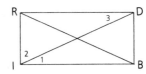

9 Given: $\overline{JF} \cong \overline{JG}$;
    F and G trisect $\overline{EH}$.
    ∠EFJ ≅ ∠HGJ

Conclusion: △EHJ is isosceles.

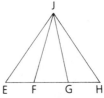

10 If we assume any pair of sides are ≅, then x = 1 and each side = 8.

10 In △RST, RS = x + 7, RT = 3x + 5, and ST = 9 − x. If △RST is isosceles, is it also equilateral? Yes

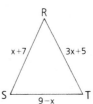

11 If △VSY is isosceles and its perimeter is less than 45, which side of △VSY is the base? $\overline{VY}$

12 Given: AB = x + 3,
AC = 3x + 2,
BC = 2x + 3;
Perimeter of △ABC = 20.
Show that △ABC is scalene.

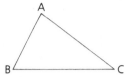

13 The average of the lengths of the sides of △ABC is 14. How much longer than the average is the longest side? 4

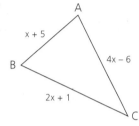

Problem Set C

14 Given: $\overline{AB}$ and $\overline{AC}$ are the legs of isosceles △ABC.
m∠1 = 5x,
m∠3 = 2x + 12
Find: m∠2 60

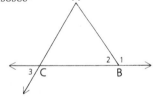

15 Draw an obtuse triangle PQR with longest side $\overline{PR}$. Then draw equilateral triangles APQ and BQR lying outside the given triangle. Assuming that the measure of each angle of an equilateral triangle is 60, prove that $\overline{AR} \cong \overline{PB}$.

16 How many different isosceles triangles can you find that have sides that are whole-number lengths and that have a perimeter of 18? 4

Problem-Set Notes and Additional Answers, continued

■ See *Solution Manual* for answer to problem **12.**

11 If 10 = x + 7, SY is negative. If x + 7 = 2x − 8, P = 54.
∴ 10 = 2x − 8, and $\overline{VY}$ is the base.

14 (This problem uses the results of Section **3.7.**)
Since $\overline{AB} \cong \overline{AC}$, ∠2 ≅ ∠ACB and ∠ACB ≅ ∠3.
Thus, ∠3 + ∠1 = 180.
5x + 2x + 12 = 180
x = 24
m∠1 = 120 ∴ m∠2 = 60

15 There are three cases to consider.
Case 1: 90° < ∠Q < 120°

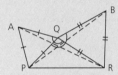

$\overline{AQ} \cong \overline{PQ}$ and $\overline{BQ} \cong \overline{RQ}$.
∠AQR ≅ ∠PQB by addition.
∴ △AQR ≅ △PQB by SAS.
∴ $\overline{AR} \cong \overline{PB}$ by CPCTC.
Case 2: ∠Q = 120°

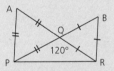

A, Q, and R are collinear, and B, Q, and P are collinear.
∴ $\overline{AR} \cong \overline{PB}$ by addition.
Case 3: 120° < ∠Q < 180°

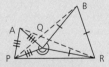

The proof is the same as the first case.

16 Triangles with sides of lengths 5, 5, 8; 6, 6, 6; 7, 7, 4; 8, 8, 2; ∴ 4 triangles

T 147

ANGLE-SIDE THEOREMS

Class Planning

Time Schedule
All levels: 2 days

Resource References
Teacher's Resource Book
    Class Opener 3.7A, 3.7B
    Additional Practice Worksheet 6
Evaluation
    Tests and Quizzes
        Quiz 3, Series 1

Class Opener

Given: $\overline{AM} \cong \overline{AN}$;
    $\overline{AT}$ is a median of
    $\triangle AMN$.
Prove: $\angle M \cong \angle N$

Statements	Reasons
1 $\overline{AM} \cong \overline{AN}$	1 Given
2 $\overline{AT} \cong \overline{AT}$	2 Reflexive Prop.
3 $\overline{AT}$ is a median of $\triangle AMN$.	3 Given
4 $\overline{MT} \cong \overline{NT}$	4 Median divides a side of a $\triangle$ into 2 $\cong$ segments.
5 $\triangle ATM \cong \triangle ATN$	5 SSS
6 $\angle M \cong \angle N$	6 CPCTC

You may want to discuss the possibility of proving the isosceles triangle theorem and its converse this way.

Lesson Notes

■ Students sometimes misunderstand Theorem 20 and try to apply it when the congruent segments are in different triangles.

Objective
After studying this section, you will be able to
■ Apply theorems relating the angle measures and side lengths of triangles

Part One: Introduction

It can be shown that the base angles of any isosceles triangle are congruent.

Theorem 20   *If two sides of a triangle are congruent, the angles opposite the sides are congruent. (If △, then △.)*

Given: $\overline{AB} \cong \overline{AC}$
Prove: $\angle B \cong \angle C$

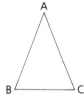

Proof:

Statements	Reasons
1 $\overline{AB} \cong \overline{AC}$	1 Given
2 $\overline{BC} \cong \overline{BC}$	2 Reflexive Property
3 $\triangle ABC \cong \triangle ACB$	3 SSS (1, 2, 1)
4 $\angle B \cong \angle C$	4 CPCTC

You should be accustomed to proving that one triangle is congruent to another triangle. But notice that to prove the preceding theorem, we proved that a triangle is congruent to itself (its mirror image). We shall use the same type of proof to show that the converse of Theorem 20 is also true.

148   Chapter 3   Congruent Triangles

Theorem 21   *If two angles of a triangle are congruent, the sides opposite the angles are congruent. (If △, then △.)*

Given: ∠D ≅ ∠E
Conclusion: $\overline{DF} \cong \overline{EF}$

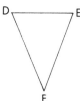

Proof:

Statements	Reasons
1 ∠D ≅ ∠E	1 Given
2 $\overline{DE} \cong \overline{DE}$	2 Reflexive Property
3 △DEF ≅ △EDF	3 ASA (1, 2, 1)
4 $\overline{DF} \cong \overline{EF}$	4 CPCTC

Theorem 21 tells us that a triangle is isosceles if two or more of its angles are congruent. We now have two ways of proving that a triangle is isosceles.

Ways to Prove That a Triangle Is Isosceles

1 If at least two sides of a triangle are congruent, the triangle is isosceles.
2 If at least two angles of a triangle are congruent, the triangle is isosceles.

The inverses of Theorems 20 and 21 are also true. (Recall that the inverse of "If *p*, then *q*" is "If not *p*, then not *q*.") In fact, it can be proved that inequalities of sides and angles are related as shown in the diagram.

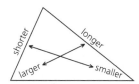

Theorem   *If two sides of a triangle are not congruent, then the angles opposite them are not congruent, and the larger angle is opposite the longer side. (If △, then △.)*

Theorem   *If two angles of a triangle are not congruent, then the sides opposite them are not congruent, and the longer side is opposite the larger angle. (If △, then △.)*

These theorems will be restated and proved in Chapter 15.

Section 3.7   Angle-Side Theorems   **149**

Cooperative Learning

Give students different examples of isosceles triangles. Have each student measure the lengths of the three sides of his or her triangle, and the three angles. Then have students work in small groups to come up with as many generalizations about isosceles triangles as they can.
Students should recognize that the base angles of an isosceles triangle are congruent.
Using the same triangle, each student should draw the median for the vertex angle of his or her triangle, then measure each of the resulting six angles. Again, pooling their information, students should draw as many conclusions as they can.
Students should recognize that the median from the vertex angle is both an angle bisector and an altitude.

Let us now consider a question we raised in Section 3.6: Is an equilateral triangle also equiangular?

Given: $\overline{GH} \cong \overline{HJ} \cong \overline{GJ}$
Is $\angle H \cong \angle J \cong \angle G$?

If $\overline{GH} \cong \overline{HJ}$, which two angles must be congruent? If $\overline{HJ} \cong \overline{GJ}$, which two angles must be congruent? Do we therefore know that $\triangle GHJ$ is equiangular? Can we also prove that an equiangular triangle is equilateral?

Because of their equivalence, the terms *equilateral triangle* and *equiangular triangle* will be used interchangeably throughout this book. We cannot, however, use the words *equilateral* and *equiangular* interchangeably when we apply them to other types of figures. For example, figure ABCD is equilateral but not equiangular. Figure EFGH, on the other hand, is equiangular but not equilateral.

Part Two: Sample Problems

Problem 1   Given: AC > AB,
             m$\angle$B + m$\angle$C < 180,
             m$\angle$B = 6x − 45,
             m$\angle$C = 15 + x

What are the restrictions on the value of x?

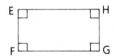

Solution   Since AC > AB, m$\angle$B > m$\angle$C.

$$6x - 45 > 15 + x$$
$$5x > 60$$
$$x > 12$$

We also know that m$\angle$B + m$\angle$C < 180.

$$6x - 45 + 15 + x < 180$$
$$7x < 210$$
$$x < 30$$

Therefore, x must be between 12 and 30.

Problem 2   Prove: *The bisector of the vertex angle of an isosceles triangle is also the median to the base.*

Proof   For a problem like this, we must set up the proof and supply the diagram.

Given: △JOM is isosceles, with
∠JOM the vertex angle.
$\overrightarrow{OK}$ bisects ∠JOM.

Conclusion: $\overline{OK}$ is the median to the base.

Lesson Notes, continued

■ Refer to Sample Problem **2.**
The entire Section **4.2**
(pp. 176–179) is used to present problems in verbal form, without diagrams.

Statements	Reasons
1 △JOM is isosceles, with ∠JOM the vertex angle.	1 Given
2 $\overline{OJ} \cong \overline{OM}$	2 The legs of an isosceles △ are ≅.
3 $\overrightarrow{OK}$ bisects ∠JOM.	3 Given
4 ∠JOK ≅ ∠MOK	4 If a ray bisects an ∠, it divides the ∠ into two ≅ ∠s.
5 $\overline{OK} \cong \overline{OK}$	5 Reflexive Property
6 △JOK ≅ △MOK	6 SAS (2, 4, 5)
7 $\overline{JK} \cong \overline{MK}$	7 CPCTC
8 $\overline{OK}$ is the median to the base.	8 If a segment from a vertex of a △ divides the opposite side into two ≅ segments, it is a median.

Problem 3   Given: ∠3 ≅ ∠4,
$\overline{BX} \cong \overline{AY}$,
$\overline{BW} \cong \overline{AZ}$

Conclusion: △WTZ is isosceles.

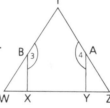

Proof

Statements	Reasons
1 ∠3 ≅ ∠4	1 Given
2 ∠3 is supp. to ∠WBX.	2 If two ∠s form a straight ∠, they are supplementary.
3 ∠4 is supp. to ∠YAZ.	3 Same as 2
4 ∠WBX ≅ ∠YAZ	4 ∠s supp. to ≅ ∠s, are ≅.
5 $\overline{BX} \cong \overline{AY}$	5 Given
6 $\overline{BW} \cong \overline{AZ}$	6 Given
7 △BWX ≅ △AZY	7 SAS (5, 4, 6)
8 ∠W ≅ ∠Z	8 CPCTC
9 △WTZ is isosceles.	9 If at least two ∠s of a △ are ≅, the △ is isosceles.

Problem 4   Given: ∠E ≅ ∠H,
                        $\overline{EF} ≅ \overline{GH}$

Conclusion: $\overline{DF} ≅ \overline{DG}$

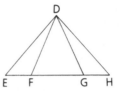

Proof

Statements	Reasons
1 ∠E ≅ ∠H	1 Given
2 $\overline{DE} ≅ \overline{DH}$	2 If △, then △.
3 $\overline{EF} ≅ \overline{GH}$	3 Given
4 △DEF ≅ △DHG	4 SAS (2, 1, 3)
5 $\overline{DF} ≅ \overline{DG}$	5 CPCTC

Part Three: Problem Sets

Problem Set A

1 Given: $\overline{AB} ≅ \overline{AC}$
    Conclusion: ∠1 ≅ ∠2

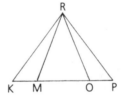

2 Given: ∠KRM ≅ ∠PRO,
        $\overline{KR} ≅ \overline{PR}$
    Prove: $\overline{RM} ≅ \overline{RO}$

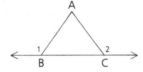

3 Given: $\overline{SX} ≅ \overline{TY}$,
        $\overline{WX} ≅ \overline{YZ}$,
        $\overline{SW} ≅ \overline{TZ}$
    Prove: $\overline{RW} ≅ \overline{RZ}$

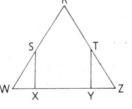

4 Given: ∠3 ≅ ∠6;
        ∠3 is comp. to ∠4.
        ∠6 is comp. to ∠5.
    Prove: △EBC is isosceles.

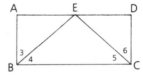

Assignment Guide

Basic

| Day 1 | 1–5, 8, 12 |
| Day 2 | 6, 7, 9–11, 16 |

Average

| Day 1 | 2–4, 8, 10–12, 15 |
| Day 2 | 17–19, 20, 21, 24 |

Advanced

| Day 1 | 3, 4, 6–12, 15 |
| Day 2 | 17, 18, 20, 21, 24, 25 |

**Problem-Set Notes
and Additional Answers**

■ See *Solution Manual* for answers to problems **1–4.**

5 Given: $\overline{FH} \cong \overline{GJ}$;
   $\triangle FKJ$ is isosceles, with $\overline{FK} \cong \overline{JK}$.
   Prove: $\triangle FKH \cong \triangle JKG$

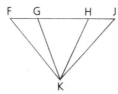

Problem-Set Notes and
Additional Answers, continued

■ See *Solution Manual* for answers to problems **5, 6, 9,** and
   10.

6 Given: $\angle 5 \cong \angle 6$;
   $\overline{JG}$ is the altitude to $\overline{FH}$.
   Prove: $\triangle FJH$ is isosceles.

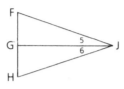

7 In $\triangle ABC$, $AC > BC > AB$. List the three
   angles in order of size, from largest to
   smallest. $\angle B$, $\angle A$, $\angle C$

8 Given: $m\angle P + m\angle R < 180$.
   $PQ < QR$

   Write an inequality to describe the restrictions on x. $6 < x < 18$

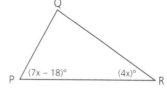

9 Given: $\overline{OP} \cong \overline{RS}$,
   $\overline{KO} \cong \overline{KS}$;
   M is the midpt. of $\overline{OK}$.
   T is the midpt. of $\overline{KS}$.
   Prove: $\overline{MP} \cong \overline{TR}$

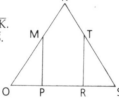

10 Given: $\odot O$,
   $\overline{OX} \cong \overline{XW}$
   Prove: $\triangle XOW$ is equilateral.

11 Given: $\overline{AC} \perp \overline{BC}$,
   $\angle C = (3x)°$,
   $BC = x + 20$,
   $AC = 2x - 20$
   Is $\triangle ABC$ isosceles? No

11 Students should realize that
   AB > either leg, although
   they cannot prove it.
   $3x = 90$
   $x = 30$
   $BC = x + 20 = 50$
   $AC = 2x - 20 = 40$

Problem-Set Notes and Additional Answers, continued

■ See *Solution Manual* for answers to problems **13** and **15–17**.

14

△ABC is isosceles, with $\overline{AB} \cong \overline{AC}$. $\overline{AD}$ is a median, so $\overline{BD} \cong \overline{CD}$. $\overline{AD} \cong \overline{AD}$, so △ABD ≅ △ACD by SSS. Then ∠BAD ≅ ∠CAD by CPCTC. Therefore, $\overline{AD}$ bisects ∠BAC, the vertex angle.

18 The base of the pyramid is a square, and since all four sides of a square are ≅, the four bases of the △ are ≅. Since each △ is isosceles and each △ shares a side with two other △, all four legs involved must be ≅. ∴ All four △ are ≅ by SSS.

Problem Set B

12 Given: ⊙Q,
$\overline{PS} \perp \overline{SR}$,
∠P = 36°

Find: **a** ∠PSQ 36°
b ∠R 54°

13 Given: $\overline{BE} \cong \overline{BD}$,
$\overline{BE} \perp \overline{AE}$,
∠BDC = 90°

Prove: ∠AED ≅ ∠CDE

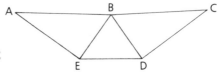

14 Prove: The median to the base of an isosceles triangle bisects the vertex angle.

15 Given: $\overline{HK} \cong \overline{JM}$,
$\overline{GJ} \cong \overline{JK}$,
$\overline{OK} \cong \overline{JK}$;
$\overline{GJ}$ and $\overline{OK}$ are ⊥ to $\overline{HM}$.

Prove: △FHM is isosceles.

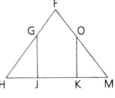

16 Given: $\overline{PR} \cong \overline{ST}$,
$\overline{NP} \cong \overline{VT}$,
∠P ≅ ∠T

Prove: △WRS is isosceles.

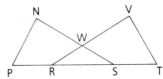

17 Given: $\overline{YZ}$ is the base of an isosceles triangle.
∠2 ≅ ∠Z,
∠1 ≅ ∠Y

Prove: $\overrightarrow{XA}$ bisects ∠BXZ.

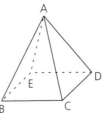

18 The pyramid shown has four isosceles triangular faces, and its base is a square. Explain why the four triangles are congruent.

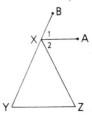

19 Given: $\overline{HJ} \cong \overline{MK}$,
  $\angle HJK \cong \angle MKJ$

Conclusion: $\triangle JOK$ is isosceles.

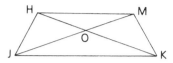

20 Given: $\angle A$ is the vertex of an isosceles $\triangle$.
  The number of degrees in $\angle B$ is *twice* the number of centimeters in $\overline{BC}$.
  The number of degrees in $\angle C$ is *three* times the number of centimeters in $\overline{AB}$.
  $m\angle B = x + 6$,
  $m\angle C = 2x - 54$

Find: The perimeter of $\triangle ABC$ 77 cm

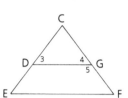

21 Given: $\overline{CE} \cong \overline{CF}$,
  $\angle F \cong \angle 3$;
  $\angle E$ is supp. to $\angle 5$.

Prove: $\triangle CDG$ is isosceles.

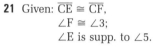

Problem Set C

22 Given: $\overline{FG} \cong \overline{JH}$,
  $\angle FGH \cong \angle JHG$

Conclusion: $\triangle FKJ$ is isosceles.

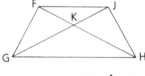

23 Given: $\odot O$,
  $\odot P$;
  $\overleftrightarrow{AB}$ bisects $\angle$s OAP and OBP.

Prove: Figure AOBP is equilateral.

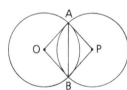

24 Given: Figure XSTOW is equilateral and equiangular.

Prove: $\triangle YTO$ is isosceles.

25 Given: $\triangle FED$ is equilateral.
  $\overline{GE} \perp \overline{DE}$,
  $m\angle FEG = x + y$,
  $m\angle D = 3x - 6$,
  $m\angle F = 6y + 12$

Find: x, y, and $\angle F$ 22; 8; 60°

Problem-Set Notes and Additional Answers, continued

- See Solution Manual for answers to problems **19** and **21**.

20 $x + 6 = 2x - 54$
    $60 = x$
  $\therefore m\angle B = m\angle C = 66$
  $AB = AC = 22$, and $BC = 33$, so $P = 77$.
  (With trigonometry we can show that this $\triangle$ is actually overdetermined.)

22 Prove $\triangle FGH \cong \triangle JHG$ by SAS, so $\overline{FH} \cong \overline{JG}$ by CPCTC. Then $\triangle FJG \cong \triangle JFH$ by SSS. Thus, $\angle FJG \cong \angle JFH$ by CPCTC, and $\triangle FKJ$ is isosceles.

23 $\overline{AP} \cong \overline{BP}$, so $\angle PBA \cong \angle PAB$. $\overline{AB}$ bisects $\angle OAP$, so $\angle PAB \cong \angle OAB$. $\overline{OA} \cong \overline{OB}$, so $\angle OAB \cong \angle OBA$. $\therefore$ All the $\angle$s are $\cong$ by transitivity. Proving the $\triangle$ $\cong$ by ASA makes the figure equilateral.

24 Prove $\triangle STO \cong \triangle WOT$ by SAS. Then $\angle YOT \cong \angle YTO$ by CPCTC.

25 Solve any two of the equations.
  $3x - 6 = 6y + 12$
  $(3x - 6) + (x + y) = 90$
  $(6y + 12) + (x + y) = 90$

Communicating Mathematics

Have students verify that solving any two of the equations in problem **25** will give the same answers.

THE HL POSTULATE

Class Planning

Time Schedule
All levels: 1 day

Resource References
Teacher's Resource Book
    Class Opener 3.8A, 3.8B
Transparency 6

Class Opener

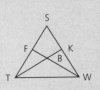

Given: ∠SWT ≅ ∠STW,
    ∠BTW ≅ ∠BWT
Prove: $\overline{SK} \cong \overline{SF}$

Statements	Reasons
1 ∠SWT ≅ ∠STW	1 Given
2 ∠BTW ≅ ∠BWT	2 Given
3 ∠STK ≅ ∠SWF	3 Subtraction Prop.
4 $\overline{ST} \cong \overline{SW}$	4 If two ∠s of a △ are ≅, then sides opposite them are ≅.
5 ∠TSK ≅ ∠WSF	5 Reflexive Prop.
6 △KST ≅ △FSW	6 ASA
7 $\overline{SK} \cong \overline{SF}$	7 CPCTC

Objective
After studying this section, you will be able to
- Use the HL postulate to prove right triangles congruent

Part One: Introduction

The two right triangles below, △ABC and △DEF, can be shown to be congruent by a method that we shall call HL. Although HL congruence can be proved, we shall treat it as a postulate.

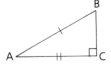

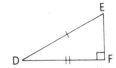

Postulate   *If there exists a correspondence between the vertices of two right triangles such that the hypotenuse and a leg of one triangle are congruent to the corresponding parts of the other triangle, the two right triangles are congruent.* **(HL)**

It is important to note that the HL postulate applies only to right triangles. When we use it in proofs, therefore, we must establish that the triangles that we are dealing with are right triangles. We do this by inserting steps showing that each triangle contains a right angle. Clearly, any triangle containing a right angle is a right triangle.
    Did you notice that once again three conditions are involved in proving that two triangles are congruent?

Part Two: Sample Problems

Problem 1

Given: $\overline{BC} \perp \overline{AC}$,
$\overline{BD} \perp \overline{AD}$,
$\overline{AC} \cong \overline{AD}$

Prove: $\overrightarrow{AB}$ bisects $\angle CAD$.

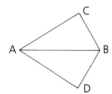

Proof

Statements	Reasons
1 $\overline{BC} \perp \overline{AC}$	1 Given
2 $\angle ACB$ is a right $\angle$.	2 If two segments are $\perp$, they form right $\angle$s.
3 $\overline{BD} \perp \overline{AD}$	3 Given
4 $\angle BDA$ is a right $\angle$.	4 Same as 2
5 $\overline{AC} \cong \overline{AD}$	5 Given
6 $\overline{AB} \cong \overline{AB}$	6 Reflexive Property
7 $\triangle ACB \cong \triangle ADB$	7 HL (2, 4, 6, 5)
8 $\angle CAB \cong \angle DAB$	8 CPCTC
9 $\overrightarrow{AB}$ bisects $\angle CAD$.	9 A ray that divides an $\angle$ into two $\cong \angle$s bisects the $\angle$.

Problem 2

Prove: Corresponding angle bisectors of congruent triangles are congruent.

Proof

Once again we must set up the proof and draw the figure. (Although this may look like a simple two-step proof based on CPCTC, it isn't. *Corresponding parts of congruent triangles* refers only to corresponding sides and angles.)

Given: $\triangle KPR \cong \triangle SAW$;
$\overrightarrow{RM}$ bisects $\angle KRP$.
$\overrightarrow{WT}$ bisects $\angle SWA$.

Prove: $\overline{RM} \cong \overline{WT}$

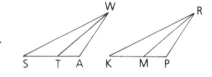

Statements	Reasons
1 $\triangle KPR \cong \triangle SAW$	1 Given
2 $\overline{KR} \cong \overline{SW}$	2 CPCTC
3 $\angle K \cong \angle S$	3 CPCTC
4 $\angle KRP \cong \angle SWA$	4 CPCTC
5 $\overrightarrow{RM}$ bisects $\angle KRP$.	5 Given
6 $\overrightarrow{WT}$ bisects $\angle SWA$.	6 Given
7 $\angle KRM \cong \angle SWT$	7 Division Property
8 $\triangle KRM \cong \triangle SWT$	8 ASA (3, 2, 7)
9 $\overline{RM} \cong \overline{WT}$	9 CPCTC

Lesson Notes

- There are many proofs of HL congruences. One straightforward proof uses complementary angles.

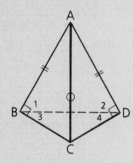

Given $\overline{AB} \cong \overline{AD}$, so $\angle 1 \cong \angle 2$. Thus, $\angle 3 \cong \angle 4$, so $\overline{BC} \cong \overline{CD}$. Then $\triangle ABC \cong \triangle ADC$ by SSS.

The proof "assumed" that figure BDC is a triangle. Students can explore how the given congruence of hypotenuses is subtly used by separating the two right triangles and then letting $\angle C_1AC_2$ "close up." It is because $\overline{AC_1} \cong \overline{AC_2}$ that C_1 and C_2 will coincide.

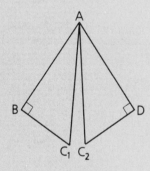

Problem 3 Given: $\overline{OF}$ is an altitude.
             ⊙O

Conclusion: $\overline{EF} \cong \overline{FG}$

Proof

	Statements		Reasons
1	$\overline{OF}$ is an altitude	1	Given
2	∠EFO and ∠GFO are right ∠s	2	An altitude of a △ forms right ∠s with the side to which it is drawn
3	$\overline{OF} \cong \overline{OF}$	3	Reflexive Property
4	⊙O	4	Given
5	$\overline{OE} \cong \overline{OG}$	5	All radii of a circle are ≅
6	△OEF ≅ △OGF	6	HL (2, 5, 3)
7	$\overline{EF} \cong \overline{FG}$	7	CPCTC

Lesson Notes, continued

The proof in Sample Problem **3** must establish right angles before using HL. The parenthesis in reason 6 indicates that step 2 established right angles.

Assignment Guide

Basic

1–5, 15

Average

3, 6, 7, 10, 12, 15

Advanced

7, 13–17

Problem-Set Notes and Additional Answers

■ See *Solution Manual* for answers to problems **1–5**.

Communicating Mathematics

Discuss why two right triangles that are congruent by HL are not also congruent by SAS.

Part Three: Problem Sets

Problem Set A

1 Given: $\overline{GJ}$ is the altitude to $\overline{HK}$.
     $\overline{HG} \cong \overline{KG}$

  Prove: △HGJ ≅ △KGJ

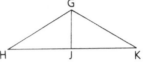

2 Given: $\overline{MO} \perp \overline{OP}$,
     $\overline{RP} \perp \overline{OP}$,
     $\overline{MP} \cong \overline{RO}$

  Prove: △MOP ≅ △RPO

3 Given: ⊙O,
     $\overline{YO} \perp \overline{YX}$,
     $\overline{ZO} \perp \overline{ZX}$

  Conclusion: $\overline{YX} \cong \overline{ZX}$

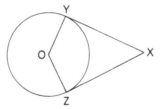

4 Given: $\overline{AE} \cong \overline{CF}$,
     $\overline{AB} \cong \overline{CD}$;
     ∠BFA is a right angle.
     ∠DEC is a right angle.

  Prove: ∠CDE ≅ ∠ABF

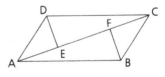

5 Set up and prove: The altitude to the base of an isosceles triangle divides the triangle into two congruent triangles.

6 Given: $\overline{GH} \cong \overline{GK}$;
   $\overline{GJ}$ is an altitude.

   Prove: $\overrightarrow{GJ}$ bisects $\angle HGK$.

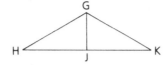

Problem Set B

7 Prove: An altitude of an equilateral triangle is also a median of the triangle.

8 Given: $\overline{BD} \perp \overline{CF}$,
   $\overline{GE} \perp \overline{CF}$,
   $\overline{CE} \cong \overline{DF}$,
   $\overline{BC} \cong \overline{GF}$

   Prove: $\triangle ACF$ is isosceles.

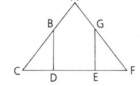

9 Given: $\overline{RK} \perp \overline{HR}$,
   $\overline{JO} \perp \overline{PM}$,
   $\overline{PH} \cong \overline{PM}$,
   $\overline{PR} \cong \overline{PO}$

   Conclusion: $\overline{RK} \cong \overline{JO}$

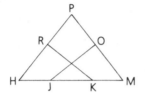

10 Given: $\odot P$,
   $\overline{ST} \cong \overline{VT}$

   Prove: $\angle PST \cong \angle PVT$

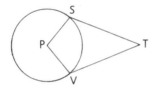

11 Prove: Corresponding medians of congruent triangles are congruent.

12 Given: $\overline{CD} \cong \overline{EF}$,
   $\overline{JF} \perp \overline{JD}$,
   $\overline{CH} \perp \overline{HE}$,
   $\overline{CH} \cong \overline{JF}$

   Prove: $\overline{JD} \cong \overline{HE}$

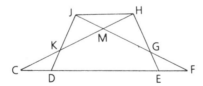

13 Given: $\triangle ABC$ and $\triangle ABD$ standing on plane p.
   $\overline{AB} \perp \overline{BC}$, $\overline{AB} \perp \overline{BD}$,
   $\overline{AC} \cong \overline{AD}$

   Prove: If $\overline{CD}$ is drawn, $\triangle BCD$ will be isosceles.

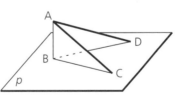

Problem-Set Notes and Additional Answers, continued

- See *Solution Manual* for answers to problems **6–13**.

Problem Set B, *continued*

14 Given: m∠A > m∠C

Find the restrictions on the value
of x. −12 < x < −8

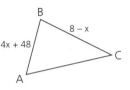

15 In the diagram, $\overline{PQ}$ is congruent to $\overline{QR}$.

a Find the coordinates of S. (4, 0)

b Explain why PS = SR. CPCTC

c Find the coordinates of R. (10, 0)

d Find the area of △PQR. 54

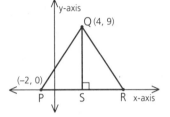

Problem Set C

16 Given: $\overline{BE} \perp \overline{AD}$, $\overline{AC} \perp \overline{BD}$,
$\overline{AC} \cong \overline{BE}$, $\overline{DE} \cong \overline{EC}$

Prove: △DEC is equilateral.

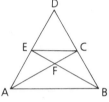

17 Given: ∠R and ∠W are right ∠s.
$\overline{RX} \cong \overline{WX}$;
S is $\frac{3}{7}$ of the way from R to T.
V is $\frac{4}{7}$ of the way from T to W.

Prove: $\overline{ST} \cong \overline{TV}$

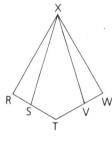

Problem Set D

18 a Which of the triangles below are congruent? A, B, C, and E

b If two of the triangles are selected at random, what is the
probability that they are congruent? $\frac{3}{5}$

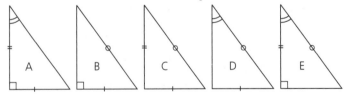

*Problem-Set Notes and
Additional Answers, continued*

16 △EAB ≅ △CBA by HL, so
$\overline{EA} \cong \overline{CB}$ and ∠EAB ≅
∠CBA by CPCTC.
∴ $\overline{AD} \cong \overline{DB}$, so $\overline{DE} \cong \overline{DC}$ by
subtraction. Using the
Transitive Property, the △ is
equilateral.

17 Draw $\overline{XT}$ and prove △XRT ≅
△XWT by HL. Then $\overline{RT} \cong$
$\overline{WT}$ and $\overline{ST} \cong \overline{TV}$ by
division.

Cooperative Learning

Problem **18** involves subtle concepts. Have students work in
small groups to solve this problem.

CHAPTER SUMMARY

CONCEPTS AND PROCEDURES

After studying this chapter, you should be able to
- Understand the concept of congruent figures (3.1)
- Accurately identify the corresponding parts of figures (3.1)
- Identify included angles and included sides (3.2)
- Apply the SSS postulate (3.2)
- Apply the SAS postulate (3.2)
- Apply the ASA postulate (3.2)
- Apply the principle of CPCTC (3.3)
- Recognize some basic properties of circles (3.3)
- Apply the formulas for the area and the circumference of a circle (3.3)
- Identify medians of triangles (3.4)
- Identify altitudes of triangles (3.4)
- Understand why auxiliary lines are used in some proofs (3.4)
- Write proofs involving steps beyond CPCTC (3.4)
- Use overlapping triangles in proofs (3.5)
- Name the various types of triangles and their parts (3.6)
- Apply theorems relating the angle measures and side lengths of triangles (3.7)
- Use the HL postulate to prove right triangles congruent (3.8)

VOCABULARY

acute triangle (3.6)	isosceles triangle (3.6)
altitude (3.4)	leg (3.6)
auxiliary line (3.4)	median (3.4)
base (3.6)	obtuse triangle (3.6)
base angles (3.6)	reflection (3.1)
congruent polygons (3.1)	Reflexive Property (3.1)
congruent triangles (3.1)	right triangle (3.6)
equiangular triangle (3.6)	rotate (3.1)
equilateral triangle (3.6)	scalene triangle (3.6)
hypotenuse (3.6)	slide (3.1)
included (3.2)	vertex angle (3.6)

3 REVIEW PROBLEMS

Problem Set A

1 For each of the following statements, write
A if the statement is always true
S if the statement is sometimes true
N if the statement is never true

  a Two triangles are congruent if two sides and an angle of one are congruent to the corresponding parts of the other. S

  b If two sides of a right triangle are congruent to the corresponding parts of another right triangle, the triangles are congruent. A

  c All three altitudes of a triangle fall outside the triangle. N

  d A median of a triangle does not contain the midpoint of the side to which it is drawn. N

  e A right triangle is congruent to an obtuse triangle. N

2 Given: $\overline{AB} \perp \overline{BC}$,
     $\overline{DC} \perp \overline{BC}$,
     $\angle 1 \cong \angle 2$
   Conclusion: $\overline{AC} \cong \overline{DB}$

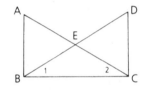

3 Given: $\odot O$,
     $\overline{OG} \perp \overline{FH}$
   Conclusion: $\overline{FG} \cong \overline{GH}$

4 Given: $\overline{PK}$ and $\overline{JM}$ bisect each other at R.
   Prove: $\overline{PJ} \cong \overline{MK}$

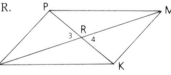

5 Given: $\odot P$;
     $\overrightarrow{PR}$ bisects $\angle KPM$.
   Conclusion: $\overline{PR}$ is a median.

6 Given: $\overline{DG} \cong \overline{JF}$,
$\overline{DE} \cong \overline{JH}$,
$\overline{EG} \cong \overline{HF}$

Prove: $\triangle HCE$ is isosceles.

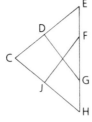

7 $\triangle HGF$ is equilateral.

a If $\angle F = (x + 32)°$ and $\angle H = (2x + 4)°$,
solve for x and find $m\angle G$. 28; 60

b If the perimeter of $\triangle HGF = 6y + 24$
and $HG = 3y - 7$, find the perimeter
of $\triangle HGF$. 114

8 Given: $\triangle RST \cong \triangle DFE$, $\angle R = 50°$, $\angle T = 40°$, $\angle E = (y + 10)°$,
$\angle S = 90°$, $\angle D = (x + 20)°$, $\angle F = (z - 30)°$

Find: The values of x, y, and z (Draw your own diagram for this
problem.) $x = 30$; $y = 30$; $z = 120$

9 Find the area of $\triangle ABC$. 8

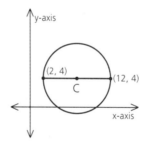

10 Find the area and the circumference of
$\odot C$ to the nearest tenth. $A \approx 78.5$; $C \approx 31.4$

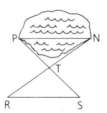

Problem Set B

11 Kate and Jaclyn wished to find the distance
from N on one side of a lake to P on the
other side. They put stakes at N, P, and T,
then extended $\overline{PT}$ to S, making sure that $\overline{PT}$
was congruent to $\overline{TS}$. They followed a similar
process in extending $\overline{NT}$ to R. They then
measured $\overline{SR}$ and found it to be 70 m long.
They concluded that NP was 70 m.
Prove that they were correct.

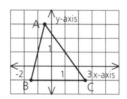

Review Problems **163**

3

Problem-Set Notes and
Additional Answers, continued

■ See *Solution Manual* for answers to problem **6**.

7a $x + 32 = 2x + 4$
$x = 28$
$m\angle G = m\angle F = 28 + 32 = 60$
b $6y + 24 = 3(3y - 7)$
$y = 15$
$P = 6y + 24 = 114$

8
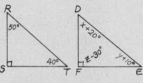

$\angle R \cong \angle D$, so $50 = x + 20$.
∴ $x = 30$.
$\angle S \cong \angle F$, so $90 = z - 30$.
∴ $z = 120$.
$\angle T \cong \angle E$, so $40 = y + 10$.
∴ $y = 30$.

11
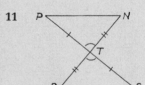

The △ are ≅ by SAS.

T 163

Problem-Set Notes and Additional Answers, continued

■ See *Solution Manual* for answers to problems **12** and **13**.

14

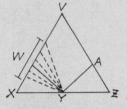

$\overline{WY}$ and $\overline{YA}$ are not "fixed" by the givens.

16

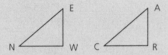

$\overline{EN} \cong \overline{AC}$, so $11 = 4x + y$.
$\overline{EW} \cong \overline{AR}$, so $10 = 2x - 4y$.
$\overline{NW} \cong \overline{CR}$, so $x + y = CR$.
Using the first equation,
$y = 11 - 4x$.
$10 = 2x - 4(11 - 4x)$
$x = 3$
$y = 11 - 4x = -1$
$CR = x + y = 2$

17 $2x + 3 = 5x - 9$
$x = 4$
∴ The sides are 22, 22, and 16.

18 Use $\angle C \cong \angle C$ and $\overline{DC} \cong \overline{EC}$ to prove $\triangle ACE \cong \triangle BDC$, so subtracting $\angle 1$ and $\angle 3$ gives $\angle FDE \cong \angle FED$, and the $\triangle$ is isosceles.

Review Problem Set B, *continued*

12 Given: $\overline{AD} \cong \overline{BC}$,
$\angle DAB \cong \angle CBA$
Prove: $\triangle ABE$ is isosceles.

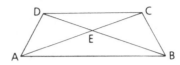

13 Given: $\overline{FJ}$ is the base of an isosceles $\triangle$.
$\overline{FG} \cong \overline{JH}$;
O is the midpt. of $\overline{MF}$.
K is the midpt. of $\overline{MJ}$.
Conclusion: $\overline{OH} \cong \overline{KG}$

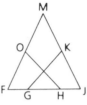

14 Given: $\overline{VX} \cong \overline{VZ}$;
Y is the midpt of $\overline{XZ}$.
Prove: $\overline{WY} \cong \overline{YA}$ Cannot be proved

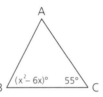

15 In the diagram, $\overline{AB} \cong \overline{AC}$. Solve for x.
$x = -5$ or $x = 11$

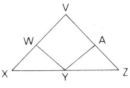

16 Given: $\triangle NEW \cong \triangle CAR$, $EN = 11$, $AR = 2x - 4y$, $NW = x + y$,
$CA = 4x + y$, $EW = 10$
Draw the triangles and find CR. 2

Problem Set C

17 Given: $\triangle FJH$ is isosceles, with base $\overline{JH}$.
K and G are midpoints.
$FK = 2x + 3$,
$GH = 5x - 9$,
$JH = 4x$
Find: The perimeter of $\triangle FHJ$ 60

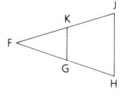

18 Given: $\overline{AC} \cong \overline{BC}$,
$\angle 1 \cong \angle 3$
Prove: $\triangle DFE$ is isosceles.

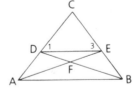

CUMULATIVE REVIEW
CHAPTERS 1–3

Problem Set A

1 a $\overline{BC} \cap \overline{CD} = \underline{\ \ C\ \ }$.
  b $\overline{BG} \cap \overline{EJ} = \underline{\ \ \overline{GJ}\ \ }$.
  c $\overrightarrow{AF} \cup \overrightarrow{AB} = \underline{\angle FAB}$.
  d $\overleftrightarrow{BC} \cap \overleftrightarrow{ED} = \underline{\ \ D\ \ }$.
  e $\overline{BC} \cap \overline{ED} = \underline{\ \ \varnothing\ \ }$.

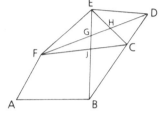

2 Three fifths of a degree is equivalent to how many minutes? 36

3 Find the complement of 43°17′51″. 46°42′09″

4 How large is the angle formed by the hands of a clock at 11:20? 140°

5 One of two supplementary angles is 8 degrees larger than the other. Find the measure of the larger angle. 94

6 Given: AB = 2r + 7,
   CD = 3r − 1,
   BC = 6;
   C is the midpt. of $\overline{AD}$.

Find: AC 41

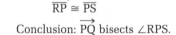

7 Given: ∠A is comp. to ∠BCA.
   ∠D is comp. to ∠DBC.
   ∠D ≅ ∠BCA
Prove: ∠A ≅ ∠DBC

8 Given: ⊙Q,
   $\overline{RP} \cong \overline{PS}$

Conclusion: $\overrightarrow{PQ}$ bisects ∠RPS.

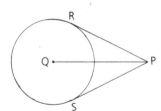

Cumulative Review
Chapters 1–3

Time Schedule
All levels: 1 day

Resource References
Evaluation
   Tests and Quizzes
      Cumulative Review Test
      Chapters 1–3
      Series 1, 2, 3

Assignment Guide

Basic
1–6, 9, 10, 12, 13, 15
Average
3–6, 17–19, 22, 23
Advanced
4, 16–19, 22–24

Problem-Set Notes
and Additional Answers

5 x + (x + 8) = 180
          x = 86
      x + 8 = 94
■ See *Solution Manual* for answers to problems **7** and **8**.

9 Given: $\overline{PS} \perp \overline{SR}$;
  $\angle QRP$ is comp. to $\angle PRS$.
  Prove: $\angle S \cong \angle QRS$

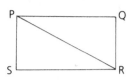

*Problem-Set Notes and
Additional Answers, continued*

■ See *Solution Manual* for answers to problems **9–14**.

10 Given: $\angle 1 \cong \angle 2$,
  $\angle 1 \cong \angle 3$
  Conclusion: $\overrightarrow{FH}$ bisects $\angle EFG$.

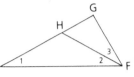

11 Given: Diagram as shown, with $\angle 6$ supp.
  to $\angle 7$
  Conclusion: $\angle 6 \cong \angle 8$

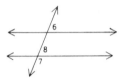

Problem Set B

12 Given: $\angle T \cong \angle W$,
  $\angle TSW \cong \angle XSV$,
  $\overline{ST} \cong \overline{SW}$
  Conclusion: $\overline{SX} \cong \overline{SV}$

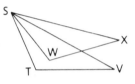

13 Given: $\overline{CE} \cong \overline{DF}$, $\overline{BD} \cong \overline{GE}$,
  $\overline{BD} \perp \overline{CF}$,
  $\overline{GE} \perp \overline{CF}$
  Conclusion: $\triangle ACF$ is isosceles.

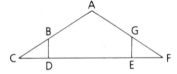

15 $6x + 5x + 4x = 180$
  $x = 12$; $6x = 72$; $5x = 60$;
  $4x = 48$

16

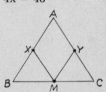

Given: $\triangle ABC$ is isosceles
with $\overline{AB} \cong \overline{AC}$. X, Y, and M
are midpts. of $\overline{AB}$, $\overline{AC}$, and
$\overline{BC}$, respectively.
Prove: $\overline{XM} \cong \overline{YM}$
$BX = \frac{1}{2}(AB)$, and $CY = \frac{1}{2}(AC)$, so $\overline{BX} \cong \overline{CY}$. M is
midpt., so $\overline{BM} \cong \overline{CM}$.
$\angle B \cong \angle C$. Thus, $\triangle XMB \cong \triangle YMC$ by SAS, and $\overline{XM} \cong \overline{YM}$ by CPCTC.

14 Given: $\triangle ZWX$ is isosceles, with base $\overline{WX}$.
  $\overrightarrow{WR}$ bisects $\angle XWZ$.
  $\overrightarrow{XR}$ bisects $\angle ZXW$.
  Prove: $\angle XWR \cong \angle RXW$

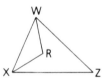

15 If angles 1, 2, and 3 are in the ratio 6:5:4,
  find their measures. 72; 60; 48

16 Prove: The segments drawn from the midpoint of the base of an
  isosceles triangle to the midpoints of the legs are congruent.

17 Q is the midpoint of $\overline{PR}$. The ratio of PQ to QS is 2:5. What are the locations of P and S? −14; 28

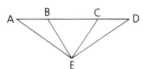

18 The measure of the supplement of an angle exceeds twice the measure of the complement of the angle by 40. Find half the measure of the complement. 25

19 The lengths of two segments are in the ratio of 5:3, and the longer segment exceeds the shorter by 14 m. Find the length of the longer segment. 35 m

20 Given: ∠AEC ≅ ∠BED,
    $\overline{AE} \cong \overline{ED}$
Conclusion: $\overline{AB} \cong \overline{CD}$

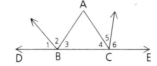

21 Given: ∠1 ≅ ∠5,
    ∠2 ≅ ∠6
Conclusion: △ABC is isosceles.

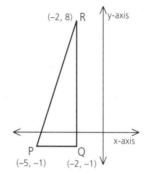

22 Copy the diagram and reflect each point of △PQR over the y-axis to produce △P′Q′R′.
  a Find the coordinates of P′, Q′, and R′.
  b Justify that △PQR ≅ △P′Q′R′.
  c Find the area of △P′Q′R′.
    a (5, −1); (2, −1); (2, 8); **c** 13.5

Problem Set C

23 Given: ⊙O,
    $\overline{OD} \cong \overline{OE}$,
    ∠DOB ≅ ∠EOA
Conclusion: $\overline{CD} \cong \overline{CE}$

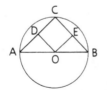

24 Given: ∠NOT ≅ ∠POV,
    O is a midpoint.
    ∠N ≅ ∠P
Prove: $\overline{ST} \cong \overline{RV}$

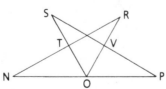

Problem-Set Notes and Additional Answers, continued

18 (180 − x) = 2(90 − x) + 40
          x = 40
      $\frac{1}{2}$ comp. = 25°

19 5x = 3x + 14 m
        x = 7 m
      5x = 35 m

20 Either subtract to get ∠AEB ≅ ∠DEC and prove △AEB ≅ △DEC or prove △AEC ≅ △DEB and then subtract to get $\overline{AB} \cong \overline{CD}$.

21 ∠DBA ≅ ∠ECA by addition, so ∠3 ≅ ∠4, since ∠s supp. to ≅ ∠s are ≅, and △ABC is isosceles.

■ See *Solution Manual* for answer to problem **22b**.

23 Use radii and angle subtraction to prove △ADO ≅ △BEO by SAS. Then $\overline{AD} \cong \overline{EB}$ and ∠A ≅ ∠B, so $\overline{AC} \cong \overline{BC}$. $\overline{CD} \cong \overline{CE}$ by subtraction.

24 △NOT ≅ △POV by ASA, so $\overline{TO} \cong \overline{VO}$. Also ∠NOV ≅ ∠POT by addition yields △NOR ≅ △POS by ASA, so $\overline{SO} \cong \overline{RO}$. Subtraction gives our conclusion.

Chapter 4 Schedule

Basic
Problem Sets:	10 days
Review:	1 day
Test:	1 day

Average
Problem Sets:	9 days
Review:	1 day
Test:	1 day

Advanced
Problem Sets:	7 days
Review:	1 day
Test:	1 day

*T*HIS CHAPTER MAY challenge many students, but if they proceed slowly, follow the text, and study the sample problems carefully, they can become more proficient in writing proofs.

4 LINES IN THE PLANE

This replica of Canada's *Silver Dart*, displayed during Expo '86 in Vancouver, British Columbia, is a "takeoff" of the geometric interpretation of lines in a plane.

DETOURS AND MIDPOINTS

Objectives

After studying this section, you will be able to
- Use detours in proofs
- Apply the midpoint formula

Part One: Introduction

Detour Proofs

To solve some problems, it is necessary to prove more than one pair of triangles congruent. We call the proofs we use in such cases *detour proofs*.

Analyze carefully the following example.

Example   Given: $\overline{AB} \cong \overline{AD}$,
$\overline{BC} \cong \overline{CD}$

Prove: $\triangle ABE \cong \triangle ADE$

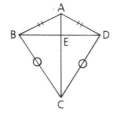

Notice that of the given information only $\overline{AB} \cong \overline{AD}$ seems to be usable. There does not seem to be enough information to prove that $\triangle ABE \cong \triangle ADE$. We must therefore prove something else first, taking a little detour to pick up the congruent parts we need.

	Statements	Reasons
1	$\overline{AB} \cong \overline{AD}$	1 Given
2	$\overline{BC} \cong \overline{CD}$	2 Given
3	$\overline{AC} \cong \overline{AC}$	3 Reflexive Property
4	$\triangle ABC \cong \triangle ADC$	4 SSS (1, 2, 3)
5	$\angle BAE \cong \angle DAE$	5 CPCTC
6	$\overline{AE} \cong \overline{AE}$	6 Reflexive Property
7	$\triangle ABE \cong \triangle ADE$	7 SAS (1, 5, 6)

DETOUR

Whenever you are asked to prove that triangles or parts of triangles are congruent and you suspect that a detour may be needed, use the following procedure.

Section 4.1   Detours and Midpoints   **169**

Class Planning

Time Schedule
Basic: 2 days
Average: $1\frac{1}{2}$ days
Advanced: $1\frac{1}{2}$ days

Resource References
Teacher's Resource Book
    Class Opener 4.1A, 4.1B
    Additional Practice Worksheet 7
Transparency 7

Class Opener

Given: $\overline{BC} \cong \overline{DC}$,
$\overline{CG} \cong \overline{CF}$,
$\angle AGB \cong \angle EFD$
Prove: $\overline{AG} \cong \overline{FE}$

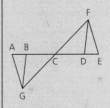

Statements	Reasons
1 $\overline{BC} \cong \overline{DC}$	1 Given
2 $\overline{CG} \cong \overline{CF}$	2 Given
3 $\angle BCG \cong \angle DCF$	3 Vertical $\angle$s are $\cong$.
4 $\triangle BCG \cong \triangle DCF$	4 SAS
5 $\angle CBG \cong \angle CDF$	5 CPCTC
6 $\angle ABG \cong \angle EDF$	6 If two $\angle$s are $\cong$, then their supplements are $\cong$.
7 $\overline{BG} \cong \overline{DF}$	7 CPCTC
8 $\angle AGB \cong \angle EFD$	8 Given
9 $\triangle ABG \cong \triangle EDF$	9 ASA
10 $\overline{AG} \cong \overline{EF}$	10 CPCTC

This is a detour problem without overlapping triangles.

Vocabulary
detour proof
midpoint formula

Lesson Notes

■ The detour is a device to describe a proof-within-a-proof. Occasionally, proofs done by students will contain several detours and thus be long and complex.
■ Many students need to see several examples of proofs with detours.
■ As noted in the text, steps 3–5 of this procedure constitute the detour. Some students choose to put detours as the first steps, as in Sample Problem **1**, whereas others put them in the middle of the proof, as in the example on page 169.

Communicating Mathematics

Ask students to assume that a close friend was home sick and phoned after school to ask what happened in class. Have students write out the way they would describe what was presented in class.

Procedure for Detour Proofs

1 Determine which triangles you must prove to be congruent to reach the required conclusion. (In the preceding example, these are △ABE and △ADE.)

2 Attempt to prove that these triangles are congruent. If you cannot do so for lack of enough given information, take a detour (steps 3–5 below).

3 Identify the parts that you must prove to be congruent to establish the congruence of the triangles. (Remember that there are many ways to prove that triangles are congruent. Consider them all.)

4 Find a pair of triangles that
   (a) You can readily prove to be congruent
   (b) Contain a pair of parts needed for the main proof (parts identified in step 3)

5 Prove that the triangles found in step 4 are congruent.

6 Use CPCTC and complete the proof planned in step 1.

The Midpoint Formula

In some coordinate-geometry problems, you will need to locate the midpoint of a line segment. A method of doing so is suggested by the following example.

Example   *On the number line below, the coordinate of A is 2 and the coordinate of B is 14. Find the coordinate of M, the midpoint of* $\overline{AB}$.

There are several ways of solving this problem. One of these is the *averaging* process (the average of two numbers is equal to half their sum). We will use x_m (read "x sub m") to represent the coordinate of M.

$$x_m = \frac{2 + 14}{2}$$

$$= \frac{16}{2} = 8$$

Check: AM = 8 − 2 = 6
       MB = 14 − 8 = 6

Therefore, 8 is the coordinate of M.

We can apply the averaging process to develop a formula, called the *midpoint formula*, that can be used to find the coordinates of the midpoint of any segment in the coordinate plane. The proof of this theorem is left to you.

170   Chapter 4   Lines in the Plane

Theorem 22 *If A = (x_1, y_1) and B = (x_2, y_2), then the midpoint M = (x_m, y_m) of $\overline{AB}$ can be found by using the midpoint formula:*

$$M = (x_m, y_m) = \left(\frac{x_1 + x_2}{2}, \frac{y_1 + y_2}{2}\right)$$

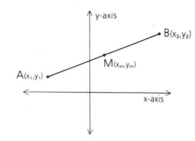

Part Two: Sample Problems

Problem 1 Given: $\overleftrightarrow{PQ}$ bisects $\overline{YZ}$.
Q is the midpt. of $\overline{WX}$.
∠Y ≅ ∠Z, $\overline{WZ}$ ≅ $\overline{XY}$

Conclusion: ∠WQP ≅ ∠XQP

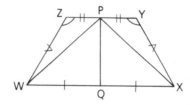

Proof To reach the required conclusion, we must prove that △WQP ≅ △XQP, but the given information is not sufficient to prove these triangles congruent. Therefore, we must detour through another pair of triangles. Can you see which pair of triangles we should use? Check your choice against the following proof.

Statements	Reasons
1 $\overleftrightarrow{PQ}$ bisects $\overline{YZ}$.	1 Given
2 $\overline{ZP}$ ≅ $\overline{PY}$	2 If a line bisects a segment, then it divides the segment into two ≅ segments.
3 ∠Z ≅ ∠Y	3 Given
4 $\overline{WZ}$ ≅ $\overline{XY}$	4 Given
5 △ZWP ≅ △YXP	5 SAS (2, 3, 4)
6 $\overline{WP}$ ≅ $\overline{PX}$	6 CPCTC
7 Q is the midpt. of $\overline{WX}$.	7 Given
8 $\overline{WQ}$ ≅ $\overline{QX}$	8 The midpoint of a segment divides the segment into two ≅ segments.
9 $\overline{PQ}$ ≅ $\overline{PQ}$	9 Reflexive Property
10 △WQP ≅ △XQP	10 SSS (6, 8, 9)
11 ∠WQP ≅ ∠XQP	11 CPCTC

Section 4.1 Detours and Midpoints **171**

Problem 2 *Find the coordinates of M, the midpoint of $\overline{AB}$.*

Solution Use the midpoint formula.

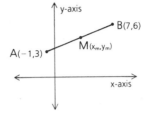

$$x_m = \frac{x_1 + x_2}{2} \qquad y_m = \frac{y_1 + y_2}{2}$$

$$= \frac{-1 + 7}{2} \qquad = \frac{3 + 6}{2}$$

$$= 3 \qquad = 4\frac{1}{2}$$

Thus, $M = (x_m, y_m) = \left(3, 4\frac{1}{2}\right)$.

Problem 3 *In $\triangle ABC$, find the coordinates of the point at which the median from A intersects $\overline{BC}$.*

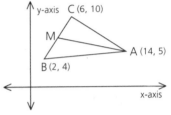

Solution Since a median is drawn to a midpoint, use the midpoint formula to find the midpoint M of $\overline{BC}$.

$$x_m = \frac{x_1 + x_2}{2} \qquad y_m = \frac{y_1 + y_2}{2}$$

$$= \frac{2 + 6}{2} \qquad = \frac{4 + 10}{2}$$

$$= 4 \qquad = 7$$

Thus, the coordinates are (4, 7).

Part Three: Problem Sets

Problem Set A

1 Copy this problem and proof and fill in the missing statements and reasons.

Given: $\overline{WX} \cong \overline{WZ}$, $\overline{XY} \cong \overline{ZY}$

Prove: $\triangle XAY \cong \triangle ZAY$

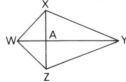

Statements	Reasons
1 $\overline{WX} \cong \overline{WZ}$	1 Given
2 $\overline{XY} \cong \overline{ZY}$	2 Given
3 $\overline{WY} \cong \overline{WY}$	3 Reflexive Property
4 $\triangle WXY \cong \triangle WZY$	4 SSS (1, 2, 3)
5 $\angle XYW \cong \angle ZYW$	5 CPCTC
6 $\overline{AY} \cong \overline{AY}$	6 Reflexive Property
7 $\triangle XAY \cong \triangle ZAY$	7 SAS (2, 5, 6)

---

2 Given: $\overline{MN} \cong \overline{NS}$,
$\overline{MP} \cong \overline{PS}$

Prove: $\angle MQP \cong \angle SQP$

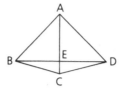

3 Given: A is *equidistant from* B and D
(that is, AB = AD).
$\overrightarrow{AC}$ bisects $\angle BAD$.

Prove: $\overline{AC}$ bisects $\overline{BD}$.

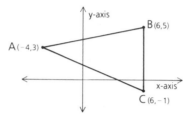

4 Find the coordinates of the midpoint of
each side of $\triangle ABC$. (1, 4); (6, 2); (1, 1)

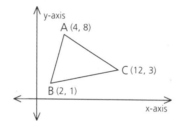

5 Find the coordinates of the point where
the median from A intersects $\overline{BC}$. (7, 2)

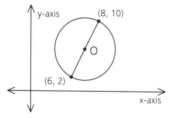

6 A circle with center at O ($\odot O$) has the
diameter shown. Find the coordinates of
O. (7, 6)

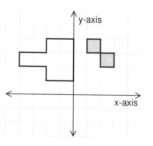

7 If the figure graphed in blue is reflected
across the y-axis and the reflection is to
be shaded, how many additional small
squares must be shaded? 6

**Problem-Set Notes
and Additional Answers**

- See *Solution Manual* for answers to problems **2** and **3**.

Cooperative Learning

Have students work in small
groups to extend problem **6**. Ask
them to draw another circle that
contains point (8, 10) but has its
center at (6, 2). Have them determine the other endpoint of the
diameter through (8, 10) and
(6, 2). Ask whether they can find
the ratio of the area of the larger
circle to the area of the smaller
circle without actually computing either diameter.
(4, −6); 4:1

Problem Set A, *continued*

8 If the shaded square has center at C and
an area of $A_\square$, find C and $A_\square$. $(-2, 2)$; 16

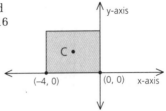

Problem Set B

9 Given: △ABC is isosceles, with base $\overline{BC}$.
$\overline{AD} \perp \overline{BC}$

Prove: △BEC is isosceles.

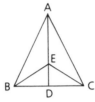

10 Given: ⊙O, $\overline{WX} \cong \overline{WY}$

Prove: $\overleftrightarrow{WZ}$ bisects $\overline{XY}$.

11 Given: $\overline{AD} \cong \overline{BC}$, $\overline{AF} \cong \overline{EC}$,
$\overleftrightarrow{BD} \perp \overleftrightarrow{AF}$, $\overleftrightarrow{BD} \perp \overleftrightarrow{EC}$

Conclusion: $\overline{AB} \cong \overline{DC}$

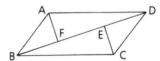

12 Given: $\overline{PR} \cong \overline{PU}$,
$\overline{QR} \cong \overline{QU}$,
$\overline{RS} \cong \overline{UT}$

Conclusion: ∠1 ≅ ∠2

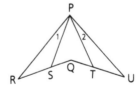

13 Given: $\overline{AB} \cong \overline{AC}$,
$\overline{AD} \cong \overline{AE}$

Prove: △FBC is isosceles.

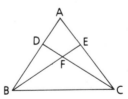

14 Given: T is the midpt. of $\overline{MN}$.
$\angle PMT$ and $\angle QNT$ are right $\angle$s.
$\overline{MR} \cong \overline{SN}$, $\angle 1 \cong \angle 2$

Conclusion: $\angle P \cong \angle Q$

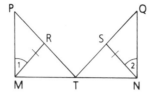

15 Given: $\odot O$, $\angle B \cong \angle C$
Prove: $\overline{AO}$ bisects $\overline{BC}$.

Problem Set C

16 Given: $\overline{AB} \cong \overline{AC}$;
$\overrightarrow{BD}$ bisects $\angle ABE$.
$\overrightarrow{CD}$ bisects $\angle ACE$.

Conclusion: $\overline{AE}$ bisects $\overline{BC}$.

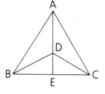

17 Given: $\overline{PT} \cong \overline{PU}$,
$\overline{PR} \cong \overline{PS}$

Prove: $\overrightarrow{PQ}$ bisects $\angle RPS$.

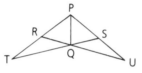

18 Given: $\overline{AD} \cong \overline{DB}$,
$\overline{AE} \cong \overline{BC}$,
$\overline{CD} \cong \overline{ED}$

Prove: $\triangle AFB$ is isosceles.

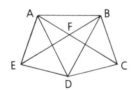

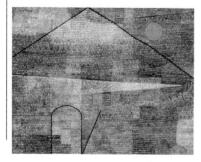

Problem-Set Notes and Additional Answers, continued

■ See *Solution Manual* for answers to problems **14** and **15**.

16 By division, $\angle DBA \cong \angle DCA$, and then $\overline{BD} \cong \overline{CD}$. Thus, $\triangle ADB \cong \triangle ADC$ by SAS. Then $\triangle BAE \cong \triangle CAE$ by ASA.

17 With $\angle TPS \cong \angle UPR$ by Reflexive Prop., $\triangle PRU \cong \triangle PST$ by SAS. Then $\angle PRU \cong \angle PST$ and $\angle T \cong \angle U$ by CPCTC. $\angle TRQ \cong \angle USQ$ because, if two $\angle$s are $\cong$, then their supplements are $\cong$; and $\overline{TR} \cong \overline{US}$ by subtraction. Thus, $\triangle TRQ \cong \triangle USQ$, and $\overline{RQ} \cong \overline{SQ}$ by CPCTC. Then $\triangle PRQ \cong \triangle PSQ$ by SAS, and $\angle RPQ \cong \angle SPQ$ by CPCTC.

18 $\triangle ADE \cong \triangle BDC$ by SSS. Then $\angle ADE \cong \angle BDC$ by CPCTC. By addition, $\angle EDB \cong \angle CDA$, and thus, $\triangle EDB \cong \triangle CDA$ by SAS. $\overline{EB} \cong \overline{CA}$ by CPCTC, and then $\triangle ABE \cong \triangle BAC$ by SSS. $\therefore$ $\angle ABF \cong \angle BAF$, and $\triangle AFB$ is isosceles.

4.2 THE CASE OF THE MISSING DIAGRAM

Class Planning

Time Schedule
Basic: 2 days
Average: $1\frac{1}{2}$ days
Advanced: $1\frac{1}{2}$ days

Resource References
Teacher's Resource Book
    Class Opener 4.2A, 4.2B
Evaluation
    Tests and Quizzes
        Quiz 1, Series 2

Class Opener

Two circles intersect at two points. Prove that the segment joining the centers of the circles bisects the segment joining the points of intersection.
The diagram below shows that the segments might not intersect, so the statement cannot be proved as a theorem.

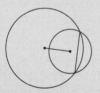

If the segments do intersect, as in the diagram below, then a detour proof can show that △AOP ≅ △BOP by SSS. Then △APM ≅ △BPM by SAS, and $\overline{AM} \cong \overline{BM}$ by CPCTC.

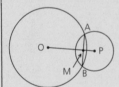

Objective
After studying this section, you will be able to
- Organize the information in, and draw diagrams for, problems presented in words

Part One: Introduction

Some of the geometry problems you encounter will not be accompanied by diagrams. When you are faced with such a problem, it is important for you to be able to "set up" the problem—that is, to draw a diagram that accurately represents the problem and to express the given information and the conclusion you must reach in terms of that diagram. The following examples show some useful techniques for setting up problems.

Example 1   Set up a proof of the statement, *"If two altitudes of a triangle are congruent, then the triangle is isosceles."*

The statement in this problem is in "If . . ., then . . ." form; it is a conditional statement. In such statements the given information is usually to be found in the hypothesis (the *if* clause) and what we are to prove is stated in the conclusion (the *then* clause).
    The diagram we draw should represent the given information but otherwise should be as general as possible. For instance, in the setup below we have not drawn the altitudes so that they bisect the sides, because bisections were not given. To draw bisectors would *overdetermine* the problem.

Setup for Example 1:

Given: $\overline{BD}$ and $\overline{CE}$ are altitudes to
$\overline{AC}$ and $\overline{AD}$ of △ACD.
$\overline{BD} \cong \overline{CE}$

Prove: △ACD is isosceles.

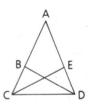

Sometimes the word *then* is left out of a conditional statement or the conclusion comes before the hypothesis. But the hypothesis always follows the word *if* and always contains given conditions. Occasionally, however, some of the given conditions appear in the conclusion, as in the next example.

Example 2   *Set up a proof of the statement, "The medians of a triangle are congruent if the triangle is equilateral."*

In the *if* clause, we are given an equilateral triangle, so we draw one. The conclusion tells us that we are to prove something about the medians, so the medians are also given. We draw them. We letter our diagram any way we wish and write our "Given:" and "Prove:" statements in terms of the diagram.

Setup for Example 2:

Given: △XYZ is equilateral.
    $\overline{PZ}$, $\overline{RY}$, and $\overline{QX}$ are medians.
Prove: $\overline{PZ} \cong \overline{RY} \cong \overline{QX}$

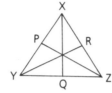

Example 3   *Set up a proof of the statement, "The altitude to the base of an isosceles triangle bisects the vertex angle."*

The statement in this example is a conditional statement with *if* and *then* left out. The main clue is that the sentence begins with given information and ends with a conclusion.

First, we are given an altitude to a base. Then we are given an isosceles triangle. We must prove that the altitude bisects the vertex angle of the triangle.

Setup for Example 3:

Given: △PQR is isosceles, with base $\overline{QR}$.
    $\overline{PM}$ is an altitude.
Prove: $\overline{PM}$ bisects ∠QPR.

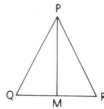

Why was it necessary to specify in the "Given:" statement that $\overline{QR}$ is the base of △PQR? Why was it not necessary to specify that ∠QPR is the vertex angle?

Lesson Notes

- Word problems in geometry, as in other mathematics courses, trouble many students. Even after overcoming initial difficulties in translating the words into a labeled diagram, some students tend to include the conclusion with the given conditions.

Communicating Mathematics

Have students write a paragraph that explains how they would set up the proof of a problem that is presented in words.

Assignment Guide

Basic

Day 1	1–5
Day 2	6–9

Average

Day 1	($\frac{1}{2}$ day) Section **4.1** 12, 15
	($\frac{1}{2}$ day) Section **4.2** 3–5
Day 2	7–9, 11

Advanced

Day 1	($\frac{1}{2}$ day) Section **4.1** 15, 17
	($\frac{1}{2}$ day) Section **4.2** 8, 9
Day 2	11–14

Problem-Set Notes and Additional Answers

- See *Solution Manual* for answers to problems 1–5.

13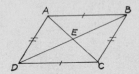

Given: $\overline{AB} \cong \overline{CD}$,
$\overline{AD} \cong \overline{BC}$
Prove: $\overline{AC}$ bisects $\overline{BD}$.
$\overline{BD}$ bisects $\overline{AC}$.
$\triangle ABC \cong \triangle CDA$ by SSS, and thus, $\angle BAC \cong \angle DCA$.
$\triangle BAD \cong \triangle DCB$ by SSS, and thus, $\angle ABD \cong \angle CDB$. Thus, $\triangle ABE \cong \triangle CDE$ by ASA, and then $\overline{AE} \cong \overline{EC}$ and $\overline{DE} \cong \overline{EB}$.

Cooperative Learning

Assign each group of students the same problem chosen from the problem set. Ask each group to draw a diagram, list the givens, and tell what the conclusion should be. Have the groups compare their results and discuss any differences. Each individual in a group should be able to explain the group's work to the class. An answer can be found in the *Solution Manual*.

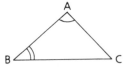

Part Two: Sample Problem

Problem   Set up a proof of the statement, *"If two angles of one triangle are congruent to two angles of another triangle, the remaining pair of angles are also congruent."*

Solution   We draw scalene triangles, since we are not told that the triangles are isosceles or equilateral. Also, we draw triangles of different sizes, since the triangles do not need to be congruent for the angles to be congruent.

Given: $\angle A \cong \angle D$,
$\quad\quad\;\; \angle B \cong \angle E$
Prove: $\angle C \cong \angle F$

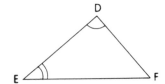

Part Three: Problem Sets

Problem Set A

In problems 1–4, draw your own diagram and write "Given:" and "Prove:" statements in terms of your diagram. Do *not* write a proof.

1 Given: An isosceles triangle and the median to the base

Prove: The median is the *perpendicular bisector* of the base. (This sentence contains *two* conclusions—"the median is perpendicular to the base" and "the median bisects the base.")

2 Given: A four-sided polygon with all four sides congruent (This figure is called a *rhombus*.)

Conclusion: The lines joining opposite vertices are perpendicular.

3 Given: Segments drawn perpendicular to each side of an angle from a point on the bisector of the angle

Conclusion: These two segments are congruent.

4 The bisector of the vertex angle of an isosceles triangle is perpendicular to the base.

In problems 5–7, set up each problem and supply a proof of the statement.

5 The altitude to a side of a scalene triangle forms two congruent angles with that side of the triangle.

6 The median to the base of an isosceles triangle divides the triangle into two congruent triangles.

7 If the base of an isosceles triangle is extended in both directions, then the exterior angles formed are congruent.

Problem Set B

In problems 8–12, set up and complete a proof of each statement.

8 If the median to a side of a triangle is also an altitude to that side, then the triangle is isosceles.

9 The line segments joining the vertex angle of an isosceles triangle to the trisection points of the base are congruent.

10 If the line joining a pair of opposite vertices of a four-sided polygon bisects both angles, then the remaining two angles are congruent.

11 If two triangles are congruent, then any pair of corresponding medians are congruent.

12 If a triangle is isosceles, the triangle formed by its base and the angle bisectors of its base angles is also isosceles.

Problem Set C

In problems 13–15, set up and complete a proof of each statement.

13 If each pair of opposite sides of a four-sided figure are congruent, then the segments joining opposite vertices bisect each other.

14 If a point on the base of an isosceles triangle is equidistant from the midpoints of the legs, then that point is the midpoint of the base.

15 If a point in the interior of an angle (between the sides) is equidistant from the sides of the angle, then the ray joining the vertex of the angle to this point bisects the angle. (Hint: The distance from a point to a line is defined as the length of the *perpendicular* segment from the point to the line.)

Problem-Set Notes and Additional Answers, continued

■ See *Solution Manual* for answers to problems **6–12**.

14

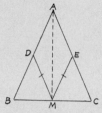

Given: $\overline{AB} \cong \overline{AC}$;
  D is the midpt. of $\overline{AB}$.
  E is the midpt. of $\overline{AC}$.
  $\overline{MD} \cong \overline{ME}$
Prove: M is the midpt. of $\overline{BC}$.
$\overline{AD} \cong \overline{AE}$ by division. Draw $\overline{AM}$. Then $\triangle ADM \cong \triangle AEM$ by SSS. Then $\angle DAM \cong \angle EAM$ by CPCTC. Thus, $\triangle BAM \cong \triangle CAM$ by SAS, and $\overline{BM} \cong \overline{CM}$ by CPCTC.

15

Given: $\overline{DE} \perp \overline{AB}$,
  $\overline{DF} \perp \overline{BC}$,
  $\overline{DE} = \overline{DF}$
Prove: $\overrightarrow{BD}$ bisects $\angle ABC$.
$\triangle BED \cong \triangle BFD$ by HL. Then $\angle DBE \cong \angle DBF$ by CPCTC, and $\overrightarrow{BD}$ bisects $\angle ABC$.

A RIGHT-ANGLE THEOREM

Class Planning

Time Schedule
All levels: 1 day

Resource References
Teacher's Resource Book
   Class Opener 4.3A, 4.3B
   Additional Practice Worksheet 8
Evaluation
   Test and Quizzes
      Quiz 1, Series 1
      Quiz 2, Series 2

Class Opener

Prove that the median to the base of an isosceles triangle is an altitude to the base.

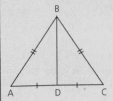

From the diagram, △ABD ≅ △CBD by SSS. Thus, ∠BDA ≅ ∠BDC. Many students will see that, because the two congruent adjacent angles at D form a straight angle, each must be a right angle.

Therefore, the median to the base of an isosceles triangle is also an altitude to the base. This leads to the theorem of the section.

Lesson Notes

■ Writing the paragraph proof of Theorem 23 in two-column form might help many students.

Objective
After studying this section, you will be able to
■ Apply one way of proving that two angles are right angles

Part One: Introduction

Proving that lines are perpendicular depends on proving that they form right angles. For this reason, it is useful to know some ways of proving that angles are right angles. The following theorem will provide you with one such way.

Theorem 23   *If two angles are both supplementary and congruent, then they are right angles.*

Given: ∠1 ≅ ∠2
Prove: ∠1 and ∠2 are right angles.

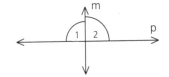

Proof: Since ∠1 and ∠2 form a straight angle (line p), they are supplementary. Therefore, m∠1 + m∠2 = 180. Since ∠1 ≅ ∠2, we can use substitution to get the equation m∠1 + m∠1 = 180, or m∠1 = 90. Thus, ∠1 is a right angle, and so is ∠2.

   In the rest of this book, we shall assume that whenever two angles (such as ∠1 and ∠2 in the diagram for Theorem 23) form a straight angle, the two angles are supplementary. No formal statement of this fact will be necessary.

Part Two: Sample Problems

By this time, you should be familiar with the format used in two-column proofs. Therefore, we shall no longer include the headings "Statements" and "Reasons" in such proofs.

Problem 1　Given: $\overline{AB} \cong \overline{AC}$,
　　　　　　　　$\overline{BD} \cong \overline{CD}$

Conclusion: $\overline{AD}$ is an altitude.

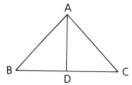

Proof

1　$\overline{AB} \cong \overline{AC}$	1　Given
2　$\overline{BD} \cong \overline{CD}$	2　Given
3　$\overline{AD} \cong \overline{AD}$	3　Reflexive Property
4　$\triangle ABD \cong \triangle ACD$	4　SSS (1, 2, 3)
5　$\angle ADB \cong \angle ADC$	5　CPCTC
6　$\angle ADB$ and $\angle ADC$ are right $\angle$s.	6　If two $\angle$s are both supp. and $\cong$, then they are right $\angle$s.
7　$\overline{AD}$ is an altitude	7　If a segment from a vertex of a $\triangle$ is $\perp$ to the opposite side, it is an altitude of the $\triangle$.

Problem 2　Given: $\overline{AB} \cong \overline{AD}$, $\overline{BC} \cong \overline{CD}$

Prove: $\overleftrightarrow{AC}$ is the $\perp$ bisector of $\overline{BD}$.

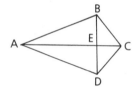

Proof

1　$\overline{AB} \cong \overline{AD}$	1　Given
2　$\overline{BC} \cong \overline{CD}$	2　Given
DETOUR　3　$\overline{AC} \cong \overline{AC}$	3　Reflexive Property
4　$\triangle ABC \cong \triangle ADC$	4　SSS (1, 2, 3)
5　$\angle BAC \cong \angle DAC$	5　CPCTC
6　$\overline{AE} \cong \overline{AE}$	6　Reflexive Property
7　$\triangle ABE \cong \triangle ADE$	7　SAS (1, 5, 6)
8　$\overline{BE} \cong \overline{ED}$	8　CPCTC
9　$\overleftrightarrow{AC}$ bisects $\overline{BD}$.	9　If a line divides a segment into two $\cong$ segments, it bisects the segment.
10　$\angle AEB \cong \angle AED$	10　CPCTC (step 7)
11　$\angle AED$ and $\angle AEB$ are right $\angle$s.	11　If two $\angle$s are both supp. and $\cong$, then they are right $\angle$s.
12　$\overleftrightarrow{AC} \perp \overleftrightarrow{BD}$	12　If two lines intersect to form right $\angle$s, they are $\perp$.
13　$\overleftrightarrow{AC}$ is the $\perp$ bisector of $\overline{BD}$.	13　Combination of steps 9 and 12

Communicating Mathematics

Ask students to write a summary, in paragraph form, of the proofs for Sample Problems **1** and **2**.

Assignment Guide

Basic

1–7

Average

1–8, 10, 13

Advanced

4–10, 13, 14

Problem-Set Notes
and Additional Answers

■ See *Solution Manual* for answers to problems **1–4**.

Part Three: Problem Sets

Problem Set A

1 Given: ⊙P;
      S is the midpt. of $\overline{QR}$.
   Prove: $\overline{PS} \perp \overline{QR}$

2 Prove: The angle bisector of the vertex angle of an isosceles triangle is perpendicular to the base.

3 Given: $\overline{AB} \cong \overline{BC} \cong \overline{CD} \cong \overline{AD}$
      (that is, ABCD is a rhombus)
   Conclusion: $\overline{AC} \perp \overline{BD}$
      (Hint: Use a detour.)

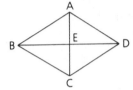

4 Given: $\overrightarrow{XR}$ bisects ∠YXZ.
      ∠Y ≅ ∠Z
   Conclusion: $\overline{XR}$ is an altitude.

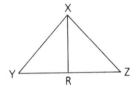

5 A diameter of a circle has endpoints with coordinates (2, 6) and (−4, 10). Find the coordinates of the center of the circle. (−1, 8)

6 If squares A and C are folded across the dotted segments onto B, find the area of B that will *not* be covered by either square. 8 sq units

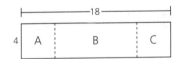

7 Find, to the nearest tenth, the area of the circle. ≈12.6

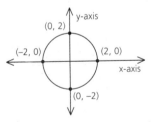

Problem Set B

8 If $\overline{CD}$ is the hypotenuse of a right triangle CAD and A has integral coordinates, find all possible values of the coordinates of A. (1, 3) or (2, 1)

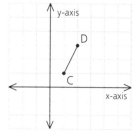

Cooperative Learning

Have students copy the diagram from problem **8,** and ask them to draw a circle that has D as the center and contains point C. Have them try to list all other points on the circle that have integral coordinates.

Seven points: (0, 2); (0, 4); (1, 5); (3, 5); (4, 4); (4, 2); (3, 1)

■ See *Solution Manual* for answers to problems **9–13.**

9 Given: $\odot$O,
$$\angle B \cong \angle C$$
Conclusion: $\overline{AO} \perp \overline{BC}$

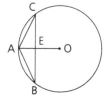

10 Prove that the median to the base of an isosceles triangle is also an altitude to the base.

11 Given: $\overleftrightarrow{PR}$ bisects $\overline{QS}$.
$$\angle RQT \cong \angle RST$$
Prove: $\overline{QS} \perp \overline{PR}$

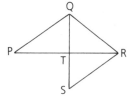

12 Prove that if two circles intersect at two points, A and B, then the line joining the circles' centers is perpendicular to $\overline{AB}$.

13 Prove that the supplement of a right angle is a right angle.

Problem Set C

14 Is b perpendicular to a? Justify your answer. Yes

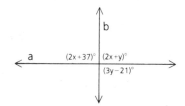

15 The ratio of the complements of two angles is 3:2, and the ratio of their supplements is 9:8. Find the two original angles. 45°; 60°

16 To the nearest second, what is the first time after 7:00 that the hands of a clock form a right angle? ≈7:21:49

14 $\begin{cases} 2x + 37 + 2x + y = 180 \\ 2x + y + 3y - 21 = 180 \end{cases}$
$x = 26\frac{1}{2}$; $y = 37$

15 If the angles have measures x and y, then $\frac{90 - x}{90 - y} = \frac{3}{2}$ and $\frac{180 - x}{180 - y} = \frac{9}{8}$, so $3y - 2x = 90$ and $9y - 8x = 180$.
∴ $x = 45$ and $y = 60$.

16

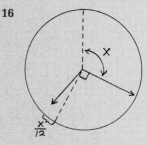

Let the minute hand be at x minutes; then the hour hand is at $35 + \frac{x}{12}$ minutes.
$$35 + \frac{x}{12} - x = 15$$
$$20 = \frac{11}{12}x$$
$$x = \frac{240}{11}$$
$$= 21\frac{9}{11} \text{ min}$$
$$\approx 21 \text{ min } 49 \text{ sec}$$

T 183

THE EQUIDISTANCE THEOREMS

Class Planning

Time Schedule
Basic: 2 days
Average: 2 days
Advanced: $1\frac{1}{2}$ days

Resource References
Teacher's Resource Book
   Class Opener 4.4A, 4.4B
   Supposer Worksheet 5
Evaluation
   Tests and Quizzes
      Quiz 2, Series 1
      Quiz 1, Series 3

Class Opener

This is an excellent opportunity for students to prove today's theorem and to review work from the past few days.
Given: $\overline{PE} \cong \overline{PR}$,
$\overline{EB} \cong \overline{RB}$
Prove: $\overleftrightarrow{PB}$ is the $\perp$ bisector of $\overline{ER}$.

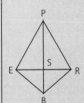

Objective
After studying this section, you will be able to
- Recognize the relationship between equidistance and perpendicular bisection

Part One: Introduction

In geometry, the term *distance* has a special meaning.

Definition   The **distance** between two objects is the length of the shortest path joining them.

Postulate   *A line segment is the shortest path between two points.*

The distance between points R and S is the length of $\overline{RS}$, or RS.

R•————————————•S

If two points P and Q are the same distance from a third point X, then X is said to be **equidistant** from P and Q.

$\overline{PX} \cong \overline{XQ}$
means that
X is equidistant from P and Q.

184   Chapter 4   Lines in the Plane

Vocabulary
distance
equidistant
perpendicular bisector

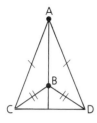 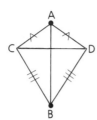 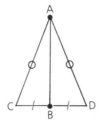

You should recall many problems with diagrams resembling those above. These diagrams have something in common. In each, both point A and point B are equidistant from the endpoints C and D of $\overline{CD}$. In each case, you could prove that $\overleftrightarrow{AB}$ is the **perpendicular bisector** of $\overline{CD}$ just by using the following definition and theorem.

Definition   The **perpendicular bisector** of a segment is the line that bisects and is perpendicular to the segment.

Theorem 24   *If two points are each equidistant from the endpoints of a segment, then the two points determine the perpendicular bisector of that segment.*

Given: $\overline{PA} \cong \overline{PB}$,
$\overline{QA} \cong \overline{QB}$

Prove: $\overleftrightarrow{PQ}$ is the ⊥ bisector of $\overline{AB}$.

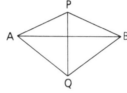

For a proof of Theorem 24, see sample problem 2 in Section 4.3.

Theorem 25   *If a point is on the perpendicular bisector of a segment, then it is equidistant from the endpoints of that segment.*

Given: $\overleftrightarrow{PQ}$ is the ⊥ bisector of $\overline{AB}$.

Prove: $\overline{PA} \cong \overline{PB}$

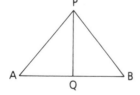

You can easily prove this theorem by using the definition of *perpendicular bisector* and some congruent triangles.

Section 4.4   The Equidistance Theorems   **185**

Class Opener, continued

Statements	Reasons
1 $\overline{PE} \cong \overline{PR}$	1 Given
2 $\overline{EB} \cong \overline{RB}$	2 Given
3 $\overline{PB} \cong \overline{PB}$	3 Reflex. Prop.
4 △PEB ≅ △PRB	4 SSS
5 ∠EPS ≅ ∠RPS	5 CPCTC
6 $\overline{PS} \cong \overline{PS}$	6 Reflex. Prop.
7 △EPS ≅ △RPS	7 SAS
8 $\overline{ES} \cong \overline{RS}$	8 CPCTC
9 $\overleftrightarrow{PB}$ bisects $\overline{ER}$.	9 A line that ÷ a seg. into two ≅ segs. bisects the seg.
10 ∠PSE ≅ ∠PSR	10 CPCTC
11 ∠s PSE and PSR are rt. ∠s	11 If 2 ∠s are both supp. and ≅, they are rt. ∠s
12 $\overleftrightarrow{PB}$ ⊥ $\overline{ER}$	12 If 2 lines intersect to form rt. ∠s, they are ⊥.
13 $\overleftrightarrow{PB}$ is the ⊥ bisector of $\overline{ER}$.	13 The ⊥ bisector of a seg. is a line that bisects and is ⊥ to the seg.

Lesson Notes

- Basic courses might omit this section. These two powerful theorems are often difficult for students to understand and apply.
- Theorem 24 justifies Construction 4, Perpendicular Bisector, pages 668–669. All six basic constructions can now be justified.
- The term *equidistant* may be difficult for some students to understand in the context of the given diagram. Several other diagrams should help students to identify points A and B as the endpoints of any given segment and points P and Q as points equidistant from these endpoints.
- Students may appreciate the power of Theorem 24 by using it in the proofs for problems **3**, **9**, and **12** of Section **4.3**.

- Emphasize that any two points on $\overleftrightarrow{AE}$ determine the perpendicular bisector of $\overline{BD}$. Students need to understand that A and E are not the only points that can be used.

The Geometric Supposer

This activity would be most effective as a whole-class activity.

An interesting way of illustrating the first equidistance theorem (Theorem 24, page 185) is to draw an isosceles triangle and then draw its reflection over its base. The result will be two congruent isosceles triangles with a common base. Ask students what can be said about the vertices of these isosceles triangles with respect to the endpoints of the common base. (The two vertices are equidistant from the endpoints.) What, then, should be true of the segment connecting the vertices of the reflected isosceles triangles? (It should be the perpendicular bisector of the common base.) Have students measure to see that this is true.

Using *The Geometric preSupposer: Points and Lines, The Geometric Supposer: Triangles,* or *The Geometric Supposer: Circles,* you can illustrate the second equidistance theorem (Theorem 25, page 185) as follows. Draw any segment and its perpendicular bisector. Label any point on the perpendicular bisector, drawing and measuring the segments connecting that point to the two endpoints of the segment.

Part Two: Sample Problems

Problem 1 Given: $\angle 1 \cong \angle 2$,
$\angle 3 \cong \angle 4$
Prove: $\overleftrightarrow{AE} \perp$ bis. $\overline{BD}$

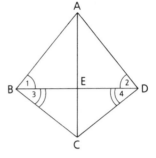

Proof

1 $\angle 1 \cong \angle 2$	1 Given
• 2 $\overline{AB} \cong \overline{AD}$	2 If $\triangle$, then $\triangle$.
3 $\angle 3 \cong \angle 4$	3 Given
• 4 $\overline{BC} \cong \overline{CD}$	4 Same as step 2
5 $\overleftrightarrow{AE} \perp$ bis. $\overline{BD}$	5 If two points are each equidistant from the endpoints of a segment, they determine the $\perp$ bisector of the segment.

Note Since we must identify two "equidistant" points to determine a perpendicular bisector, we have placed a dot before each of the statements in which we identified such a point. We proved that both A and C were equidistant from B and D. Why did we not need to use point E?

Problem 2 Prove: The line joining the vertex of an isosceles triangle to the midpoint of the base is perpendicular to the base.

Given: $\triangle ABC$ is isosceles, with $\overline{AB} \cong \overline{AC}$.
E is the midpoint of $\overline{BC}$.
Prove: $\overleftrightarrow{AE} \perp \overline{BC}$

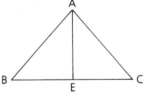

Proof

• 1 $\triangle ABC$ is isosceles, with $\overline{AB} \cong \overline{AC}$.	1 Given
2 E is the midpt. of $\overline{BC}$.	2 Given
• 3 $\overline{BE} \cong \overline{EC}$	3 The midpoint of a segment divides the segment into two $\cong$ segments.
4 $\overleftrightarrow{AE} \perp \overline{BC}$	4 Two points each equidistant from the endpoints of a segment determine the $\perp$ bisector of the segment.

Problem 3   Given: $\overline{AB} \cong \overline{AD}$,
$\overline{BC} \cong \overline{CD}$
Conclusion: $\overline{BE} \cong \overline{ED}$

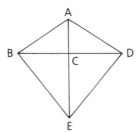

Proof

• 1 $\overline{AB} \cong \overline{AD}$	1 Given
• 2 $\overline{BC} \cong \overline{CD}$	2 Given
3 $\overleftrightarrow{AC} \perp$ bis. $\overline{BD}$	3 Two points each equidistant from the endpoints of a segment determine the $\perp$ bisector of the segment.
4 $\overline{BE} \cong \overline{ED}$	4 A point on the $\perp$ bisector of a segment is equidistant from the endpoints of the segment.

These sample problems could have been solved without the use of Theorems 24 and 25, but the proofs would have been harder and longer.

Part Three: Problem Sets

Problem Set A

As you work on these proofs, see if the equidistance theorems apply; they can save you a lot of work.

1 Given: ⊙O; M is the midpt. of $\overline{AB}$.
Conclusion: $\overline{OM} \perp \overline{AB}$ (Hint: Draw two auxiliary lines.)

2 Given: $\overleftrightarrow{WZ} \perp$ bis. $\overline{XY}$
Prove: △WXY is isosceles. (Hint: This proof can be written in three steps by using Theorem 25.)

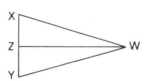

3 Given: ⊙O, $\overline{AB} \cong \overline{AC}$
Conclusion: $\overleftrightarrow{AD} \perp$ bis. $\overline{BC}$ (Hint: Show that A and O are each equidistant from B and C.)

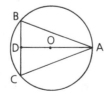

Cooperative Learning

When starting this section, give groups of students Sample Problem **3** to prove without using Theorems 24 and 25. After the theorems have been covered, have the groups compare their proofs with the one given in the text, which uses the theorems. The one in the text is obviously shorter. Ask whether it is "better" or "more correct."

Assignment Guide

Problem-Set Notes
and Additional Answers

■ See *Solution Manual* for answers to problems **1–3**.

Problem Set A, *continued*

4 Given: ⓈⓈ P and Q

Prove: $\overleftrightarrow{PQ} \perp$ bis. $\overline{RS}$

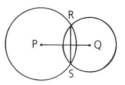

5 Given: $\overleftrightarrow{AD} \perp$ bis. $\overline{BC}$

Prove: $\triangle ABE \cong \triangle ACE$

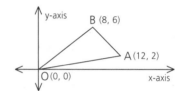

6 Given: $\overleftrightarrow{AG} \perp$ bis. $\overline{BC}$,

$\overleftrightarrow{AG} \perp$ bis. $\overline{DE}$

Conclusion: $\overline{BD} \cong \overline{CE}$

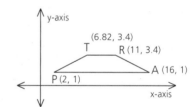

7 How much greater than the x-coordinate of the midpoint of $\overline{OA}$ is the x-coordinate of the midpoint of $\overline{AB}$? 4

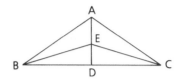

8 In the graph, if a perpendicular is drawn from T to $\overleftrightarrow{PA}$, what will the coordinates of the point where the perpendicular intersects $\overleftrightarrow{PA}$ be? (6.82, 1)

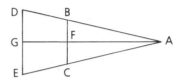

9 If △CAP is slid along the x-axis until C is at (11, 0), what will the new coordinates of P be? (15, 3)

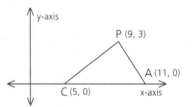

10 A fifth point, E, is located on the diagram so that m∠EBC = $\sqrt{x}$ + 83.

a Is $\overleftrightarrow{AB}$ perpendicular to $\overleftrightarrow{DC}$? Yes

b What do we know about $\overleftrightarrow{AB}$ and $\overleftrightarrow{BE}$? Same line

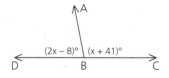

Problem Set B

Remember, the equidistance theorems will help you write a concise proof.

11 Draw isosceles △PQR, with P the vertex. Draw the bisectors of the base angles and label their point of intersection S. Prove that $\overleftrightarrow{PS} \perp \overleftrightarrow{QR}$. (Hint: Use Theorem 24.)

12 Given: $\overline{AB} \cong \overline{BC}$,
    $\overline{AE} \cong \overline{EC}$

Prove: $\overline{AD} \cong \overline{DC}$ (Hint: This can be done in four steps.)

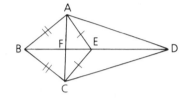

13 Given: $\overline{WY}$ and $\overline{XZ} \perp$ bis. each other.

Prove: $\overline{WX} \cong \overline{XY} \cong \overline{YZ} \cong \overline{ZW}$ (that is, WXYZ is a rhombus)

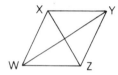

14 Given: $\overline{WX} \cong \overline{WZ}$, $\overline{XY} \cong \overline{YZ}$
    (WXYZ is a kite.)

Prove: △WPZ is a right △.

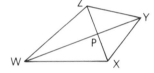

15 Given: ∠ADC and ∠ABC are right ∠s.
    $\overline{AB} \cong \overline{AD}$

Conclusion: $\overleftrightarrow{AC} \perp$ bis. $\overline{BD}$

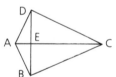

16 Prove: The median to the base of an isosceles triangle is also an altitude. (Prove this without using congruent triangles.)

17 Given: F is the midpt. of $\overline{BC}$.
    $\overline{DB} \cong \overline{EC}$,
    $\overline{DB} \perp \overline{DF}$,
    $\overline{EC} \perp \overline{EF}$

Conclusion: $\overleftrightarrow{AF} \perp \overleftrightarrow{BC}$

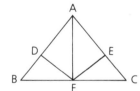

Problem-Set Notes and Additional Answers, continued

■ See *Solution Manual* for answers to problems **11–17**.

Problem-Set Notes and
Additional Answers, continued

■ See *Solution Manual* for answer
to problem **18a.**

20 From the given, it follows
that $\overleftrightarrow{BE}$ is the ⊥ bisector of
$\overline{AC}$. Since D is on $\overleftrightarrow{BE}$, it is
equidistant from the
endpoints of $\overline{AC}$.

21 Draw $\overline{BF}$ and $\overline{DF}$. △BAF ≅
△DEF by SAS. Then $\overline{BF}$ ≅
$\overline{DF}$ by CPCTC, and so F is
equidistant from B and D.
Thus, $\overleftrightarrow{FC}$ is the ⊥ bisector
of $\overline{BD}$.

22

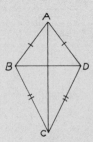

Given: $\overline{AB}$ ≅ $\overline{AD}$,
$\overline{CB}$ ≅ $\overline{CD}$
Prove: $\overleftrightarrow{AC}$ is the ⊥ bisector
of $\overline{BD}$.
Proof follows from
Theorem 24.

Problem Set B, *continued*

18 Given: $\overline{PS}$ ≅ $\overline{SR}$,
$\overline{PQ}$ ≅ $\overline{QR}$

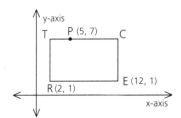

  a Prove that $\overline{QS}$ is an altitude.

  b If RS = 9, QS = 40, and QR = 41, find
the area of triangle PQR. 360

  c What relationship exists among the
numbers 9, 40, and 41, the lengths of
the sides of right triangle QRS? $9^2 + 40^2 = 41^2$

19 **a** On the rectangle shown, how much
farther is the trip from P to T to R to
E than the trip from P to C to E? 6 units

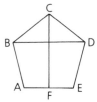

  b If rectangle RECT is rotated 90°
clockwise about point R, what will
the coordinates of the new location of
P be? (8, −2)

Problem Set C

20 Given: $\overline{AB}$ ≅ $\overline{BC}$,
$\overline{AE}$ ≅ $\overline{EC}$
Conclusion: $\overline{AD}$ ≅ $\overline{DC}$

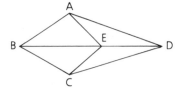

21 Given: ABCDE is equilateral and equian-
gular.
F is the midpt. of $\overline{AE}$.
Prove: $\overleftrightarrow{FC}$ ⊥ bis. $\overline{BD}$

22 A four-sided figure with two disjoint pairs of consecutive sides
congruent is called a *kite*. The two segments joining opposite
vertices are its diagonals. Prove that one of these diagonals is the
perpendicular bisector of the other diagonal.

23 Prove that if each of the three altitudes of a triangle bisects the
side to which it is drawn, then the triangle is equilateral.

24 a If two of the points A, B, C, D, E, and M are chosen at random, what is the probability that the two points determine the perpendicular bisector of $\overline{AB}$? $\frac{1}{5}$

b If three of the six points are chosen at random, what is the probability that the three points are collinear? $\frac{1}{10}$

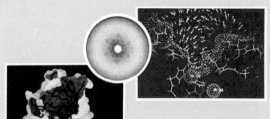

Problem-Set Notes and Additional Answers, *continued*

23

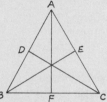

Given: $\overline{AF}$ is the altitude to $\overline{BC}$.

$\overline{BE}$ is the altitude to $\overline{AC}$.

$\overline{CD}$ is the altitude to $\overline{AB}$.

$\overline{AF}$ bisects $\overline{BC}$.

$\overline{BE}$ bisects $\overline{AC}$.

$\overline{CD}$ bisects $\overline{AB}$.

Prove: $\triangle ABC$ is equilateral.

Since $\overline{AF}$ is the $\perp$ bisector of $\overline{BC}$, $\overline{AB} \cong \overline{AC}$. Similarly, $\overline{AB} \cong \overline{CB}$ and $\overline{CA} \cong \overline{CB}$. Thus, $\overline{AB} \cong \overline{BC} \cong \overline{CA}$.

24a

AB AC AD AE AM
BC BD BE BM
CD (CE) (CM)
ED (EM)
DM $\frac{3}{15} = \frac{1}{5}$

b

ABC ABD ABE (ABM)
ACD ACE ACM
ADE ADM
AEM
BCD BCE BCM
BDE BDM
BEM
CDE CDM
(CEM)
DEM $\frac{2}{20} = \frac{1}{10}$

PLOTTING THE STRUCTURE OF A MOLECULE
Elizabeth Getzoff deduces the locations of atoms

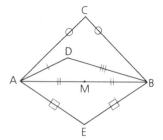

If you sight along the plane of a phonograph record toward a source of light, you'll notice a *diffraction spectrum*, a rainbow created by the separation of the light into its component colors by the parallel grooves. Crystallographers exploit a similar, though far more complex, phenomenon to create *X-ray diffraction* patterns, a powerful tool for determining the atomic structure of molecules.

Crystallographer Elizabeth Getzoff specializes in decoding the structures of protein molecules. She begins by growing crystals made of the protein she wishes to map. Then she bombards the crystals with X-rays. "The X-rays are diffracted by the parallel planes of atoms within the crystal," she explains. "The diffracted rays interfere with each other, producing an array of spots on a photographic film. I can measure the angles and spacings of the diffraction spots and deduce the arrangement and packing of molecules in the crystal." Once she understands the structure of the crystal, she can analyze the structure of the protein molecule from the intensities of the diffraction spots. "My goal is to find the x-, y-, and z-coordinates of the atoms that make up the protein. Then I can plot them in a three-dimensional coordinate system." Computer graphics, another field in which Getzoff has made a series of important contributions, help simplify the task of plotting.

As a high school student in Whippany, New Jersey, Getzoff participated in a National Science Foundation summer program in inorganic chemistry and computer science. She attended Duke University, where she earned a bachelor's degree in chemistry and a doctorate in X-ray crystallography. Since 1985 she has been an assistant member of the molecular biology department at Scripps Clinic in La Jolla, California. There, she runs a research group in molecular structure. Her work, she says, could not proceed without the use of mathematics, especially geometry. For example, to aid in her analysis of the effect of protein upon its function, she is currently developing computer graphic visualizations based on fractal geometry. However difficult the challenge, her reason for taking it on is simple: "I'm finding out how molecules work," she says. "How they work is how life works."

Class Planning

Time Schedule
Basic: 1 day
Average: 1 day
Advanced: $\frac{1}{2}$ day

Resource References
Teacher's Resource Book
   Class Opener 4.5A, 4.5B
Evaluation
   Tests and Quizzes
      Quiz 3, Series 1–2

Class Opener

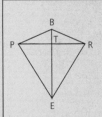

$m\angle PTB = 5x + 30$
$m\angle BTR = 7x + 6$
$\overline{PT} = 2x$
$\overline{TR} = x + 12$
$\overline{PB} = 6y + 2$
$\overline{BR} = 7y - 2$
$\overline{PE} = 4z$
$\overline{RE} = 2z + 20$

1   Find x.   12
2   Find y.   4
3   Find z.   10
4   Find the perimeter of PBRE.
     132

Lesson Notes

■ Students should read through
   Section **4.5** at the beginning of
   the class and then practice
   drawing and identifying examples of pairs of angles that are
   alternate interior, alternate exterior, and corresponding.

4.5 INTRODUCTION TO PARALLEL LINES

Objectives
After studying this section, you will be able to
■ Recognize planes
■ Recognize transversals
■ Identify the pairs of angles formed by a transversal
■ Recognize parallel lines

Part One: Introduction

Planes
In order to explain parallel lines adequately, we must first acquaint
you with the meaning of **plane**.

> **Definition**   A **plane** is a surface such that if any two points on
> the surface are connected by a line, all points of the
> line are also on the surface.

A plane has only two dimensions—length and width. Both the
length and the width are infinite. A plane has no thickness.

> **Definition**   If points, lines, segments, and so forth, lie in the
> same plane, we call them **coplanar**. Points, lines,
> segments, and so forth, that do not lie in the same
> plane are called **noncoplanar**.

Planes are discussed more fully in Chapter 6.

Transversals
In the figure, line t is a **transversal** of lines
a and b.

> **Definition**   A **transversal** is a line that intersects two coplanar
> lines in two distinct points.

192   Chapter 4   Lines in the Plane

Vocabulary
alternate exterior angles
alternate interior angles
coplanar
corresponding angles

exterior
interior
noncoplanar
parallel lines
plane
transversal

The region between lines d and e is the *interior* of the figure. The rest of the plane is the *exterior*.

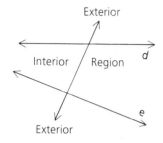

The diagram of lines f and g cut by transversal h provides another illustration of the regions formed by two lines and a transversal.

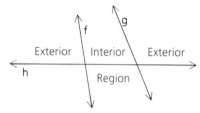

Angle Pairs Formed by Transversals

$\overleftrightarrow{AB}$ and $\overleftrightarrow{CD}$ are cut by transversal $\overleftrightarrow{EF}$.

The two pairs of *alternate interior angles* are 3 and 6, 4 and 5.
The two pairs of *alternate exterior angles* are 1 and 8, 2 and 7.
The four pairs of *corresponding angles* are 1 and 5, 2 and 6, 3 and 7, 4 and 8.

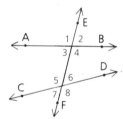

$\overleftrightarrow{GH}$ and $\overleftrightarrow{JK}$ are cut by transversal $\overleftrightarrow{MO}$.

The alternate interior angles are b and g, f and c.
The alternate exterior angles are a and h, e and d.
The corresponding angles are a and c, b and d, e and g, f and h.

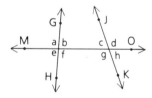

Definition *Alternate interior angles* are a pair of angles formed by two lines and a transversal. The angles must both lie in the interior of the figure, must lie on alternate sides of the transversal, and must have different vertices.

Look for an N or Z shape.

Lesson Notes, continued

- Identifying the N or Z shape can help students locate pairs of alternate interior angles.

Definition   *Alternate exterior angles* are a pair of angles formed by two lines and a transversal. The angles must both lie in the exterior of the figure, must lie on alternate sides of the transversal, and must have different vertices.

Definition   *Corresponding angles* are a pair of angles formed by two lines and a transversal. One angle must lie in the interior of the figure, and the other must lie in the exterior. The angles must lie on the same side of the transversal but have different vertices.

Look for an F shape.

It is important to be able to recognize these pairs of angles when they appear in figures made up of a number of segments. In each of the following examples, the segment corresponding to the transversal is shown in red, and the segments corresponding to the lines it cuts are shown in blue.

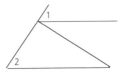 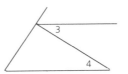 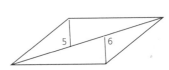

∠1 and ∠2 are corresponding ∠s.

∠3 and ∠4 are alternate interior ∠s.

∠5 and ∠6 are alternate exterior ∠s.

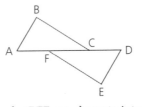 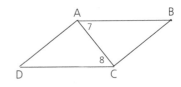

∠BCA and ∠DFE are alternate interior ∠s.

∠BCD and ∠EFA are alternate exterior ∠s.

∠7 and ∠8 are alternate interior ∠s.

Can you find a pair of alternate interior ∠s formed by $\overleftrightarrow{AD}$ and $\overleftrightarrow{BC}$ with transversal $\overleftrightarrow{AC}$?

Parallel Lines

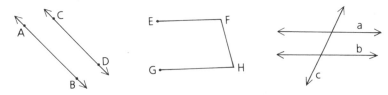

Above are three illustrations of **_parallel_** (‖) lines. We write $\overleftrightarrow{AB} \parallel \overleftrightarrow{CD}$, $\overleftrightarrow{EF} \parallel \overleftrightarrow{GH}$, and a ‖ b.

Definition	**_Parallel lines_** are two coplanar lines that do not intersect.

    We shall also call segments or rays parallel if they are parts of parallel lines. For example, we can say that in the preceding diagrams $\overline{AB} \parallel \overline{CD}$ and $\overline{EF} \parallel \overline{GH}$.

    There are many lines that do not intersect yet are not parallel. _To be parallel, lines must be coplanar._ In Chapter 6, lines that are noncoplanar and nonintersecting are defined as _skew_ lines.

Cooperative Learning

Have groups of students find three examples of pairs of parallel lines in the classroom and also pairs of lines that are not parallel but that do not intersect. Have students in each group identify the physical planes in which the pairs of parallel lines lie.　Answers will vary.

Part Two: Sample Problem

Problem

  a Which of the lines in the figure at the right is the transversal?

  b Name all pairs of alternate interior angles.

  c Name all pairs of alternate exterior angles.

  d Name all pairs of corresponding angles.

  e Name all pairs of interior angles on the same side of the transversal.

  f Name all pairs of exterior angles on the same side of the transversal.

Answers

  a $\overleftrightarrow{AD}$

  b ∠GBC and ∠FCB, ∠HBC and ∠ECB

  c ∠ABG and ∠DCF, ∠ABH and ∠DCE

  d ∠ABG and ∠BCE, ∠GBC and ∠ECD, ∠ABH and ∠BCF, ∠HBC and ∠FCD

  e ∠GBC and ∠ECB, ∠HBC and ∠FCB

  f ∠ABG and ∠DCE, ∠ABH and ∠DCF

Part Three: Problem Set

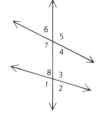

1 a Name all pairs of alternate interior angles.
∠3 and ∠7, ∠4 and ∠8
b Name all pairs of alternate exterior angles.
∠1 and ∠5, ∠2 and ∠6
c Name all pairs of corresponding angles.
∠1 and ∠7, ∠2 and ∠4, ∠3 and ∠5, ∠8 and ∠6
d Name all pairs of interior angles on the same side of the transversal.
∠7 and ∠8, ∠3 and ∠4
e Name all pairs of exterior angles on the same side of the transversal.
∠1 and ∠6, ∠2 and ∠5

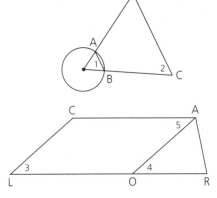

2 a What name is given to ∠1 and ∠2 for
$\overleftrightarrow{AB}$ and $\overleftrightarrow{CD}$? What is the transversal?
Corresponding ∠s; $\overleftrightarrow{BC}$
b What type of angles are 3 and 4?
Which lines and transversal form them?
Corresponding; $\overleftrightarrow{CL}$ and $\overleftrightarrow{AO}$, $\overleftrightarrow{LR}$
c What type of angles are 4 and 5?
Which lines and transversal form them?
Alternate interior; $\overleftrightarrow{CA}$ and $\overleftrightarrow{LR}$, $\overleftrightarrow{AO}$

3 Copy the diagram.

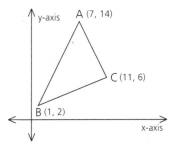

a Find the coordinates of M, the midpoint of $\overline{AB}$. (4, 8)
b Find the coordinates of N, the midpoint of $\overline{AC}$. (9, 10)
c Draw $\overleftrightarrow{MN}$. What *appears* to be true about $\overleftrightarrow{MN}$ and $\overleftrightarrow{BC}$? Parallel
d What *appears* to be true about ∠AMN and ∠ABC? Congruent
e Name a pair of corresponding angles formed by $\overleftrightarrow{MN}$ and $\overleftrightarrow{BC}$ with transversal $\overleftrightarrow{AC}$. ∠ANM and ∠ACB

4 a For which pair of lines are angles 1 and 4 a pair of alternate interior angles? $\overleftrightarrow{JK}$ and $\overleftrightarrow{OM}$

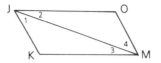

b For which pair of lines are angles 2 and 3 a pair of alternate interior angles? $\overleftrightarrow{JO}$ and $\overleftrightarrow{KM}$

c How many transversals of $\overleftrightarrow{JO}$ and $\overleftrightarrow{KM}$ are shown? 3

5 Locate the following points on a graph: $(x_1, y_1) = (0, 0)$, $(x_2, y_2) = (4, 5)$, $(x_3, y_3) = (0, 3)$ and $(x_4, y_4) = (4, 8)$.

a Find $\frac{y_2 - y_1}{x_2 - x_1}$. $\frac{5}{4}$

b Find $\frac{y_4 - y_3}{x_4 - x_3}$. $\frac{5}{4}$

c Draw a line through the first two points and a line through the second two points. What *appears* to be true about these lines? Parallel

HISTORICAL SNAPSHOT

FROM MUD TO THE STARS
The reach of geometry

Since before recorded history, human beings have used basic geometric principles in building and surveying. But with the rise of civilization, people came gradually to recognize the power of geometry as a means of controlling and explaining the world around them. As the encyclopedist Isidore of Seville (A.D. 560–636) tells us,

The science of geometry is said to have been discovered by the Egyptians, for after the Nile would flood, covering all their property with mud, they would mark off their landholdings with boundaries and measurements, thus giving geometry its name [from Greek gē, "earth," and metra, "measurements"]. Later, when this study had been further perfected by the ingenuity of the wise, it was also used to measure the expanses of sea and stars and air. For after investigating the dimensions of the earth by geometry, people began to investigate even the extent of the heavens—how far the moon is from the earth, and the sun from the moon, all the way to the limits of the universe.

The development of geometric thought from its beginnings to the present day, when it guides scientists' explorations of realms of space stretching from the subatomic to the intergalactic, makes for a fascinating story. The Historical Snapshots in this book will give you a few brief glimpses into that story. If you find them interesting, you may wish to look further into the history of geometry.

Class Planning

Time Schedule
Basic: 2 days
Average: 2 days
Advanced: 1 day

Resource References
Teacher's Resource Book
   Class Opener 4.6A, 4.6B
Transparency 8

Class Opener

The answers for problems **1, 2,** and **4** should be drawn on the same coordinate plane.

1   Draw the line through (3, 5) and (8, 7) on a graph.   See graph.
2   Draw a line through (3, 12) that is parallel to the line through (3, 5) and (8, 7). See graph.
3   Find the coordinates of at least one more point on the line you drew for problem **2.** Answers may vary. Possible answer: (8, 14)
4   Draw a line through (3, 12) that is perpendicular to the line that you drew for problem **2.**   See graph.
5   Find the coordinates of at least one more point on the line through (3, 12) that you drew for problem **4.**   Answers may vary. Possible answer: (5, 7)

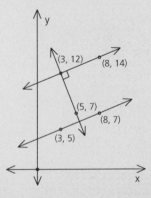

4.6 | SLOPE

Objectives
After studying this section, you will be able to
■ Understand the concept of slope
■ Relate the slope of a line to its orientation in the coordinate plane
■ Recognize the relationships between the slopes of parallel and perpendicular lines

Part One: Introduction

Definition of *Slope*

To understand how the principles of coordinate geometry can be applied to the study of parallelism and perpendicularity, you need to be familiar with the concept of **slope**.

Definition   The **slope** m of a nonvertical line, segment, or ray containing (x_1, y_1) and (x_2, y_2) is defined by the formula

$$m = \frac{y_2 - y_1}{x_2 - x_1} \text{ or } m = \frac{y_1 - y_2}{x_1 - x_2} \text{ or } m = \frac{\Delta y}{\Delta x}$$

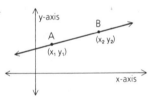

Note   In more-advanced mathematics classes, it is common to use Δy (read "delta y") instead of $y_2 - y_1$ and Δx ("delta x") instead of $x_2 - x_1$. The symbol Δ is used to indicate change, so that Δy, for example, means "the change in y-coordinates between two points."

Example   *Find the slope of the segment joining* $(-2, 3)$ *and* $(6, 5)$.

$$m = \frac{y_2 - y_1}{x_2 - x_1} \text{ or } m = \frac{y_1 - y_2}{x_1 - x_2}$$

$$= \frac{5 - 3}{6 - (-2)} \qquad = \frac{3 - 5}{-2 - 6}$$

$$= \frac{2}{8} = \frac{1}{4} \qquad = \frac{-2}{-8} = \frac{1}{4}$$

Notice that it does not matter which point is chosen as (x_1, y_1).

198   Chapter 4   Lines in the Plane

Vocabulary
opposite reciprocal
slope

When the slope formula is applied to a vertical line, such as $\overleftrightarrow{CD}$, the denominator is zero. Division by zero is undefined, so a vertical line has no slope.

$$m = \frac{y_2 - y_1}{x_2 - x_1}$$

$$= \frac{12 - 2}{6 - 6}$$

$$= \frac{10}{0} \text{ (An undefined expression)}$$

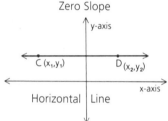

Do not confuse no slope with a slope of zero. On a horizontal line, $y_2 = y_1$, but $x_2 \neq x_1$. Therefore, the numerator is zero, while the denominator is not. Hence, a horizontal line has zero slope.

Visual Interpretation of Slope

The numerical value of a slope gives us a clue to the direction a line is taking. The following diagrams illustrate this notion.

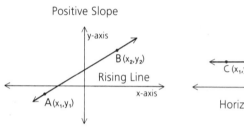

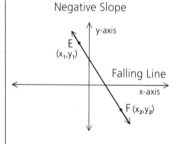

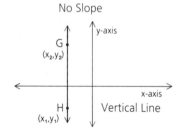

In summary,
- Rising line $\Leftrightarrow$ positive slope
- Horizontal line $\Leftrightarrow$ zero slope
- Falling line $\Leftrightarrow$ negative slope
- Vertical line $\Leftrightarrow$ no slope

Lesson Notes

- Remind students that the symbol $\Leftrightarrow$ used in this summary is usually read "is equivalent to." In this case, a better phrase might be "corresponds to."

Section 4.6   Slope   **199**

Slopes of Parallel and Perpendicular Lines

The proofs of the following four theorems require a knowledge of the properties of similar triangles and will be omitted here.

Theorem 26   *If two nonvertical lines are parallel, then their slopes are equal.*

Given: $\overleftrightarrow{AB} \parallel \overleftrightarrow{CD}$

Prove: Slope $\overleftrightarrow{AB}$ = slope $\overleftrightarrow{CD}$

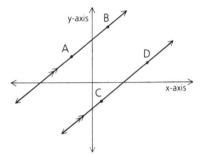

The next theorem is the converse of Theorem 26, with the statements in the *if* clause and the *then* clause reversed.

Theorem 27   *If the slopes of two nonvertical lines are equal, then the lines are parallel.*

It can also be shown that there is a relationship between the slopes of two perpendicular lines—they are **opposite reciprocals** of each other. For example, if the slope of a line is $\frac{2}{5}$, the slope of any line perpendicular to it is $-\frac{5}{2}$. As with parallel lines, we can develop two converse theorems.

Theorem 28   *If two lines are perpendicular and neither is vertical, each line's slope is the opposite reciprocal of the other's.*

Theorem 29   *If a line's slope is the opposite reciprocal of another line's slope, the two lines are perpendicular.*

Part Two: Sample Problems

Problem 1 If $A = (4, -6)$ and $B = (-2, -8)$, find the slope of $\overleftrightarrow{AB}$.

Solution By the slope formula,

$$m = \frac{y_2 - y_1}{x_2 - x_1}$$

$$= \frac{-8 - (-6)}{-2 - 4}$$

$$= \frac{-8 + 6}{-6} = \frac{1}{3}$$

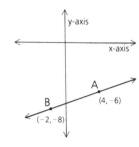

Note The line is rising, so the slope is positive. Drawing a diagram helps prevent careless errors.

Problem 2 Show that CEF is a right triangle.

Solution Find the slopes of the sides.

Slope of $\overleftrightarrow{CE} = \dfrac{\Delta y}{\Delta x} = \dfrac{4 - 3}{8 - 1} = \dfrac{1}{7}$

Slope of $\overleftrightarrow{FE} = \dfrac{\Delta y}{\Delta x} = \dfrac{7 - 4}{4 - 8} = \dfrac{3}{-4} = -\dfrac{3}{4}$

Slope of $\overleftrightarrow{FC} = \dfrac{\Delta y}{\Delta x} = \dfrac{3 - 7}{1 - 4} = \dfrac{-4}{-3} = \dfrac{4}{3}$

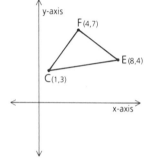

Since the slopes of $\overleftrightarrow{FE}$ and $\overleftrightarrow{FC}$ are opposite reciprocals, $\angle F$ is a right angle. Therefore, $\triangle CEF$ is a right triangle.

Problem 3 Given: $\triangle ABE$ as shown

Find: **a** The slope of altitude $\overline{AC}$
     b The slope of median $\overline{AD}$

Solution **a** Slope of $\overleftrightarrow{BE} = \dfrac{\Delta y}{\Delta x} = \dfrac{5 - 3}{6 - (-4)} = \dfrac{1}{5}$

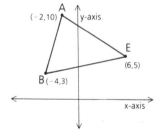

Since the slope of the altitude to $\overline{BE}$ is the opposite reciprocal of the slope of $\overleftrightarrow{BE}$, the slope of $\overleftrightarrow{AC} = -5$.

b By the midpoint formula, $D = (1, 4)$. Since $A = (-2, 10)$, the slope of $\overleftrightarrow{AD} = \dfrac{\Delta y}{\Delta x} = \dfrac{4 - 10}{1 - (-2)} = -2$.

4.6

Problem 4 *Find the slope of $\overleftrightarrow{AB}$ to the nearest hundredth.*

Solution By the slope formula,

$$m = \frac{y_2 - y_1}{x_2 - x_1}$$

$$= \frac{2\sqrt{5} - \sqrt{5}}{6 - 3}$$

$$= \frac{\sqrt{5}}{3}$$

To approximate, use a calculator.

$$\frac{\sqrt{5}}{3} \approx \frac{2.236067977}{3}$$

$$\approx 0.75$$

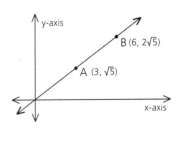

Part Three: Problem Sets

Problem Set A

1 Find the slope of the line determined by each pair of points.

a (1, 7) and (10, 15) $\frac{8}{9}$

b (−2, 6) and (5, 7) $\frac{1}{7}$

c (−8, −7) and (−2, 4) $\frac{11}{6}$

d (5, 4) and (−2, 4) 0

e ($\sqrt{3}$, 7) and ($\sqrt{3}$, −9) No slope

f (5a, 6c) and (2a, −9c) $\frac{5c}{a}$

2 $\overleftrightarrow{AB}$ has a slope of $1\frac{2}{3}$ and $\overleftrightarrow{CD} \perp \overleftrightarrow{AB}$. What is the slope of $\overleftrightarrow{CD}$? $-\frac{3}{5}$

3 If $\overleftrightarrow{EF} \parallel \overleftrightarrow{GH}$ and $\overleftrightarrow{EF}$ has a slope of −4, what is the slope of $\overleftrightarrow{GH}$? −4

4 If $\angle F$ is a right angle, find the slope of $\overleftrightarrow{FH}$. $\frac{1}{2}$

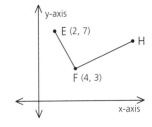

5 Given the diagram as marked, with $\overline{AC}$ an altitude and $\overline{AD}$ a median, find the slope of each line.

a $\overleftrightarrow{BE}$ $\frac{1}{4}$ b $\overleftrightarrow{AC}$ −4   c $\overleftrightarrow{AD}$ $-\frac{7}{2}$

d A line through A and parallel to $\overline{BE}$ $\frac{1}{4}$

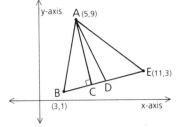

6 $\overleftrightarrow{AB}$ has a slope of $2\frac{1}{2}$. If A = (2, 7) and B = (12, k), what is the value of k? 32

7 Show that $\overleftrightarrow{FH} \parallel \overleftrightarrow{JK}$ and $\overleftrightarrow{FK} \parallel \overleftrightarrow{HJ}$. (Since both pairs of opposite sides of FHJK are parallel, we call the figure a *parallelogram*.)

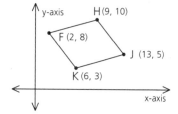

8 a Is $\overleftrightarrow{RE}$ parallel to $\overleftrightarrow{TC}$? Yes
 b Is $\overleftrightarrow{TR}$ parallel to $\overleftrightarrow{CE}$? Yes
 c Show that $\angle R$ is a right angle.

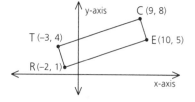

9 a Find the slope of $\overleftrightarrow{PT}$. $\frac{1}{2}$
 b Find the slope of $\overleftrightarrow{TV}$. $\frac{1}{2}$
 c Are P, T, and V collinear or noncollinear? Collinear

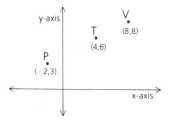

10 a Are $(-6, 5)$, $(1, 7)$, and $(15, 10)$ collinear? No
 b Are $(74, 20)$, $(50, 16)$, and $(2, 8)$ collinear? Yes

11 Complete each of the following statements.
 a For $\overleftrightarrow{EC}$ and $\overleftrightarrow{AB}$, a pair of corresponding angles are $\angle ABC$ and ___?___. $\angle ECD$
 b For $\overleftrightarrow{EC}$ and $\overleftrightarrow{AB}$, a pair of alternate interior angles are $\angle ABE$ and ___?___. $\angle CEB$

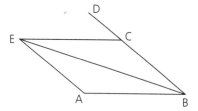

Problem Set B

12 Write an argument to show that $\overline{CM}$ is not the median to $\overline{AB}$.

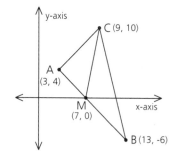

Problem-Set Notes and Additional Answers

■ See *Solution Manual* for answers to problems **7, 8c,** and **12.**

Cooperative Learning

Have students find the midpoints of the diagonals of the parallelograms pictured in problems **7** and **8.** Ask what they can conclude about the diagonals of parallelograms. Have them graph a rectangle of their own choosing to test their hypothesis.
The diagonals of parallelograms bisect each other.

- Problems **12, 14,** and **17** are informal applications of indirect proof.
- See *Solution Manual* for answers to problems **13** and **14.**

17 The midpoint M of $\overline{AB}$ is (7, 2) from the midpoint formula. The slope of $\overline{CM}$ is −5, and the slope of $\overline{AB}$ is $\frac{1}{5}$. Hence, $\overline{CM}$ is the altitude from C to $\overline{AB}$. Since the median and the altitude are the same segment, they have equal length. Thus, the median is not longer than the altitude.

18 Apply the given Pythagorean Theorem to find
$$(CA)^2 + (AT)^2 = (CT)^2$$
$$= 16$$

Problem Set B, *continued*

13 If A = (6, 11), B = (1, 5), and C = (7, 0), show by means of slopes that △ABC is a right triangle. Name the hypotenuse. $\overline{AC}$

14 Suppose that point H is rotated 90° in a clockwise direction about the origin to point J.

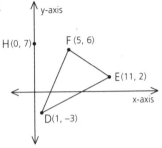

  a Does J lie on $\overleftrightarrow{DE}$? Show why or why not. Yes

  b Write an argument to show that $\overline{FJ}$ is not the altitude to $\overline{DE}$.

15 If square OABC is rotated 180° clockwise about its center, what will the new coordinates of O be? (7, 7)

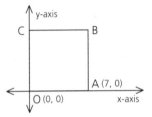

16 Goofy Guff wanted to reflect the outline figure (the figure to the right of the y-axis) across the dashed line. Goofy shaded what he thought was the reflected figure as shown, but Goofy had goofed.

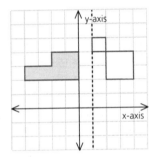

  a How many 1-unit squares did Goofy shade that he shouldn't have? 2

  b How many additional 1-unit squares should Goofy have shaded? 1

Problem Set C

17 △ABC has vertices at A = (2, 1), B = (12, 3), and C = (6, 7). Write an argument to show that the median from C to $\overline{AB}$ is not longer than the altitude from C to $\overline{AB}$.

18 In any right triangle, if *a* and *b* are the lengths of the legs and *c* is the length of the hypotenuse, $a^2 + b^2 = c^2$. Given △CAT as shown, find $(CA)^2 + (AT)^2$. 16

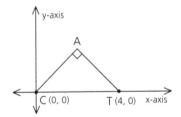

CHAPTER SUMMARY

CONCEPTS AND PROCEDURES

After studying this chapter, you should be able to
- Use detours in proofs (4.1)
- Apply the midpoint formula (4.1)
- Organize the information in, and draw diagrams for, problems presented in words (4.2)
- Apply one way of proving that two angles are right angles (4.3)
- Recognize the relationship between equidistance and perpendicular bisection (4.4)
- Recognize planes (4.5)
- Recognize transversals (4.5)
- Identify the pairs of angles formed by a transversal (4.5)
- Recognize parallel lines (4.5)
- Understand the concept of slope (4.6)
- Relate the slope of a line to its orientation in the coordinate plane (4.6)
- Recognize the relationships between the slopes of parallel and perpendicular lines (4.6)

VOCABULARY

alternate exterior angles (4.5)
alternate interior angles (4.5)
coplanar (4.5)
corresponding angles (4.5)
detour proof (4.1)
distance (4.4)
equidistant (4.4)
exterior (4.5)
interior (4.5)

midpoint formula (4.1)
noncoplaner (4.5)
opposite reciprocal (4.6)
parallel lines (4.5)
perpendicular bisector (4.4)
plane (4.5)
slope (4.6)
transversal (4.5)

4

REVIEW PROBLEMS

Problem Set A

1 Copy the problem and proof, filling in the blanks with the correct statements and reasons.

Given: P is the midpt. of $\overline{XZ}$.
    $\angle 1 \cong \angle 2$

Conclusion: $\overline{XY} \cong \overline{YZ}$

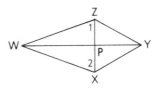

1 P is the midpt. of $\overline{XZ}$.	1 Given
2 _____	2 _____
3 $\angle 1 \cong \angle 2$	3 Given
4 _____	4 _____
5 $\overleftrightarrow{WY} \perp$ bis. $\overline{XZ}$	5 _____
6 $\overline{XY} \cong \overline{YZ}$	6 _____

2 a Identify a pair of corresponding angles formed by $\overleftrightarrow{BE}$ and $\overleftrightarrow{CD}$ with transversal $\overleftrightarrow{BC}$. $\angle C$ and $\angle ABE$

b Identify a pair of alternate interior angles formed by $\overleftrightarrow{BE}$ and $\overleftrightarrow{CD}$ with transversal $\overleftrightarrow{BD}$. $\angle EBD$ and $\angle CDB$

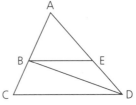

3 Given: $\angle ADB \cong \angle CDB$, $\overline{AD} \cong \overline{DB}$
Prove: $\overline{BD}$ is an altitude.

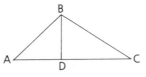

4 Given: $\angle 1 \cong \angle 4$;
    $\overrightarrow{FC}$ bisects $\angle BFD$.
Conclusion: $\overleftrightarrow{CF} \perp \overleftrightarrow{AE}$

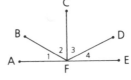

5 Set up a proof for the following information. Then complete the proof.

Given: Two isosceles triangles with the same base

Prove: The line joining the vertices of the vertex ∠s of the △ is the ⊥ bisector of the base.

6 Given: △ABC is isosceles, with base $\overline{AC}$.
∠1 ≅ ∠2
Conclusion: $\overleftrightarrow{BD} \perp \overleftrightarrow{AC}$

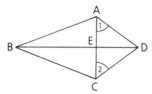

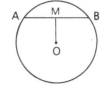

7 Given: ⊙O;
M is the midpt. of $\overline{AB}$.
Conclusion: $\overline{OM} \perp \overline{AB}$

8 Set up a proof of, but do not prove, the statement, "If two chords of a circle are congruent, then the segments joining the midpoints of the chords to the center of the circle are congruent." (A *chord* is a segment whose endpoints are on the circle.)

9 a If the median from A intersects $\overline{BC}$ at M, what are the coordinates of M? (9, 4)

b Find the slope of $\overleftrightarrow{BC}$. $\frac{1}{2}$

c Is $\overleftrightarrow{AR}$ parallel to $\overleftrightarrow{BC}$? Why or why not?

d Find the slope of the altitude from A to $\overline{BC}$. −2

e If Rhonda Right walked from A to M, how far did she walk? 7 units

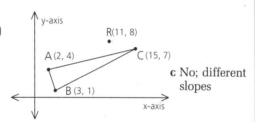

c No; different slopes

Problem Set B

10 a Point A = (1, 4) is reflected across the y-axis to point B. Find the coordinates of B. (−1, 4)

b Point A is rotated, with respect to the origin, 90° clockwise to point C. Find the coordinates of C. (4, −1)

c If A is slid two units up and then seven units to the right to point D, what are the coordinates of D? (This "sliding" procedure is called a *translation*.) (8, 6)

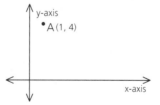

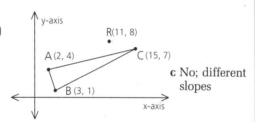

Problem-Set Notes and Additional Answers, continued

■ See *Solution Manual* for answers to problems **5–8**.

Problem-Set Notes and
Additional Answers, continued

■ See *Solution Manual* for answers to problems **11–16.**

Review Problem Set B, *continued*

11 Given: ⊙O,
∠1 ≅ ∠2
Conclusion: $\overline{OY} \perp \overline{WX}$

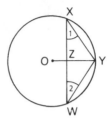

12 Given: ∠1 ≅ ∠2 ≅ ∠3 ≅ ∠4,
$\overline{BE} \cong \overline{BF}$
Conclusion: △ABE ≅ △CBF

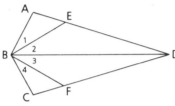

13 Given: ⊙O,
$\overline{DX} \cong \overline{DY}$
Conclusion: $\overleftrightarrow{DZ}$ bisects $\overline{XY}$.

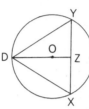

14 Given: ∠WXY ≅ ∠ZYX,
$\overline{WX} \cong \overline{ZY}$
Conclusion: $\overline{WR} \cong \overline{RZ}$

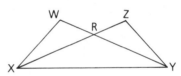

15 Given: $\overline{AB} \cong \overline{AF}$,
$\overline{BD} \cong \overline{DF}$,
∠1 ≅ ∠2
Conclusion: $\overline{AD} \perp \overline{CE}$

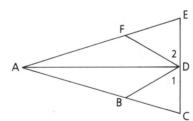

16 Given: $\overleftrightarrow{AD} \perp$ bis. $\overline{BC}$
Conclusion: ∠1 ≅ ∠2

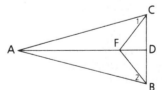

17 Use slopes to show that EFHJ is a parallelogram.

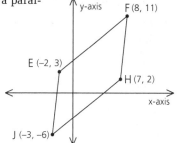

Problem Set C

18 ∠F is a right angle. Explain why (9, 6) could not be the coordinates of H.

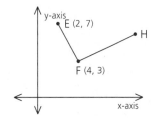

19 Given △PQR, with P = (3, 6), Q = (4, 1), and R = (14, 3), find the measure of the largest angle of △PQR. Explain your reasoning. m∠Q = 90

20 Given: $\overline{AB} \cong \overline{AF}$,
$\overline{BC} \cong \overline{FE}$
Conclusion: $\overline{CD} \cong \overline{DE}$

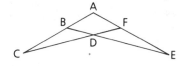

21 Prove: If the bisector of an angle whose vertex lies on a circle passes through the center of the circle, then it is the perpendicular bisector of the segment joining the points where the sides of the angle intersect the circle.

22 Given: $\overline{AB} \cong \overline{AC}$,
$\overline{BF} \cong \overline{FC}$,
∠BAE ≅ ∠CAD
Prove: $\overleftrightarrow{AF} \perp \overleftrightarrow{DE}$

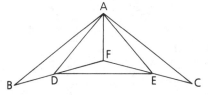

Problem-Set Notes and Additional Answers, continued

■ See *Solution Manual* for answers to problems **17** and **18**.

19 $\overline{PQ} \perp \overline{QR}$ by computing slopes.

20 △ACF ≅ △AEB by SAS. Thus, ∠C ≅ ∠E by CPCTC, and ∠AFC ≅ ∠ABE by CPCTC. Thus, ∠CBD ≅ ∠EFD because, if two ∠s are ≅, then their supplements are ≅. Thus, △BCD ≅ △FED by ASA, and then $\overline{CD} \cong \overline{ED}$ by CPCTC.

21

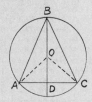

Given: $\overrightarrow{BD}$ bisects ∠ABC.
Prove: $\overleftrightarrow{BD}$ is the ⊥ bisector of $\overline{AC}$.
$\overline{OA} \cong \overline{OB}$; thus, ∠OAB ≅ ∠OBA. $\overline{OC} \cong \overline{OB}$; thus, ∠OCB ≅ ∠OBC. Since $\overrightarrow{BD}$ bisects ∠ABC, ∠OBC ≅ ∠OBA, and thus, ∠OAB ≅ ∠OCB by the Transitive Prop. $\overline{OA} \cong \overline{OC}$; thus, ∠OAD ≅ ∠OCD. Thus, ∠BAD ≅ ∠BCA by addition. Thus, $\overline{BA} \cong \overline{BC}$ because sides opposite ≅ ∠s are ≅. Since $\overline{OA} \cong \overline{OC}$, $\overleftrightarrow{BD}$ is the ⊥ bisector of $\overline{AC}$.

22 △ABF ≅ △ACF by SSS; thus, ∠B ≅ ∠C and ∠BAF ≅ ∠CAF by CPCTC. ∠BAD ≅ ∠CAE by subtraction. Thus, △ABD ≅ △ACE by ASA. Then $\overline{AD} \cong \overline{AE}$ and $\overline{BD} \cong \overline{CE}$ by CPCTC. Thus, $\overline{DF} \cong \overline{EF}$ by subtraction. ∴ $\overline{AF}$ is the ⊥ bisector of $\overline{DE}$.

CHAPTER 5

PARALLEL LINES AND RELATED FIGURES

THE FIRST SECTION of this chapter introduces indirect proofs. The remainder of the chapter can be considered as two parts: parallel lines (Sections **5.2** and **5.3**) and special quadrilaterals (Sections **5.4** to **5.7**).

The properties developed for parallel lines and for quadrilaterals are used in many subsequent chapters, both in proofs and in application problems.

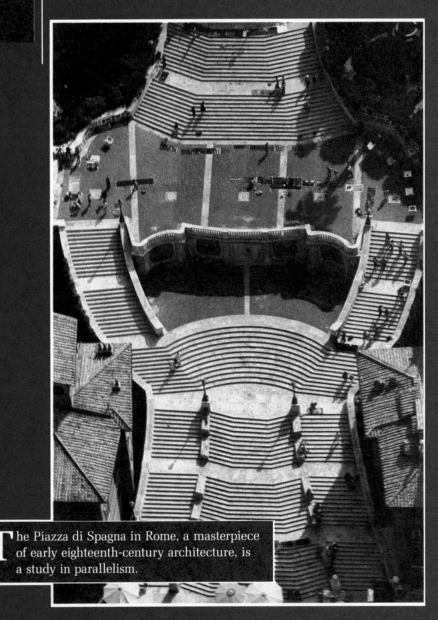

The Piazza di Spagna in Rome, a masterpiece of early eighteenth-century architecture, is a study in parallelism.

5.1 INDIRECT PROOF

Objective

After studying this section, you will be able to
- Write indirect proofs

Part One: Introduction

At the beginning of this book, we mentioned that mathematicians believe that today's students should be familiar with a variety of proof styles. This is why we have provided you with several alternatives to the two-column proof. To give you an efficient way to work certain problems, we now introduce the concept of ***indirect proof.***

An indirect proof may be useful in a problem where a direct proof would be difficult to apply. Study the following example of an indirect proof.

Example

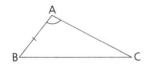 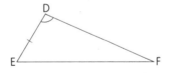

Given: $\angle A \cong \angle D$, $\overline{AB} \cong \overline{DE}$, $\overline{AC} \not\cong \overline{DF}$

Prove: $\angle B \not\cong \angle E$

Proof: Either $\angle B \cong \angle E$ or $\angle B \not\cong \angle E$.
Assume $\angle B \cong \angle E$.
From the given information, $\angle A \cong \angle D$ and $\overline{AB} \cong \overline{DE}$.
Thus, $\triangle ABC \cong \triangle DEF$ by ASA.
$\therefore \overline{AC} \cong \overline{DF}$

But this is impossible, since $\overline{AC} \not\cong \overline{DF}$ is given.
Thus, our assumption was false and $\angle B \not\cong \angle E$, because this is the only other possibility.

Vocabulary
indirect proof

Class Planning

Time Schedule
All levels: 1 day

Resource References
Teacher's Resource Book
  Class Opener 5.1A, 5.1B

Class Opener

Given: $\overline{DB} \perp \overline{AC}$;
   M is the midpoint of $\overline{AC}$.
Prove: $\overline{AD} \not\cong \overline{CD}$

In order for sides $\overline{AD}$ and $\overline{CD}$ to be congruent, $\triangle ADC$ must be isosceles. But then the foot (point B) of the altitude from vertex D and the midpoint (M) of the side opposite vertex D would have to coincide. Therefore, $\overline{AD} \not\cong \overline{CD}$ unless point B = point M.

This leads into indirect proof.

The following procedure will help you to write indirect proofs.

Indirect-Proof Procedure

1 List the possibilities for the conclusion.
2 Assume that the *negation* of the desired conclusion is correct.
3 Write a chain of reasons until you reach an impossibility.
 This will be a contradiction of either
  (a) given information or
  (b) a theorem, definition, or other known fact.
4 State the remaining possibility as the desired conclusion.

Part Two: Sample Problems

Note   Remember to start by looking at the conclusion.

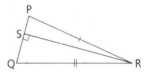

Problem 1   Given: $\overline{RS} \perp \overline{PQ}$,
  $\overline{PR} \neq \overline{QR}$
  Prove: $\overrightarrow{RS}$ does not bisect $\angle PRQ$.

Proof   Either $\overrightarrow{RS}$ bisects $\angle PRQ$ or $\overrightarrow{RS}$ does not bisect $\angle PRQ$.
  Assume $\overrightarrow{RS}$ bisects $\angle PRQ$.
  Then we can say that $\angle PRS \cong \angle QRS$.
  Since $\overline{RS} \perp \overline{PQ}$, we know that $\angle PSR \cong \angle QSR$.
  Thus, $\triangle PSR \cong \triangle QSR$ by ASA ($\overline{SR} \cong \overline{SR}$).
  Hence, $\overline{PR} \cong \overline{QR}$ by CPCTC.

  But this is impossible because it contradicts the given fact that $\overline{PR} \neq \overline{QR}$. Consequently, the assumption must be false. $\therefore \overrightarrow{RS}$ does not bisect $\angle PRQ$, the only other possibility.

Problem 2   Given: $\odot O$, $\overline{AB} \neq \overline{BC}$
  Prove: $\angle AOB \neq \angle COB$

Proof   Either $\angle AOB \cong \angle COB$ or $\angle AOB \neq \angle COB$. We will assume that $\angle AOB \cong \angle COB$. Since O is the center of the circle, $\overline{AO} \cong \overline{CO}$. By the Reflexive Property, $\overline{BO} \cong \overline{BO}$. Thus, $\triangle AOB \cong \triangle COB$ by SAS, which means that $\overline{AB} \cong \overline{CB}$ by CPCTC.
   This is impossible because it contradicts the given fact that $\overline{AB} \neq \overline{BC}$. Consequently, our assumption ($\angle AOB \cong \angle COB$) is false. $\therefore \angle AOB \neq \angle COB$, because that is the only other possibility.

Chapter 5   Parallel Lines and Related Figures

Part Three: Problem Sets

Problem Set A

1 Given: $\overline{AB} \cong \overline{AD}$, $\angle BAC \not\cong \angle DAC$
Prove: $\overline{BC} \not\cong \overline{DC}$

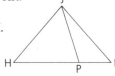

2 Given: P is not the midpoint of $\overline{HK}$.
$\overline{HJ} \cong \overline{JK}$
Prove: $\overrightarrow{JP}$ does not bisect $\angle HJK$.

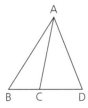

3 Given: $\overline{AC} \perp \overline{BD}$,
$\overline{BC} \cong \overline{EC}$,
$\overline{AB} \not\cong \overline{ED}$
Prove: $\angle B \not\cong \angle CED$

4 Given: $\angle H \not\cong \angle K$
Prove: $\overline{JH} \not\cong \overline{JK}$

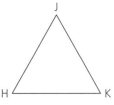

5 Given: $\odot O$;
$\overline{OB}$ is not an altitude.
Prove: $\overrightarrow{OB}$ does not bisect $\angle AOC$.

6 ODEF is a square.
In terms of a, find E($2a$, $2a$), F(0, $2a$)
  a The coordinates of points E and F
  b The area of the square $4a^2$
  c The midpoint of $\overline{FD}$ (a, a)
  d The midpoint of $\overline{OE}$ (a, a)

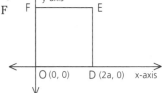

Assignment Guide

Basic
1, 2, 5–9
Average
1, 2, 5–10
Advanced
6–8, 11–15

Problem-Set Notes and Additional Answers

■ See *Solution Manual* for answers to problems **1–5**.

Problem Set A, *continued*

7 RECT is a rectangle.

  a In terms of a and b, find the coordinates of C. $(2a, 2b)$

  b Does $\overline{RC}$ appear to be congruent to $\overline{ET}$? Yes

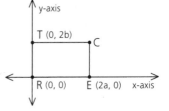

8 With respect to the origin, point A is rotated 90° clockwise to point C, and point B is rotated 180° clockwise to point D. Find the slope of $\overleftrightarrow{CD}$. $\frac{11}{7}$

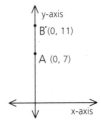

9 Identify each of the following pairs of angles as alternate interior, alternate exterior, or corresponding.

  a For $\overleftrightarrow{BE}$ and $\overleftrightarrow{CD}$ with transversal $\overleftrightarrow{BC}$, ∠1 and ∠C

  b For $\overleftrightarrow{AE}$ and $\overleftrightarrow{BD}$ with transversal $\overleftrightarrow{BE}$, ∠2 and ∠4

  a Corr.; **b** Alt. int.

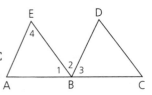

Problem-Set Notes and Additional Answers, continued

■ See Solution Manual for answers to problems **10–12**.

Problem Set B

10 Given: $\overline{PA} \perp \overline{AB}$,
      $\overline{PA} \perp \overline{AC}$,
      ∠B ≇ ∠C

  Prove: $\overline{AB} \not\cong \overline{AC}$

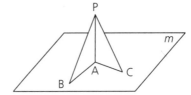

11 Given: $\overrightarrow{BD}$ bisects ∠ABC.
       ∠ADB is acute.

  Prove: $\overline{AB} \not\cong \overline{BC}$

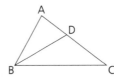

12 Given: ⊙O; $\overline{HE}$ is not the perpendicular bisector of $\overline{DF}$.

  Prove: $\overline{DE} \not\cong \overline{EF}$

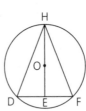

13 Prove that if △ABC is isosceles, with base $\overline{BC}$, and if P is a point on $\overline{BC}$ that is not the midpoint, then $\overrightarrow{AP}$ does not bisect ∠BAC.

Problem Set C

14 Prove that if no two medians of a triangle are congruent, then the triangle is scalene.

15 a Show that the diagonals, $\overline{CS}$ and $\overline{OI}$, of the given isosceles trapezoid do not bisect each other.

b Are the diagonals of this isosceles trapezoid perpendicular? Yes

c Do you think that the diagonals of every isosceles trapezoid are perpendicular? No

d Can you figure out what to draw so that you could use the formula $a^2 + b^2 = c^2$ (see Section 4.6, problem 18) to find that OS ≈ 8.25?

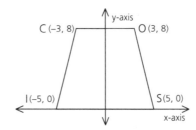

C (–3, 8)   y-axis   O (3, 8)

I (–5, 0)   S (5, 0)   x-axis

A LINE TO THE STARS

Geometry guides navigator Paul Wotherspoon

Perhaps the primary task of a navigator is to fix, or identify exactly, the ship's position. To appreciate the problem this presents, imagine yourself on the ocean with no land in sight. How do you tell where you are? The most reliable way, says Senior Chief Quartermaster Paul Wotherspoon—an assistant navigator in the United States Coast Guard—is to get out your sextant.

A sextant is a hand-held instrument used to measure the angle between a star and the horizon. Its effectiveness is based on the fact that all lines of sight to a star from anywhere on earth are parallel. Because of the earth's curvature, the angle between the lines of sight to the star and to the horizon changes as your position on earth changes.

"With the sextant," explains Wotherspoon, "we find the position of a known star in the sky by measuring its angle above the horizon.

Horizon
EP
Earthly position of star

Next, [based on the time of night, charts, and a complicated procedure called sight reduction] we identify the point on the earth that is directly beneath the star. Using this information we can draw a *line of position*, an arc the ship must lie on." The navigator repeats the process for a second star. "Your location is the point where the two lines of position intersect. Since there are two such places on the earth's surface, it's best to take a third sighting to confirm your position."

Line of position 2
Line of position 1

Wotherspoon attended high school in Vernon, Connecticut. Following graduation he joined the Coast Guard. He attended quartermaster school. During his nineteen years in the Coast Guard, he has served on five ships. Today he is stationed in Boston. His many duties as an assistant navigator include planning trips, giving directions on the bridge, securing tide and current information, steering the ship in close quarters, and taking official deck logs.

Section 5.1   Indirect Proof   **215**

Proving That Lines Are Parallel

Class Planning

Time Schedule
Basic: 2 days
Average: 2 days
Advanced: 1 day

Resource References
Teacher's Resource Book
  Class Opener 5.2A, 5.2B
  Using Manipulatives 7

Class Opener

Given: M is the midpoint of $\overline{NL}$.
$\overline{KN} \not\cong \overline{KL}$
Prove: $\overline{NW} \not\cong \overline{LW}$

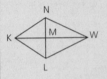

Proof: Assume that $\overline{NW} \cong \overline{LW}$.
Since M is the midpoint of $\overline{NL}$, it
follows that $\overline{NM} \cong \overline{LM}$. So, $\overleftrightarrow{WM}$
is the ⊥ bisector of $\overline{NL}$ by the
equidistance theorem. ∴ $\overline{KN} \cong$
$\overline{KL}$ because a point on the ⊥
bisector of a segment is equidistant from the endpoints of the
segment. But this contradicts
the given that $\overline{KN} \not\cong \overline{KL}$.
So $\overline{NW} \not\cong \overline{LW}$.

Objectives

After studying this section, you will be able to
- Apply the Exterior Angle Inequality Theorem
- Use various methods to prove lines parallel

Part One: Introduction

The Exterior Angle Inequality Theorem

An exterior angle of a triangle is formed
whenever a side of the triangle is extended
to form an angle supplementary to the adjacent interior angle.

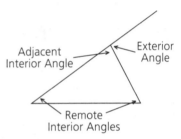

Adjacent
Interior Angle

Exterior
Angle

Remote
Interior Angles

Theorem 30 *The measure of an exterior angle of a triangle is
greater than the measure of either remote interior
angle.*

Given: Exterior angle BCD
Prove: m∠BCD > m∠B,
  m∠BCD > m∠BAC

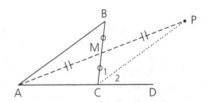

Proof: Locate the midpoint, M, of $\overline{BC}$. Draw $\overline{AP}$ so that AM = MP.
Draw $\overline{CP}$. $\overline{MB} \cong \overline{MC}$, $\overline{AM} \cong \overline{MP}$, and vertical angles are congruent. Thus, △ABM ≅ △PCM and∠1 ≅ ∠B. Since m∠BCD
> m∠1, we know that m∠BCD > m∠B. The second part of
the theorem is proved by extending $\overline{BC}$ to form the other
exterior angle, a vertical angle to ∠BCD. The result follows.

Identifying Parallel Lines

When two lines are cut by a transversal, eight angles are formed. You can use several pairs of angles to prove that the lines are parallel.

Theorem 31   *If two lines are cut by a transversal such that two alternate interior angles are congruent, the lines are parallel.* (**Short form: Alt. int. ∠s ≅ ⇒ ∥ lines**)

Given: ∠3 ≅ ∠6

Prove: a ∥ b

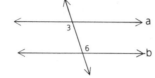

Proof: (Indirect proof) *Assume that the lines are not parallel.* Then a and b must intersect at some point P. ∠3 is an exterior angle of the triangle formed, so by the Exterior Angle Inequality Theorem, m∠3 > m∠6. But this contradicts the given: ∠3 ≅ ∠6. Thus, our assumption was false; the lines *are* parallel.

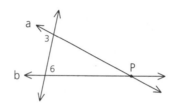

Theorem 32   *If two lines are cut by a transversal such that two alternate exterior angles are congruent, the lines are parallel.* (*Alt. ext. ∠s ≅ ⇒ ∥ lines.*)

Given: ∠1 ≅ ∠8

Prove: a ∥ b

This can be proved by use of alt. int. ∠s ≅ ⇒ ∥ lines.

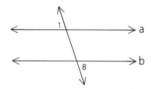

Theorem 33   *If two lines are cut by a transversal such that two corresponding angles are congruent, the lines are parallel.* (*Corr. ∠s ≅ ⇒ ∥ lines*)

Given: ∠2 ≅ ∠6

Prove: a ∥ b

This can be proved by use of alt. int. ∠s ≅ ⇒ ∥ lines.

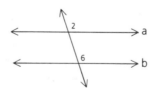

5.2

Lesson Notes

- Some teachers use AIP as a short form for Theorem 31.
- Some students may appreciate the five-step chain of reasoning for Theorem 31.

1   Suppose a ∦ b; then they intersect at P.

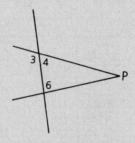

2   From the Parallel Postulate (see Section **5.3**, page 224), it follows that m∠4 + m∠6 + m∠P = 180. (This is proved as Theorem 50, p. 295.)

3   m∠3 = m∠6 + m∠P (This is proved as Theorem 51, p. 296.)

4   m∠3 > m∠6 (This is proved as Theorem 30, p. 216.)

5   But it is given that m∠3 = m∠6. Thus, a ∥ b.

Cooperative Learning

Using indirect proofs, have one group of students prove each of Theorems 32–36 on pages 217 and 218. Choose one student from each group to present the theorem and proof to the class. Answers may vary. A possible proof for Theorem 34 is given.

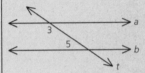

Given: Lines a and b and
      transversal t,
      ∠3 supp. ∠5
Prove: a ∥ b
Assume a ∦ b. Then a and b must intersect at P.

∠9 is supp. to ∠3 because they form a straight ∠. ∴ ∠9 ≅ ∠5 because ∠s supp. to ≅ ∠s are ≅. But ∠5 is an exterior ∠ to the triangle formed and must have a measure greater than m∠9 by the Exterior Angle Inequality Theorem. Since ∠5 cannot be ≅ ∠9, ∠5 cannot be supp. to ∠3. But this contradicts the given. ∴ a ∥ b.

Note: Indirect proofs may not use hints given in the text.

Theorem 34 *If two lines are cut by a transversal such that two interior angles on the same side of the transversal are supplementary, the lines are parallel.*

Given: ∠3 supp. ∠5
Prove: a ∥ b

This can be proved by use of alt. int. ∠s ≅ ⇒ ∥ lines.

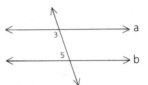

Theorem 35 *If two lines are cut by a transversal such that two exterior angles on the same side of the transversal are supplementary, the lines are parallel.*

Given: ∠1 supp. ∠7
Prove: a ∥ b

This can be proved by use of alt. int. ∠s ≅ ⇒ ∥ lines.

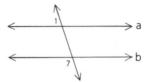

Theorem 36 *If two coplanar lines are perpendicular to a third line, they are parallel.*

Given: a ⊥ c and b ⊥ c
Prove: a ∥ b

This can be proved by use of corr. ∠s ≅ ⇒ ∥ lines.

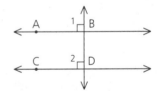

Part Two: Sample Problems

Problem 1 Prove Theorem 36.
  Given: $\overleftrightarrow{AB} \perp \overleftrightarrow{BD}$ and $\overleftrightarrow{CD} \perp \overleftrightarrow{BD}$
  Prove: $\overleftrightarrow{AB} \parallel \overleftrightarrow{CD}$

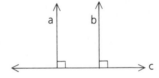

Proof

1 $\overleftrightarrow{BD} \perp \overleftrightarrow{AB}$	1 Given
2 ∠1 is a right ∠.	2 ⊥ lines form right ∠s.
3 $\overleftrightarrow{BD} \perp \overleftrightarrow{CD}$	3 Given
4 ∠2 is a right ∠.	4 Same as 2
5 ∠1 ≅ ∠2	5 Right ∠s are ≅.
6 $\overleftrightarrow{AB} \parallel \overleftrightarrow{CD}$	6 Corr. ∠s ≅ ⇒ ∥ lines

218   Chapter 5   Parallel Lines and Related Figures

Problem 2   A parallelogram is a four-sided figure with both pairs of opposite sides parallel.

Given: ∠1 ≅ ∠2,
   ∠PQR ≅ ∠RSP
Prove: PQRS is a parallelogram.

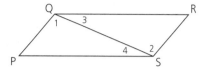

Proof

1 ∠1 ≅ ∠2	1 Given
2 $\overline{PQ} \parallel \overline{RS}$	2 Alt. int. ∠s ≅ ⇒ ∥ lines
3 ∠PQR ≅ ∠RSP	3 Given
4 ∠3 ≅ ∠4	4 Subtraction Property
5 $\overline{QR} \parallel \overline{PS}$	5 Same as 2
6 PQRS is a parallelogram.	6 A four-sided figure with both pairs of opposite sides parallel is a parallelogram.

Problem 3   A trapezoid is a four-sided figure with exactly one pair of parallel sides.

Given: ∠1 supp. ∠3,
   ∠2 ≅ ∠3
Prove: TRAP is a trapezoid.

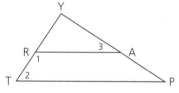

Proof   We can use a flow diagram.

Reasons ① Given
   ② Substitution
   ③ Int. ∠s on same side supp. ⇒ ∥ lines

Part Three: Problem Sets

Problem Set A

1 In each case, state the theorem that proves a ∥ b.

a
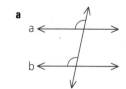
Corr.
∠s ≅ ⇒ ∥ lines

b
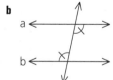
Alt. int.
∠s ≅ ⇒ ∥ lines

c
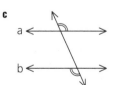
Alt. ext.
∠s ≅ ⇒ ∥ lines

Section 5.2   Proving That Lines Are Parallel   **219**

Communicating Mathematics

Have students write a paragraph that summarizes the various conditions under which two lines that are cut by a transversal will be parallel.

Assignment Guide

Basic
Day 1 1, 2, 4–9, 12, 13
Day 2 14–17, 19, 20, 24
Average
Day 1 1, 2, 4–9, 12–14
Day 2 15–17, 19, 24–26, 28
Advanced
9, 12–14, 19, 22, 24, 26, 27

2a Ext. ∠s on same side supp. ⇒ ∥ lines

b 2 lines ⊥ to same line are ∥.

c Int. ∠s on same side supp. ⇒ ∥ lines

■ See *Solution Manual* for answer to problem **6**.

Problem Set A, *continued*

2 In each case, state the theorem that proves c ∥ d.

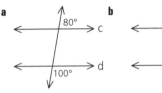

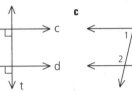

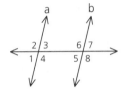

∠1 supp. ∠2

3 If certain pairs of angles in the diagram are given to be congruent, we can prove that a ∥ b. List all such pairs. (1, 5), (1, 7), (2, 6), (2, 8), (3, 7), (3, 5), (4, 8), (4, 6)

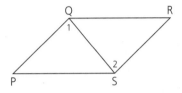

4 If ∠1 ≅ ∠2, which lines are parallel? Write the theorem that justifies your answer.
$\overleftrightarrow{PQ} \parallel \overleftrightarrow{SR}$; alt. int. ∠s ≅ ⇒ ∥ lines

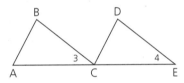

5 If ∠3 ≅ ∠4, which lines are parallel? Write the theorem that justifies your answer.
$\overleftrightarrow{BC} \parallel \overleftrightarrow{DE}$; corr. ∠s ≅ ⇒ ∥ lines

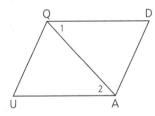

6 Given: $\overleftrightarrow{QD} \nparallel \overleftrightarrow{UA}$
Prove: ∠1 ≇ ∠2

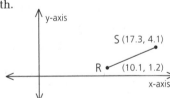

7 Find the slope of $\overleftrightarrow{RS}$ to the nearest tenth.
≈ 0.4

y-axis

S (17.3, 4.1)

R • (10.1, 1.2)

x-axis

8 Which two lines are parallel? Write the theorem that justifies your answer.
$\overleftrightarrow{RA} \parallel \overleftrightarrow{TP}$; int. ∠s on same side supp. ⇒ ∥ lines

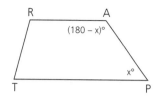

9 If ∠1 ≅ ∠2, which two lines are parallel? Write the theorem that justifies your answer.
$\overleftrightarrow{BE} \parallel \overleftrightarrow{DF}$; alt. ext. ∠s ≅ ⇒ ∥ lines

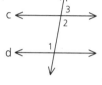

10 If exactly two of the three labeled angles are congruent, what is the probability that one can prove that c ∥ d?
$\frac{1}{3}$

11 Complete the inequality that shows the restrictions on x.

$$\frac{?}{\ \ } < x < \frac{?}{\ \ }$$

0 < x < 110

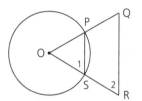

12 Given: ⊙O,
∠1 ≅ ∠2
Prove: $\overline{PS} \parallel \overline{QR}$

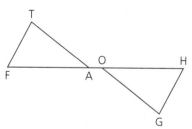

13 Given: ∠FAT ≅ ∠HOG
Prove: $\overline{AT} \parallel \overline{GO}$

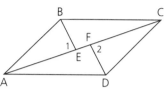

14 Given: $\overline{AB} \cong \overline{CD}$,
$\overline{BC} \cong \overline{AD}$
Prove: $\overline{AB} \parallel \overline{CD}$

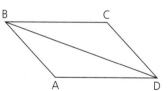

Problem-Set Notes and Additional Answers, continued

■ See *Solution Manual* for answers to problems **12–14**.

Section 5.2   Proving That Lines Are Parallel   **221**

Problem Set A, *continued*

15 If P = (−3, 1), Q = (2, 4), R = (1, −2), and S = (7, 2), are $\overline{PQ}$ and $\overline{RS}$ parallel? Explain your answer. No, because the slopes are not equal.

16 Solve for x and justify that m ∥ n.
x = 26; x + 84 = 110,
so m ∥ n by corr. ∠s ≅ ⇒ ∥ lines.

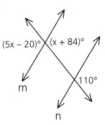

(5x − 20)° (x + 84)°
m 110°
n

17 Write a valid inequality and find the restrictions on x.
16 < x < 66

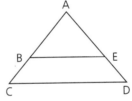

30° (3x − 18)°

18 If x is 14.5, are p and q parallel? Explain. Yes, if x = 14.5, then 8x + (5x − 8.5) = 180. Thus, p ∥ q because int. ∠s on same side of transv. supp. ⇒ ∥ lines.

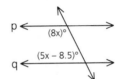

p (8x)°
q (5x − 8.5)°

Problem Set B

19 Given: ∠D ≅ ∠ABE,
$\overline{BE}$ ∦ $\overline{CD}$
Prove: $\overline{AC}$ ≇ $\overline{AD}$

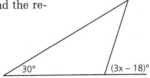

A
B E
C D

20 Given: ∠1 comp. ∠2,
∠3 comp. ∠2
Prove: $\overline{QT}$ ∥ $\overline{RS}$

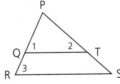

P
Q 1 2 T
R 3 S

21 Given: ∠MOP is a right angle.
$\overline{RP}$ ⊥ $\overline{OP}$
Prove: $\overline{MO}$ ∥ $\overline{RP}$

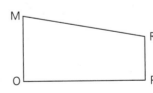

M R
O P

22 Find the restrictions on x. (Point P may be anywhere between K and M.) $81 < x < 143$

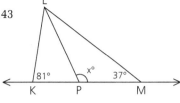

23 If $\overleftrightarrow{PQ} \parallel \overleftrightarrow{RS}$, can x be 25? Explain.
No, because if x = 25, then ∠QBD
supp. ∠RCD and $\overleftrightarrow{PQ} \parallel \overleftrightarrow{RS}$, a contradiction.

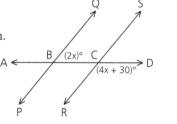

24 Given: ∠1 supp. ∠2,
∠3 supp. ∠2
Prove: ABCD is a parallelogram. (See sample problem 2 for a definition of *parallelogram*.)

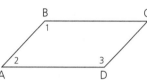

Problem Set C

25 Show that the quadrilateral formed by joining consecutive midpoints of RECT is a parallelogram. (See sample problem 2 for a definition of *parallelogram*.)

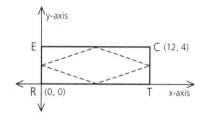

26 Find the value(s) of x (to the nearest tenth) that will allow you to prove that m ∥ n. (Hint: You may wish to review the quadratic formula.)

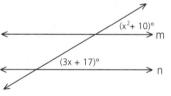

27 Use a coordinate proof to prove that the quadrilateral determined by the mid-points of PQRS is a parallelogram.

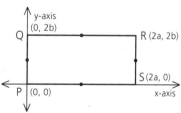

28 Prove that the diagonals $\overline{PR}$ and $\overline{QS}$ in problem 27 bisect each other.

Problem-Set Notes and Additional Answers, continued

■ See *Solution Manual* for answer to problem **24**.

25

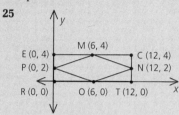

Find the coordinates of all points as shown.
Slope $\overline{PM}$ = slope $\overline{ON}$ = $\frac{1}{3}$
Slope $\overline{MN}$ = slope $\overline{PO}$ = $-\frac{1}{3}$
∴ PMNO is a ▱.

26 To be ∥, corr. ∠s must be ≅, so
$$x^2 + 10 = 3x + 17.$$
$$x^2 - 3x - 7 = 0$$
$$x = \frac{3 \pm \sqrt{37}}{2}$$
x ≈ 4.5 or x ≈ −1.5

27

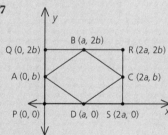

Find the coordinates of all points as shown.
Slope $\overline{AB}$ = slope $\overline{DC}$ = $\frac{b}{a}$.
∴ $\overline{AB} \parallel \overline{DC}$.
Slope $\overline{AD}$ = slope $\overline{BC}$ = $-\frac{b}{a}$.
∴ $\overline{AD} \parallel \overline{BC}$.
∴ ABCD is a ▱.

28 Use the diagram in problem **27**.
Midpt. of $\overline{PR}$ is (a, b).
Midpt. of $\overline{QS}$ is (a, b).
Since $\overline{PR}$ and $\overline{QS}$ have the same midpt., they must bisect each other.

Class Planning

Time Schedule
All levels: 2 days

Resource References
Teacher's Resource Book
  Class Opener 5.3A, 5.3B
  Additional Practice
  Worksheet 9
Transparency 9
Evaluation
  Tests and Quizzes
    Quizzes 1–2, Series 1
    Quiz 1, Series 2

Class Opener

Given: a ∥ b
Find: ∠1

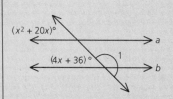

Since corresponding ∠s are ≅,
$x^2 + 20x = 4x + 36.$
$x^2 + 16x - 36 = 0$
$(x - 2)(x + 18) = 0$
$x = 2$ or $x = -18$
If $x = 2$, then
$4x + 36 = 8 + 36 = 44.$
If $x = -18$, then
$4x + 36 = -72 + 36 = -36.$
The last solution is not possible.
Therefore, $\angle 1 = 180 - 44 = 136.$

This provides a good review of solving quadratic equations. The need to consider both solutions above should be discussed.

CONGRUENT ANGLES ASSOCIATED WITH PARALLEL LINES

Objectives

After studying this section, you will be able to
- Apply the Parallel Postulate
- Identify the pairs of angles formed by a transversal cutting parallel lines
- Apply six theorems about parallel lines

Part One: Introduction

The Parallel Postulate

In this section we shall see that the converses of many of the theorems in Section 5.2 are also true.

A fundamental postulate for parallel lines in the plane is the *Parallel Postulate*.

Postulate	*Through a point not on a line there is exactly one parallel to the given line.*

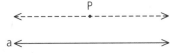

Although this idea may seem reasonable, mathematicians argued for centuries over the truth of the Parallel Postulate. Some even created their own geometries based on the assumptions that there may be *more than one parallel* to a given line at a given point or *no parallels* to the given line (a geometry in which *any* two lines intersect at some point). If you are interested in learning more about these *non-Euclidean* geometries, see your teacher for a list of sources. We, however, will assume that the Parallel Postulate is true.

Angles Formed When Parallel Lines Are Cut by a Transversal

In Section 5.2 we saw that when alternate interior angles are congruent, lines are parallel. In this section you will learn that the converse is true—that is, if we start with parallel lines, then we can conclude that alternate interior angles are congruent. In fact, many pairs of congruent angles are determined by parallel lines cut by a transversal.

Vocabulary
Parallel Postulate

Theorem 37 *If two parallel lines are cut by a transversal, each pair of alternate interior angles are congruent. (Short form: ∥ lines ⇒ alt. int. ∠s ≅)*

Given: Lines a and b are parallel.

Prove: ∠1 ≅ ∠2

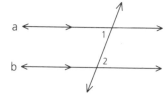

Notice the special tick marks (➤➤) used to designate parallel lines.

Proof: This theorem can be verified by indirect proof.

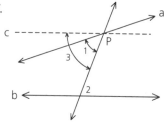

We are given a ∥ b. Assume that ∠1 is not congruent to ∠2. Then there must be another line, c, that intersects the transversal at P to form an angle, ∠3, that is congruent to ∠2. But in Section 5.2 we observed that congruent alternate interior angles lead to parallel lines. Thus, c ∥ b.

This means that line b is parallel to *two* lines in the plane at point P. This violates the Parallel Postulate. So we can conclude that our assumption is false. Therefore, ∠1 ≅ ∠2. You may be surprised to learn the following.

Theorem 38 *If two parallel lines are cut by a transversal, then any pair of the angles formed are either congruent or supplementary.*

The proof of this may be developed algebraically by letting x be the measure of any one of the angles. Follow the steps below. In each diagram, a ∥ b.

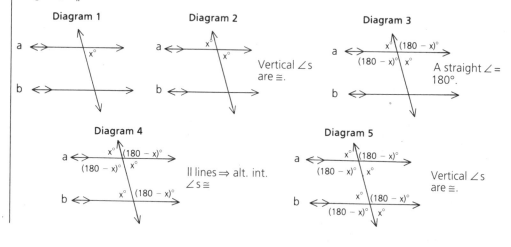

Section 5.3   Congruent Angles Associated with Parallel Lines   **225**

Lesson Notes

■ Now that we have the Parallel Postulate, we can prove that the sum of the measures of the angles of a triangle is 180. A few students may wish to pursue the topic of non-Euclidean geometry by researching János Bolyai, Nicolay Ivanovich Lobachevsky, Bernhard Riemann, and others who worked in this field.

■ List of sources for non-Euclidean geometries
Beck, A., M. Bleicher, and D. Crowe. *Excursions into Mathematics.* New York: Worth, 1969.
Bonola, R. *Non-Euclidean Geometry.* New York: Dover, 1955.
Coxeter, H.S.M., and S.L. Greitzer. *Geometry Revisited.* New York: Random House, 1967.
Eves, H. *A Survey of Geometry,* 2 vols. Boston: Allyn and Bacon, 1963–1965.
Greenberg, Marvin Jay. *Euclidean and Non-Euclidean Geometries.* San Francisco: Freeman, 1973.
Heath, T.L. *Euclid's Elements,* 3 vols. New York: Dover, 1956.
Kasner, E. and J. Newman. *Mathematics and the Imagination.* New York: Simon & Schuster, 1965.
Kline, M. *Mathematics in Western Culture.* New York: Oxford University Press, 1964.
Moise, E.E. *Elementary Geometry from an Advanced Standpoint.* Reading, Mass.: Addison-Wesley, 1963.
Sawyer, W.W. *Prelude to Mathematics.* Baltimore, Md.: Penguin Books, 1955.
Wolfe, H.E. *Introduction to Non-Euclidean Geometry.* New York: Holt, Rinehart and Winston, 1945.

■ Some teachers use PAI as a short form for Theorem 37.

Six Theorems About Parallel Lines

Diagram 5 on the preceding page is the basis for each of the following five theorems.

Theorem 39   *If two parallel lines are cut by a transversal, each pair of alternate exterior angles are congruent.*
(‖ *lines* ⇒ *alt. ext.* ∠s ≅)

Given: a ‖ b
Prove: ∠1 ≅ ∠8
Proof: See Diagram 5.

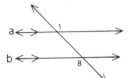

Theorem 40   *If two parallel lines are cut by a transversal, each pair of corresponding angles are congruent.*
(‖ *lines* ⇒ *corr.* ∠s ≅)

Given: a ‖ b
Prove: ∠1 ≅ ∠5
Proof: See Diagram 5.

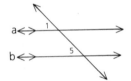

Lesson Notes, continued

■ Theorem 41 is used later in this chapter (p. 241) for consecutive angles of a parallelogram.

Theorem 41   *If two parallel lines are cut by a transversal, each pair of interior angles on the same side of the transversal are supplementary.*

Given: a ‖ b
Prove: ∠3 supp. ∠5
Proof: See Diagram 5.

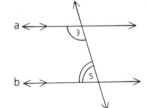

Theorem 42   *If two parallel lines are cut by a transversal, each pair of exterior angles on the same side of the transversal are supplementary.*

Given: a ‖ b
Prove: ∠1 supp. ∠7
Proof: See Diagram 5.

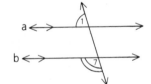

Theorem 43 *In a plane, if a line is perpendicular to one of two parallel lines, it is perpendicular to the other.*

Given: a ∥ b,
     c ⊥ a

Prove: c ⊥ b

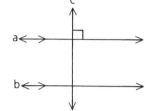

Proof: See Diagram 5 and let x = 90.

The following is another useful theorem about parallel lines.

Theorem 44 *If two lines are parallel to a third line, they are parallel to each other. (Transitive Property of Parallel Lines)*

Given: a ∥ b, b ∥ c

Prove: a ∥ c

    a ⟷

    c ⟷

    b ⟷

By using "∥ lines ⇒ alt. int. ∠s ≅" and "alt. int. ∠s ≅ ⇒ ∥ lines," you can prove that Theorem 44 is true when all three lines lie in the same plane. It also can be shown that the theorem holds for lines in three-dimensional space.

In summary, if two parallel lines are cut by a transversal, then
- Each pair of alternate interior angles are congruent
- Each pair of alternate exterior angles are congruent
- Each pair of corresponding angles are congruent
- Each pair of interior angles on the same side of the transversal are supplementary
- Each pair of exterior angles on the same side of the transversal are supplementary

Section 5.3   Congruent Angles Associated with Parallel Lines   **227**

Lesson Notes, continued

- Perpendicularity to a plane is explored in Chapter **6**.

Cooperative Learning

Divide students into six groups. Have each of the groups prove one of Theorems 39–44. Have one member of each group present the proof to the class.

Part Two: Sample Problems

Problem 1   If c ∥ d, find m∠1.

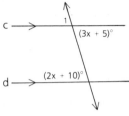

Solution   Since alt. int. ∠s are ≅,

$$3x + 5 = 2x + 10$$
$$x + 5 = 10$$
$$x = 5$$
$$3x + 5 = 20$$

Because vertical angles are ≅, m∠1 = 20.

Problem 2   Given: $\overline{FA} \parallel \overline{DE}$,
$\overline{FA} \cong \overline{DE}$,
$\overline{AB} \cong \overline{CD}$

Prove: ∠F ≅ ∠E

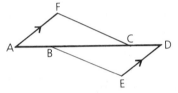

Proof

1 $\overline{FA} \parallel \overline{DE}$	1 Given
2 ∠A ≅ ∠D	2 ∥ lines ⇒ alt. int. ∠s ≅
3 $\overline{FA} \cong \overline{DE}$	3 Given
4 $\overline{AB} \cong \overline{CD}$	4 Given
5 $\overline{AC} \cong \overline{BD}$	5 Addition Property ($\overline{BC}$ to step 4)
6 △FAC ≅ △EDB	6 SAS (3, 2, 5)
7 ∠F ≅ ∠E	7 CPCTC

Problem 3   Given: g ∥ h

Prove: ∠1 supp. ∠2

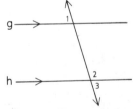

Proof

1 g ∥ h	1 Given
2 ∠2 supp. ∠3	2 If two angles form a straight angle, they are supplementary.
3 ∠1 ≅ ∠3	3 ∥ lines ⇒ alt. ext. ∠s ≅
4 ∠1 supp. ∠2	4 Substitution

Lesson Notes, continued

■ For step 5 of Sample Problem **2,** students may need to review the algebraic properties from Chapter **2.**

Problem 4   (*A crook problem*)
If a ∥ b, find m∠1.

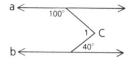

Solution   Using the Parallel Postulate, draw m parallel to a. By three theorems about ∥ lines, it can be proved that m∠1 = 120.

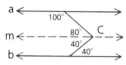

Problem 5   Given: Figure ABCD, with $\overline{AD} \parallel \overline{BC}$,
$\overline{AB} \cong \overline{DC}$, and $\overline{AB} \not\parallel \overline{DC}$
Prove: ∠B ≅ ∠C

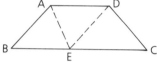

Note   Figure ABCD is called an isosceles trapezoid.

Proof

1 Figure ABCD, with $\overline{AD} \parallel \overline{BC}$	1 Given
2 $\overline{AB} \not\parallel \overline{DC}$	2 Given
3 Draw $\overline{DE} \parallel \overline{AB}$.	3 Parallel Postulate
4 Draw $\overline{AE}$.	4 Two points determine a line.
5 ∠DAE ≅ ∠BEA	5 ∥ lines ⇒ alt. int. ∠s ≅
6 ∠BAE ≅ ∠DEA	6 Same as 5
7 $\overline{AE} \cong \overline{AE}$	7 Reflexive Property
8 △AEB ≅ △EAD	8 ASA (5, 7, 6)
9 $\overline{AB} \cong \overline{DE}$	9 CPCTC
10 $\overline{AB} \cong \overline{DC}$	10 Given
11 $\overline{DE} \cong \overline{DC}$	11 Transitive Property
12 ∠DEC ≅ ∠C	12 If △, then △.
13 ∠B ≅ ∠DEC	13 ∥ lines ⇒ corr. ∠s ≅
14 ∠B ≅ ∠C	14 Transitive Property

Part Three: Problem Sets

Problem Set A

1   Given: $\overline{AB} \cong \overline{DC}$,
$\overline{AB} \parallel \overline{DC}$
Conclusion: $\overline{AD} \cong \overline{BC}$

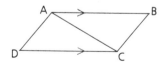

2   Given: $\overline{EF} \parallel \overline{GH}$,
$\overline{EF} \cong \overline{GH}$
Conclusion: $\overline{EJ} \cong \overline{JH}$

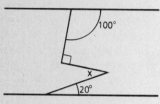

Lesson Notes, continued

■ "Crook" problems (the diagram resembles a shepherd's crook) can be solved in several ways; the method in Sample Problem **4** illustrates an immediate application of the Parallel Postulate. Some students may be challenged by extensions of crook problems.

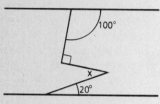

x = 30°

■ The result in Sample Problem **5** is used later in this chapter (p. 242) as a property of isosceles trapezoids.

Communicating Mathematics

Describe how mathematicians created non-Euclidean geometries based on assumptions about parallel lines.

Assignment Guide

Basic	
Day 1	1–6
Day 2	7, 9, 10, 13, 15, 21
Average	
Day 1	1–7
Day 2	9, 10, 15, 17–19
Advanced	
Day 1	3, 5, 6, 10, 12, 13
Day 2	15, 16, 19, 21, 23, 24

Problem-Set Notes and Additional Answers

■ See *Solution Manual* for answers to problems **1** and **2**.

Problem Set A, *continued*

3 Given: a ∥ b,
   30° angle as shown

Copy the diagram and fill in the measures of the seven remaining angles.

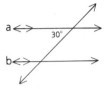

4 Given: ∠5 ≅ ∠6,
   $\overline{RS}$ ∥ $\overline{NP}$

Prove: △NPR is isosceles.

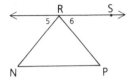

5 Given: a ∥ b
   Find m∠1. 41

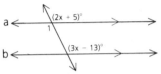

6 Given: $\overline{TE}$ ∥ $\overline{XW}$,
   $\overline{TE}$ ≅ $\overline{XW}$

Conclusion: $\overline{TX}$ ∥ $\overline{EW}$ (Hint: Draw an auxiliary segment and prove that some ▲ are ≅.)

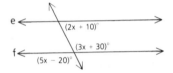

7 Are e and f parallel? No

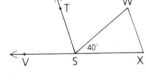

8 Given: $\overrightarrow{ST}$ ∥ $\overline{XW}$;
   $\overrightarrow{ST}$ bisects ∠VSW.

Find: m∠X and 70
   m∠W 70

What do you notice about △WSX? Isosceles

9 Given: $\overline{EA}$ ∥ $\overline{DB}$ and $\overline{EA}$ ≅ $\overline{DB}$;
   B is the midpt. of $\overline{AC}$.

Prove: $\overline{EB}$ ∥ $\overline{DC}$

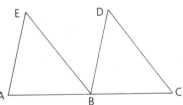

10 (*A crook problem*)
If f ∥ g, find m∠8. 150

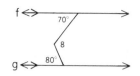
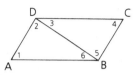

11 Given: $\overline{AD} \parallel \overline{BC}$
Name all pairs of angles that must be congruent. ∠2 ≅ ∠5

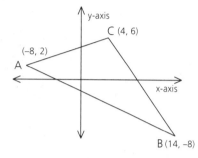

12 One of the sides of △ABC has a midpoint whose x-coordinate is negative. Find the coordinates of that midpoint. (−2, 4)

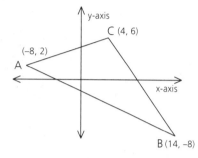

13 If a ∥ b, solve for x and find m∠1. x = 19.72
m∠1 = 42.2008

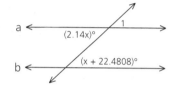

Problem Set B

14 Given: $\overline{FJ} \parallel\!\!\!/ \overline{KO}$,
$\overline{FH} \parallel \overline{MO}$,
$\overline{HK} \cong \overline{MJ}$
Prove: $\overline{FH} \neq \overline{MO}$

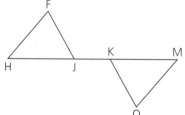

15 Given: $\overleftrightarrow{FH} \parallel \overleftrightarrow{JM}$,
∠1 ≅ ∠2,
$\overline{FH} \cong \overline{JM}$
Prove: $\overline{GJ} \cong \overline{HK}$

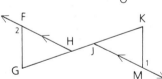

Problem-Set Notes and Additional Answers, continued

■ See *Solution Manual* for answers to problems **14** and **15**.

■ See *Solution Manual* for answers to problems **16** and **19–
21.**

17 2x + y = 180
x = y
∴ x = 60

Problem Set B, *continued*

16 Given: ⊙O,
$\overline{NR} \parallel \overline{PS}$

Prove: △OSP is isosceles.

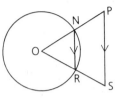

17 If $\overleftrightarrow{DA} \parallel \overleftrightarrow{BC}$, is △ABC equilateral? Yes
Find m∠DAB. 60

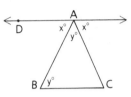

18 Explain why lines m and n are parallel.
Same slope

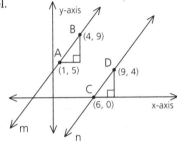

19 Given: ∠C supp. ∠D
Prove: ∠A supp. ∠B

20 Given: $\overline{CY} \cong \overline{AY}$,
$\overline{YZ} \parallel \overline{CA}$
Prove: $\overrightarrow{YZ}$ bisects ∠AYB.

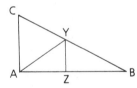

21 Prove that bisectors of a pair of alternate exterior angles formed
by a transversal cutting parallel lines are parallel.

22 Given: a ∥ b,
∠1 = (x + 3y)°,
∠2 = (2x + 30)°,
∠3 = (5y + 20)°

Find: m∠1 70

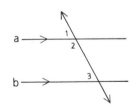

23 A line is described by the equation $y = 5x - 7$. Points (x, y) that solve the equation must lie on the line. Which of the points $(3, 8)$, $(-2, -17)$, $(0, -7)$, $(4, 11)$ and $(1, -2)$ are on the line? All but $(4, 11)$

24 Prove that the opposite sides of a parallelogram are congruent. (Recall that a parallelogram is a four-sided figure in which both pairs of opposite sides are parallel.)

25 Prove that the opposite angles of a parallelogram are congruent.

Problem Set C

26 If two parallel lines are cut by a transversal, eight angles are formed (not counting the straight angles).

  a How many pairs of angles are formed? 28
  b If one of these pairs is chosen at random, what is the probability that the angles will be alternate interior angles *or* alternate exterior angles *or* corresponding angles? $\frac{2}{7}$
  c If one of the pairs is chosen at random, what is the probability that the angles are supplementary? $\frac{4}{7}$ (unless the transv. intersects the lines at rt. ∠s)

27 Given: ⊙O,
    $\overline{DC} \parallel \overline{AB}$
  Prove: $\overline{AD} \cong \overline{BC}$

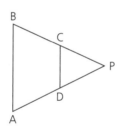

28 Given: $\overline{BC} \cong \overline{AD}$
  Prove: $\overline{AB} \ne \overline{CD}$ (Hint: Draw $\overline{AC}$.)

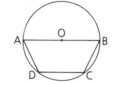

Problem Set D

29 Given: $\overleftrightarrow{BE} \parallel \overleftrightarrow{CD}$ and $\overline{BE} \cong \overline{EF}$;
    B is the midpt. of $\overline{AC}$.
    E is the midpt. of $\overline{AD}$.
  Prove: $BE = \frac{1}{2}(CD)$

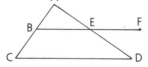

30 Write a paragraph proof that shows that the sum of the three angles of a triangle is 180°. (Hint: Draw a triangle and use the Parallel Postulate.)

30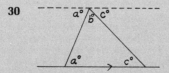

(See pp. 295–296.)
$a° + b° + c° = 180°$

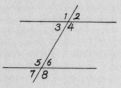
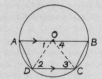

Class Planning

Time Schedule
All levels: 1 day

Resource References
Teacher's Resource Book
  Class Opener 5.4A, 5.4B
  Using Manipulatives 8

Class Opener

Given: $\overline{SX} \parallel \overline{NO}$,
  $\angle NSX \cong \angle NOX$
Prove: $\overline{SN} \parallel \overline{XO}$

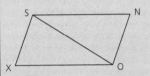

Statements	Reasons
1 $\overline{SX} \parallel \overline{NO}$	1 Given
2 $\angle XSO \cong \angle NOS$	2 $\parallel$ lines $\Rightarrow$ alt. int. $\angle$s $\cong$
3 $\angle NSX \cong \angle NOX$	3 Given
4 $\angle NSO \cong \angle XOS$	4 Subtraction Prop.
5 $\overline{SN} \parallel \overline{XO}$	5 Alt. int. $\angle$s $\cong \Rightarrow$ $\parallel$ lines

This problem requires students to distinguish between a theorem and its converse and to focus on the correct angles.

Lesson Notes

- This section serves as an introduction and a preview to the material in Chapter 7.

FOUR-SIDED POLYGONS

Objectives

After studying this section, you will be able to
- Recognize polygons
- Understand how polygons are named
- Recognize convex polygons
- Recognize diagonals of polygons
- Identify special types of quadrilaterals

Part One: Introduction

Polygons

Polygons are plane figures. The following are examples of polygons.

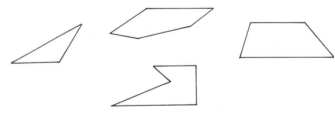

The following are examples of figures that are not polygons.

EFGH is not a polygon, because a polygon consists entirely of segments.

ABCDE is not a polygon. In a polygon, consecutive sides intersect only at endpoints. Nonconsecutive sides do not intersect.

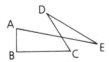

PKMO, PKMOR, and POR are polygons, but PKMOPRO is not, because each vertex must belong to exactly two sides. (Vertex P belongs to three sides in PKMOPRO.)

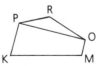

Vocabulary

convex polygon	parallelogram	rhombus
diagonal	polygon	square
isosceles trapezoid	quadrilateral	trapezoid
kite	rectangle	upper base angles
lower base angles		

SVTY is a polygon, but SVTXY is not, because consecutive sides must be noncollinear.

Why is PLAN not a polygon?

Naming Polygons

We name a polygon by starting at any vertex and then proceeding either clockwise or couterclockwise. If we start at A, we can call this polygon ABCDEF or AFEDCB. Can you start at B and name the polygon in two different ways?

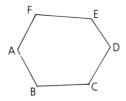

Convex Polygons

Many of the polygons you encounter in your geometry studies will be *convex.*

Definition   A *convex polygon* is a polygon in which each interior angle has a measure less than 180.

Polygon ABCDE is not convex because the angle that lies in the interior of the polygon at E has a measure greater than 180.
    In the rest of this book, unless it is expressly stated otherwise, assume that all polygons are convex.

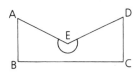

Diagonals of Polygons

In the two following figures, the dashed segments are *diagonals* of the polygons.

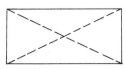
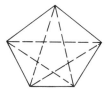

Definition   A *diagonal* of a polygon is any segment that connects two nonconsecutive (nonadjacent) vertices of the polygon.

Communicating Mathematics

Write a definition for the term *nonconvex* (or *concave*) *polygon.*

Section 5.4   Four-Sided Polygons   **235**

■ These definitions allow a rhombus to be thought of as a special case of a kite. They do not allow a parallelogram to be thought of as a special trapezoid.

Quadrilaterals

A **quadrilateral** is a four-sided polygon.

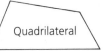
Quadrilateral

The following are special quadrilaterals.

A **parallelogram** is a quadrilateral in which both pairs of opposite sides are parallel.

Parallelogram

A **rectangle** is a parallelogram in which at least one angle is a right angle.

Rectangle

A **rhombus** is a parallelogram in which at least two consecutive sides are congruent.

Rhombus

A **kite** is a quadrilateral in which two *disjoint* pairs of consecutive sides are congruent.

Kite

A **square** is a parallelogram that is both a rectangle and a rhombus.

Square

A **trapezoid** is a quadrilateral with exactly one pair of parallel sides. The parallel sides are called *bases* of the trapezoid.

Trapezoid

An **isosceles trapezoid** is a trapezoid in which the nonparallel sides (legs) are congruent. In the figure, $\angle A$ and $\angle B$ are called the **lower base angles,** and $\angle C$ and $\angle D$ are called the **upper base angles**.

Isosceles Trapezoid

236 Chapter 5   Parallel Lines and Related Figures

We have given the meaning (definition) of each of the previous figures in as simple a manner as possible. Each special quadrilateral will have further properties associated with it. Those properties are discussed in the next section.

Part Two: Sample Problem

Solve the Quadrilateral Mystery!

No solution is provided for the following problem. It is intended to help you understand how mathematicians go about testing ideas that they *think* are true but which they have not yet *proved*. As you work through the problem, think carefully about the ideas you formulate and the ways you test them. (A computer with exploratory geometry software—such as *The Geometric Supposer*, by Sunburst—is an excellent tool for testing ideas. If you do not have access to such resources, try making careful drawings and using a ruler and a protractor to test your ideas.)

Problem   *What truths can you discover about a parallelogram and a rectangle?*

    a *Draw a parallelogram ABCD.*
      i *What true statements do you think you might be able to make about the parallelogram? Test your ideas and discuss your results in class.*
      ii *Draw diagonals $\overline{AC}$ and $\overline{BD}$. What true statements can be made about the diagonals? Again, test your ideas and discuss your results in class.*
    b *Draw a rectangle PQRS.*
      i *What true statements can be made about the rectangle? Test your ideas and discuss your results in class.*
      ii *Draw diagonals $\overline{PR}$ and $\overline{QS}$. What true statements can be made about the diagonals? Again, test your ideas and discuss your results in class.*

Part Three: Problem Sets

Problem Set A

A computer and exploratory geometry software may be used for problems 1–5.

  1 Examine the rhombus. Which of the following statements appear to be true? **a–d**

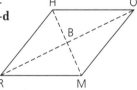

    a All four sides are congruent.
    b The diagonals are perpendicular.
    c The diagonals bisect the angles.
    d The diagonals bisect each other.
    e The diagonals are congruent.

Problem Set A, *continued*

2 Examine the isosceles trapezoid. Which of the following statements appear to be true? **e**

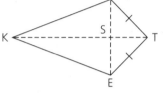

  a The opposite sides are congruent.
  b Opposite sides are parallel.
  c The diagonals bisect the angles.
  d The diagonals bisect each other.
  e The diagonals are congruent.

3 Examine the kite. Which of the following statements appear to be true? **f**

  a The opposite sides are congruent.
  b Opposite sides are parallel.
  c The diagonals bisect the angles.
  d The diagonals bisect each other.
  e The diagonals are congruent.
  f The diagonals are perpendicular.

**Problem-Set Notes
and Additional Answers**

■ See *Solution Manual* for answers to problems **4–6.**

4 List all the properties that a nonisosceles trapezoid appears to have.

5 List all the properties that a square appears to have.

6 a Draw an equilateral quadrilateral that is not equiangular.
  b Draw an equiangular quadrilateral that is not equilateral.

7 In the isosceles trapezoid shown, $\overline{ST} \parallel \overline{RV}$.

Name: **a** The bases $\overline{ST}$ and $\overline{RV}$
       b The diagonals $\overline{SV}$ and $\overline{RT}$
       c The legs $\overline{RS}$ and $\overline{VT}$
       d The lower base angles $\angle SRV$ and $\angle TVR$
       e The upper base angles $\angle RST$ and $\angle VTS$
       f All pairs of congruent alternate interior angles $\angle STR \cong \angle VRT$, $\angle TSV \cong \angle RVS$

8 Examine each statement below. If the statement is always true, write A; if sometimes true, write S; if never true, write N.

  a A square is a rhombus. A
  b A rhombus is a square. S
  c A kite is a parallelogram. S
  d A rectangle is a polygon. A
  e A polygon has the same number of vertices as sides. A
  f A parallelogram has three diagonals. N
  g A trapezoid has three bases. N

9 Why is a circle not a polygon? A polygon consists entirely of segments.

10 Using the diagram, explain how the formula for the area of a parallelogram can be the same as that for the area of a rectangle.

11 If the sum of the measures of the angles of a triangle is 180, what is the sum of the measures of the angles in
   a A quadrilateral? 360
   b A pentagon (five-sided polygon)? 540

12 Find the area of a square whose perimeter is 65 feet. ≈264 sq ft

Problem Set B

13 Prove that in a parallelogram each pair of consecutive angles are supplementary.

14 Prove that in a parallelogram each pair of opposite sides are congruent.

15 Prove that the diagonals of a rectangle are congruent.

16 Given: ABCD is a kite.
      $AB = x + 3,$
      $BC = x + 4,$
      $CD = 2x - 1,$
      $AD = 3x - y$

   a Solve for x and y. x = 5; y = 7
   b What is the perimeter of the kite? 34
   c Is it possible for $\overline{AC}$ to be 19 units long? Why or why not?
      No, since AC must be less than AB + BC.

17 a PQRS is a kite and also a rectangle. What else do we know about PQRS? It is a square.
   b Draw a quadrilateral that is not convex and still satisfies the definition of a kite.

18 What is the area of a triangle whose vertices are (−4, −3), (8, 7), and (8, −3)? 60 sq units

19 The trapezoidal region is actually the union of two triangles and a rectangle. Find the area of the trapezoid. 112 sq units

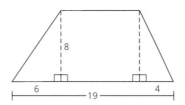

Problem-Set Notes and Additional Answers, continued

- Some students may find a paper model of the diagram helpful in answering problem **10**.

10 The area of a parallelogram can be rearranged to form a rectangle.

13 In a parallelogram, each pair of consecutive ∠s are also int. ∠s on the same side of a transv.

14

Draw $\overline{AC}$ and prove △ACD ≅ △CAB by ASA.

15

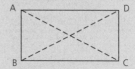

Draw both diagonals and prove △ABC ≅ △DCB by SAS.

16a $\begin{cases} x + 3 = 3x - y \\ x + 4 = 2x - 1 \end{cases}$

17b

21 From each of the n vertices you can draw $n - 3$ diagonals. Thus, there are $n(n - 3)$ diagonals. Since this method counts each diagonal twice (once from each endpoint), you must divide by 2.

$\therefore d = \frac{n(n - 3)}{2}$

22 Number the figures consecutively from 1 through 7. Possibilities include:

1,2
1,3 2,3
1,4 2,4 3,4
1,5 2,5 3,5 4,5
1,6 2,6 3,6 4,6 5,6
1,7 2,7 3,7 4,7 5,7 6,7

$\therefore$ Prob. $= \frac{10}{21}$

Problem Set B, *continued*

20 How many rectangles are shown in the figure at the right, in which all of the angles are right angles? 10

Problem Set C

21 a How many diagonals does a triangle have? 0
   b How many diagonals does a quadrilateral have? 2
   c How many diagonals does a five-sided polygon have? 5
   d How many diagonals does a six-sided polygon have? 9
   e How many diagonals meet at one vertex of a polygon with n sides? $n - 3$
   f How many vertices does an n-sided polygon have? n
   g How many diagonals does an n-sided polygon have? $\frac{n(n - 3)}{2}$

22 Refer to the seven special quadrilaterals on page 236. What is the probability that if two are picked at random, each will have a pair of congruent opposite sides? $\frac{10}{21}$

HISTORICAL SNAPSHOT

A NEW KIND OF PROOF
The computer and the four-color conjecture

How many colors does it take to color any map so that no two adjacent regions will be the same color? (Regions that touch only at a single point are not considered to be adjacent.) In 1852 it was suggested that four colors are enough for any possible map. Although no one ever succeeded in constructing a map that needed more than four colors, for over 100 years no one was able to furnish a satisfactory proof that such a map could not exist.

Then, in 1976, it was announced that a group of mathematicians led by Kenneth Appel and Wolfgang Haken at the University of Illinois had proved the four-color conjecture. Having determined that all possible maps could be represented by a set of 1936 particular configurations of regions, they programmed a computer to test each of these cases for four-colorability.

The computer found no instance in which more than four colors were required.

Traditionally, however, a proof has been considered a way of presenting mathematical reasoning that can be understood and verified by other people. The four-color proof is so complex that it would take lifetimes to verify it by hand. It is one of the first examples of a proof that can be produced and checked only using a computer.

5.5 PROPERTIES OF QUADRILATERALS

Objectives

After studying this section, you will be able to
- Identify some properties of parallelograms
- Identify some properties of rectangles
- Identify some properties of kites
- Identify some properties of rhombuses
- Identify some properties of squares
- Identify some properties of isosceles trapezoids

Part One: Introduction

Properties of Parallelograms

In this section, we will list some of the properties of special quadrilaterals, beginning with parallelograms. (You should be able to prove many of these properties.) Read the properties carefully and learn them. They will be used often in the sections to follow.

Learning so many properties may seem overwhelming at first, but most are concepts that you already know or that you discovered in Section 5.4. With some effort you will soon learn them all.

In a parallelogram,

1 The opposite sides are parallel by definition ($\overline{PL} \parallel \overline{AR}$, $\overline{AP} \parallel \overline{RL}$)
2 The opposite sides are congruent ($\overline{PL} \cong \overline{AR}$, $\overline{AP} \cong \overline{RL}$)
3 The opposite angles are congruent ($\angle PAR \cong \angle PLR$, $\angle ARL \cong \angle APL$)
4 The diagonals bisect each other ($\overline{AL}$ bis. $\overline{PR}$, $\overline{PR}$ bis. $\overline{AL}$)
5 Any pair of consecutive angles are supplementary ($\angle PAR$ supp. $\angle ARL$, etc.)

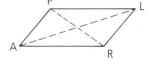

Properties of Rectangles

In a rectangle,

1 All the properties of a parallelogram apply by definition
2 All angles are right angles ($\angle REC$ is a right angle, etc.)
3 The diagonals are congruent ($\overline{ET} \cong \overline{RC}$)

Class Planning

Time Schedule
Basic: 3 days
Average: 3 days
Advanced: 2 days

Resource References
Teacher's Resource Book
  Class Opener 5.5A, 5.5B
  Using Manipulatives 9
  Additional Practice
  Worksheet 10
  Supposer Worksheet 6
Evaluation
  Tests and Quizzes
    Quiz 3, Series 1
    Quizzes 2–3, Series 2
    Quizzes 1–2, Series 3

Class Opener

Elena used a rectangle, a square, a kite, a rhombus, and an isosceles trapezoid as part of a computer game she was creating. The player selects two of these shapes at random. If each of the selected shapes has at least one pair of opposite sides parallel, the player can use these shapes as keys to a higher level of the game. What is the probability of selecting a pair of keys?

Represent each shape with a letter.

A: rectangle; B: square; C: kite; D: rhombus; E: isosceles trapezoid. A, B, D, and E are keys.

List all possible twosomes, and ring those that are successes.

AE

The probability of getting a pair of keys is $\frac{9}{10}$.

Properties of Kites

In a kite,

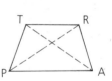

1 Two disjoint pairs of consecutive sides are congruent by definition ($\overline{IT} \cong \overline{ET}$, $\overline{IK} \cong \overline{EK}$)
2 The diagonals are perpendicular ($\overline{TK} \perp \overline{IE}$)
3 One diagonal is the perpendicular bisector of the other ($\overline{TK} \perp$ bis. $\overline{IE}$)
4 One of the diagonals bisects a pair of opposite angles ($\overrightarrow{TK}$ bis. $\angle ITE$, $\overrightarrow{TK}$ bis. $\angle IKE$)
5 One pair of opposite angles are congruent ($\angle TIK \cong \angle TEK$)

Properties 3–5 are sometimes called the *half properties* of kites.

Properties of Rhombuses

In a rhombus,

1 All the properties of a parallelogram apply by definition
2 All the properties of a kite apply (In fact, the half properties become full properties)
3 All sides are congruent—that is, a rhombus is equilateral ($\overline{RH} \cong \overline{HO} \cong \overline{OM} \cong \overline{MR}$)
4 The diagonals bisect the angles ($\overrightarrow{RO}$ bis. $\angle MRH$, $\overrightarrow{RO}$ bis. $\angle MOH$, etc.)
5 The diagonals are perpendicular bisectors of each other ($\overline{RO} \perp$ bis. $\overline{MH}$, $\overline{MH} \perp$ bis. $\overline{RO}$)
6 The diagonals divide the rhombus into four congruent right triangles

Properties of Squares

In a square,

1 All the properties of a rectangle apply by definition
2 All the properties of a rhombus apply by definition
3 The diagonals form four isosceles right triangles (45°-45°-90° triangles)

Properties of Isosceles Trapezoids

In an isosceles trapezoid,

1 The legs are congruent by definition ($\overline{TP} \cong \overline{RA}$)
2 The bases are parallel (by definition of *trapezoid*) ($\overline{TR} \parallel \overline{PA}$)
3 The lower base angles are congruent ($\angle RAP \cong \angle TPA$)
4 The upper base angles are congruent ($\angle PTR \cong \angle ART$)
5 The diagonals are congruent ($\overline{PR} \cong \overline{AT}$)
6 Any lower base angle is supplementary to any upper base angle ($\angle PAR$ supp. $\angle PTR$, etc.)

In the problems that follow, you will be asked to prove some of these properties. You may use any prior property to help in the proof of a later property. For example, if you are asked to prove property 5 of parallelograms, you may use properties 1–4 to help you in the proof.

Part Two: Sample Problems

Problem 1   Given: ABCD is a ▱ (parallelogram).
∠GHA ≅ ∠FEC,
$\overline{HB} \cong \overline{DE}$

Conclusion: $\overline{GH} \cong \overline{EF}$

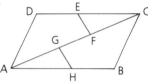

Proof

1 ABCD is a ▱.	1 Given
2 $\overline{DC} \parallel \overline{AB}$	2 Opposite sides of a ▱ are ∥.
3 ∠ECF ≅ ∠HAG	3 ∥ lines ⇒ alt. int. ∠s ≅
4 $\overline{AB} \cong \overline{DC}$	4 Opposite sides of a ▱ are ≅.
5 $\overline{HB} \cong \overline{DE}$	5 Given
6 $\overline{HA} \cong \overline{EC}$	6 Subtraction Property
7 ∠GHA ≅ ∠FEC	7 Given
8 △GAH ≅ △FCE	8 ASA (3, 6, 7)
9 $\overline{GH} \cong \overline{EF}$	9 CPCTC

Problem 2   Given: VRZA is a ▱.
AV = 2x − 4,
VR = 3y + 5,
RZ = $\frac{1}{2}$x + 8
ZA = y + 12

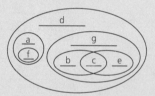

Find: The perimeter of VRZA

Solution   The opposite sides of a ▱ are congruent, so we can write two equations.

$2x - 4 = \frac{1}{2}x + 8$ $3y + 5 = y + 12$
$1\frac{1}{2}x - 4 = 8$ $2y + 5 = 12$
$1\frac{1}{2}x = 12$ $2y = 7$
$x = 8$ $y = 3\frac{1}{2}$

AV = 12 and RZ = 12   VR = $15\frac{1}{2}$ and ZA = $15\frac{1}{2}$

Adding the measures of the four sides, we find that the perimeter is 55.

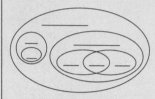

Lesson Notes, continued

Each list of properties in Section **5.5** is ordered; students should realize that they can use parallelogram property 2 to prove property 4 but that violation of the order may involve circular reasoning.

Assignment Guide

Basic

Day 1	1, 2, 4–6, 12, 24
Day 2	3, 7, 8, 14, 23
Day 3	10, 11, 15, 17, 25, 26

Average

Day 1	1, 2, 4–6, 8, 12, 24
Day 2	3, 7, 10, 11, 14, 15
Day 3	17–19, 21, 23, 25, 26

Advanced

| Day 1 | 5–9, 11–15, 23 |
| Day 2 | 17, 19, 21, 22, 25, 26, 28, 30 |

Problem-Set Notes and Additional Answers

See *Solution Manual* for answers to problems **1** and **3**.

Problem 3   Prove property 4 of parallelograms.

Given: ▱ ABCD

Prove: $\overline{AC}$ and $\overline{BD}$ bisect each other.

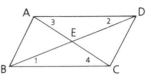

Proof

1 ▱ ABCD	1 Given
2 $\overline{AD} \parallel \overline{BC}$	2 Opposite sides of a ▱ are $\parallel$.
3 ∠1 ≅ ∠2	3 $\parallel$ lines ⇒ alt. int. ∠s ≅
4 ∠3 ≅ ∠4	4 $\parallel$ lines ⇒ alt. int. ∠s ≅
5 $\overline{AD} \cong \overline{BC}$	5 Opposite sides of a ▱ are ≅.
6 △BEC ≅ △DEA	6 ASA (3, 5, 4)
7 $\overline{BE} \cong \overline{DE}$	7 CPCTC
8 $\overline{AE} \cong \overline{EC}$	8 CPCTC
9 $\overline{AC}$ and $\overline{BD}$ bisect each other.	9 If two segments divide each other into ≅ segments, they bisect each other.

Part Three: Problem Sets

Problem Set A

1 Given: ▱ ABCD (ABCD is a ▱.)

Conclusion: △ABC ≅ △CDA

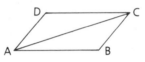

2 Given: ▱ EFHJ,

∠1 ≅ ∠2

Conclusion: $\overline{KH} \cong \overline{EG}$

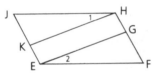

Supply each missing reason.

1 ▱ EFHJ	1 Given
2 ∠J ≅ ∠F	2 Opp. ∠s of a ▱ are ≅.
3 $\overline{JH} \cong \overline{EF}$	3 Opp. sides of a ▱ are ≅.
4 ∠1 ≅ ∠2	4 Given
5 △KJH ≅ △GFE	5 ASA (2, 3, 4)
6 $\overline{KH} \cong \overline{EG}$	6 CPCTC

3 Given: Rectangle MPRS,

$\overline{MO} \cong \overline{PO}$

Prove: △ROS is isosceles.

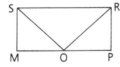

4 Given: ▱ABCD,
$\overline{AE} \cong \overline{CF}$

Conclusion: $\overline{DE} \cong \overline{BF}$

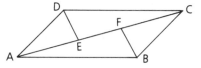

Problem-Set Notes and
Additional Answers, continued

■ See *Solution Manual* for answers to problems **4** and **8–10**.

5 Given: ▱ WSTV,
WS = x + 5,
WV = x + 9,
VT = 2x + 1

Find the perimeter of WSTV. 44

6 Given: ▱ ABCD,
∠A = (x)°,
∠D = (3x − 4)°

Find: m∠D and m∠C 134; 46

7 Given: EFGH is an isosceles trapezoid,
with legs $\overline{HE}$ and $\overline{GF}$.

EJ = x + 5,
JG = 2x − 1,
HF = 13

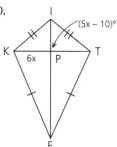

Find: EJ, JG, and HJ 8; 5; 5

8 Prove property 3 of parallelograms.

9 Prove property 4 of rhombuses.

10 Prove property 5 of isosceles trapezoids.

11 Given: m∠IPT = 5x − 10,
KP = 6x

Find: KT 240

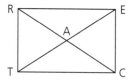

12 Given: RECT is a rectangle.
RA = 43x,
AC = 214x − 742

Find: The length of $\overline{ET}$ to the nearest
tenth ≈373.2

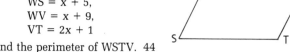

Problem Set A, *continued*

13 Which of the dotted lines represent an *axis of symmetry* of the figure? (One side of a figure is a reflection of the other side over an axis of symmetry.)

a Rectangle   ℓ   m

b Kite   n   p

c Rhombus   q and r

14 Given: ∠AFB ≅ ∠DCE,
△AFB ≇ △DCE
Prove: ACDF is not a parallelogram.

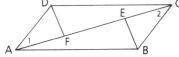

Problem Set B

15 Given: ABCD is a ▱.
$\overline{AF} \cong \overline{CE}$
Prove: $\overline{DF} \parallel \overline{EB}$

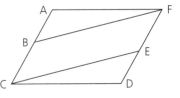

16 Given: PHJM is a rectangle.
$\overline{PG} \cong \overline{MK}$
Prove: △OGK is isosceles.

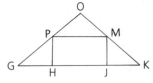

Problem-Set Notes and Additional Answers, continued

■ See *Solution Manual* for answers to problems **14–16.**

17 Given: VRST is an isosceles trapezoid, with legs $\overline{VR}$ and $\overline{TS}$.
Prove: $\triangle ARS$ is isosceles.

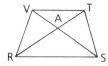

18 Prove that the diagonals of a rhombus divide the rhombus into four $\cong$ $\triangle$.

19 Given: $\square$ KMOP,
$\angle M = (x + 3y)°$,
$\angle O = (x - 4)°$,
$\angle P = (4y - 8)°$
Find: $m\angle K$ 28

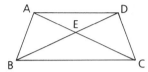

20 ABCD is an isosceles trapezoid with upper base $\overline{AD}$.
$BE = x + 7$, $CE = y - 3$,
$AE = x + 5$, $BD = y + 4$
Find AC. 16

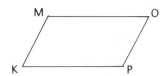

21 An author wrote a problem involving kite KITE but forgot to say which pairs of sides were congruent. Work the problem twice to see which pairs of sides are congruent.
$\overline{KI} \cong \overline{KE}$,
$\overline{IT} \cong \overline{ET}$

22 Prove, in paragraph form, that one diagonal of a kite divides it into two congruent triangles, while the other diagonal divides it into two isosceles triangles.

23 RHOM is a rhombus.
 a Find the coordinates of point O.
 b Find the slopes of $\overleftrightarrow{HM}$ and $\overleftrightarrow{RO}$.
 c What does the result in part **b** verify?
  a (19, 15); **b** Slope $\overleftrightarrow{HM} = -\frac{3}{2}$; slope $\overleftrightarrow{RO} = \frac{2}{3}$; **c** The diagonals of a rhombus are $\perp$.

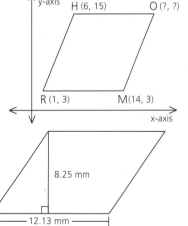

24 The area of a parallelogram is equal to the product of its base and its height. Find the area of the parallelogram.
≈ 100.07 mm^2

Section 5.5   Properties of Quadrilaterals   **247**

Problem-Set Notes and
Additional Answers, *continued*

- See *Solution Manual* for answers to problems **17, 18, 21,** and **22.**

19 Solve any two of the three equations.
$x + 3y = 4y - 8$
$x + 3y + x - 4 = 180$
$x - 4 + 4y - 8 = 180$

20 $AE + CE = AC = BD$ and $BE = CE$, so $(x + 5) + (y - 3) = (y + 4)$ and $(x + 7) = (y - 3)$.

Communicating Mathematics

Have students write a paragraph explaining why all the properties of a kite apply to a rhombus but not every property of a rhombus applies to a kite.

24 Many students will answer 100.0725. Point out that if the measurements are in error by 0.01 mm (e.g., 12.14 mm instead of 12.13 mm), the computed area changes by at least 0.08 sq mm. A general rule: use the same number of decimal places in the answer as was used in the problem.

Problem Set B, *continued*

25 SQUA is a square. If one of the triangles shown in the figure is chosen at random, what is the probability that it is isosceles? 1

26 $\overline{CM}$ is a median.

  a Find the coordinates of M. $(-2, 0)$
  b Is $\overline{CM}$ an altitude? Yes
  c What type of triangle is △ABC?
    Isosceles

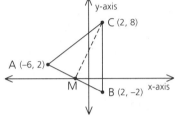

27 RHOM is a rhombus.
  m∠MBR = 21x + 13,
  MR = 6.2x

  Find the length of $\overline{RH}$ to the nearest
  tenth. ≈22.7

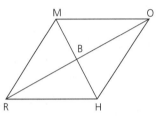

Problem Set C

28 TRAP is an isosceles trapezoid. The measure of one of its angles is 2.43 greater than 5.12 times the measure of another. If m∠T is less than m∠R, find ∠A to the nearest second. ≈150°59′7″

29 Given: m ∥ n

  a Solve for a in terms of x and y.
  b If $a > 90$, what must be true of $y - x$?
    a $a = 180 - y + x$; **b** $y - x < 90$

30 In the solid box,
  ▱ ABCD ≅ ▱ EFGH.
  Prove: $\overline{HF} \cong \overline{DB}$

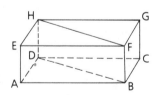

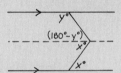
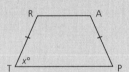

5.6 PROVING THAT A QUADRILATERAL IS A PARALLELOGRAM

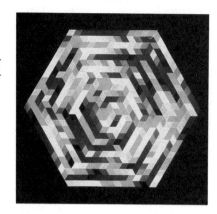

Objective

After studying this section, you will be able to
- Prove that a quadrilateral is a parallelogram

Part One: Introduction

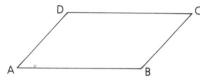

Any one of the following methods might be used to prove that quadrilateral ABCD is a parallelogram.

1 If both pairs of opposite sides of a quadrilateral are parallel, then the quadrilateral is a parallelogram (reverse of the definition).

2 If both pairs of opposite sides of a quadrilateral are congruent, then the quadrilateral is a parallelogram (converse of a property).

3 If one pair of opposite sides of a quadrilateral are both parallel and congruent, then the quadrilateral is a parallelogram.

4 If the diagonals of a quadrilateral bisect each other, then the quadrilateral is a parallelogram (converse of a property).

5 If both pairs of opposite angles of a quadrilateral are congruent, then the quadrilateral is a parallelogram (converse of a property).

- The proof of method 5 is postponed until it is established that the sum of the angles of a quadrilateral is 360° (Theorem 55, p. 308). Method 5 follows readily from that result.

Class Planning

Time Schedule
Basic: 2 days
Average: 2 days
Advanced: $1\frac{1}{2}$ days

Resource References
Teacher's Resource Book
  Class Opener 5.6A, 5.6B

Class Opener

Given: BCDF is a kite.
  BC = 3x + 4y,
  CD = 20,
  BF = 12,
  FD = x + 2y
Find: The perimeter of BCDF
BCDF is a kite, so it must have two sides with length 12 and two sides with length 20.
∴ $P = 2(12) + 2(20) = 64$
This problem should remind students to think before rushing into a solution. There is a longer algebraic solution.

Lesson Notes

- To prove methods 2–4, show that two triangles are congruent by

T 249

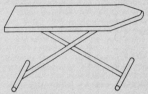
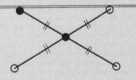
Part Two: Sample Problems

Problem 1

Given: ACDF is a ▱.
∠AFB ≅ ∠ECD

Prove: FBCE is a ▱.

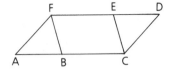

Proof

1 ACDF is a ▱.	1 Given
2 ∠A ≅ ∠D	2 Opposite ∠s of a ▱ are ≅.
3 $\overline{AF} \cong \overline{DC}$	3 Opposite sides of a ▱ are ≅.
4 ∠AFB ≅ ∠ECD	4 Given
5 △AFB ≅ △DCE	5 ASA (2, 3, 4)
6 $\overline{FB} \cong \overline{EC}$	6 CPCTC
7 $\overline{AB} \cong \overline{ED}$	7 CPCTC
8 $\overline{AC} \cong \overline{FD}$	8 Same as 3
9 $\overline{BC} \cong \overline{FE}$	9 Subtraction Property
10 FBCE is a ▱.	10 If both pairs of opposite sides of a quadrilateral are ≅, it is a ▱.

Problem 2

Given: △CAR is isosceles, with base $\overline{CR}$.
$\overline{AC} \cong \overline{BK}$,
∠C ≅ ∠K

Prove: BARK is a ▱.

Proof

1 △CAR is isos., with base $\overline{CR}$.	1 Given
2 $\overline{AC} \cong \overline{AR}$	2 Legs of an isos. △ are ≅.
3 $\overline{AC} \cong \overline{BK}$	3 Given
4 $\overline{AR} \cong \overline{BK}$	4 Transitive Property
5 ∠C ≅ ∠ARC	5 Base ∠s of an isos. △ are ≅.
6 ∠C ≅ ∠K	6 Given
7 ∠ARC ≅ ∠K	7 Transitive Property
8 $\overleftrightarrow{AR} \parallel \overleftrightarrow{BK}$	8 Corr. ∠s ≅ ⇒ ∥ lines
9 BARK is a ▱.	9 If one pair of opposite sides of a quadrilateral are both ∥ and ≅, it is a ▱.

Problem 3

Given: Quadrilateral QUAD, with angles as shown

Show that QUAD is a ▱.

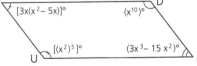

Solution

By the Distributive Property of Multiplication over Subtraction, $3x(x^2 - 5x) = 3x^3 - 15x^2$; and $(x^2)^5 = x^{10}$ by the rules of exponents. This means that ∠Q ≅ ∠A and ∠U ≅ ∠D. Thus, QUAD is a parallelogram, since both pairs of opposite angles are congruent.

Problem 4 Given: NRTW is a ▱.
$\overline{NX} \cong \overline{TS}$,
$\overline{WV} \cong \overline{PR}$

Prove: XPSV is a ▱.

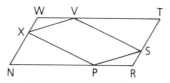

Proof

1 NRTW is a ▱.	1 Given
2 ∠N ≅ ∠T	2 Opposite ∠s of a ▱ are ≅.
3 $\overline{NX} \cong \overline{TS}$	3 Given
4 $\overline{NR} \cong \overline{WT}$	4 Opposite sides of a ▱ are ≅.
5 $\overline{WV} \cong \overline{PR}$	5 Given
6 $\overline{NP} \cong \overline{VT}$	6 Subtraction Property
7 △NXP ≅ △TSV	7 SAS (3, 2, 6)
8 $\overline{XP} \cong \overline{VS}$	8 CPCTC
9 In a similar manner, △WXV ≅ △RSP and $\overline{XV} \cong \overline{PS}$.	9 Steps 1–8
10 XPSV is a ▱.	10 If both pairs of opposite sides of a quadrilateral are ≅, it is a ▱.

Part Three: Problem Sets

Problem Set A

1 For each quadrilateral QUAD, state the property or definition (if there is one) that proves that QUAD is a parallelogram.

a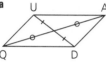

c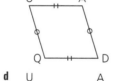

e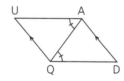

b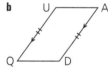

d

2 Given: ∠XRV ≅ ∠RST,
∠RSV ≅ ∠TVS

Conclusion: RSTV is a ▱.

3 Given: ⊙O
Conclusion: SMPR is a ▱.

Assignment Guide

Basic

Day 1	1–5, 9
Day 2	8, 10, 13, 14, 17

Average

Day 1	1, 3–6, 8, 10
Day 2	11–14, 17

Advanced

Day 1	2, 3, 5, 11–13, 17
Day 2	($\frac{1}{2}$ day) Section 5.6 14
	($\frac{1}{2}$ day) Section 5.7 8, 12, 13, 21, 22, 25, 28

Problem-Set Notes and Additional Answers

1a If the diagonals of a quadrilateral bisect each other, then it is a ▱.

b If one pair of opposite sides of a quadrilateral are ‖ and ≅, then it is a ▱.

c If the opposite sides of a quadrilateral are ≅, then it is a ▱.

d Cannot be proved to be a ▱

e If the opposite sides of a quadrilateral are ‖, then it is a ▱.

■ See *Solution Manual* for answers to problems **2** and **3**.

Problem Set A, *continued*

4 Given: RKMP is a ▱.
∠JRK ≅ ∠PMO

Prove: RJMO is a ▱.

Supply each missing reason.

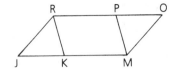

1 RKMP is a ▱.	1 Given
2 ⃡RO ∥ ⃡JM	2 Definition of ▱
3 RK ≅ PM	3 Opp. sides of a ▱ are ≅.
4 ∠RKM ≅ ∠MPR	4 Opp. ∠s of a ▱ are ≅.
5 ∠JKR supp. ∠RKM	5 Definition of supp.
6 ∠OPM supp. ∠MPR	6 Same as 5
7 ∠JKR ≅ ∠OPM	7 ∠s supp. to ≅ ∠s are ≅.
8 ∠JRK ≅ ∠PMO	8 Given
9 △JRK ≅ △OMP	9 ASA (7, 3, 8)
10 JK ≅ PO	10 CPCTC
11 RP ≅ KM	11 Opp. sides of a ▱ are ≅.
12 RO ≅ JM	12 Addition Property
13 RJMO is a ▱.	13 If one pair of opp. sides of a quad are both ∥ and ≅, it is a ▱.

5 Show that PQRS is a parallelogram.

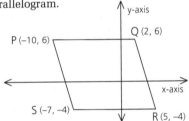

6 Given: CD ∥ AB,
∠EDA ≅ ∠CBF

Prove: ABCD is a parallelogram.

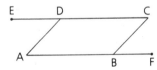

7 Given: RSTU is a square.
VR ≅ SW

a Is VWTU an isosceles trapezoid? Yes
b Is △ VWX an isosceles triangle? Yes
c Is △ UTX an isosceles triangle? Yes

8 In ▱ ABCD, the ratio of AB to BC is 5:3. If the perimeter of ABCD is 32, find AB. 10

Problem-Set Notes and Additional Answers, continued

■ See *Solution Manual* for answers to problems **5** and **6**.

Communicating Mathematics

Have students write a summary of ways to prove that a quadrilateral is a parallelogram.

9 JKMO is a $\square$.

$\overleftrightarrow{JM}$ bisects $\angle OJK$ and $\angle OMK$.

$OJ = x + 5$, $KM = y - 3$,

$JK = 2x - 4$

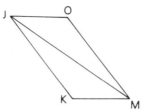

a Solve for x. 9
b Solve for y. 17
c Find the perimeter of OJKM. 56

10 The measure of one angle of a parallelogram is 40 more than 3 times another. Find the measure of each angle. 35; 145

Problem Set B

11 Answer Always, Sometimes, or Never: A quadrilateral is a parallelogram if

a Diagonals are congruent S
b One pair of opposite sides are congruent and one pair of opposite sides are parallel S
c Each pair of consecutive angles are supplementary A
d All angles are right angles A

12 Given: Quadrilateral PQRS,

$P = (-10, 7)$, $Q = (4, 3)$,

$R = (-2, -5)$, $S = (-16, 1)$

a Prove that quadrilateral PQRS is not a parallelogram.
b Prove that the quadrilateral formed by joining consecutive midpoints of the sides of PQRS is a parallelogram.

13 Prove that the quadrilateral is a parallelogram.

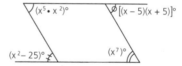

14 Given: RSOT is a $\square$.

$\overline{MS} \cong \overline{TP}$

Conclusion: MOPR is a $\square$.

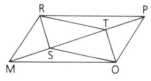

15 Prove: If both pairs of opposite sides of a quadrilateral are $\cong$, the quadrilateral is a $\square$ (method 2 of proving that a quadrilateral is a $\square$). (Hint: Use method 1.)

16 Prove: If two sides of a quadrilateral are both $\parallel$ and $\cong$, the quadrilateral is a $\square$ (method 3 of proving that a quadrilateral is a $\square$).

Problem-Set Notes and Additional Answers, continued

■ See *Solution Manual* for answers to problems **12–16**.

■ Problem **17** is a Crook problem.

18 1 $\angle 1 \cong \angle 4$ by $\parallel$ lines $\Rightarrow$ alt. ext. $\angle$s $\cong$
  2   $\overline{BF} \cong \overline{DE}$ by subtraction
  3   SAS
  4   $\angle 2 \cong \angle 3$ by CPCTC
      $\overline{AB} \cong \overline{CD}$ by CPCTC
  5   Alt. int. $\angle$s $\cong$ $\Rightarrow$ $\parallel$ lines
  6   Quad. has one pair of opp. sides both $\parallel$ and $\cong$ $\Rightarrow$ $\square$.

Answers for chart are as follows:
  1     $\angle 1 \cong \angle 4$
  2     $\overline{BF} \cong \overline{DE}$
  4a $\angle 2 \cong \angle 3$
  b $\overline{AB} \cong \overline{CD}$

19   Use the given $\square$s to prove $\overleftrightarrow{JM} \parallel \overleftrightarrow{RO}$ and $\overleftrightarrow{JP} \parallel \overleftrightarrow{KO}$. Since opposite sides of $\square$s are $\cong$, all of the short segments in the figure can be proved $\cong$. The conclusion follows by addition.

Problem-Set Notes and Additional Answers, continued

20

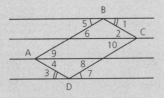

The Parallel Postulate allows you to draw lines through A and C that are ∥ to the lines in the diagram. Then by ∥ lines ⇒ alt. int. ∠s ≅, ∠1 ≅ ∠2 and ∠3 ≅ ∠4, and since ∠1 ≅ ∠3 then ∠2 ≅ ∠4. Similarly, ∠5 ≅ ∠6 and ∠7 ≅ ∠8 by ∥ lines ⇒ alt. int. ∠s ≅, and since ∠5 ≅ ∠7, then ∠6 ≅ ∠8. We also know ∠9 ≅ ∠6 and ∠10 ≅ ∠8 by ∥ lines ⇒ corr. ∠s ≅, and thus, ∠9 ≅ ∠10. ∴ ∠BCD ≅ ∠DAB by addition. Finally, ∠ABC ≅ ∠ADC by subtraction, and thus, ABCD is a ▱ because its opposite ∠s are ≅.

21a

Possibilities

(1,2)
(1,3)(2,3)
(1,4)(2,4)(3,4)
1,5 2,5 3,5 4,5
1,6 2,6 3,6 4,6 5,6
1,7 2,7 3,7 4,7 (5,7)6,7
1,8 2,8 3,8 4,8 5,8 (6,8)7,8

Prob. $= \frac{8}{28} = \frac{2}{7}$

b Prob. (neither are winners)
$= \frac{5}{7} \cdot \frac{5}{7} = \frac{25}{49}$

Prob. (at least 1 winner)
= 1 − Prob. (neither are winners)
$= 1 - \frac{25}{49} = \frac{24}{49}$

Problem Set B, *continued*

17 Find the value of x. 145

18 Given: $\overline{AF} \parallel \overline{EC}$,
 $\overline{AF} \cong \overline{EC}$,
 $\overline{BE} \cong \overline{FD}$

Prove ABCD is a ▱.

Copy and complete the flow diagram for the proof. Be sure to list reasons 1–6.

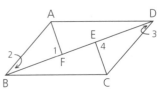

Problem Set C

19 Given: △KOR is equilateral.
 KOPR is a ▱.
 KMOR is a ▱.

Prove: △JMP is equilateral.

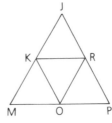

20 Given two parallel lines with a quadrilateral ABCD forming congruent angles as shown, prove that ABCD is a parallelogram (paragraph proof).

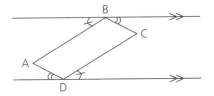

Problem Set D

21 The angles of a rectangle and a parallelogram that is not a rectangle are in a box.

a If two of the eight angles are selected at random, what is the probability that the angles are congruent? $\frac{2}{7}$

b A man offers to let you have two tries at getting a pair of congruent angles. In other words, you would draw a pair of angles at random, then replace the pair, and then draw a pair again. The man is willing to bet you $20 that you won't draw a congruent pair either time. Should you take the bet?
No. You will win $\frac{24}{49}$ of the time and lose $\frac{25}{49}$ of the time.

PROVING THAT FIGURES ARE SPECIAL QUADRILATERALS

Objectives

After studying this section, you will be able to
- Prove that a quadrilateral is a rectangle
- Prove that a quadrilateral is a kite
- Prove that a quadrilateral is a rhombus
- Prove that a quadrilateral is a square
- Prove that a quadrilateral is an isosceles trapezoid

Part One: Introduction

Proving That a Quadrilateral Is a Rectangle

When you want to prove that a figure is one of the special quadrilaterals, you must be sure that you prove sufficient properties to establish the quadrilateral's identity.

You can prove that quadrilateral EFGH is a rectangle by first showing that the quadrilateral is a parallelogram and then using either of the following methods to complete the proof.

1 If a parallelogram contains at least one right angle, then it is a rectangle (reverse of the definition).

2 If the diagonals of a parallelogram are congruent, then the parallelogram is a rectangle.

You can also prove that a quadrilateral is a rectangle without first showing that it is a parallelogram.

3 If all four angles of a quadrilateral are right angles, then it is a rectangle.

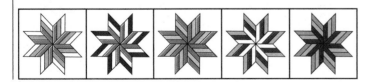

Section 5.7 Proving That Figures Are Special Quadrilaterals **255**

Class Planning

Time Schedule
Basic: 1 day
Average: 1 day
Advanced: $1\frac{1}{2}$ days

Resource References
Teacher's Resource Book
   Class Opener 5.7A, 5.7B
   Using Manipulatives 10

Class Opener

Given: $\overline{AB} \parallel \overline{CD}$,
   $\angle ABC \cong \angle ADC$,
   $\overline{AB} \cong \overline{AD}$
Prove: ABCD is a rhombus.

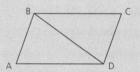

Statements	Reasons
1 $\overline{AB} \parallel \overline{CD}$	1 Given
2 $\angle ABD \cong \angle CDB$	2 $\parallel$ lines $\Rightarrow$ alt. int. $\angle$s $\cong$
3 $\angle ABC \cong \angle ADC$	3 Given
4 $\angle CBD \cong \angle ADB$	4 Subtraction Prop.
5 $\overline{BC} \parallel \overline{AD}$	5 Alt. int. $\angle$s $\cong \Rightarrow \parallel$ lines
6 ABCD is a $\square$.	6 If a quad has 2 pairs of $\parallel$ sides, it is a $\square$.
7 $\overline{AB} \cong \overline{AD}$	7 Given
8 ABCD is a rhombus.	8 $\square$ with at least 1 pair of consecutive sides $\cong$ is a rhombus.

This problem is very similar to the class opener for Section **5.4** except that it extends to the definition of rhombus. Building on previous problems helps students to make connections, reviews important ideas, and previews concepts to come later.

Lesson Notes

- Rectangle method 2 on page 255:
  △EFG ≅ △FEH by SSS.
  ∠FEG and ∠EFH are ≅ and supp.

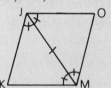

- Rhombus method 2:
  △JKM ≅ △JOM by ASA,
  so $\overline{JK} \cong \overline{JO}$.

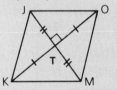

- Rhombus method 3:
  △JKT ≅ △JOT by SAS,
  so $\overline{JK} \cong \overline{JO}$.

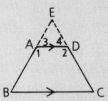

- Isosceles trapezoid method 2:

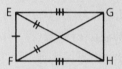

Extend the nonparallel sides to their intersection at E. Given ∠1 ≅ ∠2 or ∠B ≅ ∠C. In either case, ∠3 ≅ ∠4.
Thus, $\overline{EA} \cong \overline{ED}$ and $\overline{EB} \cong \overline{EC}$.
By subtraction, $\overline{AB} \cong \overline{DC}$.

- Isosceles trapezoid method 3:

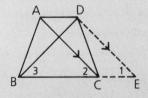

Proving That a Quadrilateral Is a Kite

To prove that a quadrilateral is a kite, either of the following methods can be used.

1. If two disjoint pairs of consecutive sides of a quadrilateral are congruent, then it is a kite (reverse of the definition).

2. If one of the diagonals of a quadrilateral is the perpendicular bisector of the other diagonal, then the quadrilateral is a kite.

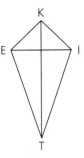

Proving That a Quadrilateral Is a Rhombus

To prove that quadrilateral KMOJ is a rhombus, you may first show that it is a parallelogram and then apply either of the following methods.

1. If a parallelogram contains a pair of consecutive sides that are congruent, then it is a rhombus (reverse of the definition).

2. If either diagonal of a parallelogram bisects two angles of the parallelogram, then it is a rhombus.

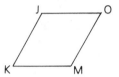

You can also prove that a quadrilateral is a rhombus without first showing that it is a parallelogram.

3. If the diagonals of a quadrilateral are perpendicular bisectors of each other, then the quadrilateral is a rhombus.

Proving That a Quadrilateral is a Square

The following method can be used to prove that NPRS is a square:

- If a quadrilateral is both a rectangle and a rhombus, then it is a square (reverse of the definition).

Proving That a Trapezoid Is Isosceles

Any one of the following methods can be used to prove that a trapezoid is isosceles.

1. If the nonparallel sides of a trapezoid are congruent, then it is isosceles (reverse of the definition).

2. If the lower or the upper base angles of a trapezoid are congruent, then it is isosceles.

3. If the diagonals of a trapezoid are congruent, then it is isosceles.

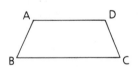

256 | Chapter 5 Parallel Lines and Related Figures

Draw $\overline{DE} \parallel \overline{AC}$. Since $\overline{AD} \parallel \overline{CE}$, ADEC is a ▱ and $\overline{DE} \cong \overline{AC}$. $\overline{AC} \cong \overline{BD}$, thus $\overline{DE} \cong \overline{DB}$ and ∠3 ≅ ∠1. ∠2 ≅ ∠1 by ∥ lines ⇒ corr. ∠s ≅. Thus, ∠3 ≅ ∠2. △ACB ≅ △DBC by SAS, and $\overline{AB} \cong \overline{DC}$.

Part Two: Sample Problems

Problem 1   *What is the most descriptive name for quadrilateral ABCD with vertices A = (−3, −7), B = (−9, 1), C = (3, 9), and D = (9, 1)?*

Solution   We must check every detail to see if sides are parallel or perpendicular, and we must also check diagonals. We must be careful to identify what we are finding with each calculation. A graph may prove helpful in directing our work.

Slope of $\overleftrightarrow{AB} = \dfrac{1-(-7)}{-9-(-3)} = \dfrac{8}{-6} = -\dfrac{4}{3}$

Slope of $\overleftrightarrow{BC} = \dfrac{9-1}{3-(-9)} = \dfrac{8}{12} = \dfrac{2}{3}$

Slope of $\overleftrightarrow{CD} = \dfrac{1-9}{9-3} = \dfrac{-8}{6} = -\dfrac{4}{3}$

Slope of $\overleftrightarrow{AD} = \dfrac{1-(-7)}{9-(-3)} = \dfrac{8}{12} = \dfrac{2}{3}$

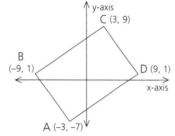

Since the slopes of $\overleftrightarrow{AB}$ and $\overleftrightarrow{CD}$ are equal, $\overline{AB} \parallel \overline{CD}$. Similarly, slope $\overleftrightarrow{BC}$ = slope $\overleftrightarrow{AD}$, so $\overline{BC} \parallel \overline{AD}$. Thus, ABCD is at least a parallelogram.

Is it a rectangle or a rhombus? Since the slopes of $\overleftrightarrow{AB}$ and $\overleftrightarrow{BC}$ are not opposite reciprocals of each other, $\angle ABC$ is not a right angle. ABCD is not a rectangle.

For the figure to be a rhombus, the diagonals must be perpendicular.

Slope of $\overleftrightarrow{AC} = \dfrac{9-(-7)}{3-(-3)} = \dfrac{16}{6} = \dfrac{8}{3}$

Slope of $\overleftrightarrow{BD} = \dfrac{1-1}{9-(-9)} = \dfrac{0}{18} = 0$

The slopes are not opposite reciprocals, so $\overline{AC} \not\perp \overline{BD}$. We conclude that ABCD is a parallelogram.

Problem 2   *Prove that if either diagonal bisects two angles of a ▱, the ▱ is a rhombus (method 2 of proving that a quadrilateral is a rhombus).*

Given: ABCD is a ▱.
  $\overleftrightarrow{BD}$ bisects $\angle ADC$ and $\angle ABC$.

Prove: ABCD is a rhombus.

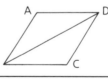

Proof

1 ABCD is a ▱.	1 Given
2 $\angle ADC \cong \angle ABC$	2 Opposite ∠s of a ▱ are ≅.
3 $\overleftrightarrow{BD}$ bis. $\angle ADC$ and $\angle ABC$	3 Given
4 $\angle ABD \cong \angle ADB$	4 Division Property
5 $\overline{AB} \cong \overline{AD}$	5 If △, then △.
6 ABCD is a rhombus.	6 If a ▱ contains a consecutive pair of sides that are ≅, it is a rhombus.

Cooperative Learning

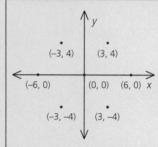

Have students work in small groups. If four of the seven points in the diagram above are chosen at random and joined with line segments, find the probability that the resulting line segments form

a   A quadrilateral $\frac{23}{35}$
b   A rectangle $\frac{1}{35}$
c   A rhombus $\frac{2}{35}$
d   A square   0
e   A trapezoid $\frac{6}{35}$
f   An isosceles trapezoid $\frac{2}{35}$
g   A kite $\frac{2}{35}$
h   A parallelogram $\frac{9}{35}$

Note   The probabilities above are based on the assumptions that the points are joined in the manner most conducive to forming a quadrilateral (if possible) and that the resulting figure is named with as many names as apply.

Problem 3   Given: GJMO is a ▱.
$\overline{OH} \perp \overline{GK}$;
$\overline{MK}$ is an altitude of △MKJ.

Prove: OHKM is a rectangle.

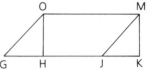

Proof

1 GJMO is a ▱.	1 Given
2 $\overleftrightarrow{OM} \parallel \overleftrightarrow{GK}$	2 Opposite sides of a ▱ are ∥.
3 $\overline{OH} \perp \overline{GK}$	3 Given
4 $\overline{MK}$ is an alt. of △MKJ.	4 Given
5 $\overline{MK} \perp \overline{GK}$	5 An altitude of a △ is ⊥ to the side to which it is drawn.
6 $\overline{OH} \parallel \overline{MK}$	6 If two coplanar lines are ⊥ to a third line, they are ∥.
7 OHKM is a ▱.	7 If both pairs of opposite sides of a quadrilateral are ∥, it is a ▱.
8 ∠OHK is a right ∠.	8 ⊥ segments form a right ∠.
9 OHKM is a rectangle.	9 If a ▱ contains at least one right ∠, it is a rectangle.

Assignment Guide

Basic

1–3, 5, 6, 10, 21, 25

Average

2, 3, 5, 8, 13, 14, 21, 22, 24, 25

Advanced

Day 1   ($\frac{1}{2}$ day) Section 5.6 14
          ($\frac{1}{2}$ day) Section 5.7 8,
          12, 13, 21, 22, 25, 28
Day 2   15, 17–20, 24, 26, 27

Problem-Set Notes
and Additional Answers

- See *Solution Manual* for answer to problem **2**.

Part Three: Problem Sets

Problem Set A

1 Locate points Q = (2, 4), U = (2, 7), A = (10, 7), and D = (10, 4) on a graph. Then give the most descriptive name for QUAD.   Rectangle

2 If $\overline{AB} \cong \overline{DC}$, show that ABCD is not a rhombus.

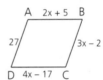

3 In order for RECT to be a rectangle, what must the value of x be?   21

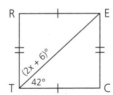

4 What is the most descriptive name for a quadrilateral with vertices (−11, 5), (7, 5), (7, −13), and (−11, −13)?   Square

5 a Write an expression for the area of the rectangle.
  b Write an expression for the perimeter of the rectangle.
  c Evaluate each when x is 4.2.
      a $18x^2 - 45x$; **b** $22x - 10$; **c** A = 128.5; P = 82.4

258   Chapter 5   Parallel Lines and Related Figures

6 Give the most descriptive name for

a A quadrilateral with diagonals that are perpendicular bisectors of each other Rhombus

b A rectangle that is also a kite Square

c A quadrilateral with opposite angles supplementary and consecutive angles supplementary Rectangle

d A quadrilateral with one pair of opposite sides congruent and the other pair of opposite sides parallel Isos. trapezoid

7 Given: $\overrightarrow{AC}$ bisects $\angle BAD$.
$\overline{AB} \cong \overline{BC}$,
$\overline{AB} \not\parallel \overline{CD}$

Prove: ABCD is a trapezoid.

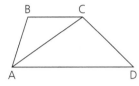

8 Given: YTWX is a $\square$.
$\overline{YP} \perp \overline{TW}$,
$\overline{ZW} \perp \overline{TY}$,
$\overline{TP} \cong \overline{TZ}$

Conclusion: TWXY is a rhombus.

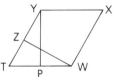

9 Given: Right $\triangle PQR$, with hypotenuse $\overline{PR}$. M is the midpoint of $\overline{PR}$. Through M, lines are drawn parallel to the legs.

Prove: The quadrilateral formed is a rectangle.

10 Given: $\overline{ID}$ bisects $\overline{RB}$,
$\overline{BI} \cong \overline{IR}$

Prove: BIRD is a kite.

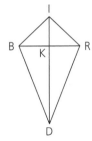

Problem Set B

11 Given: ABDE is a $\square$.
$\overline{BC}$ is the base of isosceles $\triangle BCD$.

Prove: ACDE is an isosceles trapezoid.

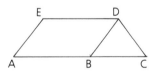

12 ABCD is a parallelogram with perimeter 52. The perimeter of ABNM is 36, and $\overline{NC} \cong \overline{AM}$. Find NM. 10

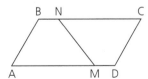

Problem-Set Notes and Additional Answers, continued

■ See *Solution Manual* for answers to problems **7–11**.

Communicating Mathematics

Have students write a description of each of three special quadrilaterals without using the names of the quadrilaterals. Each description should include sufficient properties to establish the quadrilateral's identity.

Problem Set B, *continued*

13 What is the most descriptive name for each quadrilateral below?

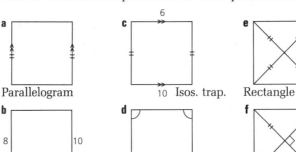

a Parallelogram

c 6 / 10 Isos. trap.

e Rectangle

g Kite

b 8 / 10 Trapezoid

d Rectangle

f Rhombus

h Quadrilateral

14 What is the most descriptive name for a quadrilateral with vertices $(-7, 2)$, $(2, 8)$, $(6, 2)$, and $(-3, -4)$? Justify your conclusion. Rectangle

15 Given: ▱ PQRS;
  A is the midpoint of $\overline{QR}$.
  $\overrightarrow{PA}$ bisects ∠QPS.
Prove: $\overrightarrow{SA}$ bisects ∠PSR.

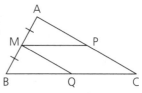

16 Find the area of the parallelogram. (Hint: Area = base · height.) 108

(−5, 9) y-axis
(4, 5)
x-axis
(−5, −3)
(4, −7)

17 a If a quadrilateral is symmetrical across both diagonals, it is a ___rhombus___.
  b If a quadrilateral is symmetrical across exactly one diagonal, it is a ___kite___.
  c Which quadrilateral has four axes of symmetry? Square

18 Given: △ABC;
  M is the midpoint of $\overline{AB}$.
  Segments are drawn from M parallel to $\overline{AC}$ and $\overline{BC}$.
Prove: **a** PMQC is a ▱.
  b △MAP ≅ △BMQ.

Problem-Set Notes and
Additional Answers, continued

■ See *Solution Manual* for answers to problems **15** and **18**.

19 Write a quadratic equation to represent the area of the rectangle. If the area is 160 square meters, find the perimeter. $A = x^2 + 6x$; $P = 52$ m

x

x + 6

20 The rectangle has a vertex on the line. The equation of the line is $y = -2x + 17$. Find the area of the rectangle.   33

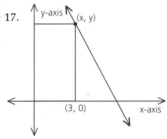

y-axis   (x, y)

(3, 0)   x-axis

21 ⊙P just touches (is tangent to) the x-axis. Find the area of ⊙P to the nearest hundredth.   ≈29.61

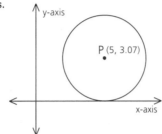

y-axis

P (5, 3.07)

x-axis

22 M, N, O, and P are midpoints of the sides of ABCD. Make up your own coordinates for A, B, C, and D.

a Find the coordinates of M, N, O, and P.

b Find the slopes of $\overleftrightarrow{MN}$ and $\overleftrightarrow{PO}$.

c What is true about $\overleftrightarrow{MN}$ and $\overleftrightarrow{PO}$?
  a Answers will vary; **b** Answers will vary but slopes will be equal; **c** Parallel

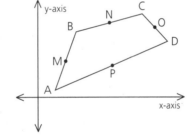

y-axis

B   N   C

O

M   D

A   P

x-axis

23 R, H, O, and M are midpoints. Find the slopes of the diagonals of RHOM. $-\frac{1}{4}$; 4

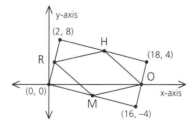

y-axis

(2, 8)

H

R   (18, 4)

O

(0, 0)

M   x-axis

(16, −4)

5.7

26 SQUA is a square with sides 7 units each. Q' = (−3, 10) and A' = (10, −3). Slope $\overline{QA}$ = slope $\overline{Q'A'}$ = −1, and Q'Q = 6 while AA' = 6. Q'QAA' is an isos. trapezoid.

27

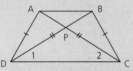

$\triangle ADC \cong \triangle BCD$ by SSS. Then $\angle 1 \cong \angle 2$ by CPCTC. Since the base $\angle$s of $\triangle DPC$ are $\cong$, $\overline{DP} \cong \overline{CP}$. $\overline{PA} \cong \overline{PB}$ by subtraction, and so $\triangle PAB$ is isos.

28

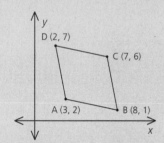

Slope $\overline{AB}$ = slope $\overline{CD}$ = $-\frac{1}{5}$; slope $\overline{AD}$ = slope $\overline{BC}$ = −5. ∴ ABCD is a $\square$. Slope $\overline{AC}$ = 1; slope $\overline{BD}$ = −1. ∴ The diagonals are $\perp$, and ABCD is a rhombus. It is not a square, because $\overline{AB} \not\perp \overline{BC}$.

29a Shaded area = area of rectangle − area of square = $324 - x^2$

b The side of the square must be less than 12. ∴ $0 < \text{area} < 144$

Problem Set B, *continued*

24 D is the reflection of E = (x, 14) across the y-axis. If $\overleftrightarrow{CD} \parallel \overleftrightarrow{AB}$, solve for x. −3

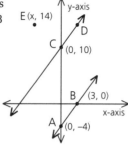

25 ABCD is a parallelogram. If two of the following conclusions are selected at random, what is the probability that both conclusions are true? $\frac{1}{2}$

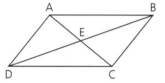

a $\overline{AB} \cong \overline{DC}$

b E is the midpoint of $\overline{AC}$.

c $\angle ADC$ is supplementary to $\angle DAB$.

d $\overrightarrow{AC}$ bisects $\angle DAB$.

Problem Set C

26 What is the most descriptive name for the quadrilaterals SQUA and Q'QAA', where Q' is the reflection of Q over the y-axis and A' is the reflection of A over the x-axis? Justify your conclusions. SQUA is a square; Q'QAA' is an isos. trapezoid.

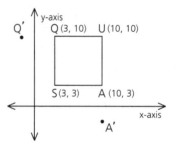

27 The diagonals of a quadrilateral are congruent. Exactly one pair of opposite sides are congruent. Prove that two of the triangles formed are isosceles.

28 What is the most descriptive name for the quadrilateral with vertices (3, 2), (8, 1), (7, 6), and (2, 7)? Rhombus

29 a Write an expression to represent the shaded area. $324 - x^2$

b Write an inequality that gives the limits of the area of the square. $0 < \text{area} < 144$

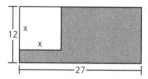

CHAPTER SUMMARY

CONCEPTS AND PROCEDURES

After studying this chapter, you should be able to
- Write indirect proofs (5.1)
- Apply the Exterior Angle Inequality Theorem (5.2)
- Use various methods to prove lines parallel (5.2)
- Apply the Parallel Postulate (5.3)
- Identify the pairs of angles formed by a transversal cutting parallel lines (5.3)
- Apply six theorems about parallel lines (5.3)
- Solve crook problems (5.3)
- Recognize polygons (5.4)
- Understand how polygons are named (5.4)
- Recognize convex polygons (5.4)
- Recognize diagonals of polygons (5.4)
- Identify special types of quadrilaterals (5.4)
- Identify some properties of parallelograms, rectangles, kites, rhombuses, squares, and isosceles trapezoids (5.5)
- Prove that a quadrilateral is a parallelogram (5.6)
- Prove that a quadrilateral is a rectangle (5.7)
- Prove that a quadrilateral is a kite (5.7)
- Prove that a quadrilateral is a rhombus (5.7)
- Prove that a quadrilateral is a square (5.7)
- Prove that a quadrilateral is an isosceles trapezoid (5.7)

VOCABULARY

convex polygon (5.4)
diagonal (5.4)
indirect proof (5.1)
isosceles trapezoid (5.4)
kite (5.4)
lower base angles (5.4)
parallelogram (5.4)
Parallel Postulate (5.3)

polygon (5.4)
quadrilateral (5.4)
rectangle (5.4)
rhombus (5.4)
square (5.4)
trapezoid (5.4)
upper base angles (5.4)

5

REVIEW PROBLEMS

Problem Set A

1 Give the most descriptive name for

  a A quadrilateral whose consecutive sides measure 15, 18, 15, and 18 Parallelogram

  b A quadrilateral whose consecutive sides measure 15, 18, 18, and 15 Kite

  c A quadrilateral with consecutive angles of 30°, 150°, 110°, and 70° Trapezoid

  d A quadrilateral whose diagonals are perpendicular and congruent and bisect each other Square

  e A quadrilateral whose congruent diagonals bisect each other and bisect the angles Square

2 ABCD is a ▱.
AB = 2x + 6,
BC = 8,
CD = x + 8
Find the perimeter of ABCD. 36

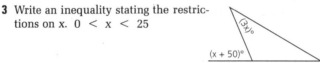

3 Write an inequality stating the restrictions on x. $0 < x < 25$

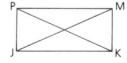

4 JKMP is a rectangle.
PK = 0.2x,
JM = x − 12
Find PK. 3

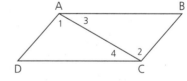

5 Given: ∠1 ≅ ∠2;
      ABCD is not a ▱.
Prove: ∠3 ≇ ∠4

6 In a parallelogram, the measure of one of the angles is twice that of another. Are these opposite angles or consecutive angles? Find the measure of each angle of the parallelogram. Consecutive; 60, 120, 60, 120

7 In each of these diagrams, is m ∥ p?

a
Yes

b
No

c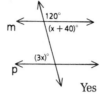
Yes

8 Name five properties of a parallelogram.

9 Given: $\overline{AB} \cong \overline{CD}$, $\overline{AG} \cong \overline{BE}$,
$\overline{AG} \parallel \overline{BE}$
Conclusion: $\overline{GC} \parallel \overline{ED}$

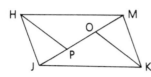

10 Given: HJKM is a ▱.
∠JHP ≅ ∠MKO
Conclusion: $\overline{MP} \cong \overline{JO}$

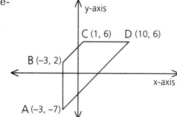

11 Show that ABCD is an isosceles trapezoid.

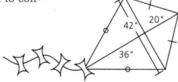

12 Find the area of the paper used to construct the kite. **1176 sq in.**

13 Two polygons are selected at random from a group consisting of a nonisosceles trapezoid, an isosceles trapezoid, and a parallelogram. Find the probability that both polygons have two pairs of congruent angles. $\frac{1}{3}$

14 a How many squares appear to be in the figure at the right? **6**
  b How many rectangles? **19**

Problem-Set Notes and Additional Answers, continued

■ See *Solution Manual* for answers to problems **8–11.**

Problem-Set Notes and
Additional Answers, continued

■ See Solution Manual for answers to problems **15, 20,** and
21.

Review Problem Set A, *continued*

15 Given: $\overline{KR}$ is a median to $\overline{JO}$.
$\overline{RP} \cong \overline{KM}$,
$\overline{RM} \cong \overline{KM}$

Prove: JMOP is a ▱.

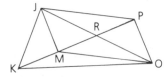

16 In ▱ ABCD, $\angle A = (2x + 6)°$ and $\angle B = (x + 24)°$. Find m$\angle$C. 106

17 $\overline{TP}$ (a diameter) passes through the center of ⊙O. Find the area of the circle to
three decimal places. ≈78.540 sq units

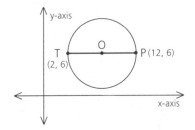

Problem Set B

18 Given: $\overleftrightarrow{AD} \parallel \overleftrightarrow{BC}$,
m$\angle$1 = 5.63x + 2.42,
m$\angle$2 = 2.1x,
m$\angle$3 = 6x − 5.1,
m$\angle$4 = 42

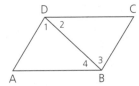

a Find m$\angle$1. ≈116.85
b Is $\overleftrightarrow{DC} \parallel \overleftrightarrow{AB}$? No

19 If the statement is always true, write A; if sometimes true, write
S; if never true, write N.

a If the diagonals of a quadrilateral are congruent, the quadrilateral is an isosceles trapezoid. S
b If the diagonals of a quadrilateral divide each angle into two
45-degree angles, the quadrilateral is a square. A
c If a parallelogram is equilateral, it is equiangular. S
d If two of the angles of a trapezoid are congruent, the trapezoid
is isosceles. S

20 Prove: The figure produced by joining the consecutive midpoints
of a parallelogram is a parallelogram.

21 Prove: If the bisector of an exterior angle formed by extending
one of the sides of a triangle is parallel to a side of the triangle,
the triangle is isosceles.

22 What is the most descriptive name for the quadrilateral with
vertices (0, − 6), (− 4, 2), (4, 6), and (8, − 2)? Justify your
conclusion. Square

23 Given: EFGH is a ▱.
$\overline{AE} \cong \overline{BF} \cong \overline{CG} \cong \overline{DH}$

Prove: ABCD is a ▱.

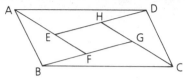

24 Given: P is the midpt. of $\overline{RO}$.
K is the midpt. of $\overline{JM}$.
$\angle 1 \cong \angle 2$,
$\overline{PS} \cong \overline{KS}$

Prove: RJMO is a ▱.

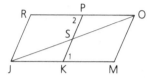

25 Find the value of x. 70

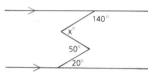

26 Given: △TWX is isosceles, with base $\overline{WX}$.
$\overline{RY} \parallel \overline{WX}$

Prove: RWXY is an isosceles trapezoid.

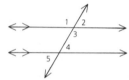

27 If two of the five labeled angles are chosen at random, what is the probability that they are supplementary? $\frac{3}{5}$

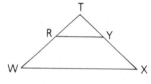

Problem Set C

28 Given: $\overline{AB} \cong \overline{DC}$,
$\overline{AB} \perp \overline{BC}, \overline{DC} \perp \overline{BC}$

Prove: △DEC is isosceles.

29 Given: △AED and △BEC are isosceles,
with congruent bases $\overline{AD}$ and $\overline{BC}$.

Prove: ABCD is a rectangle.

30 Given: Kite KITE

Find: The three possible values of the
perimeter of KITE 60; 4; 102

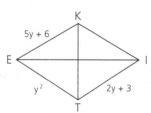

Problem-Set Notes and
Additional Answers, continued

■ See *Solution Manual* for answers to problems **23, 24,**
and **26.**

25

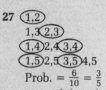

27 (1,2)
1,3 (2,3)
(1,4) 2,4 (3,4)
(1,5) 2,5 (3,5) 4,5
Prob. $= \frac{6}{10} = \frac{3}{5}$

28 Prove $\overleftrightarrow{AB} \parallel \overleftrightarrow{DC}$, since both
are $\perp$ to $\overline{BC}$. Then ABCD is
a ▱ and a rectangle, so
diagonals are $\cong$ and bisect
each other. By division,
$\overline{DE} \cong \overline{CE}$.

29

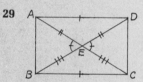

(**Note** The given does not
say that $\overline{AE} \cong \overline{EC}$ or
$\overline{BE} \cong \overline{ED}$.)
Prove △AEB $\cong$ △DEC by
SAS, so $\overline{AB} \cong \overline{DC}$ and
ABCD is a ▱. Use addition
to prove $\overline{AC} \cong \overline{BD}$.

30 Either KE = ET or ET = TI,
so $y^2 = 5y + 6$ or
$y^2 = 2y + 3$. y = 6 or −1, or
y = 3 or −1. Thus, P = 60,
4, or 102.

CHAPTER

6

LINES AND PLANES IN SPACE

*T*HREE-DIMENSIONAL GEOMETRY is introduced in this chapter and is integrated throughout the rest of the text. Many properties of triangles and quadrilaterals are reinforced and extended by considering them in space.

The remainder of the text stresses the interrelationships among geometric concepts rather than the technical rigor of proof. Students in advanced courses should be encouraged to derive theorems, but other students should concentrate on applying the concepts to problem solving.

All students should study Sections **6.1** and **6.2**, which are necessary for Chapters **11** and **12**. In Section **6.1**, students learn the basic properties of planes, identify four methods by which a plane can be determined, and apply postulates relating lines and planes. Section **6.2** defines perpendicularity of a line and a plane and establishes a theorem related to this concept.

Basic and average courses might omit Section **6.3**. The concepts presented here are not required in later chapters.

I n this model, the Ruck-A-Chucky Bridge in Auburn, California, appears suspended in space.

6.1 RELATING LINES TO PLANES

Objectives

After studying this chapter, you will be able to
- Understand basic concepts relating to planes
- Identify four methods of determining a plane
- Apply two postulates concerning lines and planes

Part One: Introduction

Preliminary Concepts

Recall the definition of *plane* from Section 4.5: *A plane is a surface such that if any two points on the surface are connected by a line, all points of the line are also on the surface.* Because a surface has no thickness, a plane must be "flat" if it is to contain the straight lines determined by all pairs of points on it. It must also be infinitely long and wide. Thus, a plane has only two dimensions, length and width.

A surface that is not a plane   Plane surface

A plane is frequently drawn as shown in the right-hand diagram above. In this case, the diagram represents part of a horizontal plane, with the edges nearest to you darkened. A plane can be named by placing a single lowercase letter in one of the corners.

It is important to understand that although our picture of a plane has edges and corners, an actual plane has neither and should be thought of as infinite in length and width.

You may recall the following definitions from Section 4.5: *If points, lines, segments, and so forth, lie in the same plane, we call them coplanar. Points, lines, segments, and so forth, that do not lie in the same plane are called noncoplanar.* In the diagram on the next page, $\overleftrightarrow{AB}$ and $\overleftrightarrow{ST}$ lie in plane m. $\overleftrightarrow{RP}$ does not lie in the plane but intersects m at V.

Vocabulary

foot

6.1

Class Planning

Time Schedule
Basic: 2 days
Average: 1 day
Advanced: 1 day

Resource References
Teacher's Resource Book
    Class Opener 6.1A, 6.1B
Transparency 10

Class Opener

Given: Plane m,
    ∠ABC ≅ ∠ABD,
    $\overline{BD} \cong \overline{BC}$
Prove: $\overline{AD} \cong \overline{AC}$

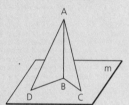

Statements	Reasons
1 ∠ABC ≅ ∠ABD	1 Given
2 $\overline{BD} \cong \overline{BC}$	2 Given
3 $\overline{AB} \cong \overline{AB}$	3 Reflexive Prop.
4 △ABD ≅ △ABC	4 SAS
5 $\overline{AD} \cong \overline{AC}$	5 CPCTC

This is not a three-dimensional proof. To incorporate three dimensions, ask students to prove ∠ADC ≅ ∠ACD using the same given.

Lesson Notes

- After students have read the introduction, they will find classroom demonstrations using three-dimensional models (either commercial or made by hand) of intersecting lines and planes helpful in visualizing the concepts.

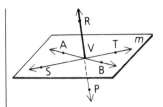

A, B, S, T, and V are *coplanar* points.
$\overleftrightarrow{AB}$ and $\overleftrightarrow{ST}$ are *coplanar* lines.
$\overline{AB}$ and $\overline{ST}$ are *coplanar* segments.

A, B, S, T, and R are *noncoplanar* points.
$\overleftrightarrow{AB}$, $\overleftrightarrow{ST}$, and $\overleftrightarrow{RP}$ are *noncoplanar* lines.
$\overline{AB}$, $\overline{ST}$, and $\overline{RP}$ are *noncoplanar* segments.

Definition The point of intersection of a line and a plane is called the ***foot*** of the line.

In the preceding diagram, V is the foot of $\overleftrightarrow{RP}$ in plane m.

Four Ways to Determine a Plane

In Chapter 3, you learned that two points determine a line. We would now like to find conditions under which a plane is determined.

One point obviously does not determine a plane, since infinitely many planes pass through a single point.

The diagram at the right shows that two points also do not determine a unique plane. It shows two different planes, m and n, each of which contains both point A and point B. The same diagram shows that three points—A, B, and C—do not determine a plane if the three points are collinear.

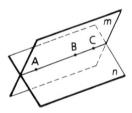

If the three points are noncollinear, however, they do determine a plane.

There is one and only one plane that contains the three noncollinear points A, B, and C. This plane can be called either plane ABC or plane k.

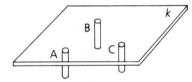

The preceding observations suggest an important postulate.

Postulate ***Three noncollinear points determine a plane.***

There are other ways of determining a plane. The following three are stated as theorems.

Lesson Notes, continued

■ Ask students to compare the stability of a three-legged stool with that of a four-legged chair when both are placed on an uneven surface.

270 Chapter 6 Lines and Planes in Space

6.1

T 270

Theorem 45   *A line and a point not on the line determine a plane.*

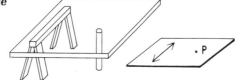

Theorem 46   *Two intersecting lines determine a plane.*

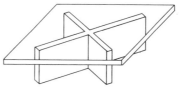

The proofs of Theorems 45 and 46 are asked for in Problem Set B.

Theorem 47   *Two parallel lines determine a plane.*

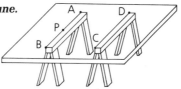

Proof: If $\overleftrightarrow{AB}$ and $\overleftrightarrow{CD}$ are parallel, then according to the definition of *parallel lines*, they lie in a plane. We need to show that they lie in *only one* plane. If P is any point on $\overleftrightarrow{AB}$, then according to Theorem 45, there is only one plane containing P and $\overleftrightarrow{CD}$. Thus, there is only one plane that contains $\overleftrightarrow{AB}$ and $\overleftrightarrow{CD}$, because every plane containing $\overleftrightarrow{AB}$ contains P.

Two Postulates Concerning Lines and Planes

We shall assume the following two statements.

Postulate   *If a line intersects a plane not containing it, then the intersection is exactly one point.*

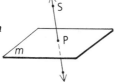

Postulate   *If two planes intersect, their intersection is exactly one line.*

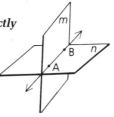

Lesson Notes, continued

■ Three students, two with meter sticks and one with a large piece of cardboard, can demonstrate that any two intersecting or parallel lines are coplanar. By definition, any two skew lines are not coplanar.

Lesson Notes, continued

■ These sample problems should
be read and discussed in class
so that students can have some
help in visualizing lines in
space.

Communicating Mathematics

Ask students to use three-
dimensional models made of
cardboard to demonstrate the information conveyed by the diagram in Sample Problem **1**.

Part Two: Sample Problems

Problem 1

a m ∩ n = __?__

b A, B, and V determine plane __?__.

c Name the foot of $\overleftrightarrow{RS}$ in m.

d $\overleftrightarrow{AB}$ and $\overleftrightarrow{RS}$ determine plane __?__.

e $\overleftrightarrow{AB}$ and point __?__ determine plane n.

f Does W lie in plane n?

g Line AB and line __?__ determine plane m.

h A, B, V, and __?__ are coplanar points.

i A, B, V, and __?__ are noncoplanar points.

j If R and S lie in plane n, what can be said about $\overleftrightarrow{RS}$?

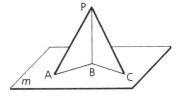

Answers

a $\overleftrightarrow{AB}$	d n	g VW	j $\overleftrightarrow{RS}$ lies in
b m	e R or s	h W or P	plane n.
c P	f No	i R or S	

Note In this problem, other planes are determined besides the two
shown in the diagram. For example, the noncollinear points R, P,
and V determine a plane.

Problem 2

Given: A, B, and C lie in plane m.
  $\overline{PB} \perp \overline{AB}$,
  $\overline{PB} \perp \overline{BC}$,
  $\overline{AB} \cong \overline{BC}$

Prove: ∠APB ≅ ∠CPB

Proof

1 $\overline{PB} \perp \overline{AB}$, $\overline{PB} \perp \overline{BC}$	1 Given
2 ∠PBA and ∠PBC are right ∠s.	2 ⊥ lines form right ∠s.
3 ∠PBA ≅ ∠PBC	3 Right ∠s are ≅.
4 $\overline{AB} \cong \overline{BC}$	4 Given
5 $\overline{PB} \cong \overline{PB}$	5 Reflexive Property
6 △PBA ≅ △PBC	6 SAS (4, 3, 5)
7 ∠APB ≅ ∠CPB	7 CPCTC

Assignment Guide

Basic

Day 1	1–6
Day 2	7, 8, 13, 14

Average

7, 8, 13, 14

Advanced

7–9, 14, 15, 18

Part Three: Problem Sets

Problem Set A

1 Consider a spherical object, such as an orange or a globe. If two
points are marked on it and a straight line is drawn through the
two points, does the line lie on the surface? Is it possible to draw
a straight line that will lie entirely on the surface? No; no

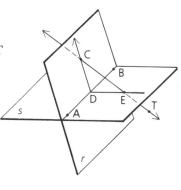

2 a $r \cap s = \underline{\quad ? \quad}$ $\overleftrightarrow{AB}$

b $\overleftrightarrow{AB} \cap s = \underline{\quad ? \quad}$ $\overleftrightarrow{AB}$

c Name three collinear points. A, D, B or C, E, T

d Name four noncoplanar points.

e What plane do points A, B, and E determine? s

f What plane do $\overleftrightarrow{AB}$ and $\overleftrightarrow{ED}$ determine? s

g Name the foot of $\overleftrightarrow{TC}$ in plane s. E

h Name the foot of $\overleftrightarrow{TC}$ in plane r. C

i Do $\overleftrightarrow{CD}$ and $\overleftrightarrow{ED}$ determine a plane? Yes

j If $\overleftrightarrow{CD} \perp \overleftrightarrow{AB}$, name the right angles formed.
∠CDB and ∠CDA

3 Consider two points on a cylindrical surface, such as the curved surface of a tin can. Does the line connecting two such points *always* lie on the surface? Does it *ever* lie on the surface? No; yes

4 Make freehand sketches of a horizontal plane, a vertical plane, and two intersecting planes.

5 A three-legged stool will not rock, even if the legs are of different lengths. Many four-legged stools wobble. Explain.

6 What theorem or assumption in this chapter provides the best explanation for the fact that when you saw a board, the edge of the cut is a straight line?

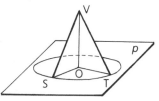

Problem Set B

7 Given: A, P, and B lie in plane m.
$\overleftrightarrow{CP} \perp \overleftrightarrow{AP}$, $\overleftrightarrow{CP} \perp \overleftrightarrow{PB}$,
$\overline{PA} \cong \overline{PB}$
Prove: $\overline{CA} \cong \overline{CB}$

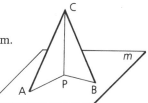

8 Given: ⊙O lies in plane p.
$\overleftrightarrow{VO} \perp \overleftrightarrow{OS}$,
$\overleftrightarrow{VO} \perp \overleftrightarrow{OT}$
Prove: ∠VSO ≅ ∠VTO

9 Prove Theorem 45: A line and a point not on the line determine a plane. (Write a paragraph proof.)

Cooperative Learning

Have small groups of students use spherical and cylindrical objects to demonstrate their responses to problems **1, 3, 11,** and **13.** Each student should be prepared to explain each model. When all groups have completed making their models, have one student from each group display and explain the group's model.

Problem-Set Notes and Additional Answers

2d Answers will vary. Possible answer: A, C, D, and E.

■ See *Solution Manual* for answers to problems **4** and **7–9.**

5 A three-legged stool has three points, and three points are on one plane. A four-legged stool has four points, and four points may lie on one plane, but usually form two planes.

6 If two planes intersect, their intersection is exactly one line.

Problem-Set Notes and
Additional Answers, continued

■ See *Solution Manual* for answers to problems **10** and **14**.

13 If the points are in space
and not in a plane, the line
joining them may not go
through the segment.

16 It is given that $\overleftrightarrow{AB}$ is the ⊥
bisector of $\overleftrightarrow{ST}$ and $\overleftrightarrow{BC}$ is the
⊥ bisector of $\overleftrightarrow{ST}$. Thus,
$\overline{AS} \cong \overline{AT}$ and $\overline{CS} \cong \overline{CT}$
because if a point is on the
⊥ bisector of a segment, it is
equidistant from the
endpoints of the segment.
Thus, △SAC ≅ △TAC by
SSS. ∠SAD ≅ ∠TAD by
CPCTC, and △SAD ≅ △TAD
by SAS. Thus, $\overline{DS} \cong \overline{DT}$
by CPCTC, and so $\overline{DB}$ is the
⊥ bisector of $\overline{ST}$.

Problem Set B, *continued*

10 Prove Theorem 46: Two intersecting lines determine a plane.
(Write a paragraph proof.)

11 Can you hold two pencils so that they do not intersect and are
not parallel? Are they coplanar? (Lines that do not intersect and
that are not coplanar are called *skew* lines.) Yes; no

12 Cut a quadrilateral out of paper and fold
it along a diagonal as shown in the figure. Is every four-sided figure a plane
figure? No

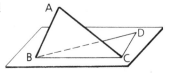

13 If two points in space are equidistant from the endpoints of a
segment, will the line joining them be the perpendicular bisector
of the segment? Explain. Not necessarily

14 Given: Planes m and n
intersect in $\overleftrightarrow{RS}$.
m contains R, S, and V.
n contains R, S, and T.
$\overline{TS} \cong \overline{VR}$,
$\overline{TR} \perp \overline{RS}$,
$\overline{VS} \perp \overline{RS}$
Prove: $\overline{TR} \cong \overline{VS}$

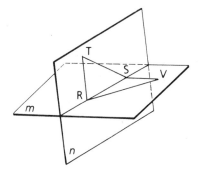

Problem Set C

15 The figure at the right is a square pyramid. How many planes are determined
by its vertices? (There are more than
five.) Name them.
7; ABC, ABP, BCP, CDP, DAP, ACP, BDP

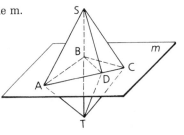

16 Given: A, B, C, and D lie in plane m.
$\overleftrightarrow{ST}$ intersects m at B.
D is any point on $\overline{AC}$.
$\overleftrightarrow{ST} \perp \overleftrightarrow{AB}$, $\overleftrightarrow{ST} \perp \overleftrightarrow{BC}$,
$\overline{SB} \cong \overline{TB}$
Prove: $\overleftrightarrow{ST} \perp \overleftrightarrow{BD}$

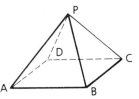

17 Given: A, B, and X lie in plane m.
X is on $\overline{AB}$. P and Q are above m.
B is equidistant from P and Q.
A is equidistant from P and Q.
Prove: X is equidistant from P and Q.

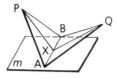

18 Given: △ABC ≅ △DBC
Prove: △AXD is isosceles.

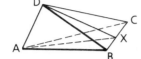

17 Since B is equidistant from P and Q, $\overline{PB} \cong \overline{BQ}$. Since A is equidistant from P and Q, $\overline{PA} \cong \overline{AQ}$. Thus, △PAB ≅ △QAB by SSS. Then ∠PAB ≅ ∠QAB by CPCTC. Thus, △PAX ≅ △QAX by SAS, and $\overline{PX} \cong \overline{QX}$ by CPCTC.

18 ∠ABC ≅ ∠DBC and $\overline{AB} \cong \overline{DB}$ by CPCTC. $\overline{BX} \cong \overline{BX}$ by the Reflexive Property. ∴ △ABX ≅ △DBX by SAS. Thus, $\overline{AX} \cong \overline{DX}$ by CPCTC.

CAREER PROFILE

THE GEOMETRY OF ARCHITECTURE
Thalia and Steve Lubin organize space

The geometry of a building can express itself in a multitude of ways. On a technical level there are the angles and dimensions of the hallways and rooms that compose the building. At the creative level the architect who designs the building must be able to see it in abstract geometrical terms. Explains architect Steve Lubin: "When we look at an empty lot we envision volumes of space where there are none now. It's all geometry, imagining a progression of interlocking spaces that will ultimately become a building."

Then there is the geometry of scale. According to architect Thalia Lubin: "When you enter a space you relate it to yourself. That's why a house cannot be restful and orderly unless everything in it relates to people and their sense of proportion and scale."

In designing a building, an architect must take into consideration the client's wishes, legal requirements, and environmental constraints dictated by the building site. The purest expression of geometry in a building is one of logic. "The final design is a bundle of compromises," says Thalia Lubin. "The architect's job is to impose a sense of logic on all of the competing forces, to find the natural order of things."

Both members of this unusual husband-and-wife team of architects took five-year degrees in architecture from the University of Oregon at Eugene. After working briefly for others, they decided to go into business together in Woodside, California. In designing a building, Thalia works with the client, while Steve oversees the technical aspects of the project. The system works, though Thalia admits, "We take a lot of chaos wherever we go." Steve says, "Every job is completely different. You need incredible patience to be an architect, but you dream of achieving poetry in the end."

Computer-generated renderings courtesy of Skidmore, Owings & Merrill.

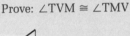
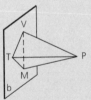
6.2   PERPENDICULARITY OF A LINE AND A PLANE

Objectives

After studying this section, you will be able to
- Recognize when a line is perpendicular to a plane
- Apply the basic theorem concerning the perpendicularity of a line and a plane

Part One: Introduction

A Line Perpendicular to a Plane

What does it mean to say that a line is perpendicular to a plane? Think about that for a moment and then read the following formal definition.

Definition   A line is perpendicular to a plane if it is perpendicular to *every one* of the lines in the plane that pass through its foot.

Observe that we now have two kinds of perpendicularity:

1 Between two lines ($\overleftrightarrow{AB} \perp \overleftrightarrow{BD}$)
2 Between a line and a plane ($\overleftrightarrow{AB} \perp m$)

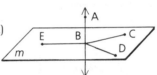

The definition above is a very powerful statement because of the words *every one*. If we are given that $\overleftrightarrow{AB} \perp m$ (in the diagram above), we can draw *three* conclusions:

$$\overleftrightarrow{AB} \perp \overleftrightarrow{BC} \qquad \overleftrightarrow{AB} \perp \overleftrightarrow{BD} \qquad \overleftrightarrow{AB} \perp \overleftrightarrow{BE}$$

The Basic Theorem Concerning the Perpendicularity of a Line and a Plane

You have just seen that a number of conclusions can be drawn from the information that a line is perpendicular to a plane. What about the reverse situation? How can we prove that a given line is perpendicular to a plane? To apply the preceding definition in reverse, we would have to show that the line is perpendicular to *every* line that passes through its foot. Considering the infinite number of lines one by one would be an endless process.

　　If a line is perpendicular to only one line that lies in the plane and passes through its foot, is it perpendicular to the plane? Or must it be perpendicular to two, three, or four lines in order to be perpendicular to the plane? The following theorem answers that question.

Theorem 48　*If a line is perpendicular to two distinct lines that lie in a plane and that pass through its foot, then it is perpendicular to the plane.*

Given: $\overleftrightarrow{BF}$ and $\overleftrightarrow{CF}$ lie in plane m.
　　　$\overleftrightarrow{AF} \perp \overleftrightarrow{FB}$,
　　　$\overleftrightarrow{AF} \perp \overleftrightarrow{FC}$
Prove: $\overleftrightarrow{AF} \perp m$

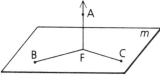

The proof is left as a challenge. (You may already have written part of the proof in Section 6.1, problem 16.)

Part Two: Sample Problems

Problem 1　If ∠STR is a right angle, can we conclude that $\overleftrightarrow{ST} \perp m$? Why or why not?

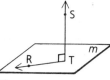

Solution　No. To be perpendicular to plane m, $\overleftrightarrow{ST}$ must be perpendicular to at least *two* lines that lie in m and pass through T, the foot of $\overleftrightarrow{ST}$.

Problem 2　Given: $\overline{PF} \perp k$,
　　　　　　$\overline{PG} \cong \overline{PH}$

　　　　　Prove: ∠G ≅ ∠H

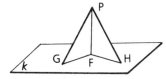

Proof

1 $\overline{PF} \perp k$	1 Given
2 $\overline{PF} \perp \overline{FG}$, $\overline{PF} \perp \overline{FH}$	2 If a line is ⊥ to a plane, it is ⊥ to every line in the plane that passes through its foot.
3 ∠PFG is a right ∠. ∠PFH is a right ∠.	3 ⊥ lines form right ∠s.
4 $\overline{PG} \cong \overline{PH}$	4 Given
5 $\overline{PF} \cong \overline{PF}$	5 Reflexive Property
6 △PFG ≅ △PFH	6 HL (3, 4, 5)
7 ∠G ≅ ∠H	7 CPCTC

Cooperative Learning

Have students work in small groups and compare their answers to problem **1**, making a list of all the right angles. Then, without formally proving the congruences, have them list sets of triangles that appear to be congruent. Ask what conjecture they might make about $\overline{PA}$, $\overline{PB}$, $\overline{PC}$, and $\overline{PD}$.
Answers will vary. Students should conclude that $\overline{PA}$, $\overline{PB}$, $\overline{PC}$, and $\overline{PD}$ appear to be congruent. Some might generalize that if points are equidistant from the foot of a line perpendicular to the plane in which the points lie, then those points are also equidistant from a point on that line.

Problem-Set Notes
and Additional Answers

■ See *Solution Manual* for answers to problems **2** and **3**.

Problem 3   Given: B, C, D, and E lie in plane n.
$\overline{AB} \perp n$,
$\overleftrightarrow{BE} \perp$ bis. $\overline{CD}$

Prove: △ADC *is isosceles.*

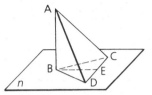

Proof

1 $\overline{AB} \perp n$	1 Given
2 $\overline{AB} \perp \overline{BD}$, $\overline{AB} \perp \overline{BC}$	2 If a line is $\perp$ to a plane, it is $\perp$ to every line in the plane that passes through its foot.
3 $\angle ABC$ is a right $\angle$. $\angle ABD$ is a right $\angle$.	3 $\perp$ lines form right $\angle$s.
4 $\angle ABC \cong \angle ABD$	4 All right $\angle$s are $\cong$.
5 $\overleftrightarrow{BE} \perp$ bis. $\overline{CD}$	5 Given
6 $\overline{BC} \cong \overline{BD}$	6 If a point is on the $\perp$ bis. of a segment, it is equidistant from the segment's endpoints.
7 $\overline{AB} \cong \overline{AB}$	7 Reflexive Property
8 △ABC $\cong$ △ABD	8 SAS (6, 4, 7)
9 $\overline{AD} \cong \overline{AC}$	9 CPCTC
10 △ADC is isosceles.	10 A △ with two $\cong$ sides is isosceles.

Part Three: Problem Sets

Problem Set A

1 If ABCD is a square that lies in plane t and $\overleftrightarrow{PF} \perp t$, how many right angles can be found in the figure? 12

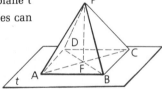

2 Given: $\overleftrightarrow{PB} \perp m$,
$\angle APB \cong \angle CPB$

Prove: $\overline{AB} \cong \overline{CB}$

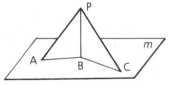

3 Given: ⊙O lies in plane n.
$\overline{RO} \perp n$

Prove: $\overline{RS} \cong \overline{RT}$

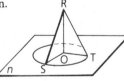

4 Given: $\overleftrightarrow{TS} \perp m$;
   $\overline{PV}$ bisects $\overline{TS}$.
   Prove: $\overrightarrow{PV}$ bisects $\angle TPS$.

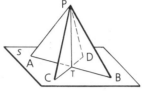

*Problem-Set Notes and
Additional Answers, continued*

■ See *Solution Manual* for answers to problems **4, 5, 7, 9,** and **10.**

5 Given: $\overleftrightarrow{AB}$ and $\overleftrightarrow{CD}$ lie in plane s.
   $\overleftrightarrow{PT} \perp s$,
   $\overline{PC} \cong \overline{PD}$,
   $\overline{PA} \cong \overline{PB}$

   Prove: T is the midpt. of $\overline{AB}$ and $\overline{CD}$.

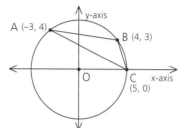

6 A *chord* of a circle is a segment joining two points on the circle. In the figure shown, $\overline{AB}$ and $\overline{AC}$ are chords of $\odot O$.
   a Find the slope of $\overline{AB}$. $-\frac{1}{7}$
   b Find the slope of $\overline{AC}$. $-\frac{1}{2}$

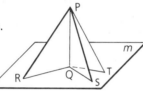

Problem Set B

7 Given: Q, R, S, and T lie in plane m.
   $\angle PQR$ and $\angle PQT$ are right $\angle$s.
   Prove: $\angle PQS$ is a right $\angle$.

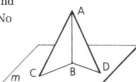

8 If $\overline{AB} \perp \overline{BD}$, $m\angle ABD = \frac{2}{3}x + 56$, and $m\angle ABC = 2x - 10$, is $\overline{AB} \perp m$? No

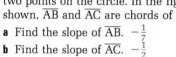

9 Given: $\overline{PA} \perp s$;
   P is equidistant from B and C.
   Prove: A is equidistant from B and C.

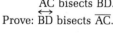

10 Given: $\overline{AB} \perp n$,
   $\overline{CD} \perp n$;
   $\overleftrightarrow{AC}$ bisects $\overline{BD}$.
   Prove: $\overleftrightarrow{BD}$ bisects $\overline{AC}$.

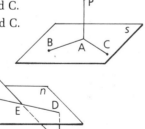

Problem-Set Notes and
Additional Answers, continued

■ See *Solution Manual* for answers to problems **11–13**.

14 Since ∠FCD ≅ ∠FDC, $\overline{CF}$ ≅ $\overline{DF}$ by sides opp. ≅ ∠s are ≅. Thus, △EFD ≅ △EFC by SSS, so ∠EFD ≅ ∠EFC by CPCTC. $\overline{EF}$ ⊥ $\overline{CF}$, so ∠EFC must be a rt. ∠, and thus, ∠EFD must also be a rt. ∠. This means $\overline{EF}$ ⊥ $\overline{FD}$. Thus, $\overline{EF}$ ⊥ m because if a line is ⊥ to two distinct lines that lie in a plane and that pass through its foot, then it is ⊥ to the plane.

15 ABCD can be shown to be a parallelogram and then a rectangle. Then $\overline{AD}$ ≅ $\overline{BC}$ because the diagonals of a rectangle are ≅.

17 Use the diagram for problem **3**. △ROS ≅ △ROT by SAS. Then $\overline{RS}$ ≅ $\overline{RT}$ by CPCTC.

Problem Set B, *continued*

11 Given: $\overline{AB}$ ⊥ m;
equilateral △ DBC lies in plane m.
Prove: △ACD is isosceles.

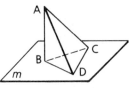

12 Given: $\overleftrightarrow{PB}$ ⊥ m;
D is the midpt. of $\overline{AC}$.
△PAC is isosceles, with base $\overline{AC}$.
Prove: $\overleftrightarrow{BD}$ ⊥ bis. $\overline{AC}$

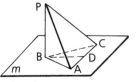

13 From any point on a line perpendicular to a plane, two lines are drawn oblique to the plane. If the foot of the perpendicular is equidistant from the feet of the oblique lines, prove that the oblique segments are congruent.

Problem Set C

14 Given: $\overline{EF}$ ⊥ $\overline{CF}$,
$\overline{CE}$ ≅ $\overline{DE}$,
∠FCD ≅ ∠FDC
Prove: $\overline{EF}$ ⊥ m

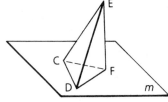

15 Given: $\overleftrightarrow{AD}$ and $\overleftrightarrow{BC}$ intersect at E.
$\overleftrightarrow{AC}$ ⊥ m, $\overleftrightarrow{AC}$ ⊥ n,
$\overleftrightarrow{BD}$ ⊥ m, $\overleftrightarrow{BD}$ ⊥ n
Prove: $\overline{AD}$ ≅ $\overline{BC}$

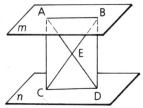

16 Given: A, B, C, and D lie in m.
$\overleftrightarrow{ED}$ ⊥ $\overleftrightarrow{BC}$,
$\overleftrightarrow{AD}$ ⊥ bis. $\overline{BC}$

a Which segment is ⊥ to which plane?

b How many planes are determined in this figure? 5

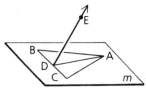

$\overline{BC}$ ⊥ plane ADE

17 Prove that if a line is perpendicular to the plane of a circle and passes through the circle's center, any point on the line is equidistant from any two points of the circle.

18 Given: ABCD is an isosceles trapezoid in
plane t.
$\overline{BC} \parallel \overline{AD}$,
$\overline{PF} \perp t$;
$\overline{PF}$ bisects $\overline{AD}$.

Prove: $\triangle PAB \cong \triangle PDC$

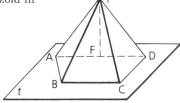
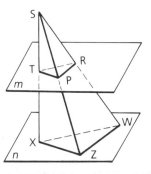

19 Given: $\overleftrightarrow{SX} \perp m$,
$\overleftrightarrow{SX} \perp n$,
$\overline{TP} \cong \overline{TR}$

Prove: $\triangle SZW$ is isosceles.

*Problem-Set Notes and
Additional Answers, continued*

18 $\triangle PFA \cong \triangle PFD$ by SAS, so
$\overline{PA} \cong \overline{PD}$ by CPCTC. Since
ABCD is an isos. trap. with
$\overline{AB} \cong \overline{CD}$, $\angle BAD \cong CDA$.
Thus, $\triangle AFB \cong \triangle DFC$ by
SAS, and $\overline{PB} \cong \overline{PC}$ by
CPCTC. Thus, $\triangle PAB \cong$
$\triangle PDC$ by SSS.

19 $\triangle STP \cong \triangle STR$ by SAS.
Then $\angle TSP \cong \angle TSR$ by
CPCTC. $\triangle SXZ \cong \triangle SXW$ by
ASA. Then $\overline{SZ} \cong \overline{SW}$ by
CPCTC.

HISTORICAL SNAPSHOT

PROBABILITY AND PI
The ubiquity of a geometric constant

Georges-Louis Leclerc, comte de Buffon
(1707–1788), one of the most celebrated
naturalists of all time and a pioneer in
such fields as ecology and paleontology,
was also extremely interested in mathematics. Besides translating Isaac
Newton's work on the calculus into
French, he was among the first to deal
with probability in a geometrical fashion.

Imagine a tabletop ruled with equally spaced
parallel lines. If you toss a needle at random
onto the table, what is the probability that it
will land across one of the ruled lines? Buffon
found that if the length of the needle is less
than or equal to the distance between the lines,
this probability can be expressed as $\frac{2\ell}{\pi d}$, where ℓ
is the needle's length and *d* is the distance
between the lines.

It is somewhat surprising that the answer to
a probability problem that does not involve circles should involve π, the ratio between a cir-

cle's circumference and its diameter.
But π has a habit of popping up in
unlikely places, even ones entirely
unconnected with geometry, as in
these amazing infinite sums:

$$\frac{1}{1} - \frac{1}{3} + \frac{1}{5} - \frac{1}{7} + \frac{1}{9} - \frac{1}{11} + \dots = \frac{\pi}{4}$$

$$\frac{1}{1^2} + \frac{1}{2^2} + \frac{1}{3^2} + \frac{1}{4^2} + \frac{1}{5^2} + \dots = \frac{\pi^2}{6}$$

BASIC FACTS ABOUT PARALLEL PLANES

Class Planning

Time Schedule
Basic: 2 days
Average: 2 days
Advanced: 1 day

Resource References
Teacher's Resource Book
 Class Opener 6.3A, 6.3B
 Additional Practice Worksheet 12
Evaluation
 Tests and Quizzes
 Quiz 1, Series 3

Class Opener

Given: $\overleftrightarrow{AB} \perp \overleftrightarrow{BC}$,
 $\overleftrightarrow{AB} \perp \overleftrightarrow{BE}$,
 $\overline{BC} \cong \overline{BD}$
Prove: $\angle CAB \cong \angle DAB$

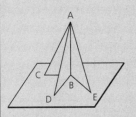

Statements	Reasons
1 $\overleftrightarrow{AB} \perp \overleftrightarrow{BC}$	1 Given
2 $\overleftrightarrow{AB} \perp \overleftrightarrow{BE}$	2 Given
3 $\overleftrightarrow{AB} \perp$ plane BCDE	3 If a line is $\perp$ to two lines in a plane, it is $\perp$ to the plane.
4 $\angle ABC \cong \angle ABD$	4 $\perp$ lines determine $\cong \angle$s.
5 $\overline{BC} \cong \overline{BD}$	5 Given
6 $\overline{AB} \cong \overline{AB}$	6 Reflexive Prop.
7 $\triangle ABC \cong \triangle ABD$	7 SAS
8 $\angle CAB \cong \angle DAB$	8 CPCTC

Lesson Notes

- Basic and average courses may omit this section, since mastery of it is not essential for subsequent chapters.

Objectives
After studying this section, you will be able to
- Recognize lines parallel to planes, parallel planes, and skew lines
- Use properties relating parallel lines and planes

Part One: Introduction

Lines Parallel to Planes, Parallel Planes, Skew Lines
Since we examined the concept of parallel lines in Chapter 4, it seems logical now to investigate the possibilities of a line being parallel to a plane and of two planes being parallel to each other.

Definition A line and a plane are parallel if they do not intersect.

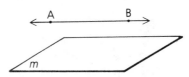

Definition Two planes are parallel if they do not intersect.

The diagram at the right shows two lines located in two parallel planes. Although the planes are parallel, the lines are not parallel, because A, B, C, and D do not determine a plane. Such lines are said to be *skew.*

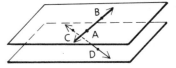

Vocabulary
skew

Definition   Two lines are **skew** if they are not coplanar.

You will see that parallelism in space is very similar to parallelism in a plane. There are, however, a few notable differences. For example, there are no skew planes. Planes are either intersecting or parallel.

The following theorem is basic to the understanding of parallelism in space.

Theorem 49   *If a plane intersects two parallel planes, the lines of intersection are parallel.*

Given: m ∥ n;
    s intersects
    m and n in lines
    $\overleftrightarrow{AB}$ and $\overleftrightarrow{CD}$.
Prove: $\overleftrightarrow{AB}$ ∥ $\overleftrightarrow{CD}$

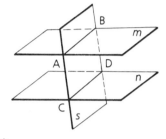

Proof: We know that $\overleftrightarrow{AB}$ and $\overleftrightarrow{CD}$ are coplanar, since they both lie in plane s. Also, they cannot intersect each other, because one lies in plane m and the other lies in plane n—two planes that, being parallel, have no intersection. Thus, $\overleftrightarrow{AB}$ ∥ $\overleftrightarrow{CD}$ by the definition of *parallel lines.*

Properties Relating Parallel Lines and Planes

There are numerous properties relating lines and planes in space, many of which are similar to the theorems about parallel lines you have already seen. We will present some of these properties without their proofs.

Parallelism of Lines and Planes

1 If two planes are perpendicular to the same line, they are parallel to each other.
2 If a line is perpendicular to one of two parallel planes, it is perpendicular to the other plane as well.
3 If two planes are parallel to the same plane, they are parallel to each other.
4 If two lines are perpendicular to the same plane, they are parallel to each other.
5 If a plane is perpendicular to one of two parallel lines, it is perpendicular to the other line as well.

Communicating Mathematics

Have students write a lesson plan that explains the difference between parallel and skew lines. Encourage them to use physical models. Then choose one or two students to present their lesson to the class.

Lesson Notes, continued

■ If you do cover this section in basic or average courses, you will need to emphasize two points:
  1 *When* Theorem 49 can be applied. The wording is simple, but it is not always clear when the theorem can be applied.
  2 *How* Theorem 49 can be applied. The theorem is applicable only when the proof using it has steps that establish the existence of the plane containing the two lines.

Students may need several examples to understand when and how to apply the theorem.

Cooperative Learning

Have students work in small groups to make a physical model of each of the properties relating parallel lines and planes on page 283. Each student should be able to model the property for the class. Then have students find physical examples of each property.
Answers will vary. For property 2, students might observe that the line formed by the intersection of two walls is perpendicular to both the floor and the ceiling.

Part Two: Sample Problem

Problem

Given: m ∥ n;
$\overleftrightarrow{AB}$ lies in m.
$\overleftrightarrow{CD}$ lies in n.
$\overleftrightarrow{AC} \parallel \overleftrightarrow{BD}$

Prove: $\overline{AD}$ bisects $\overline{BC}$.

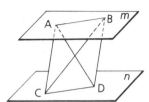

Proof

1 m ∥ n	1 Given
2 $\overleftrightarrow{AB}$ lies in m. $\overleftrightarrow{CD}$ lies in n.	2 Given
3 $\overleftrightarrow{AC} \parallel \overleftrightarrow{BD}$	3 Given
4 $\overleftrightarrow{AC}$ and $\overleftrightarrow{BD}$ determine a plane, ACDB.	4 Two ∥ lines determine a plane.
5 $\overleftrightarrow{AB} \parallel \overleftrightarrow{CD}$	5 If a plane intersects two ∥ planes, the lines of intersection are ∥.
6 ACDB is a ▱.	6 If both pairs of opposite sides of a quadrilateral are ∥, it is a ▱.
7 $\overline{AD}$ bisects $\overline{BC}$.	7 The diagonals of a ▱ bisect each other.

Note Before making statement 6, we had to show that ABDC is a plane figure. See Section 6.1, problem 12, for an example of a four-sided figure that is not a plane figure.

Part Three: Problem Sets

Problem Set A

1 Indicate whether each statement is True (T) or False (F).

  a If a plane contains one of two skew lines, it contains the other. F

  b If a line and a plane never meet, they are parallel. T

  c If two parallel lines lie in different planes, the planes are parallel. F

  d If a line is perpendicular to two planes, the planes are parallel. T

  e If a plane and a line not in the plane are each perpendicular to the same line, then they are parallel to each other. T

2 By substituting 3 for x and 4 for y, verify that point D is on the circle that is the graph of the equation $x^2 + y^2 = 25$.
$3^2 + 4^2 = 9 + 16 = 25$

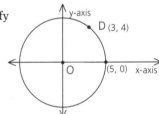

Assignment Guide

Basic

Day 1   1, 3, 4
Day 2   2, 5, 7

Average

Day 1   1–5
Day 2   6–9

Advanced

2–5, 9

3 Given: r ∥ s,
       s ∥ t,
       $\overleftrightarrow{AE}$ ∥ $\overleftrightarrow{BF}$

Prove: **a** r ∥ t
       b ABFE is a plane figure.
       c $\overleftrightarrow{AB}$ ∥ $\overleftrightarrow{EF}$
       d $\overline{AB}$ ≅ $\overline{EF}$

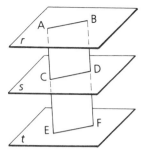

4 Given: $\overline{GJ}$ and $\overline{KH}$ bisect each other at P.

  a Is GHJK a plane figure? Yes

  b Are $\overline{GH}$ and $\overline{KJ}$ parallel? Yes

  c Are $\overline{GH}$ and $\overline{KJ}$ congruent? Yes

  d Are plane e and plane f parallel? Not necessarily

  e What is the most descriptive name for
     GHJK? Parallelogram

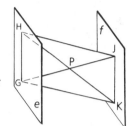

5 In the figure shown, find the slope of
chord $\overline{EF}$. Then find the slope of chord
$\overline{FG}$. What type of triangle is △EFG?
Why? $\frac{1}{2}$; −2; right; the slopes are
opposite reciprocals, so $\overline{EF}$ ⊥ $\overline{FG}$.

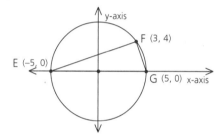

Problem Set B

6 Given: m ∥ n,
       $\overleftrightarrow{AB}$ ∥ $\overleftrightarrow{CD}$
Prove: $\overline{AB}$ ≅ $\overline{CD}$

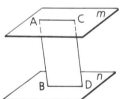

7 Given: e ∥ f,
       $\overline{RT}$ ∩ $\overline{VS}$ = P,
       $\overline{RS}$ ≅ $\overline{VT}$
Prove: $\overline{RV}$ ≅ $\overline{ST}$

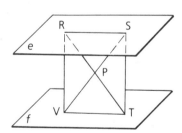

**Problem-Set Notes
and Additional Answers**

3 Once a plane has been
established, assume that
points lie in that plane.
Planes r, s, and t are given, so
assume that A and B lie in
plane r, C and D in plane s,
and E and F in plane t. After
establishing that A, B, F, and
E determine a plane, assume
that C and D lie in that
plane.
Students may want to add
this to their own list of
assumptions from diagrams,
page 19.

■ See *Solution Manual* for answers to problems **6** and **7**.

*Problem-Set Notes and
Additional Answers, continued*

■ See *Solution Manual* for answers to problems **8–11.**

Problem Set B, *continued*

8 If a slide projector is set up so that the slide is parallel to the
screen,

a Prove that a segment on the slide is parallel to its image on the
screen

b Prove that the angles marked 1 and 2 in the diagram are
congruent

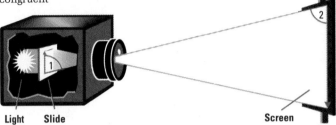

Light   Slide   Screen

9 Given: f ∥ g;
RTW is an isosceles △,
with base $\overline{TW}$.

Prove: △RSV is isosceles.

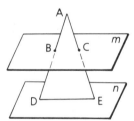

Problem Set C

10 Given: m ∥ n,
$\overline{BD} \cong \overline{CE}$

Prove: △ADE is isosceles.

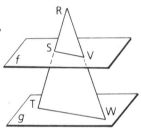

11 Given: p ∥ q,
$\overleftrightarrow{AD} \parallel \overleftrightarrow{BE}, \overleftrightarrow{CF} \parallel \overleftrightarrow{BE}$

Prove: ∠BAC ≅ ∠EDF

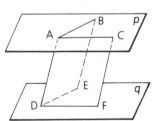

CHAPTER SUMMARY

CONCEPTS AND PROCEDURES

After studying this chapter, you should be able to
- Understand basic concepts relating to planes (6.1)
- Identify four methods of determining a plane (6.1)
- Apply two postulates concerning lines and planes (6.1)
- Recognize when a line is perpendicular to a plane (6.2)
- Apply the basic theorem concerning the perpendicularity of a line and a plane (6.2)
- Recognize lines parallel to planes, parallel planes, and skew lines (6.3)
- Use properties relating parallel lines and planes (6.3)

VOCABULARY

foot (6.1)
skew (6.3)

REVIEW PROBLEMS

Chapter 6 Review
Class Planning

Time Schedule
All levels: 1 day

Resource References
Evaluation
    Tests and Quizzes
        Test 6, Series 1, 2, 3

Assignment Guide

Basic
Chapter Review   1–3
Cumulative Review   1–4

Average
1–4, 6, 7

Advanced
Section **6.2**   17, 18
Chapter Review   7, 10, 11

Problem-Set Notes and Additional Answers

■ See *Solution Manual* for answers to problems **2** and **3**.

Problem Set A

1 Indicate whether each statement is True or False. Be prepared to defend your choice.

a Two lines must either intersect or be parallel. F

b In a plane, two lines perpendicular to the same line are parallel. T

c In space, two lines perpendicular to the same line are parallel. F

d If a line is perpendicular to a plane, it is perpendicular to every line in the plane. F

e It is possible for two planes to intersect at one point. F

f If a line is perpendicular to a line in a plane, it is perpendicular to the plane. F

g Two lines perpendicular to the same line are parallel. F

h A triangle is a plane figure. T

i A line that is perpendicular to a horizontal line is vertical. F

j Three parallel lines must be coplanar. F

k Every four-sided figure is a plane figure. F

2 Given: $\overline{PB} \perp m$,
$\overline{PA} \cong \overline{PC}$

Prove: △ABC is isosceles.

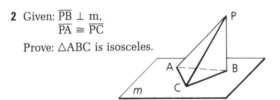

3 Given: $\overline{AB} \cong \overline{AC}$,
∠DAB ≅ ∠DAC

Prove: $\overline{DB} \cong \overline{DC}$

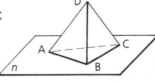

Problem Set B

4 How many planes are determined by a set of four noncoplanar points if no three of the points are collinear? 4

5 From the top of a flagpole 48 ft in height, two 60-ft ropes reach two points on the ground, each of which is 36 ft from the pole. If the ground is level, is the pole perpendicular to the ground? **Not necessarily**

6 a At a given point on a line, how many lines can be drawn perpendicular to the given line? **Infinitely many**

  b At a given point on a plane, how many lines can be drawn perpendicular to the plane? **1**

7 Given: $\angle ADC = (x + 88)°,$
  $\angle ADE = (74 - 8x)°,$
  $\angle BDE = (2x + 94)°$

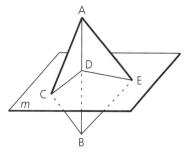

  a Are $\overline{AD}$ and m perpendicular? **No**

  b Are $\overline{AD}$ and $\overline{CD}$ perpendicular? **No**

  c Are $\overline{AD}$ and $\overline{DE}$ perpendicular? **Yes**

8 ⊙P lies in plane m. If A and B are points on ⊙P and if $\overleftrightarrow{QP} \perp$ m, which of the following must be true? **a and b**

  a $\angle APQ \cong \angle BPQ$

  b $\overline{AP} \cong \overline{PB}$

  c $\overline{QP} \perp \overline{AB}$

9 $\overleftrightarrow{AB}$ is parallel to plane m and perpendicular to plane r. $\overleftrightarrow{CD}$ lies in r. Which of the following must be true? **a**

  a $r \perp m$ **c** $\overleftrightarrow{CD} \perp$ m   **e** $\overleftrightarrow{AB}$ and $\overleftrightarrow{CD}$ are skew.

  b $r \parallel m$ **d** $\overleftrightarrow{AB} \parallel \overleftrightarrow{CD}$

10 Given: △BDC is isosceles, with $\overline{BD} \cong \overline{CD}$.
  $\angle ADB \cong \angle ADC$

  Prove: △BAC is isosceles.

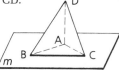

11 Given: $\overline{BP} \perp \overline{PQ}$,
  $\overline{AP} \perp \overline{PQ}$;
  A and B are equidistant from P.

  Prove: $\angle ABQ \cong \angle BAQ$

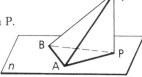

12 A line is drawn perpendicular to the plane of a square at the point of intersection of the square's diagonals. Prove that any point on the perpendicular is equidistant from the vertices of the square.

Problem-Set Notes and Additional Answers, continued
■ See *Solution Manual* for answers to problems **10–12**.

13 ∠PAB ≅ ∠PBA, so that
PA ≅ PB by sides opp. ≅ ∠s
are ≅. Then △PQA ≅ △PQB
by HL, and ∠APR ≅ ∠BPR
by CPCTC. △APR ≅ △BPR
by SAS. ∴ ∠PAR ≅ ∠PBR
by CPCTC.

14 △PAB ≅ △PCB by SSS.
Then ∠BPA ≅ ∠BPC by
CPCTC. PT ≅ PS by
division, and thus, △RPT ≅
△RPS by SAS. ∴ RT ≅ RS
by CPCTC.

15 △PCB ≅ △QCB by SAS, so
BP ≅ BQ by CPCTC.
Combining this with AP ≅
AQ gives ⟷BA is the ⊥ bisector
of ⟷PQ by Theorem 24.

16 By the Transitive Property,
m ∥ p. ABDC is a plane
because two intersecting
lines determine a plane.
Then AB ∥ CD by Theorem
49. Then ∠3 ≅ ∠4 by ∥
lines ⇒ alt. int. ∠s ≅. BE ≅
CE and ∠1 ≅ ∠2. Then
△ABE ≅ △DCE by ASA,
and AE ≅ ED by CPCTC, so
BC bisects AD.

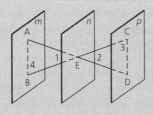

Problem Set C

13 Given: ⟷PR ⊥ m,
∠PAB ≅ ∠PBA
Prove: ∠PAR ≅ ∠PBR

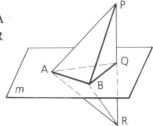

14 Given: △ABC lies in n.
PA ≅ PC,
AB ≅ BC;
T and S are midpoints.
Prove: RT ≅ RS

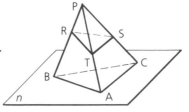

15 Given: PC ≅ QC;
A is the midpoint of PQ.
∠PCB ≅ ∠QCB
Prove: ⟷BA ⊥ ⟷PQ

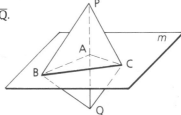

16 Given: m ∥ n,
p ∥ n;
AD bisects BC.
Prove: BC bisects AD.

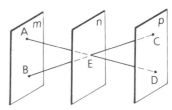

CUMULATIVE REVIEW
CHAPTERS 1–6

Problem Set A

1 Write the most descriptive name for each figure.

a A four-sided figure in which the diagonals are perpendicular bisectors of each other Rhombus

b A four-sided figure in which the diagonals bisect each other Parallelogram

c A triangle in which there is a hypotenuse Right triangle

d A four-sided figure in which the diagonals are congruent and all sides are congruent Square

2 Find the angle formed by the hands of a clock at 9:30. 105°

3 If one of two supplementary angles is 16° smaller than three times the other, find the measure of the larger. 131

4 Given: ∠OMP ≅ ∠OPM,
     ∠PMR ≅ ∠MPR
Prove: $\overleftrightarrow{OR}$ ⊥ bis. $\overline{PM}$

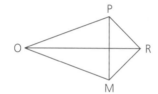

5 Given: TVAX is a rectangle.
Conclusion: ∠TXV ≅ ∠VAT

6 Two consecutive angles of a parallelogram are in a ratio of 7 to 5. Find the measure of the larger. 105

7 Given: NPRS is a ▱, with diagonals $\overline{SP}$
     and $\overline{NR}$ intersecting at O.
     $\overline{TO}$ is perpendicular to the plane
     of ▱NPRS.
Prove: △STP is isosceles.

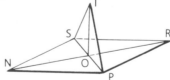

Cumulative Review
Chapters 1–6

Time Schedule
All levels: 1 day

Resource References
Evaluation
   Tests and Quizzes
      Cumulative Review Test
      Chapters 1–6
      Series 1, 2, 3

Assignment Guide

Basic
5–7, 9–13
Average
2, 4, 6, 7, 10, 14, 15, 17, 18
Advanced

Chapter **6** Review   8, 9, 15
Cumulative Review   14–16,
                     18, 19

Problem-Set Notes
and Additional Answers

■ See *Solution Manual* for answers to problems **4, 5,** and **7.**

Cumulative Review Problem Set A, *continued*

8 Find m∠1. 110

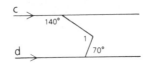

9 Given: ⊙P;
   M is the midpoint of $\overline{AB}$.
   Prove: $\overleftrightarrow{PQ} \perp \overleftrightarrow{AB}$

10 Indicate whether each statement is true Always, Sometimes, or Never (A, S, or N).

   a If a triangle is obtuse, it is isosceles. S

   b The bisector of the vertex angle of a scalene triangle is perpendicular to the base. N

   c If one of the diagonals of a quadrilateral is the perpendicular bisector of the other, the quadrilateral is a kite. A

   d If A, B, C, and D are noncoplanar, $\overline{AB} \perp \overline{BC}$, and $\overline{AB} \perp \overline{BD}$, then $\overline{AB}$ is perpendicular to the plane determined by B, C, and D. A

   e Two parallel lines determine a plane. A

   f Planes that contain two skew lines are parallel. S

   g Supplements of complementary angles are congruent. S

11 Given: FGHJ is a ▱.
   FG = x + 5, GH = 2x + 3,
   ∠G = 40°, ∠J = (4x + 12)°

   Find: **a** m∠F 140
      b The perimeter of FGHJ 58

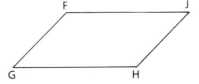

12 In the figure shown, find the slope of chord $\overline{AC}$. Then find the slope of chord $\overline{AT}$. What type of triangle is △CAT? Why?
   $\frac{1}{3}$, −3; right; the slopes are opposite reciprocals, so $\overline{AC} \perp \overline{AT}$.

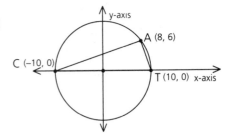

Problem Set B

13 Given: $\overline{AB} \cong \overline{AC}$,
$\overline{BD} \cong \overline{DC}$

Conclusion: $\angle B \cong \angle C$

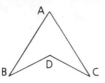
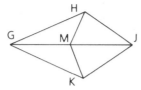

14 Given: $\overline{GH} \cong \overline{GK}$,
$\overline{HM} \cong \overline{KM}$

Conclusion: HMKJ is a kite.

15 Given: ABCD is a ▱.
$\angle A = (3x + y)°$,
$\angle D = (5x + 10)°$,
$\angle C = (5y + 20)°$

Find: m∠B 110

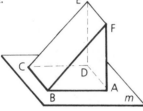

16 Given: A, B, C, and D lie in m.
FBCE is a ▱.
$\overline{FE} \parallel \overline{AD}$,
$\overline{AD} \cong \overline{BC}$

Prove: ABCD is a ▱.

17 Prove: If segments drawn from the midpoint of one side of a triangle perpendicular to the other two sides are congruent, then the triangle is isosceles.

18 The measure of the supplement of an angle exceeds three times the measure of the complement of the angle by 12. Find the measure of half of the supplement. $64\frac{1}{2}$

Problem Set C

19 Given: $\overline{AC} \cong \overline{BD}$,
$\overline{AB} \cong \overline{CD}$

Prove: $\angle B \cong \angle C$

20 Given: ⊙A lies in m.
$\overline{PA} \perp m$,
$\overline{PD} \cong \overline{PE}$

Prove: $\overline{BE} \cong \overline{CD}$

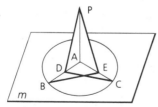

Cumulative Review Problems **293**

Chapter 7 Schedule

Basic

Problem Sets:	8 days
Review:	2 days
Test:	1 day

Average

Problem Sets:	8 days
Review:	1 day
Test:	1 day

Advanced

Problem Sets:	6 days
Review:	1 day
Test:	1 day

*C*HAPTERS **7–13** EMPHASIZE the application of geometric concepts to solve problems and the use of algebraic language to describe geometric relationships.

The Geometric Supposer

Use *The Geometric Supposer: Triangles* to explore angle relationships in a triangle before beginning work on this chapter.

Have students use *The Geometric Supposer* to draw three or four triangles and measure the interior angles of each. Do they notice a pattern in the data? Ask whether they can make a conjecture. (The sum of the angles of any triangle is 180°.)

Remind students that one feature of the program is that they can calculate using anything they measure. Guide students to have the program add the angles as they are measured.

Go a step farther by asking students what the relationship is between the angle formed by extending one of the sides of the triangle and the three interior angles of the triangle.

CHAPTER

7 POLYGONS

Polygons of various shapes and sizes are interwoven to create this intricate ceiling design of inlaid wood in the Hall of the Blessing, the Alhambra, Granada, Spain.

7.1 TRIANGLE APPLICATION THEOREMS

Class Planning

Time Schedule
Basic: 2 days
Average: 2 days
Advanced: $1\frac{1}{2}$ days

Resource References
Teacher's Resource Book
  Class Opener 7.1A, 7.1B
  Using Manipulatives 11
  Supposer Worksheet 7

Objective

After studying this section, you will be able to
- Apply theorems about the interior angles, the exterior angles, and the midlines of triangles.

Part One: Introduction

In elementary school you probably learned that the sum of the measures of the angles of a triangle is 180°. This property of triangles has a number of important applications.

Theorem 50 *The sum of the measures of the three angles of a triangle is 180.*

Given: △ABC
Prove: $m\angle A + m\angle B + m\angle C = 180$

Proof: According to the Parallel Postulate, there exists exactly one line through point A parallel to $\overleftrightarrow{BC}$, so the figure at the right can be drawn.

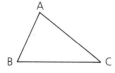

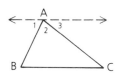

Because of the straight angle, we know that
$\angle 1 + \angle 2 + \angle 3 = 180°$. Since $\angle 1 \cong \angle B$ and $\angle 3 \cong \angle C$ (by ∥ lines ⇒ alt. int. ∠s ≅), we may substitute to obtain $\angle B + \angle 2 + \angle C = 180°$. Hence, $m\angle A + m\angle B + m\angle C = 180$.

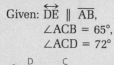
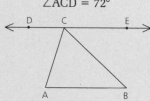
Vocabulary
exterior angle

Before proving the next theorem, we need to explain what an *exterior angle* of a polygon is. In each of the figures below, ∠1 is an exterior angle of a polygon.

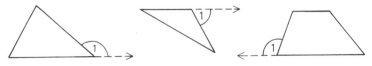

You can see that an exterior angle of a polygon is formed by extending one of the sides of the polygon. The following definition puts this idea in a form that is much more useful in proofs and problems.

Definition   An *exterior angle* of a polygon is an angle that is adjacent to and supplementary to an interior angle of the polygon.

The next theorem applies only to triangles.

Theorem 51   *The measure of an exterior angle of a triangle is equal to the sum of the measures of the remote interior angles.*

Given: △DEF, with exterior angle 1 at F
Prove: m∠1 = m∠D + m∠E

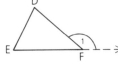

Do you see how Theorem 50 and the definition of exterior angle are the keys to a proof of Theorem 51?
    The following theorem could have been presented in the chapter on parallelograms, but you may find it more useful now.

Theorem 52   *A segment joining the midpoints of two sides of a triangle is parallel to the third side, and its length is one-half the length of the third side. (Midline Theorem)*

Given: H is a midpoint.
    M is a midpoint.
Prove: a $\overline{HM} \parallel \overline{JK}$
    b HM = $\frac{1}{2}$(JK)

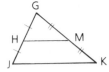

Proof: Extend $\overleftrightarrow{HM}$ through M to a point P so that $\overline{MP} \cong \overline{HM}$. P is now established, so P and K determine $\overleftrightarrow{PK}$.

We know that $\overline{GM} \cong \overline{KM}$ (by the definition of midpoint) and that $\angle GMH \cong \angle KMP$ (vertical $\angle$s are $\cong$). Thus, $\triangle GMH \cong \triangle KMP$ by SAS.

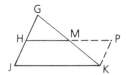

Since $\angle G = \angle PKM$ by CPCTC, $\overleftrightarrow{PK} \parallel \overleftrightarrow{HJ}$ by alt. int. $\angle$s $\Rightarrow \parallel$ lines. Also, $\overline{GH} \cong \overline{PK}$ by CPCTC, and $\overline{GH} \cong \overline{HJ}$ (by the definition of midpoint). By transitivity, then, $\overline{PK} \cong \overline{HJ}$.

Two sides, $\overline{PK}$ and $\overline{HJ}$, are parallel and congruent, so PKJH is a parallelogram. Therefore, $\overleftrightarrow{HP} \parallel \overleftrightarrow{JK}$.

Opposite sides of a parallelogram are congruent, so HP = JK.

Also, since we made MP = HM, HM = $\frac{1}{2}$(HP) and, by substitution, HM = $\frac{1}{2}$(JK).

Part Two: Sample Problems

Problem 1
Given: Diagram as marked
Find: x, y, and z

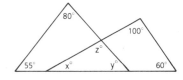

Solution
Since the sum of the measures of the angles of a triangle is 180,

x + 100 + 60 = 180	55 + 80 + y = 180	x + y + z = 180
x + 160 = 180	135 + y = 180	20 + 45 + z = 180
x = 20	y = 45	z = 115

Substitution

Substitution

Problem 2
The measures of the three angles of a triangle are in the ratio 3:4:5. Find the measure of the largest angle.

Solution
Let the measures of the three angles be 3x, 4x, and 5x. Since the sum of the measures of the three angles of a triangle is 180,

$3x + 4x + 5x = 180$
$12x = 180$
$x = 15$

Therefore, the measure of the largest angle is 5(15), or 75.

Lesson Notes

■ Reading and discussing Sample Problem **1** in class can help students shift focus among the three triangles.

Lesson Notes, *continued*

- Sample Problem **3** calls for students to keep the geometry problem in mind while they go through the algebra. Although they cannot solve for x or for y, they need only find the sum to solve the problem.

Problem 3 *If one of the angles of a triangle is 80°, find the measure of the angle formed by the bisectors of the other two angles.*

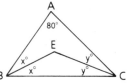

Solution The bisectors, $\overrightarrow{BE}$ and $\overrightarrow{CE}$, meet at E, so we want to find m∠E. Let ∠ABC = (2x)° and ∠ACB = (2y)°.

In △ABC,

$$2x + 2y + 80 = 180$$
$$2x + 2y = 100$$
$$x + y = 50$$

In △EBC,

$$x + y + m∠E = 180$$
$$50 + m∠E = 180 \quad \text{(Substitution)}$$
$$m∠E = 130$$

Problem 4 *∠1 = 150°, and the measure of ∠D is twice that of ∠E. Find the measure of each angle of the triangle.*

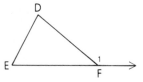

Solution Let ∠E = x° and ∠D = (2x)°.

Since the measure of an exterior angle of a triangle is equal to the sum of the measures of the remote interior angles,

$$150 = x + 2x$$
$$150 = 3x$$
$$50 = x$$

Hence, ∠E = 50°, ∠D = 100°, and ∠DFE = 30°.

Part Three: Problem Sets

Problem Set A

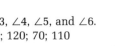

1 Given: Diagram as marked

Find: m∠B 70

2 Given: ∠1 = 130°,
        ∠7 = 70°

Find the measures of ∠2, ∠3, ∠4, ∠5, and ∠6.
        50; 60; 120; 70; 110

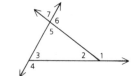

Assignment Guide

Basic

Day 1	1–6, 10
Day 2	7–9, 11–13, 15

Average

Day 1	1–7, 10
Day 2	8, 9, 11–13, 15, 16, 18

Advanced

Day 1	9, 12, 14–18
Day 2	($\frac{1}{2}$ day) Section **7.1** 19, 20, 23
	($\frac{1}{2}$ day) Section **7.2** 17, 18

3 Given: ∠CAB = 80°,
∠CBA = 60°,
$\overline{AE}$ and $\overline{BD}$ are altitudes.
Find: m∠C and m∠AFB **40; 140**

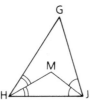

4 In the diagram as marked,
if m∠G = 50, find m∠M.
(Hint: See sample problem 3.) **115**

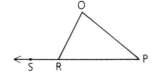

5 The measures of the three angles of a triangle are in the ratio
4:5:6. Find the measure of each. **48; 60; 72**

6 Given: ∠ORS = (4x + 6)°,
∠P = (x + 24)°,
∠O = (2x + 4)°
Find: m∠O **48**

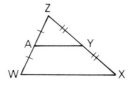

7 In the diagram as marked,
if WX = 18, find AY. **9**

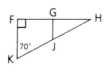

8 Given: Diagram as marked;
G and J are midpoints.
Find: m∠H, m∠HGJ, and m∠HJG
20; 90; 70

9 Tell whether each statement is true Always, Sometimes, or Never
(A, S, or N).

a The acute angles of a right triangle are complementary. **A**

b The supplement of one of the angles of a triangle is equal in
measure to the sum of the other two angles of the triangle. **A**

c A triangle contains two obtuse angles. **N**

d If one of the angles of an isosceles triangle is 60°, the triangle
is equilateral. **A**

e If the sides of one triangle are doubled to form another triangle, each angle of the second triangle is twice as large as the
corresponding angle of the first triangle. **N**

Cooperative Learning

Have small groups of students
compare answers to problem **9**.
Then have each group write a
clear, concise sentence to justify
why each statement is always,
sometimes, or never true. Encourage students to draw diagrams to support their answers.
Answers may vary. For example,
students might explain part **c** as
follows.
The measure of one obtuse angle
is greater than 90. The sum of
the measures of the angles of a
triangle is 180. If a triangle had
two obtuse angles, the sum of
their measures would be greater
than 180. Therefore, a triangle
never contains two obtuse angles.

Problem Set A, *continued*

10 The vertex angle of an isosceles triangle is twice as large as one of the base angles. Find the measure of the vertex angle. 90

11 Given: ∠P = 10°;
$\overrightarrow{RO}$ bisects ∠MRP.

Find: m∠ORP and m∠MOR 40; 50

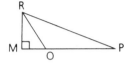

Problem Set B

12 In △DEF, the sum of the measures of ∠D and ∠E is 110. The sum of the measures of ∠E and ∠F is 150. Find the sum of the measures of ∠D and ∠F. 100

13 Prove, in paragraph form, that the acute angles of a right triangle are complementary.

14 Prove, in paragraph form, that if a right triangle is isosceles, it must be a 45°-45°-90° triangle.

15 The measures of two angles of a triangle are in the ratio 2:3. If the third angle is 4 degrees larger than the larger of the other two angles, find the measure of an exterior angle at the third vertex. 110

16 Given: ∠A = 30°, $\overline{AB} \cong \overline{AC}$;
$\overrightarrow{CD}$ bisects ∠ACB.
$\overrightarrow{BD}$ is one of the trisectors of ∠ABC.

Find: m∠D 117.5 or 92.5

17 Given: EFGH is a rectangle.
FH = 20;
J, K, M, and O are midpoints.

a Find the perimeter of JKMO. 40

b What is the most descriptive name for JKMO? Rhombus

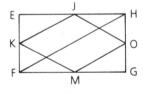

18 Given: ∠PST = (x + 3y)°,
∠P = 45°, ∠R = (2y)°,
∠PSR = (5x)°

Find: m∠PST 105

Problem-Set Notes and Additional Answers

■ See *Solution Manual* for answers to problems **13** and **14**.

■ In solving problem **16**, students might need to be cautioned that $\overrightarrow{BD}$ can be either trisector; two answers are possible.

19

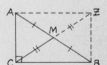

Extend $\overline{CM}$ to Z so that $\overline{CM} \cong \overline{MZ}$. Then ACBZ is a parallelogram, since the diagonals bisect each other, and a rectangle, because it has a right angle. Since the diagonals of a rectangle are ≅, $\overline{AM} \cong \overline{CM} \cong \overline{MB}$.

Problem Set C

19 Prove that the midpoint of the hypotenuse of a right triangle is equidistant from all three vertices. (Hint: See the method used to prove the Midline Theorem, page 296.)

20 Prove that if the midpoints of a quadrilateral are joined in order, the figure formed is a parallelogram.

21 Given: $\overline{AB} \cong \overline{AC}$,
$\overline{AE} \cong \overline{DE} \cong \overline{DB} \cong \overline{BC}$
Find: $m\angle A$ $25\frac{5}{7}$

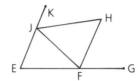

22 Given: $\angle E = 70°$;
$\overrightarrow{JH}$ and $\overrightarrow{FH}$ bisect the exterior angles of $\triangle JEF$ at J and F.

a Find $m\angle H$. 55

b Can you find a formula that expresses $m\angle H$ in terms of $m\angle E$?
$m\angle H = 90 - \frac{1}{2}(m\angle E)$

23 Show that $a + e_1 + c_1 = d + e_2 + b_2$.

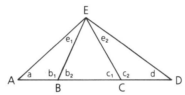

Problem-Set Notes and Additional Answers, continued

20

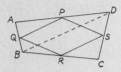

Draw $\overline{BD}$. Consider $\triangle CDB$. $\overline{RS}$ is a midline, and thus, $\overline{RS} \parallel \overline{BD}$ and $RS = \frac{1}{2}BD$. Similarly, consider $\triangle ABD$. $\overline{QP}$ is a midline, and thus, $\overline{QP} \parallel \overline{BD}$ and $QP = \frac{1}{2}BD$. Thus, $\overline{QP} \parallel \overline{RS}$, $QP = RS$, and PQRS is a parallelogram.

21 Let $m\angle A = x$. Then $m\angle ADE = x$ and $m\angle DEB = 2x$ because the measure of an exterior angle of a $\triangle$ is = to the sum of the measures of the remote interior angles. Then $m\angle DBE = 2x$, also. Thus, $m\angle BDE = 180 - 4x$. $\angle BDC$ is supp. to $\angle BDA$ and so is $3x$. Thus, $m\angle C = 3x$ and so is $m\angle CBA$. By subtraction, $m\angle CBD = x$. Thus, in $\triangle BCD$, $x + 3x + 3x = 180$, or $x = 25\frac{5}{7}$.

22b Let $m\angle EJF = x$. Then $m\angle EFJ = 180 - x - m\angle E$. $m\angle HJF = \frac{180 - x}{2}$ and $m\angle JFH = \frac{x + m\angle E}{2}$. Thus, $m\angle H = 180 - \left(\frac{180 - x}{2} + \frac{x + m\angle E}{2}\right) = 90 - \frac{1}{2}m\angle E$.

23 $\begin{cases} b_2 = a + e_1 \\ d + e_2 = c_1 \end{cases}$ to the sum of the measures of the remote interior angles.
The measure of an exterior angle of a $\triangle$ is = $\therefore a + e_1 + c_2 = d + e_2 + b_2$

TWO PROOF-ORIENTED TRIANGLE THEOREMS

Class Planning

Time Schedule
Basic: 2 days
Average: 2 days
Advanced: $1\frac{1}{2}$ days

Resource References
Teacher's Resource Book
    Class Opener 7.2A, 7.2B
    Additional Practice Worksheet 13
Evaluation
    Tests and Quizzes
        Quiz 1, Series 1–2

Class Opener

Given: $\overline{PR} \cong \overline{RT}$,
     $\angle Q \cong \angle S$
Prove: $\overline{PQ} \cong \overline{TS}$

Statements	Reasons
1 $\angle PRQ \cong \angle TRS$	1 Vert. $\angle$s are $\cong$.
2 $\overline{PR} \cong \overline{RT}$	2 Given
3 $\angle Q \cong \angle S$	3 Given
4 $\angle P \cong \angle T$	4 If two pairs of $\angle$s of a $\triangle$ are $\cong$, the third pair must be $\cong$ because the sum of the measures of the $\angle$s in a $\triangle$ is 180.
5 $\triangle PRQ \cong \triangle TRS$	5 ASA
6 $\overline{PQ} \cong \overline{TS}$	6 CPCTC

Reason 4 leads into the No-Choice and AAS theorems.

Objective

After studying this section, you will be able to
- Apply the No-Choice Theorem and the AAS theorem

Part One: Introduction

We shall refer to the following theorem as the No-Choice Theorem, since it shows that two angles "have no choice" but to be congruent.

Theorem 53   ***If two angles of one triangle are congruent to two angles of a second triangle, then the third angles are congruent. (No-Choice Theorem)***

Given: $\angle A \cong \angle D$,
     $\angle B \cong \angle E$
Conclusion: $\angle C \cong \angle F$

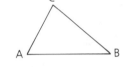

Proof: Since the sum of the angles in each triangle is 180°, the sums may be set equal. If we then apply the Subtraction Property, we see that $\angle C$ and $\angle F$ must be congruent.

Note   The two triangles need not be congruent for us to apply the No-Choice Theorem.

Theorem 54   ***If there exists a correspondence between the vertices of two triangles such that two angles and a nonincluded side of one are congruent to the corresponding parts of the other, then the triangles are congruent. (AAS)***

Given: ∠G ≅ ∠K,
  ∠H ≅ ∠M,
  $\overline{JH} ≅ \overline{OM}$

Prove: △GHJ ≅ △KMO

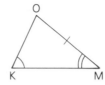

Proof:

1 ∠G ≅ ∠K	1 Given
2 ∠H ≅ ∠M	2 Given
3 ∠J ≅ ∠O	3 No-Choice Theorem
4 $\overline{JH} ≅ \overline{OM}$	4 Given
5 △GHJ ≅ △KMO	5 ASA (2, 4, 3)

Part Two: Sample Problems

Problem 1 Given: ∠A ≅ ∠D
  Prove: ∠E ≅ ∠C

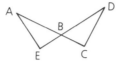

Proof

1 ∠A ≅ ∠D	1 Given
2 ∠ABE ≅ ∠DBC	2 Vertical ∠s are ≅.
3 ∠E ≅ ∠C	3 No-Choice Theorem

Problem 2 Given: ∠N ≅ ∠R, ∠NTR ≅ ∠P,
  $\overline{TO} \perp \overline{NP}$, $\overline{TS} \perp \overline{PR}$,
  $\overline{TO} ≅ \overline{TS}$

  Prove: NPRT is a rhombus.

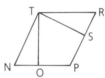

Proof

1 ∠N ≅ ∠R	1 Given
2 ∠NTR ≅ ∠P	2 Given
3 NPRT is a ▱.	3 If both pairs of opposite ∠s of a quadrilateral are ≅, it is a ▱.
4 $\overline{TO} \perp \overline{NP}$	4 Given
5 ∠TON is a right ∠.	5 ⊥ segments form right ∠s.
6 $\overline{TS} \perp \overline{PR}$	6 Given
7 ∠TSR is a right ∠.	7 Same as 5
8 ∠TON ≅ ∠TSR	8 Right ∠s are ≅.
9 $\overline{TO} ≅ \overline{TS}$	9 Given
10 △TON ≅ △TSR	10 AAS (1, 8, 9)
11 $\overline{TN} ≅ \overline{TR}$	11 CPCTC
12 NPRT is a rhombus.	12 If two consecutive sides of a ▱ are ≅, it is a rhombus.

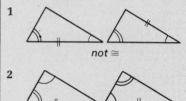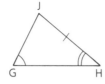

Part Three: Problem Sets

Problem Set A

1 Given: $\overline{JM} \perp \overline{GM}$,
$\overline{GK} \perp \overline{KJ}$

Conclusion: $\angle G \cong \angle J$

2 Given: $\overline{CB} \perp \overline{AB}$,
$\overleftrightarrow{DE} \parallel \overleftrightarrow{AB}$,
$\angle CDE = 40°$

Find: m$\angle A$, m$\angle C$, and m$\angle CED$
40; 50; 90

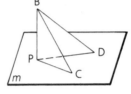

3 Given: $\overline{PD}$ and $\overline{PC}$ lie in plane m.
$\overline{BP} \perp$ m,
$\angle C \cong \angle D$

Prove: $\angle PBC \cong \angle PBD$

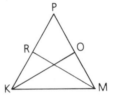

4 Given: $\overline{MR} \perp \overline{KP}$,
$\overline{KO} \perp \overline{PM}$,
$\angle RKM \cong \angle OMK$

Prove: $\triangle RKM \cong \triangle OMK$

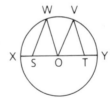

5 Given: $\odot$O,
$\angle SOV \cong \angle TOW$,
$\angle WSO \cong \angle VTO$

Prove: $\overline{SO} \cong \overline{TO}$

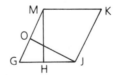

6 Given: GJKM is a rhombus.
$\overline{OJ} \perp \overline{GM}$,
$\overline{MH} \perp \overline{GJ}$

Conclusion: $\overline{MH} \cong \overline{JO}$

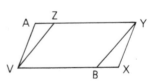

7 Given: $\angle A \cong \angle X$,
$\angle AVZ \cong \angle XYB$,
$\angle ZVB \cong \angle YBX$

Prove: VBYZ is a $\square$.

8 The measures of the angles of a triangle are in the ratio 3:4:8. Find the measure of the supplement of the largest angle. 84

9 Given: Triangle as marked
Find: m∠1 50

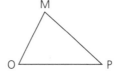

10 Given: ∠J ≅ ∠O,
 $\overline{JK} \cong \overline{OP}$,
 $\overline{HK} \not\cong \overline{MP}$
Prove: ∠H ≇ ∠M

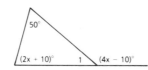

Problem-Set Notes and
Additional Answers, continued

■ See *Solution Manual* for answers to problems **10–15**.

Problem Set B

11 Prove that the altitude to the base of an isosceles triangle is also a median to the base.

12 Prove that segments drawn from the midpoint of the base of an isosceles triangle and perpendicular to the legs are congruent if they terminate at the legs.

13 Given: OHJM is an isosceles trapezoid, with bases $\overline{HJ}$ and $\overline{OM}$.
 ∠HPJ ≅ ∠JKH

Prove: **a** △HRJ is isosceles.

 b $\overline{HP} \cong \overline{JK}$

 c R is equidistant from O and M.

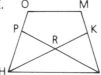

Cooperative Learning

Have students work in small groups to solve problem **17** on page 306. Suggest that each group draw each of the quadrilaterals on a coordinate plane, find the midpoint of each side of each figure, and then connect the midpoints.

14 Given: $\overleftrightarrow{AC} \parallel \overleftrightarrow{XY}$,
 $\overleftrightarrow{AB} \parallel \overleftrightarrow{CY}$,
 ∠ZAC ≅ ∠XAB
Prove: ∠X ≅ ∠Z

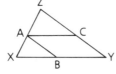

15 Prove the HL postulate.

Given: $\overline{TW} \cong \overline{GR}$,
 $\overline{WE} \cong \overline{AR}$;
 ∠E and ∠A are rt. ∠s.
Conclusion: △WET ≅ △RAG

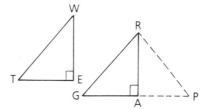

(Hint: Extend $\overrightarrow{GA}$ to P so that $\overline{AP} \cong \overline{ET}$. Use SAS to prove that △WET ≅ △RAP. Prove that △RGP is isosceles. Use AAS to prove that △RAG ≅ △RAP. What does it mean that two triangles are congruent to △RAP?)

16b If the midpoints of the sides of a triangle are joined, the new triangle has a perimeter that is half the perimeter of the original triangle.

17a Rectangle
  b Rectangle
  c Square
  d Rhombus
  e Parallelogram
  f Parallelogram
  g Rhombus

18 Since $\angle CBD \cong \angle ADB$, $\overline{AD} \parallel \overline{BC}$ by alt. int. $\angle s \cong$ $\Rightarrow \parallel$ lines. Thus, $\angle DAF \cong \angle FCB$ by $\parallel$ lines $\Rightarrow$ alt. int. $\angle s \cong$. Since $\overline{EF}$ is the median to $\overline{AC}$, $\overline{AF} \cong \overline{FC}$. Since $\triangle FDC$ is isosceles, $\overline{FC} \cong \overline{FD}$. Thus, $\overline{AF} \cong \overline{DF}$ and $\angle FAD \cong \angle FDA$. By the Transitive Property, $\angle FBC \cong \angle FCB$, and thus, $\overline{FB} \cong \overline{FC}$ and $\overline{FA} \cong \overline{FD} \cong \overline{FC} \cong \overline{FB}$. So ABCD is a $\square$ with $\cong$ diagonals and thus is a rectangle.

19 $\angle TNR \cong \angle TSR$ will force $\overline{NR} \cong \overline{SR}$; $\angle TNP \cong \angle TSP$ will force $\overline{SP} \cong \overline{NP}$. Combining this with $\overline{PR} \cong \overline{PR}$, $\triangle NPR \cong \triangle SPR$ by SSS.

20 Let $m\angle HGK = x$,
  $m\angle GKH = x$,
  $m\angle GHK = 180 - 2x$,
  $m\angle KHT = 2x - a$,
  $m\angle SGH = 180 - x - b$.
In $\triangle GSH$,
$a + 60 + 180 - x - b = 180$,
so $x + b - a = 60$ and
$2x + 2b - 2a = 120$.
In $\triangle KHT$,
$c + 60 + 2x - a = 180$,
so $c + 2x - a = 120$.
$\therefore 2x + 2b - 2a = c + 2x - a$
and $2b - c = a$.

Problem Set B, *continued*

16 a If the perimeter of $\triangle DEF$ is 145, find the perimeter of $\triangle GHJ$. 72.5

  b Can you state a generalization based on your solution to part **a**?

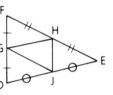

Problem Set C

17 Give the most descriptive name to the figure formed by connecting consecutive midpoints of each of the following figures. Be prepared to defend your answer in each case.

  a Rhombus   **c** Square   **e** Parallelogram   **g** Isosceles trapezoid

  b Kite   **d** Rectangle   **f** Quadrilateral

18 Given: $\overline{EF}$ is the median to $\overline{AC}$.
    $\angle CBD \cong \angle ADB$;
    $\overline{CD}$ is the base of isosceles $\triangle FDC$.

  Prove: ABCD is a rectangle.

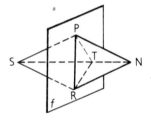

19 Given: P, T, and R lie in plane f.
    $\angle TNR \cong \angle TSR$, $\overline{NS} \perp f$,
    $\angle TNP \cong \angle TSP$

  Conclusion: $\triangle NPR \cong \triangle SPR$

Problem Set D

20 Given: $\triangle RST$ is equiangular.
    $\overline{GH} \cong \overline{KH}$

  Solve for a in terms of b and c.
  $a = 2b - c$

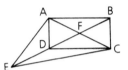

EUCLIDEAN DOG

7.3 FORMULAS INVOLVING POLYGONS

Objective

After studying this section, you will be able to

- Use some important formulas that apply to polygons

Part One: Introduction

A polygon with three sides can be called a 3-gon. Similarly, a polygon with seven sides can be called a 7-gon. Most of the polygons you will encounter have special names, like those given in the following chart.

No. of Sides (or Vertices)	Polygon	No. of Sides (or Vertices)	Polygon
3	Triangle	8	Octagon
4	Quadrilateral	9	Nonagon
5	Pentagon	10	Decagon
6	Hexagon	12	Dodecagon
7	Heptagon	15	Pentadecagon
		n	n-gon

What is the sum of the measures of the five angles in the figure? To answer that question, start at any vertex and draw diagonals. Three triangles are formed. By adding the measures of the angles of the three triangles, we can obtain the sum of the measures of the five original angles. In this case, the sum of the measures of the angles of pentagon ABCDE is 3(180), or 540.

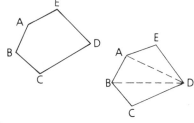

We follow a similar process with the next figure.

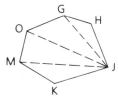

Since there are four triangles, the sum of the measures of the angles of figure GHJKMO is 4(180), or 720.

These two examples suggest the following theorem, which we present without formal proof.

Vocabulary

decagon	hexagon	octagon
dodecagon	interior angle	pentadecagon
heptagon	nonagon	pentagon

Class Planning

Time Schedule
All levels: 2 days

Resource References
Teacher's Resource Book
   Class Opener 7.3A, 7.3B
   Using Manipulatives 12, 13
   Additional Practice
   Worksheet 14
Evaluation
   Tests and Quizzes
      Quiz 2, Series 2
      Quizzes 1–2, Series 3

Class Opener

Given: $\overrightarrow{BD}$ bisects ∠PBQ.
   $\overline{PD} \perp \overline{PB}$, $\overline{QD} \perp \overline{QB}$
Prove: $\overleftrightarrow{BD}$ is the ⊥ bisector of $\overline{PQ}$.

Statements	Reasons
1 $\overrightarrow{BD}$ bisects ∠PBQ.	1 Given
2 ∠PBD ≅ ∠QBD	2 If a ray bisects an ∠, the ray divides the ∠ into 2 ≅ ∠s.
3 $\overline{PD} \perp \overline{PB}$	3 Given
4 $\overline{QD} \perp \overline{QB}$	4 Given
5 ∠BPD ≅ ∠BQD	5 ⊥ lines form ≅ rt. ∠s.
6 $\overline{BD} \cong \overline{BD}$	6 Reflex. Prop.
7 △DPB ≅ △DQB	7 AAS
8 $\overline{PB} \cong \overline{QB}$	8 CPCTC
9 $\overline{PD} \cong \overline{QD}$	9 CPCTC
10 $\overleftrightarrow{BD}$ is the ⊥ bisector of $\overline{PQ}$.	10 Two points equidistant from the endpoints of a segment determine the ⊥ bisector of the segment.

Theorem 55   *The sum S_i of the measures of the angles of a polygon with n sides is given by the formula $S_i = (n - 2)180$.*

On occasion, we may refer to the angles of a polygon as the *interior angles* of the polygon.

In the following diagram, we have formed an exterior angle at each vertex by extending one of the sides of the polygon.

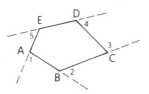

At vertex A, $m\angle 1 + m\angle EAB = 180$. In a similar manner, we can add each exterior angle to its adjacent interior angle, getting a sum of 180 at each vertex. Since there are five vertices, the total is 5(180), or 900.

According to Theorem 55, the sum of the measures of the angles of polygon ABCDE is 540. Since $900 - 540 = 360$, we may conclude that $m\angle 1 + m\angle 2 + m\angle 3 + m\angle 4 + m\angle 5 = 360$.

What is the sum of the measures of exterior angles 1, 2, 3, 4, 5, and 6 in this figure?

Again, the sum of the interior and the exterior angle is 180 at each of the six vertices, for a total measure of 6(180), or 1080. Moreover, according to Theorem 55, the sum of the measures of the angles of polygon GHJKMO is 720.

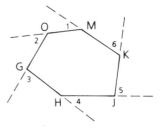

Because $1080 - 720 = 360$, we may conclude that in this figure, too, $m\angle 1 + m\angle 2 + m\angle 3 + m\angle 4 + m\angle 5 + m\angle 6 = 360$.

These two examples suggest the next theorem, which we present without formal proof.

Theorem 56   *If one exterior angle is taken at each vertex, the sum S_e of the measures of the exterior angles of a polygon is given by the formula $S_e = 360$.*

The following theorem is presented without proof. Problem 21 in Problem Set C asks you to explain this formula.

Theorem 57   *The number d of diagonals that can be drawn in a polygon of n sides is given by the formula*

$$d = \frac{n(n - 3)}{2}.$$

Part Two: Sample Problems

Problem 1 Find the sum of the measures of the angles of the figure to the right.

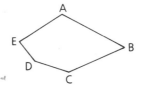

Solution The figure has five sides and five vertices.
The formula is $S_i = (n - 2)180$.
By substituting 5 for n, we find that
$S_i = (5 - 2)180$, or 540.

Problem 2 Find the number of diagonals that can be drawn in a pentadecagon.

Solution We use the formula in Theorem 57.

$$d = \frac{n(n - 3)}{2}$$
$$= \frac{15(15 - 3)}{2}$$
$$= 90$$

Problem 3 What is the name of a polygon if the sum of the measures of its angles is 1080?

Solution We use the formula in Theorem 55.

$$S_i = (n - 2)180$$
$$1080 = (n - 2)180$$
$$1080 = 180n - 360$$
$$1440 = 180n$$
$$8 = n$$

Since it has eight sides, the polygon is an octagon.

Part Three: Problem Sets

Problem Set A

1 Find the sum of the measures of the angles of

  a A quadrilateral 360   **c** An octagon 1080   **e** A 93-gon 16,380

  b A heptagon 900   **d** A dodecagon 1800

2 Given: $m\angle A = 160$, $m\angle B = 50$,
       $m\angle C = 140$, $m\angle D = 150$

  Find: $m\angle E$ 40

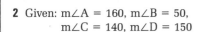

3 How many diagonals can be drawn in each figure below?

  a 5   b 9   c 2   d 0

Assignment Guide

Basic

Day 1	1a,c,e, 2, 3a,c, 4, 5, 6a,c, 7
Day 2	1b,d, 3b,d, 6b,d, 8, 10a,c,e, 13a, 14, 15, 17

Average

Day 1	1a,c,e, 2, 3a,c, 4–8
Day 2	1d, 3b, 10a,c,e, 11, 12, 13a, 14–17, 19

Advanced

Day 1	1a,c,e, 2, 3, 6–8, 10a,c,e, 15, 16
Day 2	Section **7.1** 21
	Section **7.3** 17–19, 21, 23

Problem Set A, *continued*

4 Given: m∠F = 110,
    m∠G = 80,
    m∠H = 74

Find: m∠1 84

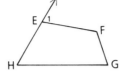

5 Given: K is a midpoint.
    P is a midpoint.
    m∠M = 70,
    m∠JKP = y + 15,
    m∠JPK = y − 10

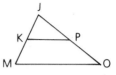

Find: **a** m∠JKP   **b** m∠JPK   **c** m∠J
        70   45   65

6 Find the sum of the measures of the exterior angles, one per vertex, of each of these polygons.

  a A triangle   **b** A heptagon   **c** A nonagon   **d** A 1984-gon
     360   360   360   360

7 What is the fewest number of sides a polygon can have? 3

Problem Set B

8 On a clock a segment is drawn connecting the mark at the 12 and the mark at the 1; then another segment connecting the mark at the 1 and the mark at the 2; and so forth, all the way around the clock.

  a What is the sum of the measures of the angles of the polygon formed? 1800

  b What is the sum of the measures of the exterior angles, one per vertex, of the polygon? 360

9 Prove that corresponding altitudes of congruent triangles are congruent.

10 How many sides does a polygon have if the sum of the measures of its angles is

  a 900? 7   **c** 2880? 18   **e** 436? Impossible

  b 1440? 10   **d** 180x − 720? x − 2   **f** Six right angles? 5

11 a In what polygon is the sum of the measures of the exterior angles, one per vertex, equal to the sum of the measures of the angles of the polygon? Quadrilateral

  b In what polygon is the sum of the measures of the angles of the polygon equal to twice the sum of the measures of the exterior angles, one per vertex? Hexagon

Problem-Set Notes and Additional Answers

■ See *Solution Manual* for answer to problem **9**.

12 If the sum of the measures of the angles of a polygon is increased by 900, how many sides will have been added to the polygon? **5**

13 What are the names of the polygons that contain the following numbers of diagonals?

a 14 **b** 35 **c** 209
Heptagon Decagon 22-gon

14 Tell whether each statement is true Always, Sometimes, or Never (A, S, or N).

a As the number of sides of a polygon increases, the number of exterior angles increases. **A**

b As the number of sides of a polygon increases, the sum of the measures of the exterior angles increases. **N**

c The sum of the lengths of the diagonals of a polygon is greater than the perimeter of the polygon. **S**

d The sum of the measures of the angles of a polygon formed by joining consecutive midpoints of a polygon's sides is equal to the sum of the measures of the angles of the original polygon. **A**

15 Find the restrictions on x. **18 < x < 36**

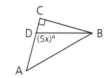

16 If AB > BC, find the restrictions on point B's

a x-coordinate **6 < x < 11**

b y-coordinate **7 < y < 10**

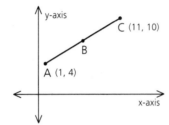

17 Find the area of a rectangle with vertices at $(-5, 2)$, $(3, 2)$, $(3, 8)$, and $(-5, 8)$. **48**

18 Using the diagram, write a coordinate proof of the Midline Theorem.

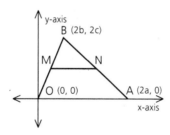

Cooperative Learning

Have students work in small groups to solve problem **14.** Have each group write a clear, concise sentence to justify why each statement is always, sometimes, or never true. Encourage students to draw diagrams to support their answers.

Answers will vary. For example, students might explain part **b** as follows.

This is never true because the sum of the measures of the exterior angles is always 360.

■ See *Solution Manual* for answer to problem **18.**

Problem Set B, continued

19 If three of the following four statements are chosen at random as given information, what is the probability that the fourth statement can be proved? $\frac{3}{4}$

a $\angle C \cong \angle D$ **c** $\angle A \cong \angle F$

b $\overline{AC} \cong \overline{DF}$ **d** $\overline{AB} \cong \overline{EF}$

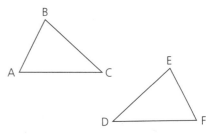

Problem Set C

20 In Chapter 5, we noted that one of the ways to show that a quadrilateral is a parallelogram is to prove that both pairs of opposite angles are congruent. Without the information presented in this chapter, the proof of that method would be extremely long and involved. Use your new knowledge to prove it now.

Given: $\angle B \cong \angle D$,
     $\angle A \cong \angle C$

Prove: ABCD is a ▱. (Hint: Let m∠B = x and m∠C = y.)

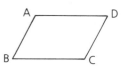

21 Explain why each of the three ingredients in the formula of Theorem 57 (the n, the n − 3, and the 2) is needed.

22 We have stated that in this text the word *polygon* will mean a *convex polygon* and that angles greater than 180° will not be considered. Ignore those rules for this problem.

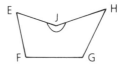

a Consider the nonconvex polygon EFGHJ, whose interior angle at J is greater than 180°. Can you demonstrate that the sum of the measures of the angles of this nonconvex polygon is 540?

b Can you demonstrate that the sum of the measures of the angles of the nonconvex octagon at the right is 1080?

c Is the sum of the measures of the angles of a nonconvex polygon of n sides (n − 2)180? Yes

d Is the sum of the measures of the exterior angles, one per vertex, of a nonconvex polygon equal to 360? Explain.

23 Seven of the angles of a decagon have measures whose sum is 1220. Of the remaining three angles, exactly two are complementary and exactly two are supplementary. Find the measures of these three angles. 40; 50; 130

Problem Set D

24 Find the set of polygons in which the number of diagonals is greater than the sum of the measures of the angles.

{All polygons with 362 or more sides}

Problem-Set Notes and Additional Answers, continued

24 $\dfrac{n(n-3)}{2} > (n-2)180$

Students must complete the square (or use the quadratic formula) and then approximate the square root.

CAREER PROFILE

PRECISE ANGLES PAY OFF
John C. Buchholz cuts a solid with 58 facets

Make a mistake drawing a 34° angle with your pencil and protractor, and the consequences will probably be minimal. Make a similar mistake cutting a facet on a diamond, and the consequences may be disastrous. A flawless, beautifully colored 1 carat (200 milligram or $\frac{1}{142}$ ounce) diamond may be worth $25,000, according to Denver, Colorado, diamond cutter John C. Buchholz. That is the size Buchholz typically works on, and an error in cutting a diamond cannot be corrected.

Diamond is the hardest, and one of the rarest, naturally occurring substances. Diamond crystals often occur as octahedra. To turn a rough diamond into a brilliant gem requires precise and painstaking work. Buchholz describes the cutting of the fifty-eight *facets* (faces or planes) that characterize the familiar *round brilliant-cut* diamond: "Since only diamond can cut diamond, I use a 3000-rpm wheel impregnated with diamond powder. As I cut, I aim for maximum brilliance. I use gauges to cut the first

eight facets: four in the crown at $32\frac{1}{2}°$ and four in the pavilion at $40\frac{3}{4}°$ The other fifty facets I cut by eye." As he works on the tiny facets he must keep his eye on the overall proportions of the diamond. For example, the table, or top facet, must be uniform and centered, with a diameter 53 percent to 57 percent of the stone's diameter.

Buchholz was born in Iola, Wisconsin. Following his discharge from the army he undertook a three-year apprenticeship at a diamond-cutting school in Gardnerville, Nevada. Says Buchholz: "American cutters are the most skilled and the best paid in the world today." Unlike many cutters, he has refused to specialize, remaining proficient in all facets of cutting. He takes as his motto the words of Michelangelo: "Only human genius enlivens a rough stone into a masterpiece."

Describe the plane figures that form the facets of a round brilliant-cut diamond.

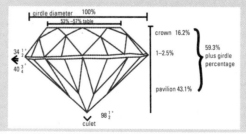

The part of the diamond we can see in the illustration consists of kites and triangles.

7.4 REGULAR POLYGONS

Objectives

After studying this section, you will be able to
- Recognize regular polygons
- Use a formula to find the measure of an exterior angle of an equiangular polygon

Part One: Introduction

Regular Polygons

The figures below are examples of ***regular polygons***.

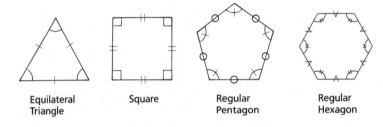

Equilateral Triangle   Square   Regular Pentagon   Regular Hexagon

Definition   A ***regular polygon*** is a polygon that is both equilateral and equiangular.

A Special Formula for Equiangular Polygons

Can you find m∠1 in the equiangular pentagon below?

In Section 7.3, you learned that the sum of the measures of the exterior angles, one per vertex, of any polygon is 360. Since each of the five exterior angles has the same measure, we can find m∠1 by dividing 360 by 5.

$$m\angle 1 = \frac{360}{5} = 72$$

This result suggests the next theorem, which we present without formal proof.

Vocabulary
regular polygon

Theorem 58   *The measure E of each exterior angle of an equiangular polygon of n sides is given by the formula*

$$E = \frac{360}{n}$$

You will see several applications of this theorem in the problems that follow.

Part Two: Sample Problems

Problem 1   *How many degrees are there in each exterior angle of an equiangular heptagon?*

Solution   Using $E = \frac{360}{n}$, we find that $E = \frac{360}{7}$, or $51\frac{3}{7}$.

Problem 2   *If each exterior angle of a polygon is 18°, how many sides does the polygon have?*

Solution   We can use the formula $E = \frac{360}{n}$.

$$18 = \frac{360}{n}$$
$$18n = 360$$
$$n = 20$$

Problem 3   *If each angle of a polygon is 108°, how many sides does the polygon have?*

Solution   First, we find the measure of an exterior angle. Since an angle of a polygon and its adjacent exterior angle are supplementary, an exterior angle of this polygon has a measure of 180 − 108, or 72. Now we can substitute 72 for E in the formula $E = \frac{360}{n}$.

$$72 = \frac{360}{n}$$
$$72n = 360$$
$$n = 5$$

Problem 4   *Find the measure of each angle of a regular octagon.*

Solution   We use the formula $E = \frac{360}{n}$, finding that $E = \frac{360}{8}$, or 45. Thus, the measure of each interior angle is 180 − 45, or 135.

Problem 5   *Find the measure of each exterior angle of an equilateral quadrilateral.*

Solution   An equilateral quadrilateral is not necessarily equiangular, so there is no answer.

Section 7.4   Regular Polygons   **315**

Part Three: Problem Sets

Problem Set A

1 Find the measure of an exterior angle of each of the following equiangular polygons.

  a A triangle **120**   **c** An octagon **45**   **e** A 23-gon $15\frac{15}{23}$

  b A quadrilateral **90**   **d** A pentadecagon **24**

2 Find the measure of an angle of each of the following equiangular polygons.

  a A pentagon **108**   **c** A nonagon **140**   **e** A 21-gon $162\frac{6}{7}$

  b A hexagon **120**   **d** A dodecagon **150**

3 Find the number of sides an equiangular polygon has if each of its exterior angles is

  a 60° **6**   **b** 40° **9**   **c** 36° **10**   **d** 2° **180**   **e** $7\frac{1}{2}$° **48**

4 Find the number of sides an equiangular polygon has if each of its angles is

  a 144° **10** **b** 120° **6** **c** 156° **15** **d** 162° **20** **e** $172\frac{4}{5}$° **50**

5 Given: PENTA is a regular pentagon.
Prove: △PNT is isosceles.

6 In the stop sign shown, is △NTE scalene, isosceles, equilateral, or undetermined? **Isosceles**

7 In an equiangular polygon, the measure of each exterior angle is 25% of the measure of each interior angle. What is the name of the polygon? **Equiangular decagon**

Problem Set B

8 a Prove that the perpendicular bisector of a side of a regular pentagon passes through the opposite vertex.

  b Can you generalize about the perpendicular bisectors of the sides of regular polygons?

9 Given: $\overline{AB} \cong \overline{AD}$,
$\overline{FC} \perp \overline{BD}$

Conclusion: △AEF is isosceles.

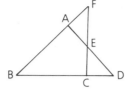

10 The sum of the measures of the angles of a regular polygon is 5040. Find the measure of each angle. 168

11 The sum of a polygon's angle measures is nine times the measure of an exterior angle of a regular hexagon. What is the polygon's name? Pentagon

12 What is the name of an equiangular polygon if the ratio of the measure of an interior angle to the measure of an exterior angle is 7:2? Equiangular nonagon

13 Tell whether each statement is true Always, Sometimes, or Never (A, S, or N).

a If the number of sides of an equiangular polygon is doubled, the measure of each exterior angle is halved. A

b The measure of an exterior angle of a decagon is greater than the measure of an exterior angle of a quadrilateral. S

c A regular polygon is equilateral. A

d An equilateral polygon is regular. S

e If the midpoints of the sides of a scalene quadrilateral are joined in order, the figure formed is equilateral. S

f If the midpoints of the sides of a rhombus are joined in order, the figure formed is equilateral but not equiangular. N

Problem Set C

14 Given: ABCDEF is a regular hexagon.
Prove: ACDF is a rectangle.

15 Given: $\overrightarrow{RO} \perp$ plane GHJ;
O, M, and K are coplanar.
GHJKMP is a regular hexagon.
$\overrightarrow{HO}$ bisects ∠GHJ.
$\overline{RH} \cong \overline{RJ}$

Prove: △HOJ is regular.

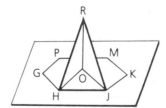

*Problem-Set Notes and
Additional Answers, continued*

■ See *Solution Manual* for answer to problem **9**.

14 ∠B = 120°, and since △BAC is isosceles, ∠BAC = 30°. Thus, ∠FAC = 90°. In a similar manner, ∠ACD = ∠CDF = ∠DFA = 90°.

15 ∠GHJ = 120°, and ∠OHJ = 60°. △ROH ≅ △ROJ by HL. Thus, $\overline{HO} \cong \overline{JO}$, and △HOJ is equiangular.

Problem-Set Notes and
Additional Answers, continued

16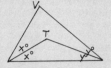

$$\begin{cases} m\angle V + 2x + 2y = 180 \\ m\angle T + x + y = 180 \end{cases}$$
$\therefore m\angle V = 2m\angle T - 180$
The set of possibilities for
$m\angle T$ is $\{108, 120, 128\frac{4}{7},$
$135, 140, 144\}$. Substituting
the individual values we
see that $m\angle V$ can be
$\{36, 60, 77\frac{1}{7}, 90, 100, 108\}$.

17 $\begin{cases} 3x + 3y + 9 = 135 \\ 2x + y - 4\frac{1}{2} = 67\frac{1}{2} \end{cases}$
$x = 30; y = 12$

Problem Set C, *continued*

16 Given: $105 < m\angle T < 145$;
an equiangular polygon
can be drawn with $\angle T$
as one of the angles.

Find: The set of possible values of $m\angle V$
$\{36, 60, 77\frac{1}{7}, 90, 100, 108\}$

17 We shall call the figure to the right a
regular semioctagon. (What do you think
that means?)
If $m\angle E = 3x + 3y + 9$ and $m\angle A = 2x + y - 4\frac{1}{2}$,
what are the values of x and y? $x = 30; y = 12$

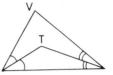

MATHEMATICAL EXCURSION

POLYGONS IN THE NORTH COUNTRY
The Vegreville Egg

Polygons can be tiled in three dimensions as well as two. One result: an aluminum sculpture of an egg—31 feet long, three and a half stories high, weighing 2.5 tons, and decorated in the intricate Ukrainian style—in the town of Vegreville, Alberta, Canada.

To make a long story short, the town had got a grant to build a huge Ukrainian-style egg to celebrate the centennial of the Royal Canadian Mounted Police. The project, however, was more than most architects and engineers were willing to take on. Their reluctance arose from the fact that the surface of an egg cannot be defined mathematically.

Fortunately, true to the spirit of the Mounties, there was one computer science professor from Utah who would not give up until he had cracked the problem and who finally hatched a plan. After much computer analysis of the structures of various birds' eggs, he designed an egg that could be built using very thin aluminum tiles.

He tiled the egg using more than two thousand congruent equilateral triangles and more than five hundred hexagons in the shapes of stars, as shown in the illustration. The tiles, ranging from $\frac{1}{16}$ inch to $\frac{1}{8}$ inch thick, are joined at angles ranging from less than 1° near the middle of the egg to about 7° at its tip. They are held together by an internal structure consisting of a central shaft from which radiate spokes that connect it with the egg's "shell."

How can flat tiles be used to simulate a curved surface such as an egg's? Are the stars equilateral hexagons? Are they regular hexagons? Why or why not?

Using tiles that are relatively small compared with the overall surface and placing them at suffic
slight angles allows us to simulate curvature. Yes, because all their sides are the same length.
No, because the angles within each hexagon vary.

CHAPTER SUMMARY

CONCEPTS AND PROCEDURES

After studying this chapter, you should be able to
- Apply theorems about the interior angles, the exterior angles, and the midlines of triangles. (7.1)
- Apply the No-Choice Theorem and the AAS theorem (7.2)
- Use some important formulas that apply to polygons (7.3)
- Recognize regular polygons (7.4)
- Use a formula to find the measure of an exterior angle of an equiangular polygon (7.4)

VOCABULARY

decagon (7.3)
dodecagon (7.3)
exterior angle (7.1)
heptagon (7.3)
hexagon (7.3)
interior angle (7.3)

octagon (7.3)
pentadecagon (7.3)
pentagon (7.3)
nonagon (7.3)
regular polygon (7.4)

REVIEW PROBLEMS

Chapter 7 Review
Class Planning

Time Schedule
Basic: 2 days
Average: 1 day
Advanced: 1 day

Resource References
Evaluation
Tests and Quizzes
Test 7, Series 1, 2, 3

Assignment Guide

Basic

Day 1   1, 2, 5–12
Day 2   13–16, 18, 21, 23

Average

1, 4–6, 9–16, 18–21, 24, 28

Advanced

Section 7.4 15–17
Chapter 7 Review 18, 19, 21, 27–29

Problem-Set Notes and Additional Answers

- See *Solution Manual* for answers to problems **1–5**.

Problem Set A

1 Given: ∠DBC ≅ ∠E
Conclusion: ∠A ≅ ∠BDC

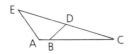

2 Given: ⊙O,
∠W ≅ ∠V,
∠ORW ≅ ∠OSV
Prove: $\overline{PR} \cong \overline{ST}$

3 Given: $\overline{AC} \cong \overline{AE}$,
∠CBD ≅ ∠EFD
Prove: ∠BDC ≅ ∠FDE

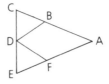

4 Given: ∠S ≅ ∠ROP,
∠ROS ≅ ∠P
Prove: ∠SRO ≅ ∠PRO (Hint: Why can't you use AAS to prove that the triangles are congruent?)

5 Given: $\overline{SV}$ lies in plane m.
$\overline{VX}$ lies in plane m.
∠S ≅ ∠X,
$\overleftrightarrow{TV} \perp$ plane m
Prove: $\overline{TS} \cong \overline{TX}$

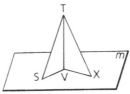

6 The measures of three of the angles of a quadrilateral are 40, 70, and 130. What is the measure of the fourth angle? 120

7 The measures of the angles of a triangle are in the ratio 1:2:3. Find half the measure of the largest angle. 45

8 Given: Diagram as marked
Find: m∠1 and m∠2 50; 50

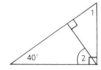

9 Given: Diagram as marked
Find: m∠YZA 50

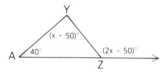

10 Given: C is the midpt. of $\overline{BD}$.
E is the midpt. of $\overline{BF}$.
DF = 12,
m∠D = 80, m∠B = 60

Find: CE, m∠BCE, and m∠BEC
6; 80; 40

11 Find m∠3 in the diagram as marked. 20

12 If the measure of an exterior angle of a regular polygon is 15,
how many sides does the polygon have? 24

13 If a polygon has 33 sides, what is

a The sum of the measures of the angles of the polygon? 5580

b The sum of the measures of the exterior angles, one per vertex,
of the polygon? 360

14 The sum of the measures of the angles of a polygon is 1620. Find
the number of sides of the polygon. 11

15 Find the number of diagonals that can be drawn in a
pentadecagon. 90

16 The measure of an exterior angle of an equiangular polygon is
twice that of an interior angle. What is the name of the polygon? Equilateral triangle

Problem Set B

17 Prove that any two diagonals of a regular pentagon are congruent. Are any two diagonals congruent in any regular polygon?

*Problem-Set Notes and
Additional Answers, continued*

■ See *Solution Manual* for answer
to problem **17**.

Review Problem Set B, *continued*

18 The measure of one of the angles of a right triangle is five times the measure of another angle of the triangle. What are the possible values of the measure of the second largest angle? 72 or 75

19 Given: △ABC is isosceles,
with base $\overline{BC}$.
$\overrightarrow{BE}$ bisects ∠ABC.
$\overrightarrow{CE}$ bisects ∠FCD.
∠A = 50°

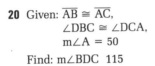

Find: **a** m∠ABF   **b** m∠BCE   **c** m∠E
        32.5   122.5   25

20 Given: $\overline{AB} \cong \overline{AC}$,
∠DBC ≅ ∠DCA,
m∠A = 50
Find: m∠BDC 115

21 Tell whether each statement is true Always, Sometimes, or Never (A, S, or N).

a An equiangular triangle is isosceles. A

b The number of diagonals in a polygon is the same as the number of sides. S

c An exterior angle of a triangle is larger in measure than any angle of a triangle. S

d One of the base angles of an isosceles triangle has a measure greater than that of one of the exterior angles of the triangle. N

22 The sum of the measures of five of the angles of an "octagon" is 540. What conclusion can you draw about the "octagon"? It is nonconvex.

23 An *arithmetic progression* is a sequence of terms in which the difference between any two consecutive terms is always the same. (For example, 1, 5, 9, 13 is an arithmetic progression because the difference between any two consecutive terms is 4.) Do the numbers of diagonals in a triangle, a quadrilateral, a pentagon, and a hexagon form an arithmetic progression? No

24 The measure of an angle of an equiangular polygon exceeds four times the measure of one of the polygon's exterior angles by 30. What is the name of the polygon? Equiangular dodecagon

Problem Set C

25 Given: $\overline{BC} \cong \overline{FE}$,
   $\angle C \cong \angle E$
   Prove: $\triangle ABF$ is isosceles.

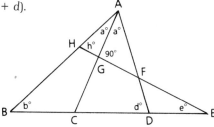

26 Show that $h = \frac{1}{2}(b + d)$.

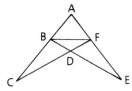

27 Given: $PR = PS$;
   $\overrightarrow{RV}$ bisects $\angle PRS$.
   $\overrightarrow{SV}$ bisects $\angle PST$.
   Prove: $m\angle V = \frac{1}{2}(m\angle P)$
   (Hint: Let $m\angle P = 4x$.)

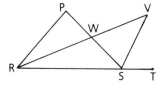

28 In a drawer there is a regular triangle, a regular quadrilateral, a regular pentagon, and a regular hexagon. The drawer is opened, and an angle from one of the polygons is selected at random. What is the probability that the measure of the angle is an integral multiple of 30? $\frac{13}{18}$

29 A square has vertices $A = (-4, 0)$, $B = (-4, 4)$, $C = (0, 4)$, and $O = (0, 0)$. When the square is rotated 90° counterclockwise about the origin, points A, B, and C are rotated to points E, F, and G respectively. Find the area of the polygon with vertices at A, B, F, and G. **32**

Problem Set D

30 Show that the number of diagonals in a polygon is never the same as the sum of the measures of the exterior angles, one per vertex, of the polygon.

Problem-Set Notes and Additional Answers, continued

25 $\triangle BCD \cong \triangle FED$ by AAS. $\overline{BD} \cong \overline{FD}$ and $\overline{CD} \cong \overline{ED}$ by CPCTC. By addition, $\overline{CF} \cong \overline{EB}$. $\triangle AFC \cong \triangle ABE$ by AAS. Then $\overline{AB} \cong \overline{AF}$ by CPCTC.

26 $\triangle AGH \cong \triangle AGF$ by ASA. $\therefore \angle AFG = h°$. Thus, $\angle DFE = h°$. Consider $\triangle HBE$ and $\angle AHE$; $h = b + e$. Consider $\triangle FDE$ and $\angle ADB$; $d = h + e$. $\therefore e = d - h$. Thus, $h = b + (d - h)$, or $h = \frac{1}{2}(b + d)$.

27 $m\angle WRS = 45 - x$
   $m\angle PSR = 90 - 2x$
   $m\angle WSV = 45 + x$
   $m\angle WRS + m\angle RSV = 180 - 2x$
   Thus, $m\angle V = 2x$.
   $m\angle V = \frac{1}{2}m\angle P$

28 ⬭60 ⬭60 ⬭60
   ⬭90 ⬭90 ⬭90 ⬭90
   108 108 108 108 108
   ⬭120 ⬭120 ⬭120 ⬭120 ⬭120
   ⬭120 $\frac{13}{18}$

29

The rotation sends A to C, B to F, and C to G. The area of ABFG = $bh = 8(4) = 32$.

30 Assume there is such a polygon. $\frac{n(n - 3)}{2} = 360$, where n is a positive integer greater than 2. $n^2 - 3n = 720$; $n^2 - 3n - 720 = 0$. Thus, $b^2 - 4ac$ (the discriminant) is 2889. Since 2889 is not a perfect square, there are no rational factors for n, and hence no integral solution for n is possible.

Chapter 8 Schedule

Basic

Problem Sets:	9 days
Review:	1 day
Test:	1 day

Average

Problem Sets:	8 days
Review:	1 day
Test:	1 day

Advanced

Problem Sets:	6 days
Review:	1 day
Test:	1 day

*A*LGEBRA IS USED extensively in this chapter for problems of ratio and proportion.

Many average courses may mark the end of half the school year during this chapter.

CHAPTER

8 | SIMILAR POLYGONS

L inda MacDonald used similar polygons to create a geometric design in her quilt, *Titus Canyon*.

8.1 | RATIO and PROPORTION

Objectives

After studying this section, you will be able to
- Recognize and work with ratios
- Recognize and work with proportions
- Apply the product and ratio theorems
- Calculate geometric means

Part One: Introduction

Ratio

You may recall the following definition from your previous mathematics studies.

Definition   A **ratio** is a quotient of two numbers.

The ratio of 5 meters to 3 meters can be written in any of the following ways:

$$\frac{5}{3} \qquad 5{:}3 \qquad 5 \text{ to } 3 \qquad 5 \div 3$$

Notice that the first number, 5, is the numerator and the second number, 3, is the denominator.

Unless otherwise specified, a ratio is given in lowest terms. For example, the ratio of 15 to 6, or $\frac{15}{6}$, when reduced to lowest terms is $\frac{5}{2}$.

The slope of a line is the ratio of the **rise** between any two points on the line to the **run** between the two points.

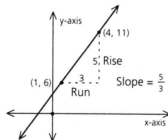

Section 8.1   Ratio and Proportion   **325**

Class Planning

Time Schedule
All levels: 2 days

Resource References
Teacher's Resource Book
   Class Opener 8.1A, 8.1B

Class Opener

Two supplementary angles are in the ratio of 3 to 5. Find the measure of the larger angle.
112.5°
This problem reviews terminology and algebra and introduces ratio.

Lesson Notes

- The terminology in this section and the difference between Theorems 59 and 60 are fundamental to the entire chapter.

Vocabulary

arithmetic mean	means
extremes	proportion
geometric mean	ratio
mean proportion	rise
mean proportional	run

On a map, the scale gives the ratio of the map distance to the actual distance. The distance from A-ville to B-ville on the map is 2.5 centimeters. The scale indicates that 1 centimeter represents 30 miles. We can conclude that the distance from A-ville to B-ville is 2.5(30), or 75, miles.

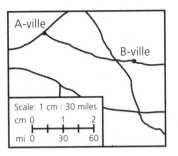

Proportion

Proportions are related to ratios.

Definition A **proportion** is an equation stating that two or more ratios are equal. Here are three examples of proportions.

$$\frac{1}{2} = \frac{5}{10} \qquad 5{:}15 = 15{:}45 \qquad \frac{4}{6} = \frac{10}{15} = \frac{x}{y} = \frac{2}{3}$$

Most proportions you encounter, however, will contain only two ratios and will be written in one of the following equivalent forms.

$$\frac{a}{b} = \frac{c}{d} \qquad a{:}b = c{:}d$$

In both of these forms,
a is called the first term c is called the third term
b is called the second term d is called the fourth term

The equation $y = \frac{2}{3}x + 4$ relates the x- and y-coordinates of points on the graph of the equation.

If $x = -3$, then $y = \frac{2}{3}(-3) + 4 = 2$, so $(-3, 2)$ is on the line.

If $x = 3$, then $y = \frac{2}{3}(3) + 4 = 6$, so $(3, 6)$ is on the line.

If $x = 6$, then $y = 8$, so $(6, 8)$ is on the line.

If $x = 15$, then $y = 14$, so $(15, 14)$ is on the line.

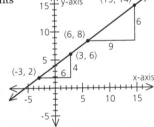

The slope of the segment joining $(-3, 2)$ and $(3, 6)$ is the ratio

$$\frac{6 - 2}{3 - (-3)} = \frac{4}{6}$$

The slope of the segment joining $(6, 8)$ and $(15, 14)$ is the ratio

$$\frac{14 - 8}{15 - 6} = \frac{6}{9}$$

No matter what pair of points on the line we choose, the slope should be the same. The proportion $\frac{4}{6} = \frac{6}{9}$ is a true statement, since both ratios reduce to $\frac{2}{3}$.

The Product and Ratio Theorems

In a proportion containing four terms,
- The first and fourth terms are called the *extremes*
- The second and third terms are called the *means*

Theorem 59 *In a proportion, the product of the means is equal to the product of the extremes. (Means-Extremes Products Theorem)*

This theorem allows us to "multiply out" a proportion.

$$\text{If } \frac{a}{b} = \frac{c}{d}, \text{ then } ad = bc.$$

Theorem 60 *If the product of a pair of nonzero numbers is equal to the product of another pair of nonzero numbers, then either pair of numbers may be made the extremes, and the other pair the means, of a proportion. (Means-Extremes Ratio Theorem)*

This theorem is harder to state than to use. Given that $pq = rs$, we can create proportions such as $\frac{p}{r} = \frac{s}{q}$, $\frac{p}{s} = \frac{r}{q}$, and $\frac{r}{p} = \frac{q}{s}$. All these proportions are equivalent forms, since multiplying them out yields equivalent equations.

The Geometric Mean

In a *mean proportion*, the means are the same.

$$\frac{1}{4} = \frac{4}{16} \qquad \frac{a}{x} = \frac{x}{r}$$

Definition If the means in a proportion are equal, either mean is called a *geometric mean*, or *mean proportional*, between the extremes.

In the first example above, 4 is a geometric mean between 1 and 16. What is the mean proportional (geometric mean) in the second example?

Lesson Notes, continued
- Proof of Theorem 59, Means-Extremes Products Theorem:
  Given: $\frac{a}{b} = \frac{c}{d}$
  Prove: $ad = bc$

Statements	Reasons
1 $\frac{a}{b} = \frac{c}{d}$	1 Given
2 $bd\left(\frac{a}{b}\right) = bd\left(\frac{c}{d}\right)$	2 Multiplication Prop.
3 $ad = bc$	3 Algebra

- Proof of Theorem 60, Means-Extremes Ratio Theorem:
  Given: $ad = bc$, where a, b, c, $d \neq 0$
  Prove: $\frac{a}{b} = \frac{c}{d}$

Statements	Reasons
1 $ad = bc$	1 Given
2 $\frac{ad}{bd} = \frac{bc}{bd}$	2 Division Prop.
3 $\frac{a}{b} = \frac{c}{d}$	3 Algebra

In other mathematics classes, you have probably had to calculate averages. The average of two numbers is another kind of mean between the numbers, called the **arithmetic mean**.

Example   *Find the geometric and arithmetic means between 3 and 27.*

Arithmetic Mean:

$$\text{Average} = \frac{3 + 27}{2}$$
$$= 15$$

Geometric Mean:

Write a proportion, using 3 and 27 as the extremes and x as both means.

$$\frac{3}{x} = \frac{x}{27}$$
$$x^2 = 81$$
$$x = \pm 9$$

The arithmetic mean is 15. There are two possible values of the geometric mean. The geometric mean is either 9 or -9.

Part Two: Sample Problems

Problem 1   *If $\frac{3}{x} = \frac{7}{14}$, solve for x.*

Solution
$$\frac{3}{x} = \frac{7}{14}$$
$$\frac{3}{x} = \frac{1}{2} \qquad \text{$\frac{7}{14}$ reduces to $\frac{1}{2}$.}$$
$$1 \cdot x = 3 \cdot 2 \qquad \text{Means-Extremes Products Theorem}$$
$$x = 6$$

Problem 2   *Find the fourth term (sometimes called the fourth proportional) of a proportion if the first three terms are 2, 3, and 4.*

Solution
$$\frac{2}{3} = \frac{4}{x}$$
$$\frac{2}{x} = 3 \cdot 4$$
$$x = 6$$

Problem 3   *Find the mean proportional(s) between 4 and 16.*

Solution
$$\frac{4}{x} = \frac{x}{16}$$
$$x \cdot x = 4 \cdot 16$$
$$x^2 = 64$$
$$x = \pm 8 \text{ (Two answers)}$$

Note   There are two mean proportionals (or geometric means) between the numbers. In certain geometry problems, we reject one of these algebraic answers. For example, a segment cannot have a length of -8.

Problem 4   If $3x = 4y$, *find the ratio of* x *to* y.

Solution   Use Theorem 60 to write a proportion, making x and 3 the extremes and y and 4 the means.

$$3x = 4y$$
$$\frac{x}{y} = \frac{4}{3}$$

Problem 5   Is $\frac{x}{y} = \frac{a}{b}$ equivalent to $\frac{x - 2y}{y} = \frac{a - 2b}{b}$?

Solution

$$\frac{x}{y} = \frac{a}{b} \qquad \frac{x - 2y}{y} = \frac{a - 2b}{b}$$
$$xb = ya \qquad (x - 2y)b = (a - 2b)y$$
$$xb = ay \qquad xb - 2by = ay - 2by$$
$$\qquad\qquad\qquad xb = ay$$

The answer is yes. The Means-Extremes Products Theorem reveals that the two proportions are equivalent forms.

Problem 6   *Show that* $\frac{a}{b} = \frac{c}{d}$ *and* $\frac{a + b}{b} = \frac{c + d}{d}$ *are equivalent proportions.*

Solution   Start with the first proportion and add 1 to each side.

$$\frac{a}{b} = \frac{c}{d}$$
$$\frac{a}{b} + 1 = \frac{c}{d} + 1$$
$$\frac{a}{b} + \frac{b}{b} = \frac{c}{d} + \frac{d}{d} \qquad \text{Substitute fractions equal to 1.}$$
$$\frac{a + b}{b} = \frac{c + d}{d}$$

Part Three: Problem Sets

Problem Set A

1 a In $\frac{3}{4} = \frac{9}{12}$, what is the third term? 9

  b Name the means and the extremes of the proportion in part **a**.
   4 and 9; 3 and 12

2 Is $\frac{p}{q} = \frac{r}{s}$ equivalent to $\frac{r}{p} = \frac{s}{q}$? Yes

3 Solve each proportion for x.

  a $\frac{3}{x} = \frac{12}{16}$ 4   **b** $\frac{x}{18} = \frac{3}{7}$ $\frac{54}{7}$ **c** $\frac{7}{x - 4} = \frac{3}{5}$ $\frac{47}{3}$

4 Find the fourth proportional for each set of three terms.

  a 1, 2, 3 6   **b** $\frac{1}{2}$, 3, 4 24   **c** a, b, 5 $\frac{5b}{a}$

Section 8.1   Ratio and Proportion   **329**

Lesson Notes, continued
- Many students are surprised by the result of Sample Problem **4**; they expect the reciprocal.

Communicating Mathematics

See Sample Problem **6**. Have students show that if $\frac{a}{b} = \frac{c}{d}$, then $\frac{a + 2b}{b} = \frac{c + 2d}{d}$. Challenge students to write a generalization for adding a whole number to both sides of a proportion.

$$\frac{a}{b} + 2 = \frac{c}{d} + 2$$
$$\frac{a}{b} + \frac{2b}{b} = \frac{c}{d} + \frac{2d}{d}$$
$$\frac{a + 2b}{b} = \frac{c + 2d}{d}$$
$$\frac{a + nb}{b} = \frac{c + nd}{d}$$

Assignment Guide

Basic	
Day 1	1, 2, 3a,b, 4a,b, 5, 6a,b, 7, 8
Day 2	3c, 4c, 6c, 9–16

Average	
Day 1	2, 3a,c, 4, 5, 6a,b, 7–9, 11–13
Day 2	14–18, 20–22

Advanced	
Day 1	4b,c, 7–9, 11a,c, 12, 13, 15, 17
Day 2	14, 18, 20–22, 24–27

Problem Set A, *continued*

5 a Use the coordinates of points A and B to find the slope of $\overleftrightarrow{AC}$. $-\frac{1}{2}$

b Use the coordinates of points B and C to find the slope of $\overleftrightarrow{AC}$. $-\frac{1}{2}$

c Should your answers in parts **a** and **b** be the same? Yes

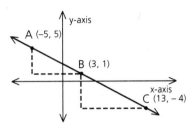

6 Find the ratio of x to y if

a $2x = 3y$ $\frac{3}{2}$ **b** $6(y + 3) = 2(x + 9)$ $\frac{3}{1}$ **c** $\dfrac{3}{x + 5} = \dfrac{9}{y + 15}$ $\frac{1}{3}$

7 What is the ratio of the number of diagonals in a pentagon to the measure of each exterior angle of a regular decagon? $\frac{5}{36}$

8 Given two squares with sides 5 and 7,

a What is the ratio of their perimeters? **5:7**

b What is the ratio of their areas? **25:49**

9 If the ratio of the measures of a pair of sides of a parallelogram is 2:3 and the ratio of the measures of the diagonals is 1:1, what is the most descriptive name of the parallelogram? **Rectangle**

10 a What is the ratio of AB to BC? **1:1**

b What is AB:AC? **1:2**

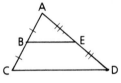

11 Find the geometric mean(s) between each pair of extremes.

a 4 and 25 ± 10 **b** 3 and 5 $\pm\sqrt{15}$ **c** a and b $\pm\sqrt{ab}$

12 A 60-m steel pole is cut into two parts in the ratio of 11 to 4. How much longer is the longer part than the shorter? **28 m**

13 The ratio of the measures of the sides of a quadrilateral is 2:3:5:7. If the figure's perimeter is 68, find the length of each side. **8; 12; 20; 28**

Problem Set B

14 Find the positive arithmetic and geometric means between each pair of numbers. Note which mean is greater in each case.

a 8 and 50 **29; 20; A.M. is greater.**   **b** 6 and 12 **9; $6\sqrt{2}$; A.M. is greater.**

15 If 4 is a mean proportional between 6 and a number, what is the number? $\frac{8}{3}$

Cooperative Learning

Challenge students to find two different-sized squares such that the ratio of their perimeters is the same as the ratio of their areas.

Have students work in small groups to prove algebraically that there are or are not two such squares.

There are no two such squares.

In order for $\frac{4x}{4y} = \frac{x^2}{y^2}$ to be true, x must equal y. Therefore, the squares will not be different sizes.

16 Copy the number line and locate the arithmetic mean and the positive geometric mean between the two numbers. 10; 5√3

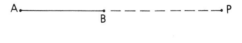

17 The ratio of the measure of the supplement of an angle to the measure of the complement of the angle is 5:2. Find the measure of the supplement. 150

18 Is $\frac{x-5}{4} = \frac{c}{3}$ equivalent to $\frac{x-1}{4} = \frac{c+3}{3}$? (Hint: Use what was proved in sample problem 6 as a theorem.) Yes

19 If $x(a+b) = y(c+d)$, find the ratio of x to y. $\frac{c+d}{a+b}$

20 If $ex - fy = gx + hy$, find the ratio of x to y. $\frac{f+h}{e-g}$

21 Reduce the ratio $\frac{x^2-7x+12}{x^2-16}$ to lowest terms. $\frac{x-3}{x+4}$

22 The length of a model plane is $10\frac{1}{2}$ in. The scale of the model is 1:72.

   a What is the length of the real plane? 756 in., or 63 ft

   b If the real plane has a wingspan of $43\frac{1}{2}$ ft, find the wingspan of the model. 7.25 in.

   c If another model of the same plane has a scale of 1:48, find the length of that model. 15.75 in.

Problem Set C

23 Show that no polygon exists in which the ratio of the number of diagonals to the sum of the measures of the polygon's angles is 1 to 18.

24 If $\frac{a}{b} = \frac{c}{d}$, show that $\frac{a-b}{b} = \frac{c-d}{d}$.

25 In the figure, P is said to *divide* $\overline{AB}$ *externally* into two segments, $\overline{AP}$ and $\overline{PB}$. If AB = 30 and $\frac{AP}{AB} = \frac{5}{2}$, find AP. 75

26 The equation $y = \frac{5}{2}x - 3$ relates the x- and y-coordinates of points on a line. Find the points on the line whose x-coordinates are 6 and 10. Then use these points to find the slope of the line. (6, 12), (10, 22); slope = $\frac{5}{2}$

Problem Set D

27 If two ratios are formed at random from the four numbers 1, 2, 4, and 8, what is the probability that the ratios are equal? $\frac{1}{3}$

Problem-Set Notes and Additional Answers

23 $\frac{\frac{n(n-3)}{2}}{180(n-2)} = \frac{1}{18}$ becomes $n^2 - 23n + 40 = 0$. There is no integral solution for n.

24 $\frac{a}{b} = \frac{c}{d}$
$\frac{a}{b} - 1 = \frac{c}{d} - 1$
$\frac{a}{b} - \frac{b}{b} = \frac{c}{d} - \frac{d}{d}$
$\frac{a-b}{b} = \frac{c-d}{d}$

25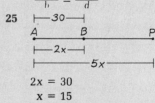
$2x = 30$
$x = 15$
$AP = 5x = 5(15) = 75$

26 Substitute 6 and 10 into the equation to find corresponding y values.
slope $= \frac{22-12}{10-6} = \frac{5}{2}$

27 $\frac{8}{4 \cdot 3 \cdot 2 \cdot 1} = \frac{1}{3}$

8.2 SIMILARITY

Objective

After studying this section, you will be able to
- Identify the characteristics of similar figures

Part One: Introduction

Below are three pairs of *similar* figures—figures that have the same shape but not necessarily the same size.

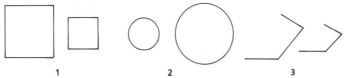

1   2   3

You need only look around you to find examples of similar figures. Whenever you use a pair of binoculars, look at a photograph, or read a map, you are dealing with similar figures. A knowledge of similarity and proportion is also useful in the building of model planes and automobiles and in the construction of electric-train layouts.

One way in which a figure similar to another figure can be produced is called *dilation*, or enlargement. The opposite of dilation, called *reduction*, also produces similar figures.

Example 1   *A pinhole camera produces a reduced image of a candle. The size of the image is proportional to the distance of the candle from the camera. Given the measurements shown in the diagram, find the height of the candle.*

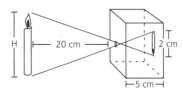

To find the height, we write and solve a proportion.

$\frac{H}{2} = \frac{20}{5}$

$H = 8$

The candle is 8 cm tall.

332 | Chapter 8 Similar Polygons

Vocabulary

dilation	similar
reduction	similar polygon

In this book, except for a few problems, we shall limit our study of similar figures to similar polygons.

Definition   ***Similar polygons*** are polygons in which
> 1 The ratios of the measures of corresponding sides are equal
> 2 Corresponding angles are congruent

Communicating Mathematics

Have students write a paragraph that explains the difference between similar and congruent polygons.

The triangles below are similar triangles. They have the same shape, although they differ in size.

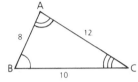
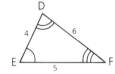

We write △ABC ~ △DEF ("triangle ABC is similar to triangle DEF"), which means that A corresponds to D, B corresponds to E, and C corresponds to F.

As you can see,

> 1 The ratios of the measures of all pairs of corresponding sides are equal

$$\frac{AB}{DE} = \frac{2}{1} \qquad \frac{AC}{DF} = \frac{2}{1} \qquad \frac{BC}{EF} = \frac{2}{1}$$

> 2 Each pair of corresponding angles are congruent

$$\angle B \cong \angle E \qquad \angle A \cong \angle D \qquad \angle C \cong \angle F$$

Example 2   *△MCN is a dilation of △MED, with an enlargement ratio of 2:1 for each pair of corresponding sides. Find the lengths of the sides of △MCN.*

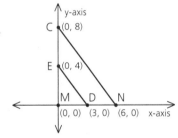

Since each side of △MCN is twice as long as the corresponding side of △MED, MC = 8 and MN = 6. To find the length of $\overline{CN}$, we can use the fact that in any right triangle with legs a and b and hypotenuse c, $a^2 + b^2 = c^2$.

$$(CN)^2 = 8^2 + 6^2$$
$$= 100$$
$$CN = 10$$

Example 3   Given: ABCD ~ EFGH, *with measures as shown.*

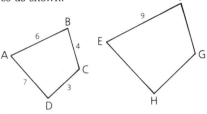

a *Find FG, GH, and EH.*

Since the quadrilaterals are similar, the ratios of the measures of their corresponding sides are equal. We begin with one ratio of measures of corresponding sides, preferably one we can simplify.

Thus, $\frac{AB}{EF} = \frac{6}{9} = \frac{2}{3}$.

$\frac{AB}{EF} = \frac{BC}{FG}$	$\frac{AB}{EF} = \frac{CD}{GH}$	AB:EF = AD:EH
$\frac{2}{3} = \frac{4}{FG}$	$\frac{2}{3} = \frac{3}{GH}$	2:3 = 7:EH
2(FG) = 12	2(GH) = 9	2(EH) = 21
FG = 6	GH = $4\frac{1}{2}$	EH = $10\frac{1}{2}$

b *Find the ratio of the perimeter of ABCD to the perimeter of EFGH.*

Perimeter of ABCD = 6 + 4 + 3 + 7 = 20

Perimeter of EFGH = 9 + 6 + $4\frac{1}{2}$ + $10\frac{1}{2}$ = 30

$$\frac{P_{ABCD}}{P_{EFGH}} = \frac{20}{30} = \frac{2}{3}$$

Notice that in the preceding example the ratio of perimeters was equal to the ratio of sides. This result suggests the following theorem.

Theorem 61   ***The ratio of the perimeters of two similar polygons equals the ratio of any pair of corresponding sides.***

Part Two: Sample Problems

Problem 1   *Given that △JHK ~ △POM, ∠H = 90°, ∠J = 40°, m∠M = x + 5, and m∠O = $\frac{1}{2}y$, find the values of x and y.*

Solution   First draw triangles JHK and POM so that ∠H = 90°, ∠J = 40°, and the corresponding angles are congruent.

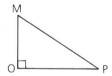

Lesson Notes, continued

■ Proof of Theorem 61

Given: △GHJ ~ △RST

Prove: $\frac{GH + HJ + GJ}{RS + ST + RT} = \frac{HJ}{ST}$

Proof: Suppose $\frac{HJ}{ST} = k$. Then

HJ = k · ST. Similarly, GH = k · RS and GJ = k · RT. Adding, GH + HJ + GJ = k · RS + k · ST + k · RT = k(RS + ST + RT).

Thus, $\frac{GH + HJ + GJ}{RS + ST + RT} = k$.

$$\left.\begin{array}{l}\angle J \text{ comp. } \angle K \\ \angle J = 40°\end{array}\right\} \Rightarrow$$

$$\angle K = 50°$$
$$\angle K \cong \angle M \qquad \angle H \cong \angle O$$
$$50 = x + 5 \qquad 90 = \tfrac{1}{2}y$$
$$45 = x \qquad 180 = y$$

Problem 2

Given: △BAT ~ △DOT,
OT = 15, BT = 12, TD = 9

Find the value of x (AO).

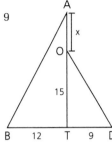

Solution

Since △BAT ~ △DOT, the ratios of the measures of corresponding sides are equal.

$$\frac{AT}{OT} = \frac{BT}{TD}$$
$$\frac{x + 15}{15} = \frac{12}{9}$$
$$\frac{x + 15}{15} = \frac{4}{3}$$
$$3(x + 15) = 4(15) \qquad \text{Means-Extremes Products Theorem}$$
$$3x + 45 = 60$$
$$3x = 15$$
$$x = 5$$

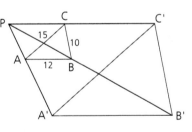

Problem 3

In the diagram, segments PA, PB, and PC are drawn to the vertices of △ABC from an external point P, then extended to three times their original lengths to points A', B', and C'. What are the lengths of the sides of △A'B'C'?

Solution

It appears that △A'B'C' ~ △ABC. (In the next section we will develop some theorems that will allow you to prove that the triangles are similar.) In fact, △A'B'C' is a dilation of △ABC, with a dilation ratio of 3:1 for each pair of corresponding sides.

A'B' = 3(AB) = 3(12) = 36
B'C' = 3(BC) = 3(10) = 30
A'C' = 3(AC) = 3(15) = 45

Assignment Guide

Basic
_____
1, 4–10, 15
Average
_____
3, 6, 8–16
Advanced
_____
6, 8–18

Cooperative Learning

1 Have students work in small groups to compare their answers to problems **1** and **2**. Then have them determine what is the *least* amount of information necessary to prove each of the following.
   a Two rectangles are similar.
   Lengths of adjacent sides of both rectangles
   b Two circles are similar.
   No information—all circles are similar
   c Two triangles are similar.
   Two angle measures of both triangles
   d Two nonregular hexagons
   Similar lengths of six sides, or measures of five angles
   e Two regular hexagons
   All are similar.
2 What generalization can be drawn about similarity and regular polygons from the above information? Students should generalize that regular polygons with the same number of sides are similar.

Part Three: Problem Sets

Problem Set A

1 Which pairs of figures *appear* to be similar? *b, c,* and *d*

a

c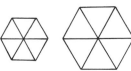

b

d

2 Which pairs of polygons *can be proved* to be similar? *b* and *c*

a

c

b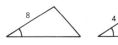

d

3 Given: △NPR ~ △STV,
m∠P = 90, m∠R = 60,
SV = 15, NR = 20, RP = 10
Find: m∠T, m∠S, and VT 90; 30; 7.5

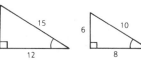

4 Given: △ABC ~ △DEF,
with lengths as shown
Find: EF 5

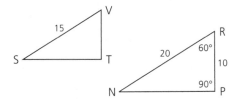

5 Given: ⊙O, ⊙P, △AOB ~ △RPS,
OA = 2, AB = 3, PR = 6
Find: PS and RS 6; 9

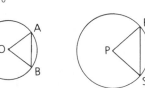

6 Find the mean proportionals between each pair of extremes.
   a 4 and 25 ±10   **b** 2 and 5 ±√10

7 If 3x = 5y, find the ratio of x to y. $\frac{5}{3}$

8 △OKM is a dilation of △OHJ, with a dilation ratio of 3:1 for each pair of corresponding sides.

a Find the coordinates of K.

b Find the lengths of the sides of △OHJ.

c Find the lengths of the sides of △OKM.

a (0, 9); **b** OH = 3, OJ = 4, HJ = 5;
c OK = 9, OM = 12, KM = 15

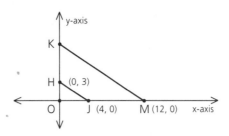

Problem Set B

9 Given: △SVT ~ △WYX,
with measures as shown

Find: WY and VT 5.6; 7.5

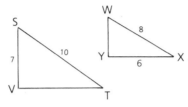

10 Given: Quad ABCD ~ quad HGFE,
with measures as shown

Find: **a** The ratio of lengths of corresponding sides $\frac{3}{2}$

b EF 8

c The perimeter of EFGH 32

d The ratio of the perimeters $\frac{3}{2}$

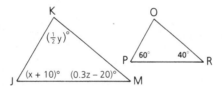

11 Given: △KJM ~ △OPR,
with angles as shown

Find: $\dfrac{x + y + z}{2}$ 205

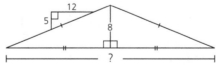

12 Find the ratio of the fourth proportional of 1, 2, and 3 to the fourth proportional of 4, 5, and 6. $\frac{4}{5}$

13 If $\dfrac{8}{2x - 3y} = \dfrac{7}{6x - 4y}$, find the ratio of x to y. $\frac{11}{34}$

14 The roof of a house has a slope of $\frac{5}{12}$. What is the width of the house if the height of the roof is 8 ft? 38.4 ft

15 Hammond R. looked at the plans for the new house he was building. The plans were drawn to a scale of $\frac{1}{4}$ in. = 1 ft. He measured the size of a room on the plans and found it to be 2.75 in. by 3.5 in. About how large is the room? 11 ft by 14 ft

16 Answers will vary.

18
$$\begin{cases} 2x + 5y = 50 \\ 180 - (2x + 5y + 5x + y) \\ = 102 - x \end{cases}$$
$$x = 5$$
$$y = 8$$
$$\therefore \angle F = 5x + y = 33$$

19 Use the Pythagorean
Theorem.
$$10^2 + (NP)^2 = 20^2$$
$$NP = 10\sqrt{3}$$
From similar triangles,
$ST = 7.5\sqrt{3}$.
Note: This problem is an
excellent preview of radicals
and of 30°-60°-90° triangles.
It is suggested that you do
this problem in class or
assign it for extra credit.

Problem Set B, *continued*

16 Draw a triangle. Using some point P in the interior of the triangle
as the point of dilation, draw a triangle twice the size of the
original triangle.

17 The projector shown uses a
slide in which the rectangular
transparency measures 3 cm by
4 cm. The slide is 5 cm behind
the lens. How large is the rectangular image on the screen?
180 cm by 240 cm

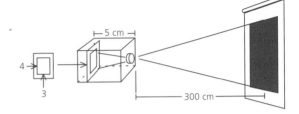

Problem Set C

18 Given: △ABC ~ △DEF,
m∠A = 50, m∠D = 2x + 5y,
m∠F = 5x + y, m∠B = 102 − x

Find: m∠F 33

19 Look again at problem 3. Find the length of $\overline{NP}$ in simplified
form. Then quickly find ST. $10\sqrt{3}$; $7.5\sqrt{3}$

8.3 METHODS OF PROVING TRIANGLES SIMILAR

Objective

After studying this section, you will be able to
- Use several methods to prove that triangles are similar

Part One: Introduction

In this section, we will present ways to prove that triangles are similar. We start by accepting one method as a postulate.

Postulate | *If there exists a correspondence between the vertices of two triangles such that the three angles of one triangle are congruent to the corresponding angles of the other triangle, then the triangles are similar. (AAA)*

The following three theorems will be used in proofs much as SSS, SAS, ASA, HL, and AAS were used in proofs to establish congruency.

Theorem 62 | *If there exists a correspondence between the vertices of two triangles such that two angles of one triangle are congruent to the corresponding angles of the other, then the triangles are similar. (AA)*

Given: ∠A ≅ ∠D,
          ∠B ≅ ∠E
Conclusion: △ABC ~ △DEF

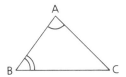
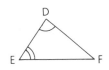

The proof of Theorem 62 follows from the No-Choice Theorem (p. 302).

We also present, without proof, two additional methods of proving that two triangles are similar. You will discover, however, that AA is the most frequently used of the three methods.

Class Planning

Time Schedule
Basic: 2 days
Average: $1\frac{1}{2}$ days
Advanced: 1 day

Resource References
Teacher's Resource Book
    Class Opener 8.3A, 8.3B
    Additional Practice Worksheet 15
Transparency 13

Class Opener

Given: △PRB ~ △WNM,
          PR = 20, PB = 18,
          RB = 22, WN = 12
Find: NM and WM.
$\frac{PR}{WN} = \frac{PB}{WM} = \frac{RB}{NM}$,
so $\frac{20}{12} = \frac{18}{WM} = \frac{22}{NM}$.
∴ WM = 10.8 and NM = 13.2
This problem emphasizes the importance of corresponding vertices in similar figures. It also shows that it is unrealistic to expect whole-number answers.

Theorem 63 *If there exists a correspondence between the vertices of two triangles such that the ratios of the measures of corresponding sides are equal, then the triangles are similar. (SSS~)*

Given: $\dfrac{AB}{DE} = \dfrac{BC}{EF} = \dfrac{AC}{DF}$

Prove: $\triangle ABC \sim \triangle DEF$

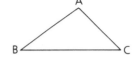

Theorem 64 *If there exists a correspondence between the vertices of two triangles such that the ratios of the measures of two pairs of corresponding sides are equal and the included angles are congruent, then the triangles are similar. (SAS~)*

Given: $\dfrac{AB}{DE} = \dfrac{BC}{EF}$,
$\quad\ \angle B \cong \angle E$

Prove: $\triangle ABC \sim \triangle DEF$

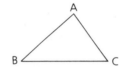

Part Two: Sample Problems

Problem 1 Given: ABCD is a ▱.

Prove: $\triangle BFE \sim \triangle CFD$

Proof

1 ABCD is a ▱.	1 Given
2 $\overleftrightarrow{AB} \parallel \overleftrightarrow{DC}$	2 Opposite sides of a ▱ are ∥.
3 $\angle CDF \cong \angle E$	3 ∥ lines ⟹ alt. int. ∠s ≅
4 $\angle DFC \cong \angle EFB$	4 Vertical angles are ≅.
5 $\triangle BFE \sim \triangle CFD$	5 AA (3, 4)

Problem 2 Given: $\overline{LP} \perp \overline{EA}$;
$\qquad\qquad$ N is the midpoint of $\overline{LP}$.
$\qquad\qquad$ P and R trisect $\overline{EA}$.

Prove: $\triangle PEN \sim \triangle PAL$

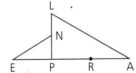

Proof Since $\overline{LP} \perp \overline{EA}$, $\angle NPE$ and $\angle LPA$ are congruent right angles. If N is the midpoint of $\overline{LP}$, $\frac{NP}{LP} = \frac{1}{2}$. But P and R trisect $\overline{EA}$, so $\frac{EP}{PA} = \frac{1}{2}$. Therefore, $\triangle PEN \sim \triangle PAL$ by SAS~.

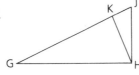

Problem 3   Given: $\overline{KH}$ *is the altitude to*
hypotenuse $\overline{GJ}$ *of right* $\triangle GHJ$.

Prove: $\triangle KHJ \sim \triangle HGJ$

Proof

1 $\overline{KH}$ is the altitude to hypotenuse $\overline{GJ}$ of $\triangle GHJ$.	1 Given
2 $\angle HKJ$ is a right angle.	2 An altitude of a $\triangle$ is drawn from a vertex and forms right $\angle$s with the opposite side.
3 $\angle JHG$ is a right angle.	3 The hypotenuse is opposite the right $\angle$.
4 $\angle HKJ \cong \angle JHG$	4 Right $\angle$s are $\cong$.
5 $\angle J \cong \angle J$	5 Reflexive Property
6 $\triangle KHJ \sim \triangle HGJ$	6 AA (4, 5)

Problem 4   *The sides of one triangle are 8, 14, and 12, and the sides of another triangle are 18, 21, and 12. Prove that the triangles are similar.*

Proof   We can determine the ratios of corresponding sides to see whether the ratios are equal.

Shortest sides: $\dfrac{8}{12} = \dfrac{2}{3}$

Longest sides: $\dfrac{14}{21} = \dfrac{2}{3}$

Other sides: $\dfrac{12}{18} = \dfrac{2}{3}$

Since the ratio is the same for each pair of corresponding sides, the two triangles are similar by SSS~.

Part Three: Problem Sets

Problem Set A

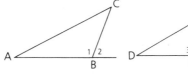

1 Given: $\angle A \cong \angle D$,
$\angle 2 \cong \angle 4$

Prove: $\triangle ABC \sim \triangle DEF$

2 Draw a triangle GJK. Then indicate a point H on $\overline{GJ}$ and a point M on $\overline{GK}$ such that $\overline{HM} \parallel \overline{JK}$. Prove that $\triangle GHM \sim \triangle GJK$.

3 Given: NPRV is a $\square$.

Conclusion: $\triangle NWO \sim \triangle SWT$

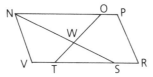

Assignment Guide

Basic

Day 1	1, 3, 6, 8, 9
Day 2	4, 7, 11, 12, 16, 19

Average

Day 1	2, 3, 6–10
Day 2	($\frac{1}{2}$ day) Section **8.3** 16–18
	($\frac{1}{2}$ day) Section **8.4** 1, 3, 5, 9

Advanced

3, 7, 11, 17–20, 22

Problem-Set Notes and Additional Answers

■ See *Solution Manual* for answers to problems **1–3**.

Communicating Mathematics

Have students do problem **2** with an acute, a right, and an obtuse triangle. Will the result be different for each type of triangle? No. Explain. All triangles will be similar by AA~.

T 341

Problem-Set Notes and
Additional Answers, continued

■ See Solution Manual for answers to problems **4–6** and **9**.

Problem Set A, continued

4 Given: $\overline{AC} \cong \overline{AE}$,
       $\angle CBD \cong \angle EFD$
  Prove: $\triangle BCD \sim \triangle FED$

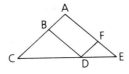

5 Given: TVPX is a trapezoid, with bases
    $\overline{TV}$ and $\overline{XP}$.
  Conclusion: $\triangle TVY \sim \triangle PXY$

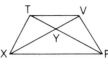

6 Find the coordinates of B if $\triangle OAC \sim$
$\triangle OBD$. Then write a paragraph proof to
show that $\triangle OAC \sim \triangle OBD$. Challenge:
Can you find the length of $\overline{BD}$? (0, 10); BD = 26

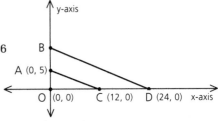

7 Given: $\angle G$ is a right $\angle$.
      $\angle K$ is a right $\angle$.
      $HJ = \frac{1}{2}(MO)$
  Prove: $\triangle GHJ \sim \triangle KMO$
Cannot be proved

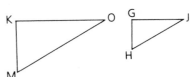

8 In $\triangle FGH$, FG = 6, GH = 8, and FH = 12. $\triangle FGH$ is projected
onto a wall, and the image, $\triangle F'G'H'$, has sides $F'G' = 15$,
$G'H' = 20$, and $F'H' = 30$. Is $\triangle FGH$ similar to $\triangle F'G'H'$?
Explain. **Yes, by SSS~**

9 Given: $\angle PTA \cong \angle B$
  Prove: $\triangle PAT \cong \triangle PTB$

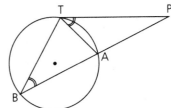

10 The slope of line PQ is $-\frac{3}{2}$. Find the
coordinates of P and Q. (0, 3); (2, 0)

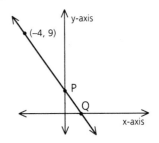

11 Given: △A'B'C' is not a dilation of
△ABC.

Prove: A'C' ≠ 12.3

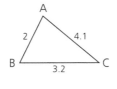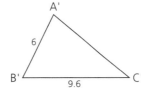

Problem Set B

12 Given: $\overline{SP}$ is the altitude from S to $\overline{NR}$.
$\overline{RT}$ is the altitude from R to $\overline{NS}$.

Conclusion: △NRT ~ △NSP

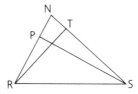

13 Prove that if an acute angle of one right triangle is congruent to
an acute angle of another right triangle, the triangles are similar.

14 Prove that if the vertex angle of one isosceles triangle is congruent to the vertex angle of a second isosceles triangle, the triangles
are similar.

15 Given: $\dfrac{GJ}{HK} = \dfrac{GK}{GM}$,
∠1 ≅ ∠G

Conclusion: $\overleftrightarrow{HM} \parallel \overleftrightarrow{JK}$

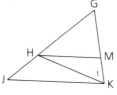

16 Indicate whether the statement is true Always, Sometimes, or
Never (A, S, or N).

a If two triangles are similar, then they are congruent. S

b If two triangles are congruent, then they are similar. A

c An obtuse triangle is similar to an acute triangle. N

d Two right triangles are similar. S

e Two equilateral polygons are similar. S

f Two equilateral triangles are similar. A

g Two rectangles are similar if neither is a square. S

17 From two points, one on each leg of an isosceles triangle, perpen-
diculars are drawn to the base. Prove that the triangles formed
are similar.

18 Given: A = (1, 2), B = (9, 8), C = (1, 8),
P = (5, −3), Q = (−7, 6), R = (−7, −3),
AB = 10, PQ = 15

By which theorem is △ABC ~ △QPR? SAS~ or SSS~

*Problem-Set Notes and
Additional Answers, continued*

■ See *Solution Manual* for answers to problems **11–15** and
17.

Cooperative Learning

Have students work in small
groups to solve problem **16.** Students should be prepared to justify their answers. Choose a
student at random from each
group to present the solution to
each part of problem **16.**

Problem Set B, *continued*

19 Given: Figure as shown

a Is △PQT ~ △PRS? Justify your reasoning. Yes, by SAS~

b Is $\overline{QT}$ parallel to $\overline{RS}$? Justify your reasoning. Yes, by corr. ∠s ⇒ ‖ lines

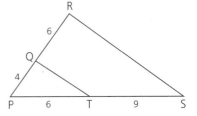

20 A line is graphed at the right.

a What is the slope of the line? $-\frac{1}{2}$

b As the x values of points on the line increase by 3, by how much do the y values increase or decrease? Decrease by 1.5

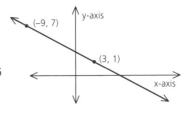

Problem Set C

21 Prove that two triangles similar to a third triangle are similar to each other (the transitive property of similar triangles). Do you think the transitive property could be applied to other similar polygons?

22 If two of the six triangles below are selected at random, what is the probability that the two triangles are similar? $\frac{4}{15}$

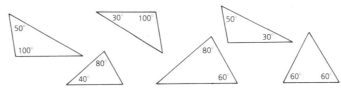

Problem-Set Notes and Additional Answers, continued

21 The transitive property of similar △ follows directly from the transitive property of congruent ∠s.
Yes. For similar polygons, draw the diagonals from a set of corresponding vertices. The corresponding △ will be similar, so the corresponding sides of the polygons will be proportional and their corresponding ∠s will be ≅.

22 I,II II,III III,IV IV,V V,VI
Ⓘ,Ⓘ Ⓘ,Ⓘ Ⓘ,Ⓥ IV,VI
I,IV II,V III,VI
Ⓘ,Ⓥ II,VI $\frac{4}{15}$
I,VI

CONGRUENCES AND PROPORTIONS IN SIMILAR TRIANGLES

Objective

After studying this section, you will be able to
- Use the concept of similarity to establish the congruence of angles and the proportionality of segments

Part One: Introduction

As you have seen, if we know that two triangles are congruent, we can use the definition of *congruent triangles* (CPCTC) to prove that pairs of angles and sides are congruent. In like fashion, once we know that two triangles are similar, we can use the definition of *similar polygons* to prove that

1 Corresponding sides of the triangles are proportional (The ratios of the measures of corresponding sides are equal.)
2 Corresponding angles of the triangles are congruent

If a problem asks you to prove that *products* of the measures of sides are equal, try using the Means-Extremes Products Theorem.

Example 1   *Given:* $\triangle ABC \sim \triangle DEF$
Prove: $\angle A \cong \angle D$

| 1 $\triangle ABC \sim \triangle DEF$ | 1 Given |
| 2 $\angle A \cong \angle D$ | 2 Corresponding $\angle$s of $\sim$ ⚠ are $\cong$. |

Example 2   *Given:* $\triangle ABC \sim \triangle DEF$
Prove: $\dfrac{AB}{DE} = \dfrac{AC}{DF}$

| 1 $\triangle ABC \sim \triangle DEF$ | 1 Given |
| 2 $\dfrac{AB}{DE} = \dfrac{AC}{DF}$ | 2 Corresponding sides of $\sim$ ⚠ are proportional. |

Note   We may also write $\dfrac{AB}{AC} = \dfrac{DE}{DF}$, since this proportion is equivalent to $\dfrac{AB}{DE} = \dfrac{AC}{DF}$.

Class Planning

Time Schedule
Basic: 2 days
Average: $1\frac{1}{2}$ days
Advanced: 1 day

Resource References
Teacher's Resource Book
  Class Opener 8.4A, 8.4B
  Using Manipulatives 14
Evaluation
  Tests and Quizzes
    Quizzes 1–3, Series 1
    Quizzes 1–4, Series 2

Class Opener

Given: N is the midpoint of $\overline{ML}$.
    F is the midpoint of $\overline{TL}$.
Prove: $\triangle NFL \sim \triangle MTL$
Discussion should revolve around two possible logical solutions.

1   Using the fact that a midpoint divides a segment in half, SAS $\sim$ follows.
2   Using the midline theorem, SSS $\sim$ follows.

Lesson Notes

- Students should realize that in a statement of triangle similarity, as in a statement of congruence, the order of the vertices is important. If the vertices are correctly ordered, students can read the proportions directly from the statement of similarity.

Example 3 Given: △ABC ~ △DEF
Prove: AB · DF = AC · DE

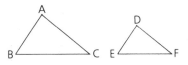

1 △ABC ~ △DEF	1 Given
2 $\dfrac{AB}{DE} = \dfrac{AC}{DF}$ (or $\dfrac{AB}{AC} = \dfrac{DE}{DF}$)	2 Corresponding sides of ~ △ are proportional.
3 AB · DF = AC · DE	3 Means-Extremes Products Theorem

Part Two: Sample Problems

Problem 1 Given: $\overleftrightarrow{BD} \parallel \overleftrightarrow{CE}$
Prove: AB · CE = AC · BD

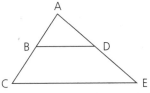

Proof

1 $\overleftrightarrow{BD} \parallel \overleftrightarrow{CE}$	1 Given
2 ∠ABD ≅ ∠C	2 ∥ lines ⇒ corr. ∠s ≅
3 ∠ADB ≅ ∠E	3 Same as 2
4 △ABD ≅ △ACE	4 AA (2, 3)
5 $\dfrac{AB}{AC} = \dfrac{BD}{CE}$	5 Corresponding sides of ~ △ are proportional.
6 AB · CE = AC · BD	6 Means-Extremes Products Theorem

Note In sample problem 1, we worked backwards. In order to conclude that AB · CE = AC · BD, we looked for a proportion involving AB, AC, CE, and BD, the lengths of sides of a pair of similar triangles. Working backwards helped us to think through the logical steps that we would need.

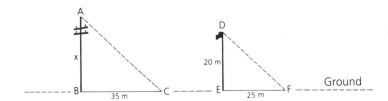

Problem 2 *While strolling one morning to get a little sun, Judy noticed that a 20-m flagpole cast a 25-m shadow. Nearby was a telephone pole that cast a 35-m shadow. How tall was the telephone pole? (A shadow problem)*

Solution

Because the sun is very far from us, its rays are nearly parallel. △ABC ~ △DEF by AA, so we can write a proportion.

$$\frac{x}{35} = \frac{20}{25}$$

$$\frac{x}{35} = \frac{4}{5}$$

$$5x = 140$$

$$x = 28$$

The pole was 28 m high.

Problem 3

Given: ▱YSTW,
SX ⊥ YW,
SV ⊥ WT

Prove: SX · YW = SV · WT

Proof

1 ▱YSTW	1 Given
2 ∠Y ≅ ∠T	2 Opposite ∠s of a ▱ are ≅.
3 SX ⊥ YW	3 Given
4 ∠SXY is a right ∠.	4 ⊥ segments form right ∠s.
5 SV ⊥ WT	5 Given
6 ∠SVT is a right ∠.	6 Same as 4
7 ∠SXY ≅ ∠SVT	7 Right ∠s are ≅.
8 △SXY ~ △SVT	8 AA (2, 7)
9 $\frac{SX}{SV} = \frac{SY}{ST}$	9 Corresponding sides of ~ △ are proportional.
10 SX · ST = SV · SY	10 Means-Extremes Products Theorem
11 ST ≅ YW	11 Opposite sides of a ▱ are ≅.
12 SY ≅ WT	12 Same as 11
13 SX · YW = SV · WT	13 Substitution (11 and 12 in 10)

In this proof, we again found it useful to work backwards. This time, lengths YW and WT were not sides of similar triangles. But since SYTW is a parallelogram, we were able to substitute these lengths for the lengths of the opposite sides.

Part Three: Problem Sets

Problem Set A

1 Given: ∠C ≅ ∠F,
AB ⊥ BC,
DE ⊥ EF

Prove: $\frac{AB}{BC} = \frac{DE}{EF}$

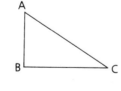

Lesson Notes, continued

■ Students should draw a diagram and mark the congruent angles in it. This step will help them set up the proper correspondence and hence the proper proportions.

Assignment Guide

Basic

Day 1	1–5, 7
Day 2	6, 8–11, 14, 18

Average

Day 1	($\frac{1}{2}$ day) Section **8.3** 16–18
	($\frac{1}{2}$ day) Section **8.4** 1, 3, 5, 9
Day 2	2, 8, 10–12, 14, 17, 20

Advanced

6, 8, 10, 12, 14, 18, 20, 24

**Problem-Set Notes
and Additional Answers**

■ See *Solution Manual* for answer to problem **1**.

Communicating Mathematics

Have students use the given condition in problem **6** to write two
other multiplication facts that
could be proved.
SW · SY = SX · SZ
SZ · XY = SY · WZ

Problem Set A, *continued*

2 Given: $\angle X \cong \angle ZBA$

Conclusion: $\dfrac{AZ}{AB} = \dfrac{ZY}{XY}$

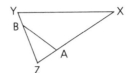

3 Given: $\angle D \cong \angle G$

Conclusion: $\dfrac{CD}{FG} = \dfrac{DE}{EG}$

4 Given: $\angle HJK$ is a right $\angle$.
$\overline{JM}$ is an altitude.

Prove: $\dfrac{JM}{MK} = \dfrac{HJ}{JK}$

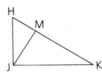

5 Given: $\triangle NOP \sim \triangle RST$
Prove: NO · RT = RS · NP

6 Given: $\overleftrightarrow{WZ} \parallel \overleftrightarrow{XY}$

Conclusion: WS · XY = XS · WZ

7 $\triangle ABC \sim \triangle DEF$
Find: AC and EF 4; 14

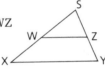

8 Given: GJKL ~ MOPR
Find: OP, PR, and MR 8; 4; $5\frac{1}{3}$

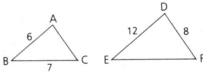

9 *A shadow problem:* Mannertink observed that a tree was casting
a 30-m shadow. A nearby flagpole was casting a 24-m shadow. If
the flagpole was 20 m high, how tall was the tree? 25 m

10 If two similar kites have perimeters of 21 and 28, what is the
ratio of the measures of two corresponding sides? $\frac{3}{4}$

11 Using the diagram at the right, show that $\overline{AB} \parallel \overline{DE}$.

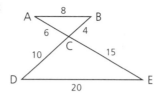

Problem Set B

12 Given: $\square ACEG$, with F the
midpoint of EG,
$\angle ABH \cong \angle EFD$

Prove: $AB \cdot FD = HB \cdot GF$

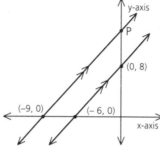

13 Prove that the ratio of corresponding altitudes of similar triangles
is equal to the ratio of any pair of corresponding sides of the
triangles.

14 Find the coordinates of point P in the
diagram. (0, 12)

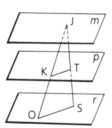

15 Given: $m \parallel p \parallel r$;
J lies in m.
$\overline{KT}$ lies in p.
$\overline{OS}$ lies in r.

Prove: $\dfrac{JK}{JO} = \dfrac{JT}{JS}$

16 Given: Trapezoid ABCD, with bases $\overline{AB}$ and $\overline{CD}$
Prove: $AE \cdot CD = EC \cdot AB$

17 Given: $\angle M \cong \angle S$,
MP = 8,
PR = 6,
SP = 7

Find: PO $6\frac{6}{7}$

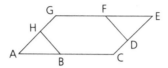

8.4

Problem-Set Notes and
Additional Answers, continued

■ See *Solution Manual* for answers to problems **11–13, 15,**
and **16.**

Cooperative Learning

Have students work in small
groups to compare their proofs of
problem **11.** Is there more than
one way to approach this problem? Is there any information
given in the problem that is
unnecessary?

It is not necessary to know the
lengths of $\overline{AB}$ and $\overline{DE}$. Similar
triangles can be proven by
SAS ~ by using segments $\overline{BC}$,
$\overline{DC}$, $\overline{AC}$, $\overline{EC}$, $\angle ACB$, and $\angle DCE$.

Problem Set B, *continued*

18 If $\triangle TVK \sim \triangle XZY$, TV = 8, VK = 9, TK = 10, and ZY = 4, find
XY. $4\frac{4}{9}$

19 Given: $\overleftrightarrow{BE} \parallel \overleftrightarrow{CD}$,
    AB = 6, BC = 2, BE = 9
Find: CD 12

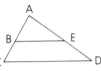

20 Shad is 3 ft from a lamppost that is 12 ft
high. Shad is $5\frac{1}{2}$ ft tall. How long is
Shad's shadow? $2\frac{7}{13}$ ft

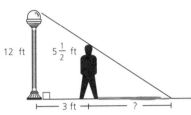

21 Given: $\overline{AD} \parallel \overline{BC}$,
    AB = 24, BC = 9,
    AD = 16, DB = 12

a How can you show that the two triangles are similar? SAS~

b Which angle is congruent to $\angle A$? $\angle BDC$

c Find CD. 18

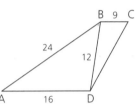

Problem Set C

22 Given: $\angle ACB$ is a right $\angle$.
    $\overline{CD}$ is an altitude.
Prove: $(CD)^2 = (AD)(DB)$

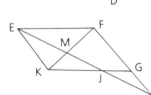

23 Given: EFGK is a $\square$.
    MJ = 4,
    JH = 5
Find: EM 6

24 When an object is placed on a ramp, part
of its weight w (which is a downward
force) is directed along the ramp as a
sliding force f. In physics, these forces
are represented by vectors with lengths
proportional to w and f.

a Find the angle between the two vectors. 67°

b If w is 50, what is f? 20

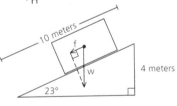

---

*Problem-Set Notes and
Additional Answers, continued*

22 $\angle A$ is comp. to $\angle ACD$.
    $\angle BCD$ is comp. to $\angle ACD$.
    $\therefore \angle A \cong \angle BCD$. $\angle ADC \cong$
    $\angle BDC$, and so $\triangle ADC \sim$
    $\triangle CDB$ by AA ~. Thus,
    $\frac{AD}{CD} = \frac{CD}{DB}$.
    $\therefore CD^2 = AD \cdot DB$

23 $\triangle EMF \sim \triangle JMK$, so
    $\frac{EM}{MJ} = \frac{MF}{MK}$.
    $\triangle EMK \sim \triangle HMF$, so
    $\frac{MF}{MK} = \frac{MH}{EM}$.
    Thus, $\frac{EM}{MJ} = \frac{MH}{EM}$,
    or $\frac{x}{4} = \frac{9}{x}$.
    $\therefore x = 6$

24a

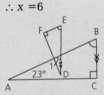

Since $\overline{BC} \parallel \overline{ED}$, $\angle B \cong \angle 1$
by corresponding $\angle$s.
Since $\overline{AB} \parallel \overline{FE}$, $\angle 1 \cong \angle E$
by corresponding $\angle$s.
$\therefore \angle B \cong \angle E$, and thus,
$\angle E = 90° - 23° = 67°$.

b From part **a**, $\triangle ABC \sim$
$\triangle DEF$.
Thus, $\frac{EF}{BC} = \frac{ED}{AB}$, or
    $\frac{EF}{4} = \frac{50}{10}$.
    $\therefore EF = 20$

THREE THEOREMS INVOLVING PROPORTIONS

Objective

After studying this section, you will be able to
- Apply three theorems frequently used to establish proportionality

Part One: Introduction

You will find the theorems presented in this section useful in a number of applications.

Theorem 65   *If a line is parallel to one side of a triangle and intersects the other two sides, it divides those two sides proportionally. (Side-Splitter Theorem)*

Given: $\overleftrightarrow{BE} \parallel \overleftrightarrow{CD}$

Prove: $\dfrac{AB}{BC} = \dfrac{AE}{ED}$

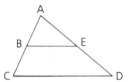

Theorem 66   *If three or more parallel lines are intersected by two transversals, the parallel lines divide the transversals proportionally.*

Given: $\overleftrightarrow{AB} \parallel \overleftrightarrow{CD} \parallel \overleftrightarrow{EF}$

Conclusion: $\dfrac{AC}{CE} = \dfrac{BD}{DF}$

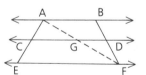

If you wish to prove this theorem, draw auxiliary segment AF and think about two opportunities of using the Side-Splitter Theorem. You may also find it a challenge to prove Theorem 66 for transversals intersecting four parallel lines.

Class Planning

Time Schedule
Basic: 2 days
Average: 2 days
Advanced: 1 day

Resource References
Teacher's Resource Book
  Class Opener 8.5A, 8.5B
  Additional Practice
  Worksheet 16
  Supposer Worksheets 8–10
Evaluation
  Tests and Quizzes
    Quiz 4, Series 1
    Quiz 1, Series 3

Class Opener

Given: $\overline{AT} \parallel \overline{BN}$
      FA = 12, FT = 15,
      AT = 14, AB = 8
Find: The length of BN and
      TN.   BN = $23\frac{1}{3}$, TN = 10

This problem is important because it leads into the Side-Splitter Theorem.

Lesson Notes

- Steps in proof of Theorem 65:
  $\triangle ABE \sim \triangle ACD$, so
  $$\frac{AB}{AC} = \frac{AE}{AD}$$
  $$\frac{AB}{AB + BC} = \frac{AE}{AE + ED}$$
  $$AB(AE + ED) = AE(AB + BC)$$
  $$AB \cdot AE + AB \cdot ED = AE \cdot AB + AE \cdot BC$$
  $$AB \cdot ED = AE \cdot BC$$
  $$\frac{AB}{BC} = \frac{AE}{ED}$$

T 351

Lesson Notes, continued

■ Some classes may need several days to grasp the relationships in Theorems 65, 66, and 67.
■ In the last days of covering Chapter **8**, all classes can include some review of radicals and quadratic equations to prepare students for their extensive use in Chapter **9**.

Cooperative Learning

Give students a sheet with three parallel lines cut by a transversal. The transversal should be cut in a common ratio (e.g., 2 to 5). Have students first individually measure the segments of the transversal and verify the ratio formed. Then have them draw another transversal across the parallel lines and measure the resulting line segments. Have students work in small groups to compare their measurements and try to form a generalization based on their results.
Students should recognize that the transversals drawn are divided proportionally by the parallel lines.

Another useful statement that can be made about parallel lines and their transversals is the following: *If parallel lines cut off (intercept) congruent segments on one transversal, they cut off congruent segments on any transversal.* Do you see how this statement is a consequence of Theorem 66? What is the ratio of lengths in such a case?

Theorem 67 *If a ray bisects an angle of a triangle, it divides the opposite side into segments that are proportional to the adjacent sides. (Angle Bisector Theorem)*

Given: $\triangle ABD$;
$\overrightarrow{AC}$ bisects $\angle BAD$.

Prove: $\dfrac{BC}{CD} = \dfrac{AB}{AD}$

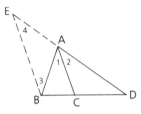

Proof:

1 $\triangle ABD$	1 Given
2 $\overrightarrow{AC}$ bisects $\angle BAD$.	2 Given
3 $\angle 1 \cong \angle 2$	3 If a ray bisects an $\angle$, it divides the $\angle$ into two $\cong$ $\angle$s.
4 Draw through B the line that is $\parallel$ to $\overline{AC}$.	4 Parallel Postulate
5 Extend $\overleftrightarrow{DA}$ to intersect the $\parallel$ line at some point E.	5 A line can be extended as far as desired
6 $\dfrac{BC}{CD} = \dfrac{EA}{AD}$	6 Side-Splitter Theorem
7 $\angle 1 \cong \angle 3$	7 $\parallel$ lines $\Rightarrow$ alt. int. $\angle$s $\cong$
8 $\angle 2 \cong \angle 4$	8 $\parallel$ lines $\Rightarrow$ corr. $\angle$s $\cong$
9 $\angle 3 \cong \angle 4$	9 Transitive Property (3, 7, 8)
10 $\overline{EA} \cong \overline{AB}$	10 If $\triangle$, then $\triangle$.
11 $\dfrac{BC}{CD} = \dfrac{AB}{AD}$	11 Substitution (10 in 6)

Part Two: Sample Problems

Problem 1   Given: $\overleftrightarrow{BE} \parallel \overleftrightarrow{CD}$,
          lengths as shown

Find: **a** ED
    b CD

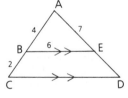

Solution Be alert. In problems involving this type of figure, you may need to use both the Side-Splitter Theorem and the properties of similar triangles.

a By the Side-Splitter Theorem,

$$\frac{AB}{BC} = \frac{AE}{ED}$$

$$\frac{4}{2} = \frac{7}{ED}$$

$$\frac{2}{1} = \frac{7}{ED}$$

$$2(ED) = 7$$

$$ED = 3\tfrac{1}{2}$$

b Since the parallel segments are involved, use the fact that △ABE ~ △ACD to write a proportion.

$$\frac{AB}{AC} = \frac{BE}{CD}$$

$$\frac{4}{4+2} = \frac{6}{CD}$$

$$\frac{2}{3} = \frac{6}{CD}$$

$$2(CD) = 18$$

$$CD = 9$$

Problem 2 *Given:* a ∥ b ∥ c ∥ d,
lengths as shown,
KP = 24

Find: KM

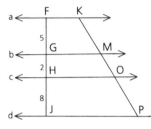

Solution According to Theorem 66, the ratio KM:MO:OP is equal to 5:2:8. Therefore, we let KM = 5x, MO = 2x, and OP = 8x. Since KP = 24,

$$5x + 2x + 8x = 24$$

$$15x = 24$$

$$x = \frac{24}{15} = \frac{8}{5}$$

Thus, KM = $5\left(\frac{8}{5}\right)$ = 8

Problem 3 *Given:* ∠RVS ≅ ∠SVT,
lengths as shown

Find: ST

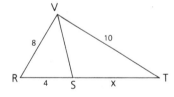

Solution By Theorem 67,

$$\frac{VR}{VT} = \frac{RS}{ST}$$

$$\frac{8}{10} = \frac{4}{ST}$$

$$\frac{4}{5} = \frac{4}{ST}$$

$$4(ST) = 20$$

$$ST = 5$$

Problem 4   Given: $\overleftrightarrow{XA} \parallel \overleftrightarrow{YZ}$,
$\angle XAY \cong \angle XYA$

Conclusion: $\dfrac{WX}{XA} = \dfrac{WA}{AZ}$

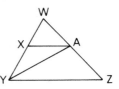

Proof

1 $\overleftrightarrow{XA} \parallel \overleftrightarrow{YZ}$	1 Given
2 $\dfrac{WX}{XY} = \dfrac{WA}{AZ}$	2 Side-Splitter Theorem
3 $\angle XAY \cong \angle XYA$	3 Given
4 $\overline{XA} \cong \overline{XY}$	4 If $\triangle$, then $\triangle$.
5 $\dfrac{WX}{XA} = \dfrac{WA}{AZ}$	5 Substitution (4 in 2)

Problem 5   Given: $\overleftrightarrow{DH} \parallel \overleftrightarrow{BC}$,
$\overleftrightarrow{HF} \parallel \overleftrightarrow{BG}$

Prove: $\dfrac{CD}{DE} = \dfrac{GF}{FE}$

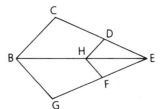

Proof

1 $\overleftrightarrow{DH} \parallel \overleftrightarrow{BC}$	1 Given
2 $\dfrac{CD}{DE} = \dfrac{BH}{HE}$	2 Side-Splitter Theorem
3 $\overleftrightarrow{HF} \parallel \overleftrightarrow{BG}$	3 Given
4 $\dfrac{BH}{HE} = \dfrac{GF}{FE}$	4 Same as 2
5 $\dfrac{CD}{DE} = \dfrac{GF}{FE}$	5 Transitive Property (2, 4)

Part Three: Problem Sets

Problem Set A

For problems 1–3, see sample problem 1.

1 Given: $\overline{BE} \parallel \overline{CD}$,
lengths as shown

Find: **a** ED **8**
b CD $\dfrac{35}{3}$

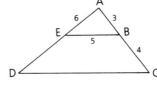

2 Solve for x and y in the figure shown.
6; $\frac{72}{5}$

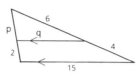

3 Solve for p and q in the figure shown.
3; 9

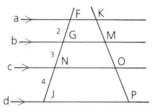

For problems 4 and 5, see sample problem 2.

4 Given: a ∥ b ∥ c ∥ d,
      lengths as shown,
      KP = 15
Find: KM, MO, and OP $\frac{10}{3}$; 5; $\frac{20}{3}$

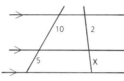

5 Solve for x in the diagram shown. 1

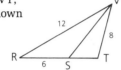

For problems 6 and 7, see sample problem 3.

6 Given: ∠RVS ≅ ∠SVT,
      lengths as shown
Find: ST 4

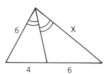

7 Given the diagram as marked, solve
for x. 9

8 A 60-m tower casts a 50-m shadow, while one-half block away a
telephone pole casts a 20-m shadow. How tall is the telephone
pole? 24 m

**Problem-Set Notes
and Additional Answers**

3 This is a "degenerative
triangle"; it violates the
Triangle Inequality Postulate.

Communicating Mathematics

Have students write any assumptions that they think were made
in problem **8** that might affect
the answer.
Typical response: A building is
not blocking the sun, so a
shadow can be cast.

The Geometric Supposer

You can use *The Geometric pre-Supposer: Points and Lines* or *The Geometric Supposer: Triangles* or *Circles* to explore proportions in triangles and parallel lines. Before beginning Section **8.5**, have students do the following.

Draw or construct a triangle ABC. Label a point D other than the midpoint on $\overline{AB}$. Draw a line parallel to $\overline{AC}$ through D. Label the point of intersection of that line with $\overline{BC}$ as E. Using the Measure option, find the ratios of the longer to the shorter segments formed on the two sides of the triangle: that is, find AD:DB and BE:EC. What is true about these two ratios? Repeat the activity and describe the pattern that emerges.

Draw three lines that are all parallel but different distances apart. Draw two nonparallel transversals through these lines. Using the Measure option, find the ratio between the lengths of each pair of segments formed on the transversal. Ask students to predict what will happen, given a new set of parallel lines and transversals. Have students repeat the activity to test their predictions.

Draw or construct any triangle. Draw the bisector of one of its angles. Define the bisector as a segment ending at the opposite side of the triangle. Using Measure, find the two ratios: one between the lengths of the two sides adjacent to the angle bisected, and the second between the two segments formed by the angle bisector on the side opposite the bisected angle. Ask students to make conjectures and to explore their conjectures by repeating the activity with a variety of different triangles.

Problem Set A, *continued*

9 Given: $\angle J \cong \angle MKO$,
MK = 12, KO = 8,
MO = 10, JK = 3
Find: PO and JP $\frac{5}{2}$; 10

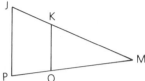

10 Given: $\overleftrightarrow{SV} \parallel \overleftrightarrow{RW}$,
RW = 15, RS = 10,
ST = 3, WV = 8
Find: SV and VT $\frac{45}{13}$, $\frac{12}{5}$

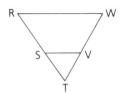

11 Given: $\overleftrightarrow{CD} \parallel \overleftrightarrow{BE}$,
AC = 18, AB = 12,
AE = 10, CD = 24
Find: The perimeter of trapezoid BEDC 51

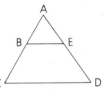

12 Given: $\overrightarrow{GJ}$ bisects $\angle FGH$.
FG = 10, GH = 8,
FJ = 7
Find: JH $\frac{28}{5}$

13 Given: r $\parallel$ s $\parallel$ t,
WV = 3,
WX = 8,
QY = 9
Find: QZ and ZY $\frac{27}{11}$, $\frac{72}{11}$

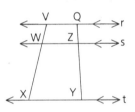

14 Given: $\overrightarrow{PT}$ bisects $\angle RPQ$.
PR = 30, PQ = 24
Find: The coordinates of point R (0, 18)

Problem Set B

15 Given: $\angle 1 \cong \angle 2$

Conclusion: $\dfrac{KM}{JK} = \dfrac{MO}{OP}$

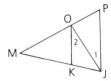

Problem-Set Notes and
Additional Answers, continued

■ See *Solution Manual* for answers to problems **15–17**.

16 Given: $\angle 3 \cong \angle 5$

Prove: $\dfrac{RV}{VT} = \dfrac{RS}{ST}$

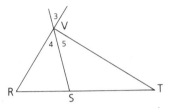

17 Given: WYZB is a trapezoid with bases $\overline{WB}$ and $\overline{YZ}$.
$\overleftrightarrow{XA} \parallel \overleftrightarrow{YZ}$

Prove: $\dfrac{WX}{XY} = \dfrac{WC}{CZ} = \dfrac{BA}{AZ}$

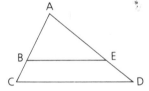

18 Given: $\overleftrightarrow{BE} \parallel \overleftrightarrow{CD}$,
    $AB = 4x$, $BC = x$,
    $AD = 8x$, $BE = 5x$

Find: AE and CD (in terms of x)
$\dfrac{32x}{5}, \dfrac{25x}{4}$

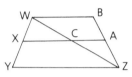

19 a One side of a triangle is 4 cm longer than another side. The ray bisecting the angle formed by these sides divides the opposite side into 5-cm and 3-cm segments. Find the perimeter of the triangle. 24 cm

   b If the first side of the triangle in part **a** were x cm longer than the second side and the other information were unchanged, find the triangle's perimeter in terms of x. 4x + 8

20 Given: $\overleftrightarrow{GK} \parallel \overleftrightarrow{HJ}$,
    lengths as shown

Find: The perimeter of △HJF 40

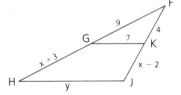

Section 8.5   Three Theorems Involving Proportions   **357**

Problem-Set Notes and
Additional Answers, continued

■ See *Solution Manual* for answers to problems **23–25**.

Problem Set B, *continued*

21 Sketch a triangle ABC, and locate a point P on $\overline{BC}$ such that $\overrightarrow{AP}$ bisects ∠BAC. If the perimeter of △ABC is 44, BP = 6, and PC = 10, find AB and AC. 10.5; 17.5

22 Given: $\overleftrightarrow{VS} \parallel \overleftrightarrow{MR}$,
   TV = 12, VM = 8, TS = 15,
   SR = TW = TX
Find: XP $\frac{20}{3}$

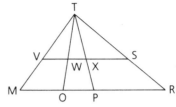

23 Given: $\overleftrightarrow{AC} \perp \overline{BE}$,
   ∠1 ≅ ∠4
   Conclusion: $\dfrac{BD}{BF} = \dfrac{DE}{EF}$

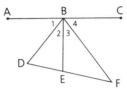

24 Given: $\overleftrightarrow{GD} \parallel \overleftrightarrow{FE}$,
   $\overleftrightarrow{BD} \parallel \overleftrightarrow{CE}$
   Prove: $\dfrac{AB}{AC} = \dfrac{AG}{AF}$

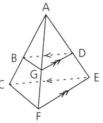

25 Prove that if a line bisects one side of a triangle and is parallel to a second side, it bisects the third side.

Problem Set C

26 Given: $\overleftrightarrow{GK} \parallel \overleftrightarrow{HJ}$,
   lengths as shown
   Find: The perimeter of △HJF $42\frac{1}{4}$

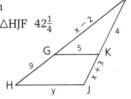

27 If two flagpoles are 10 m and 70 m tall and are 100 m apart, find the height of the point where a line from the top of the first to the bottom of the second intersects a line from the bottom of the first to the top of the second. $8\frac{3}{4}$ m

26 $\dfrac{x-2}{9} = \dfrac{4}{x+3}$
∴ x = 6
$\dfrac{FG}{FH} = \dfrac{GK}{HJ}$
$\dfrac{4}{13} = \dfrac{5}{y}$
∴ $y = 16\frac{1}{4}$
Thus, the perimeter of
△HJF = $42\frac{1}{4}$.

27
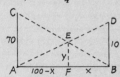

Consider △BAC.
$\dfrac{x}{100} = \dfrac{y}{70}$
Consider △ABD.
$\dfrac{100-x}{100} = \dfrac{y}{10}$
Solving simultaneously,
$y = 8\frac{3}{4}$.

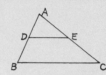

28 Prove that a line that divides two sides of a triangle proportionally is parallel to the third side.

29 Given: $\overrightarrow{RW}$ bisects ∠SRT.
$\overrightarrow{TV}$ bisects ∠RTS.
RV = 4, SV = 5,
SW = 6, WT = 7

Show that the given information is impossible.

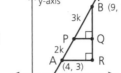

30 In the diagram, $\overline{BR}$ ∥ y-axis, $\overline{AR}$ ∥ x-axis, and point P divides $\overline{AB}$ in the ratio 2:3. Find the coordinates of points R and Q. (Hint: Find BQ and QR.)   (9, 3); (9, 7)

Problem-Set Notes and Additional Answers, continued

28

Given: $\frac{AD}{DB} = \frac{AE}{EC}$
Prove: $\overline{DE}$ ∥ $\overline{BC}$
From the given it can be shown that $\frac{AD}{AB} = \frac{AE}{AC}$ and then △ADE ~ △ABC by SAS ~. Thus, ∠ADE ≅ ∠ABC and $\overline{DE}$ ∥ $\overline{BC}$ by corr. ∠s ≅ ⟹ ∥ lines.

29 By applying Theorem 67 to △RST twice,
$\frac{4}{5} = \frac{RT}{13}$ and $\frac{6}{7} = \frac{9}{RT}$.
RT = $10\frac{2}{5}$, RT = $10\frac{1}{2}$
It is impossible for $\overline{RT}$ to have two different lengths.

30 By observation, R = (9, 3). Since ∠B ≅ ∠B and ∠BQP ≅ ∠BRA, we have △BQP ~ △BRA.
∴ $\frac{BQ}{BR} = \frac{BP}{BA}$, or
$\frac{BQ}{10} = \frac{3k}{5k}$.
Solving, BQ = 6.
Thus, Q = (9,7).

PUTTING QUILTS IN PERSPECTIVE

The patchwork world of Linda MacDonald, quilt maker

Quilt patterns traditionally have been based on two-dimensional geometric shapes. Linda MacDonald, a California artist, has taken this tradition and raised it to a new dimension.

"I'm interested in creating a three-dimensional space instead of a flat pattern," she explains, "a window you can move through—a fantasy landscape." Her quilt *Salmon Ladders*, for example, consists of hundreds of delicately colored polygons arranged in an intricate lattice of interlocking planes.

To create perspective, MacDonald designs a set of similar polygons and arranges them from largest to smallest moving toward a horizon. She designs these figures freehand, which gives her the artistic flexibility that precisely constructed figures might not allow. She hand-dyes her fabrics and stitches them by hand.

MacDonald received a bachelor of arts degree in painting from San Francisco State University. She began quilt making in 1974. "It's such a rich art form," she says. "Traditionally, quilt

patterns have told the story of American history. With the themes that I choose, I'm trying to tell my own history."

Project: Use one or more sets of hand-drawn similar polygons to create a sense of space in a rectangular area.

8 | CHAPTER SUMMARY

CONCEPTS AND PROCEDURES

After studying this chapter, you should be able to
- Recognize and work with ratios (8.1)
- Recognize and work with proportions (8.1)
- Apply the product and ratio theorems (8.1)
- Calculate geometric means (8.1)
- Identify the characteristics of similar figures (8.2)
- Use several methods to prove that triangles are similar (8.3)
- Use the concept of similarity to establish the congruence of angles and the proportionality of segments (8.4)
- Solve shadow problems (8.4)
- Apply three theorems frequently used to establish proportionality (8.5)

VOCABULARY

arithmetic mean (8.1)
dilation (8.2)
extremes (8.1)
geometric mean (8.1)
mean proportion (8.1)
mean proportional (8.1)
means (8.1)

proportion (8.1)
ratio (8.1)
reduction (8.2)
rise (8.1)
run (8.1)
similar (8.2)
similar polygons (8.2)

REVIEW PROBLEMS

Problem Set A

1 Identify the means and the extremes in the proportion $\frac{a}{b} = \frac{c}{d}$. *b and c; a and d*

2 Find the fourth proportional to 4, 6, and 8. **12**

3 Find the mean proportionals between 5 and 20. **±10**

4 Find the geometric means between 3 and 6. **±3√2**

5 If $9x = 4y$, find the ratio of x to y. $\frac{4}{9}$

6 Given: △ABC ~ DEF, with lengths as shown
Find: DF and EF 9; $\frac{15}{2}$

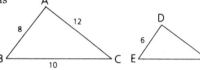

7 Pentagon ABCDE is similar to pentagon A′B′C′D′E′. The pentagons' respective perimeters are 24 and 30. If AB = 6, find A′B′. $\frac{15}{2}$

8 If $\frac{GH}{HJ} = \frac{3}{4}$ and GJ = 56, find HJ. **32**

G •————————— H •————————————— J

9 If $\frac{r}{3x} = \frac{a}{2b}$, what is the value of x in terms of a, b, and r? $\frac{2br}{3a}$

10 A radio antenna that is 100 m tall casts an 80-m shadow. At the same time, a nearby telephone pole casts a 16-m shadow. Find the height of the telephone pole. **20 m**

11 Given: $\overleftrightarrow{MR} \parallel \overleftrightarrow{OP}$,
lengths as shown
Find: RP and OP 4; 15

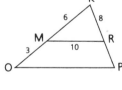

12 Given: $\overleftrightarrow{TP} \parallel \overleftrightarrow{VW}$,
lengths as shown
Find: ST, TV, and PT $\frac{90}{13}, \frac{40}{13}, \frac{72}{13}$

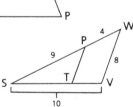

Chapter 8 Review
Class Planning

Time Schedule
All levels: 1 day

Resource References
Evaluation
    Tests and Quizzes
        Test 8, Series 1, 2, 3

Assignment Guide

Basic
1–15, 17, 18

Average
2–4, 7, 8, 10, 11, 14, 18, 19, 25, 26, 28, 29

Advanced
Section **8.5:**   28
Chapter **8** Review: 2–4, 12, 19, 20, 25–30, 34

To integrate constructions, Construction 8 on page 674 and Construction 10 on page 675 can be included at this point.

13 Given: $\overrightarrow{CD}$ bisects $\angle ACB$.
   $AC = 8$, $BC = 6$, $BD = 5$
   Find: AD $\frac{20}{3}$

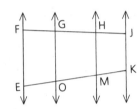

14 Given: Diagram as shown
   Find: x 12

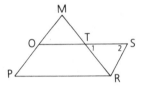

15 A scale model of the *Titanic* is $18\frac{1}{2}$ in. long. The scale is 1:570. To the nearest foot, how long was the *Titanic*? $\approx$879 ft

16 Given: $\overleftrightarrow{EF} \parallel \overleftrightarrow{GO} \parallel \overleftrightarrow{HM} \parallel \overleftrightarrow{JK}$,
   $FG = 2$, $GH = 8$,
   $HJ = 5$, $EM = 6$
   Find: EO and EK $\frac{6}{5}$; 9

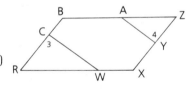

Problem-Set Notes and Additional Answers

■ See *Solution Manual* for answers to problems **17** and **18**.

17 Given: $\overline{OS} \parallel \overline{PR}$,
   $\angle 1 \cong \angle 2$
   Prove: $\dfrac{MO}{OP} \cong \dfrac{MT}{SR}$

18 Given: BRXZ is a ▱.
   $\angle 3 \cong \angle 4$
   Prove: (RC) (ZA) = (ZY) (RW)

19 If PQ = 30, find the coordinates of Q in the diagram. **(−13, 0)**

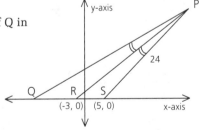

Problem Set B

20 Indicate whether the statement is true Always, Sometimes, or Never (A, S, or N)

a Two isosceles triangles are similar if a base angle of one is congruent to a base angle of the other. A

b Two isosceles triangles are similar if the vertex angle of one is congruent to the vertex angle of the other. A

c An equilateral triangle is similar to a scalene triangle. N

d If two sides of one triangle are proportional to two sides of another triangle, the triangles are similar. S

e In △ABC, ∠A = 40°, AB = 6, and BC = 8.
In △RST, RS = 12, ST = 16, and ∠R = 80°.
Therefore, △ABC ~ △RST. N

f If a line intersects a side of a triangle at one of its trisection points and is parallel to a second side, then it intersects the third side at one of its trisection points. A

g Two right triangles are similar if the legs of one are proportional to the legs of the other. A

h If the ratio of the measures of a pair of corresponding sides of two polygons is 3:4, then the ratio of the polygons' perimeters is 5:6. S

Problem-Set Notes and Additional Answers, continued

■ See *Solution Manual* for answers to problems **21–23.**

21 Given: ABDF is a ▱.
Conclusion: △CBD ~ △DFE

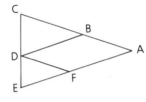

22 Given: $\overleftrightarrow{HR} \parallel \overleftrightarrow{GM}$

Prove: $\dfrac{PR}{OM} = \dfrac{PH}{OG}$

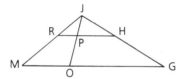

23 Prove that diagonals of a trapezoid divide each other proportionally.

24 If 78 is divided into three parts in the ratio 3:5:7, what is the sum of the smallest and the largest part? 52

25 One side of a triangle is 4 cm shorter than a second side. The ray bisecting the angle formed by these sides divides the opposite side into 4-cm and 6-cm segments. Find the perimeter of the triangle. 30

Problem-Set Notes and
Additional Answers, continued

30

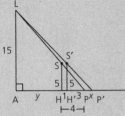

$\triangle PHS \sim \triangle PAL$

$\therefore \frac{PH}{PA} = \frac{HS}{AL}$

$\frac{4}{4 + y} = \frac{5}{15}$

$y = 8$

$\triangle P'H'S' \sim \triangle P'AL$

$\therefore \frac{P'H'}{P'A} = \frac{H'S'}{AL}$

$\frac{x + 3}{x + 12} = \frac{5}{15}$

$x = \frac{3}{2}$

Original shadow is PH = 4.
New shadow is P'H' = $4\frac{1}{2}$.

$4\frac{1}{2} - 4 = \frac{1}{2}$

31 Let the numbers be 50x,
75x, 60x, 72x. Then 50x +
75x + 60x + 72x = 771
$x = 3$
$\therefore 75x = 225$

32 Consider quadrilateral
GFAB. Since $\angle F$ and $\angle B$ are
both rt. $\angle$s, $\angle A$ is supp. to
$\angle FGB$. But $\angle HGJ$ is supp. to
$\angle FGB$. $\therefore \angle A \cong \angle HGJ$.
Similarly, $\angle C \cong \angle JHG$.
$\therefore \triangle ACE \sim \triangle GHJ$ by AA $\sim$.

$\frac{AC}{GH} = \frac{AE}{GJ} = \frac{CE}{HJ}$

$\frac{20}{5} = \frac{10}{GJ} = \frac{15}{HJ}$

$\therefore GJ = \frac{5}{2}, HJ = \frac{15}{4}$

33

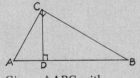

Given: $\triangle ABC$ with
hypotenuse $\overline{AB}$;
$\overline{CD}$ is altitude to $\overline{AB}$.
Prove: $CD \cdot AB = AC \cdot BC$
$\triangle ADC \sim \triangle CDB$, and thus
$\frac{AD}{CD} = \frac{CD}{DB}$, or $CD^2 = AD \cdot DB$.
$\triangle ABC \sim \triangle ACD$ and
$\triangle ABC \sim \triangle CBD$, so $\frac{AB}{AC} = \frac{AC}{AD}$

Review Problem Set B, *continued*

26 If $\frac{7}{x + 4y} = \frac{9}{2x - y}$, find the ratio of x to y. $\frac{43}{5}$

27 Given: $\overleftrightarrow{PT} \parallel \overleftrightarrow{RS}$,
NP = 5x − 21, PR = 5,
NT = x, TS = 8
Find: NR + NS $\frac{104}{5}$

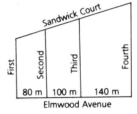

28 The diagram shows a part of the town
of Oola, La. First, Second, Third, and
Fourth streets are each perpendicular
to Elmwood Avenue. If the total frontage on Sandwick Court is 400 m, find
the length of each block of Sandwick
Court. 100 m; 125 m; 175 m

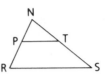

Problem Set C

29 Given: $\overrightarrow{TQ}$ bisects $\angle RTP$.
$\overleftrightarrow{QS} \parallel \overleftrightarrow{PT}$
Find: QP, RS, and QS 8, 9, 12

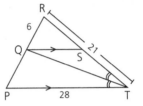

30 Llong is 5 ft tall and is standing in the light of a 15-ft lamppost.
Her shadow is 4 ft long. If she walks 1 ft farther away from the
lampost, by how much will her shadow lengthen? $\frac{1}{2}$ ft

31 The sum of four numbers is 771. The ratio of the first to the
second is 2:3. The ratio of the second to the third is 5:4. The
ratio of the third to the fourth is 5:6. Find the second number. 225

32 Given: $\overline{GB} \perp \overline{AC}, \overline{HD} \perp \overline{EC}$,
$\overline{JF} \perp \overline{AE}$,
AB = 8,
BC = 12,
EC = 15,
AE = 10,
GH = 5
Find: GJ and HJ $\frac{5}{2}; \frac{15}{4}$

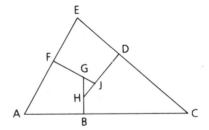

33 Prove that if an altitude is drawn to the hypotenuse of a right
triangle, then the product of the measures of the altitude and the
hypotenuse is equal to the product of the measures of the legs of
the right triangle.

364 Chapter 8 Similar Polygons

---

and $\frac{AB}{BC} = \frac{BC}{BD}$, or $AC^2 =$
$AB \cdot AD$ and $BC^2 = AB \cdot BD$.
Thus, $AC^2 \cdot BC^2 = AB \cdot$

$AD \cdot AB \cdot BD = AB^2 \cdot$
$AD \cdot BD = AB^2 \cdot CD^2$
$\therefore AC \cdot BC = AB \cdot CD$

34 Filbert knows that the two triangles ABC and XYZ are similar, but he cannot remember what the correct correspondence of vertices should be. He guesses that △ABC ~ △XYZ.

a What is the probability that his guess is correct? $\frac{1}{6}$

b If Filbert finds out that the triangles are isosceles, what will the probability be then? $\frac{1}{3}$

c If the triangles are equilateral, what are his chances of guessing a correct correspondence? 1

35 Given: CARD and GAME are parallelograms.
The perimeter of GAME is 48.
AD:DE = 2:1

Find: The perimeter of CARD 32

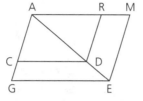

Problem-Set Notes and Additional Answers, continued

34a △ABC ↔ △XYZ
△ABC ↔ △XZY
△ABC ↔ △YXZ
△ABC ↔ △YZX
△ABC ↔ △ZXY
△ABC ↔ △ZYX

b △ABC ↔ △XYZ
△ABC ↔ △XZY
△ABC ↔ △ZXY

c All equilateral triangles are similar.

35 The angles of CARD can be shown to be congruent to the angles of GAME, respectively, through an analysis of several sets of parallel lines.
Since △ARD ~ △AME, it can be shown that $\frac{AR}{AM} = \frac{RD}{ME} = \frac{AD}{AE}$. Since △ACD ~ △AGE, it can be shown that $\frac{AC}{AG} = \frac{CD}{GE} = \frac{AD}{AE}$. Thus, $\frac{AR}{AM} = \frac{RD}{ME} = \frac{AC}{AG} = \frac{CD}{GE}$.
∴ CARD ~ GAME
Thus, the ratio of their perimeters will be the same as the ratio of any linear dimension, namely, their diagonals.
$\frac{P(CARD)}{P(GAME)} = \frac{2}{3}$
$\frac{P(CARD)}{48} = \frac{2}{3}$
∴ P(CARD) = 32

HISTORICAL SNAPSHOT

A MASTER TECHNOLOGIST
The sketchbook of Villard de Honnecourt

In the Middle Ages, master architects were much more than merely designers of buildings. Because they had to supervise every aspect of the planning and construction of many types of structures, they needed to be adept in all the arts and sciences of their times. The wide-ranging interests and expertise of these men are strikingly illustrated by the surviving sketchbook of one of them, the thirteenth-century French architect Villard de Honnecourt.

Villard seems to have used his sketchbook as a sort of technological diary and as a way of sharing his ideas and discoveries with his colleagues. In addition to drawings of interesting architectural features and procedures that Villard noticed in his travels, the book includes such diverse material as plans for a variety of mechanical devices, including a powerful catapult and a water-driven saw; notes on lion taming; recipes for medicines; and more than a hundred drawings of different animals.

Among the most curious pages, however, are those devoted to sketches of people, beasts, and birds on which Villard has superimposed various geometric figures. Some of these sketches may have been intended to demonstrate the proper proportions to use in drawing, but it is thought that their main purpose was to exemplify a method of reproducing drawings in any desired size. By associating a sketch with a geometric diagram and then drawing a dilation of the diagram on a wall or on a block of stone that was to be carved, an artist could create a basic framework that would serve as a guide for the accurate enlargement of the sketch itself.

Chapter 9 Schedule

Basic
Problem Sets:	13 days
Review:	2 days
Test:	1 day

Average
Problem Sets:	13 (17) days
Trig (optional):	4 days
Review:	2 days
Test:	1 day

Advanced
Problem Sets:	12 days
Review:	2 days
Test:	1 day

*S*TUDENTS WILL WORK with radicals and quadratic equations in this chapter. A review of the related algebra skills prior to beginning the chapter can be helpful.

CHAPTER

9 THE PYTHAGOREAN THEOREM

There are many ways to prove that the square of the length of the hypotenuse of a right triangle is equal to the sum of the squares of the lengths of its legs, but Euclid's proof of the Pythagorean Theorem deals directly with squares and their areas.

9.1 REVIEW OF RADICALS AND QUADRATIC EQUATIONS

Objective

After studying this section, you will be able to
- Simplify radical expressions and solve quadratic equations

Part One: Introduction

Some of the problems in the next three chapters will involve radicals and quadratic equations. Although you have already completed a course in algebra, you may have forgotten some algebraic techniques. Carefully read the following sample problems, which review these two concepts.

Part Two: Sample Problems

Problem 1 *Simplify* $\sqrt{48}$.

Solution
$$\sqrt{48} = \sqrt{16 \cdot 3} \quad \text{(16 is a perfect square.)}$$
$$= \sqrt{16} \cdot \sqrt{3}$$
$$= 4\sqrt{3}$$

Problem 2 *Simplify* $\sqrt{18} + \sqrt{32} + \sqrt{75}$.

Solution
$$\sqrt{18} + \sqrt{32} + \sqrt{75} = \sqrt{9 \cdot 2} + \sqrt{16 \cdot 2} + \sqrt{25 \cdot 3}$$
$$= 3\sqrt{2} + 4\sqrt{2} + 5\sqrt{3}$$
$$= 5\sqrt{3} + 7\sqrt{2}$$

Problem 3 *Simplify* $\sqrt{\frac{5}{3}}$.

Solution
$$\sqrt{\frac{5}{3}} = \frac{\sqrt{5}}{\sqrt{3}}$$
$$= \frac{\sqrt{5}}{\sqrt{3}} \cdot \frac{\sqrt{3}}{\sqrt{3}}$$
$$= \frac{\sqrt{15}}{3} \text{ or } \frac{1}{3}\sqrt{15} \quad \text{(The two answers are equivalent simplifications.)}$$

Class Planning

Time Schedule
All levels: 1 day

Resource References
Teacher's Resource Book
  Class Opener 9.1A, 9.1B

Class Opener

Write each number as the product of two whole-number factors where one of the factors is a perfect square.

1 12 $4 \cdot 3$
2 27 $9 \cdot 3$
3 50 $25 \cdot 2$
4 48 $16 \cdot 3$ or $4 \cdot 12$
5 72 $36 \cdot 2$ or $9 \cdot 8$
6 54 $9 \cdot 6$

Complete the following.

7 If $5^2 = 25$, then $\sqrt{25} =$ ___. 5

8 If $(\sqrt{7})^2 = \sqrt{7} \cdot \sqrt{7} = 7$, then $(3\sqrt{7})^2 =$ ___ $\cdot$ ___ $=$ ___. $(3\sqrt{7})^2 = 3\sqrt{7} \cdot 3\sqrt{7} = 63$

Lesson Notes

- Some classes may need numerous examples of working with radicals. All students should read the sample problems before starting the problem sets.

Problem 4   *Solve $x^2 + 9 = 25$ for x.*

Solution   *Method 1:*

$$x^2 + 9 = 25$$
$$x^2 = 16$$
$$x = \pm 4$$

Method 2 (factoring):

$$x^2 + 9 = 25$$
$$x^2 - 16 = 0$$
$$(x - 4)(x + 4) = 0$$
$$x - 4 = 0 \text{ or } x + 4 = 0$$
$$x = 4 \text{ or } x = -4$$

Problem 5   *Solve $(3\sqrt{5})^2 + (3\sqrt{2})^2 = x^2$ for x.*

Solution

$$(3\sqrt{5})^2 + (3\sqrt{2})^2 = x^2$$
$$9 \cdot 5 + 9 \cdot 2 = x^2$$
$$45 + 18 = x^2$$
$$63 = x^2$$
$$\pm\sqrt{63} = x$$
$$\pm\sqrt{9 \cdot 7} = x$$
$$\pm 3\sqrt{7} = x$$

Problem 6   *Solve for x.*   **a** $x^2 - 10x = -16$ **b** $x^2 + 5x = 0$

Solution   **a**

$$x^2 - 10x = -16$$
$$x^2 - 10x + 16 = 0$$
$$(x - 8)(x - 2) = 0$$
$$x - 8 = 0 \text{ or } x - 2 = 0$$
$$x = 8 \text{ or } x = 2$$

b $x^2 + 5x = 0$

$$x(x + 5) = 0$$
$$x = 0 \text{ or } x + 5 = 0$$
$$x = 0 \text{ or } x = -5$$

Part Three: Problem Sets

Problem Set A

1   Simplify.
   a $\sqrt{4}$ 2
   b $\sqrt{27}$ $3\sqrt{3}$
   c $\sqrt{72}$ $6\sqrt{2}$
   d $\sqrt{32}$ $4\sqrt{2}$
   e $\sqrt{98}$ $7\sqrt{2}$
   f $\sqrt{200}$ $10\sqrt{2}$
   g $\sqrt{20}$ $2\sqrt{5}$
   h $\sqrt{24}$ $2\sqrt{6}$

2   Simplify.
   a $5\sqrt{18}$ $15\sqrt{2}$
   b $\sqrt{4 + 9}$ $\sqrt{13}$
   c $\sqrt{3^2 + 4^2}$ 5
   d $\sqrt{5^2 + 12^2}$ 13
   e $\frac{1}{6}\sqrt{48}$ $\frac{2\sqrt{3}}{3}$
   f $\sqrt{49 \cdot 3}$ $7\sqrt{3}$

3   Simplify.
   a $\dfrac{1}{\sqrt{2}}$ $\dfrac{\sqrt{2}}{2}$
   b $\dfrac{1}{\sqrt{5}}$ $\dfrac{\sqrt{5}}{5}$
   c $\dfrac{4}{\sqrt{2}}$ $2\sqrt{2}$
   d $\dfrac{6}{\sqrt{3}}$ $2\sqrt{3}$

4   Simplify.
   a $4\sqrt{3} + 7\sqrt{3}$ $11\sqrt{3}$
   b $7\sqrt{2} + \sqrt{3} + 6\sqrt{3} + \sqrt{2}$ $8\sqrt{2} + 7\sqrt{3}$
   c $\sqrt{12} + \sqrt{27}$ $5\sqrt{3}$
   d $\sqrt{72} + \sqrt{75} - \sqrt{48}$ $6\sqrt{2} + \sqrt{3}$

5 Solve for x.

a $x^2 = 25$ ±5 **c** $x^2 = 169$ ±13 **e** $x^2 = 12$ ±2$\sqrt{3}$

b $x^2 = 144$ ±12 **d** $x^2 = \frac{1}{4}$ ±$\frac{1}{2}$ **f** $x^2 = 18$ ±3$\sqrt{2}$

6 Solve for x.

a $x^2 + 16 = 25$ ±3 **c** $12^2 + x^2 = 13^2$ ±5 **e** $(\sqrt{5})^2 + (\sqrt{11})^2 = x^2$ ±4

b $x^2 + 6^2 = 100$ ±8 **d** $x^2 + (3\sqrt{3})^2 = 36$ ±3 **f** $x^2 = (5\sqrt{3})^2 + (\sqrt{5})^2$ ±4$\sqrt{5}$

7 Solve for x.

a $x^2 - 5x - 6 = 0$ {6, −1} **c** $x^2 - 8x + 15 = 0$ {5, 3} **e** $x^2 - 36 = 9x$ {12, −3}

b $x^2 + 4x - 12 = 0$ {2, −6} **d** $x^2 - 18 - 3x = 0$ {6, −3} **f** $-x^2 + 5x + 36 = 0$ {9, −4}

8 Solve for x.

a $x^2 - 4x = 0$ {0, 4} **c** $x^2 - 2x = 11x$ {0, 13}

b $x^2 = 10x$ {0, 10} **d** $5x = x^2 - 3x$ {0, 8}

9 If, in the given figure, $x^2 + y^2 = r^2$,

a Find x if y = 21 and r = 29 20

b Find y, in simplified radical form, if x = 2 and r = 4 2$\sqrt{3}$

c Find r to the nearest tenth if x = 4.1 and y = 7.1 ≈8.2

Problem Set B

10 Solve for x.

a $3x^2 + 5x - 7 = x^2 + 8x + 28$ $\left\{5, -\frac{7}{2}\right\}$ **c** $8x^2 - 7x + 9 = 2x^2 + 6x + 7$ $\left\{\frac{1}{6}, 2\right\}$

b $12x^2 - 15 = -11x$ $\left\{-\frac{5}{3}, \frac{3}{4}\right\}$

11 Solve $\frac{7}{x+1} = \frac{2x+4}{3x-3}$ for x. $\left\{\frac{5}{2}, 5\right\}$

12 Find AB to the nearest tenth. ≈12.2

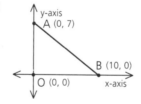

Problem Set C

13 Simplify.

a $\sqrt{h^2}$, if h represents a negative number −h

b $\sqrt{(x-3)^2}$, if x < 3 3 − x

c $\sqrt{p^2q^2}$, if p and q both represent negative numbers pq

d $\sqrt{x^3y^2}$, if x > 0 and y < 0 −xy$\sqrt{x}$

Section 9.1 Review of Radicals and Quadratic Equations **369**

Cooperative Learning

Have each student work independently with part **c** in problems **1–8**. Then have students work in small groups to compare answers. To correct any errors, each group member should explain what caused the error and how to avoid making the same mistake. Each group should then work some of the remaining problems in a cooperative manner. Each group member should be able to explain each problem at the end of this group activity.

Problem-Set Notes and Additional Answers

13a $\sqrt{h^2} = |h| = -h$, since h < 0.

b $\sqrt{(x-3)^2} = |x-3| = 3-x$, since x < 3.

c $\sqrt{p^2q^2} = |p| \cdot |q| = (-p)(-q) = pq$, since p < 0 and q < 0.

d $\sqrt{x^3y^2} = |x| \cdot |y|\sqrt{x} = -xy\sqrt{x}$, since x > 0 and y < 0.

INTRODUCTION TO CIRCLES

Class Planning

Time Schedule
Basic: 1 day
Average: 1 day
Advanced: Omit

Resource References
Teacher's Resource Book
  Class Opener 9.2A, 9.2B
  Using Manipulatives 15

Class Opener

This circle is divided into equal pieces.

The pieces are then rearranged to form a figure that is nearly a parallelogram.

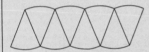

1 What is the circumference of a circle with radius r?
   $2 \cdot \pi \cdot r$
2 What is half the circumference of a circle with radius r? $\pi \cdot r$
3 What is the formula for the area of a parallelogram?
   $A = b \cdot h$
4 Use the answers to problems 1–3 to help you derive the formula for the area of a circle.

$$\square\; A = \quad b \quad \cdot h$$
$$\odot\; A = \tfrac{1}{2} \cdot \overbrace{\quad C \quad} \cdot r$$
$$A = \tfrac{1}{2} \cdot \overbrace{2 \cdot \pi \cdot r} \cdot r$$
$$A = \pi \cdot r \cdot r,\ \text{or}\ A = \pi r^2$$

Objective
After studying this section, you will be able to
■ Begin solving problems involving circles

Part One: Introduction

Because of the unfamiliar terms and concepts involved, many students find working with circles the most difficult part of their geometry studies. To help you deal with circles more effectively, this section will informally introduce you to some of the basic concepts used in circle problems. If you study this section carefully and solve the circle problems presented in the problem sets of this chapter, you will be better prepared for the formal study of circles in Chapter 10.

You have already encountered some problems that have asked you to find the *circumferences* (perimeters) and the areas of circles, so you should be familiar with the relevant formulas.

Example 1   *Find the circumference and the area of ⊙O.*

The circumference is found with the formula $C = \pi d$, where d is the diameter of the circle.
$$C = \pi d$$
$$ = 14\pi$$
The area is found with the formula $A = \pi r^2$, where r is the circle's radius.
$$A = \pi r^2$$
$$ = \pi(7^2)$$
$$ = 49\pi$$
The circle's circumference is 14π, or about 43.98, inches and its area is 49π, or about 153.94, square inches.

An *arc* is made up of two points on a circle and all the points of the circle needed to connect those two points by a single path. The blue portion of the figure at the right is called arc CD (symbolized $\overset{\frown}{CD}$).

The *measure* of an arc is equivalent to the number of degrees it occupies. (A complete circle occupies 360°.) The *length* of an arc is a fraction of a circle's circumference, so it is expressed in linear units, such as feet, centimeters, or inches.

Example 2   *Find the measure and the length of* $\overset{\frown}{AB}$.

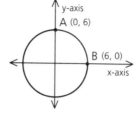

Since the arc is one fourth of the circle, its measure is $\frac{1}{4}(360)$, or 90. The arc's length (ℓ) can be expressed as a part of the circle's circumference.

$$\ell = \tfrac{90}{360}C$$
$$= \tfrac{1}{4}\pi d$$
$$= \tfrac{1}{4}(\pi \cdot 12)$$
$$= 3\pi, \text{ or } \approx 9.42$$

A *sector* of a circle is a region bounded by two radii and an arc of the circle. The figure at the right shows sector CAT of ⊙A.

Since we know that $\overset{\frown}{CT}$ has a measure of 90, we can calculate the area of sector CAT as a fraction of the area of ⊙A.

$$\text{Area of sector CAT} = \tfrac{90}{360}(\text{area of } \odot A)$$
$$= \tfrac{1}{4}(\pi \cdot 6^2)$$
$$= 9\pi, \text{ or } \approx 28.27$$

A *chord* is a line segment joining two points on a circle. (A diameter is a chord that passes through the center of its circle.) An *inscribed angle* is an angle whose vertex is on a circle and whose sides are determined by two chords of the circle.

In the figure at the right, $\overline{AB}$ and $\overline{AC}$ are chords, and ∠BAC is an inscribed angle.

∠BAC is said to *intercept* $\overset{\frown}{BC}$. (An intercepted arc is an arc whose endpoints are on the sides of an angle and whose other points all lie within the angle. Although ∠BAC intercepts only one arc, in Chapter 10 you will deal with some angles that intercept two arcs of a circle.)

Lesson Notes

■ Students might have difficulty assimilating the new terms and concepts in this section. After presenting the lesson, ask students to draw a picture that describes the meaning of each term.

Communicating Mathematics

Have students write a paragraph that explains why every diameter of a circle is a chord but every chord is not a diameter.

Part Two: Sample Problems

Problem 1   Find the circumference and the area
of ⊙P.

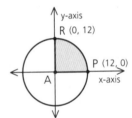

2.7

P

Solution

$C = \pi d$ $A = \pi r^2$
$\quad = \pi(5.4)$ $\quad = \pi(2.7^2)$
$\quad = 5.4\pi$, or ≈ 16.96 $\quad = 7.29\pi$, or ≈ 22.90

Problem 2   Given: Diagram as marked
Find: **a** m$\widehat{AB}$
b The length of $\widehat{AB}$

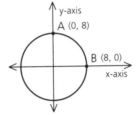

y-axis
A (0, 8)

B (8, 0)
x-axis

Solution   The circle's radius is 8, and $\widehat{AB}$ is one fourth of the circle.

a m$\widehat{AB} = \frac{1}{4}(360)$

$\quad = 90$

b Length of $\widehat{AB} = \frac{90}{360}C$

$\quad = \frac{1}{4}(\pi \cdot 16)$
$\quad = 4\pi$, or ≈ 12.57

Problem 3   Find the area of the shaded region
(sector PAR).

y-axis
R (0, 12)

P (12, 0)
A
x-axis

Solution   The radius of ⊙A is 12, and m$\widehat{RP} = 90$.

Area of sector PAR $= \frac{90}{360}$(area of ⊙A)
$\quad = \frac{1}{4}(\pi \cdot 12^2)$
$\quad = 36\pi$, or ≈ 113.10

Problem 4   Harry Halph looked ahead to Chapter 10 and discovered that the measure of an inscribed angle is half
the measure of its intercepted arc. Use this information
to find the measure of inscribed angle DEF.

E

D

80°

F

Solution $\widehat{DF}$ is the arc intercepted by ∠DEF.

m∠DEF $= \frac{1}{2}$(m$\widehat{DF}$)
$\quad = \frac{1}{2}(80)$
$\quad = 40$

372   Chapter 9   The Pythagorean Theorem

Problem 5 *If $\overline{HJ}$ is a diameter of $\odot O$, what are the coordinates of point O?*

J (8, 5)

O

H (−4, −3)

Solution We can use the midpoint formula.

$$x_m = \frac{x_1 + x_2}{2} \qquad y_m = \frac{y_1 + y_2}{2}$$

$$= \frac{8 + (-4)}{2} \qquad = \frac{5 + (-3)}{2}$$

$$= 2 \qquad = 1$$

The coordinates of point O are (2, 1).

Problem 6 *Show that $\triangle INS$ is a right triangle by*
  a *Finding $m\angle INS$*
  b *Finding the slopes of $\overleftrightarrow{IN}$ and $\overleftrightarrow{NS}$*

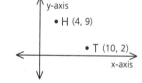

y-axis

N (6, 8)

S (10, 0)

I (−10, 0)

x-axis

C

Solution **a** $\overparen{ICS}$ is one-half the circle, so $m\overparen{ICS} = 180$.
  Since $\overparen{ICS}$ is intercepted by inscribed angle INS,

$$m\angle INS = \tfrac{1}{2}(m\overparen{ICS})$$

$$= \tfrac{1}{2}(180)$$

$$= 90$$

Therefore, $\angle INS$ is a right angle, and $\triangle INS$ is a right triangle.

  b Recall that two lines are perpendicular if their slopes are opposite reciprocals.

$$\text{Slope of } \overleftrightarrow{IN} = \frac{8 - 0}{6 - (-10)} = \tfrac{1}{2}$$

$$\text{Slope of } \overleftrightarrow{NS} = \frac{0 - 8}{10 - 6} = -2$$

Since $\overleftrightarrow{IN} \perp \overleftrightarrow{NS}$, $\triangle INS$ is a right triangle.

Problem 7 *Reflect point H over the y-axis to H′. Then find the slope of $\overleftrightarrow{TH'}$.*

y-axis

• H (4, 9)

• T (10, 2)

x-axis

Solution Since H is four units to the right of the y-axis, H′ must be four units to the left of the y-axis. Therefore, H′ = (−4, 9).

$$\text{Slope of } \overleftrightarrow{TH'} = \frac{9 - 2}{-4 - 10}$$

$$= \frac{7}{-14}$$

$$= -\frac{1}{2}$$

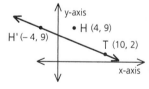

y-axis

H′ (−4, 9)

• H (4, 9)

T (10, 2)

x-axis

Assignment Guide

Basic

1–11, 14, 17

Average

1, 3, 4, 7, 8, 10–12, 14–17

Advanced

Advanced students do not need to spend a day on this section. It is suggested that students do some of these problems in class during the latter part of Chapter 8 and the early part of Chapter 9.

Part Three: Problem Sets

Problem Set A

1 Find the circumference and the area of ⊙O. $C = 19.6\pi$, or ≈ 61.58; $A = 96.04\pi$, or ≈ 301.72

2 Given: Diagram as marked
Find: **a** The measure of the arc from C to D (m$\overset{\frown}{CD}$) 90
 b The length of $\overset{\frown}{CD}$ 2.5π, or ≈ 7.85

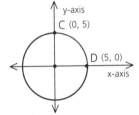

3 Given: Diagram as marked
Find: **a** m$\overset{\frown}{EF}$ 180
 b The length of $\overset{\frown}{EF}$ to the nearest tenth ≈ 12.6

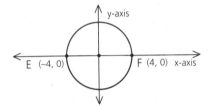

4 Given: Diagram as marked
Find: **a** The coordinates of D (0, 4)
 b The area of the shaded region (sector DOG) 4π, or ≈ 12.57

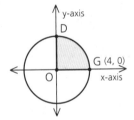

5 If AB = 10, what is the area of the shaded region (sector AOB)? 12.5π, or ≈ 39.27

6 Find m∠F. 45

7 Given: Diagram as marked
 Find: **a** m∠A 40
   b m∠D 40

8 In ⊙O, mÂB = 50. Find mB̂C and
 m∠BCA. 130; 25

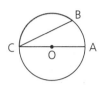

9 In the figure shown, $\overline{AB}$ is a diameter.
 Find the coordinates of point O, the cen-
 ter of the circle. (3, 3)

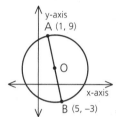

10 Find the coordinates of Q, the center of
 the circle. Then use slopes to show that
 △DEF is a right triangle.
 (4, −1); slope of DE = $-\frac{1}{7}$ and slope of
 EF = 7 (opp. reciprocals), so $\overleftrightarrow{DE} \perp \overleftrightarrow{EF}$
 and △DEF is a rt. △.

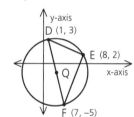

11 Copy the diagram, reflecting H across
 the y-axis to H′. Then find
 a The coordinates of H′ (−3, 14)
 b The slope of $\overleftrightarrow{TH'}$ −1

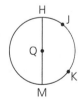

Problem Set B

12 In ⊙Q, mĤJ = 20 and mM̂K = 40. The
 circumference of ⊙Q is 27π.
 a Find mĴK. 120
 b Find the length of ĴK. 9π
 c Find HM (the length of $\overline{HM}$). 27

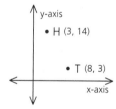

13 Use the diagram of ⊙O to find the coordinates of H. Then find the coordinates of G', the reflection of G over the y-axis. (7, 1); (−3, 7)

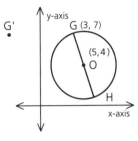

Problem-Set Notes
and Additional Answers

14 Vertical angles are congruent. ∠B ≅ ∠C (or ∠A ≅ ∠D) because they intercept the same arc. △ABE ~ △DCE by AA.

14 Write a convincing argument to show that △ABE ~ △DCE.

15 Given: AB = 4, BE = 5, AE = 6, CE = 3

Find: CD $2\frac{2}{5}$

16 In the diagram of ⊙O at the right, ∠JHK = 45°.
  a Find m$\widehat{JK}$. 90
  b Find the length of $\widehat{JK}$. 3π, or ≈9.42

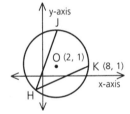

17 Replace x with 5 and y with 8 in $(x − 2)^2 + (y − 4)^2 = 25$.
$(5 − 2)^2 + (8 − 4)^2 = 25$;
$3^2 + 4^2 = 25$; $9 + 16 = 25$

17 Verify by substitution that point A = (5, 8) is on the circle that is the graph of the equation $(x − 2)^2 + (y − 4)^2 = 25$.

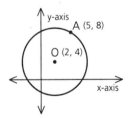

ALTITUDE-ON-HYPOTENUSE THEOREMS

Objective

After studying this section, you will be able to
- Identify the relationships between the parts of a right triangle when an altitude is drawn to the hypotenuse

Part One: Introduction

When altitude $\overline{CD}$ is drawn to the hypotenuse of $\triangle ABC$, three similar triangles are formed.

$$\triangle ABC \sim \triangle ACD \sim \triangle CBD$$

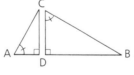

$\triangle ABC \sim \triangle ACD$ by AA, and we notice that

$$\frac{AB}{AC} = \frac{AC}{AD}, \text{ or } (AC)^2 = (AB)(AD)$$

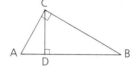

Therefore, AC is the mean proportional between AB and AD.

$\triangle ABC \sim \triangle CBD$ by AA, and we notice that

$$\frac{AB}{CB} = \frac{CB}{DB}, \text{ or } (CB)^2 = (AB)(DB)$$

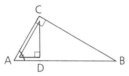

Therefore, CB is the mean proportional between AB and DB.

$\triangle ACD \sim \triangle CBD$ by transitivity of similar triangles, and we notice that

$$\frac{AD}{CD} = \frac{CD}{DB}, \text{ or } (CD)^2 = (AD)(DB)$$

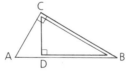

Therefore, CD is the mean proportional between AD and DB.

These illustrations prove three closely related theorems, which we will present as one theorem.

Class Planning

Time Schedule
All levels: 2 days

Resource References
Teacher's Resource Book
  Class Opener 9.3A, 9.3B
  Additional Practice Worksheet 17
Transparencies 14, 15

Class Opener

Given: $\angle ATS$ is a right angle.
  $\overline{TU}$ is an altitude of $\triangle ATS$.
  $\triangle AUT \sim \triangle ATS$,
  $AS = 12$, $AU = 8$
Find: AT
Since $\triangle AUT \sim \triangle ATS$, $\frac{AT}{AS} = \frac{AU}{AT}$.
So $\frac{AT}{12} = \frac{8}{AT}$ and $AT^2 = 96$.
$\therefore AT = \sqrt{96}$, or $AT \approx 9.8$.

Lesson Notes

- If students learn the three formulas in words (e.g., "The altitude to the hypotenuse is the mean proportional to the legs"), they can take any labeled diagram and readily write the proportions.

Communicating Mathematics

Have students write an explanation of the term *mean proportional*.

Lesson Notes, continued

- The proofs of the parts of Theorem 68 are brief. For part **a**, the two smaller right triangles are each similar to the larger one by AA and hence are similar to each other. Part **b** follows from the similarity between the two smaller triangles, and part **c**, from the similarity of each smaller triangle to the larger one.

- Students should learn the formulas as phrases.
   (height)^2 = product of parts of the hypotenuse
   (leg)^2 = product of the hypotenuse and the projection (shadow) of the leg on the hypotenuse

This will help students to readily write the relationships for any labeled diagram.

Theorem 68 *If an altitude is drawn to the hypotenuse of a right triangle, then*

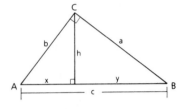

a *The two triangles formed are similar to the given right triangle and to each other*
$$\triangle ADC \sim \triangle ACB \sim \triangle CDB$$

b *The altitude to the hypotenuse is the mean proportional between the segments of the hypotenuse*

$$\frac{x}{h} = \frac{h}{y}, \text{ or } h^2 = xy$$

c *Either leg of the given right triangle is the mean proportional between the hypotenuse of the given right triangle and the segment of the hypotenuse adjacent to that leg (i.e., the projection of that leg on the hypotenuse)*

$$\frac{y}{a} = \frac{a}{c}, \text{ or } a^2 = yc; \text{ and } \frac{x}{b} = \frac{b}{c}, \text{ or } b^2 = xc$$

Parts **b** and **c** of Theorem 68 can be summarized as follows.

$$h^2 = x \cdot y$$
$$b^2 = x \cdot c$$
$$a^2 = y \cdot c$$

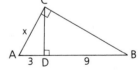

Part Two: Sample Problems

Problem 1   If AD = 3 and DB = 9, find CD.

Solution $(CD)^2 = AD \cdot DB$
$\qquad x^2 = 3 \cdot 9$
$\qquad x = \pm\sqrt{3}\sqrt{9}$
$\qquad x = \pm 3\sqrt{3}$
$\qquad CD = 3\sqrt{3}$ (CD cannot be negative, so reject $-3\sqrt{3}$.)

Problem 2   If AD = 3 and DB = 9, find AC.

Solution $(AC)^2 = AD \cdot AB$
$\qquad x^2 = 3 \cdot 12$
$\qquad x^2 = 36$
$\qquad x = \pm 6$
$\qquad AC = 6$ (Reject -6, since AC cannot be negative.)

378 | Chapter 9   The Pythagorean Theorem

Problem 3 *If* DB = 21 *and* AC = 10, *find* AD.

Solution

$$(AC)^2 = AD \cdot AB$$
$$10^2 = x(x + 21)$$
$$x(x + 21) = 10 \cdot 10$$
$$x^2 + 21x = 100$$
$$x^2 + 21x - 100 = 0$$
$$(x + 25)(x - 4) = 0$$
$$x + 25 = 0 \text{ or } x - 4 = 0$$
$$x = -25 \text{ or } x = 4$$

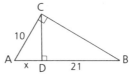

Since AD cannot be negative, AD = 4.

Problem 4 *Given:* $\overline{PK} \perp \overline{JM}$, $\overline{RK} \perp \overline{JP}$, $\overline{KO} \perp \overline{PM}$

Prove: (PO)(PM) = (PR)(PJ)

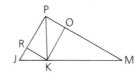

Proof

1 $\overline{PK} \perp \overline{JM}$	1 Given
2 $\angle$PKJ is a right $\angle$.	2 $\perp$ segments form right $\angle$s.
3 $\angle$PKM is a right $\angle$.	3 Same as 2
4 $\overline{RK} \perp \overline{JP}$	4 Given
5 $\overline{RK}$ is an altitude.	5 A segment drawn from a vertex of a $\triangle$ $\perp$ to the opposite side is an altitude.
6 $(PK)^2 = (PR)(PJ)$	6 If the altitude is drawn to the hypotenuse of a right $\triangle$, then either leg of the given right $\triangle$ is the mean proportional between the hypotenuse and the segment of the hypotenuse adjacent to that leg.
7 Similarly, $(PK)^2 = (PO)(PM)$	7 Reasons 1–6
8 (PO)(PM) = (PR)(PJ)	8 Transitive Property

Part Three: Problem Sets

Problem Set A

1 a If EH = 7 and HG = 3, find HF. $\sqrt{21}$
  b If EH = 7 and HG = 4, find EF. $\sqrt{77}$
  c If GF = 6 and EG = 9, find HG. 4

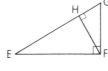

2 a Find 2x.   **b** Find $\frac{1}{2}$y.   **c** Find z + 8.
    $4\sqrt{5}$ 3 $3\sqrt{5} + 8$

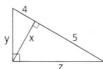

Assignment Guide

Basic

Day 1	1, 3a,b,c, 4a,d, 8, 9, 11
Day 2	3d,e, 5, 10, 12–14

Average

Day 1	1, 3a,b,c, 4a,d, 8–11
Day 2	12–14, 16, 17a, 19

Advanced

Day 1	1a,c, 2, 3a,c,e, 4a,c, 5, 9, 11
Day 2	12–14, 16, 17a,c, 19, 24

Problem-Set Notes and Additional Answers

- In problems **1**, **3**, and **4**, students should draw a separate diagram for each part.

Problem Set A, *continued*

3 Given: $\overline{AC} \perp \overline{CB}$, $\overline{CD} \perp \overline{AB}$

a If AD = 4 and BD = 9, find CD. 6

b If AD = 4 and AB = 16, find AC. 8

c If BD = 6 and AB = 8, find BC. $4\sqrt{3}$

d If CD = 8 and BD = 16, find AD. 4

e If AD = 3 and BD = 24, find AC. 9

f If BC = 8 and BD = 20, find AB. Impossible

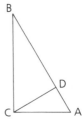

4 Given: $\angle$JOM = 90°; $\overline{OK}$ is an altitude.

a If JK = 12 and KM = 5, find OK. $2\sqrt{15}$

b If OK = $3\sqrt{5}$ and JK = 9, find KM. 5

c If JO = $3\sqrt{2}$ and JK = 3, find JM. 6

d If KM = 5 and JK = 6, find OM. $\sqrt{55}$

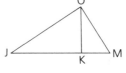

5 a Find a. 9

b Find ab. 54

c Find $a + b + c$. $15 + 6\sqrt{3}$

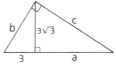

6 Given: $\overline{RT}$ is an altitude. $\angle$PRS is a right $\angle$.

Conclusion: $\dfrac{PR}{RS} = \dfrac{RT}{ST}$

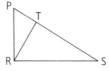

7 Given: $\overline{SY}$ is an altitude. $\angle$VSX is a right $\angle$.

Prove: XY · SV = XS · YS

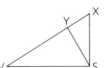

8 Find the coordinates of P, the center of the circle. (3, 1)

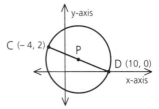

9 Given: Diagram as marked

Find: m$\angle$HJP, m$\angle$HKP, and m$\angle$HMP

25; 25; 25

Problem-Set Notes and Additional Answers, continued

6 Show that △PRT ~ △RST.

7 Show that △XYS ~ △SYV.

10 Find the measure of $\overset{\frown}{RH}$. 180

11 Find the area of sector MOG.
25π, or ≈ 78.54

Problem Set B

12 a Find the coordinates of point C. $(0, -12)$
  b Find the measure of the arc from A to
     B to C $(m\overset{\frown}{ABC})$. 270
  c Find the length of $\overset{\frown}{ABC}$. 18π, or ≈ 56.55

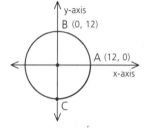

13 In $\odot P$, $m\overset{\frown}{FG} = 80$ and $m\overset{\frown}{DE} = 40$. Find
   $m\overset{\frown}{EF}$ and $m\angle EDF$. 60; 30

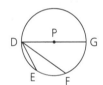

14 As Slarpy stood at B, the foot of a 6-m
   pole, he asked Carpy how far it was
   across the pond from B to C. Carpy got
   his carpenter's square and climbed the
   pole. Using his lines of sight, he set up
   the figure shown. When Slarpy found
   that AB = 3 m, Carpy knew the answer.
   What was it? 12 m

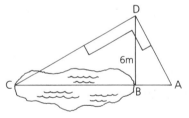

15 Given: $\odot O$, $\overline{CD} \perp \overline{AB}$;
     $\angle ACB$ is a right $\angle$.

   Conclusions: **a** $\dfrac{AD}{CD} = \dfrac{CD}{BD}$

   b $\dfrac{AD}{ED} = \dfrac{ED}{BD}$

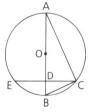

*Problem-Set Notes and
Additional Answers, continued*

14   Use Theorem 68b.

15a   Use Theorem 68b.
   b   Draw $\overline{OE}$, $\overline{OC}$. Then
        $\triangle OED \cong \triangle OCD$ by HL.
        Thus, $\overline{ED} \cong \overline{CD}$. Substitute
        into part **a.**

Problem Set B, continued

16 a If HG = 4 and EF = $3\sqrt{5}$, find EH. 5
  b If GF = 6 and EH = 9, find EG. 12

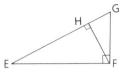

17 a If AD = 7 and AB = 11, find CD. $2\sqrt{7}$
  b If CD = 8 and AD = 6, find AB. $16\frac{2}{3}$
  c If AB = 12 and AD = 4, find BC. $4\sqrt{6}$
  d If AC = 7 and AB = 12, find BD. $7\frac{11}{12}$

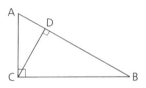

18 $\overline{CD}$ is the altitude to hypotenuse $\overline{AB}$. If the lengths AD, CD, CD, and BD are written down at random to form two ratios, what is the probability that the ratios are equal? $\frac{1}{3}$

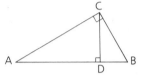

Problem Set C

19 If $\sqrt{5} \approx 2.236$, find DE to the nearest tenth. (The symbol $\approx$ means "approximately equals.") 1.8

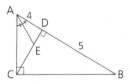

20 Prove: The product of the measures of the legs of a right triangle is equal to the product of the measures of the hypotenuse and the altitude to the hypotenuse.

21 Given: $\overline{AD} \perp \overline{CD}$,
  $\overline{BD} \perp \overline{AC}$,
  BC = 5, AD = 6
Find: BD $2\sqrt{5}$

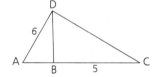

22 Given: $\overrightarrow{FG} \perp \overline{GH}$;
  ∠1 is comp. to ∠3.
Prove: $\dfrac{JH}{GH} = \dfrac{GH}{HF}$

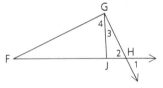

Problem-Set Notes and Additional Answers, continued

18 $\frac{AD}{CD} \overset{?}{=} \frac{BD}{CD}$, no; $\frac{AD}{CD} \overset{?}{=} \frac{CD}{BD}$, yes; $\frac{AD}{BD} \overset{?}{=} \frac{CD}{CD}$, no

19 AC = 6 by Theorem 68c
  $\frac{4}{6} = \frac{DE}{EC}$
  $\frac{2}{3} = \frac{x}{2\sqrt{5} - x}$
  $x = \frac{4\sqrt{5}}{5} \approx 1.8$

20

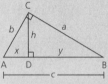

  $a^2 = cy$ and $b^2 = cx$.
  ∴ $a^2b^2 = c^2xy$. $a^2b^2 = c^2h^2$
  ∴ $ab = ch$.
  The equation $a \cdot b = h \cdot c$ can be presented as another formula relating the parts of a right triangle.

21

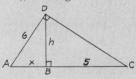

  $6^2 = x(x + 5)$
  ∴ $x = 4$
  $h^2 = 4 \cdot 5$
  $h = 2\sqrt{5}$

22 ∠1 is comp. to ∠3, so ∠1 ≅ ∠2. ∴ ∠2 is comp. to ∠3. Thus, $\overline{GJ} \perp \overline{FH}$. Then apply Theorem 68c.

Problem Set D

23 Given: HKMO is a rectangle.
$\overline{PK} \perp \overline{HM}$,
$\overline{PJ} \perp \overline{HK}$

Prove: $ab = fx$

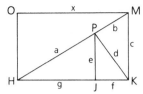

24 In the figure, CD is the mean proportional (or geometric mean) between AD and BD.

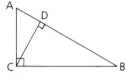

For any two numbers a and b, the arithmetic mean is $\frac{1}{2}(a + b)$.

For any two numbers a and b, the harmonic mean is $\dfrac{2}{\frac{1}{a} + \frac{1}{b}}$.

a Find the arithmetic mean, the geometric mean, and the harmonic mean between each pair of lengths.

   i $AD = 2$, $BD = 8$ A.M. = 5; G.M. = 4; H.M. = 3.2
   ii $AD = 3$, $BD = 12$ A.M. = 7.5; G.M. = 6; H.M. = 4.8
   iii $AD = 4$, $BD = 25$ A.M. = $14\frac{1}{2}$; G.M. = 10; H.M. = $6\frac{26}{29}$

b Given two positive numbers a and b, prove that their arithmetic mean, $\frac{1}{2}(a + b)$, is always greater than or equal to their positive geometric mean, $\sqrt{ab}$.

MATHEMATICAL EXCURSION

THE PYTHAGOREAN THEOREM AND TRIGONOMETRIC RATIOS

The magnifying properties of gravity

  Using sophisticated technology, astronomers have recently observed a phenomenon called *Einstein rings*, which occur when three objects, such as a galaxy, a quasar, and the earth, are collinear. Einstein rings are multiple images of the farther object—for example, the quasar—as its light or energy curves around the intervening object —the galaxy. In this case, the galaxy acts as a gravitational lens. It helps astronomers see more of the distant object than they could by observing a single image.

  Astronomers can calculate the distance to a flaring quasar by applying the Pythagorean The-

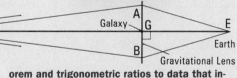

orem and trigonometric ratios to data that include: the difference in arrival times of the light from the flare by different paths it takes around the galaxy, the angles separating the images, the red-shift velocities of light from the quasar and from the galaxy, and the mass of the galaxy.

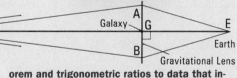

Problem-Set Notes and Additional Answers, continued

23 $d^2 = f(f + g)$
      $d^2 = fx$
      $d^2 = ab$
      $ab = fx$

24a In general, the harmonic mean of n numbers $a_1, a_2, \ldots, a_n$ is

$$\dfrac{n}{\frac{1}{a_1} + \frac{1}{a_2} + \ldots + \frac{1}{a_n}}$$

The harmonic mean is the average speed of individual speeds, each over the same distance. An important application of harmonic mean is in music.

b Two positive numbers are unequal in the same order as their squares.

$\left(\dfrac{a + b}{2}\right)^2 ? (\sqrt{ab})^2$

$\dfrac{a^2 + 2ab + b^2}{4} ? ab$

$a^2 + 2ab + b^2 ? 4ab$

$a^2 - 2ab + b^2 ? 0$

$(a - b)^2 ? 0$

But $(a - b)^2 \geq 0$ for all real a and b.

So $\left(\dfrac{a + b}{2}\right)^2 \geq (\sqrt{ab})^2$.

Cooperative Learning

Have students work in small groups to solve problem **24**.

Quasars are giant, brilliant objects that might be in the process of collapsing. Quasars give off bursts of luminosity, or flares, from their centers. The time elapsed between two observations (through a gravitational lens) of a single flare as its light travels by different paths around an intervening object can be used in calculating the distance to the quasar. It would not be unusual for the two observations of a single flare to be two years apart!

GEOMETRY'S MOST ELEGANT THEOREM

Objective

After studying this section, you will be able to
■ Use the Pythagorean Theorem and its converse

Part One: Introduction

As the plays of Shakespeare are to literature, as the Constitution is to the United States, so is the **Pythagorean Theorem** to geometry. First, it is basic, for it is the rule for solving right triangles. Second, it is widely applied, because every polygon can be divided into right triangles by diagonals and altitudes. Third, it enables many ideas (and objects) to fit together very simply. Indeed, it is elegant in concept and extremely powerful.

Theorem 69   *The square of the measure of the hypotenuse of a right triangle is equal to the sum of the squares of the measures of the legs.* **(Pythagorean Theorem)**

Given: △ACB is a right △
         with right ∠ACB.
Prove: $a^2 + b^2 = c^2$

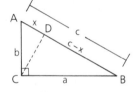

Proof:

1 ∠ACB is a right ∠.	1 Given
2 Draw $\overline{CD} \perp$ to $\overline{AB}$.	2 From a point outside a line, only one ⊥ can be drawn to the line.
3 $\overline{CD}$ is an altitude.	3 A segment drawn from a vertex of a △ ⊥ to the opposite side is an altitude.
4 $a^2 = (c - x)c$	4 In a right △ with an altitude drawn to the hypotenuse, $(\text{leg})^2 = (\text{adjacent seg.}) (\text{hypot.})$.
5 $a^2 = c^2 - cx$	5 Distributive Property
6 $b^2 = xc$	6 Same as 4
7 $a^2 + b^2 = c^2 - cx + cx$	7 Addition Property
8 $a^2 + b^2 = c^2$	8 Algebra

384 | Chapter 9   The Pythagorean Theorem

The Pythagorean Theorem was known to the ancient Egyptians and Greeks. The first proof is attributed to Pythagoras, a Greek mathematician who lived about 500 B.C. There are now more than 300 proofs of the theorem, and a book has been published consisting solely of such proofs. (Different sets of postulates and theorems lead to different proofs.)

One of the simplest ways to know that two lines are perpendicular is to find out if they form a right angle in a triangle. To use this method, we need the converse of the Pythagorean Theorem, given next.

Theorem 70 *If the square of the measure of one side of a triangle equals the sum of the squares of the measures of the other two sides, then the angle opposite the longest side is a right angle.*

If $a^2 + b^2 = c^2$,
then $\triangle ACB$ is a right $\triangle$
and $\angle C$ is the right $\angle$.

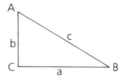

If, in the diagram above, we increased c while keeping a and b the same, $\angle C$ would become larger. Try it. Thus, a valuable extension of Theorem 70 can be stated:

If c is the length of the longest side of a triangle, and
- $a^2 + b^2 > c^2$, then the triangle is acute
- $a^2 + b^2 = c^2$, then the triangle is right
- $a^2 + b^2 < c^2$, then the triangle is obtuse

Part Two: Sample Problems

Problem 1 *Solve for x.*

Solution We use the Pythagorean Theorem.
$$6^2 + 8^2 = x^2$$
$$36 + 64 = x^2$$
$$100 = x^2$$
$$\pm 10 = x \quad \text{(Reject } -10.)$$
$$x = 10$$

Since $a^2 + b^2 = c^2$ and $a^2 + b^2 = EF^2$ in $\triangle EDF$ (Pythagorean Theorem), $c^2 = EF^2$ by substitution and $c = EF$. The $\triangle$ are $\cong$ by SSS, and $\angle C \cong \angle D$ by CPCTC. Since $\angle D$ is a rt. $\angle$, $\angle C$ is also a rt. $\angle$.

Lesson Notes

- One simple proof of the Pythagorean Theorem uses two elementary area formulas.

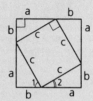

Total area = $(a + b)^2$
We can show that $\angle 1$ is complementary to $\angle 2$, and that the interior figure is a square.
The figure can be rearranged.

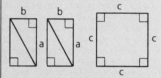

Total area = $2ab + c^2$
Thus, $(a + b)^2 = 2ab + c^2$;
$a^2 + 2ab + b^2 = 2ab + c^2$;
$a^2 + b^2 = c^2$.

References:

The Pythagorean Proposition by Elisha S. Loomis (Reston, Va.: National Council of Teachers of Mathematics, 1968).
Was Pythagoras Chinese? by Frank J. Swetz and T. I. Kao (Reston, Va.: National Council of Teachers of Mathematics, 1977).

- Proof of Theorem 70
  Given: $a^2 + b^2 = c^2$
  Prove: $\angle C$ is a right angle.

Construct rt. $\triangle EDF$ with rt. $\angle$ at D, legs a and b units long.

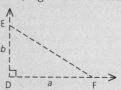

T 385

- The students should study all the sample problems; each problem illustrates a different situation for applying the Pythagorean Theorem.
- Sample Problem **3** represents a type of problem that appears on many standardized tests.

- The technique of considering a trapezoid as a rectangle and two right triangles is used frequently in the text.

Problem 2   *Find the perimeter of the rectangle shown.*

Solution   We use the Pythagorean Theorem.

$$x^2 + 5^2 = 13^2$$
$$x^2 + 25 = 169$$
$$x^2 = 144$$
$$x = \pm 12 \quad \text{(Reject } -12.\text{)}$$

Perimeter = 5 + 12 + 5 + 12 = 34

Problem 3   *Find the perimeter of a rhombus with diagonals of 6 and 10.*

Solution   Remember that the diagonals of a rhombus are perpendicular bisectors of each other.

$$3^2 + 5^2 = x^2$$
$$9 + 25 = x^2$$
$$34 = x^2$$
$$\pm\sqrt{34} = x \quad \text{(Reject } -\sqrt{34}.\text{)}$$

Since all sides of a rhombus are congruent, the perimeter is $4\sqrt{34}$.

Problem 4   *Nadia skipped 3 m north, 2 m east, 4 m north, 13 m east, and 1 m north. How far is Nadia from where she started?*

Solution   Since Nadia started at S and ended at E, we are looking for the hypotenuse of $\triangle SAE$. She has gone a total of 8 m north and 15 m east.

$$8^2 + 15^2 = x^2$$
$$64 + 225 = x^2$$
$$289 = x^2$$
$$\pm 17 = x \quad \text{(Reject } -17.\text{)}$$
$$SE = 17 \text{ m}$$

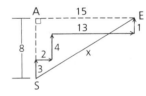

Problem 5   *Find the altitude of an isosceles trapezoid whose sides have lengths of 10, 30, 10, and 20.*

Solution   An altitude of a trapezoid is a segment, such as $\overline{AE}$, perpendicular to both bases. We often draw two altitudes, such as $\overline{AE}$ and $\overline{BD}$, to obtain a rectangle, AEDB. Thus, ED = 20, right $\triangle AEF$ is congruent to right $\triangle BDC$, and FE = DC = $\frac{1}{2}(30 - 20) = 5$.

In $\triangle AEF$,
$$x^2 + 5^2 = 10^2$$
$$x^2 + 25 = 100$$
$$x^2 = 75$$
$$x = \pm\sqrt{75}$$
$$= \pm\sqrt{25 \cdot 3}$$
$$= \pm 5\sqrt{3} \quad \text{(Reject } -5\sqrt{3}.\text{)}$$

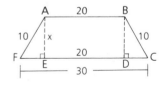

Altitude = $5\sqrt{3}$

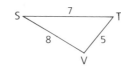

Problem 6   *Classify the triangle shown.*

Solution   If $5^2 + 7^2 > 8^2$, the triangle is acute.
If $5^2 + 7^2 = 8^2$, the triangle is right.
If $5^2 + 7^2 < 8^2$, the triangle is obtuse.

$$5^2 + 7^2 \; ? \; 8^2$$
$$25 + 49 \; ? \; 64$$
$$74 > 64$$

Therefore, the triangle is acute.

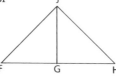

Part Three: Problem Sets

Problem Set A

1 Solve for the third side.

 a $x = 4$, $y = 5$ $\sqrt{41}$

 b $x = 15$, $r = 17$ 8

 c $y = 9$, $r = 15$ 12

 d $x = 12$, $r = 13$ 5

 e $x = 5$, $y = 5\sqrt{3}$ 10

 f $x = 5$, $r = \sqrt{29}$ 2

 g $x = 2\sqrt{5}$, $r = \sqrt{38}$
  $3\sqrt{2}$

2 Find the length of the diagonal of a square with perimeter 12 cm. $3\sqrt{2}$ cm

3 Find the perimeter of a rhombus with diagonals 12 km and 16 km. 40 km

4 Find the perimeter of a rectangle whose diagonal is 17 mm long and whose base is 15 mm long. 46

5 Given: $\overline{JG}$ is the altitude to base $\overline{FH}$ of isosceles triangle JFH.
 $FJ = 15$, $FH = 24$

 Find: JG 9

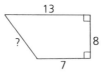

6 $\overline{PM}$ is an altitude of equilateral triangle PKO. If $PK = 4$, find PM. $2\sqrt{3}$

7 Find the missing length in the trapezoid. 10

Cooperative Learning

After presenting Theorem 70, have small groups of students draw a large acute triangle, right triangle, and obtuse triangle using a protractor and a straightedge. The longest side of each triangle is to be labeled "c" and the other two sides labeled "a" and "b." Each student should record the measure of each side and then compute and record $a^2 + b^2$ and c^2 for each triangle. Each student's measurements and computations should be checked by other members of the group. Ask each group to compare $a^2 + b^2$ with c^2 for each triangle and to write generalizations that are suggested by the comparisons.

These generalizations should be equivalent to the converse extension of Theorem 70.

Communicating Mathematics

Have students write a paragraph that explains the Pythagorean Theorem.

Assignment Guide

Basic

Day 1	1a,c,e,g, 2–8
Day 2	9, 10, 12a,c, 13, 14, 17, 20, 22

Average

Day 1	1a,c,e,g, 2–10
Day 2	12a,c, 13–15, 17, 22–24

Advanced

Day 1	2–10, 12a,c, 13–17
Day 2	$\left(\frac{1}{2}\text{ day}\right)$ Section **9.4** 23–28, 31
	$\left(\frac{1}{2}\text{ day}\right)$ Section **9.5** 6, 9, 11, 12

Problem Set A, *continued*

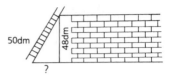

8 How far is the foot of the ladder from the wall? **14 dm**

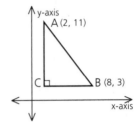

9 a Find the coordinates of C. **(2, 3)**
   b Find AC and CB. **8; 6**
   c Find AB. **10**
   d Is $AB = \sqrt{(8-2)^2 + (11-3)^2}$? **Yes**

10 Use the method suggested by part **d** of problem 9 to find PQ. **13**

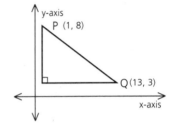

Problem Set B

11 Find the missing length in terms of the variable(s) provided.

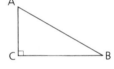

   a $AC = x$, $BC = y$, $AB = \underline{\quad?\quad}$ $\sqrt{x^2 + y^2}$
   b $AC = 2$, $BC = x$, $AB = \underline{\quad?\quad}$ $\sqrt{4 + x^2}$
   c $AC = 3a$, $BC = 4a$, $AB = \underline{\quad?\quad}$ $5a$
   d $AB = 13c$, $AC = 5c$, $BC = \underline{\quad?\quad}$ $12c$

12 $\angle ACB$ is a right angle and $\overline{CD} \perp \overline{AB}$.

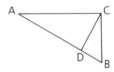

   a If $AD = 7$ and $BD = 4$, find CD. $2\sqrt{7}$
   b If $CD = 8$ and $DB = 6$, find CB. **10**
   c If $BC = 8$ and $BD = 2$, find AB. **32**
   d If $AC = 21$ and $AB = 29$, find CB. **20**

13 Al Capone walked 2 km north, 6 km west, 4 km north, and 2 km west. If Big Al decides to "go straight," how far must he walk across the fields to his starting point? **10 km**

14 Find the altitude (length of a segment perpendicular to both bases) of the isosceles trapezoid shown. $6\sqrt{2}$

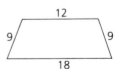

15 A piece broke off rectangle ABDF, leaving trapezoid ACDF. If BD = 16, BC = 7, FD = 24, and E is the midpoint of $\overline{FD}$, what is the perimeter of △ACE? **60**

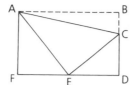

16 Given: Diagram as shown
Find: CD **4.8**

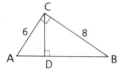

17 Solve for x in the partial spiral to the right. $\sqrt{5}$

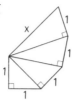

18 If the perimeter of a rhombus is $8\sqrt{5}$ and one diagonal has a length of $4\sqrt{2}$, find the length of the other diagonal. $4\sqrt{3}$

19 Woody Woodpecker pecked at a 17-m wooden pole until it cracked and the upper part fell, with the top hitting the ground 10 m from the foot of the pole. Since the upper part had not completely broken off, Woody pecked away where the pole had cracked. How far was Woody above the ground? ≈5.56 m

20 Find the perimeter of an isosceles right triangle with a 6-cm hypotenuse. $(6 + 6\sqrt{2})$ cm

21 The lengths of the diagonals of a rhombus are in the ratio 2:1. If the perimeter of the rhombus is 20, find the sum of the lengths of the diagonals. $6\sqrt{5}$

22 Classify the triangles.

Obtuse
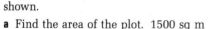

Impossible; 3 + 5 ≯ 10

23 George and Diane bought a plot of land along Richard Road with the dimensions shown.

a Find the area of the plot. **1500 sq m**

b Find, to the nearest meter, the length of frontage on Richard Road. ≈36 m

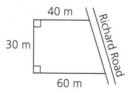

Section 9.4 Geometry's Most Elegant Theorem   **389**

**Problem-Set Notes
and Additional Answers**

19

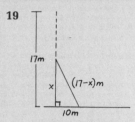

$$x^2 + 10^2 = (17 - x)^2$$
$$x^2 + 100 = 289 - 34x + x^2$$
$$34x = 189$$
$$x = \frac{189}{34} = 5\frac{19}{34}$$

27 PO = x, JK = 2x,
KM = 15 − 2x.
In △PKM,
$(15 − 2x)^2 + (20 + x)^2 = 25^2$.
x = 0 or x = 4. Reject 0.
∴ KM = 7.

28 The solution involves a system of equations.

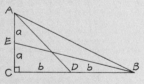

In △ACD, $4a^2 + b^2 = 52$.
In △BCE, $a^2 + 4b^2 = 73$.
$a = 3$. $b = 4$; ∴ AB = 10.

29

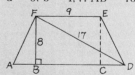

In △FBD, $8^2 + (BD)^2 = 17^2$.
BD = 15, CD = 6, and
AB = 6.
In △ECD, $8^2 + 6^2 = (ED)^2$.
So ED = 10. ∴ P = 50.

30a

$x^2 + y^2 = (x + 1)^2$
$x^2 + y^2 = x^2 + 2x + 1$
$\therefore y^2 = 2x + 1$

b Counterexamples have
nonintegral values for two
sides of the △; the sides 4,
$7\frac{1}{2}$, and $8\frac{1}{2}$ do form a rt. △
(as shown in the △ below),
but $7\frac{1}{2}$ and $8\frac{1}{2}$ are not
consecutive integers.

31a Slope of $\overleftrightarrow{QU}$ and $\overleftrightarrow{DA}$ is $\frac{15}{8}$;
slope of $\overleftrightarrow{QD}$ and $\overleftrightarrow{UA}$ is $-\frac{3}{4}$;
so the opposite sides are ∥.
b From the distance formula
previewed in **9d,**
QU = AD = 17 and
QD = UA = 10. P = 54.

32

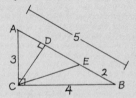

$(AC)^2 = AD \cdot AB$
$9 = 5x$
$x = \frac{9}{5}$
$AD = \frac{9}{5}, DE = \frac{6}{5}, BD = \frac{16}{5}$
$(CD)^2 = AD \cdot DB$
$CD^2 = \frac{9}{5} \cdot \frac{16}{5}$
$CD = \frac{12}{5}$
$(CD)^2 + (DE)^2 = (CE)^2$
$\left(\frac{12}{5}\right)^2 + \left(\frac{6}{5}\right)^2 = (CE)^2$
$CE = \frac{6\sqrt{5}}{5}$

Problem Set C

24 Find the perimeter of △DBC. $18 + 6\sqrt{5}$

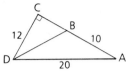

25 a Find HF. **8**
b Is △EHF similar to △HGF? **No**

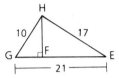

26 The perimeter of an isosceles triangle is 32, and the length of the
altitude to its base is 8. Find the length of a leg. **10**

27 A ladder 25 ft long (JO) is leaning against
a wall, reaching a point 20 ft above the
ground (MO). The ladder is then moved
so that JK = 2(PO). Find KM. **7 ft**

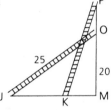

28 The medians of a right triangle that are drawn from the vertices
of the acute angles have lengths of $2\sqrt{13}$ and $\sqrt{73}$. Find the
length of the hypotenuse. **10**

29 The diagonals of an isosceles trapezoid are each 17, the altitude
is 8, and the upper base is 9. Find the perimeter of the trapezoid. **50**

30 a Show that if the lengths of one leg of a right triangle and the
hypotenuse are consecutive integers, then the square of the
length of the second leg is equal to the sum of the lengths of
the first leg and the hypotenuse.

b Show by counterexample that the converse of the statement in
part **a** is not necessarily true. (The converse is, "If the square
of the length of one of the legs of a right triangle is equal to
the sum of the lengths of the other leg and the hypotenuse,
then the lengths of the second leg and the hypotenuse are
consecutive integers.")

31 Quadrilateral QUAD has vertices at Q = (−7, 1), U = (1, 16),
A = (9, 10), and D = (1, −5).

a Plot the figure and indicate what type of quadrilateral
QUAD is. **Parallelogram**

b Find the perimeter of QUAD. **54**

Problem Set D

32 The legs of a right triangle have lengths of 3 m and 4 m. A point on the hypotenuse is 2 m from the intersection of the hypotenuse with the longer leg. How far is the point from the vertex of the right angle? $\frac{6\sqrt{5}}{5}$ m

33 RSTV is an isosceles trapezoid with RS = 9, RV = 12, and ST = 18.

Find the length of the perpendicular segment from T to $\overline{SW}$. $12\sqrt{2}$

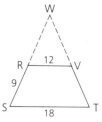

34 Given: $\overline{PR} \perp \overline{RT}$, PT = 25, PR = 15,
PS = ST + 12
Find: SR $\frac{799}{64}$

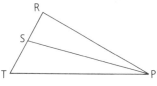

35 Abigail Adventuresome took a shortcut along the diagonal of a rectangular field and saved a distance equal to $\frac{1}{3}$ the length of the longer side. Find the ratio of the length of the shorter side of the rectangle to that of the longer side. $\frac{5}{12}$

36 a Given: P is any point in the interior of rectangle ABCD.

Show: $(BP)^2 + (PD)^2 = (AP)^2 + (CP)^2$

b Is the result the same when P is in the exterior of the rectangle? Yes

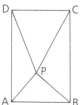

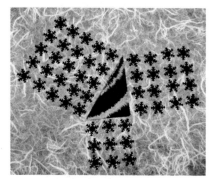

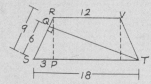

*Problem-Set Notes and
Additional Answers, continued*

33

By isos. trap. procedure,
SP = 3 in △SPR.
△SPR ~ △SQT;
∴ $\frac{9}{18} = \frac{3}{SQ}$ and SQ = 6.
Then by the Pythagorean
Theorem, in △SQT, $6^2 + (QT)^2 = 18^2$; QT = $12\sqrt{2}$.

34 In △PRT, $15^2 + (RT)^2 = 25^2$;
RT = 20. Let ST = x,
SR = 20 − x, and
PS = x + 12. In △PSR,
$(20 − x)^2 + 15^2 = (x + 12)^2$;
$x = \frac{481}{64}$.
∴ SR = $\frac{1280}{64} - \frac{481}{64} = \frac{799}{64}$.

35

$x^2 + y^2 = \left(x + \frac{2}{3}y\right)^2$
$x^2 + y^2 = x^2 + \frac{4}{3}xy + \frac{4}{9}y^2$
$\frac{5}{9}y^2 = \frac{4}{3}xy$
$\frac{5}{9}y = \frac{4}{3}x$
$\frac{x}{y} = \frac{\frac{5}{9}}{\frac{4}{3}} = \frac{5}{12}$

36
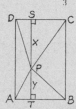

$(BP)^2 = y^2 + (TB)^2$
$(PD)^2 = x^2 + (DS)^2$
$(BP)^2 + (PD)^2 = x^2 + y^2 + (TB)^2 + (DS)^2$
$(AP)^2 = y^2 + (AT)^2$
$(CP)^2 = x^2 + (SC)^2$
$(AP)^2 + (CP)^2 = x^2 + y^2 + (AT)^2 + (SC)^2$
But TB = SC and DS = AT.
Thus, $(BP)^2 + (PD)^2 = (AP)^2 + (CP)^2$.

THE DISTANCE FORMULA

Class Planning

Time Schedule
Basic: 2 days
Average: 2 days
Advanced: $1\frac{1}{2}$ days

Resource References
Teacher's Resource Book
   Class Opener 9.5A, 9.5B
Evaluation
   Tests and Quizzes
      Quizzes 6–8, Series 1
      Quiz 4, Series 2
      Quiz 4, Series 3

Class Opener

Plot A(3, 7) and B(11, 10). Draw $\overline{AB}$. Draw a horizontal segment from A to the right, and draw a vertical segment downward from B.

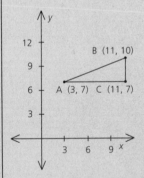

1 What is the coordinate of the point where the two segments intersect?   (11, 7)
    Label this point C.
2 What is the length of $\overline{AC}$?   8
    Of $\overline{BC}$?   3
3 How would you find the length of $\overline{AB}$?   Use Pythagorean Theorem
4 Find the length of $\overline{AB}$.
$$(AB)^2 = 8^2 + 3^2$$
$$= 64 + 9$$
$$= 73$$
$$AB = \sqrt{73}, \text{ or } AB \approx 8.5$$

Objective

After studying this section, you will be able to
- Use the distance formula to compute lengths of segments in the coordinate plane

Part One: Introduction

In $\triangle AOB$, AO = 4, since we can count the 4 spaces from O to A. OB = 3, since we can count the 3 spaces from O to B.

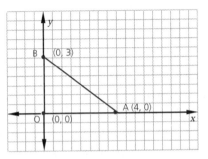

When a segment in the coordinate plane is either horizontal or vertical, its length is easily computed. To compute the length of $\overline{AB}$, we must find a new method. Since $\triangle AOB$ is a right triangle, we can apply the Pythagorean Theorem.

$$(OA)^2 + (OB)^2 = (BA)^2$$
$$3^2 + 4^2 = (BA)^2$$
$$25 = (BA)^2$$
$$5 = BA$$

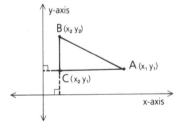

To compute any nonvertical, nonhorizontal length, we could draw a right triangle and use the Pythagorean Theorem.

$$(AB)^2 = (CA)^2 + (BC)^2$$
$$(AB)^2 = (x_2 - x_1)^2 + (y_2 - y_1)^2$$
$$AB = \sqrt{(x_2 - x_1)^2 + (y_2 - y_1)^2}$$
$$\text{or } AB = \sqrt{(\Delta x)^2 + (\Delta y)^2}$$

However, it is easier to use the **distance formula,** which is derived from the Pythagorean Theorem.

Vocabulary
distance formula

Theorem 71   *If P = (x₁, y₁) and Q = (x₂, y₂) are any two points, then the distance between them can be found with the formula*

$$PQ = \sqrt{(x_2 - x_1)^2 + (y_2 - y_1)^2} \text{ or}$$
$$PQ = \sqrt{(\Delta x)^2 + (\Delta y)^2}$$

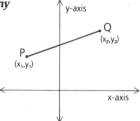

When doing coordinate proofs (sometimes called analytic proofs), you may select any convenient position in the coordinate plane for the figure *as long as complete generality is preserved.* Here are some convenient locations for a right triangle, an isosceles triangle, and a parallelogram.

Right Triangle	Isosceles Triangle	Parallelogram

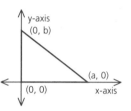 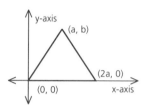 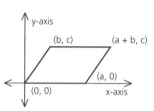

When midpoints are involved in a problem, it is helpful to use coordinates that make computations easier. For example, you could locate a rectangle as shown at the right.

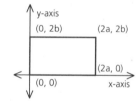

Part Two: Sample Problems

Problem 1   If A = (2, 3) and B = (7, 15), find AB.

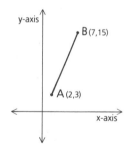

Solution   By the distance formula,

$$AB = \sqrt{(\Delta x)^2 + (\Delta y)^2}$$
$$= \sqrt{(7 - 2)^2 + (15 - 3)^2}$$
$$= \sqrt{5^2 + 12^2}$$
$$= \sqrt{169}$$
$$= 13$$

Lesson Notes

■ The distance formula is another instance in coordinate geometry of treating the coordinates separately and then combining them in the Pythagorean Theorem.

Cooperative Learning

Have students work in small groups to do the following problems. For each of the points, state whether Theorem 71 is needed to find the distance. Give an explanation for each answer.

1 (2, −4); (2, −9)   No, they lie on a vertical segment.
2 (−3, −5); (8, −5)   No, they lie on a horizontal segment.
3 (−1, −5); (3, 4)   Yes, they do not lie on a vertical or horizontal segment.
4 (a, b); (6, 6)   No, if a = 6 or b = 6; otherwise, yes

Communicating Mathematics

Have students write a paragraph that explains how the distance formula is derived from the Pythagorean Theorem.

Problem 2   If D = (7, 1), E = (9, −5), and F = (6, −4), find the length of the median from F to $\overline{DE}$.

Solution   By the midpoint formula, the midpoint M of $\overline{DE}$ is (8, −2).

By the distance formula,

$$FM = \sqrt{(\Delta x)^2 + (\Delta y)^2}$$
$$= \sqrt{(6-8)^2 + [-4-(-2)]^2}$$
$$= \sqrt{(-2)^2 + (-2)^2}$$
$$= \sqrt{8} = 2\sqrt{2}$$

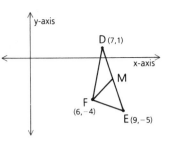

Problem 3   Prove: The medians to the legs of an isosceles triangle are congruent.

Proof   Use the general isosceles △ABC as shown.

By the midpoint formula, M = (a, b) and N = (3a, b).

By the distance formula,

$$MB = \sqrt{(4a-a)^2 + (0-b)^2} = \sqrt{9a^2 + b^2}$$
$$NA = \sqrt{(3a-0)^2 + (b-0)^2} = \sqrt{9a^2 + b^2}$$

Thus, $\overline{MB} \cong \overline{NA}$.

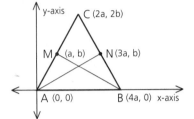

Part Three: Problem Sets

Problem Set A

1 Find the distance between each pair of points.

a (4, 0) and (6, 0)   2

b (2, 3) and (2, −1)   4

c (4, 1) and (7, 5)   5

d (−2, −4) and (−8, 4)   10

e The origin and (2, 5) $\sqrt{29}$

f (2, 1) and (6, 3) $2\sqrt{5}$

2 Find, to the nearest tenth, the perimeter of △ABC if A = (2, 6), B = (5, 10), and C = (0, 13).   ≈18.1

3 Show that the triangle with vertices at (8, 4), (3, 5), and (4, 10) is a right triangle by using

a The distance formula   **b** Slopes

4 Use the distance formula to show that △DOG is equilateral if D = (6, 0), O = (0, 0), and G = (3, 3$\sqrt{3}$).

5 Find the area of the circle that passes through (9, −4) and whose center is (−3, 5).   225π

Basic

Day 1   1a,c,e, 2, 3, 5, 10, 11
Day 2   6–9, 13, 15, 18

Average

Day 1   1a,c,e, 2, 3, 5–8, 10, 11
Day 2   9, 13–16, 22, 23

Advanced

Day 1   ($\frac{1}{2}$ day) Section **9.4**
          23–28, 31
          ($\frac{1}{2}$ day) Section **9.5**   6,
          9, 11, 12
Day 2   13–16, 19, 22–24

Problem-Set Notes and Additional Answers

■ See *Solution Manual* for answers to problems **3** and **4**.

6 Given: △RTV as shown

Find: **a** The length of the median from T $\sqrt{85}$

b The length of the segment joining the midpoints of $\overline{RT}$ and $\overline{TV}$ $3\sqrt{5}$

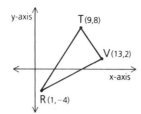

7 Find AD and BC. $\frac{9}{2}$; 10

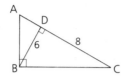

8 Given: Rectangle ABCO

a Find the coordinates of A, B, C, and O. (0, 2b); (2a, 2b); (2a, 0); (0, 0)

b Find the coordinates of M, N, P, and Q, the midpoints of the sides. (0, b); (a, 2b); (2a, b); (a, 0)

c Find the slopes of $\overline{MN}$, $\overline{QP}$, $\overline{MQ}$, and $\overline{NP}$. What can we conclude about MNPQ? $\frac{b}{a}$; $\frac{b}{a}$; $-\frac{b}{a}$; $-\frac{b}{a}$; MNPQ is a parallelogram.

d Find the lengths of $\overline{MN}$, $\overline{QP}$, $\overline{MQ}$, and $\overline{NP}$. What can we now conclude about MNPQ? All have length $\sqrt{b^2 + a^2}$; thus, MNQP is a rhombus.

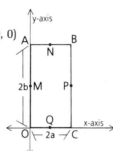

Problem-Set Notes and Additional Answers, continued

■ See *Solution Manual* for answer to problem **9**.

9 Given: Trapezoid PQRS

a Find PQ and SR and verify that PQRS is an isosceles trapezoid.

b Prove that the diagonals $\overline{PR}$ and $\overline{QS}$ are congruent.

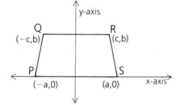

10 In the figure at the right, RECT is a rectangle. Is $\overline{RC}$ a diameter? Why or why not? Yes; since m∠E = 90, m$\overset{\frown}{RTC}$ = 180, and therefore $\overline{RC}$ is a diameter.

11 In rectangle RECT, RE = 5 and EC = 12.

a Find the circumference of the circle. 13π

b Find the area of the circle to the nearest tenth. ≈ 132.7

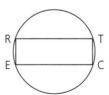

*Problem-Set Notes and
Additional Answers, continued*

■ See *Solution Manual* for answers to problems **12, 14, 16,
  17,** and **19.**

Problem Set A, *continued*

12 Prove that the diagonals of a square are congruent and
perpendicular.

Problem Set B

13 Given: ⊙R, m$\overset{\frown}{VW}$ = 120,
      RW = 9

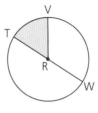

   Find: **a** The area of sector TRV to the
        nearest tenth ≈42.4

       b The difference, to the nearest
        tenth, between the length of $\overline{TW}$
        and the length of $\overset{\frown}{VW}$ ≈0.8

14 Show that (7, 11), (7, −13), and (14, 4) lie on a circle with its
center at (2, −1).

15 Find, to the nearest tenth, the perimeter of a quadrilateral with
vertices A = (2, 1), B = (7, 3), C = (12, 1), and D = (7, −4), and
give the figure's most descriptive name. ≈24.9; kite

16 Show that the parallelogram whose vertices are (−1, −3), (2, 1),
(3, −2), and (−2, 0) is not a rhombus.

17 Show that the triangle with vertices (−2, 1), (5, 5), and (−1, −7)
is isosceles.

18 The vertices of a rectangle are (0, 0) (8, 0), (0, 6), and (8, 6). Find
the sum of the lengths of the two diagonals. 20

19 Show that (1, 2) (4, 6), and (10, 14) are collinear by using
   a The distance formula (Hint: What is true about the lengths of
     the three segments joining three collinear points?)
   b Slopes

20 The point (5, y) is equidistant from (1, 4) and (10, −3). Find y. $-\frac{1}{7}$

21 Find the altitude of a trapezoid with sides having the respective
lengths 2, 41, 20, and 41. 40

22 A model rocket shot up to a point 20 m
above the ground, hitting a smokestack,
and then dropped straight down to a
point 11 m from its launch site. Find, to
the nearest meter, the total distance traveled from launch to touchdown. ≈43 m

23 Prove that the midpoint of the hypotenuse of a right triangle is equidistant from the three vertices.

24 Prove that the sum of the squares of the sides of a parallelogram is equal to the sum of the squares of the diagonals.

Problem Set C

25 Prove that in any quadrilateral the sum of the squares of the sides is equal to the sum of the squares of the diagonals plus four times the square of the segment joining the midpoints of the diagonals.

26 In isosceles trapezoid ABCD, A = $(-2a, 0)$ and B = $(2a, 0)$, where $a > 0$. The altitude of the trapezoid is $2h$, and the upper base, $\overline{CD}$, has a length of $4p$.

Find: **a** The coordinates of C and D $(2p, 2h); (-2p, 2h)$

 b The length of the lower base $4a$

 c The length of the segment joining the midpoints of $\overline{AD}$ and $\overline{BC}$ $2a + 2p$

 d The length of the segment joining the midpoints of the diagonals of the trapezoid $2a - 2p$

27 Two of the vertices of an equilateral triangle are (2, 1) and (6, 5). Find the possible coordinates of the remaining vertex.
$\left(4 + 2\sqrt{3}, 3 - 2\sqrt{3}\right)$ or $\left(4 - 2\sqrt{3}, 3 + 2\sqrt{3}\right)$

CAREER PROFILE

FINDING DISTANCE WITH LASERS
Don Milligan draws the contours of the land

Don Milligan, whose work as a surveyor is heavily dependent on mathematics, admits that he hated geometry in high school. "But I enjoyed trigonometry," he says. "Then I found that trigonometric identities are related to SAS similarity. That changed my mind about geometry."

A surveyor's job is to determine the exact size, shape, and location of a plot of land. A survey can help establish boundary lines and compute the areas of irregularly shaped lots.

Angles are measured using a tool called a *transit*, a small telescope on a tripod. Transits in use today often employ lasers. Surveyors apply trigonometry, triangle proportions, and triangle similarity in their work.

Born in Salt Lake City, Utah, Milligan ob-

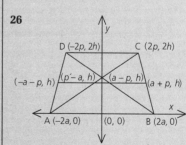

tained a bachelor's degree in forestry and wildlife management at Utah State University. After doing some survey work during the summer, he was hired by the Utah State Fish and Game Department. He left the department to work for a private surveying company, which he bought three years later.

Section 9.5   The Distance Formula   **397**

27 Equation of ⊥ bis. of given side is $y = -x + 7$. The length of a side is $4\sqrt{2}$.
∴ If the remaining vertex is

(x, y), $\sqrt{(x - 2)^2 - (y - 1)^2} = 4\sqrt{2}$; $y = -x + 7$.
Solving, $x = 4 \pm 2\sqrt{3}$.

9.5

Problem-Set Notes and Additional Answers, continued

■ See *Solution Manual* for answers to problems **23** and **24**.

25

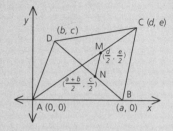

Prove:
$AB^2 + BC^2 + CD^2 + DA^2 = AC^2 + BD^2 + 4MN^2$
$AB^2 = a^2$
$BC^2 = (a - d)^2 + (e - 0)^2$
$\quad = a^2 - 2ad + d^2 + e^2$
$CD^2 = (b - d)^2 + (c - e)^2$
$\quad = b^2 - 2bd + d^2 + c^2 - 2ce + e^2$
$DA^2 = b^2 + c^2$
$AC^2 = d^2 + e^2$
$BD^2 = (a - b)^2 + (0 - c)^2$
$\quad = a^2 - 2ab + b^2 + c^2$
$MN^2 = \left(\frac{a + b}{2} - \frac{d}{2}\right)^2 + \left(\frac{c}{2} - \frac{e}{2}\right)^2$
$\quad = \frac{a^2 + b^2 + d^2 + 2ab - 2ad - 2bd}{4}$
$\quad + \frac{c^2 - 2ce + e^2}{4}$
$4MN^2 = a^2 + b^2 + d^2 + 2ab - 2ad - 2bd + c^2 - 2ce + e^2$
$AB^2 + BC^2 + CD^2 + DA^2 = 2a^2 + 2b^2 + 2c^2 + 2d^2 + 2e^2 - 2ad - 2bd - 2ce$
$AC^2 + BD^2 + 4MN^2 = 2a^2 + 2b^2 + 2c^2 + 2d^2 + 2e^2 - 2ad - 2bd - 2ce$

26

c $(a + p) - (-a - p) = 2a + 2p$
d $(p - a) - (a - p) = 2a - 2p$

T 397

Class Planning

Time Schedule
Basic: 2 days
Average: 2 days
Advanced: 1 day

Resource References
Teacher's Resource Book
  Class Opener 9.6A, 9.6B
Transparency 17
Evaluation
  Tests and Quizzes
    Quizzes 7–8, Series 1
    Quiz 5, Series 2

Class Opener

1 Use these formulas,
   $a = 2uv$, $b = u^2 - v^2$, and
   $c = u^2 + v^2$, to find a, b, and
   c for these values.
   a $u = 2$ and $v = 1$
      $a = 4$; $b = 3$; $c = 5$
   b $u = 3$ and $v = 2$
      $a = 12$; $b = 5$; $c = 13$
   c $u = 4$ and $v = 1$
      $a = 8$; $b = 15$; $c = 17$
2 Find a^2, b^2, and c^2 for each
   part of problem **1**.
   a $a^2 = 16$; $b^2 = 9$; $c^2 = 25$
   b $a^2 = 144$; $b^2 = 25$; $c^2 = 169$
   c $a^2 = 64$; $b^2 = 225$; $c^2 = 289$
3 What relationship do you
   notice about a^2, b^2, and c^2?
   $a^2 + b^2 = c^2$

You can use the three formulas
in problem **1** to generate all
Pythagorean triples.

Lesson Notes

- The ability to solve problems
  quickly and accurately will require students to drill, drill,
  drill on triples and reducing
  triangles.
- Problems involving Pythagorean triples frequently appear on
  standardized tests.

9.6 FAMILIES OF RIGHT TRIANGLES

Objectives

After studying this section, you will be able to
- Recognize groups of whole numbers known as Pythagorean triples
- Apply the Principle of the Reduced Triangle

Part One: Introduction

Pythagorean Triples

In this section we consider some combinations of whole numbers
that satisfy the Pythagorean Theorem. Knowing these combinations
is not essential, but knowing some of them can save you appreciable
time and effort.

Definition   Any three whole numbers that satisfy the equation
$a^2 + b^2 = c^2$ form a ***Pythagorean triple.***

Below is a set of right triangles you have encountered many
times in this chapter. Do you see how the triangles are related?

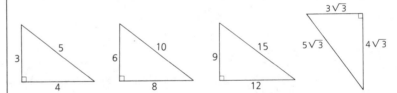

These four triangles are all members of the (3, 4, 5) family. For
example, the triple (6, 8, 10) is $(3 \cdot 2, 4 \cdot 2, 5 \cdot 2)$.
   Even though the last triangle, $\left(3\sqrt{3}, 4\sqrt{3}, 5\sqrt{3}\right)$, is a member of
the (3, 4, 5) family, the measures of its sides are not a Pythagorean
triple because they are not whole numbers.
   Other common families are

(5, 12, 13), of which (15, 36, 39) is another member

(7, 24, 25), of which (14, 48, 50) is another member

(8, 15, 17), of which $\left(4, 7\frac{1}{2}, 8\frac{1}{2}\right)$ is another member

   There are infinitely many families, including (9, 40, 41), (11, 60, 61),
(20, 21, 29), and (12, 35, 37), but most are not used very often.

Vocabulary
Pythagorean triple

The Principle of the Reduced Triangle

The following problem shows how a knowledge of Pythagorean triples can be useful even in situations where their applicability is not immediately apparent.

Example 1   *Given: The right triangle shown*
Find: x

The fraction may complicate our work, and we may not wish to complete a long calculation to solve $4^2 + \left(7\frac{1}{2}\right)^2 = x^2$.

An alternative is to find a more easily recognized member of the same family. We multiply each side by the denominator of the fraction, 2. Clearly, the family is (8, 15, 17). Thus, $2x = 17$ and $x = 8\frac{1}{2}$ (in the original triangle).

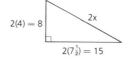

Principle of the Reduced Triangle

1 Reduce the difficulty of the problem by multiplying or dividing the three lengths by the same number to obtain a similar, but simpler, triangle in the same family.
2 Solve for the missing side of this easier triangle.
3 Convert back to the original problem.

The next example shows that the method may save time even if the sides of the "reduced" triangle are not a proper Pythagorean triple.

Example 2   *Find the value of x.*

First, notice that both 55 and 77 are multiples of 11. Then reduce the problem to an easier problem as shown below.

   is in the family

$$\text{where } 5^2 + y^2 = 7^2$$
$$25 + y^2 = 49$$
$$y^2 = 24$$
$$y^2 = \pm 2\sqrt{6} \quad \text{(Reject } -2\sqrt{6}.\text{)}$$

Thus, $x = 11 \cdot 2\sqrt{6} = 22\sqrt{6}$.

Lesson Notes, continued

- (6, 8, 10) is a Pythagorean triple. Students should realize that the greatest number (not automatically the third one) must be the hypotenuse.
- The other primitive triples with hypotenuses less than 100 are (13, 84, 85), (16, 63, 65), (28, 45, 53), (33, 56, 65), (36, 77, 85), (39, 80, 89), (48, 55, 73), and (65, 72, 97).
- There are two types of problems involving the Principle of the Reduced Triangle.

1 Sides with fractional (non-integral) measures

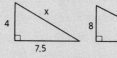

$2x = 17$, so $x = 8\frac{1}{2}$.
Multiply each measure by the least common denominator.

2 Sides whose measures have a common factor

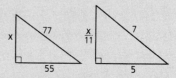

$$\frac{x}{11} = \sqrt{49 - 25}$$
$$= \sqrt{24} = 2\sqrt{6}$$
$$x = 22\sqrt{6}$$

Divide each measure by the greatest common factor.

- The Principle of the Reduced Triangle can also be used, with slight modifications, for sides with irrational measures. (See Sample Problem **3**, p. 400.)
- Students should note that the Principle of the Reduced Triangle can be applied even if the sides of the triangle are not a Pythagorean triple.

Communicating Mathematics

Ask students to explain how they could use the Principle of the Reduced Triangle to solve Sample Problem **1**.
Divide 24 and 10 by 2, and solve using the Pythagorean Theorem. Then multiply this answer by 2.

Cooperative Learning

Use this small-group activity prior to presenting the Principle of the Reduced Triangle.

1

Refer to △ABC to find x, the length of the hypotenuse in each problem below.

a 　**b**

$x = 2\sqrt{2}$　　$x = 38\sqrt{2}$

2

Refer to △DEF to find x, the length of the hypotenuse in each problem below.

a 　**b**

$x = 10$　　$x = 2\frac{1}{2}$

The students in each group should be able to explain how they found each answer without doing paper-and-pencil computations.

Part Two: Sample Problems

Problem 1　Find AB.

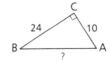

Solution

Method One:

(10, 24, ?) belongs to the (5, 12, 13) family.

$$10 = 5 \cdot 2$$
$$24 = 12 \cdot 2$$
So AB = 13 · 2 = 26

Method Two:

$$10^2 + 24^2 = (AB)^2$$
$$100 + 576 = (AB)^2$$
$$676 = (AB)^2$$
$$\pm\sqrt{676} = AB \quad (\text{Reject } -\sqrt{676}.)$$
$$26 = AB$$

Problem 2　Find x.

Solution

You may think that 5 is the answer, but in a (3, 4, 5) triangle the 5 must represent the length of the hypotenuse. Therefore, we are stuck with the long way.

$$3^2 + x^2 = 4^2$$
$$x^2 = 7$$
$$x = \pm\sqrt{7} \quad (\text{Reject } -\sqrt{7}.)$$
$$x = \sqrt{7}$$

Problem 3　Find the hypotenuse of the right triangle.

Solution

Method One:
Reduced-Triangle Principle

Divide each given length by 6 to obtain the reduced similar triangle.
$$1^2 + (3\sqrt{3})^2 = y^2$$
$$1 + 27 = y^2$$
$$\pm\sqrt{28} = y$$
$$\pm2\sqrt{7} = y \quad (\text{Reject } -2\sqrt{7}.)$$
Now multiply by 6 to convert back to the original triangle.
$$x = 6(2\sqrt{7}) = 12\sqrt{7}$$

Method Two:
Pythagorean Theorem

$$6^2 + (18\sqrt{3})^2 = x^2$$
$$36 + 972 = x^2$$
$$1008 = x^2$$
$$\sqrt{1008} = x$$
$$\pm\sqrt{144 \cdot 7} = x$$
(Would you have discovered those factors?)
Reject the negative root.
$$12\sqrt{7} = x$$

Part Three: Problem Sets

Problem Set A

In problems 1–5, find the missing side in each triangle.

1 (3, 4, 5)

a 25, 20, ?, 15

b ?, 45, 36, 27

c 28, 35, ? 21

d $\frac{5}{3}$, $1\frac{1}{3}$, ?, 1

e ?, 75, 60, 45

2 (5, 12, 13)

a 24, 26, ? 10

b 72, 78, ?, 30

c ?, 39, 36, 15

d 65, 60, ?, 25

e 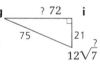 6, ? $6\frac{1}{2}$, $2\frac{1}{2}$

3 (7, 24, 25)

a 250, 240, ?, 70

b ?, 50, 48, 14

c 96, 100, ? 28

d ?, 2.5, 2.4, 0.7

e ?, 275, 264, 77

4 (8, 15, 17)

a 45, ? 51, 24

b 3.4, 3, ?, 1.6

c ?, 85, 75, 40

d $1\frac{1}{2}$, $1\frac{7}{10}$, ? $\frac{4}{5}$

e 150, 170, ?, 80

5 Mixed

a 9, 15, ? 12

b 6, 8, ? $2\sqrt{7}$

c 24, ?, 10, 26

d 1.3, 1.2, ? .5

e 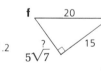 16, 30, ? 34

f 20, ?, 15, $5\sqrt{7}$

g 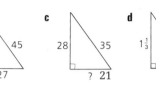 ? 72, 75, 21, $12\sqrt{7}$, ?

h 51, 24, ? 45

i 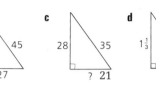 $5\sqrt{7}$, $13\sqrt{7}$

Section 9.6 Families of Right Triangles **401**

Lesson Notes, continued

- Sample Problem **2** is included to warn students against indiscriminate application of Pythagorean triples.
- Sample Problem **3** illustrates that the Principle of the Reduced Triangle can also be used, with slight modifications, for sides with irrational measures. Method 2 does not use the principle. It applies the Pythagorean Theorem to the given numbers.

Assignment Guide

Basic

| Day 1 | 1–5 |
| Day 2 | 6–14 |

Average

| Day 1 | 1–8 |
| Day 2 | 9–15, 17, 18 |

Advanced

9–11, 13, 14, 17, 18, 22, 23

Problem-Set Notes and Additional Answers

- After this assignment, a "speed" quiz (e.g., 20 problems in ten minutes) of mixed problems can encourage students to use triples and the Principle of the Reduced Triangle.

Problem Set A, *continued*

6 Find the diagonal of a rectangle whose sides are 20 and 48. 52

7 Find the perimeter of an isosceles triangle whose base is 16 dm and whose height is 15 dm. 50 dm

8 Find the length of the upper base of the isosceles trapezoid. 17

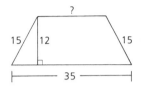

9 Use the reduced-triangle principle to find each missing side.

a

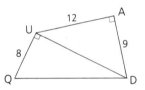

b

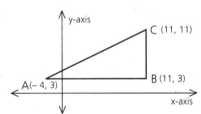

10 Find QD. 17

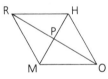

11 Find the perimeter and the area of △ABC. 40; 60

12 RHOM is a rhombus with diagonals RO = 48 and HM = 14. Find the perimeter of the rhombus. 100

Problem Set B

13 Mary and Larry left the riding stable at 10 A.M. Mary trotted south at 10 kph while Larry galloped east at 16 kph. To the nearest kilometer, how far apart were they at 11:30? ≈28 km

14 Write a coordinate proof to show that the diagonals of a rectangle are congruent.

402   Chapter 9   The Pythagorean Theorem

Problem-Set Notes and Additional Answers, continued

■ See *Solution Manual* for answer to problem **14**.

19a If n is odd, then n^2 is odd, and thus, $n^2 + 1$ and $n^2 - 1$ are even. Thus, n, $\frac{n^2 + 1}{2}$, and $\frac{n^2 - 1}{2}$ are integers.

$$n^2 + \left(\frac{n^2 - 1}{2}\right)^2 \stackrel{?}{=} \left(\frac{n^2 + 1}{2}\right)^2$$

$$n^2 + \frac{n^4 - 2n^2 + 1}{4} \stackrel{?}{=} \frac{n^4 + 2n^2 + 1}{4}$$

$$\frac{n^4 + 2n^2 + 1}{4} = \frac{n^4 + 2n^2 + 1}{4}$$

Parts **b, c**, and **d** are done in a similar manner.

Note that the Rule of Euclid in part **c** always generates values that represent lengths of sides of right triangles, but generates Pythagorean triples only when mn is a perfect square.

20

$$x^2 + (x + 1)^2 = (x + 2)^2$$
$$x^2 + x^2 + 2x + 1 = x^2 + 4x + 4$$
$$x^2 - 2x - 3 = 0$$
$$(x - 3)(x + 1) = 0$$

x = 3 or x = −1
Reject −1.
∴ 3, 4, 5

15 Find the missing side of each triangle.

a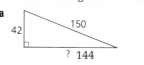
42, 150, ? 144

b
$\frac{3}{8}$, ?, $\frac{1}{2}$, $\frac{5}{8}$

c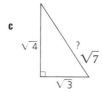
$\sqrt{4}$, ?, $\sqrt{7}$, $\sqrt{3}$

16 a Find x. 7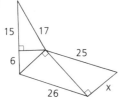
15, 17, 6, 25, 26, x

b Find x and y.
20; 20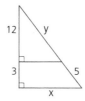
12, y, 3, 5, x

17 a What is the most descriptive name for quadrilateral PQRS? Isosceles trapezoid

b Find the area of PQRS. 48

c Find PR and QS. $4\sqrt{10}$

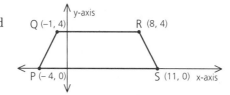
y-axis
Q (−1, 4)   R (8, 4)
P (−4, 0)   S (11, 0) x-axis

18 A submarine travels an evasive course, trying to outrun a destroyer. It travels 1 km north, then 1 km west, then 1 km north, then 1 km west, and so forth, until it has traveled a total of 41 km. How many kilometers is the sub from the point at which it started? 29 km

Problem Set C

19 Each of the following is a method for generating sets of whole numbers that represent the sides of a right triangle. Prove that each rule does indeed generate Pythagorean triples.

a Rule of Pythagoras
(*n* is any odd number.)

$\frac{n^2+1}{2}$, $\frac{n^2-1}{2}$, n

c Rule of Euclid
(*m* and *n* are both odd or both even.)

$\frac{m+n}{2}$, $\sqrt{mn}$, $\frac{m-n}{2}$

b Rule of Plato
(*m* is any even number.)

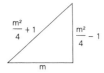
$\frac{m^2}{4}+1$, $\frac{m^2}{4}-1$, m

d Rule of Masères
(*m* and *n* are any two integers.)

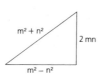
$\frac{m^2+n^2}{}$, 2mn, m^2-n^2

Section 9.6 Families of Right Triangles | **403**

Problem-Set Notes and Additional Answers, continued

21 x = 390 − 250 = 140

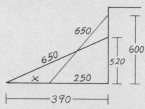
650, 650, 600, 520, x, 250, 390

22 $x^2 + (3x + y)^2 = (4x − y)^2$
$x^2 + 9x^2 + 6xy + y^2 = 16x^2 − 8xy + y^2$
$14xy = 6x^2$
$14y = 6x$
$\therefore \frac{x}{y} = \frac{14}{6} = \frac{7}{3}$

23a 3,4 4,5 5,6 6,8 8,10
3,5 4,6 5,8 6,10
3,6 4,8 5,10
3,8 4,10
3,10
$\frac{6}{15} = \frac{2}{5}$

b 3,4 4,5 5,6 6,8 8,10
3,5 4,6 5,8 6,10
3,6 4,8 5,10
3,8 4,10
3,10
$\frac{4}{15}$

24

16, c, b

$16^2 + b^2 = c^2$, so $16^2 = c^2 − b^2$.
$\therefore 256 = (c + b)(c − b)$.
Since *b* and *c* are integers, examine integral pairs of factors of 256, 16^2, until we find the largest pair that works.
c + b = 256
c − b =   1
2c   = 257
Since *c* will be a fraction, this pair doesn't work.
c + b = 128
c − b =   2
2c   = 130; c = 65.
$\therefore 65 + b = 128$ and $b = 63$.
The length of the hypotenuse of this (16, 65, 63) triangle is 65.

Problem-Set Notes and
Additional Answers, continued

25

Case 1: 20 is a leg.
$20^2 + b^2 = c^2$, so $20^2 = c^2 - b^2$.
$\therefore 400 = (c + b)(c - b)$.
Since b and c are integers,
examine integral pairs of
factors of 400.

$$c + b = 400 \qquad c + b = 200$$
$$\underline{c - b = 1} \qquad \underline{c - b = 2}$$
$$2c = 401 \qquad 2c = 202$$
$$c = 200\tfrac{1}{2} \qquad c = 101$$

$\therefore$ Impossible. $\quad \therefore b = 99$

$$c + b = 100 \qquad c + b = 80$$
$$\underline{c - b = 4} \qquad \underline{c - b = 5}$$
$$2c = 104 \qquad 2c = 85$$
$$c = 52 \qquad c = 42\tfrac{1}{2}$$

$\therefore b = 48 \qquad$ Impossible.

$$c + b = 50 \qquad c + b = 40$$
$$\underline{c - b = 8} \qquad \underline{c - b = 10}$$
$$2c = 58 \qquad 2c = 50$$
$$c = 29 \qquad c = 25$$
$$\therefore b = 21 \qquad \therefore b = 15$$

$$c + b = 25$$
$$\underline{c - b = 16}$$
$$2c = 41, \ c = 20\tfrac{1}{2} \ \text{Imposs.}$$

Case 2: 20 is the hypotenuse.
$a^2 + b^2 = 20^2$, $a \le b$.
$\therefore a^2 = 400 - b^2$.
Perfect squares between 200
and 400 are 225, 256, 289,
324, and 361. Let b^2 equal
each of these perfect squares.
Since a^2 must be a perfect
square, then $400 - b^2$ must
be a perfect square. The only
value of b^2 that works is 256;
that is, $400 - 256 = 144$, and
$\therefore a = 12$. So the triangles are
(20, 99, 101); (20, 48, 52);
(20, 21, 29); (15, 20, 25); and
(12, 16, 20).

Problem Set C, *continued*

20 Show that the only right triangle in which the lengths of the sides are consecutive integers is the (3, 4, 5) triangle.

21 If a 650-cm ladder is placed against a building at a certain angle, it just reaches a point on the building that is 520 cm above the ground. If the ladder is moved to reach a point 80 cm higher up, how much closer will the foot of the ladder be to the building? 140 cm

22 The lengths of the legs of a right triangle are x and 3x + y. The length of the hypotenuse is 4x − y. Find the ratio of x to y. $\frac{7}{3}$

23 Six slips of paper, each containing a different one of the numbers 3, 4, 5, 6, 8, and 10, are placed in a hat. Then two of the slips are drawn at random.

a What is the probability that the numbers drawn are the lengths of two of the sides of a triangle of the (3, 4, 5) family? $\frac{2}{5}$

b What is the probability that the numbers drawn are lengths of a leg and hypotenuse of a triangle of the (3, 4, 5) family? $\frac{4}{15}$

Problem Set D

24 Find the length of the hypotenuse of the largest Pythagorean-triple triangle in which 16 is the measure of a leg. 65

25 Find all right triangles in which one side is 20 and other sides are integral.
(20, 99, 101); (20, 48, 52); (20, 21, 29); (15, 20, 25); (12, 16, 20)

9.7 | SPECIAL RIGHT TRIANGLES

Objectives

After studying this section, you will be able to
- Identify the ratio of side lengths in a 30°-60°-90° triangle
- Identify the ratio of side lengths in a 45°-45°-90° triangle

Part One: Introduction

30°-60°-90° Triangles

You will find it useful to know the ratio of the sides of a triangle with angles of 30°, 60°, and 90°.

Theorem 72 **In a triangle whose angles have the measures 30, 60, and 90, the lengths of the sides opposite these angles can be represented by x, $x\sqrt{3}$, and $2x$ respectively. (30°-60°-90°-Triangle Theorem)**

Given: △ABC is equilateral.
$\overrightarrow{CD}$ bisects ∠ACB.
Prove: AD:DC:AC = $x:x\sqrt{3}:2x$

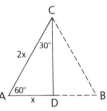

Proof: Since △ABC is equilateral, ∠ACD = 30°, ∠A = 60°, ∠ADC = 90°, and AD = $\frac{1}{2}$(AC).
By the Pythagorean Theorem, in △ADC,

$$x^2 + (DC)^2 = (2x)^2$$
$$x^2 + (DC)^2 = 4x^2$$
$$(DC)^2 = 3x^2$$
$$DC = x\sqrt{3}$$

Thus, AD:DC:AC = $x:x\sqrt{3}:2x$

Class Planning

Time Schedule
All levels: 2 days

Resource References
Teacher's Resource Book
    Class Opener 9.7A, 9.7B
    Additional Practice
    Worksheet 18
Evaluation
    Tests and Quizzes
        Quizzes 9–10, Series 1
        Quizzes 6–9, Series 2
        Quiz 5, Series 3

Class Opener

In the diagram, △ABC is equilateral, $\overline{CD}$ is an altitude, and $\overline{AC}$ = 12.

1   What relationship exists between △ADC and △BDC?
    △ADC ≅ △BDC
2   Find AD and BD.
    AD = 6; BD = 6
3   Find CD. $6\sqrt{3}$
4   What relationship exists between AD and AC?
    AC = 2AD
5   What relationship exists between CD and AD?
    CD = $\sqrt{3}$AD

Lesson Notes

- Basic courses often gain from spending one day on only the 30°-60°-90° triangle, and a second day on the rest of the section.

Lesson Notes, continued

- Theorems 72 and 73 are applicable in many polygons. Every equilateral triangle and regular hexagon break into 30°-60°-90° triangles, and every square and regular octagon break into 45°-45°-90° triangles.
- All students should be able to apply each of these six families readily and accurately.

Communicating Mathematics

Ask students to describe each of the following in their own words.

1 The relationship between the lengths of the sides of a 30°-60°-90° triangle
The side opposite the 30° angle (the shorter leg) is one-half of the hypotenuse. The side opposite the 60° angle is $\sqrt{3}$ times the shorter leg.

2 The relationship between the lengths of the sides of a 45°-45°-90° triangle
The hypotenuse is $\sqrt{2}$ times the leg.

- Labeling the diagrams with x, $x\sqrt{3}$, and $2x$ in Sample Problems **1** and **2** helps students track the correspondence between the sides of the triangle and the results of the algebraic steps.

45°-45°-90° Triangles

The sides of a triangle with angles of 45°, 45°, and 90° are also in an easily remembered ratio.

Theorem 73 *In a triangle whose angles have the measures 45, 45, and 90, the lengths of the sides opposite these angles can be represented by x, x, and x√2, respectively. (45°-45°-90°-Triangle Theorem)*

Given: △ACB, with ∠A = 45° and ∠B = 45°.
Prove: AC:CB:AB = $x:x:x\sqrt{2}$
The proof of this theorem is left to you.

You will see 30°-60°-90° and 45°-45°-90° triangles frequently in this book and in other mathematics courses. Their ratios are worth memorizing now.

Six Common Families of Right Triangles

30°-60°-90° $\Longleftrightarrow$ $(x, x\sqrt{3}, 2x)$ (5, 12, 13)
45°-45°-90° $\Longleftrightarrow$ $(x, x, x\sqrt{2})$ (7, 24, 25)
(3, 4, 5)   (8, 15, 17)

Part Two: Sample Problems

Problems 1 and 2 involve 30°-60°-90° triangles. In each, start by placing x on the side opposite (across from) the 30° angle, $x\sqrt{3}$ on the side opposite the 60° angle, and $2x$ on the hypotenuse.

Problem 1 *Type: Hypotenuse (2x) known*
Find BC and AC.

Solution Place x, $x\sqrt{3}$, and $2x$ on a copy of the diagram.

$$2x = 10$$
$$x = 5$$
Hence, BC = 5, and
$$AC = 5\sqrt{3}$$

Problem 2 Type: Longer leg $(x\sqrt{3})$ known
Find JK and HK.

Lesson Notes, continued
- Sample Problem **2** is usually the most difficult case. Several additional examples may be necessary.
- Labeling the sides of the triangle x, 2x, and x√3 is important (especially so when the hypotenuse is given as a whole number).

Solution Place x, $x\sqrt{3}$, and 2x on the figure as shown.

$$x\sqrt{3} = 6$$

$$x = \frac{6}{\sqrt{3}}$$

$$= \frac{6}{\sqrt{3}} \cdot \frac{\sqrt{3}}{\sqrt{3}}$$

$$= \frac{6\sqrt{3}}{3} = 2\sqrt{3}$$

Hence, JK $= 2\sqrt{3}$, and HK $= 2(2\sqrt{3}) = 4\sqrt{3}$.

Problems 3 and 4 involve 45°-45°-90° triangles. In each, start by placing x on each leg and $x\sqrt{2}$ on the hypotenuse.

Problem 3 Type: Leg (x) known
MOPR is a square.
Find MP.

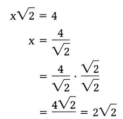

Solution A diagonal divides a square into two 45°-45°-90° triangles.
Place x, x, and $x\sqrt{2}$ as shown.
Since x = 9, MP = $9\sqrt{2}$.

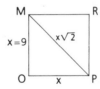

Problem 4 Type: Hypotenuse $(x\sqrt{2})$ known
Find ST and TV.

Solution Place x, x, and $x\sqrt{2}$ as shown.

$$x\sqrt{2} = 4$$

$$x = \frac{4}{\sqrt{2}}$$

$$= \frac{4}{\sqrt{2}} \cdot \frac{\sqrt{2}}{\sqrt{2}}$$

$$= \frac{4\sqrt{2}}{2} = 2\sqrt{2}$$

Hence, ST = TV = $2\sqrt{2}$.

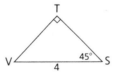

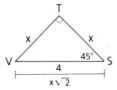

Part Three: Problem Sets

Problem Set A

1 Find the two missing sides in each 30°-60°-90° triangle. Try to do the calculations in your head.

a 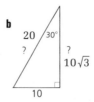   **b** 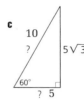   **c** 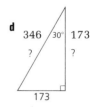   **d** 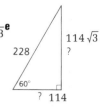   **e**

2 Find the two missing sides of each triangle. (Hint: These are a bit harder, and you may want to put x, x√3, and 2x on the proper sides as shown in the sample problems.)

a 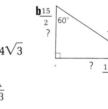   **b** 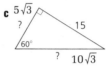   **c** 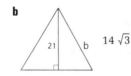   **d**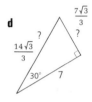

3 Solve for the variable in each of these equilateral triangles.

a   **b**   **c**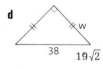

4 Solve for the variable in each of these 45°-45°-90° triangles.

a   **b**   **c**   **d**

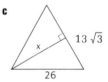

5 The perimeter of a square is 44. Find the length of a diagonal. $11\sqrt{2}$

6 Find the length of the diagonal of the rectangle. 16m

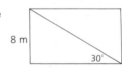

7 Find the altitude of an equilateral triangle if a side is 6 mm long. $3\sqrt{3}$ mm

8 Given: $\overline{AC} \perp \overline{BC}$, $\overline{CD} \perp \overline{AB}$,
   $\angle B = 30°$, $BC = 8\sqrt{3}$
   Find: CD $4\sqrt{3}$

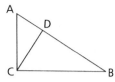

9 Given: TRWX is a kite ($\overline{TR} \cong \overline{WR}$ and $\overline{TX} \cong \overline{XW}$).
   RY = 5, TW = 10, YX = 12
   Find: **a** TR $5\sqrt{2}$
   b WX 13

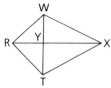

10 a Find the ratio of the longer leg to the hypotenuse in a
   30°-60°-90° triangle. $\frac{\sqrt{3}}{2}$
   b Find the ratio of one of the legs to the hypotenuse in a
   45°-45°-90° triangle. $\frac{\sqrt{2}}{2}$

11 Plato is alleged to have said that the 30°-60°-90° triangle was the
   most beautiful right triangle in the world. Grunts Giraffe, a
   sophomore student at Animal High, is alleged to have said that
   the 30°-60°-90° triangle didn't look very pretty to him. Who was
   Plato, and what do you think he meant by *beautiful*?

*Problem-Set Notes and
Additional Answers, continued*

■ See *Solution Manual* for answers to problems **11** and **14**.

Problem Set B

12 a Find the coordinates of B. $(1, \sqrt{3})$
   b Find the slope of $\overleftrightarrow{OB}$. $\sqrt{3}$
   c Find $\frac{AB}{OA}$. (In a trigonometry class, this
   ratio is called the *tangent* of angle
   BOA.) $\sqrt{3}$

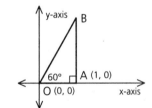

13 a Find the coordinates of D. (1, 1)
   b Find the slope of $\overleftrightarrow{OD}$. 1
   c Find the tangent of 45°. 1

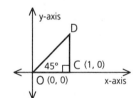

14 Show that in a 30°-60°-90° triangle the
   altitude to the hypotenuse divides the
   hypotenuse in the ratio 1:3. (Hint: Let
   DB = x. Then CD = $x\sqrt{3}$. Now solve
   for AD.)

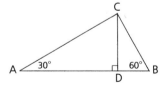

Problem Set B, *continued*

15 Find the perimeter of the isosceles trapezoid EFGH. (Hint: Drop altitudes of the
trapezoid from E and H.) 38

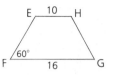

16 Given: $\overline{PK}$ is an altitude of isosceles trap-
ezoid JMOP.
PK = 6, PO = 8, ∠J = 45°
Find: The perimeter of JMOP $28 + 12\sqrt{2}$

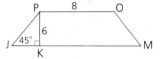

17 Using the figure, find
a VS $3\sqrt{3}$
b ST 9
c VT $6\sqrt{3}$
d The ratio of the perimeter of △VSR to
the perimeter of △VRT 1:2

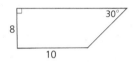

18 One of the angles of a rhombus has a measure of 120. If the
perimeter of the rhombus is 24, find the length of each diagonal. 6; $6\sqrt{3}$

19 Find, to the nearest tenth, the perimeter
of the trapezoid. ≈57.9

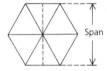

20 Any regular hexagon can be divided into
six equilateral triangles by drawing the
three diagonals shown. Find the span of
a regular hexagon with sides 12 dm long.
$12\sqrt{3}$ dm

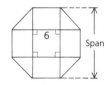

21 Any regular octagon can be divided into
rectangles and right triangles. Here, a
side of the central square is 6 units long.
a Find the perimeter of the octagon. 48
b Find the span of the octagon. $6 + 6\sqrt{2}$

22 Find the altitude to the base of the isosceles triangle shown. $3\sqrt{3}$

23 If rectangle RECT is rotated about the origin until E lies on the positive y-axis, what will the new coordinates of E be? $(0, 2\sqrt{5})$

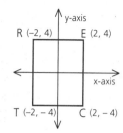

Problem Set C

24 Find x and y. $3\sqrt{3}; \frac{3}{2}\sqrt{3}$

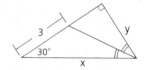

25 Given: ABCD is a trapezoid ($\overline{DC} \parallel \overline{AB}$).
$AB = AD = 4$,
$\angle A = 60°, \angle C = 45°$
Find: **a** DC $2 + 2\sqrt{3}$
   b BC $2\sqrt{6}$

26 If the area of rectangle ABCE is eight times that of △BCD, how far is D from the origin? $\sqrt{146}$, or ≈ 12.08

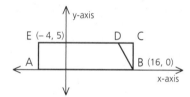

Problem Set D

27 Given: $\angle ACB$ is a right angle.
   $\overrightarrow{CD}$ and $\overrightarrow{CE}$ trisect $\angle ACB$.
   $AC = 5, BC = 12$
Find: CE (Hint: Draw a perpendicular from E to $\overline{CB}$.) $\frac{40(12 - 5\sqrt{3})}{23}$

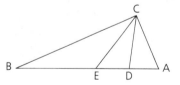

24

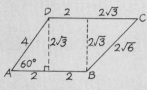

25

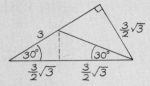

26 $A_{rect} = 100; A_\triangle = \frac{1}{2} \cdot CD \cdot BC = \frac{1}{2} \cdot CD \cdot 5$. Thus, $100 = 8 \cdot \frac{1}{2} \cdot 5 \cdot CD; 100 = 20CD; 5 = CD$. So D = (11, 5). From the distance formula, $d = \sqrt{11^2 + 5^2} = \sqrt{146}; d \approx 12.08$.

27

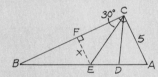

Let $\overline{EF}$ be the ⊥ from E to $\overline{CB}$. Let EF = x. Since △CEF is a 30°-60°-90° △, $CF = x\sqrt{3}, CE = 2x, BF = 12 - x\sqrt{3}$.
△BEF ~ △BAC, so $\frac{EF}{AC} = \frac{BF}{BC}$.
$\frac{x}{5} = \frac{12 - x\sqrt{3}}{12}$
$12x = 60 - 5x\sqrt{3}$
$12x + 5x\sqrt{3} = 60$
$x(12 + 5\sqrt{3}) = 60$
$$x = \frac{60}{12 + 5\sqrt{3}}$$
$$x = \frac{60}{12 + 5\sqrt{3}} \cdot \frac{12 - 5\sqrt{3}}{12 - 5\sqrt{3}}$$
$$x = \frac{60(12 - 5\sqrt{3})}{144 - 75}$$
$$x = \frac{20(12 - 5\sqrt{3})}{23}$$
$$CE = \frac{40(12 - 5\sqrt{3})}{23}$$

Problem Set D, continued

In solving probability problems, a tree diagram is sometimes helpful. Consider the following problem:

A bag contains seven red marbles, two blue marbles, and a white marble. A woman reaches into the bag and draws two marbles.

a What is the probability that she has drawn two red marbles?

b What is the probability that she has drawn one or more red marbles?

Solution:

a The tree diagram below shows that the probability of drawing a red marble and then another red marble is $\frac{7}{10} \cdot \frac{2}{3} = \frac{7}{15}$. So RR = $\frac{7}{15}$.

What are the probabilities of the other eight possible outcomes?

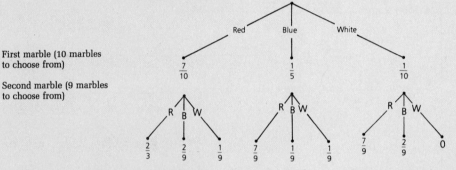

b The probability of drawing one or more red marbles is the sum of the probabilities of RR, RB, RW, BR, and WR, or $\frac{14}{15}$.

Problem-Set Notes and Additional Answers, continued

28

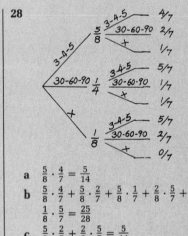

a $\frac{5}{8} \cdot \frac{4}{7} = \frac{5}{14}$

b $\frac{5}{8} \cdot \frac{4}{7} + \frac{5}{8} \cdot \frac{2}{7} + \frac{5}{8} \cdot \frac{1}{7} + \frac{2}{8} \cdot \frac{5}{7} + \frac{1}{8} \cdot \frac{5}{7} = \frac{25}{28}$

c $\frac{5}{8} \cdot \frac{2}{7} + \frac{2}{8} \cdot \frac{5}{7} = \frac{5}{14}$

28 Use a tree diagram to solve the following problem:

A bag contains eight right triangles. Five are members of the (3, 4, 5) family, and two are 30°-60°-90° triangles. A puppy falls over the bag, and two triangles fall out on the floor.

a What is the probability that both are members of the (3, 4, 5) family? $\frac{5}{14}$

b What is the probability that at least one of the triangles is a member of the (3, 4, 5) family? $\frac{25}{28}$

c What is the probability that one is a member of the (3, 4, 5) family and the other is a 30°-60°-90° triangle? $\frac{5}{14}$

THE PYTHAGOREAN THEOREM AND SPACE FIGURES

Objective

After studying this section, you will be able to
- Apply the Pythagorean Theorem to solid figures

Part One: Introduction

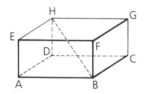

Rectangular Solid

Regular Square Pyramid

Many of the problems in this section will involve the two figures shown above.

In the rectangular solid:

ABFE is one of the 6 rectangular *faces*

$\overline{AB}$ is one of the 12 *edges*

$\overline{HB}$ is one of the 4 *diagonals* of the solid. (The others are $\overline{AG}$, $\overline{CE}$, and $\overline{DF}$.)

In the regular square pyramid:

JKMO is a square, and it is called the *base*

P is the *vertex*

$\overline{PR}$ is the *altitude* of the pyramid and is perpendicular to the base at its center.

$\overline{PS}$ is called a *slant height* and is perpendicular to a side of the base.

Note A *cube* is a rectangular solid in which all edges are congruent.

Part Two: Sample Problems

Problem 1 The dimensions of a rectangular solid are 3, 5, and 7. Find the diagonal.

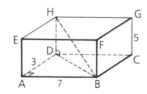

Section 9.8 The Pythagorean Theorem and Space Figures | **413**

Vocabulary

altitude	edge	regular square pyramid
base	face	slant height
cube	rectangular solid	vertex
diagonal		

Class Planning

Time Schedule
All levels: 1 day

Resource References
Teacher's Resource Book
    Class Opener 9.8A, 9.8B
Evaluation
    Test and Quizzes
        Quiz 10, Series 2
        Quizzes 6–7, Series 3

Class Opener

This box is a rectangular solid.

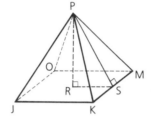

Each of the six sides (faces) is a rectangle.

1 $\overline{BD}$ is a diagonal that forms right triangle BAD.
  a If BC = 3, what is AD? 3
  b If AB = 4, what is DB? 5
2 $\overline{BH}$ is a diagonal that forms right triangle BDH.
  a If CG = 12, what is DH? 12
  b What is BH? AG? 13; 13

Lesson Notes

- The new terminology for solid figures is important and is used throughout the rest of the text.
- Students who have difficulty making three-dimensional drawings may need to reproduce the diagrams by tracing.

Cooperative Learning

The following small-group activity is designed to be used before presenting the sample problems. Each group will need a cardboard box, a piece of string, a ruler or meter stick, and two different colored markers.

1 Ask each group to attach the string to the box so that the string forms a diagonal of the rectangular solid.
2 Have students use a colored marker to draw the legs of one of the right triangles formed by an edge, a diagonal of a face, and the string. Ask students to measure the legs of this triangle and use the Pythagorean Theorem to find the length of the diagonal.
3 Using the other colored marker, students should draw the legs of a *different* right triangle that is *not* congruent to the first. They should then complete the same procedure as in step 2.
4 Ask students to compare their two computed answers and the actual measurement of the diagonal.
Answers to this group activity will vary depending on the size of the box.

Assignment Guide

Basic
1–3, 5, 7–12

Average
1–3, 5–7, 9, 11, 12, 16

Advanced
2, 5–7, 9, 11, 12, 16, 18, 21, 22

Solution It does not matter which edges are given the lengths 3, 5, and 7. Let AD = 3, AB = 7, and HD = 5, and use the Pythagorean Theorem twice.

In △ABD,
$$3^2 + 7^2 = (DB)^2$$
$$9 + 49 = (DB)^2$$
$$\sqrt{58} = DB$$

In △HDB,
$$5^2 + (\sqrt{58})^2 = (HB)^2$$
$$25 + 58 = (HB)^2$$
$$\sqrt{83} = HB$$

The measure of the diagonal is $\sqrt{83}$.

Problem 2 Given: The regular square pyramid shown, with altitude $\overline{PR}$ and slant height $\overline{PS}$, perimeter of JKMO = 40, PK = 13

Find: **a** JK **b** PS **c** PR

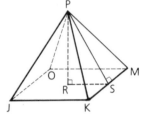

Solution **a** JK = $\frac{1}{4}(40)$ = 10
b The slant height of the pyramid is the ⊥ bis. of $\overline{MK}$, so PSK is a right △.
$$(SK)^2 + (PS)^2 = (PK)^2$$
$$5^2 + (PS)^2 = 13^2$$
$$PS = 12$$
c The altitude of a regular pyramid is perpendicular to the base at its center. Thus, RS = $\frac{1}{2}$(JK) = 5, and PRS is a right △.
$$(RS)^2 + (PR)^2 = (PS)^2$$
$$5^2 + (PR)^2 = 12^2$$
$$25 + (PR)^2 = 144$$
$$PR = \sqrt{119}$$

Part Three: Problem Sets

Problem Set A

1 Given: The rectangular solid shown, BY = 3, OB = 4, EY = 12
   Find: **a** YO, a diagonal of face BOXY 5
   b EO, a diagonal of the solid 13

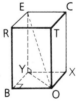

2 Find the diagonal of a rectangular solid whose dimensions are 3, 4, and 5. $5\sqrt{2}$

3 Given: Regular square pyramid ABCDE,
with slant height $\overline{AF}$, altitude $\overline{AG}$,
and base BCDE;
perimeter of BCDE = 40,
∠AFG = 60°

Find: The altitude and the slant height $5\sqrt{3}$; 10

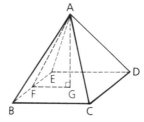

4 Given: The rectangular solid shown,
GC = 8, HG = 12, BC = 9

Find: **a** HB, a diagonal of the solid 17

  b AG, another diagonal of the solid 17

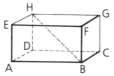

5 Given: The regular square pyramid shown, with altitude
$\overline{PY}$ and slant height $\overline{PR}$,
ID = 14, PY = 24

Find: **a** AD 14

  b YR 7

  c PR 25

  d The perimeter of base AMID 56

  e A diagonal of the base (not shown
in the diagram) $14\sqrt{2}$

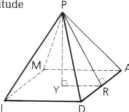

6 Find the slant height of a regular square pyramid if the altitude
is 12 and one of the sides of the square base is 10. 13

7 A line that intersects a circle at two
points is called a *secant*. Which of the
four lines in the diagram $\left(\overleftrightarrow{EF}, \overleftrightarrow{PB}, \overleftrightarrow{PD},\right.$
and $\left.\overleftrightarrow{GH}\right)$ are secants? $\overleftrightarrow{PB}$ and $\overleftrightarrow{PD}$

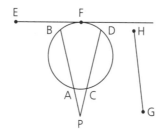

8 Given: Diagram as marked
Find: m$\widehat{AC}$ 20

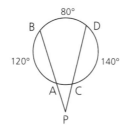

Communicating Mathematics

The following questions relate to
properties of rectangular solids.
Have students write their answers as complete sentences.

1 What is the relationship between the opposite faces of a
rectangular solid?
They are congruent and
parallel.

2 What is the relationship between the four diagonals of
the solid?
They are congruent.

3 What is the relationship between the opposite edges?
They are congruent and
parallel.

Problem Set A, *continued*

9 Daffy Difference looked ahead to Chapter 10 and found that the measure of a secant-secant angle (such as ∠BPD) is one-half the difference of its two intercepted arcs. Use this information to find m∠BPD. 30

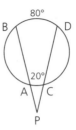

10 Given: Diagram as marked
Find: m∠EJF 35

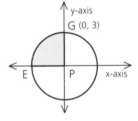

Problem Set B

11 Given: ⊙P as shown
Find: **a** The coordinates of point E (−3, 0)
   b The area of sector EPG to the nearest tenth ≈7.1
   c The length of $\overparen{GE}$ to the nearest tenth ≈4.7

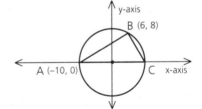

12 Given: Diagram as marked
Find: AB (the length of $\overline{AB}$)
   $8\sqrt{5}$, or ≈17.9

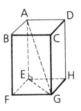

13 ABCDEFGH is a rectangular solid.
   a If face diagonal $\overline{CH}$ measures 17, edge $\overline{GH}$ measures 8, and edge $\overline{FG}$ measures 6, how long is diagonal $\overline{AG}$? $5\sqrt{13}$
   b If diagonal $\overline{AG}$ measures 50, edge $\overline{AE}$ measures 40, and edge $\overline{EF}$ measures 3, how long is edge $\overline{FG}$? $9\sqrt{11}$

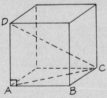

Problem-Set Notes and Additional Answers

18

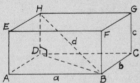

a AB = x; thus, BC = x and AC = $x\sqrt{2}$. Then in △DAC, AD² + AC² = DC². So x² + 2x² = DC²; 3x² = DC². Thus, DC = $x\sqrt{3}$.

b If AC = x, then a side of the cube is $\frac{x}{\sqrt{2}} = \frac{x\sqrt{2}}{2}$. Substituting into part **a**, DC = $\frac{x\sqrt{2}}{2} \cdot \sqrt{3}$; DC = $\frac{x\sqrt{6}}{2}$.

19

In △DAB, AB = a, DA = b, and BD = $\sqrt{a^2 + b^2}$. In △BDH, c² + ($\sqrt{a^2 + b^2}$)² = d²; c² + a² + b² = d²; $\sqrt{a^2 + b^2 + c^2}$ = d.

14 PADIM is a regular square pyramid. Slant height $\overline{PR}$ measures 10, and the base diagonals measure $12\sqrt{2}$.

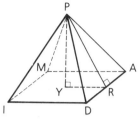

  a Find ID. 12

  b Find the altitude of the pyramid. 8

  c Find RD. 6

  d Find PD (length of a lateral edge). $2\sqrt{34}$

15 Find the diagonal of a cube if each edge is 2. $2\sqrt{3}$

16 Find the diagonal of a cube if the perimeter of a face is 20. $5\sqrt{3}$

17 The perimeter of the base of a regular square pyramid is 24. If the slant height is 5, find the altitude. 4

Problem Set C

18 In the cube, find the measure of the diagonal in terms of x if

  a AB = x $x\sqrt{3}$ **b** AC = x $\frac{x\sqrt{6}}{2}$

19 Find a formula for the length of a diagonal of a rectangular solid. (Use a, b, and c for the three dimensions.) $d = \sqrt{a^2 + b^2 + c^2}$

20 The dimensions of a rectangular solid are in the ratio 3:4:5. If the diagonal is $200\sqrt{2}$, find the three dimensions. 120; 160; 200

21 The face diagonals of a rectangular box are 2, 3, and 6. Find the diagonal of the box. Impossible

22 A pyramid is formed by assembling four equilateral triangles and a square having sides 6 cm long. Find the altitude and the slant height. $3\sqrt{2}$; $3\sqrt{3}$

Problem Set D

23 The strongest rectangular beam that can be cut from a circular log is one having a cross section in which the diagonal joining two vertices is trisected by perpendicular segments dropped from the other vertices.

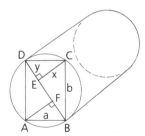

  If AB = a, BC = b, CE = x, and DE = y, show that $\frac{b}{a} = \frac{\sqrt{2}}{1}$.

Problem-Set Notes and Additional Answers, continued

20 Let the sides be $3x$, $4x$, and $5x$. Substituting into problem **19**, $200\sqrt{2} = \sqrt{(3x)^2 + (4x)^2 + (5x)^2}$; $200\sqrt{2} = \sqrt{50x^2}$; $200\sqrt{2} = 5x\sqrt{2}$; $40 = x$. Thus, sides are 120, 160, and 200.

21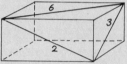

The diagonals do not satisfy the Triangle Inequality Theorem.

22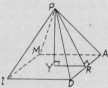

Since $\triangle$PDA is equilateral with side 6, then PR = $3\sqrt{3}$. In $\triangle$PYR, PR = $3\sqrt{3}$ and YR = 3. Thus, $PY^2 + 3^2 = (3\sqrt{3})^2$; $PY^2 + 9 = 27$
$$PY^2 = 18$$
$$PY = 3\sqrt{2}$$

23 In $\triangle$AFB, $x^2 + y^2 = a^2$. In $\triangle$CEB, $x^2 + 4y^2 = b^2$. In $\triangle$CDB, $a^2 + b^2 = 9y^2$. From $\triangle$AFB and $\triangle$CEB, $a^2 + b^2 = 2x^2 + 5y^2$. $\therefore 9y^2 = 2x^2 + 5y^2$; $4y^2 = 2x^2$; $2y^2 = x^2$. Substituting $2y^2$ for x^2,
$$\begin{cases} a^2 = 2y^2 + y^2 = 3y^2 \\ b^2 = 2y^2 + 4y^2 = 6y^2 \end{cases}$$

$$\frac{b^2}{a^2} = \frac{6y^2}{3y^2}$$

$$\frac{b^2}{a^2} = \frac{2}{1}$$

$$\frac{b}{a} = \frac{\sqrt{2}}{1}$$

9.9

INTRODUCTION TO TRIGONOMETRY

Class Planning

Time Schedule
Basic: Omit
Average: 2 days (optional)
Advanced: 1 day

Resource References
Teacher's Resource Book
  Class Opener 9.9A, 9.9B
  Supposer Worksheet 11

Class Opener

Given: Right $\triangle FMP$, $\overline{KW} \perp \overline{FM}$,
    $PF = 17$, $FM = 15$
Find: $\dfrac{KW}{FW}$

$\triangle FWK \sim \triangle FMP$ by AA.
$\therefore \dfrac{KW}{FW} = \dfrac{PM}{FM}$. $PM = 8$ by the
(8, 15, 17) family. Since
$\dfrac{PM}{FM} = \dfrac{8}{15}$, $\dfrac{KW}{FW} = \dfrac{8}{15}$.
This Class Opener previews the
trig ratios.

Lesson Notes

- This section and the rest of
  Chapter **9** can be omitted
  without losing course continu-
  ity. (The review problems, pp.
  429–433, and the cumulative
  review problems, pp. 434–437,
  do not require trigonometry.)

Objective

After studying this section, you will be able to
- Understand three basic trigonometric relationships

Part One: Introduction

This section presents the three basic trigonometric ratios **sine**, **co-sine**, and **tangent**. The concept of similar triangles and the Pythago-rean Theorem can be used to develop the **trigonometry of right triangles.**
  Consider the following 30°-60°-90° triangles.

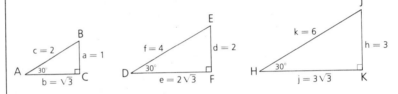

Compare the length of the leg opposite the 30° angle with the length of the hypotenuse in each triangle.

In $\triangle ABC$, $\dfrac{a}{c} = \dfrac{1}{2} = 0.5$.   In $\triangle DEF$, $\dfrac{d}{f} = \dfrac{2}{4} = 0.5$.   In $\triangle HJK$, $\dfrac{h}{k} = \dfrac{3}{6} = 0.5$.

If you think about similar triangles, you will see that in every 30°-60°-90° triangle,

$$\frac{\text{leg opposite } 30° \angle}{\text{hypotenuse}} = \frac{1}{2}$$

For each triangle shown, verify that $\dfrac{\text{leg adjacent to } 30° \angle}{\text{hypotenuse}} = \dfrac{\sqrt{3}}{2}$.

For each triangle shown, find the ratio $\dfrac{\text{leg opposite } 30° \angle}{\text{leg adjacent to } 30° \angle}$.

Vocabulary
cosine
sine
tangent
trigonometry of right triangles

In △ABC and △DEF,

$$\frac{a}{c} = \frac{d}{f} = \frac{6}{10} = \frac{3}{5}$$

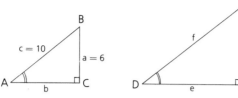

Engineers and scientists have found it convenient to formalize these relationships by naming the ratios of sides. You should memorize these three basic ratios.

Definition   **Three Trigonometric Ratios**

$$\textit{sine} \text{ of } \angle A = \sin \angle A = \frac{\text{opposite leg}}{\text{hypotenuse}}$$

$$\textit{cosine} \text{ of } \angle A = \cos \angle A = \frac{\text{adjacent leg}}{\text{hypotenuse}}$$

$$\textit{tangent} \text{ of } \angle A = \tan \angle A = \frac{\text{opposite leg}}{\text{adjacent leg}}$$

Part Two: Sample Problems

Problem 1   Find: **a** $\cos \angle A$
                      b $\tan \angle B$

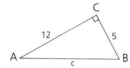

Solution   By the Pythagorean Theorem, $c = 13$.

a $\cos \angle A = \dfrac{\text{leg adjacent to } \angle A}{\text{hypotenuse}} = \dfrac{12}{13}$

b $\tan \angle B = \dfrac{\text{leg opposite } \angle B}{\text{leg adjacent to } \angle B} = \dfrac{12}{5}$

Problem 2   Find the three trigonometric ratios for $\angle A$ and $\angle B$.

Solution

$$\sin \angle A = \frac{3}{5} \qquad \sin \angle B = \frac{4}{5}$$

$$\cos \angle A = \frac{4}{5} \qquad \cos \angle B = \frac{3}{5}$$

$$\tan \angle A = \frac{3}{4} \qquad \tan \angle B = \frac{4}{3}$$

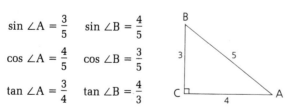

Lesson Notes, continued

■ The labeling follows the mathematical convention that side a is found opposite angle A, and so on.

Communicating Mathematics

Before presenting the sample problems, you may want to use the following exercises to determine how well students understand the meaning of sine, cosine, and tangent. Ask students to draw and label the triangle in the given circumstances.

1 $\sin \angle A = \dfrac{1}{\sqrt{2}}$

$\sin \angle B = \dfrac{1}{\sqrt{2}}$

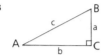

2 $\cos \angle A = \dfrac{8}{10}$

$\cos \angle B = \dfrac{6}{10}$

3 $\tan \angle A = \dfrac{8}{15}$

$\tan \angle B = \dfrac{15}{8}$

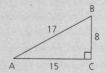

The Geometric Supposer

You can preview trigonometric ratios with your students by using *The Geometric Supposer: Triangles* to draw groups of similar right triangles, using the program to measure and calculate ratios between pairs of sides.

To draw a series of similar right triangles, use the angle-side-angle option under "New Triangle/Your Own." Have students draw a series of three similar right triangles, varying the length of the included side in each triangle. For each triangle, students should measure and calculate three ratios: the ratio of the included side to the hypotenuse, of the opposite side to the hypotenuse, and of the opposite side to the included side. Ask students to make observations about each ratio throughout the series of similar triangles.

Ask students to make a conjecture: if one of the angles they input is always a right angle, what ratios of sides will result for any acute angle in a series of similar right triangles? Have them test their conjectures by drawing and measuring three or four sets of similar triangles.

Assignment Guide

Basic	Omit

Average	Optional
Day 1	1–8
Day 2	9, 10, 13–16, 18

Advanced	
2–5, 9, 10, 13, 14, 18	

Problem 3 △ABC is an isosceles triangle as marked. Find sin ∠C.

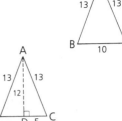

Solution We must have a right triangle, so we draw the altitude to the base.
Thus, in △ADC, sin ∠C = $\frac{12}{13}$.

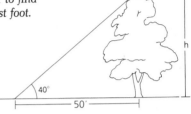

Problem 4 Use the fact that tan 40° ≈ 0.8391 to find the height of the tree to the nearest foot.

Solution

$$\tan 40° = \frac{h}{50}$$

$$0.8391 \approx \frac{h}{50}$$

$$h \approx 41.955$$

$$\approx 42 \text{ ft}$$

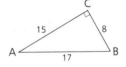

Part Three: Problem Sets

Problem Set A

1 Find each ratio.

  a sin ∠A $\frac{8}{17}$ **d** sin ∠B $\frac{15}{17}$
  b cos ∠A $\frac{15}{17}$ **e** cos ∠B $\frac{8}{17}$
  c tan ∠A $\frac{8}{15}$ **f** tan ∠B $\frac{15}{8}$

2 Find each ratio.

  a sin 30° $\frac{1}{2}$ **d** sin 60° $\frac{\sqrt{3}}{2}$
  b cos 30° $\frac{\sqrt{3}}{2}$ **e** cos 60° $\frac{1}{2}$
  c tan 30° $\frac{\sqrt{3}}{3}$ **f** tan 60° $\sqrt{3}$

3 Find each ratio.

  a sin 45° $\frac{\sqrt{2}}{2}$
  b cos 45° $\frac{\sqrt{2}}{2}$
  c tan 45° 1

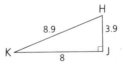

4 Find each ratio.

  a cos ∠H $\frac{39}{89}$
  b tan ∠K $\frac{39}{80}$

5 If tan $\angle M = \frac{3}{4}$, find cos $\angle M$. (Hint: Start by drawing the triangle.) $\frac{4}{5}$

6 Using the figure as marked, name each missing angle.

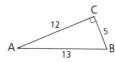

a $\frac{5}{12} = \tan \angle \underset{A}{\underline{\quad ? \quad}}$ **b** $\frac{12}{13} = \cos \angle \underset{A}{\underline{\quad ? \quad}}$ **c** $\frac{5}{13} = \sin \angle \underset{A}{\underline{\quad ? \quad}}$

7 Find each quantity.

a BC $2\sqrt{6}$ **b** sin $\angle A$ $\frac{2\sqrt{6}}{7}$ **c** tan $\angle B$ $\frac{5\sqrt{6}}{12}$

8 Given: RECT is a rectangle.
ET = 26, RT = 24
Find: **a** sin $\angle RET$ $\frac{12}{13}$ **b** cos $\angle RET$ $\frac{5}{13}$

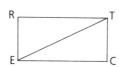

Problem Set B

9 Using the given figures, find
a cos $\angle A$ $\frac{7}{25}$
b sin $\angle E$ $\frac{8}{17}$
c sin $\angle DFG$ $\frac{4}{5}$

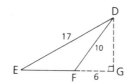

10 Use the fact that sin 40° ≈ 0.6428 to find the height of the kite to the nearest meter. ≈129 m

11 a If tan $\angle A = 1$, find m$\angle A$. **45**
b If sin $\angle P = 0.5$, find m$\angle P$. **30**

12 Given: sin $\angle P = \frac{3}{5}$, PQ = 10
Find: cos $\angle P$ $\frac{4}{5}$

13 Using the figure, find
a tan $\angle ACD$ $\frac{2}{3}$
b sin $\angle A$ $\frac{3\sqrt{13}}{13}$

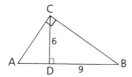

Cooperative Learning

Have students work in small groups to solve problem **16.** Each member of the group should be able to justify each answer. Answers will vary. Possible reasons are given.

a Always true; the side opposite $\angle A$ will be the side adjacent to $\angle B$.

b Never true; the hypotenuse is always greater than either leg. Therefore, the hypotenuse can never equal the adjacent leg.

c Sometimes true, as in the case of an isosceles triangle

Problem-Set Notes and Additional Answers

■ See *Solution Manual* for answer to problem **17**.

19

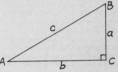

a $\sin \angle A = \frac{a}{c}$; $\cos \angle A = \frac{b}{c}$. Thus, $(\sin \angle A)^2 + (\cos \angle A)^2 = \left(\frac{a}{c}\right)^2 + \left(\frac{b}{c}\right)^2 = \frac{a^2}{c^2} + \frac{b^2}{c^2} = \frac{a^2 + b^2}{c^2}$. However, $a^2 + b^2 = 1$ by the Pythagorean Theorem. Thus, $(\sin \angle A)^2 + (\cos \angle A)^2 = 1$.

b $\frac{a}{\sin \angle A} = \frac{a}{a} = c$; $\frac{b}{\sin \angle B} = \frac{b}{b} = c$. Thus, $\frac{a}{\sin \angle A} = \frac{b}{\sin \angle B}$.

c $\frac{\sin \angle A}{\cos \angle A} = \frac{c}{b} = \frac{a}{b} = \tan \angle A$

d $\sin \angle A = \frac{a}{c}$
$\cos(90 - \angle A) = \cos(\angle B) = \frac{a}{c}$
∴ $\sin \angle A = \cos(90 - \angle A)$

20

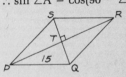

If SQ = 15, then $TQ = \frac{15}{2}$ and $PT = \frac{15}{2}\sqrt{3}$. Then sin $\angle PQS = \frac{\frac{15}{2}\sqrt{3}}{15} = \frac{\sqrt{3}}{2}$.

If PR = 15, then $PT = \frac{15}{2}$ and $\sin \angle PQS = \frac{\frac{15}{2}}{15} = \frac{1}{2}$.

21 $\frac{5}{13}, \frac{5}{12}, \frac{12}{13}, \frac{13}{12}, \boxed{\frac{12}{5}}, \frac{13}{5}$

22

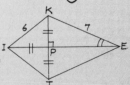

Problem Set B, *continued*

14 Given: RHOM is a rhombus.
RO = 18, HM = 24
Find: **a** cos ∠BRM $\frac{3}{5}$ **b** tan ∠BHO $\frac{3}{4}$

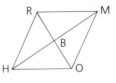

15 Given a trapezoid with sides 5, 10, 17, and 10, find the sine of one of the acute angles. $\frac{4}{5}$

16 Given △ABC with ∠C = 90°, indicate whether each statement is true Always (A), Sometimes (S), or Never (N).

a sin ∠A = cos ∠B A   **b** sin ∠A = tan ∠A N   **c** sin ∠A = cos ∠A S

17 If △EQU is equilateral and △RAT is a right triangle with RA = 2, RT = 1, and ∠T = 90°, show that sin ∠E = cos ∠A.

18 If the slope of $\overleftrightarrow{AB}$ is $\frac{5}{8}$, find the tangent of ∠BAC. $\frac{5}{8}$

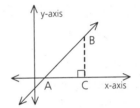

Problem Set C

19 Use the definitions of the trigonometric ratios to verify the following relationships, given △ABC in which ∠C = 90°.

a $(\sin \angle A)^2 + (\cos \angle A)^2 = 1$

c $\frac{\sin \angle A}{\cos \angle A} = \tan \angle A$

b $\frac{a}{\sin \angle A} = \frac{b}{\sin \angle B}$

d sin ∠A = cos (90° − ∠A)

20 Rhombus PQRS has a perimeter of 60 and one diagonal of 15. Find the two possible values of sin ∠PQS. $\frac{1}{2}$ or $\frac{\sqrt{3}}{2}$

21 Two sides of the triangle shown are picked at random to form a ratio. What is the probability that the ratio is the tangent of ∠A? $\frac{1}{6}$

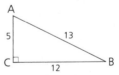

22 Given: KITE is a kite with sides as marked.
Find: tan ∠KEI $\frac{3\sqrt{62}}{31}$

Since △KPI is an isos. rt. △ with hypotenuse 6, KP = 3√2. Apply Pythagorean Theorem in

△KPE. $(3\sqrt{2})^2 + (PE)^2 = 7^2$; PE = √31. Thus, tan ∠KEI = $\frac{3\sqrt{2}}{\sqrt{31}} = \frac{3\sqrt{62}}{31}$.

9.10 TRIGONOMETRIC RATIOS

Class Planning

Time Schedule
Basic: Omit
Average: 2 days (optional)
Advanced: 1 day

Resource References
Teacher's Resource Book
  Class Opener 9.10A, 9.10B

Objective
After studying this section, you will be able to
- Use trigonometric ratios to solve right triangles

Part One: Introduction

Trigonometry is used to solve triangles other than 30°-60°-90° and 45°-45°-90° triangles. The Table of Trigonometric Ratios on the next page shows four-place decimal approximations of the ratios for other angles—for instance, sin 23° ≈ 0.3907, and the angle whose tangent is 1.5399 is approximately 57°.

Unless your teacher directs otherwise, we suggest you use a scientific calculator rather than the table to find trigonometric ratios.

For some applications of trigonometry, you need to know the meanings of **angle of elevation** and **angle of depression**.

If an observer at a point P looks upward toward an object at A, the angle the line of sight $\overrightarrow{PA}$ makes with the horizontal $\overrightarrow{PH}$ is called the **angle of elevation**.

If an observer at a point P looks downward toward an object at B, the angle the line of sight $\overrightarrow{PB}$ makes with the horizontal $\overrightarrow{PH}$ is called the **angle of depression**.

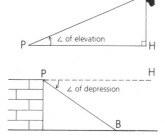

Note Do not forget that an angle of elevation or depression is an angle between a line of sight and the *horizontal*. Do not use the vertical.

Class Opener

1 Find the sine and cosine of ∠A and ∠B in △ACB.
   sin ∠A = $\frac{3}{5}$; sin ∠B = $\frac{4}{5}$;
   cos ∠A = $\frac{4}{5}$; cos ∠B = $\frac{3}{5}$

2a Compare sin ∠A with cos ∠B.   Equal

  b Compare sin ∠B with cos ∠A.   Equal

3 Write a generalization that your answers to problem **2** suggest.
   The sine of an angle equals the cosine of the complementary angle, or the cosine of an angle equals the sine of the complementary angle. Also, the tangents of complementary angles are reciprocals.

Vocabulary
angle of depression
angle of elevation

Communicating Mathematics

1 In the Table of Trigonometric Ratios, what is happening to the sin ∠A as ∠A increases? The sin ∠A increases.
2 As ∠A increases, what number is sin ∠A approaching? 1
3 Write a generalization based on your answers to problems 1 and 2.
The sin ∠A approaches 1 as the measure of ∠A approaches 90.
4 Write a generalization that describes the relationship between the cos ∠A and the measure of ∠A.
The cos ∠A approaches 1 as the measure of ∠A approaches 0.

Cooperative Learning

Have students work in small groups. One of the following measures is chosen at random: 0°, 30°, 45°, 60°, 90°. Have each group of students consider the sine, cosine, and tangent ratios. Find each probability.

1 One of the ratios related to the angle will be zero. $\frac{2}{5}$
2 All three ratios will be less than 1. $\frac{1}{5}$
3 Two of the ratios will be equal. $\frac{2}{5}$
4 At least one of the ratios will be undefined. $\frac{1}{5}$

Each member of a group should be able to explain why you get an undefined value in problem 4. When the angle is 90°, the length of the adjacent side is 0. Since division by 0 is undefined, tan 90° is undefined.

Table of Trigonometric Ratios

∠A	sin ∠A	cos ∠A	tan ∠A	∠A	sin ∠A	cos ∠A	tan ∠A
1°	.0175	.9998	.0175	46°	.7193	.6947	1.0355
2°	.0349	.9994	.0349	47°	.7314	.6820	1.0724
3°	.0523	.9986	.0524	48°	.7431	.6691	1.1106
4°	.0698	.9976	.0699	49°	.7547	.6561	1.1504
5°	.0872	.9962	.0875	50°	.7660	.6428	1.1918
6°	.1045	.9945	.1051	51°	.7771	.6293	1.2349
7°	.1219	.9925	.1228	52°	.7880	.6157	1.2799
8°	.1392	.9903	.1405	53°	.7986	.6018	1.3270
9°	.1564	.9877	.1584	54°	.8090	.5878	1.3764
10°	.1736	.9848	.1763	55°	.8192	.5736	1.4281
11°	.1908	.9816	.1944	56°	.8290	.5592	1.4826
12°	.2079	.9781	.2126	57°	.8387	.5446	1.5399
13°	.2250	.9744	.2309	58°	.8480	.5299	1.6003
14°	.2419	.9703	.2493	59°	.8572	.5150	1.6643
15°	.2588	.9659	.2679	60°	.8660	.5000	1.7321
16°	.2756	.9613	.2867	61°	.8746	.4848	1.8040
17°	.2924	.9563	.3057	62°	.8829	.4695	1.8807
18°	.3090	.9511	.3249	63°	.8910	.4540	1.9626
19°	.3256	.9455	.3443	64°	.8988	.4384	2.0503
20°	.3420	.9397	.3640	65°	.9063	.4226	2.1445
21°	.3584	.9336	.3839	66°	.9135	.4067	2.2460
22°	.3746	.9272	.4040	67°	.9205	.3907	2.3559
23°	.3907	.9205	.4245	68°	.9272	.3746	2.4751
24°	.4067	.9135	.4452	69°	.9336	.3584	2.6051
25°	.4226	.9063	.4663	70°	.9397	.3420	2.7475
26°	.4384	.8988	.4877	71°	.9455	.3256	2.9042
27°	.4540	.8910	.5095	72°	.9511	.3090	3.0777
28°	.4695	.8829	.5317	73°	.9563	.2924	3.2709
29°	.4848	.8746	.5543	74°	.9613	.2756	3.4874
30°	.5000	.8660	.5774	75°	.9659	.2588	3.7321
31°	.5150	.8572	.6009	76°	.9703	.2419	4.0108
32°	.5299	.8480	.6249	77°	.9744	.2250	4.3315
33°	.5446	.8387	.6494	78°	.9781	.2079	4.7046
34°	.5592	.8290	.6745	79°	.9816	.1908	5.1446
35°	.5736	.8192	.7002	80°	.9848	.1736	5.6713
36°	.5878	.8090	.7265	81°	.9877	.1564	6.3138
37°	.6018	.7986	.7536	82°	.9903	.1392	7.1154
38°	.6157	.7880	.7813	83°	.9925	.1219	8.1443
39°	.6293	.7771	.8098	84°	.9945	.1045	9.5144
40°	.6428	.7660	.8391	85°	.9962	.0872	11.4301
41°	.6561	.7547	.8693	86°	.9976	.0698	14.3007
42°	.6691	.7431	.9004	87°	.9986	.0523	19.0811
43°	.6820	.7314	.9325	88°	.9994	.0349	28.6363
44°	.6947	.7193	.9657	89°	.9998	.0175	57.2900
45°	.7071	.7071	1.0000				

Part Two: Sample Problems

Problem 1 Given: Right △DEF as shown
Find: **a** m∠D to the nearest degree
  b e to the nearest tenth

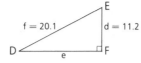

f = 20.1   d = 11.2

Solution **a** $\sin \angle D = \dfrac{11.2}{20.1}$

$\sin \angle D \approx 0.5572$

The number nearest to 0.5572 in the
sine column of the table is sin 34°,
so ∠D ≈ 34°.

b We use the result from part **a**.

$\cos 34° \approx \dfrac{e}{20.1}$

$0.8290 \approx \dfrac{e}{20.1}$

$16.7 \approx e$

Problem 2 To an observer on a cliff 360 m above sea level, the angle of depression of a ship is 28°. What is the horizontal distance between the ship and the observer?

Solution Start by drawing a diagram.
By ∥ lines ⟹ alt. int. ∠s ≅, ∠CSH = 28°.

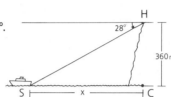

360 m

Thus, $\tan 28° = \dfrac{360}{x}$

$0.5317 \approx \dfrac{360}{x}$

$x \approx 677$

The horizontal distance is about 677 m.

Part Three: Problem Sets

Problem Set A

1 Find each of the following in the Table of Trigonometric Ratios.

  a sin 21° **b** tan 52° **c** cos 5° **d** tan 45° **e** sin 60°
   ≈0.3584   ≈1.2799   ≈0.9962   1.0000   ≈0.8660

2 Using the table, find m∠A in each case.

  a sin ∠A = 0.4067 **b** tan ∠A = 3.4874 **c** cos ∠A = .7071
   ≈24   ≈74   ≈45

3 Without using the table, find m∠A in each case.

  a tan ∠A = 1 **b** sin ∠A = $\dfrac{1}{2}$ **c** sin ∠A = $\dfrac{\sqrt{3}}{2}$
   45   30   60

4 In each case, find x to the nearest integer.

  a   **b**   **c** A   **d** A
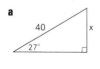   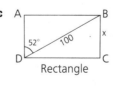

   ≈18   ≈44

   Rectangle   Rhombus
   ≈62   ≈16

Assignment Guide

Basis	Omit
Average	Optional
Day 1	1–7
Day 2	8–12, 15
Advanced	
4–8, 11, 14, 15, 20	

9.10

Problem Set A, *continued*

5 Find the height of isosceles trapezoid
ABCD. ≈15

Problem Set B

6 Solve each equation for x to the nearest integer.

a $\sin 25° = \dfrac{x}{40}$
 ≈17

b $\cos 73° = \dfrac{35}{x}$
 ≈120

c $\sin x° = \dfrac{29}{30}$
 ≈75

7 A department-store escalator is 80 ft long. If it rises 32 ft vertical-ly, find the angle it makes with the floor. ≈24°

8 Given the regular pentagon shown, with center at O and EN = 12 cm,

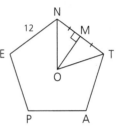

a Find m∠E 108

b Find m∠NOM 36

c Find OM to the nearest hundredth ≈8.26 cm

d Find the area of △NOT to the nearest hundredth ≈49.55 sq cm

e Explain how you could find the area of the pentagon

9 Find, to the nearest degree, the angles of a (3, 4, 5) triangle. ≈37°; ≈53°, 90°

10 A sonar operator on a cruiser detects a submarine at a distance of 500 m and an angle of depression of 37°. How deep is the sub? ≈301 m

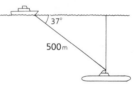

11 The legs of an isosceles triangle are each 18. The base is 14.

a Find the base angles to the nearest degree. ≈67°

b Find the exact length of the altitude to the base. $5\sqrt{11}$

12 One diagonal of a rhombus makes an angle of 27° with a side of the rhombus. If each side of the rhombus has a length of 6.2 in., find the length of each diagonal to the nearest tenth of an inch. ≈5.6 in.; ≈11.0 in.

13 Find the perimeter of trapezoid ABCD, in which $\overline{CD} \parallel \overline{AB}$, $\cos \angle A = \frac{1}{2}$, and AD = DC = CB = 2. 10

14 Find the length of the apothem of a regular pentagon that has a perimeter of 50 cm. ≈6.88 cm

■ Students with calculators can be encouraged to use them for these problem sets.
■ See *Solution Manual* for answer to problem **8e.**

16

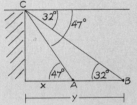

$\tan 47° = \dfrac{1000}{x}$, $\tan 32° = \dfrac{1000}{y}$
$x = \dfrac{1000}{\tan 47°}$, $y = \dfrac{1000}{\tan 32°}$
$x ≈ 932.48$, $y ≈ 600.26$
Thus, the distance between the ships is $y - x = 668$ dm.

17

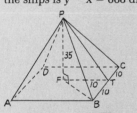

a In △PFT, PT = $5\sqrt{53} ≈ 36.4$.
b In △PTB, BP = $5\sqrt{57} ≈ 37.75$.
c In △PFT, $\tan \angle PTF = \frac{35}{10} = 3.5000$; ∠PTF ≈ 74°.
d In △PBF, $\tan \angle PBF = \frac{35}{10\sqrt{2}} ≈ 2.4752$; ∠PBF ≈ 68°

18

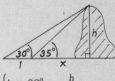

$\begin{cases} \tan 30° = \dfrac{h}{x+1} \\ \tan 35° = \dfrac{h}{x} \end{cases}$
Solving the two equations simultaneously, h ≈ 3.29 km.

T 426

15 Two buildings are 100 dm apart across a street. A sunbather at point P finds the angle of elevation of the roof of the taller building to be 25° and the angle of depression of its base to be 30°. Find the height of the taller building to the nearest decimeter. ≈104 dm

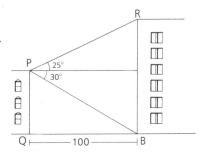

Problem Set C

16 An observer on a cliff 1000 dm above sea level sights two ships due east. The angles of depression of the ships are 47° and 32°. Find, to the nearest decimeter, the distance between the ships. ≈668 dm

17 Each side of the base of a regular square pyramid is 20 and the altitude is 35.

Find: **a** PT   **b** BP   **c** ∠PTF   **d** ∠PBF
≈36.4 ≈37.75 ≈74° ≈68°

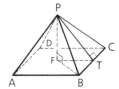

18 Find the height, PB, of a mountain whose base and peak are inaccessible. At point A the angle of elevation of the peak is 30°. One kilometer closer to the mountain, at point C, the angle of elevation is 35°. ≈3.29 km

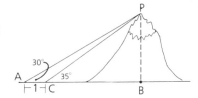

19 a Find the slope of line h. $\frac{2}{3}$
   b Find m∠1 to the nearest integer. ≈34

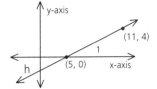

Problem Set D

20 Prove that $c^2 = a^2 + b^2 - 2ab(\cos \angle C)$ is true for any acute △ABC. (This formula is called the Law of Cosines.)

21 Given: Diagram as shown
   a Find ∠R to the nearest degree. ≈52°
   b Find QR to the nearest integer. ≈11
   c Show that $\frac{PR}{\sin \angle Q} = \frac{PQ}{\sin \angle R}$.
   d Generalize the result of part **c** for the sides and angles of any acute triangle. (The resulting formula is the Law of Sines.) $\frac{a}{\sin \angle A} = \frac{b}{\sin \angle B}$

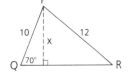

Problem-Set Notes and Additional Answers, continued

19a $m = \frac{4-0}{11-5} = \frac{4}{6} = \frac{2}{3}$

   b Since the slope is the tan ∠1, tan ∠1 = $\frac{2}{3}$ and ∠1 ≈ 33.69°, or ∠1 ≈ 34°.

20

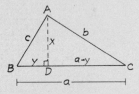

From △ACD,
$x^2 + (a - y)^2 = b^2$.
From △ABD, $x^2 + y^2 = c^2$.
$x^2 + a^2 - 2ay + y^2 = b^2$
$\underline{x^2 \qquad\quad + y^2 = c^2}$
$\quad a^2 - 2ay \qquad = b^2 - c^2$
So $c^2 = b^2 - a^2 + 2ay$,
or $c^2 = a^2 + b^2 - 2a^2 + 2ay$;
$\quad c^2 = a^2 + b^2 - 2a(a - y)$;
$\quad c^2 = a^2 + b^2 - 2ab\left(\frac{a-y}{b}\right)$.
From △ADC, however,
$\frac{a-y}{b} = \cos \angle C$. So $c^2 = a^2 + b^2 - 2ab(\cos \angle C)$.

21

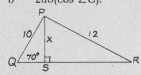

a $\sin 70° = \frac{x}{10}$; 9.397 = x
   So, $\sin \angle R = \frac{9.397}{12} = .7831$;
   ∠R ≈ 52°.

b In △PSQ, QS = 3.4 by the Pythagorean Theorem. In △PSR, SR ≈ 7.5 by the Pythagorean Theorem. Since QR = QR + SR, QR ≈ 11.

c $\frac{12}{.9397} = \frac{10}{\frac{9.397}{12}}$

d $\sin \angle Q = \frac{x}{PQ}$
   $\sin \angle R = \frac{x}{PR}$
   $(PQ)(\sin \angle Q) = (PR)(\sin \angle R)$
   $\frac{PR}{\sin \angle Q} = \frac{PR}{\sin \angle R}$

T 427

9 CHAPTER SUMMARY

CONCEPTS AND PROCEDURES

After studying this chapter, you should be able to

- Simplify radical expressions and solve quadratic equations (9.1)
- Begin solving problems involving circles (9.2)
- Identify the relationships between the parts of a right triangle when an altitude is drawn to the hypotenuse (9.3)
- Use the Pythagorean Theorem and its converse (9.4)
- Use the distance formula to compute lengths of segments in the coordinate plane (9.5)
- Recognize groups of whole numbers known as Pythagorean triples (9.6)
- Apply the Principle of the Reduced Triangle (9.6)
- Identify the ratio of side lengths in a 30°-60°-90° triangle (9.7)
- Identify the ratio of side lengths in a 45°-45°-90° triangle (9.7)
- Apply the Pythagorean Theorem to solid figures (9.8)
- Understand three basic trigonometric relationships (9.9)
- Use trigonometric ratios to solve right triangles (9.10)

VOCABULARY

altitude (9.8)
angle of depression (9.10)
angle of elevation (9.10)
base (9.8)
cosine (9.9)
cube (9.8)
diagonal (9.8)
distance formula (9.5)
edge (9.8)

face (9.8)
Pythagorean triple (9.6)
rectangular solid (9.8)
regular square pyramid (9.8)
sine (9.9)
slant height (9.8)
tangent (9.9)
trigonometry of right triangles (9.9)
vertex (9.8)

REVIEW PROBLEMS

Problem Set A

1 a Find GF if HG = 4 and EG = 6. 9
b Find EH if GH = 4 and GF = 12. 8
c Find HF if EF = $2\sqrt{5}$ and GF = 4. 5
d Find HF if EH = 2 and EF = 3. $\sqrt{13}$

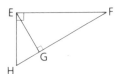

2 Identify the family of each of these special right triangles.

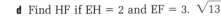

a
30°-60°-90°

b
(3, 4, 5)

c
(5, 12, 13)

d
(8, 15, 17)

e
45°-45°-90°

3 Find the missing lengths.

a

c

e

g

i

b

d

f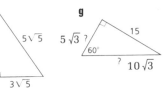

h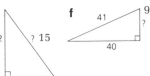

j

4 If AE = 6 and BE = 8, what is the perimeter of the rhombus shown? 40

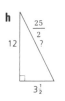

5 Find the altitude of the triangle shown. $3\sqrt{3}$

Chapter 9 Review
Class Planning

Time Schedule
All levels: 2 days

Resource References
Evaluation
   Tests and Quizzes
      Test 9, Series 1, 2, 3

Assignment Guide

Basic
1, 3–11, 13, 15, 17–20, 22
Average
3, 4, 6, 9, 12, 14, 15, 17–20, 22, 23, 28, 29, 32
Advanced
Section **9.10**   16, 17, 19
Chapter 9 Review 21, 22, 25–30

To integrate constructions, Construction 9 on page 674 can be included at this point.

Problem-Set Notes
and Additional Answers

■ A speed quiz on triples, the Principle of the Reduced Triangle, and the two special right triangles could be appropriate after the review problems.

Review Problem Set A, *continued*

6 Vail skied 2 km north, 2 km west, 1 km north, and 2 km west. How far was she from her starting point? 5 km

7 A 25-ft ladder just reaches a point on a wall 24 ft above the ground. How far is the foot of the ladder from the wall? 7 ft

8 Find, to the nearest tenth, the altitude to the base of an isosceles triangle whose sides have lengths of 8, 6, and 8. ≈7.4

9 If the altitude of an equilateral triangle is $8\sqrt{3}$, find the perimeter of the triangle. 48

10 What is the length of a diagonal of a 2-by-5 rectangle? $\sqrt{29}$

11 In the trapezoid shown, find RS. $\sqrt{85}$

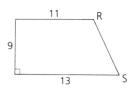

12 Given: TVWX is an isosceles trapezoid.
TX = 8, VW = 12, ∠V = 30°
Find: TV and TZ $\frac{4\sqrt{3}}{3}, \frac{2\sqrt{3}}{3}$

13 Find the diagonal of a rectangular solid whose dimensions are 4, 3, and 12. 13

14 Given: The regular square pyramid shown,
PR = 20, PS = 25
Find: The perimeter of base JKMO 120

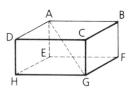

15 In the rectangular solid shown, find AG to the nearest tenth if DC = 12, CG = 7, and AD = 4. ≈14.5

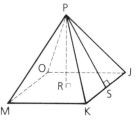

16 Given: $\overline{AC} \perp \overline{CB}$, $\overline{DE} \parallel \overline{CB}$,
AC = 15, AB = 17, DE = 4
Find: **a** CB 8   **c** AE 8.5   **e** DC 7.5
        b AD 7.5   **d** EB 8.5

430 Chapter 9   The Pythagorean Theorem

17 Find the distance from A to B if A = (1, 11) and B = (4, 15). 5

18 Given: Diagram as marked
Find: m∠M 25

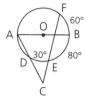

19 Given: ⊙O, mDE = 30,
    mEB = 80, mBF = 60
Find: **a** mAF 120
    b m∠C 45
    c m∠BAD 55

20 Given: RECT is a rectangle.
    RE = 6, EC = 8
Find: **a** The measure of RTC 180
    b The length of RTC 5π
    c The area of the shaded region to
       the nearest tenth ≈30.5

Problem Set B

21 a Find m∠DEF. 90
    b Find mDEF. 180
    c Find the length of DEF.
       10π, or ≈31.42

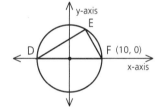

22 Given: ⊙P, ∠CAB = 30°
Find: **a** mBC 60
    b mAC 120
    c The length of BC 2π, or ≈6.28
    d The area of the shaded region
       6π, or ≈18.85

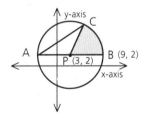

23 Two boats leave the harbor at 9:00 A.M. Boat A sails north at 20 km/hr. Boat B sails west at 15 km/hr. How far apart are the two boats at noon? 75 km

24 a Find x.   **b** Find y.

25 To swim directly would take $\frac{\sqrt{10}}{2} \approx 1\frac{1}{2}$ hr. To swim across and walk would take $\frac{1}{2}$ hr $+ \frac{3}{4}$ hr $= 1\frac{1}{4}$ hr.

Review Problem Set B, *continued*

25 A boy standing on the shore of a lake 1 mi wide wants to reach the "Golden Arches" 3 mi down the shore on the opposite side of the lake. If he swims at 2 mph and walks at 4 mph, is it quicker for him to swim directly across the lake and then walk to the Golden Arches or to swim directly to the Golden Arches? Swim directly across and walk

26 A boat is tied to a pier by a 25′ rope. The pier is 15′ above the boat. If 8′ of rope is pulled in, how many feet will the boat move forward? 12 ft

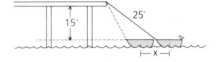

27 Find x. 7.5

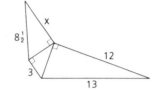

28 Follow the treasure map of Captain Zig Zag to see how far the treasure is from the old stump. 51 paces

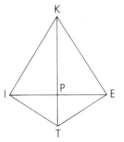

From the ol' pirate stump take ye 30 paces east, then 20 paces north, 6 paces west, and then another 25 paces north, and there ye find my treasure

29 Given: Kite KITE with right ∠s KIT and KET, KP = 9, TP = 4

Find: **a** IE 12

     b The perimeter of KITE $10\sqrt{13}$

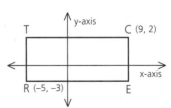

30 Given: RECT is a rectangle.
    $\overline{TC} \parallel$ x-axis.
    $\overline{RE} \parallel$ x-axis.

a Find the coordinates of E. (9, −3)

b Find the area of RECT. 70

c Find, to the nearest tenth, the length of $\overline{RC}$. ≈14.9

432   Chapter 9   The Pythagorean Theorem

31 Show that quadrilateral QUAD, with Q = (−1, −4), U = (4, 11), A = (1, 12), and D = (−4, −3), is a rectangle.

Problem Set C

32 Given: ∠C is a right angle.
E is the midpoint of $\overline{AC}$.
F is the midpoint of $\overline{BC}$.
AF = $\sqrt{41}$, BE = $2\sqrt{26}$
Find: AB $2\sqrt{29}$

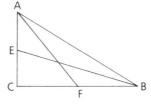

33 The altitude to the hypotenuse of a right triangle divides the hypotenuse in the ratio 4:1. What is the ratio of the legs of the triangle? 1:2

34 A 12-m rope is used to form a triangle the lengths of whose sides are integers. If one of the possible triangles is selected at random, what is the probability that the triangle is a right triangle? $\frac{1}{3}$

35 Find the edge of a cube whose diagonal is $7\sqrt{3}$. 7

36 If △PQR is a right triangle, what is the probability that tan ∠R is not a trigonometric ratio? $\frac{1}{3}$

37 Find the angle formed by
a A diagonal of a cube and a diagonal of a face of the cube ≈35°
b Two face diagonals that intersect at a vertex of a cube 60°

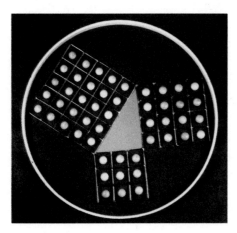

Problem-Set Notes and
Additional Answers, continued

■ See *Solution Manual* for answer to problem **31**.

32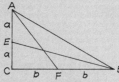

In △ACF, $(2a)^2 + b^2 = (\sqrt{41})^2$. In △BCE, $a^2 + (2b)^2 = (2\sqrt{26})^2$
$\begin{cases} 4a^2 + b^2 = 41 \\ a^2 + 4b^2 = 104 \end{cases}$
Solving the equations simultaneously, $a = 2$, $b = 5$. ∴ In △ACB, $4^2 + 10^2 = AB^2$; $2\sqrt{29} = AB$.

33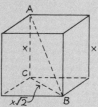

By Theorem 68c, $x^2 = 5 \cdot 1 = 5$ and $y^2 = 5 \cdot 4 = 20$.
$\frac{x^2}{y^2} = \frac{5}{20} = \frac{1}{4}$. ∴ $\frac{x}{y} = \frac{1}{2}$.

34 (2, 5, 5) (3, 4, 5) (4, 4, 4)

36 $p = \frac{1}{3}$ that ∠R is the rt. ∠ of △PQR.

37a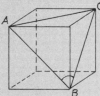

tan ∠ABC = $\frac{x}{x\sqrt{2}} = \frac{\sqrt{2}}{2} \approx .707$
∠ABC ≈ 35°

b

To find m∠ABC note that, by drawing $\overline{AC}$, an equilateral triangle △ABC is formed. ∴ ∠ABC = 60°.

**Cumulative Review
Chapters 1–9**

Class Planning

Time Schedule
All levels: 1 day

Resource References
Evaluation
  Tests and Quizzes
    Cumulative Review Test
    Chapters 1–9
    Series 1, 2, 3

Assignment Guide

Basic
1–14
Average
3, 4, 8, 13–17, 21
Advanced
Chapter 9 Review   27, 29
Cumulative Review Chapters
1–9   3, 4, 8, 17, 18, 21, 22, 24

CUMULATIVE REVIEW
CHAPTERS 1–9

Problem Set A

1 A pair of consecutive angles of a parallelogram are in the ratio 5:3. Find the measure of the smaller angle. $67\frac{1}{2}$

2 Find x. 95

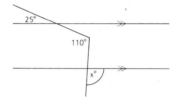

3 a Find the sum of the measures of the angles of a nonagon. 1260

b If each angle of a regular polygon is a 168° angle, how many sides does the polygon have? 30

c How many diagonals does a heptagon have? 14

4 a Find x. 20

b Is △ABC isosceles? Yes

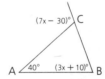

5 A boy 180 cm tall casts a 150-cm shadow. A nearby flagpole casts a shadow 12 m long. What is the length of the flagpole? 14.4 m

6 Given: ∠ABC = 60°, ∠ACB = 70°
Find: ∠BFC 130°

7 Find the perimeter of a rhombus whose diagonals are 10 and 24. 52

8 a Find BC. $20\frac{1}{4}$

b Find AB. $\frac{27}{2}$

c Is $\triangle DEC$ acute, right, or obtuse? Acute

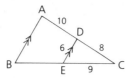

9 a Find the mean proportionals between $\frac{1}{4}$ and 49. $\pm\frac{7}{2}$

b Solve $\frac{5}{5-y} = \frac{10}{y-10}$ for y. $\frac{20}{3}$

10 Are lines a and b parallel? No

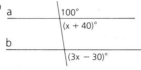

11 Given: $\overline{EB} \cong \overline{DF}$, $\overline{AG} \cong \overline{GC}$,
$\angle EAG \cong \angle FCG$

Prove: ABCD is a parallelogram.

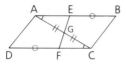

12 Given: $\odot Q$ lies in plane m.
$\overline{PQ} \perp m$

Prove: $\angle R \cong \angle S$

Problem Set B

13 Find the angle formed by the hands of a clock at each time.

a 11:50 55°　　　　**b** 12:01 $5\frac{1}{2}°$

14 The sum of an angle and four times its complement is 20° greater than the supplement of the angle. Find the angle's complement. 10°

15 The sum of the angles of an equiangular polygon is 3960°. Find the measure of each exterior angle. 15

16 Find AB. Impossible

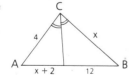

17 Points $(1, 3)$, $(-4, 7)$, and $(-29, k)$ are collinear. Find k. 27

***Problem-Set Notes
and Additional Answers***

■ See *Solution Manual* for answers to problems **11** and **12**.

13a

In ten minutes, the minute hand travels $\frac{10}{60}$ of 360° = 60°, and the hour hand travels $\frac{1}{12}$ as far as the minute hand. $\frac{1}{12} \cdot \frac{60°}{1} = 5°$; 60° − 5° = 55°.

b

In one minute, the minute hand travels $\frac{1}{60}$ of 360° = 6°, and the hour hand travels $\frac{1}{12}$ as far as the minute hand. $\frac{6}{12} = \frac{1}{2}°$; $6° - \frac{1}{2}° = 5\frac{1}{2}°$.

14　$x + 4(90 - x) = 20 + 180 - x$
$$x = 80$$
$$90 - x = 10$$

15　$(n - 2)180 = 3960$
$$n = 24$$
$$\frac{360}{24} = 15$$

16　$\frac{x+2}{12} = \frac{4}{x}$
$x \neq -8$ or $x = 6$
Thus, BC = 6 and AB = 20, but 4, 6, and 20 do not satisfy the Triangle Inequality Postulate.

T 435

Cumulative Review Problem Set B, *continued*

18 Given: ∠A = 120°;
$\overrightarrow{BD}$ and $\overrightarrow{BE}$ trisect ∠ABC.
$\overrightarrow{CD}$ and $\overrightarrow{CE}$ trisect ∠ACB.

a Find m∠D and m∠E. 140; 160

b Do m∠A, m∠D, and m∠E form an
arithmetic progression? (Hint: See
Chapter 7, review problem 23.) Yes

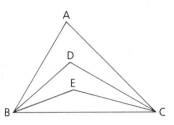

**Problem-Set Notes and
Additional Answers, continued**

■ See *Solution Manual* for answer
to problem **19**.

19 Given: $\overline{AB}$ lies in m, $\overline{CD}$ lies in n,
and m ∥ n.
$\overline{AC}$ intersects $\overline{BD}$ at P.
$\overline{AD} \perp$ n, $\overline{BC} \perp$ n

Prove: $\overline{AC} \cong \overline{DB}$

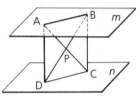

20 The shorter diagonal of a regular hexagon is 6. Find the length of
the longer diagonal. $4\sqrt{3}$

21 a Point A is rotated 127° clockwise
about the origin to point B. Through
how many degrees must B be rotated
counterclockwise to be at (0, −4)? 7°

b If A is reflected over a line parallel to
the x-axis and 1 unit below the axis,
find the coordinates of the point of
reflection. $(2\sqrt{3}, -4)$

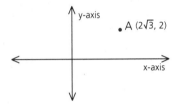

22

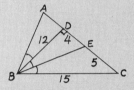

a $\frac{BC}{5} = \frac{12}{4}$
BC = 15

b From part **a**, note that
△BDC is a (9, 12, 15) rt. △,
and thus, $\overline{BD} \perp \overline{AC}$. ∴ In
△EDB, $12^2 + 4^2 = BE^2$;
BE = $4\sqrt{10}$.

c △ABD ≅ △EBD by ASA.
∴ $\overline{AD} \cong \overline{DE}$. Thus, AD = 4.

Problem Set C

22 Given: $\overrightarrow{BD}$ and $\overrightarrow{BE}$ trisect ∠ABC.
DE = 4, EC = 5, BD = 12

Find: **a** BC **b** BE **c** AD
15 $4\sqrt{10}$ 4

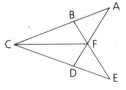

23

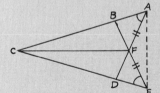

Draw $\overline{AE}$. ∠FAE ≅ ∠FEA
and ∠CAE ≅ ∠CEA by
addition. Thus, $\overline{CA} \cong \overline{CE}$
and △CAF ≅ △CEF. Thus,
∠ACF ≅ ∠ECF, and $\overrightarrow{CF}$
bisects ∠BCD.

23 Given: ∠A ≅ ∠E,
$\overline{FA} \cong \overline{FE}$
Prove: $\overrightarrow{CF}$ bisects ∠BCD.

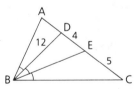

24 Given: A cube as shown, with
J the midpoint of $\overline{HF}$,
AB = 6
Find: AJ $3\sqrt{6}$

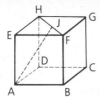

25 Given: $\overline{AD} \cong \overline{DC}$,
$\angle B + 50° = \angle DAB$
Find: m$\angle$CAB 25

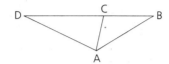

26 a Find the sum of the measures of angles A, B, C, D, and E. 180

b Does your answer depend on knowing
whether any polygons are equilateral
or equiangular? No

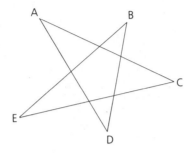

27 What is the probability that a diagonal chosen at random in a
regular decagon will be one of the shortest diagonals? $\frac{2}{7}$

28 $\overline{GH} \cong \overline{GJ}$,
$\angle 1 = (3x)°$,
$\angle 2 = x°$
Find: m$\angle$J 45

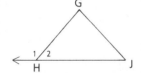

29 The consecutive sides of a quadrilateral measure $(x - 17)$,
$(24 - x)$, $(3x - 40)$, and $(x + 1)$. The perimeter is 42. Is the figure
a parallelogram? Explain. No; opposite sides are not of equal length.

30 Given: VRST is a $\square$.
V is the midpt. of $\overline{NT}$.
R is the midpt. of $\overline{PS}$.
Prove: NPST is a $\square$.

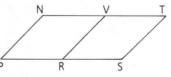

31 If one of the four angles of parallelogram
ABCD is selected at random, what is the
probability that the angle is congruent to
$\angle$C? $\frac{1}{2}$

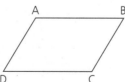

Cumulative Review Problems **437**

Problem-Set Notes and
Additional Answers, continued

24 $\triangle$HAF is equilateral with
sides $6\sqrt{2}$. $\triangle$HAJ is a
30°-60°-90° $\triangle$ with HJ =
$3\sqrt{2}$, AJ = $3\sqrt{6}$, and
HA = $6\sqrt{2}$.

25

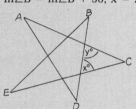

Let m$\angle$CAB = x. Then
m$\angle$DCA = x + m$\angle$B by
Exterior Angle Theorem.
Then m$\angle$DAC = x + m$\angle$B,
and thus, m$\angle$DAB = 2x +
m$\angle$B. But m$\angle$DAB =
m$\angle$B + 50. Thus, 2x +
m$\angle$B = m$\angle$B + 50; x = 25.

26

x = m$\angle$E + m$\angle$B
y = m$\angle$A + m$\angle$D
180 − x − y = m$\angle$C
180 = m$\angle$A + m$\angle$B +
m$\angle$C + m$\angle$D + m$\angle$E

27 $d = \frac{n(n-3)}{2}$

$= \frac{10(10-3)}{2}$

$= 35$

The shortest diagonal is one
that "skips" exactly one
vertex. There are ten of
these. $\frac{10}{35} = \frac{2}{7}$

■ See *Solution Manual* for answer
to problem **30**.

T 437

*T*HIS CHAPTER INTRODUCES much new vocabulary and many individual theorems that the students will need for the problem sets.

The circle theorems frequently involve perpendicular relationships and the Pythagorean Theorem.

CHAPTER

10 CIRCLES

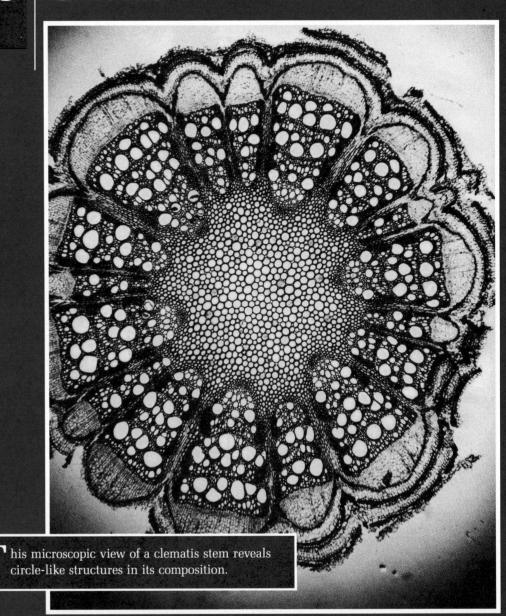

*T*his microscopic view of a clematis stem reveals circle-like structures in its composition.

10.1 THE CIRCLE

Objectives

After studying this section, you will be able to
- Identify the characteristics of circles
- Recognize chords and diameters of circles
- Recognize special relationships between radii and chords

Part One: Introduction

Basic Properties and Definitions

The following definitions will help you extend and organize what you already know about circles.

Definition	A *circle* is the set of all points in a plane that are a given distance from a given point in the plane. The given point is the *center* of the circle, and the given distance is the *radius*. A segment that joins the center to a point on the circle is also called a radius. (The plural of *radius* is *radii*.)

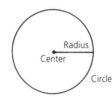

The definitions of *circle* and *radius* can be used to prove a theorem you saw in Chapter 3: *All radii of a circle are congruent* (Theorem 19).
   Although all circles have the same shape, their sizes are determined by the measures of their radii.

Definition	Two or more coplanar circles with the same center are called *concentric* circles.

Definition	Two circles are congruent if they have congruent radii.

Definition	A point is inside (in the *interior* of) a circle if its distance from the center is less than the radius.

Section 10.1   The Circle   **439**

Class Planning

Time Schedule
Basic: 2 days
Average: 2 days
Advanced: 1 day

Resource References
Teacher's Resource Book
   Class Opener 10.1A, 10.1B

Class Opener

Given: $\odot O$,
      Radius = 12,
      $AB = 12$,
      $\overline{OP} \perp \overline{AB}$
Find: OP

Draw $\overline{OA}$ and $\overline{OB}$.

$\triangle OPA \cong \triangle OPB$ by HL. So AP = BP = 6. By the Pythagorean Theorem, $144 = 36 + (OP)^2$; $108 = (OP)^2$. $\therefore OP = 6\sqrt{3}$

Lesson Notes

- Many students will need more than one class day to assimilate all the new terms in this section.

Vocabulary

center	diameter
chord	exterior
circle	interior
concentric	radius

Cooperative Learning

Have students work in small groups to formulate answers for these questions. For what circles of radius r will the ratio of the circumference of the circle to the area of the circle be less than one? Equal to one? Greater than one?   For $r > 2$; for $r = 2$; for $r < 2$

Communicating Mathematics

Have students write a paragraph that explains how to determine whether a point is on a circle, in the interior of a circle, or in the exterior of a circle.

Points O and A are in the interior of $\odot$O.

| **Definition** | A point is outside (in the **exterior** of) a circle if its distance from the center is greater than the radius. |

Point C is in the exterior of $\odot$O.

| **Definition** | A point is on a circle if its distance from the center is equal to the radius. |

Point B is on $\odot$O.

Chords and Diameters

Points on a circle can be connected by segments called **chords.**

| **Definition** | A **chord** of a circle is a segment joining any two points on the circle. |

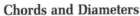

What is the longest chord of a circle? Is there a shortest chord?

| **Definition** | A **diameter** of a circle is a chord that passes through the center of the circle. |

The ideas of circumference and area of a circle are important in geometry. We now review two formulas presented in Chapter 3.

Circumference and Area of a Circle

The area of a circle can be found with the formula
$$A = \pi r^2$$
and the circumference (perimeter) of a circle can be found with the formula
$$C = \pi d$$
where r is the circle's radius, d is its diameter, and $\pi \approx 3.14$.

Radius-Chord Relationships

OP is the distance from O to chord $\overline{AB}$.

Definition	The distance from the center of a circle to a chord is the measure of the perpendicular segment from the center to the chord.	

The following three theorems are useful in establishing special relationships between radii and chords.

Theorem 74   *If a radius is perpendicular to a chord, then it bisects the chord.*

Given: $\odot$O,
     $\overline{OD} \perp \overline{AB}$
Prove: $\overline{OD}$ bisects $\overline{AB}$.

Theorem 75   *If a radius of a circle bisects a chord that is not a diameter, then it is perpendicular to that chord.*

Given: $\odot$O;
     $\overline{OH}$ bisects $\overline{EF}$.
Prove: $\overline{OH} \perp \overline{EF}$

Theorem 76   *The perpendicular bisector of a chord passes through the center of the circle.*

Given: $\overleftrightarrow{PQ}$ is the $\perp$ bisector
    of $\overline{CD}$.
Prove: $\overleftrightarrow{PQ}$ passes through O.

Lesson Notes, continued

- Point out that, in general, the distance from a point to a line is measured along a perpendicular.

- The proof of Theorem 74 must consider two cases.
  1. If the chord $\overline{AB}$ is a diameter, then the point E is at the center of the circle, and AE = EB = r.
  2. If chord $\overline{AB}$ is not a diameter, then $\triangle$ AOE $\cong$ $\triangle$ BOE by HL, so that $\overline{AE} \cong \overline{EB}$.

- For Theorem 75, remind students that if two points are each equidistant from the endpoints of a segment, the points determine the perpendicular bisector of the segment. Since O and G are each equidistant from E and F, $\overleftrightarrow{OG}$ is the perpendicular bisector of $\overline{EF}$.

- For Theorem 76, be sure students understand why point O is on $\overleftrightarrow{PQ}$. The perpendicular bisector of $\overline{CD}$ is the set of *all* points equidistant from the endpoints of $\overline{CD}$. Since $\overline{OC} \cong \overline{OD}$, O is equidistant from C and D. $\therefore$ O is on $\overleftrightarrow{PQ}$.

Part Two: Sample Problems

Problem 1

Given: ⊙Q,
$\overline{PR} \perp \overline{ST}$

Prove: $\overline{PS} \cong \overline{PT}$

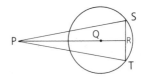

Proof

1 ⊙Q, $\overline{PR} \perp \overline{ST}$	1 Given
2 $\overline{PR}$ bisects $\overline{ST}$.	2 If a radius is ⊥ to a chord, it bisects the chord. ($\overline{QR}$ is part of a radius.)
3 $\overline{PR}$ ⊥ bis. $\overline{ST}$	3 Combination of steps 1 and 2
4 $\overline{PS} \cong \overline{PT}$	4 If a point is on the ⊥ bis. of a segment, then it is equidistant from the endpoints.

Problem 2

The radius of circle O is 13 mm. The length of chord $\overline{PQ}$ is 10 mm. Find the distance from chord $\overline{PQ}$ to the center, O.

Solution

Draw $\overline{OR}$ perpendicular to $\overline{PQ}$. Draw radius $\overline{OP}$ to complete a right △.
Since a radius perpendicular to a chord bisects the chord,

$$PR = \tfrac{1}{2}(PQ) = \tfrac{1}{2}(10) = 5$$

By the Pythagorean Theorem, $x^2 + 5^2 = 13^2$, so OR = 12.

Problem 3

Given: △ABC is isosceles $(\overline{AB} \cong \overline{AC})$.
ⓈP and Q,
$\overline{BC} \parallel \overline{PQ}$

Prove: ⊙P ≅ ⊙Q

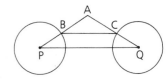

Proof

1 △ABC is isosceles $(\overline{AB} \cong \overline{AC})$.	1 Given
2 ⓈP and Q, $\overline{BC} \parallel \overline{PQ}$	2 Given
3 $\angle ABC \cong \angle P, \angle ACB \cong \angle Q$	3 ∥ lines ⟹ corr. ∠s ≅
4 $\angle ABC \cong \angle ACB$	4 If △, then △.
5 $\angle P \cong \angle Q$	5 Transitive Property
6 $\overline{AP} \cong \overline{AQ}$	6 If △, then △.
7 $\overline{PB} \cong \overline{CQ}$	7 Subtraction (1 from 6)
8 ⊙P ≅ ⊙Q	8 Ⓢ with ≅ radii are ≅.

Part Three: Problem Sets

Problem Set A

1 Given: ⊙O, chord $\overline{AB}$
  Prove: **a** △AOB is isosceles.
     b ∠A ≅ ∠B

2 Given: ⊙Q, $\overline{PR} \perp \overline{ST}$
  Prove: ∠S ≅ ∠T

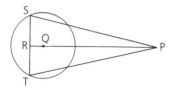

3 Given: ⊙O; $\overline{OM}$ is a median.
  Conclusion: $\overline{OM}$ is an altitude.

4 Given: ⊙Q, $\overline{QT} \perp \overline{RS}$
  Prove: $\overrightarrow{TQ}$ bisects ∠RTS.

5 Chord $\overline{AB}$ measures 12 mm and the radius of ⊙P is 10 mm. Find the distance from $\overline{AB}$ to P. 8 mm

6 Find the length of a chord that is 15 cm from the center of a circle with a radius of 17 cm. 16 cm

7 Given: PQRS is an isosceles trapezoid, with $\overleftrightarrow{SR} \parallel \overleftrightarrow{PQ}$.
  Conclusion: ⊙P ≅ ⊙Q

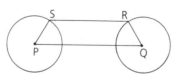

8 Find, to the nearest tenth, the circumference and the area of a circle whose diameter is 7.8 cm. C ≈ 24.5 cm; A ≈ 47.8 sq cm

---

Assignment Guide

Basic
Day 1	1–6, 8
Day 2	9–12, 16, 17

Average
Day 1	3–6, 8, 9, 11, 16
Day 2	14, 15, 17, 20, 23

Advanced
8, 10, 14, 17, 20, 22, 24

Problem-Set Notes and Additional Answers

- See *Solution Manual* for answers to problems **1–4** and **7**.

Problem-Set Notes and
Additional Answers, continued

■ See Solution Manual for answers to problems **9, 10, 13,**
and **15.**

Problem Set A, *continued*

9 Given: ⊙A ≅ ⊙B,
$\overleftrightarrow{AD} \parallel \overleftrightarrow{BC}$
Prove: ABCD is a ▱.

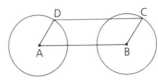

Problem Set B

10 Given: ⊙O;
$\overleftrightarrow{OR}$ bisects $\overline{PQ}$.
Prove: $\overrightarrow{RO}$ bisects ∠PRQ.

11 Find the distance from the center of a circle to a chord 30 m long
if the diameter of the circle is 34 m.   8 m

12 Find the radius of a circle if a 24-cm chord is 9 cm from the
center.   15 cm

13 Given: ⓢ A and B intersect as shown.
$\overline{DE} \parallel \overline{FC}$, ∠ADE ≅ ∠FCB,
$\overline{DE} \cong \overline{FC}$
Prove: ⊙A ≅ ⊙B

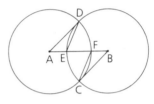

14 Two circles intersect and have a common chord 24 cm long. The
centers of the circles are 21 cm apart. The radius of one circle is
13 cm. Find the radius of the other circle.   20 cm

15 Given: ⊙P,
$\overleftrightarrow{QT} \parallel \overleftrightarrow{RS}$
Conclusion: $\overline{QR} \cong \overline{TS}$

16 $\overline{PQ}$ is a diameter of ⊙O. P = (−3, 17) and Q = (5, 2). Find the
center and the radius of ⊙O.   (1, 9.5); ≈8.5

17 ⊙P just touches (is tangent to) the x-axis. P = (15, 13) and
Q = (19, 16).

a Find the radius of ⊙P.   13

b Find PQ.   5

c Find the length of $\overline{AB}$.   24

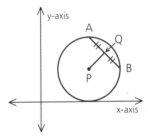

444   Chapter 10   Circles

18 Given: ⊙P;
     Z is the midpt. of $\overline{WX}$.
     △WAX is isosceles, with
     base $\overline{WX}$.
    Prove: $\overrightarrow{AZ}$ passes through P.

Problem Set C

19 Given: Two concentric circles with center P.
     Line m intersects the circles at A,
     B, C, and D.
    Conclusion: $\overline{AB} \cong \overline{CD}$

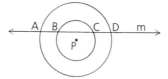

20 Given: ⊙P, $\overline{WX} \cong \overline{YZ}$
    Prove: $\overline{WQ} \cong \overline{ZR}$

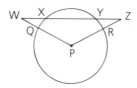

21 Given: $\overline{AB}$ is a diameter of ⊙O.
     $\overleftrightarrow{AC} \parallel \overleftrightarrow{BD}$
    Conclusion: $\overline{AC} \cong \overline{BD}$

22 Find the radius of a circle in which a 48-cm chord is 8 cm closer to the center than a 40-cm chord. 25 cm

23 In circle O, PQ = 4, RQ = 10, and PO = 15.
Find PS (the distance from P to ⊙O). 2

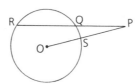

24 An isosceles triangle with each leg measuring 13 is inscribed in a circle. If the altitude to the base of the triangle is 5, find the radius of the circle. 16.9

25 Two circles intersect and have a common chord. The radii of the circles are 13 and 15. The distance between their centers is 14. Find the length of their common chord. 24

Problem-Set Notes and Additional Answers, continued

■ See Solution Manual for answer to problem **18**.

19 Let $\overline{PM}$ be the ⊥ to m from point P. Then $\overline{BM} \cong \overline{CM}$ and $\overline{AM} \cong \overline{DM}$. Thus, $\overline{AB} \cong \overline{CD}$ by subtraction.

20 △WXP ≅ △ZYP; $\overline{WP} \cong \overline{ZP}$, and $\overline{QP} \cong \overline{RP}$. Thus, $\overline{WQ} \cong \overline{ZR}$.

21 Draw $\overline{OC}$ and $\overline{OD}$. △AOC ≅ △BOD by AAS.

22

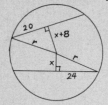

$$\begin{cases} x^2 + 24^2 = r^2 \\ (x+8)^2 + 20^2 = r^2 \end{cases}$$
$$x = 7; \therefore r = 25$$

23

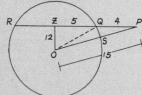

OZ = 12. ∴ Radius = 13. Thus, PS = 2.

24

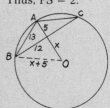

$$x^2 + 144 = (x+5)^2$$
$$x = 11.9$$
$$\therefore x + 5 = 16.9$$

25

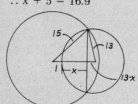

$$15^2 - (x+1)^2 = 13^2 - (13-x)^2$$
$$x = 8; \therefore \text{Common chord} = 24$$

T 445

CONGRUENT CHORDS

Objective

After studying this section, you will be able to
- Apply the relationship between congruent chords of a circle

Part One: Introduction

If two chords are the same distance from the center of a circle, what can we conclude?

Theorem 77   *If two chords of a circle are equidistant from the center, then they are congruent.*

Given: ⊙P, $\overline{PX} \perp \overline{AB}$, $\overline{PY} \perp \overline{CD}$, $\overline{PX} \cong \overline{PY}$
Prove: $\overline{AB} \cong \overline{CD}$

The proof of Theorem 77 is left for you to do. (Use four congruent triangles.) The converse of Theorem 77 can also be proved.

Theorem 78   *If two chords of a circle are congruent, then they are equidistant from the center of the circle.*

Given: ⊙O, $\overline{AB} \cong \overline{CD}$, $\overline{OE} \perp \overline{AB}$, $\overline{OF} \perp \overline{CD}$
Prove: $\overline{OE} \cong \overline{OF}$

Part Two: Sample Problems

Problem 1   Given: $\odot O$, $\overline{AB} \cong \overline{CD}$,
                      $OP = 12x - 5$, $OQ = 4x + 19$
              Find: OP

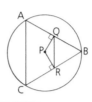

Solution   Since $\overline{AB} \cong \overline{CD}$, $OP = OQ$.

$$12x - 5 = 4x + 19$$
$$x = 3$$

Thus, $OP = 12(3) - 5 = 31$.

Problem 2   Given: $\triangle ABC$ is isosceles, with base $\overline{AC}$.
                      $\odot P$, $\overline{PQ} \perp \overline{AB}$, $\overline{PR} \perp \overline{CB}$
              Prove: $\triangle PQR$ is isosceles.

Proof

1	$\odot P$, $\overline{PQ} \perp \overline{AB}$, $\overline{PR} \perp \overline{CB}$	1	Given
2	$\triangle ABC$ is isosceles, with base $\overline{AC}$.	2	Given
3	$\overline{AB} \cong \overline{BC}$	3	An isosceles $\triangle$ has two $\cong$ sides.
4	$\overline{PQ} \cong \overline{PR}$	4	If two chords of a circle are $\cong$, then they are equidistant from the center.
5	$\triangle PQR$ is isosceles.	5	A $\triangle$ with two $\cong$ sides is isosceles.

Why do you think it was necessary to be given $\overline{PQ} \perp \overline{AB}$ and $\overline{PR} \perp \overline{CB}$, even though they did not seem to play an active role in the proof?

I FORGOT ALREADY!
I BETTER REREAD
THE LAST SECTION.

Part Three: Problem Sets

Problem Set A

1 In a circle, chord $\overline{AB}$ is 325 cm long and chord $\overline{CD}$ is $3\frac{1}{4}$ m long. Which is closer to the center? Same distance

2 Given: $\odot P$, $\overline{PQ} \cong \overline{PR}$,
            $AB = 6x + 14$,
            $CD = 4 - 4x$
      Find: AB 8

Lesson Notes, continued

- The proof of Theorem 78 (page 446) follows. If $\overline{OF}$ and $\overline{OE}$ are perpendicular to the chords, they bisect the chords by Theorem 74. Since $\overline{AB} \cong \overline{CD}$, $\overline{CF} \cong \overline{AE}$ by division. $\overline{OC}$ and $\overline{OA}$ are congruent radii, and thus, $\triangle OFC \cong \triangle OEA$ by HL. $\therefore \overline{OE} \cong \overline{OF}$ by CPCTC.

Communicating Mathematics

Have students discuss the question following the proof in Sample Problem **2**.

Answers may vary. Students should realize that distance from the center must be determined by a segment drawn perpendicular to the chord.

Assignment Guide

Basic
1–7
Average
2, 3, 6, 7, 9, 11, 12
Advanced
6, 7, 10–14

Problem Set A, *continued*

3 Given: ⊙P, $\overline{PR} \perp \overline{WX}$,
   $\overline{PS} \perp \overline{XY}$, $\overline{PR} \cong \overline{PS}$

Conclusion: $\angle W \cong \angle Y$

4 Given: Equilateral △ ABC is inscribed
   in ⊙Q.

Conclusion: $\overline{AB}$, $\overline{BC}$, and $\overline{CA}$ are equidistant from the center.

5 Given: ⊙P;
   P is the midpoint of $\overline{MN}$.
   $\overline{MN} \perp \overline{AD}$, $\overline{MN} \perp \overline{BC}$

Conclusion: ABCD is a ▱.

6 A fly is sitting at the midpoint of a
wooden chord of a circular wheel. The
wheel has a radius of 10 cm, and the
chord has a length of 12 cm.

   a How far from the hub (center) is the fly? 8 cm
   b The wheel is spun. What is the path of the fly?
      Circle

Problem Set B

7 To the nearest hundredth, find
   a The area of the circle ≈283.53 sq mm
   b The circumference of the circle
      ≈59.69 mm

19 mm

8 Given: ⊙Q, $\overline{PS} \perp \overline{RT}$,
   $\overline{MQ} \perp \overline{RP}$, $\overline{NQ} \perp \overline{PT}$

Conclusion: $\overline{MQ} \cong \overline{QN}$

9 Given: ⊙F,
   $\overline{FE} \perp \overline{BC}$, $\overline{FD} \perp \overline{AB}$;
   $\overrightarrow{BF}$ bisects $\angle ABC$.

Prove: $\overline{BC} \cong \overline{BA}$

10 Given: $\odot F$, $\overline{AB} \cong \overline{AC}$,
$\overline{DF} \perp \overline{AB}$, $\overline{EF} \perp \overline{AC}$

Prove: $\triangle ADE$ is isosceles.

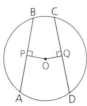

11 In circle O, $PB = 3x - 17$, $CD = 15 - x$,
and $OQ = OP = 3$.

a Find AB. 8

b Find the radius of $\odot O$. 5

12 A regular hexagon with a perimeter of
24 is inscribed in a circle. How far from
the center is each side? $2\sqrt{3}$

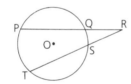

13 A 16-by-12 rectangle is inscribed in a circle. Find the radius of
the circle. 10

Problem Set C

14 Given: $\odot O$, $\overline{PQ} \cong \overline{TS}$

Prove: $\overline{RQ} \cong \overline{RS}$

15 Given: $\triangle ABC$ is isosceles, with
$\overline{AB} \cong \overline{AC}$.
$\odot E$, $\overline{AD} \perp \overline{BC}$, $\overline{EF} \perp \overline{AC}$,
$AF = 6$, $ED = 1$

Find: **a** The radius of the circle 8
b The perimeter of $\triangle ABC$ $24 + 6\sqrt{7}$

16 Two chords intersect inside a circle. Prove that if a diameter
drawn through the intersection point bisects the angle formed by
the chords, then the chords are congruent. (Hint: Prove that the
chords are equidistant from the center of the circle.)

10.2

*Problem-Set Notes and
Additional Answers, continued*

■ See Solution Manual for answer
to problem **10**.

14 Draw $\overline{OX} \perp \overline{PQ}$, $\overline{OY} \perp \overline{TS}$.
They are $\cong$ by Theorem 78.
Draw $\overline{OR}$. $\triangle OXR \cong \triangle OYR$
by HL. $\overline{XR} \cong \overline{YR}$ by CPCTC,
$\overline{XQ} \cong \overline{YS}$ by division, and
$\overline{RQ} \cong \overline{RS}$ by subtraction.

15

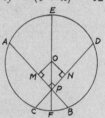

$$\begin{cases} y^2 + 1 = x^2 \\ y^2 + (1 + x)^2 = 12^2 \end{cases}$$

16

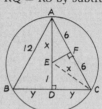

$\triangle OMP \cong \triangle ONP$ by AAS.
$\overline{OM} \cong \overline{ON}$ by CPCTC.

Cooperative Learning

Have students work in small
groups to prove that the midpoints of all congruent chords of
a circle form another circle concentric with the given circle.
Since all chords are $\cong$, they are
equidistant from the center and
their midpts. form a $\odot$.

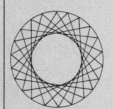

ARCS OF A CIRCLE

Class Planning

Time Schedule
Basic: 2 days
Average: $1\frac{1}{2}$ days
Advanced: $1\frac{1}{2}$ days

Resource References
Teacher's Resource Book
   Class Opener 10.3A, 10.3B
   Additional Practice
   Worksheet 19
Evaluation
   Tests and Quizzes
     Quiz 1, Series 2

Class Opener

Given: Two concentric circles
    with center O;
    ∠BOD is a right angle.
    AB = AO = 5

1 What fraction of the small
    circle is $\widehat{AFC}$? $\frac{1}{4}$

2 What fraction of the large
    circle is $\widehat{BJD}$? $\frac{1}{4}$

3 Which is longer, $\widehat{BJD}$ or
    $\widehat{AFC}$? $\widehat{BJD}$ is twice as long
    by similarity

4a Find the lengths of $\overline{AC}$ and
    $\overline{BD}$. $AC = 5\sqrt{2}$;
    $BD = 10\sqrt{2}$

b Find the ratio $\overline{BD}:\overline{AC}$.
    BD:AC = 2:1

c What is the ratio of the
    measure of $\widehat{BJD}$ to the mea-
    sure of $\widehat{AFC}$?
    $m\widehat{BJD}:m\widehat{AFC} = 2:1$

This leads into a discussion of
the measures of arcs.

Objectives

After studying this section, you will be able to
- Identify the different types of arcs
- Determine the measure of an arc
- Recognize congruent arcs
- Apply the relationships between congruent arcs, chords, and central angles

Part One: Introduction

Types of Arcs

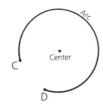
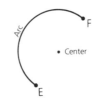

Definition   An **arc** consists of two points on a circle and all points on the circle needed to connect the points by a single path.

Definition   The center of an arc is the center of the circle of which the arc is a part.

Definition   A **central angle** is an angle whose vertex is at the center of a circle.

Radii $\overline{OA}$ and $\overline{OB}$ determine central angle AOB.

450   Chapter 10   Circles

Vocabulary

arc   minor arc
central angle   semicircle
major arc

Definition   A **minor arc** is an arc whose points are on or between the sides of a central angle.

Central angle APB determines minor arc AB.

Definition   A **major arc** is an arc whose points are on or outside of a central angle.

Central angle CQD determines major arc CD.

Definition   A **semicircle** is an arc whose endpoints are the endpoints of a diameter.

Arc EF is a semicircle.

    The symbol ⌒ is used to label arcs. The minor arc joining A and B is called $\overset{\frown}{AB}$. The major arc joining A and B is called $\overset{\frown}{AXB}$. (The extra point, X, is named to make it clear that we are referring to the arc from A to B *by way of* point X. This helps to avoid confusion when a major arc or a semicircle is being discussed.)

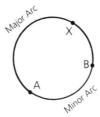

The Measure of an Arc

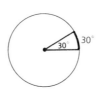

Definition   The measure of a minor arc or a semicircle is the same as the measure of the central angle that intercepts the arc.

Definition   The measure of a major arc is 360 minus the measure of the minor arc with the same endpoints.

Example

a *Given:* $m\widehat{AB} = 20$
   Find: $m\widehat{ACB}$

$$m\widehat{ACB} = 360 - 20$$
$$= 340$$

b *Given:* $m\angle XQY = 110$
   Find: $m\widehat{XDY}$

$$m\widehat{XY} = m\angle XQY = 110$$
$$\text{Therefore, } m\widehat{XDY} = 360 - 110$$
$$= 250$$

Congruent Arcs

Two arcs that have the same measure are not necessarily congruent arcs. In the con-centric circles shown, $m\widehat{AB} = 65$ and $m\widehat{CD} = 65$, but $\widehat{AB}$ and $\widehat{CD}$ are *not* congru-ent. Under what conditions, do you think, will two arcs be congruent?

Definition

Two arcs are congruent whenever they have the same measure and are parts of the same circle or congru-ent circles.

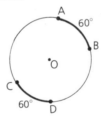

We may conclude that $\widehat{AB} \cong \widehat{CD}$.

If $\odot P \cong \odot Q$, we may conclude that $\widehat{EF} \cong \widehat{GH}$.

Relating Congruent Arcs, Chords, and Central Angles

In the diagram, points A and B determine one central angle, one chord, and two arcs (one major and one minor).

You can readily prove the following theorems.

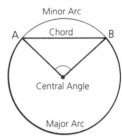

Theorem 79	*If two central angles of a circle (or of congruent circles) are congruent, then their intercepted arcs are congruent.*

Theorem 80	*If two arcs of a circle (or of congruent circles) are congruent, then the corresponding central angles are congruent.*

Theorem 81	*If two central angles of a circle (or of congruent circles) are congruent, then the corresponding chords are congruent.*

Theorem 82	*If two chords of a circle (or of congruent circles) are congruent, then the corresponding central angles are congruent.*

Theorem 83	*If two arcs of a circle (or of congruent circles) are congruent, then the corresponding chords are congruent.*

Theorem 84	*If two chords of a circle (or of congruent circles) are congruent, then the corresponding arcs are congruent.*

To summarize, in the same circle or in congruent circles, congruent chords ⟺ congruent arcs ⟺ congruent central angles.

Part Two: Sample Problems

Problem 1   Given: ⊙B;
　　　　　　　D is the midpt. of $\overset{\frown}{AC}$.
　　　　　　Conclusion: $\overrightarrow{BD}$ bisects ∠ABC.

Proof

1 ⊙B; D is the midpt. of $\overset{\frown}{AC}$.	1 Given
2 $\overset{\frown}{AD} \cong \overset{\frown}{DC}$	2 The midpoint of an arc divides the arc into two ≅ arcs.
3 ∠ABD ≅ ∠DBC	3 If two arcs of a circle are ≅, then the corresponding central ∠s are ≅.
4 $\overrightarrow{BD}$ bisects ∠ABC.	4 If a ray divides an ∠ into two ≅ ∠s, then the ray bisects the ∠.

Lesson Notes

■ Stress this statement because it summarizes the equivalence relation among congruent chords, arcs, and central angles.

Communicating Mathematics

Discuss how Theorems 79–84 are incorporated into the summary statement.

Problem 2 *If* m$\widehat{AB}$ = 102 *in* ⊙O, *find* m∠A *and* m∠B *in* △AOB.

Solution $\widehat{AB}$ = 102°, so ∠AOB = 102°.
The sum of the measures of the angles of a triangle is 180, so

$$m∠AOB + m∠A + m∠B = 180$$
$$102 + m∠A + m∠B = 180$$
$$m∠A + m∠B = 78$$

But $\overline{OA}$ ≅ $\overline{OB}$, so that ∠A ≅ ∠B.
Hence, m∠A = 39 and m∠B = 39.

Problem 3 **a** *What fractional part of a circle is an arc of 36°? Of 200°?*
b *Find the measure of an arc that is* $\frac{7}{12}$ *of its circle.*

Solution **a** 36° is $\frac{36}{360}$, or $\frac{1}{10}$, of a ⊙.
200° is $\frac{200}{360}$, or $\frac{5}{9}$, of a ⊙.

b There are 360° in a whole ⊙.
$\frac{7}{12}$ of 360 = $\frac{7}{12} \cdot \frac{360}{1}$ = 210

Problem 4 Given: ⊛ P *and* Q,
∠P ≅ ∠Q, $\overline{AR}$ ≅ $\overline{RD}$
Prove: $\widehat{AB}$ ≅ $\widehat{CD}$ (Hint: First prove
that ⊙P ≅ ⊙Q.)

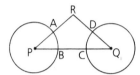

Proof

1 ⊛ P and Q	1 Given
2 ∠P ≅ ∠Q	2 Given
3 $\overline{RP}$ ≅ $\overline{RQ}$	3 If △, then △.
4 $\overline{AR}$ ≅ $\overline{RD}$	4 Given
5 $\overline{AP}$ ≅ $\overline{DQ}$	5 Subtraction Property
6 ⊙P ≅ ⊙Q	6 ⊛ with ≅ radii are ≅.
7 $\widehat{AB}$ ≅ $\widehat{CD}$	7 If two central ∠s of ≅ ⊛ are ≅, then their intercepted arcs are ≅.

Part Three: Problem Sets

Problem Set A

1 Match each item in the left column with the correct term in the right column.

a $\widehat{QRS}$ **6**	**1** Radius
b $\overline{QS}$ **2**	**2** Diameter
c $\widehat{RQS}$ **5**	**3** Chord
d $\overline{RS}$ **4**	**4** Minor arc
e $\overline{RS}$ **3**	**5** Major arc
f ∠RPQ **7**	**6** Semicircle
g $\overline{PS}$ **1**	**7** Central angle

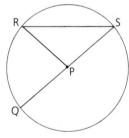

2 Given: Two concentric circles with center O;
∠BOC is acute.

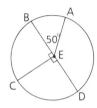

a Name a major arc of the smaller
circle. $\overset{\frown}{QRP}$ or $\overset{\frown}{QPR}$

b Name a minor arc of the larger circle. $\overset{\frown}{BC}$ or $\overset{\frown}{AB}$

c What is $m\overset{\frown}{BC} + m\overset{\frown}{PQ}$? 180

d Which is greater, $m\overset{\frown}{BC}$ or $m\overset{\frown}{PQ}$? $m\overset{\frown}{PQ}$

e Is $\overset{\frown}{BC}$ congruent to $\overset{\frown}{QR}$? No

3 In circle E, find each of the following.

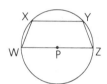

a $m\overset{\frown}{BC}$ 90 **c** $m\overset{\frown}{ACD}$ 230 **e** $m\overset{\frown}{ADC}$ 220

b $m\overset{\frown}{AD}$ 130 **d** $m\overset{\frown}{BAD}$ 180

4 Given: ⊙Q, ∠A = 25°
Find: $m\overset{\frown}{AB}$ 130

5 Given: ⊙P,
$\overset{\frown}{WY} \cong \overset{\frown}{XZ}$
Conclusion: $\overline{WX} \cong \overline{YZ}$

6 Given: ⊙D, ∠B ≅ ∠C
Conclusion: $\overset{\frown}{AB} \cong \overset{\frown}{AC}$

7 Given: $\overline{AB} \cong \overline{CD}$
Conclusion: $\overset{\frown}{AC} \cong \overset{\frown}{BD}$

8 Given: ⊙E,
$\overline{AB} \cong \overline{CD}$
Prove: $\overline{BD} \cong \overline{AC}$

***Problem-Set Notes
and Additional Answers***

■ See *Solution Manual* for answers to problems **5–8.**

Problem Set A, *continued*

9 What fractional part of a circle is an arc that measures

a $8\frac{1}{45}$

c $144\frac{2}{5}$

b $240\frac{2}{3}$

d $315\frac{7}{8}$

10 Find the measure of an arc that is

a $\frac{3}{5}$ of its circle 216

b $\frac{5}{9}$ of its circle 200

c 70% of its circle 252

Problem Set B

11 Given: $\overline{AD}$ is a diameter of $\odot E$.

C is the midpoint of $\widehat{BD}$.

$m\widehat{AB} = 9x + 30$,

$m\widehat{CD} = 54 - x$

Find: $m\angle AEC$ 132

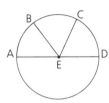

12 Find the length of a chord that cuts off an arc measuring 60 in a circle with a radius of 12. 12

13 Find the length of each arc described. (The length is a fractional part of the circumference.)

a An arc that is $\frac{5}{8}$ of the circumference of a circle with radius 12 15π

b An arc that has a measure of 270 and is part of a circle with radius 12 18π

14 $\overline{AB}$ is a chord of circle E, and C is the midpoint of $\widehat{AB}$. Prove that $\overleftrightarrow{EC}$ is the perpendicular bisector of chord $\overline{AB}$.

15 Given: $\odot Q$;

B is the midpt. of $\widehat{AC}$.

Conclusion: $\angle A \cong \angle C$

16 Given: $\odot B \cong \odot D$,

$\widehat{AE} \cong \widehat{CE}$

Prove: ABCD is a $\square$.

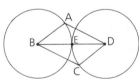

Problem-Set Notes and Additional Answers, continued

■ See *Solution Manual* for answers to problems **14–16**.

17 Given: ⊙O,
$\overline{OP} \perp \overline{AC}$, $\overline{OQ} \perp \overline{BD}$,
$\overline{OP} \cong \overline{OQ}$
Conclusion: $\overparen{AB} \cong \overparen{CD}$

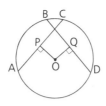

18 A polygon is *inscribed in* a ⊙ if all its vertices lie on the ⊙. Find the measure of the arc cut off by a side of each of the following inscribed polygons.

   a A regular hexagon 60

   b A regular pentagon 72

   c A regular octagon 45

19 Point P is located at $(-5, 5)$.

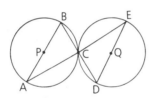

   a Find the radius of ⊙O. $5\sqrt{2}$

   b Find the measure of $\overparen{PQ}$. 135

20 Given: ⊙P ≅ ⊙Q,
$\overline{BC} \cong \overline{CD}$
Conclusion: ∠A ≅ ∠E

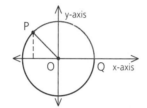

Problem Set C

21 Given: ⊙E,
$\overline{AB} \cong \overline{CD}$
Conclusion: $\overparen{FB} \cong \overparen{CG}$

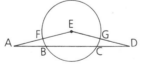

22 From point Q on circle P, an arc is drawn that contains point P. Find the measure of the arc AQB that is cut off. 120

Problem-Set Notes and Additional Answers, continued

■ See Solution Manual for answers to problems **17** and **20**.

21

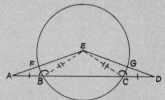

△EBA ≅ △ECD by SAS.
∠AEB ≅ ∠DEC by CPCTC.
∴ $\overparen{FB} \cong \overparen{CG}$

22

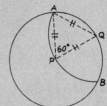

10.3

Problem-Set Notes and
Additional Answers, continued

■ In problem **23a,** from each
point, $n - 1$ chords can be
drawn, for a total of $n(n - 1)$
chords. Each chord has been
drawn twice, however, so there
are $\frac{n(n - 1)}{2}$ different chords.
In problem **23b,** for every
chord in part **a,** there will be
two arcs. Thus, there are
$n(n - 1)$ arcs.

Cooperative Learning

Have students work in small
groups to develop the formulas
for problem **23.** Have students
compare their expressions and
resolve any differences.

24

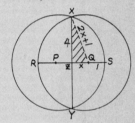

$$x^2 + 16 = (2x + 1)^2$$

25

The theorem can be shown
to be true in the triangle
above and then generalized
to any polygon.

26 If $m\overarc{PQ} = 40$, then $\angle POQ = 40°$.
$\cos 40° = \frac{x}{10}$
$\quad x = 10(\cos 40°)$
$\quad x \approx 7.7$
$\sin 40° = \frac{y}{10}$
$\quad y = 10(\sin 40°)$
$\quad y \approx 6.4$
$\quad P \approx (7.7, 6.4)$

Problem Set C, *continued*

23 If *n* points are selected on a given circle,
find a formula

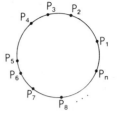

  a For the number of chords that can be
drawn between pairs of these points $\frac{n(n - 1)}{2}$

  b For the number of arcs formed—
including major and minor arcs and
semicircles (Hint: Draw circles and
count arcs for $n = 1, 2, 3, \ldots$ until
you see a number pattern.) $n(n - 1)$

  c For the measure of an arc formed by a
side of a regular *n*-gon inscribed in
the circle $\frac{360}{n}$

24 Given: $\odot P \cong \odot Q$,
       $XY = 8$,
       $RP = QS = 1$
       Find: PQ $\frac{10}{3}$

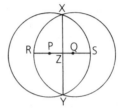

25 Prove that if an equilateral polygon is inscribed in a circle, then
it is equiangular.

26 Find, to the nearest tenth, the coordi-
nates of point P on the circle with center
O and radius 10, given that $m\overarc{PQ} = 40$.
(Hint: Use trigonometry.) ($\approx 7.7, \approx 6.4$)

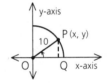

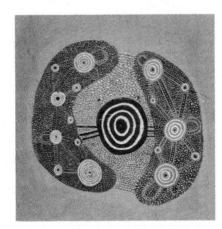

458   Chapter 10   Circles

10.4 SECANTS AND TANGENTS

Objectives

After studying this section, you will be able to
- Identify secant and tangent lines
- Identify secant and tangent segments
- Distinguish between two types of tangent circles
- Recognize common internal and common external tangents

Part One: Introduction

Secant and Tangent Lines

Some lines and circles have special relationships.

Definition   A **secant** is a line that intersects a circle at exactly two points. (Every secant contains a chord of the circle.)

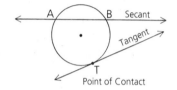

Definition   A **tangent** is a line that intersects a circle at exactly one point. This point is called the **point of tangency** or **point of contact**.

  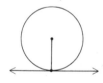

The diagrams above suggest the following postulates about tangents.

Postulate   *A tangent line is perpendicular to the radius drawn to the point of contact.*

Postulate   *If a line is perpendicular to a radius at its outer endpoint, then it is tangent to the circle.*

Section 10.4   Secants and Tangents   **459**

Class Planning

Time Schedule
Basic: 2 days
Average: $1\frac{1}{2}$ days
Advanced: $1\frac{1}{2}$ days

Resource References
Teacher's Resource Book
   Class Opener 10.4A, 10.4B
   Additional Practice Worksheet 20
   Using Manipulatives 16
Evaluation
   Tests and Quizzes
      Quiz 1, Series 1
      Quizzes 1–2, Series 3

Class Opener

Given: ⊙O with radius 10,
$\overline{AO} \perp \overline{AB}$, $m\overset{\frown}{AC} = 60$

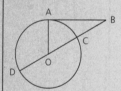

Find: BC
AO = 10 and ∠AOC = 60°, so
OB = 20. Since OC = 10,
BC = 10.

Lesson Notes

- Many students will need several days to assimilate all the new vocabulary in this section.
- Be sure students see how the two postulates relate to each of the three diagrams.

Secant and Tangent Segments

Some segments are related to circles in similar ways.

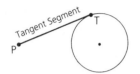

Definition   A *tangent segment* is the part of a tangent line between the point of contact and a point outside the circle.

Definition   A *secant segment* is the part of a secant line that joins a point outside the circle to the farther intersection point of the secant and the circle.

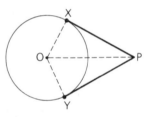

Definition   The *external part* of a secant segment is the part of a secant line that joins the outside point to the nearer intersection point.

Theorem 85   ***If two tangent segments are drawn to a circle from an external point, then those segments are congruent. (Two-Tangent Theorem)***

Given: ⊙O;
$\overline{PX}$ and $\overline{PY}$ are tangent segments.
Prove: $\overline{PX} \cong \overline{PY}$

The Two-Tangent Theorem is easily proved with congruent triangles. More theorems relating to secant segments and tangent segments are presented in Section 10.8.

Tangent Circles

Definition   *Tangent circles* are circles that intersect each other at exactly one point.

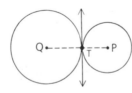

Definition   Two circles are *externally tangent* if each of the tangent circles lies outside the other. (See the left-hand figure above.)

Definition

Two circles are **internally tangent** if one of the tangent circles lies inside the other. (See the right-hand figure on the preceding page.)

Notice that in each case the tangent circles have one common tangent at their point of contact. Also, the point of contact lies on the *line of centers*, $\overleftrightarrow{PQ}$.

Common Tangents

$\overleftrightarrow{PQ}$ is the line of centers.
$\overleftrightarrow{XY}$ is a **common internal tangent**.
$\overleftrightarrow{AB}$ is a **common external tangent**.

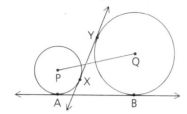

Definition

A **common tangent** is a line tangent to two circles (not necessarily at the same point). Such a tangent is a **common internal tangent** if it lies between the circles (intersects the segment joining the centers) or a **common external tangent** if it is not between the circles (does not intersect the segment joining the centers).

In practice, we will frequently refer to a *segment* as a common tangent if it lies on a common tangent and its endpoints are the tangent's points of contact. In the preceding diagram, for example, $\overline{XY}$ can be called a common internal tangent and $\overline{AB}$ can be called a common external tangent.

Part Two: Sample Problems

Problem 1

Given: $\overline{XY}$ is a common internal tangent to ⊙ P and Q at X and Y.
$\overline{XS}$ is tangent to ⊙P at S.
$\overline{YT}$ is tangent to ⊙Q at T.

Conclusion: $\overline{XS} \cong \overline{YT}$

Proof

1 $\overline{XS}$ is tangent to ⊙P. $\overline{YT}$ is tangent to ⊙Q.	1 Given
2 $\overline{XY}$ is tangent to ⊙ P and Q.	2 Given
3 $\overline{XS} \cong \overline{XY}$	3 Two-Tangent Theorem
4 $\overline{XY} \cong \overline{YT}$	4 Same as 3
5 $\overline{XS} \cong \overline{YT}$	5 Transitive Property

Communicating Mathematics

Have students write Sample Problem **1** in theorem form. They can label the theorem the "Two-Tangent Theorem."
If two tangent segments are drawn to a circle from an external point, then those segments are congruent.

Lesson Notes, continued

■ An overhead projector is a good tool to use to show the number and type of common tangents. Begin with disjoint circles and show four common tangents. Then slowly move the circles closer together. The internal tangents will begin to come together and will coincide when the circles are externally tangent, leaving three common tangents. Moving one of the circles through the other produces intersecting circles and leaves just two common external tangents. Continue moving the first circle, and notice that the external tangents eventually coincide when the circles are internally tangent, leaving just one common external tangent. Moving the circles even further produces one circle within the other, leaving no common tangents.

■ Refer to Sample Problem 3 on page 462. The common-tangent procedure is very powerful; it also works when the circles are not tangent to each other. Note that ABPQ is a trapezoid; this procedure builds on the technique for trapezoids presented in Section **9.4**.

Cooperative Learning

Have students work in small groups to make diagrams to illustrate common internal and common external tangents in each of the following cases.

1 Disjoint circles with two internal tangents and with two external tangents

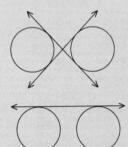

2 Externally tangent circles with one internal tangent and with two external tangents

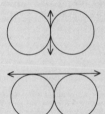

3 Concentric circles with no tangents

4 Intersecting circles with two external tangents

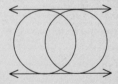

5 Internally tangent circles with one external tangent

Problem 2 $\overset{\leftrightarrow}{TP}$ is tangent to circle O at T. The radius of circle O is 8 mm. Tangent segment $\overline{TP}$ is 6 mm long. Find the length of $\overline{OP}$.

Solution Draw radius $\overline{OT}$ to form right triangle OTP.

$$(TP)^2 + (TO)^2 = (OP)^2$$
$$6^2 + 8^2 = (OP)^2$$
$$\pm 10 = OP \qquad \text{(Reject } -10.\text{)}$$

Thus, OP = 10 mm.

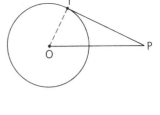

Problem 3 A circle with a radius of 8 cm is externally tangent to a circle with a radius of 18 cm. Find the length of a common external tangent.

Solution There is a standard procedure for solving a problem involving a common tangent (either internal or external).

Common-Tangent Procedure

1 Draw the segment joining the centers.

2 Draw the radii to the points of contact.

3 Through the center of the smaller circle, draw a line parallel to the common tangent.

4 Observe that this line will intersect the radius of the larger circle (extended if necessary) to form a rectangle and a right triangle.

5 Use the Pythagorean Theorem and properties of a rectangle.

In △RPQ,
$$(QR)^2 + (RP)^2 = (PQ)^2$$
$$10^2 + (RP)^2 = 26^2$$
$$RP = \pm 24$$

Thus, AB = 24 cm.

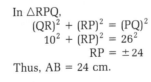

Problem 4   A *walk-around problem:*
Given: Each side of quadrilateral
ABCD is tangent to the circle.
AB = 10, BC = 15, AD = 18
Find: CD

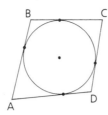

Solution   Let BE = x and "walk around" the
figure, using the given information
and the Two-Tangent Theorem.
CD = 15 − x + 18 − (10 − x)
  = 15 − x + 18 − 10 + x
  = 23

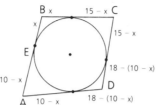

See problems 16, 21, 22, and 29 for other types of
walk-around problems.

Lesson Notes, continued

■ The walk-around problem is a good example of a technique for applying and using algebra to solve geometry problems. Together with the common-tangent procedure, such problems help students consolidate and apply the concepts of this section.

Part Three: Problem Sets

Problem Set A

1 The radius of ⊙A is 8 cm.
Tangent segment $\overline{BC}$ is 15 cm long.
Find the length of $\overline{AC}$. 17 cm

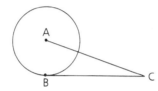

2 Concentric circles with radii 8 and 10
have center P.
$\overline{XY}$ is a tangent to the inner circle and is
a chord of the outer circle.
Find $\overline{XY}$. (Hint: Draw $\overline{PX}$ and $\overline{PY}$.) 12

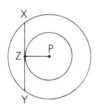

3 Given: $\overline{PR}$ and $\overline{PQ}$ are tangents to ⊙O at
R and Q.
Prove: $\overrightarrow{PO}$ bisects ∠RPQ. (Hint: Draw $\overline{RO}$
and $\overline{OQ}$.)

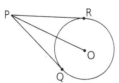

4 Given: $\overline{AC}$ is a diameter of ⊙B.
Lines s and m are tangents to the
⊙ at A and C.
Conclusion: s ∥ m

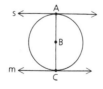

Assignment Guide

Basic
Day 1   1–3, 5, 7, 13
Day 2   6, 10–12, 14, 16, 17
Average

Day 1 $\left(\frac{1}{2}\text{ day}\right)$ Section **10.3**
          17–19
        $\left(\frac{1}{2}\text{ day}\right)$ Section **10.4**
          2, 5, 10, 13
Day 2   6, 11, 12, 14, 16, 19, 20
Advanced

Day 1 $\left(\frac{1}{2}\text{ day}\right)$ Section **10.3**
          19, 20, 23
        $\left(\frac{1}{2}\text{ day}\right)$ Section **10.4**
          5, 10, 11, 13
Day 2   12, 14, 16, 17, 19, 20,
          23

**Problem-Set Notes
and Additional Answers**

■ See *Solution Manual* for answers to problems **3** and **4**.

*Problem-Set Notes and
Additional Answers, continued*

- See Solution Manual for answers to problems 7–9.

Problem Set A, *continued*

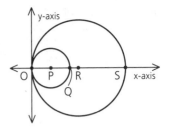

5 ⊙P and ⊙R are internally tangent at O.
P is at (8, 0) and R is at (19, 0).

   a Find the coordinates of Q and S.   Q = (16, 0);
   b Find the length of $\overline{QR}$.   3   S = (38, 0)

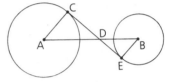

6 $\overline{AB}$ and $\overline{AC}$ are tangents to ⊙O,
and OC = 5x. Find OC.   2.5

7 Given: $\overline{CE}$ is a common internal tangent
to circles A and B at C and E.

   Prove: **a** ∠A ≅ ∠B

     b $\dfrac{AD}{BD} = \dfrac{CD}{DE}$

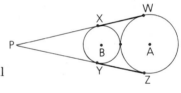

8 Given: $\overline{QR}$ and $\overline{QS}$ are tangent to ⊙P at
points R and S.

   Prove: $\overline{PQ} \perp \overline{RS}$ (Hint: This can be
proved in just a few steps.)

9 Given: $\overline{PW}$ and $\overline{PZ}$ are common tangents
to ⊚ A and B at W, X, Y, and Z.

   Prove: $\overline{WX} \cong \overline{YZ}$ (Hint: No auxiliary
lines are needed.)

   Note This is part of the proof of a useful
property: The common external tangent
segments of two circles are congruent.

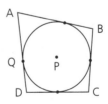

Problem Set B

10 ⊙P is tangent to each side of ABCD.
AB = 20, BC = 11, and DC = 14. Let
AQ = x and find AD.   23

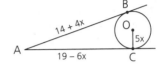

11 a Find the radius of ⊙P. $2\sqrt{13}$, or ≈ 7.21

b Find the slope of the tangent to ⊙P at point Q. $-\frac{3}{2}$

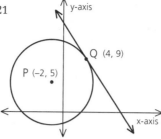

12 Two concentric circles have radii 3 and 7. Find, to the nearest hundredth, the length of a chord of the larger circle that is tangent to the smaller circle. (See problem 2 for a diagram.) ≈ 12.65

13 The centers of two circles of radii 10 cm and 5 cm are 13 cm apart.

a Find the length of a common external tangent. (Hint: Use the common-tangent procedure.) 12 cm

b Do the circles intersect? Yes

14 The centers of two circles with radii 3 and 5 are 10 units apart. Find the length of a common internal tangent. (Hint: Use the common-tangent procedure.) 6

15 Given: $\overline{PT}$ is tangent to ⊙ Q and R at points S and T.

Conclusion: $\dfrac{PQ}{PR} = \dfrac{SQ}{TR}$

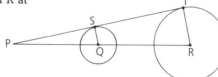

16 Given: Tangent ⊙ A, B, and C, AB = 8, BC = 13, AC = 11

Find: The radii of the three ⊙ (Hint: This is a walk-around problem.) 3; 5; 8

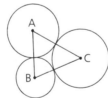

17 The radius of ⊙O is 10. The secant segment $\overline{PX}$ measures 21 and is 8 units from the center of the ⊙.

a Find the external part (PY) of the secant segment. 9

b Find OP. 17

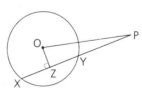

Problem-Set Notes and Additional Answers, continued

■ See *Solution Manual* for answer to problem **15**.

Problem-Set Notes and
Additional Answers, continued

- See *Solution Manual* for answer to problem **18.**
- In connection with problem **21,** students in advanced courses might be able to generalize and prove the following: If the sides of a circumscribed polygon with an even number of sides are numbered, then the sum of the measures of the odd-numbered sides is equal to the sum of the measures of the even-numbered sides.

21 By the Two-Tangent Theorem, $\overline{XQ} \cong \overline{XP}$, $\overline{YQ} \cong \overline{YR}$, $\overline{WS} \cong \overline{WP}$, and $\overline{ZS} \cong \overline{ZR}$. Then use addition.

22

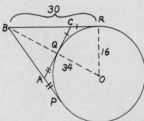

$$\begin{cases} (x + 4)^2 + (y + 4)^2 = 20^2 \\ x + y = 20 \end{cases}$$

23

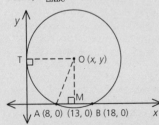

CR = CQ and AP = AQ. Thus, $P_{\triangle ABC}$ = BR + BP.

24

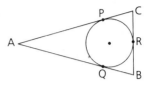

The ⊥ bis. of the chord $\overline{AB}$ passes through O. Thus, the coordinates of M will be

Problem Set B, *continued*

18 Given: △ABC is isosceles, with base $\overline{BC}$.
Conclusion: $\overline{BR} \cong \overline{RC}$

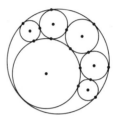

19 If two of the seven circles are chosen at random, what is the probability that the chosen pair are

a Internally tangent? $\frac{2}{7}$

b Externally tangent? $\frac{3}{7}$

c Not tangent? $\frac{2}{7}$

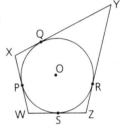

20 Find, to the nearest tenth, the distance between two circles if their radii are 1 and 4 and the length of a common external tangent is $7\frac{1}{2}$. ≈3.1

Problem Set C

21 Given: Quadrilateral WXYZ is *circumscribed about* ⊙O (that is, its sides are tangent to the ⊙).

Prove: XY + WZ = WX + YZ

22 Find the perimeter of right triangle WXY if the radius of the circle is 4 and WY = 20. **48**

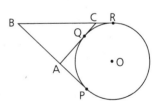

23 B is 34 mm from the center of circle O, which has radius 16 mm. $\overline{BP}$ and $\overline{BR}$ are tangent segments. $\overleftrightarrow{AC}$ is tangent to ⊙O at point Q. Find the perimeter of △ABC. **60 mm**

(13, 0). ∴ x = 13. This means OT = OA = 13. Since AM = 5, △AOM is a (15, 12, 13) rt. △, with OM = 12. Thus, the

coordinates of O will be (13, 12). In a similar manner, there will be another ⊙ satisfying the given conditions, with center at (13, −12).

24 Find the coordinates of the center of a circle that is tangent to the y-axis and intersects the x-axis at (8, 0) and (18, 0).
(13, 12) or (13, −12)

25 Given: Two concentric circles with center E,
AB = 40, CD = 24, $\overline{CD} \perp \overline{AE}$;
$\overline{AB}$ is tangent at C.

Find: AF 10

26 $\overline{BC}$ is tangent to ⊙A at B, and $\overline{BD} \cong \overline{BA}$.
Explain why $\overline{BD}$ bisects $\overline{AC}$.
△ABD is equilateral and △CAB is a
30°–60°–90° △.

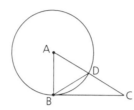

27 Given: ⓢ E and F, with $\overline{AC}$ tangent at B
and C, DE = 10, FB = 4
Find: AB $\frac{8\sqrt{10}}{3}$

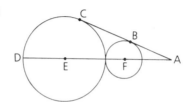

28 Circles P and Q are tangent to each other
and to the axes as shown. PQ = 26 and
AB = 24. Find the coordinates of P and Q.
(8, 42); (18, 18)

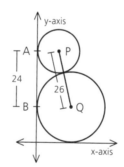

29 Given: Three tangent ⓢ, A, B, and C,
BC = a, AC = b, AB = c
Find: The radius of ⊙A in terms of a, b,
and c $\frac{c-a+b}{2}$

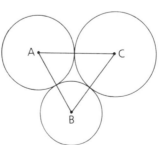

Section 10.4   Secants and Tangents   **467**

*Problem-Set Notes and
Additional Answers, continued*

25

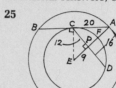

AP = 16 by the Pythagorean
Theorem. EP = 9 by
Theorem 68b. ∴ EC = 15
and AF = 10.

27

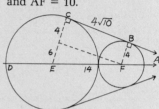

$\frac{AB}{AB + 4\sqrt{10}} = \frac{4}{10}$ by similar
triangles.

28

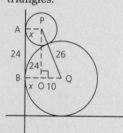

Let the radius of ⊙P =
AP = x. Drop the
perpendicular from P to $\overline{BQ}$.
Then △POQ is a (10, 24, 26)
rt. △; so the radius of the
⊙Q = $\overline{BQ}$ = x + 10. But $\overline{PQ}$
has a length equal to the
sum of the two radii, and so
x + x + 10 = 26, or x = 8.
Thus, Q = (18, 18) and
P = (8, 42).

29

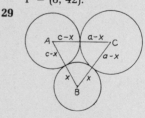

a + c − 2x = b

ANGLES RELATED TO A CIRCLE

Class Planning

Time Schedule
Basic: 3 days
Average: 2 days
Advanced: 2 days

Resource References
Teacher's Resource Book
   Class Opener 10.5A, 10.5B
   Using Manipulatives 17
Evaluation
   Tests and Quizzes
      Quizzes 2–5, Series 1
      Quizzes 2–3, Series 2
      Quizzes 3–4, Series 3

Class Opener

Given: m$\overparen{BC}$ = 48
Find: m∠A

△AOC is an isosceles triangle, and m∠AOC = 132. Therefore, m∠A + m∠C = 48. Since the base angles of an isosceles triangle are congruent, m∠A = 24.

2x = 48
  x = 24

Objectives

After studying this section, you will be able to
- Determine the measures of central angles
- Determine the measures of inscribed and tangent-chord angles
- Determine the measures of chord-chord angles
- Determine the measures of secant-secant, secant-tangent, and tangent-tangent angles

Part One: Introduction

Angles with Vertices at the Center of a Circle

The measure of an angle whose sides intersect a circle is determined by the measure of its intercepted arcs. The location of the vertex of each angle is the key to remembering how to compute the measure of the angle.

An angle with its vertex at the center of a circle is a central angle, already defined to be equal in measure to its intercepted arc (Section 10.3).

In ⊙O, $\overparen{AB}$ = 50°, so m∠AOB = 50.

Angles with Vertices on a Circle

Two important types of angles whose vertices are on a circle are shown below.

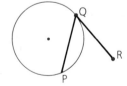

∠HKM is an *inscribed angle*.   ∠PQR is a *tangent-chord angle*.

Vocabulary
chord-chord angle
inscribed angle
secant-secant angle

secant-tangent angle
tangent-chord angle
tangent-tangent angle

Definition	An **inscribed angle** is an angle whose vertex is on a circle and whose sides are determined by two chords.

Definition	A **tangent-chord angle** is an angle whose vertex is on a circle and whose sides are determined by a tangent and a chord that intersect at the tangent's point of contact.

Theorem 86	**The measure of an inscribed angle or a tangent-chord angle (vertex on a circle) is one-half the measure of its intercepted arc.**

The proof of Theorem 86 for inscribed angles is unusual because three cases must be considered. Shown below are some key steps for each case in the proof that $m\angle BAC = \frac{1}{2}(m\widehat{BC})$.

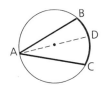

Case 1:
The center lies on a side of the angle.
1 $m\angle BOC = m\widehat{BC}$
2 $\angle BOC = \angle BAC + \angle ABO$, so $m\angle BOC = 2(m\angle BAC)$

Case 2:
The center lies inside the angle.
1 Use case 1 twice.
2 Add $\angle$s and arcs.

Case 3:
The center lies outside the angle.
1 Use case 1 twice.
2 Subtract $\angle$s and arcs.

Example 1   *Given:* $m\widehat{AC} = 112$
Find: $m\angle B$

$$m\angle B = \frac{1}{2}(m\widehat{AC})$$
$$= \frac{1}{2} \cdot 112$$
$$= 56$$

Example 2   *Given:* $\overline{FE}$ *is tangent at E.*
$m\widehat{DE} = 80$
Find: $m\angle DEF$

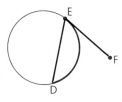

$$m\angle DEF = \frac{1}{2}(m\widehat{DE})$$
$$= \frac{1}{2} \cdot 80$$
$$= 40$$

Lesson Notes

■ The proof of Theorem 86 for tangent-chord angles is as follows.
Given: $\overleftrightarrow{BC}$ is tangent to the circle with center O.
Prove: $m\angle ABC = \frac{1}{2} m\widehat{AB}$

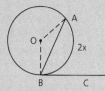

Let $m\widehat{AB} = 2x$. Thus, the central $\angle AOB = 2x$. Since $\triangle AOB$ is isos., $m\angle OBA = \frac{180 - 2x}{2} = 90 - x$. Since radius $\overline{OB} \perp$ tan $\overline{BC}$, $m\angle OBC = 90$, and thus, $m\angle ABC = x$. ∴ $m\angle ABC = \frac{1}{2}m\widehat{AB}$

Cooperative Learning

Students should work individually at the beginning. Have each one draw a circle of any size with a well-defined center. Then with a protractor and a straightedge, each student should draw and measure a central angle of the circle. Next, have each student use three colors to draw three different inscribed angles with the same intercepted arc as the central angle they drew. They should use their protractors to measure the three inscribed angles.
In small groups, students should first check the measurements of the others in their group. Have them generalize from their collected data.
Students should be able to generalize that all inscribed angles with the same intercepted arc are congruent and that the measure of each is one-half the measure of the central angle with the same intercepted arc.

T 469

Angles with Vertices Inside, but Not at the Center of, a Circle

One type of angle other than a central angle has a vertex inside a circle.

Definition   A ***chord-chord angle*** is an angle formed by two chords that intersect inside a circle but not at the center.

∠CPD is one of four chord-chord angles formed by chords $\overline{CF}$ and $\overline{DE}$ in circle O.

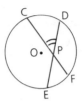

Theorem 87   *The measure of a chord-chord angle is one-half the sum of the measures of the arcs intercepted by the chord-chord angle and its vertical angle.*

Notice that one-half the sum of the arc measures is the same as the *average* of the arc measures.

Given: ∠3 is a chord-chord angle.
Prove: $m\angle 3 = \frac{1}{2}(m\widehat{AB} + m\widehat{CD})$

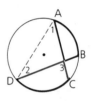

Here are two key steps in a proof of Theorem 87.
1 $m\angle 3 = m\angle 1 + m\angle 2$
2 $m\angle 3 = \frac{1}{2}(m\widehat{CD}) + \frac{1}{2}(m\widehat{AB})$

Angles with Vertices Outside a Circle

There are three types of angles having a vertex outside a circle and both sides intersecting the circle.

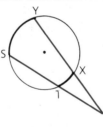
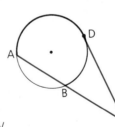
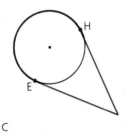

∠V is a
secant-secant angle.

∠C is a
secant-tangent angle.

∠F is a
tangent-tangent angle.

Chapter 10 Circles

Definition   A *secant-secant angle* is an angle whose vertex is outside a circle and whose sides are determined by two secants.

Definition   A *secant-tangent angle* is an angle whose vertex is outside a circle and whose sides are determined by a secant and a tangent.

Definition   A *tangent-tangent angle* is an angle whose vertex is outside a circle and whose sides are determined by two tangents.

Theorem 88   *The measure of a secant-secant angle, a secant-tangent angle, or a tangent-tangent angle (vertex outside a circle) is one-half the difference of the measures of the intercepted arcs.*

Key steps in a proof of Theorem 88 for secant-secant angles follow.

Prove: $m\angle E = \frac{1}{2}(m\widehat{SC} - m\widehat{AB})$
1 $m\angle 3 = m\angle E + m\angle 2$; solve for $m\angle E$.
2 $m\angle 2 = \frac{1}{2}(m\widehat{AB})$; $m\angle 3 = \frac{1}{2}(m\widehat{SC})$
3 Substitute and simplify.

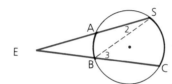

Example 1   Find $m\angle A$.
$m\angle A = \frac{1}{2}(m\widehat{CD} - m\widehat{BE})$
$\quad\;\; = \frac{1}{2}(100 - 20)$
$\quad\;\; = 40$

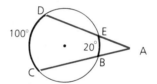

Example 2   Find $m\angle F$.
$m\widehat{JK} = 360 - 100 - 60$
$\quad\;\; = 200$
$m\angle F = \frac{1}{2}(m\widehat{JK} - m\widehat{HK})$
$\quad\;\; = \frac{1}{2}(200 - 60)$
$\quad\;\; = 70$

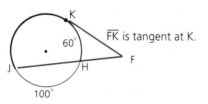

$\overline{FK}$ is tangent at K.

Example 3   Find $m\angle Q$.
$m\widehat{MAP} = 360 - 100 = 260$
$m\angle Q = \frac{1}{2}(m\widehat{MAP} - m\widehat{MP})$
$\quad\;\; = \frac{1}{2}(260 - 100)$
$\quad\;\; = 80$

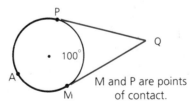

M and P are points of contact.

Section 10.5   Angles Related to a Circle   **471**

Lesson Notes, continued

■ The proof for Theorem 88 for secant-tangent angles follows.
Prove: $m\angle P = \frac{1}{2}(m\widehat{TA} - m\widehat{TB})$

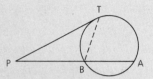

Since $\angle TBA$ is an ext. $\angle$ of $\triangle PTB$, $m\angle TBA = m\angle P + m\angle PTB$. $\therefore m\angle P = m\angle TBA - m\angle PTB$. However, inscribed $\angle TBA = \frac{1}{2}m\widehat{TA}$, and tangent-chord $\angle PTB = \frac{1}{2}m\widehat{TB}$.
$\therefore m\angle P = \frac{1}{2}m\widehat{TA} - \frac{1}{2}m\widehat{TB}$
$\qquad\;\; = \frac{1}{2}(m\widehat{TA} - m\widehat{TB})$

■ The proof for Theorem 88 for tangent-tangent angles follows.
Prove: $m\angle APB = \frac{1}{2}(m\widehat{AXB} - m\widehat{AB})$

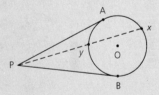

Applying the secant-tangent theorem to $\angle APX$ and $\angle BPX$ gives $m\angle APX = \frac{1}{2}(m\widehat{AX} - m\widehat{AY})$ and $m\angle BPX = \frac{1}{2}(m\widehat{XB} - m\widehat{YB})$.
Adding gives
$m\angle APX + m\angle BPX$
$= \frac{1}{2}m\widehat{AX} + \frac{1}{2}m\widehat{XB}$
$\quad - \frac{1}{2}m\widehat{AY} - \frac{1}{2}m\widehat{YB}$
$= \frac{1}{2}(m\widehat{AX} + m\widehat{XB} - m\widehat{AY}$
$\quad - m\widehat{YB})$
$= \frac{1}{2}(m\widehat{AXB} - m\widehat{AB})$

Angle-Arc Summary

Central Angle

$m\angle KOJ = m\widehat{JK}$
Vertex at center ⟹ equal

Chord-Chord Angle

$m\angle DEC = \frac{1}{2}(m\widehat{AB} + m\widehat{CD})$
Vertex inside ⟹ half the sum

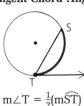

THERE IS A LOT TO REMEMBER.
BUT I SEE THE PATTERN.
THE KEY IS THE LOCATION OF THE VERTEX.

Inscribed Angle

$m\angle Q = \frac{1}{2}(m\widehat{PR})$

Tangent-Chord Angle

$m\angle T = \frac{1}{2}(m\widehat{ST})$
Vertex on circle ⟹ half the arc

Secant-Secant Angle

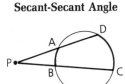

$m\angle P = \frac{1}{2}(m\widehat{CD} - m\widehat{AB})$

Tangent-Tangent Angle

$m\angle P = \frac{1}{2}(m\widehat{SXT} - m\widehat{ST})$

Secant-Tangent Angle

$m\angle P = \frac{1}{2}(m\widehat{RT} - m\widehat{QT})$

Vertex outside circle ⟹ half the difference

Part Two: Sample Problems

Problem 1

Given: $\overline{AB}$ is a diameter of $\odot P$.
$\widehat{BD} = 20°$, $\widehat{DE} = 104°$

Find: $m\angle C$

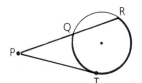

Solution

First find $m\widehat{EA}$.
$m\widehat{AEB} = 180$, so $m\widehat{EA} = 180 - (104 + 20) = 56$.
Thus, $m\angle C = \frac{1}{2}(m\widehat{EA} - m\widehat{DB}) = \frac{1}{2}(56 - 20) = 18$.

Problem 2   *Find y.*

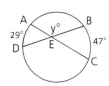

Solution

Find $m\angle BEC$ first.
$m\angle BEC = \frac{1}{2}(29 + 47) = 38$
Thus, $y = 180 - m\angle BEC = 142$.

472 Chapter 10 Circles

Problem 3 **a** Find x. **b** Find y. **c** Find z.

  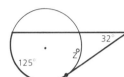

Solution

a $x = \frac{1}{2}(88 + 27)$
   $= 57\frac{1}{2}$

b $y = \frac{1}{2}(57 - 31)$
   $= 13$

c $z = \frac{1}{2}(233 - 127)$
   $= 53$

Problem 4 **a** Find y. **b** Find z. **c** Find a.

Solution

a $\frac{1}{2}(21 + y) = 72$
   $21 + y = 144$
   $y = 123$

b $\frac{1}{2}(125 - z) = 32$
   $125 - z = 64$
   $z = 61$

c $\frac{1}{2}a = 65$
   $a = 130$

Problem 5 Find m$\widehat{AB}$ and m$\widehat{CD}$.

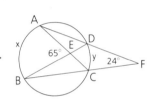

Solution Let m$\widehat{AB} = x$ and m$\widehat{CD} = y$.
Then $\frac{1}{2}(x + y) = 65$ and $\frac{1}{2}(x - y) = 24$.
So $x + y = 130$ and $x - y = 48$.

   $x + y = 130$
   $\underline{x - y = 48}$
   $\quad 2x = 178$ Add the equations.
   $\quad\; x = 89$

   $89 + y = 130$
   $\quad\quad y = 41$

Thus, m$\widehat{AB} = 89$ and m$\widehat{CD} = 41$.

Part Three: Problem Sets

Problem Set A

1 Vertex at center:
   Given: $\widehat{AB} = 62°$
   Find: m∠O 62

Lesson Notes, continued

■ Sample Problem **5** shows that
the measure of the larger arc is
the sum of the measures of the
angles, and the measure of the
smaller arc is the difference of
the measures of the angles.
Note that $\frac{1}{2}(x + y) = 65$ and
$\frac{1}{2}(x - y) = 24$ are rewritten as
$\begin{cases} x + y = 2 \cdot 65 \\ x - y = 2 \cdot 24. \end{cases}$
Then adding or subtracting,
$2x = 2(65 + 24)$
$\; x = 65 + 24$
$2y = 2(65 - 24)$
$\; y = 65 - 24$

Assignment Guide

Basic	
Day 1	1–6
Day 2	7–17
Day 3	18, 19, 21–23, 26, 27

Average	
Day 1	1–10
Day 2	12, 14–17, 19–24, 30

Advanced	
Day 1	5–17
Day 2	20, 23–30, 32, 34

Problem Set A, *continued*

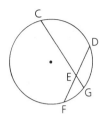

2 Vertex inside:
Given: $\overset{\frown}{CD} = 100°$, $\overset{\frown}{FG} = 30°$
Find: m∠CED 65

3 Vertex on:
a Given: $\overset{\frown}{AC} = 70°$
Find: m∠B 35

b Given: $\overline{DE}$ is tangent at E.
$\overset{\frown}{EF} = 150°$
Find: m∠DEF 75

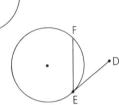

4 Vertex outside:

a

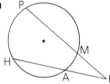

Given: $\overset{\frown}{HP} = 120°$,
$\overset{\frown}{AM} = 36°$
Find: m∠K 42

b

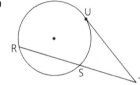

Given: $\overline{TU}$ is tangent at U.
$\overset{\frown}{RU} = 160°$,
$\overset{\frown}{SU} = 60°$
Find: m∠T 50

c

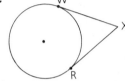

Given: W and R are points of
contact.
$\overset{\frown}{WR} = 140°$
Find: m∠X 40

5 Find the measure of each angle or arc that is labeled with a letter.

a 140

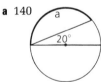

c 20 160°

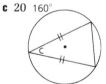

e 100

b 80

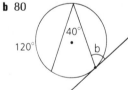

d $81\frac{1}{2}$

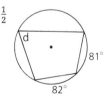

474 Chapter 10 Circles

6 Find the measure of each angle or arc that is labeled with a letter.

a 104

b 68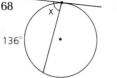

c 90

d 33

7 Given: $\overset{\frown}{AB} = 108°$, $\overset{\frown}{CD} = 62°$

Find: $\angle AXB$ and $\angle Y$ 85°; 23°

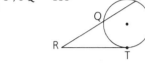

8 Given: $\overset{\frown}{TP} = 170°$, $\overset{\frown}{PQ} = 135°$

Find: $\angle R$ $57\frac{1}{2}°$

9 Given: $\angle AEB = 30°$,

$\overset{\frown}{AB} = 50°$

Find: $\overset{\frown}{CD}$ 10°

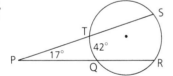

10 Given: $\angle P = 17°$,

$\overset{\frown}{TQ} = 42°$

Find: $\overset{\frown}{SR}$ 76°

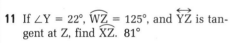

11 If $\angle Y = 22°$, $\overset{\frown}{WZ} = 125°$, and $\overleftrightarrow{YZ}$ is tangent at Z, find $\overset{\frown}{XZ}$. 81°

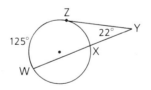

12 If $\overset{\frown}{ST} = 85°$, $\overset{\frown}{SQ} = 95°$, and $\overset{\frown}{TR} = 175°$, find $\angle P$. 40°

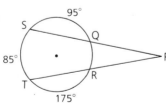

Section 10.5   Angles Related to a Circle   **475**

Problem Set A, *continued*

13 Given: $\overset{\frown}{AB} = 85°$,
$\overset{\frown}{CD} = 25°$
Find: $\angle AED$ 125°

14 Given: $\overline{WY}$ is a diameter of $\odot E$.
$\overset{\frown}{WX} = 50°$, $\angle XPY = 120°$
Find: $\overset{\frown}{WZ}$ 110°

15 A circle is divided into three arcs in the ratio of $3:4:5$. A tangent-chord angle intercepts the largest of the three arcs. Find the measure of the tangent-chord angle.   75

16 An inscribed angle intercepts an arc that is $\frac{1}{9}$ of the circle. Find the measure of the inscribed angle.   20

17 If a point is chosen at random on $\odot M$, what is the probability that it lies on
a $\overset{\frown}{IAN}$ $\frac{1}{2}$ **b** $\overset{\frown}{AN}$ $\frac{5}{18}$ **c** $\overset{\frown}{ID}$ $\frac{1}{6}$ **d** $\overset{\frown}{IE}$ $\frac{11}{36}$

Problem Set B

18 Given: $\overleftrightarrow{VQ}$ is tangent to $\odot O$ at Q.
$\overline{QS}$ is a diameter of $\odot O$.
$\overset{\frown}{PQ} = 115°$; $\angle RPS = 36°$
Find: **a** $\angle R$ $57\frac{1}{2}°$ **e** $\angle QPR$ 54° **i** $\overset{\frown}{PRQ}$ 245°
b $\angle S$ $57\frac{1}{2}°$ **f** $\angle QPS$ 90° **j** $\overset{\frown}{RSP}$ 137°
c $\overset{\frown}{SR}$ 72° **g** $\angle QTP$ $93\frac{1}{2}°$ **k** $\angle VQS$ 90°
d $\overset{\frown}{QR}$ 108° **h** $\angle PQV$ $86\frac{1}{2}°$ **l** $\angle QOP$ 115°

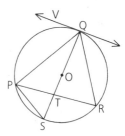

19 Given $m\angle P = 60$ and $m\overset{\frown}{PSR} = 128$, find $m\angle Q$, $m\angle R$, and $m\angle S$.   64; 120; 116

20 The major arc cut off by two tangents to a circle from an outside point is five thirds of the minor arc. Find the angle formed by the tangents.   45°

476   Chapter 10   Circles

21 Find the measure of each arc or angle labeled with a letter.

a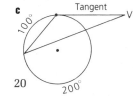

132

b

10

c

20

d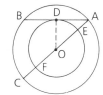

30

22 Given circles concentric at O, $\overline{AB}$ tangent to the inner circle, and $\widehat{BC} = 84°$, find the measures of $\angle A$, $\widehat{DE}$, and $\widehat{DF}$. 42; 48; 132

23 Given: $\widehat{AB} = 92°$,
$\angle AEB = 82°$
Find: $\widehat{AD}$ 98°

24 Given: $\angle AFE = 89°$,
$\angle C = 15°$
Find: $\widehat{AE}$ and $\widehat{BD}$ 104°; 74°

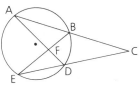

25 Given: $\widehat{SY} = 112°$,
$\widehat{DC} = 87°$
Find: $\widehat{AB}$ 25°

26 If $\widehat{DC} = (5x + 6)°$, $\widehat{AB} = (2x)°$, and $\angle AEB = 94°$, find $\widehat{AB}$. 52°

27 A secant-secant angle intercepts arcs that are $\frac{3}{5}$ and $\frac{3}{8}$ of the circle. If a chord-chord angle and its vertical angle intercept the same arcs, what is the measure of the chord-chord angle? $175\frac{1}{2}$

28 $\triangle ABC$ is inscribed in a circle (all sides are chords), AB = 12, AC = 6, and BC = $6\sqrt{3}$. Find $m\widehat{BC}$. 120

***Problem-Set Notes
and Additional Answers***

31

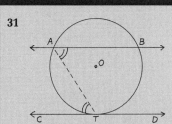

$\angle BAT \cong \angle CTA$ by $\parallel$ lines $\Rightarrow$ alt. int. $\angle$s $\cong$; $\angle BAT = \frac{1}{2} \widehat{BT}$ and $\angle CTA = \frac{1}{2} \widehat{AT}$.

32

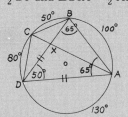

$m\angle DXC = \frac{1}{2}(80 + 100) = 90$

29 a An angle is inscribed in a circle and intercepts an arc of 140°. Find the measure of the angle. **70**

b An angle is inscribed in a 140° arc (the vertex is on the arc and the sides contain the endpoints of the arc). Find the measure of the angle. **110**

30 a Find the area and the circumference of ⊙Q to the nearest tenth. $A \approx 78.5$ sq cm; $C \approx 31.4$ cm

b Find the area of the shaded region to the nearest tenth. ≈ 19.6 sq cm

c Find the length of $\overset{\frown}{PR}$ to the nearest tenth. ≈ 7.9 cm

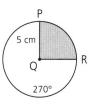

Problem Set C

31 Given: $\overset{\leftrightarrow}{AB} \parallel \overset{\leftrightarrow}{CD}$;
 $\overset{\leftrightarrow}{DC}$ is tangent to ⊙O at T.
Conclusion: $\overset{\frown}{AT} \cong \overset{\frown}{BT}$

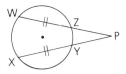

32 A quadrilateral ABCD is inscribed in a circle. Its diagonals intersect at X. If $\overset{\frown}{AB} = 100°$, $\overset{\frown}{BC} = 50°$, and $\overline{AD} \cong \overline{BD}$, find m∠DXC. **90**

33 Given: $\overline{WZ} \cong \overline{XY}$,
 $\overset{\frown}{WXY} = 200°$
Find: ∠P **20°**

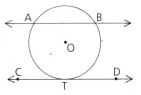

34 A secant and a tangent to a circle intersect to form an angle of 38°. If the measures of the arcs intercepted by this angle are in a ratio of 2:1, find the measure of the third arc. **132**

35 Given: ⊙ P and Q are internally tangent at T.
 Diameter $\overline{NS}$ of ⊙Q is tangent to ⊙P at A.
 $m\overset{\frown}{MR} = 42$; $\overline{TM}$ passes through A.
Find: $m\overset{\frown}{NM}$ **90**

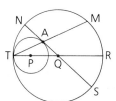

36 The two circles shown intersect at A and B. If ∠AXB = 70°, $\overset{\frown}{CD} = 20°$, and $\overset{\frown}{EF} = 160°$, find the difference between the measures of $\overset{\frown}{AB}$ of the smaller circle and $\overset{\frown}{AB}$ of the larger circle. **100**

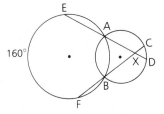

33

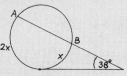

$x + 200° - x + 200° -$
$x + y = 360°$
$y - x = -40°$
$\therefore x - y = 40°$
$\frac{1}{2}(x - y) = 20° = \angle P$

34

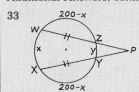

$38 = \frac{1}{2}(2x - x)$
$\quad = \frac{1}{2}x$
$76 = x$
$\therefore m\overset{\frown}{AB} = 132$

35

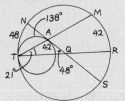

Since $m\overset{\frown}{MR} = 42$,
m∠MTR = 21. Then
$m\overset{\frown}{AQ} = 42$ and $m\overset{\frown}{AT} = 138$.
Then m∠AQT = $\frac{1}{2}(138 -$
$42) = 48$, and thus, $m\overset{\frown}{TN} =$
$48. \therefore m\overset{\frown}{NM} = 90$.

36

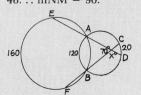

$70 = \frac{1}{2}(m\overset{\frown}{AB} + 20)$
$\therefore m\overset{\frown}{AB} = 120$ for smaller ⊙
$70 = \frac{1}{2}(160 - m\overset{\frown}{AB})$
$\therefore m\overset{\frown}{AB} = 20$ for larger ⊙

MORE ANGLE–ARC THEOREMS

Objectives

After studying this section, you will be able to
- Recognize congruent inscribed and tangent-chord angles
- Determine the measure of an angle inscribed in a semicircle
- Apply the relationship between the measures of a tangent-tangent angle and its minor arc

Part One: Introduction

Congruent Inscribed and Tangent-Chord Angles

Our knowledge of the relationships between angles and their intercepted arcs leads easily to the next two theorems.

Theorem 89   *If two inscribed or tangent-chord angles intercept the same arc, then they are congruent.*

Given: X and Y are inscribed angles intercepting arc AB.

Conclusion: ∠X ≅ ∠Y

Theorem 90   *If two inscribed or tangent-chord angles intercept congruent arcs, then they are congruent.*

If $\overleftrightarrow{ED}$ is the tangent at D and $\overset{\frown}{AB} \cong \overset{\frown}{CD}$, we may conclude that ∠P ≅ ∠CDE.

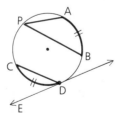

Angles Inscribed in Semicircles

All angles inscribed in semicircles have the same measure. What do you think that measure might be?

Section 10.6 More Angle-Arc Theorems   **479**

Class Planning

Time Schedule
Basic: $1\frac{1}{2}$ days
Average: $1\frac{1}{2}$ days
Advanced: 1 day

Resource References
Teacher's Resource Book
   Class Opener 10.6A, 10.6B
Transparency 18
Evaluation
   Tests and Quizzes
     Quiz 4, Series 2

Class Opener

Given: ⊙O, m$\overset{\frown}{AD}$ = 54,
     m$\overset{\frown}{BC}$ = 62, m$\overset{\frown}{DE}$ = 36,
     m$\overset{\frown}{FH}$ = 22, m$\overset{\frown}{EF}$ = 78

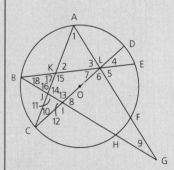

Find: The measures of the missing arcs and indicated angles

m$\overset{\frown}{CH}$ = 44, m$\overset{\frown}{AB}$ = 64,
m∠1 = 33, m∠2 = 76, m∠3 = 71,
m∠4 = 49, m∠5 = 71, m∠6 = 60,
m∠7 = 49, m∠8 = 99, m∠9 = 21,
m∠10 = 27, m∠11 = 54,
m∠12 = 99, m∠13 = 81,
m∠14 = 126, m∠15 = 104,
m∠16 = 54, m∠17 = 76,
m∠18 = 50,

- A proof of Theorem 89 follows. $m\angle X = \frac{1}{2} m\widehat{AB}$ and $m\angle Y = \frac{1}{2} m\widehat{AB}$, since the measure of an angle is one-half the measure of its arc. Thus, $m\angle X = m\angle Y$ by the Transitive Property. Therefore, $\angle X \cong \angle Y$. The proofs for two tangent-chord angles and for the combination of a tangent-chord angle and an inscribed angle would be similar.

- A proof of Theorem 91 follows. Since the measure of an inscribed angle is one-half the measure of its intercepted arc, and a semicircle is 180°, $\angle C$ is 90°.

Theorem 91   *An angle inscribed in a semicircle is a right angle.*

Given: $\overline{AB}$ is a diameter of $\odot O$.
Prove: $\angle C$ is a right angle.

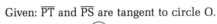

A Special Theorem About Tangent-Tangent Angles

A tangent-tangent angle has a special relationship with its minor arc.

Theorem 92   *The sum of the measures of a tangent-tangent angle and its minor arc is 180.*

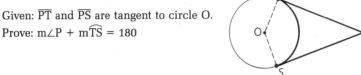

Given: $\overline{PT}$ and $\overline{PS}$ are tangent to circle O.
Prove: $m\angle P + m\widehat{TS} = 180$

Proof: Since the sum of the measures of the angles in quadrilateral SOTP is 360 and since $\angle T$ and $\angle S$ are right angles, $m\angle P + m\angle O = 180$. Therefore, $m\angle P + m\widehat{TS} = 180$.

Example $\overleftrightarrow{PT}$ and $\overleftrightarrow{PS}$ are tangents at T and S. Find $m\angle P$.

$$m\angle P + m\widehat{TS} = 180$$
$$m\angle P + 130 = 180$$
$$m\angle P = 50$$

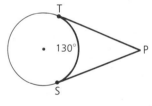

Part Two: Sample Problems

Problem 1   Given: $\odot O$

Conclusion: $\triangle LVE \sim \triangle NSE$,
$EV \cdot EN = EL \cdot SE$

Proof

1 $\odot O$	1 Given
2 $\angle V \cong \angle S$	2 If two inscribed $\angle$s intercept the same arc, they are $\cong$.
3 $\angle L \cong \angle N$	3 Same as 2
4 $\triangle LVE \sim \triangle NSE$	4 AA (2, 3)
5 $\dfrac{EV}{SE} = \dfrac{EL}{EN}$	5 Ratios of corresponding sides of $\sim$ $\triangle$ are =.
6 $EV \cdot EN = EL \cdot SE$	6 Means-Extremes Products Theorem

Problem 2   In circle O, $\overline{BC}$ is a diameter and the radius of the circle is 20.5 mm.
Chord $\overline{AC}$ has a length of 40 mm. Find AB.

Solution   Since ∠A is inscribed in a semicircle, it is a right angle. By the Pythagorean Theorem,

$$(AB)^2 + (AC)^2 = (BC)^2$$
$$(AB)^2 + 40^2 = 41^2$$
$$AB = 9 \text{ mm}$$

Problem 3   Given: ⊙O with $\overleftrightarrow{AB}$ tangent at B, $\overleftrightarrow{AB} \parallel \overleftrightarrow{CD}$
Prove: ∠C ≅ ∠BDC

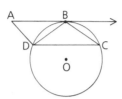

Proof

1 $\overleftrightarrow{AB}$ is tangent to ⊙O.	1 Given
2 $\overleftrightarrow{AB} \parallel \overleftrightarrow{CD}$	2 Given
3 ∠ABD ≅ ∠BDC	3 ∥ lines ⇒ alt. int. ∠s ≅
4 ∠C ≅ ∠ABD	4 If an inscribed ∠ and a tangent-chord ∠ intercept the same arc, they are ≅.
5 ∠C ≅ ∠BDC	5 Transitive Property

Part Three: Problem Sets

Problem Set A

1 Given: X is the midpt. of $\overparen{WY}$.
Prove: $\overrightarrow{ZX}$ bisects ∠WZY.

2 Given: ⊙E with diameter $\overline{AC}$, $\overline{BC} ≅ \overline{CD}$
Conclusion: △ABC ≅ △ADC

3 In ⊙P, $\overline{BC}$ is a diameter, AC = 12 mm, and BA = 16 mm. Find the radius of the circle. 10 mm

Assignment Guide

Basic

Day 1   1–9
Day 2   ($\frac{1}{2}$ day) Section **10.6**
         10–16, 18
         ($\frac{1}{2}$ day) Section **10.7**
         1–5

Average

Day 1   4, 7–15
Day 2   ($\frac{1}{2}$ day) Section **10.6**
         16, 19, 21
         ($\frac{1}{2}$ day) Section **10.7**
         1–8

Advanced

8–12, 15–17, 21, 24, 27

**Problem-Set Notes
and Additional Answers**

■ See *Solution Manual* for answers to problems **1** and **2**.

Problem Set A, *continued*

4 Given: $\overline{PQ}$ and $\overline{PR}$ are tangent segments.
$\overset{\frown}{QR} = 163°$

Find: **a** $\angle P$ **17°**
b $\angle PQR$ **$81\frac{1}{2}°$**

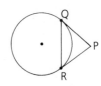

5 Given: A, B, and C are points of contact.
$\overset{\frown}{AB} = 145°$, $\angle Y = 48°$

Find: $\angle Z$ **97°**

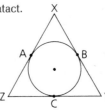

6 Given: $\overset{\frown}{BC} \cong \overset{\frown}{ED}$, AB = 8,
BC = 4, CD = 9

a Are $\overline{BE}$ and $\overline{CD}$ parallel? **Yes**
b Find BE. **6**
c Is △ACD scalene? **No**

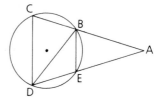

7 Given: $\overleftrightarrow{PY}$ and $\overleftrightarrow{QW}$ are tangents.
$\overset{\frown}{WZ} = 126°$, $\overset{\frown}{XY} = 40°$

Find: $\overset{\frown}{PQ}$ **137°**

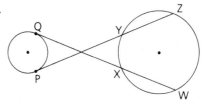

8 If △ABC is inscribed in a circle and $\overset{\frown}{AC} \cong \overset{\frown}{AB}$, tell whether each of the following must be true, could be true, or cannot be true.

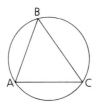

a $\overline{AB} \cong \overline{AC}$ **Must**
b $\overline{AC} \cong \overline{BC}$ **Could**
c $\overline{AB}$ and $\overline{AC}$ are equidistant from the center of the circle. **Must**

d $\angle B \cong \angle C$ **Must**
e $\angle BAC$ is a right angle. **Could**
f $\angle ABC$ is a right angle. **Cannot**

9 In the figure shown, find m$\angle P$. **42**

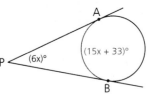

482 Chapter 10 Circles

10 If $\overline{AB}$ is a diameter of $\odot P$, $CB = 1.5$ m, and $CA = 2$ m, find the radius of $\odot P$. **1.25 m**

11 The radius of $\odot Z$ is 6 cm and $\overparen{WX} = 120°$.

Find: **a** AX **6 cm**
   b The perimeter of $\triangle WAX$
      $18 + 6\sqrt{3}$ cm

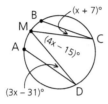

12 M is the midpoint of $\overparen{AB}$. Find $m\overparen{CD}$. **122**

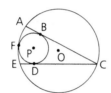

Problem Set B

13 A rectangle with dimensions 18 by 24 is inscribed in a circle. Find the radius of the circle. **15**

14 A square is inscribed in a circle with a radius of 10. Find the length of a side of the square. **$10\sqrt{2}$**

15 Quadrilateral ABCD is inscribed in circle O. $AB = 12$, $BC = 16$, $CD = 10$, and $\angle ABC$ is a right angle. Find the measure of $\overline{AD}$ in simplified radical form. **$10\sqrt{3}$**

16 Circles O and P are tangent at F. $\overline{AC}$ and $\overline{CE}$ are tangent to $\odot P$ at B and D. If $\overparen{DFB} = 223°$, find $\overparen{AE}$. **86°**

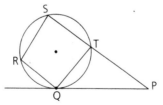

17 Given: $\angle S = 88°$, $\overparen{QT} = 104°$, $\overparen{ST} = 94°$, tangent $\overline{PQ}$

Find: **a** $\angle P$ **29°**
   b $\angle STQ$ **81°**

18 Given: $\overparen{BC} \cong \overparen{CD}$
Conclusion: $\triangle ABC \sim \triangle AED$

Problem-Set Notes and Additional Answers, continued

■ See *Solution Manual* for answer to problem **18**.

22

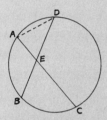

a $\angle D = \frac{1}{2}\widehat{AB}$

$\angle DAE = \frac{1}{2}\widehat{DC}$

$\angle CED = \angle D + \angle DAE$, by the Exterior Angle Theorem.

b See Sample Problem **1**.

23

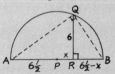

$6^2 = (6\frac{1}{2} + x)(6\frac{1}{2} - x)$

24 Let $m\widehat{HM} = x$. Then $m\angle R = 180 - x$ because it is a tangent-tangent $\angle$. However, $m\angle O = \frac{1}{2}x$ because it is an inscribed $\angle$. Since $\angle R \cong \angle O$, $180 - x = \frac{1}{2}x$. ∴ $x = 120$.

25

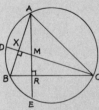

$\triangle AXM \sim \triangle CRM$, so $\angle A \cong \angle C$. Thus, $\widehat{BD} \cong \widehat{BE}$.

Problem Set B, *continued*

19 Given: $\overleftrightarrow{AC}$ is tangent at A. $\angle APR$ and $\angle AQR$ are right $\angle$s. R is the midpoint of $\widehat{AB}$.

Conclusion: $\overline{PR} \cong \overline{RQ}$ (Hint: Draw $\overline{AR}$.)

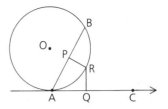

20 Given: $\triangle WXZ$ is isosceles, with $\overline{WX} \cong \overline{WZ}$. $\overline{WZ}$ is a diameter of $\odot O$.

Prove: Y is the midpoint of $\overline{XZ}$. (Hint: Draw $\overline{WY}$.)

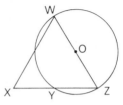

21 Given: $\overline{AC}$ is tangent to $\odot O$ at A.

Conclusion: $\triangle ADC \sim \triangle BDA$

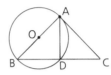

Problem Set C

22 Given: $\odot O$, with chords $\overline{AC}$ and $\overline{BD}$ intersecting at E

Prove: **a** $m\widehat{AB} + m\widehat{CD} = 2(m\angle CED)$
  b $AE \cdot EC = BE \cdot ED$

23 Given: $\overline{AB}$ is a diameter of $\odot P$. $QR = 6$, $AB = 13$, $\overline{QR} \perp \overline{AB}$

Find: RB. 4

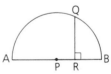

24 RHOM is a rhombus. $\overline{RH}$ and $\overline{RM}$ are tangents. Find $m\widehat{HM}$. 120

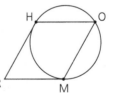

25 Given: $\triangle ABC$ is inscribed in $\odot P$. $\overline{AE}$ and $\overline{CD}$ are chords such that $\overline{AE} \perp \overline{BC}$ and $\overline{CD} \perp \overline{AB}$.

Prove: $\widehat{BD} \cong \widehat{BE}$

26 Two circles are internally tangent, and the center of the larger circle is on the smaller circle. Prove that any chord that has one endpoint at the point of tangency is bisected by the smaller circle.

27 Given: ⊙A is tangent to ⊙B at R.
$\overline{PT}$ is a common external tangent at P and T.
∠Q = 43°
Find: ∠S 47°

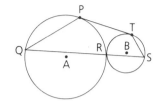

28 Given: $\overline{IT}$ is tangent to the circle.
$\overrightarrow{TS}$ bisects ∠ATM.
Prove: △SIT is isosceles.

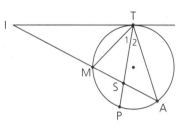

Students can work in small groups to solve problem **26**.

26

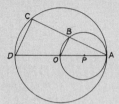

∠B and ∠C are rt. ∠s, so △AOB ~ △ADC.
$\frac{AO}{AD} = \frac{AB}{AC} = \frac{1}{2}$, so AB = BC.

27

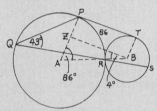

Follow common-tangent procedure. If ∠Q = 43°, then $\overset{\frown}{PR}$ = 86°, and thus, ∠PAR = 86°. Then ∠ABZ = 4°, and thus, $\overset{\frown}{TR}$ = 94°. ∴ ∠S = 47°.

Communicating Mathematics

Discuss an alternative solution to problem **27**.
In quadrilateral APTB, ∠A + ∠B = 180° = $\overset{\frown}{PR}$ + $\overset{\frown}{RT}$ = 2(∠Q + ∠S). ∴ ∠S = 47°.

28 Since ∠ITM and ∠A intercept $\overset{\frown}{TM}$, they are ≅. By the Exterior Angle Theorem, ∠MST = ∠2 + ∠A = ∠1 + ∠ITM = ∠ITS. ∴ △SIT is isos.

Class Planning

Time Schedule

Basic: $1\frac{1}{2}$ days
Average: $1\frac{1}{2}$ days
Advanced: 1 day

Resource References

Teacher's Resource Book
  Class Opener 10.7A, 10.7B
  Supposer Worksheet 12
  Using Manipulatives 18
Evaluation
  Tests and Quizzes
    Quiz 6, Series 1
    Quizzes 5–6, Series 2
    Quiz 5, Series 3

Class Opener

Given: ⊙O,
  m$\widehat{SM}$ = 80,
  m$\widehat{PS}$ = 90,
  m$\widehat{PT}$ = 70

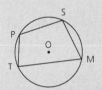

Find: Measures of $\widehat{TM}$, ∠P, ∠S,
  ∠M, ∠T, ∠S + ∠T, and
  ∠P + ∠M

m$\widehat{TM}$ = 120, so
m∠T = $\frac{90 + 80}{2}$ = 85,
m∠P = $\frac{80 + 120}{2}$ = 100,
m∠S = $\frac{70 + 120}{2}$ = 95, and
m∠M = $\frac{90 + 70}{2}$ = 80. Hence,
m∠S + m∠T = 180 and
m∠P + m∠M = 180.

10.7 INSCRIBED AND CIRCUMSCRIBED POLYGONS

Objectives

After studying this section, you will be able to
- Recognize inscribed and circumscribed polygons
- Apply the relationship between opposite angles of an inscribed quadrilateral
- Identify the characteristics of an inscribed parallelogram

Part One: Introduction

Inscribed and Circumscribed Polygons

Triangle ABC is ***inscribed in*** circle O.

| **Definition** | A polygon is ***inscribed in*** a circle if all of its vertices lie on the circle. |

Polygon PQRST is ***circumscribed about*** circle F.

| **Definition** | A polygon is ***circumscribed about*** a circle if each of its sides is tangent to the circle. |

We can also speak of a circle being circumscribed about a polygon or inscribed in a polygon.

The diagram shows that the statements "quadrilateral ABCD is inscribed in ⊙O" and "⊙O is circumscribed about quadrilateral ABCD" have the same meaning.

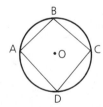

| **Definition** | The center of a circle circumscribed about a polygon is the ***circumcenter*** of the polygon. |

Vocabulary

circumcenter   incenter
circumscribed polygon   inscribed polygon

In the preceding diagram, O is the circumcenter of ABCD. Hexagon PQRSTU is circumscribed about circle F. Circle F is inscribed in hexagon PQRSTU.

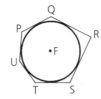

Definition   The center of a circle inscribed in a polygon is the *incenter* of the polygon.

F is the incenter of hexagon PQRSTU.

A Theorem About Inscribed Quadrilaterals

The following theorem can easily be proved by using the relationship between an inscribed angle and its intercepted arc.

Theorem 93   *If a quadrilateral is inscribed in a circle, its opposite angles are supplementary.*

Given: Quadrilateral ABCD is inscribed in circle O.

Prove: ∠A supp. ∠C, ∠B supp. ∠D

Proof: ∠A, ∠B, ∠C, and ∠D are inscribed angles, so

$m\angle A = \frac{1}{2}(m\overset{\frown}{BCD})$ and $m\angle C = \frac{1}{2}(m\overset{\frown}{BAD})$.

$m\angle A + m\angle C = \frac{1}{2}(m\overset{\frown}{BCD}) + \frac{1}{2}(m\overset{\frown}{BAD})$

$= \frac{1}{2}(m\overset{\frown}{BCD} + m\overset{\frown}{BAD})$

$= \frac{1}{2}(360) \quad (\overset{\frown}{BCD} \cup \overset{\frown}{BAD} = \text{whole} \odot)$

$= 180$

Thus, ∠A is supplementary to ∠C. Similarly, ∠B is supplementary to ∠D.

The Story of the Plain Old Parallelogram

Once there was a plain old parallelogram named Rex Tangle. Rex was always trying to fit in—into a circle, that is. One day when he awoke, he found that he had straightened out and was finally able to inscribe himself. What had the plain old parallelogram turned into?

    The following theorem shows the moral of our story.

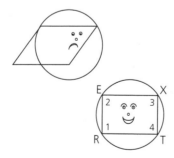

Lesson Notes

- Students should understand the relationship between *inscribed* and *circumscribed*. A given triangle can be an inscribed triangle because a circle can be circumscribed through its vertices (Theorem 128, p. 660). A circle can be inscribed in a given triangle, since a circle can be drawn tangent to its sides; the given triangle would be circumscribed about the circle (Theorem 129, p. 661).
- Not all polygons are inscribable or circumscribable. A test for circumscribing a circle about a quadrilateral is given in Theorem 93. Students can get a general feeling for inscribable polygons from the following argument: In a polygon, any three vertices determine the circumscribed circle (Theorem 128, page 660). If the remaining vertices lie on that circle, then the polygon can be inscribed in a circle.

Communicating Mathematics

Have students write an explanation, accompanied by diagrams, to help another student understand how the ideas of inscribed and circumscribed figures are interrelated. They should emphasize these ideas.

1 If the figure is inscribed, then the circle is circumscribed.
2 If the figure is circumscribed, then the circle is inscribed.

- The vertices of an inscribed polygon are called *concyclic*. This term is used in problems **22** and **23** on page 491.

Theorem 94   *If a parallelogram is inscribed in a circle, it must be a rectangle.*

Here are some of the conclusions that follow from Theorem 94.

If ABCD is an inscribed parallelogram, then
1 $\overline{BD}$ and $\overline{AC}$ are diameters
2 O is the center of the circle
3 $\overline{OA}$, $\overline{OB}$, $\overline{OC}$, and $\overline{OD}$ are radii
4 $(AB)^2 + (BC)^2 = (AC)^2$, and so forth

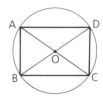

Part Two: Sample Problems

Problem 1   Given: Quadrilateral ABCD is inscribed in ⊙O.
Prove: ∠B ≅ ∠ADE

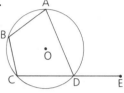

Proof

1 ABCD is inscribed in ⊙O.	1 Given
2 ∠B supp. ∠ADC	2 If a quadrilateral is inscribed in a ⊙, its opposite ∠s are supp.
3 ∠ADC supp. ∠ADE	3 Two ∠s forming a straight ∠ are supp.
4 ∠B ≅ ∠ADE	4 Two ∠s supp. to the same ∠ are ≅.

Problem 2   *Parallelogram ABCD is inscribed in a circle, and its diagonals intersect at E.*

a *Draw the figure.*   **c** *What is $\overline{BD}$?*
b *What is true about □ABCD?*   **d** *If AB = 5 and BC = 6, find AC.*

Solution   **a**

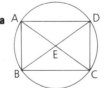

b A □ inscribed in a ⊙ must be a rectangle, so ABCD is a rectangle.

c ∠BCD is an inscribed right ∠, so $\frac{1}{2}(\overgroup{mBAD}) = 90$, making $\overgroup{BAD} = 180°$, a semicircle. Thus, $\overline{BD}$ is a diameter.

d Since △ABC is a right △, $(AB)^2 + (BC)^2 = (AC)^2$
$$5^2 + 6^2 = (AC)^2$$
$$\sqrt{61} = AC$$

Part Three: Problem Sets

Problem Set A

1 Given: ∠A = 104°, ∠B = 67°
 Find: ∠D and ∠C 113°; 76°

2 Given: $\overset{\frown}{PS}$ = 110°, $\overset{\frown}{PQ}$ = 100°
 Find: m∠R and m∠P 105; 75

3 Given: ∠A = 110°, $\overline{BC} \cong \overline{CD}$, ∠D = 95°
 Find: **a** ∠C 70°   **c** ∠B 85°
   b $\overset{\frown}{BC}$ 110°   **d** $\overset{\frown}{AB}$ 80°

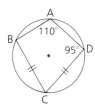

4 Given: ⊙O
 Prove: ∠Q ≅ ∠PST

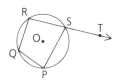

5 Can a parallelogram with a 100° angle be inscribed in a circle? No

6 Given: PQRST is a regular pentagon.
   ABCDEF is a regular hexagon.
 Find: **a** m$\overset{\frown}{PQ}$ 72   **d** m$\overset{\frown}{BD}$ 120
   b m$\overset{\frown}{RT}$ 144   **e** m$\overset{\frown}{DEA}$ 180
   c m$\overset{\frown}{AB}$ 60

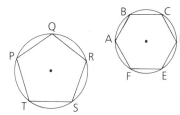

7 a If a rhombus is inscribed in a circle, what must be true about
   the rhombus? Square

 b If a trapezoid is inscribed in a circle, what must be true about
   the trapezoid? Isosceles

8 Prove: The bisector of an angle of an inscribed triangle also
 bisects the arc cut off by the opposite side.

Assignment Guide

Basic

Day 1 ($\frac{1}{2}$ day) Section **10.6**
    10–16, 18
    ($\frac{1}{2}$ day) Section **10.7**
    1–5
Day 2 6–12, 14, 17

Average

Day 1 ($\frac{1}{2}$ day) Section **10.6**
    16, 19, 21
    ($\frac{1}{2}$ day) Section **10.7**
    1–8
Day 2 11, 12, 14, 15, 17, 19,
    20

Advanced

5, 7, 9, 10, 12, 15–17, 19, 20, 23,
24

Problem-Set Notes
and Additional Answers

■ See *Solution Manual* for answers to problems **4** and **8**.

Problem Set B

9 Given: $\angle B = 115°$, $\overset{\frown}{AD} = 60°$, $\overline{BC} \parallel \overline{EF}$

Find: **a** $\angle ADC$ 65° **c** $\angle C$ 85°
b $\angle CDF$ 85° **d** $\angle A$ 95°

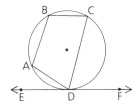

10 $PQ = 15$, $QR = 20$, $RS = 7$, and $\angle Q$ is a right angle. Find PS. 24

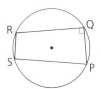

11 Trapezoid WXYZ is circumscribed about circle O. $\angle X$ and $\angle Y$ are right $\angle$s, $XW = 16$, and $YZ = 7$. Find the perimeter of WXYZ. 46

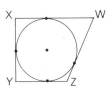

12 A circle is inscribed in a square with vertices $(-8, -3)$, $(-1, -3)$, $(-8, 4)$, and $(-1, 4)$.

a Find the coordinates of the center of the circle. $\left(-\frac{9}{2}, \frac{1}{2}\right)$
b Find the area of the circle. $\frac{49\pi}{4}$
c Find the radius of a circle circumscribed about the square. $\frac{7\sqrt{2}}{2}$

13 Prove: A trapezoid inscribed in a circle is isosceles.

14 Parallelogram RECT is inscribed in circle O. If $RE = 6$ and $EC = 8$, find the perimeter of $\triangle ECO$. 18

15 Given the figure shown, find $m\angle Q$.
116 or 80

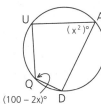

16 Given: $\odot O$; EFGH is a $\square$.
$\overset{\frown}{HG} = 120°$, $OJ = 6$

Find: The perimeter of EFGH
$12 + 12\sqrt{3}$

Problem-Set Notes and
Additional Answers, continued

■ See *Solution Manual* for answer
to problem **13**.

17 A quadrilateral can be inscribed in a circle only if a pair of opposite angles are supplementary. Which of the following quadrilaterals can be inscribed in a circle?

a

Yes

b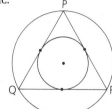

No

c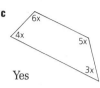

Yes

18 Prove: Any isosceles trapezoid can be inscribed in a circle. (Hint: See problem 17.)

19 Equilateral triangle PQR is inscribed in one circle and circumscribed about another circle. The circles are concentric.

a If the radius of the smaller circle is 10, find the radius of the larger circle. 20

b In general, for an equilateral triangle, what is the ratio of the radius of the inscribed circle to the radius of the circumscribed circle? $\frac{1}{2}$

20 ABCD is a kite, with $\overline{AB} \cong \overline{BC}$, $\overline{AD} \cong \overline{CD}$, and m∠B = 120. The radius of the circle is 3. Find the perimeter of ABCD. $6 + 6\sqrt{3}$

Problem Set C

21 Discuss the location of the center of a circle circumscribed about each of the following types of triangles.

a Right
  Midpoint of hypotenuse

b Acute
  Interior of △

c Obtuse
  Exterior of △

22 A set of points are *concyclic* if they all lie on the same circle. Prove that the vertices of any triangle are concyclic.

23 Are the vertices of each figure concyclic Always, Sometimes, or Never?

a A rectangle A

b A parallelogram S

c A rhombus S

d A nonisosceles trapezoid N

e An equilateral polygon S

f An equiangular polygon S

24 A right triangle has legs measuring 5 and 12. Find the ratio of the area of the inscribed circle to the area of the circumscribed circle. 16:169

Problem-Set Notes and Additional Answers, continued

■ See *Solution Manual* for answer to problem **18**.

Cooperative Learning

Have students work in small groups. Use the information given in problem **19**, but give each student in a group a different value for the radius of the smaller circle (for example, 5, 10, 12, 8, 3). Have each student find the radius of the larger circle. The group might want to collaborate to see if they can arrive at a general method that works for any radius. After students have compared answers, have them answer the question in part **b** of the problem.
They should all conclude that the ratio of the radii will always be 1:2.

22 The perpendicular bisectors of the sides of a triangle are concurrent, and this point will be the center of the circumscribed triangle.

24

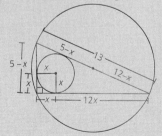

The radius of the inscribed circle can be found by a walk-around analysis to be 2. Therefore, its area is 4π. The radius of the circumscribed circle is $\frac{13}{2}$. Therefore, its area is $\frac{169\pi}{4}$. Thus, the ratio of the areas is $\frac{16}{169}$.

Problem-Set Notes and
Additional Answers, continued

25 Apply problem **21** of Section **10.4,** page 466.

26

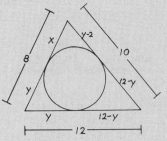

$$\begin{cases} x = y - 2 \\ x = 8 - y \end{cases}$$

27

ABC . . . N is an equiangular n-gon.
$m\angle A = \frac{1}{2}(\overarc{BC} + \overarc{CDN})$ and
$m\angle B = \frac{1}{2}(\overarc{CDN} + \overarc{AN})$.
Since $m\angle A = m\angle B$,
$\overarc{BC} \cong \overarc{AN}$ and $\overline{AN} \cong \overline{BC}$.
Similarly, each chord is $\cong$ to "its neighbor's neighbor." If n is odd, going around the circle twice will show that all the chords are $\cong$. If n is even, the figure can have two sets of $\cong$ chords.

Problem Set C, *continued*

25 Given: ⊙P is inscribed in trapezoid WXYZ.
∠W and ∠X are right ∠s.
The radius of ⊙P is 5.
YZ = 14

Find: The perimeter of WXYZ 48

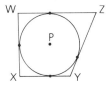

26 A circle is inscribed in a triangle with sides 8, 10, and 12. The point of tangency of the 8-unit side divides that side in the ratio $x : y$, where $x < y$. Find that ratio. $\frac{3}{5}$

27 Determine the conditions under which an equiangular polygon inscribed in a circle will be equilateral. Prove your conjecture. If the equiangular polygon has an odd number of sides, it will be equilateral.

MATHEMATICAL EXCURSION

TANGENT, SLOPE, AND LOOPS
The geometry of coasting upside down

You are on a roller coaster going 70 miles per hour. Suddenly you find yourself doing a complete loop. For an instant, you are upside-down. Why doesn't the car fall downwards from the track?

The path of a roller coaster is a series of arcs of constantly varying radii. The speed of the car at any instant is related to the slope of the tangent to the arc at that point.

A roller coaster somersault is made possible through what is called a clothoid loop, first explained by the eighteenth-century mathematician Leonhard Euler. Its name comes from that of Clotho, one of the three Fates from Greek mythology. Clotho was the spinner of the thread of human life. A clothoid loop would result from trying to draw a circle whose radius was constantly decreasing, up to a point. Because the radius near the top of the clothoid loop is relatively small, our roller coaster spends less time traveling through that part of the loop and leaves the loop before gravity can take over. The cars speed up coming out of the loop. This is similar to the "slingshot" effect observed when a comet approaches the sun, speeds up, and seems to shoot out on the other side.

10.8 THE POWER THEOREMS

Objective

After studying this section, you will be able to
- Apply the power theorems

Part One: Introduction

The following theorems involve products of the measures of segments.

Theorem 95   *If two chords of a circle intersect inside the circle, then the product of the measures of the segments of one chord is equal to the product of the measures of the segments of the other chord. (Chord-Chord Power Theorem)*

Given: Chords $\overline{VN}$ and $\overline{LS}$ intersect at point E inside circle O.

Prove: $EV \cdot EN = EL \cdot SE$

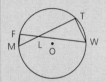

Theorem 95 was proved in Section 10.6, sample problem 1.

Theorem 96   *If a tangent segment and a secant segment are drawn from an external point to a circle, then the square of the measure of the tangent segment is equal to the product of the measures of the entire secant segment and its external part. (Tangent-Secant Power Theorem)*

Given: $\overline{PR}$ is a secant segment.
   $\overline{PT}$ is a tangent segment.

Prove: $(TP)^2 = (PR)(PQ)$

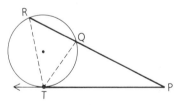

Proof: Similar triangles are formed by drawing $\overline{TQ}$ and $\overline{TR}$.
   $\angle PTQ \cong \angle R$ (why?) and $\angle P \cong \angle P$, so $\triangle PTR \sim \triangle PQT$.

   Thus, $\frac{TP}{PR} = \frac{PQ}{TP}$ and $(TP)^2 = (PQ)(PR)$.

Lesson Notes

- The term *power* indicates "product of factors." The factors are determined solely by P; for a fixed P, any line through P determines two distances to the circle. The product of the distances is constant.

Theorem 97 *If two secant segments are drawn from an external point to a circle, then the product of the measures of one secant segment and its external part is equal to the product of the measures of the other secant segment and its external part. (Secant-Secant Power Theorem)*

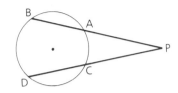

Given: Secant segments
$\overline{PB}$ and $\overline{PD}$

Prove: PB · PA = PD · PC

Part Two: Sample Problems

Problem 1   Find x, y, and z.

a   b   c

Solution   **a** By the Chord-Chord Power Theorem,

$$6 \cdot 2 = 3 \cdot x$$
$$4 = x$$

b By the Tangent-Secant Power Theorem,

$$y^2 = 2 \cdot 18$$
$$y = \pm 6 \text{ (Reject } -6.)$$
$$y = 6$$

c By the Secant-Secant Power Theorem,

$$4 \cdot (8 + 4) = 3 \cdot z$$
$$4 \cdot 12 = 3z$$
$$16 = z$$

Problem 2   Tangent segment PT measures 8 cm. The radius of the circle is 6 cm. Find the distance from P to the circle.

Solution   Draw a secant segment from P through the center R. PT = 8 and QR = RS = 6. Let x = PQ, the distance from P to the ⊙.

By the Tangent-Secant Power Theorem,

$$(PQ)(PS) = (PT)^2$$
$$x(x + 12) = 8^2$$
$$x^2 + 12x = 64$$
$$x^2 + 12x - 64 = 0$$
$$(x - 4)(x + 16) = 0$$
$$x - 4 = 0 \text{ or } x + 16 = 0$$
$$x = 4 \text{ or } x = -16$$

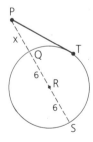

We reject the negative value, so PQ = 4 cm.

Part Three: Problem Sets

Problem Set A

1 Solve for x, y, and z.

a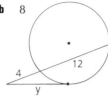

b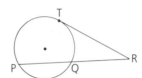

c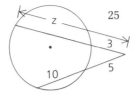

2 T is the midpoint of $\overline{QS}$, PT = 8, and QS = 40.

  a Find TR. 50

  b Find the diameter of $\odot$O. 58

3 a If TR = 10 and QR = 5, find PR. 20

  b If TR = 10 and QR = 4, find PQ. 21

  c If TR = 10 and PR = 50, find PQ. 48

4 a If AE = 6.4, AB = 8.9, and CE = 1.6, find ED. 10

  b If AE = 8, AB = 14, and ED = 16, find DC. 19

  c If CE = 2, ED = 18, and $\overline{AE} \cong \overline{EB}$, find AB. 12

5 Find the radius of $\odot$P. 3.5

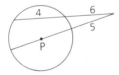

6 Given: AP = 3, PQ = 5, QB = 7, CP = 2, QD = 14

Find: PD and EQ 18; 4

Assignment Guide

Basic	
Day 1	1–5
Day 2	6–10, 13, 14

Average	
Day 1	1–8
Day 2	9–16

Advanced	
Day 1	3–10
Day 2	11–16, 18, 20

Problem Set A, *continued*

7 Given: TZ = 6, YZ = 4, SX = 3, WX = 1
Find: XT (Hint: Find SZ.) 6

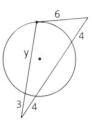

Problem Set B

8 a Find y. 9
   b Is the triangle acute, right, or obtuse?
   Acute

9 Given: AB = 7, CD = 5, ED = 2
Find: AE 1 or 6

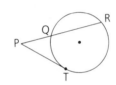

10 Given: PT = 3, QR = 8
Find: PQ 1

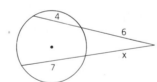

11 Solve for x. 5

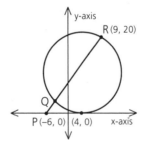

12 Find PQ. 4

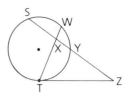

496 | Chapter 10 Circles

13 $\overline{AB}$ is a diameter of $\odot O$.
$\overline{CD}$ is tangent at D, CD = 6, and BC = 4.
Find the radius of the circle. **2.5**

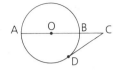

14 An arch supports a pipeline across a river 20 m wide. Midway, the suspending cable is 5 m long. Find the radius of the arch. **12.5 m**

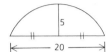

15 The diameter of the earth is approximately 8000 mi. Heavenly Helen, in a spaceship 100 mi above the earth, sights Earthy Ernest coming over the horizon. Approximately how far apart are Helen and Ernest? **900 mi**

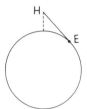

16 Solve for x. **5**

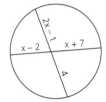

Problem Set C

17 Given concentric circles as shown, find DE and DC. **26; 39**

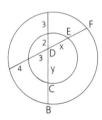

18 The radius of each circle is 3. Triangle WXY is equilateral.
  a Find WY. $12 + 6\sqrt{3}$
  b Find the ratio of the perimeters of $\triangle ABC$, $\triangle PQR$, and $\triangle WXY$. $1:2:2 + \sqrt{3}$

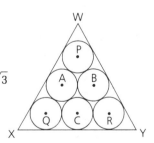

Communicating Mathematics

This can lead to an interesting classroom discussion.

Alicia and Burton were trying to decide about how far a person standing on a beach can see to a point on the horizon. Standing at sea level, they spotted a ship just appearing over the horizon. Burt's eye level is 5 feet above sea level, and the Earth's diameter is about 8000 miles. How far away is the ship?

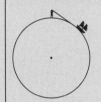

$x^2 = 5(5 + 8000 \cdot 5280)$
$x^2 = 211,200,025$
$\ x \approx 14,532.7226$ ft
$\quad \approx 2.75$ mi

Cooperative Learning

Students can work in small groups to solve problem **17**.

Problem-Set Notes and Additional Answers

17 $\begin{cases} 3x = 2y \\ 7(x + 4) = 5(y + 3) \end{cases}$

18

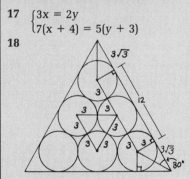

$P_{\triangle ABC} = 18;\ P_{\triangle PQR} = 36;$
$P_{\triangle WXY} = 36 + 18\sqrt{3}$

19

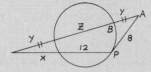

a $\begin{cases} x(x + 12) = y(y + z) \\ y(y + z) = 64 \end{cases}$

b Since $\overleftrightarrow{AB}$ is a secant, $AB < AP$, so $y < 8$. By the Triangle Inequality Theorem, $2y + z < 24$. From part **a**, $y(y + z) = 64$. By solving simultaneously, $y^2 - 24y + 64 < 0$; $12 - 4\sqrt{5} < y < 12 + 4\sqrt{5}$. $\therefore 12 - 4\sqrt{5} < y < 8$.

20

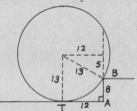

Problem Set C, *continued*

19 a Find x. 4

b What restrictions must be placed on y in this problem? $12 - 4\sqrt{5} < y < 8$

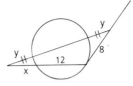

20 Tangent $\overline{AT}$ measures 12, $AB = 8$, and $\overline{AT} \perp \overline{AB}$.

a Find the diameter of the circle. 26

b How far is the circle from point A? $\sqrt{313} - 13$

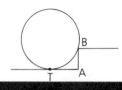

FROM ASTEROIDS TO DUST
Mathematics in space research

How does someone choose a life's profession? For physicist A.A. Jackson, it was a painting on the cover of a magazine. "It was 16 years before the first moon landing." he recalls. "*Collier's* magazine published a fantastic futuristic painting, the artist's conception of a lunar landing. It combined scientific precision with the romance and drama of space travel. From that moment I knew exactly what I wanted to do."

Pursuing his goal, Jackson majored in mathematics at North Texas State University, received a master's degree in physics from the same school, then went on to the University of Texas at Austin, where he earned his doctorate in relativistic physics. Today he is principal scientist in the solar-system exploration division of Lockheed Engineering in Houston.

Explaining the relevance of geometry to his work, Jackson refers to conic sections, the curves that result when a plane intersects a cone. "Planets move in ellipses around the sun. Satellites orbit the earth in ellipses. Some comets move in parabolic orbits. In my current research, I'm studying the motion of dust particles as they come off comets and asteroids. They move along conic sections."

Jackson's field is evolving constantly in unexpected directions. For example, recent studies

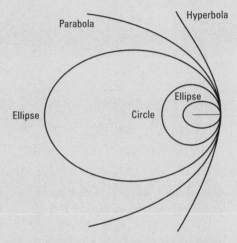

of the way three bodies interact in a plane have turned up connections with the geometry of fractals. "Every time you look at something old," says Jackson, "you see something new."

One of the things that Jackson discovered as a teenager was science fiction. He reads it avidly to this day. The best science fiction, he says, brings together provocative ideas and "super science" in a landscape that feels not fantastic, but real—"lived in." He recommends the works of Robert Heinlein especially *Starman Jones.*

10.9 CIRCUMFERENCE AND ARC LENGTH

Objectives

After studying this section, you will be able to
- Determine the circumference of a circle
- Determine the length of an arc

Part One: Introduction

Circumference

You should already know the meaning of *circumference*.

Definition   The **circumference** of a circle is its perimeter.

The formula for the circumference C of a circle of diameter d is based on the fact that, regardless of a circle's size, the ratio of its circumference to its diameter always has the same value. This value is given the special symbol π (the Greek letter pi). Its approximate value is 3.14159265.

Postulate $C = \pi d$

Example   *Find, to the nearest hundredth, the circumference of a circle whose radius is 5.37.*

The diameter is twice the radius, so $d = 2(5.37) = 10.74$.

$C = \pi d$
$= \pi(10.74) = 10.74\pi \approx 33.74$

When you are asked to find a circumference, leave the answer in terms of π unless you are asked to approximate the answer. To find an approximation, use a calculator.

Length of an Arc

Arc length is a linear measurement similar to the length of a line segment. Arc lengths are therefore expressed in terms of such units as feet, meters, and centimeters. The length of an arc depends both on the arc's measure and on the circumference of its circle.

Vocabulary
circumference

Class Planning

Time Schedule
All levels: 1 day

Resource References
Teacher's Resource Book
  Class Opener 10.9A, 10.9B
  Using Manipulatives 19
Evaluation
  Tests and Quizzes
    Quizzes 7–10, Series 1
    Quiz 8, Series 2

Class Opener

Given: AB = 20, PQ = 2
Find: The diameter of the circle

Extend $\overline{PQ}$ through Q to W.
$AQ^2 = PQ \cdot QW$
$100 = 2 \cdot QW$
$QW = 50$
$PW = 52$

Lesson Notes

- Most of the homework problems call for the answer to be expressed in terms of π. A few applications call for numerical approximations.

Communicating Mathematics

Discuss the value of the ratio of the circumference of a circle to its diameter.

Example   *Find the length of a 40° arc of a circle with an 18-cm radius.*

The circumference is 36π, and the 40° arc is $\frac{40}{360}$, or $\frac{1}{9}$, of the circle.

Length of $\widehat{AB} = \frac{1}{9}$(circumference)

$= \frac{1}{9}(36\pi)$

$= 4\pi$

Lesson Notes, continued

■ Some students find it easier to set up a proportion to solve these problems.

$$\frac{\text{arc length}}{\text{circumference}} = \frac{\text{arc measure}}{360}$$

Theorem 98   ***The length of an arc is equal to the circumference of its circle times the fractional part of the circle determined by the arc.***

$$\textbf{Length of } \widehat{PQ} = \left(\frac{m\widehat{PQ}}{360}\right)\pi d$$

where d is the diameter and $\widehat{PQ}$ is measured in degrees.

Part Two: Sample Problems

Problem 1   *Find the radius of a circle whose circumference is 50π.*

Solution
$$C = \pi d$$
$$50\pi = \pi d$$
$$50 = d$$
$$25 = r$$

Problem 2   *Find the length of each arc of a circle with a 12-cm radius.*

  a A 30° arc   **b** A 105° arc

Solution   **a** Length of arc $= \frac{30}{360}(24\pi)$ **b** Length of arc $= \frac{105}{360}(24\pi)$

$= 2\pi$ cm $= 7\pi$ cm

Problem 3   *The diameter of a bicycle wheel (including the tire) is 70 cm.*

  a *How far will the bicycle travel if the wheel rotates 1000 times? (Approximate the answer in meters.)*

  b *How many revolutions will the wheel make if the bicycle travels 15 m? (Approximate to the nearest tenth of a revolution.)*

Solution   The distance covered during one revolution is equal to the circumference of the wheel. Thus, the bicycle travels 70π, or about 220, centimeters per revolution.

  a Distance = (number of rev.)(distance per rev.)
$$\approx 1000(220)$$
$$\approx 220{,}000$$

The bicycle will travel about 220,000 cm, or 2200 m.

b The bicycle travels approximately 2.2 m per revolution. Let x be the number of revolutions.

Distance = (number of rev.)(distance per rev.)
$$15 \approx x(2.2)$$
$$6.8 \approx x$$

The wheel will revolve approximately 6.8 times.

Part Three: Problem Sets

Problem Set A

1 Find the circumference of the circle. Then approximate the circumference to the nearest hundredth.

   a A circle whose diameter is 21 mm 21π mm; ≈ 65.97 mm

   b A circle whose radius is 6 mm 12π mm; ≈ 37.70 mm

2 Find, to the nearest hundredth, the radius of a circle whose circumference is

   a 56π 28.00   **b** 314 ≈ 49.97 **c** 17π 8.50   **d** 88 ≈ 14.01

3 Find the length of each arc of a circle with a radius of 10.

   a A 72° arc 4π **b** A 90° arc 5π **c** A 60° arc $\frac{10\pi}{3}$ **d** A semicircle 10π

4 A bicycle has wheels 30 cm in diameter. Find, to the nearest tenth of a centimeter, the distance that the bicycle moves forward during

   a 1 revolution ≈ 94.2 cm   **b** 10 revolutions ≈ 942.5 cm   **c** 1000 revolutions $\approx 94{,}247.8$ cm

5 Find the complete perimeter of each figure. Leave your answers in terms of π and whole numbers.

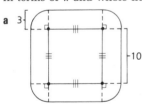
$40 + 6\pi$

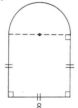
$24 + 4\pi$

$12 + 3\pi$

$6 + 7\pi$

6 **a** Find the length of $\overset{\frown}{AB}$. 3π

   b Find the perimeter of sector AOB. (The shaded region is a sector.) $12 + 3\pi$

Section 10.9   Circumference and Arc Length | **501**

Assignment Guide

Basic
1, 2a,b, 3a,b, 4b, 5, 6, 8, 11, 13
Average
1, 2a,b, 3a,b, 4b, 5, 6, 8, 10, 13, 15
Advanced
6, 8–11, 13–17

Problem Set A, *continued*

7 Find, to the nearest meter, the length of fencing needed to surround the racetrack. 138 m

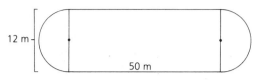
12 m
50 m

8 The radius of ⊙O is 10 mm and the length of $\overparen{AB}$ is 4π mm.

  a Find the circumference of ⊙O. 20π mm
  b Find m$\overparen{AB}$. 72

Problem Set B

9 Given arcs mounted on equilateral triangles as shown, find the length of each arc. In each case $\overline{OA}$ is a radius of $\overparen{AB}$.

a $4\pi\sqrt{3}$

b 6π

c 9π

10 There are 100 turns of thread on a spool with a diameter of 4 cm. Find the length of the thread to the nearest centimeter. $\approx$ 1257 cm

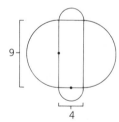

11 Awful Kanaufil plans to ride his cycle on a single-loop track. There is 100 m of straight track before the loop and 20 m after. The loop has a radius of 15 m. To the nearest meter, what is the total length of the track he must ride? $\approx$ 214 m

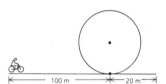
100 m   20 m

12 Find the outer perimeter of the figure, which is composed of semicircles mounted on the sides of a rectangle. 13π

9
4

13 Sandy skated on the rink shown. To the nearest tenth of a meter, how far did she travel going once around in the outside lane? In the inside lane? ≈96.5 m; ≈71.4 m

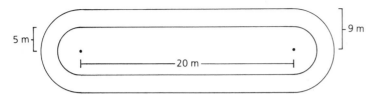

14 A belt wrapped tightly around circle O forms a right angle at P, a point outside the circle. Find the length of the belt if circle O has a radius of 6. **12 + 9π**

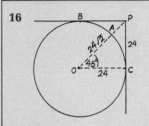

15 Find the distance traveled in one back-and-forth swing by the weight of a 12-in. pendulum that swings through a 75° angle. **10π in.**

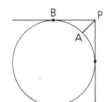

Problem Set C

16 A circular garbage can is wedged into a rectangular corner. The can has a diameter of 48 cm.

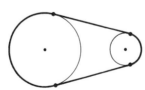

   a Find the distance from the corner point to the can (PA). **(24√2 − 24) cm**

   b Find the distance from the corner point to the point of contact of the can with the wall (PB). **24 cm**

17 Two pulleys are connected by a belt. The radii of the pulleys are 3 cm and 15 cm, and the distance between their centers is 24 cm. Find the total length of belt needed to connect the pulleys. **(24√3 + 22π) cm**

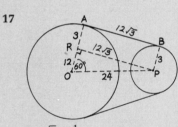

Problem-Set Notes and Additional Answers

16

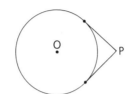

17

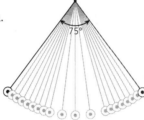

$$24\sqrt{3} + \tfrac{1}{3}(2 \cdot \pi \cdot 3) + \tfrac{2}{3}(2 \cdot \pi \cdot 15)$$
$$= 24\sqrt{3} + 22\pi$$

Cooperative Learning

Have students work in small groups to solve problem **17**.

10 CHAPTER SUMMARY

CONCEPTS AND PROCEDURES

After studying this chapter, you should be able to
- Identify the characteristics of circles, chords, and diameters (10.1)
- Recognize special relationships between radii and chords (10.1)
- Apply the relationship between congruent chords of a circle (10.2)
- Identify different types of arcs, determine the measure of an arc, and recognize congruent arcs (10.3)
- Relate congruent arcs, chords, and central angles (10.3)
- Identify secant and tangent lines and segments (10.4)
- Distinguish between two types of tangent circles (10.4)
- Recognize common internal and common external tangents (10.4)
- Determine the measures of central, inscribed, tangent-chord, chord-chord, secant-secant, secant-tangent, and tangent-tangent angles (10.5)
- Recognize congruent inscribed and tangent-chord angles (10.6)
- Determine the measure of an angle inscribed in a semicircle (10.6)
- Apply the relationship between the measures of a tangent-tangent angle and its minor arc (10.6)
- Recognize inscribed and circumscribed polygons (10.7)
- Apply the relationship between opposite angles of an inscribed quadrilateral (10.7)
- Identify the characteristics of an inscribed parallelogram (10.7)
- Apply the three power theorems (10.8)
- Determine circle circumference and arc length (10.9)

VOCABULARY

arc (10.3)
center (10.1)
central angle (10.3)
chord (10.1)
chord-chord angle (10.5)
circle (10.1)
circumcenter (10.7)
circumference (10.9)
circumscribed polygon (10.7)
common external tangent (10.4)
common internal tangent (10.4)
common tangent (10.4)
concentric (10.1)

diameter (10.1)
exterior (10.1)
externally tangent (10.4)
external part (10.4)
incenter (10.7)
inscribed angle (10.5)
inscribed polygon (10.7)
interior (10.1)
internally tangent (10.4)
line of centers (10.4)
major arc (10.3)
minor arc (10.3)
point of contact (10.4)

point of tangency (10.4)
radius (10.1)
secant (10.4)
secant-secant angle (10.5)
secant segment (10.4)
secant-tangent angle (10.5)
semicircle (10.3)
tangent (10.4)
tangent-chord angle (10.5)
tangent circles (10.4)
tangent segment (10.4)
tangent-tangent angle (10.5)

REVIEW PROBLEMS

Problem Set A

1 Find x in each case.

a

86°

x

94

b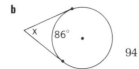

x

86°

94

c

86°

x

43

2 If $\widehat{AB} = 98°$ and $\widehat{CD} = 34°$, find x and y.
66; 32

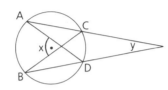

A

C

x

y

B

D

3 a Find BD. 16

A   B

6

4   E

D   8

C

b Find PT. 8

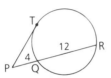

T

4

12

R

P   Q

c Find WX. 4

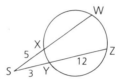

W

X

5

Z

S   3 Y   12

4 Find the radius of each circle.

a

←24→

5

13

b

15

8

8.5

c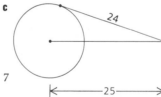

24

7

←25→

5 The circles shown are concentric at O.
$\overline{PZ}$ and $\overline{PY}$ are tangent to the inner circle
at W and X. If $\widehat{YZ} = 110°$, find the measure of $\widehat{WX}$. 125

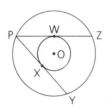

P   W   Z

•O

X

Y

Chapter 10 Review

Class Planning

Time Schedule
All levels: 1 day

Resource References
Evaluation
   Tests and Quizzes
      Test 10, Series 1, 2, 3

Assignment Guide

Basic
1–8, 10, 12, 14, 15

Average
1–8, 10, 12, 14, 15, 17, 25

Advanced
7, 8, 14–18, 20, 23, 26, 31

To integrate constructions,
assign
Basic   1, 5, 11, page 676
Average   15, page 676
         9, 12, pages 680–681
         21, page 684
Advanced   18, 19, page 681
         27, 28, 30–32, pages
         684–685

Review Problem Set A, *continued*

6 Given: △ABC is isosceles, with base $\overline{AB}$.
∠DAC = 70°, $\widehat{BC}$ = 160°
Find: $\widehat{AB}$ and $\widehat{AD}$ 40°; 20°

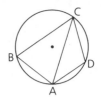

7 XOY is a sector of ⊙O.
Radius OY = 6 cm and central ∠ XOY = 45°.
Find: **a** The length of $\widehat{XY}$ $\frac{3\pi}{2}$
  b The perimeter of sector XOY 12 + 1.5π

8 Circles A, B, and C are tangent as shown.
AB = 7, BC = 10, and CA = 11.
  a Find the radius of ⊙A. 4
  b Which circle is the largest? C

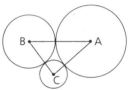

9 Given: ⊙O, $\overrightarrow{OM}$ ⊥ $\overline{AB}$
Prove: $\overrightarrow{OM}$ bisects ∠AOB.

10 Given: ⊙O, $\overline{OP}$ ⊥ $\overline{WX}$, $\overline{OQ}$ ⊥ $\overline{YZ}$;
△OPQ is isosceles, with base $\overline{PQ}$.
Conclusion: $\widehat{WX}$ ≅ $\widehat{YZ}$

11 Given: $\overline{ZX}$ and $\overline{ZY}$ are tangent at X and Y.
Prove: $\overline{WZ}$ bisects $\widehat{XY}$.

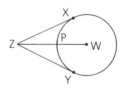

12 A parallelogram with sides 4 and 7.5 is inscribed in a circle.
Find the radius of the circle. 4.25

13 Given: TP = 8, PQ = 6
Find: RQ $4\frac{2}{3}$

14 Given: ⊚ O and P are externally tangent.
OA = 8, PB = 2

Find: The length of common external
tangent $\widehat{AB}$ 8

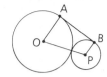

Problem Set B

15 If a point is chosen at random on ⊙P,
what is the probability that it lies on

a $\widehat{BA}$ $\frac{2}{9}$ **b** $\widehat{TUB}$ $\frac{7}{12}$

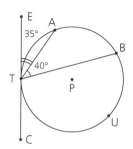

16 Jim knows that ⊙O is inscribed in isos-
celes △ ABC. He forgets which sides of
△ABC are congruent but remembers that
AB = 14 and the perimeter is 38.

a Find XC. 5

b What are the three possible lengths
of $\overline{BX}$? 7, 9, or 5

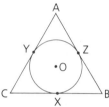

17 A quadrilateral is inscribed in a circle. Its vertices divide the
circle into four arcs in the ratio 1:2:5:4. Find the angles of the
quadrilateral. 45°, 105°, 135°, 75°

18 Given: $\widehat{AB}$ = 30°, $\widehat{BC}$ = 40°, $\widehat{CD}$ = 50°

Find: **a** ∠X 140°
      b ∠Y 30°
      c ∠Z 10°

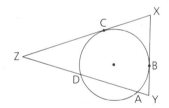

19 $\overline{TP}$ is a tangent segment, TP = 15, and
PQ = 5. Find the radius of ⊙O. 20

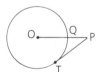

20 Given: m$\widehat{AD}$ + m$\widehat{BC}$ = 200,
       m∠P = 30
Find: m$\widehat{AB}$ and m$\widehat{CD}$ 110; 50

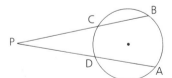

Problem-Set Notes and
Additional Answers, continued

■ See *Solution Manual* for answer
to problem **21.**

22a $\overline{WZ} \cong \overline{YZ}$ by Two-Tangent
Theorem. WXYZ is a
rhombus because a ▱ with
consecutive sides ≅ is a
rhombus.

26

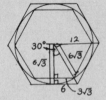

30°-60°-90° analysis for the
hexagons and triangles;
45°-45°-90° analysis for the
squares

Review Problem Set B, *continued*

21 Given: ⊙F, $\overline{EG} \perp \overline{AB}$,
   $\overline{EC} \cong \overline{ED}$

   Prove: $\overline{AD}$ and $\overline{BC}$ are equidistant
   from F.

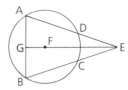

22 WXYZ is a parallelogram.
   $\overline{WZ}$ and $\overline{YZ}$ are tangent segments.

   a Show that WXYZ is a rhombus.

   b Find m∠Z. 60

   c If WY = 15, find the perimeter of
   WXYZ. 60

   d If WY = 15, find XZ. $15\sqrt{3}$

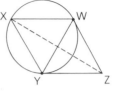

23 Find x and y. 6; 18

24 Find the area of a circle whose diameter joins the points $(10, -7)$
   and $(-2, 10)$. $\frac{433\pi}{4}$

25 Find, to the nearest centimeter, the circumference of a circle in
   which an 80-cm chord is 9 cm from the center. ≈258 cm

Problem Set C

26 Each circle below is inscribed in a regular polygon and is circumscribed about another regular polygon.

  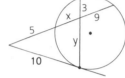

   a If the length of a side of each outer polygon is 12, find the
   length of a side of each inner polygon. $6\sqrt{3}$; $6\sqrt{2}$; 6

   b In each case, find the ratio of the sides of the smaller polygon
   to the sides of the larger polygon. $\frac{\sqrt{3}}{2}$, $\frac{\sqrt{2}}{2}$, $\frac{1}{2}$

508 | Chapter 10   Circles

27 Given: $\overline{WZ}$ is a diameter of the ⊙.

Show: $m\angle P = \dfrac{m\widehat{WX} + m\widehat{YZ}}{2}$

28 Given: ⑨ P and Q are internally tangent at T.

Prove: AC:CT = BD:DT

29 Given: $\widehat{AQ} \cong \widehat{RB}$; $\overline{PR}$ divides major and minor arcs AB in ratios of $\widehat{AQ}:\widehat{QB} = 4:3$ and $\widehat{AR}:\widehat{RB} = 7:5$.

Find: $\angle APQ:\angle BPQ$ 8:5

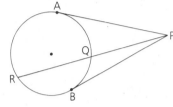

30 Three of the segments $\overline{PA}$, $\overline{PB}$, $\overline{PC}$, $\overline{PD}$, and $\overline{PE}$ are secant segments to circle O; the remaining two are tangent segments to circle O. If two of the segments are selected at random, what is the probability that a secant-tangent angle is formed? $\frac{3}{5}$

31 A flatbed truck is hauling a cylindrical container with a diameter of 6 ft. Find, to the nearest hundredth, the length of cable needed to hold down the container. ≈16.68 ft

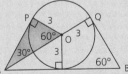

Problem-Set Notes and Additional Answers, continued

27 $\angle P = \dfrac{\widehat{WZ} - \widehat{XY}}{2}$

$= \dfrac{\widehat{WX} + \widehat{XY} + \widehat{ZY} - \widehat{XY}}{2}$

$= \dfrac{\widehat{WX} + \widehat{ZY}}{2}$

28

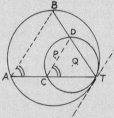

$m\widehat{DT} = m\widehat{BT}$ (draw tangent-chord ∠ if needed).
∠BAT ≅ ∠DCT, so $\overline{AB} \parallel \overline{CD}$ by corr. ∠s ≅ ⟹ ∥ lines.
AC:CT = BD:DT

29

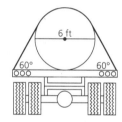

∠APR $= \frac{1}{2}(28x - 20x) = 4x$

∠BPQ $= \frac{1}{2}(20x - 15x) = \frac{5}{2}x$

30 Tangents are a, e. Secants are b, c, d.

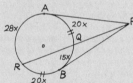

$\frac{6}{10} = \frac{3}{5}$

31

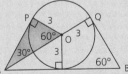

By a 30°-60°-90° analysis, AP $= 3\sqrt{3} =$ BQ and $\widehat{PQ} = \frac{120}{360} \cdot 2\pi \cdot 3 = 2\pi$.
∴ Total length of cable $= 2\pi + 6\sqrt{3} \approx 16.68$ ft.

AREA IS ONE of the most frequently used concepts of geometry. Many of the problems in the chapter use the Pythagorean Theorem and properties of similar figures, and apply them in new settings. Problems in the later sections prepare students for volumes (Chapter 12).

CHAPTER

11 AREA

This painting by Sonia Delaunay incorporates the areas of geometric shapes.

UNDERSTANDING AREA

Objectives
After studying this section, you will be able to
- Understand the concept of area
- Find the areas of rectangles and squares
- Use the basic properties of area

Part One: Introduction

The Concept of Area

When we measure lengths of line segments, we use such standard units as meters, yards, miles, centimeters, and kilometers. These are often called *linear units* because they are measures of length. The standard units of *area* are *square units*, such as square meters, square yards, and square miles. A square meter, for example, is the space enclosed by a square whose sides are each one meter in length.

One Linear Unit = One Square Unit

Definition The area of a closed region is the number of square units of space within the boundary of the region.

We can estimate the area of a region by determining the approximate number of square units it would take to fill the region.

 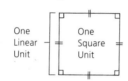

Estimated Area = 10 sq units

Estimated Area = 18 sq units

Estimated Area = 19 sq units

Counting squares, however, is neither the easiest nor the best way to find the area of a region. We will develop formulas for computing the areas of regions bounded by the common geometrical figures. Such regions are usually named by their boundaries, as when we speak of "the area of a rectangle."

Class Planning

Time Schedule
All levels: 1 day

Resource References
Teacher's Resource Book
  Class Opener 11.1A, 11.1B

Class Opener

Given ⊙O and rectangle COED. If AE = 3 and EC = 7, find the length of a radius of the circle. Since COED is a rectangle, EC = OD. OD is a radius, therefore the length of a radius of the circle is 7.

Vocabulary
area
linear unit
square unit

The Areas of Rectangles and Squares

In the figures to the right, there are two ways to find the areas:

1 The numbers of square units can be counted individually.
2 The areas can be computed by multiplying the number of columns (the measure of the base) by the number of rows (the height).

The second method suggests the following formula, which may be used to compute areas even when the lengths are fractions or irrational numbers.

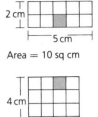

Area = 10 sq cm

Area = 16 sq cm

Postulate　　*The area of a rectangle is equal to the product of the base and the height for that base.*

$$A_{rect} = bh$$

where b is the length of the base and h is the height.

In a square, the base and the height are equal, so the following formula is used.

Theorem 99　　*The area of a square is equal to the square of a side.*

$$A_{sq} = s^2$$

where s is the length of a side.

Basic Properties of Area

We make three basic assumptions about area:

Postulate　　*Every closed region has an area.*

Postulate　　*If two closed figures are congruent, then their areas are equal.*

If ABCDEF $\cong$ PQRSTU, then the area of region I = the area of region II.

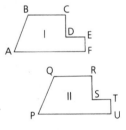

Postulate　　*If two closed regions intersect only along a common boundary, then the area of their union is equal to the sum of their individual areas.*

Part Two: Sample Problems

Problem 1 *Find the area of the rectangle.*

Solution $A_{rect} = bh$
We need to find base BZ.
$\triangle$BZY is a right $\triangle$ of the (5, 12, 13)
family, so BZ = 12.

$A_{rect} = 12(5) = 60$ sq cm

Problem 2 *Given that the area of a rectangle is 20 sq dm and the altitude is 5 dm, find the base.*

Solution Let x be the number of decimeters in the base.

$A_{rect} = bh$
$20 = x(5)$
$4 = x$
Base = 4 dm

Problem 3 *Find the area of the shaded region.*

Solution There are two methods of finding the area. One uses subtraction, and the other uses addition.

Method One:

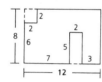

Area of large rectangle = $12 \cdot 8 = 96$
Area of square = $2^2 = 4$
Area of small rectangle = $2 \cdot 5 = 10$
Shaded area = $96 - 4 - 10 = 82$

Method Two ("*Divide and Conquer*"):

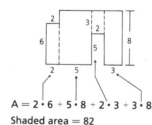

$A = 2 \cdot 6 + 5 \cdot 8 + 2 \cdot 3 + 3 \cdot 8$
Shaded area = 82

Part Three: Problem Sets

Problem Set A

1 Find the area of each figure below. (Assume right angles.)

a

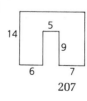

207

b

78

c

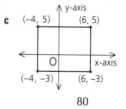

80

Section 11.1   Understanding Area   **513**

Problem Set A, *continued*

2 Find the area of a rectangle whose length and width are 12.5 cm and 6 cm respectively. 75 sq cm

3 Find the area of each rectangle.

a

120

b
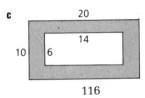
84

4 The area of a rectangle is 48 sq mm, and the altitude is 6 mm.
  a Find the length of the base. 8 mm
  b Find the length of a diagonal of the rectangle. 10 mm

5 a Find the area of a square whose side is 12. 144
  b Find the area of a square whose diagonal is 10. 50
  c Find the side of a square whose area is 49. 7
  d Find the perimeter of a square whose area is 81. 36
  e Find the area of a square whose perimeter is 36. 81

6 Find the area of each shaded region. (Assume right angles.)

a

104

b

102

c

116

7 The diagonal of a rectangle is $\sqrt{29}$, and the rectangle's base is 2.
  a Find the area of the rectangle. 10
  b Find its semiperimeter. 7

Problem Set B

8 Each rectangular garden below has an area of 100.

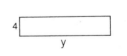

  a Find the missing dimension of each. 5; 10; 25; 12.5
  b What length of fencing is needed to surround each? 50; 40; 58; 41
  c Which figure has the shortest perimeter? The square
  d What do you think must be true about a rectangle that encloses the maximum possible area with the shortest possible perimeter? The rectangle must be a square.

9 A cross section of a steel I-beam is shown. Assume right angles and symmetry from appearances. Find the area of the cross section. **256 sq cm**

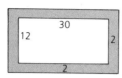

10 A rectangular picture measures 12 cm by 30 cm. It is mounted in a frame 2 cm wide. Find the area of the frame. **184 sq cm**

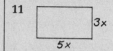

11 The sides of a rectangle are in a ratio 3:5, and the rectangle's area is 135 sq m. Find the dimensions of the rectangle. **9 m by 15 m**

12 The area of square ABCD is 64 square units. MNOP is formed by joining the midpoints of the sides of ABCD. Find the area and the perimeter of MNOP.
$A = 32$; $P = 16\sqrt{2}$

13 If the area of rectangle RCTN is six times the area of rectangle AECT, find the coordinates of A. **(18, 8)**

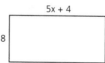

14 The dimensions of a rectangle of area 72 are whole numbers. List the dimensions of all such rectangles. If two of these rectangles are chosen at random, what is the probability that each has a perimeter greater than 40?
1 by 72, 2 by 36, 3 by 24, 4 by 18, 6 by 12, 8 by 9; $\frac{2}{5}$

15 The area of the rectangle is between 84 sq mm and 124 sq mm. What restrictions does this place on x? **1.3 < x < 2.3**

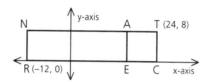

Problem Set C

16 A rectangle is formed by two diagonals of a regular hexagon as shown. Each side of the hexagon is 12. Find the area of the rectangle to the nearest tenth. **≈249.4**

17 A flag has dimensions 65 by 39. Each short stripe has a length of 39. What fractional part of the flag is red? $\frac{27}{65}$

Problem-Set Notes and Additional Answers

10 $A = 34 \cdot 16 - 30 \cdot 12$
$= 544 - 360$
$= 184$

11
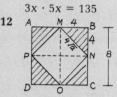
$3x \cdot 5x = 135$

12
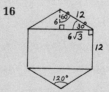
$A_{MNOP} = (4\sqrt{2})^2 = 32$
$P = 4(4\sqrt{2}) = 16\sqrt{2}$

14 Prob. $= \frac{6}{15} = \frac{2}{5}$

16

$A_{rect} = 12 \cdot 12\sqrt{3}$
$= 144\sqrt{3}$
≈ 249.4

17 Width of each strip $= \frac{39}{13} = 3$
$A_{red} = 4$ short reds + 3 long reds
$= 4(39 \cdot 3) + 3(65 \cdot 3)$
$= 468 + 585 = 1053$
$A_{flag} = 39 \cdot 65 = 2535$
$\therefore$ Fraction $= \frac{1053}{2535} = \frac{27}{65}$

AREAS OF PARALLELOGRAMS AND TRIANGLES

Class Planning

Time Schedule
All levels: 2 days

Resource References
Teacher's Resource Book
    Class Opener 11.2A, 11.2B
    Additional Practice
    Worksheet 21

Class Opener

□PQRT is inscribed in a circle
with diameter 15. If PT = 12,
find each of the following.

1 The area of PQRT   108
2 The area of △PQT   54

Lesson Notes

■ While the "cut and paste"
method can be made into a
rigorous argument, the purpose
of introducing the method is to
develop an intuitive feel for
equal areas.

Objectives

After studying this section, you will be able to
■ Find the areas of parallelograms
■ Find the areas of triangles

Part One: Introduction

The Area of a Parallelogram

Many areas can be found by a "cut and paste" method. For example,
to find the area of a parallelogram with base b and altitude h, we
may do this:

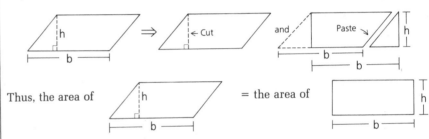

Thus, the area of ... = the area of ...

Theorem 100   *The area of a parallelogram is equal to the product
of the base and the height.*

$$A = bh$$

where b is the length of the base and h is the height.

Formal area proofs are often based on the cut-and-paste method.
For instance, the key steps in a proof of Theorem 100 could be those
below.

Given: PACT is a □.
          $\overline{RT}$ is an altitude to $\overline{PA}$.
Prove: $A_{PACT} = (PA)(RT)$

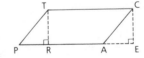

Key Steps:

1. Extend $\overleftrightarrow{PA}$ and draw altitude $\overline{CE}$ to $\overleftrightarrow{PA}$; RECT is a rectangle.
2. $A_{PRT} = A_{AEC}$ because $\triangle PRT \cong \triangle AEC$ by HL.
3. $A_{PACT} = A_{RECT}$, since $A_{CART} + A_{PRT} = A_{CART} + A_{AEC}$.
4. $A_{RECT} = (TC)(RT)$ (Why?)
5. $A_{PACT} = (PA)(RT)$, because $PA = TC$.

The Area of a Triangle

The area of any triangle can be shown to be one half of the area of a parallelogram with the same base and height.

Area of 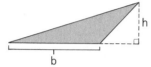 $= \frac{1}{2} \cdot$ area of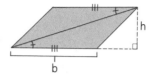

Theorem 101 *The area of a triangle is equal to one-half the product of a base and the height (or altitude) for that base.*

$$A_\triangle = \tfrac{1}{2}bh$$

where b is the length of the base and h is the altitude.

Lesson Notes, continued

- Students should realize that any pair of congruent triangles can form a parallelogram. Thus, the area of one of the triangles is half the area of the parallelogram.

- Many students have initial difficulty if the base is not on the bottom or if the triangle is obtuse.

Part Two: Sample Problems

Problem 1   *Find the area of each triangle.*

a

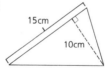

15cm

10cm

b

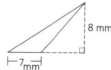

8 mm

⊢7mm⊣

Solution

a $A_\triangle = \frac{1}{2}bh$

$= \frac{1}{2}(15)(10)$

$= 75$ sq cm

Note   The base of a triangle is not always on the bottom. The 10-cm altitude is the altitude associated with the 15-cm base.

b $A_\triangle = \frac{1}{2}bh$

$= \frac{1}{2}(7)(8)$

$= 28$ sq mm

Note   The altitude of a triangle is not always inside the triangle.

Problem 2 *Find the base of a triangle with altitude 15 and area 60.*

Solution Let x be the base.

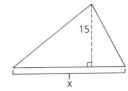

$$A_\triangle = \tfrac{1}{2}bh$$
$$60 = \tfrac{1}{2}x(15)$$
$$8 = x$$

Problem 3 *Find the area of a parallelogram whose sides are 14 and 6 and whose acute angle is 60°.*

Solution We can use 14 as the base, but we must first find the height for that base. When altitude $\overline{BE}$ is drawn, a 30°-60°-90° triangle is formed, so $h = 3\sqrt{3}$.

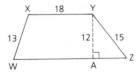

$$A_\square = bh$$
$$= 14(3\sqrt{3}) = 42\sqrt{3}$$

Problem 4 *Find the area of a trapezoid WXYZ.*

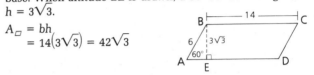

Solution Copy the diagram. Use the divide-and-conquer method. By drawing another altitude, $\overline{XB}$, you can divide the trapezoid into two right triangles and a rectangle.

Find the areas of these figures and add them.

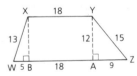

The sides of △WBX form a Pythagorean triple, so WB = 5. Similarly, in △YAZ, AZ = 9.

$$A_{\triangle WBX} = \tfrac{1}{2}bh \qquad\qquad A_{rect} = bh \qquad\qquad A_{\triangle YAZ} = \tfrac{1}{2}bh$$
$$= \tfrac{1}{2}(5)(12) \qquad\qquad = 18(12) \qquad\qquad = \tfrac{1}{2}(9)(12)$$
$$= 30 \qquad\qquad\qquad = 216 \qquad\qquad\qquad = 54$$

The sum of the three areas, 300, is the area of the trapezoid.

Part Three: Problem Sets

Problem Set A

1 Find the area of each triangle.

a  198 sq mm

b 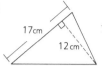 102 sq cm

c 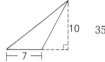 35

2 Find the area of the triangle.

 120

3 Find the total area of each figure. (In each figure the triangle is mounted on a rectangle.)

a 35

b 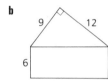 144

4 Find the altitude of a triangle if its base is 7 and its area is 21. 6

5 Find the area of an isosceles triangle with sides 10, 10, and 16. 48

6 Find the area of a parallelogram of base 17 and height 11. 187

7 Find the base of a parallelogram of height 3 and area 42. 14

8 Find the area of each obtuse triangle.

a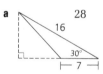

b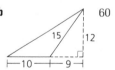

9 Find the area of each triangle.

a $18\sqrt{3}$

b 72

c $36\sqrt{3}$

Assignment Guide

Basic

Day 1   1–9
Day 2   10–13, 17, 19, 21, 23

Average

Day 1   2–12
Day 2   13–21, 23–25

Advanced

Day 1   1, 3, 5, 10, 14–19
Day 2   21, 23–26, 29, 32

Problem Set A, *continued*

10 Find the area of each parallelogram to the nearest tenth.

a ≈ 173.2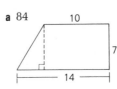

b ≈ 84.9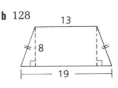

11 Find the area of each trapezoid by dividing it into a rectangle and triangle(s).

a 84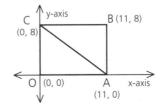

b 128

12 Find the area of △AOC. 44

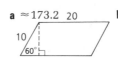

Problem Set B

13 A triangle has the same area as a 6-by-8 rectangle. The base of the triangle is 8. Find the altitude of the triangle. 12

14 Lines $\overleftrightarrow{CF}$ and $\overleftrightarrow{AB}$ are parallel and 10 mm apart. Several triangles with base $\overline{AB}$ and a vertex on $\overleftrightarrow{CF}$ have been drawn below. Which triangle has the largest area? Explain. Each has an area of 80 sq mm.

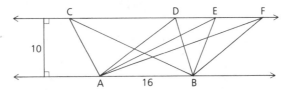

15 Find the area of the shaded region. 85

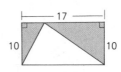

16 In a triangle, a base and its altitude are in a ratio of 3:2. The triangle's area is 48. Find the base and the altitude. 12; 8

Problem-Set Notes and Additional Answers

■ In problem **14**, the students should realize that all the triangles have equal bases and heights.

Communicating Mathematics

Refer to problem **14**. Have students write a paragraph explaining why each triangle has an area of 80 sq mm.
$\overline{AB}$ is the base of each triangle. Since $\overleftrightarrow{CF} \parallel \overleftrightarrow{AB}$, the distance between the two lines is the same. The distance between the lines is the height of each of the triangles.

■ In problem **17**, the use of diagrams can promote creative thinking.

■ In problem **25**, students should discover that the given diagram is incorrect. It should be drawn as follows.

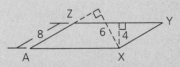

T 520

17 Find the area of the shaded triangular region. 33

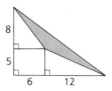

18 Given: QT = 12, PR = 15,
 PS = 10

Find: **a** The area of △PQR 90
 b RQ 18

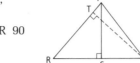

19 a Find the area of a triangle whose sides are 25, 25, and 14. 168
 b Find the area of a right triangle whose legs are 9 and 40. 180
 c Find the area of an isosceles triangle with hypotenuse 18. 81

20 Find the area of an equilateral triangle with a perimeter of 45 m. $\frac{225\sqrt{3}}{4}$ sq m

21 Find the area of each parallelogram to the nearest tenth.

a 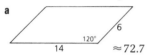 ≈72.7
b 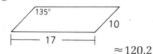 ≈120.2
c 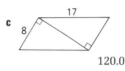 120.0

22 Find the area of each trapezoid by dividing it into other figures (rectangles and triangles or parallelograms and triangles).

a 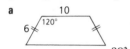 $39\sqrt{3}$
b  33
c  $128 + 32\sqrt{3}$

23 Find the area of △ABC with vertices A = (1, 3), B = (7, 3), and C = (4, −1). 12

24 The hypotenuse of a right triangle is 50, and one leg is 14.
 a Find the area of the triangle. 336
 b Find the altitude to the hypotenuse. $13\frac{11}{25}$

25 a Find m∠A in ▱AXYZ. 30
 b Find AX. 12

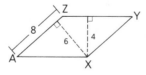

Problem-Set Notes and Additional Answers, continued

26

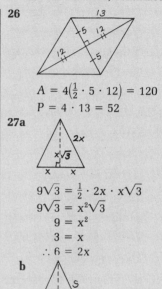

$A = 4\left(\frac{1}{2} \cdot 5 \cdot 12\right) = 120$
$P = 4 \cdot 13 = 52$

27a

$9\sqrt{3} = \frac{1}{2} \cdot 2x \cdot x\sqrt{3}$
$9\sqrt{3} = x^2\sqrt{3}$
$9 = x^2$
$3 = x$
∴ $6 = 2x$

b

$A = \frac{1}{2}s \cdot \frac{s}{2}\sqrt{3} = \frac{s^2}{4}\sqrt{3}$

28

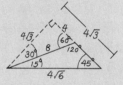

$A = A_{45°\text{-}45°\text{-}90°} - A_{30°\text{-}60°\text{-}90°}$
$= \frac{1}{2}(4\sqrt{3})(4\sqrt{3}) - \frac{1}{2}(4)(4\sqrt{3})$
$= 24 - 8\sqrt{3}$

29

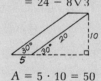

$A = 5 \cdot 10 = 50$

Cooperative Learning

Have students work in small groups to solve problem **30.** Students can choose a given perimeter, say 40 units, draw the parallelogram, then find its area. Students can then compare diagrams and areas to determine which has the greatest area.

T 521

Problem-Set Notes and
Additional Answers, continued

31

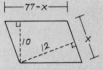

$A = 2A_\triangle$
$= 2(\frac{1}{2} \cdot 5 \cdot 24)$
$= 120$

32

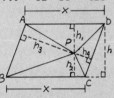

Since 10 and 12 are
altitudes and the semi-
perimeter = 77, we can
compute the area using
each altitude and then
set them equal.
$12x = 10(77 - x)$
$12x = 770 - 10x$
$22x = 770$
$x = 35$
$\therefore A = 12 \cdot 35 = 420$

33a,b

$A_{\triangle APD} + A_{\triangle BPC}$
$= \frac{1}{2}xh_1 + \frac{1}{2}xh_2$
$= \frac{1}{2}x(h_1 + h_2)$
$= \frac{1}{2}xh$
$= \frac{1}{2}A_{\square ABCD}$
$\therefore A_{\triangle APB} + A_{\triangle PDC}$ = the
other half
c This does not happen in
a trapezoid because
opposite sides need not
be congruent. Also,
altitudes to the non-
parallel sides are
noncollinear.

Problem Set C

26 If the diagonals of a rhombus are 10 and 24, find the area and
the perimeter of the rhombus. $A = 120$; $P = 52$

27 a The area of an equilateral triangle is $9\sqrt{3}$. Find the length of
one side. 6

b Find a formula for the area of an equilateral triangle with
sides s units long. $A = \frac{s^2}{4}\sqrt{3}$

28 Find the area of the triangle. $24 - 8\sqrt{3}$

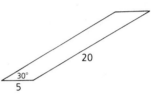

29 Find the area of the parallelogram. 50

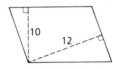

30 What is the name of the parallelogram having the greatest area
for a given perimeter? Square

31 The diagonals of a kite are 10 and 24. Find the kite's area. 120

32 The perimeter of the parallelogram is
154. Find the parallelogram's area. 420

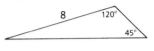

33 Let P be any point in the interior of rectangle ABCD. Four
triangles are formed by joining P to each vertex.

a Demonstrate that $A_{\triangle APD} + A_{\triangle BPC} = A_{\triangle APB} + A_{\triangle PCD}$.

b Is this equation valid if ABCD is a parallelogram? Yes

c Is the equation valid if ABCD is a trapezoid? No

$DOG^2 =$

522 | Chapter 11 Area

11.3 THE AREA OF A TRAPEZOID

Objectives

After studying this section, you will be able to
- Find the areas of trapezoids
- Use the measure of a trapezoid's median to find its area

Part One: Introduction

The Area of a Trapezoid

You have seen that the area of a trapezoid can be found by dividing the trapezoid into simpler shapes, such as triangles, rectangles, and parallelograms ("divide and conquer"). There is, however, a formula that can be used to find the area of a trapezoid.

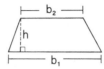

Theorem 102 *The area of a trapezoid equals one-half the product of the height and the sum of the bases.*

$$A_{trap} = \tfrac{1}{2}h(b_1 + b_2)$$

where b_1 is the length of one base, b_2 is the length of the other base, and h is the height.

The Median of a Trapezoid

We can use the Midline Theorem to find out what happens when the midpoints of the nonparallel sides of a trapezoid are joined.

Definition The line segment joining the midpoints of the non-parallel sides of a trapezoid is called the **median** of the trapezoid.

In trapezoid WXYZ, P, Q, and R are midpoints of sides of $\triangle$WXZ and $\triangle$XYZ. P, Q, and R are collinear, because $\overline{PQ}$ and $\overline{QR}$ share Q, and each segment is parallel to $\overline{WX}$ and $\overline{ZY}$. $\overline{PR}$ is the median of trapezoid WXYZ. By the Midline Theorem, PQ = $\tfrac{1}{2}$(WX) and QR = $\tfrac{1}{2}$(YZ). Thus, PR = PQ + QR = $\tfrac{1}{2}$(WX) + $\tfrac{1}{2}$(YZ) = $\tfrac{1}{2}$(WX + YZ).

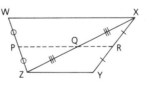

Class Planning

Time Schedule
Basic: 2 days
Average: 2 days
Advanced: 1 day

Resource References
Teacher's Resource Book
   Class Opener 11.3A, 11.3B
Evaluation
   Tests and Quizzes
      Quiz 1, Series 1–2

Class Opener

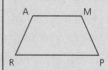

Given: $\overline{AM} \parallel \overline{RP}$,
         AR = MP = 13,
         AM = 15,
         RP = 25
Find: Area of AMPR

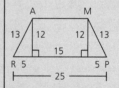

$12 \cdot 15 + 2 \cdot \tfrac{1}{2} \cdot 12 \cdot 5 =$
240 sq units
This Class Opener reviews area of rectangles and triangles and properties of trapezoids, and previews area of trapezoids.

Vocabulary
median

Lesson Notes

- Students can derive, or justify, the trapezoid formula in at least two ways.

1. Use two congruent trapezoids to form a parallelogram.

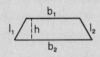

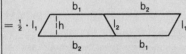

The area of the trapezoid is half the area of the parallelogram.
$$A_{trap} = \frac{1}{2}A_{\square}$$
$$= \frac{1}{2}[(b_1 + b_2)h]$$
$$= \frac{h}{2}(b_1 + b_2)$$

2. Divide the trapezoid by its median to form a parallelogram.

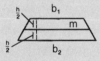

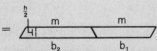

The area of the trapezoid is equal to the area of the parallelogram.
$$A_{trap} = A_{\square} = \frac{h}{2}(b_1 + b_2)$$

Theorem 103 *The measure of the median of a trapezoid equals the average of the measures of the bases.*
$$M = \frac{1}{2}(b_1 + b_2)$$
where b_1 is the length of one base and b_2 is the length of the other base.

You can now easily prove a shorter form of Theorem 102.

Theorem 104 *The area of a trapezoid is the product of the median and the height.*
$$A_{trap} = Mh$$
where M is the length of the median and h is the height.

Part Two: Sample Problems

Problem 1 Given: Trapezoid WXYZ, with height 7, lower base 18, and upper base 12

Find: The area of WXYZ

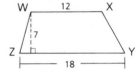

Solution $A_{trap} = \frac{1}{2}h(b_1 + b_2)$
$$= \frac{1}{2}(7)(18 + 12) = 105$$

Problem 2 Find the shorter base of a trapezoid if the trapezoid's area is 52, its altitude is 8, and its longer base is 10.

Solution Let x be the length of the shorter base.
$$A_{trap} = \frac{1}{2}h(b_1 + b_2)$$
$$52 = \frac{1}{2}(8)(10 + x)$$
$$52 = 4(10 + x)$$
$$3 = x$$

Problem 3 The height of a trapezoid is 12. The bases are 6 and 14.

a Find the median. **b** Find the area.

Solution **a** $M = \frac{1}{2}(b_1 + b_2)$ **b** $A_{trap} = Mh$
$$= \frac{1}{2}(14 + 6) \qquad \qquad = 10(12)$$
$$= 10 \qquad \qquad\qquad = 120$$

---

524 Chapter 11 Area

Part Three: Problem Sets

Problem Set A

1 A trapezoid has bases 15 and 11 and height 8.

  a Find the area. 104

  b Find the median. 13

2 Find the area of each trapezoid.

  a   75

  c   78

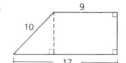

  b   72

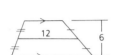

  d   312

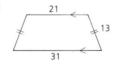

3 Given a trapezoid with bases 6 and 15 and height 7, find the median and the area. 10.5; 73.5

4 The bases of a trapezoid are 8 and 22, and the trapezoid's area is 135. Find the height. 9

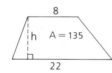

5 The height of a trapezoid is 10, and the trapezoid's area is 130. If one base is 15, find the other base. 11

6 A straight wire stretches between the tops of two poles whose heights are 30 ft and 14 ft. Find the height of a pole that is to be placed halfway between the original poles to support the wire. Assume that the poles are perpendicular to the ground. (Hint: Do you see a trapezoid and its median?) 22 ft

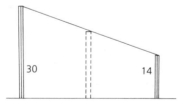

Section 11.3   The Area of a Trapezoid   **525**

Assignment Guide

Basic	
Day 1	1–6
Day 2	7–11
Average	
Day 1	2–8
Day 2	9–13, 17
Advanced	
3–6, 9–12, 14, 17, 18	

Communicating Mathematics

Refer to problem **12**. Have students explain why they do not need to know the individual lengths of the bases.
Only the sum $b_1 + b_2$ needs to be known in order to find the area.

Problem-Set Notes
and Additional Answers

14

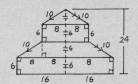

$A = 48 + 96 + 144 + 192 = 480$

15

$A_{(small \triangle)} = 4$
$A_{(large \triangle)} = 4 \cdot 4 = 16$

16

$A = \frac{1}{2}h(b_1 + b_2)$
$245 = \frac{1}{2} \cdot 4x(10x)$
$245 = 20x^2$
$\frac{49}{4} = x^2$
$\frac{7}{2} = x$
$\therefore$ Altitude $= 4x = 4 \cdot \frac{7}{2} = 14$
$P = 20x = 20 \cdot \frac{7}{2} = 70$

17a

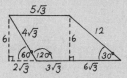

$A = \frac{1}{2} \cdot 6 \cdot (9\sqrt{3} + 5\sqrt{3})$
$= 42\sqrt{3}$

17b

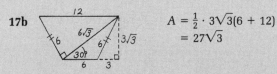

$A = \frac{1}{2} \cdot 3\sqrt{3}(6 + 12)$
$= 27\sqrt{3}$

Problem Set B

7 Find the total area of each figure.

a

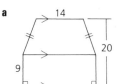

396

b

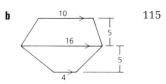

78

8 Find the total area of each figure.

a

$30\sqrt{3}$

b

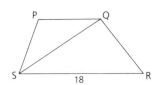

115

9 Find the lower base of a trapezoid whose upper base is 10 and whose median is 17. 24

10 The area of triangle PQS is 25.
The median of trapezoid PQRS is 14.
Base $\overline{RS}$ measures 18.

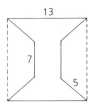

Find: **a** The length of base $\overline{PQ}$ 10
 b The height to base $\overline{PQ}$ of $\triangle PQS$ 5
 c The height of trapezoid PQRS 5
 d The area of trapezoid PQRS 70

11 Find the area of the figure shown, which was formed by cutting two identical isosceles trapezoids out of a square. 89

12 The perimeter of a trapezoid is 35. The nonparallel sides are 7 and 8. Find the trapezoid's area if its height is 5. 50

13 The consecutive sides of an isosceles trapezoid are in the ratio 2:5:10:5, and the trapezoid's perimeter is 44. Find the area of the trapezoid. 72

Problem Set C

14 The figure shown is composed of four regions of equal height. The triangle and the trapezoid are isosceles, and each side of the trapezoid is parallel to a side of the triangle. Find the total area of the figure. **480**

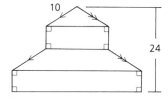

15 When an isosceles triangle is folded so that its vertex is on the midpoint of the base, a trapezoid with an area of 12 square units is formed. Find the area of the original triangle. **16 sq units**

16 The sides of a trapezoid are in the ratio $2:5:8:5$. The trapezoid's area is 245. Find the height and the perimeter of the trapezoid. **14; 70**

17 Find the area of each trapezoid.

a 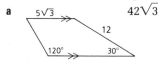 $42\sqrt{3}$

b 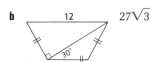 $27\sqrt{3}$

18 In trapezoid ABCD, X and Y are midpoints of sides, and P and Q are midpoints of diagonals. Develop a formula that can be used to find PQ. (Hint: See the proof of Theorem 103.) $PQ = \dfrac{b_1 - b_2}{2}$

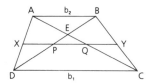

19 Prove that the area of a trapezoid is $\frac{1}{2}h(b_1 + b_2)$ by each of the following methods.

a Draw a diagonal and use the two triangles formed.

b Draw altitudes and use the rectangle and the triangles formed.

20 Write a coordinate proof that the median of a trapezoid is parallel to the bases and is equal to one-half their sum.

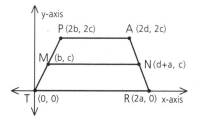

11.3

Problem-Set Notes and Additional Answers, continued

18 By the Midline Theorem, $XQ = \frac{1}{2}b_1$ and $XP = \frac{1}{2}b_2$.
∴ $PQ = XQ - XP = \frac{1}{2}b_1 - \frac{1}{2}b_2 = \dfrac{b_1 - b_2}{2}$

19a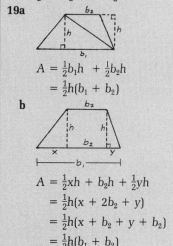

$A = \frac{1}{2}b_1h + \frac{1}{2}b_2h$
$= \frac{1}{2}h(b_1 + b_2)$

b

$A = \frac{1}{2}xh + b_2h + \frac{1}{2}yh$
$= \frac{1}{2}h(x + 2b_2 + y)$
$= \frac{1}{2}h(x + b_2 + y + b_2)$
$= \frac{1}{2}h(b_1 + b_2)$

Cooperative Learning

Have students work in small groups to solve problem **19**.

20 Slope of $\overleftrightarrow{MN} = \frac{c-c}{d+a-b} = 0$;
slope of $\overleftrightarrow{TR} = \frac{0-0}{2a-0} = 0$;
slope of $\overleftrightarrow{PA} = \frac{2c-2c}{2d-2b} = 0$.
So $\overleftrightarrow{MN} \parallel \overleftrightarrow{PA} \parallel \overleftrightarrow{TR}$.
$MN = d + a - b$.
$PA = 2d - 2b$, and
$TR = 2a$, so
$\frac{1}{2}(PA + TR) = d + a - b$.
So $MN = \frac{1}{2}(PA + TR)$.

11.4 AREAS OF KITES AND RELATED FIGURES

Objective

After studying this section, you will be able to
■ Find the areas of kites

Part One: Introduction

Remember that in a kite the diagonals are perpendicular.
   Also a kite can be divided into two isosceles triangles with a common base, so its area will equal the sum of the areas of these triangles.

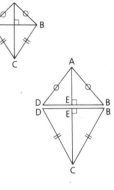

$A_{kite} = A_{\triangle ABD} + A_{\triangle DBC}$
$\quad = \frac{1}{2}(BD)(AE) + \frac{1}{2}(BD)(EC)$
$\quad = \frac{1}{2}(BD)(AE + EC)$
$\quad = \frac{1}{2}(BD)(AC)$

Notice that $\overline{BD}$ and $\overline{AC}$ are the diagonals of the kite. We have just proved the following formula.

Theorem 105   ***The area of a kite equals half the product of its diagonals.***
$$A_{kite} = \frac{1}{2}d_1 d_2$$
where d_1 is the length of one diagonal and d_2 is the length of the other diagonal.

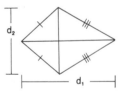

This formula can be applied to any kite, including the special cases of a rhombus and a square.

Part Two: Sample Problems

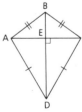

Problem 1   Find the area of a kite with diagonals 9 and 14.

Solution $A_{kite} = \frac{1}{2}d_1 d_2$ $AC = 9$
$\quad\quad = \frac{1}{2}(14)(9) = 63$ $BD = 14$

Problem 2   *Find the area of a rhombus whose perimeter is 20 and whose longer diagonal is 8.*

Solution   A rhombus is a $\square$, so its diagonals bisect each other. It is also a kite, so its diagonals are $\perp$ to each other. Thus, XZ = 8 and XP = 4.

The perimeter is 20, so XB = 5. (Why?)

$\triangle$BPX is a right triangle. Thus, BP = 3 and BY = 6.

$A_{\text{kite}} = \frac{1}{2}d_1d_2$

$\qquad = \frac{1}{2}(6)(8) = 24$

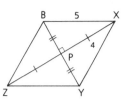

Part Three: Problem Sets

Problem Set A

1 Find the area of a kite with diagonals 6 and 20. 60

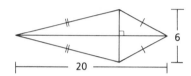

2 Find the area of each kite.

  a 56

  b 85

  c 160

3 The area of a kite is 20. The longer diagonal is 8. Find the shorter diagonal. 5

Problem Set B

4 Find the area of the kite shown. 168

5 Find the area of each rhombus.

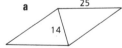 **a** 336

  b 500

6 Given: ABCD is a kite.
∠BAD is a right ∠.
BD = 10, BC = 13

Find: The area of ABCD 85

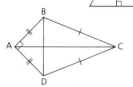

Assignment Guide

Basic
_____
1–5, 9
_____
Average
_____
2–9
_____
Advanced
_____
4–10

Problem-Set Notes and Additional Answers

■ In problem **6**, since $\triangle$ABD is isos. rt., the two small △ on the left are also isos. rt.

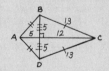

■ In problem **7**, use the "altitude on the hypotenuse" theorem (Theorem 68b) to show x = 6.

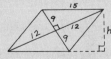

10

$A_{\text{rhom}} = bh = \frac{1}{2}d_1d_2$

$15h = \frac{1}{2} \cdot 18 \cdot 24$

$h = \frac{216}{15} = \frac{72}{5}$

11 $A = \frac{1}{2}wy + \frac{1}{2}xy + \frac{1}{2}xz + \frac{1}{2}wz$

$\qquad = \frac{1}{2}[y(w + x) + z(w + x)]$

$\qquad = \frac{1}{2}(w + x)(y + z)$

T 529

12

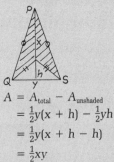

$A = A_{total} - A_{unshaded}$
$= \frac{1}{2}y(x + h) - \frac{1}{2}yh$
$= \frac{1}{2}y(x + h - h)$
$= \frac{1}{2}xy$

13

$A_{\triangle XQY} = \frac{1}{2}A_{XBCY}$
$A_{\triangle XPY} = \frac{1}{2}A_{XADY}$
$A_{XQYP} = A_{\triangle XQY} + A_{\triangle XPY}$
$\qquad = \frac{1}{2}A_{ABCD}$

Cooperative Learning

Refer to problem **13.** Have students work in small groups to compare their answers to part **a.** Then ask them to discuss the following questions.

1 How many different ways can you demonstrate the truth of your conjecture? Algebraically; with a diagram; formal proof

2 Is the ratio of the area of the quadrilateral to the area of the rectangle dependent on the quadrilateral being a kite? What has to be true for this relationship to hold? Demonstrations will vary. The quadrilateral does not have to be a kite for the ratio of the areas to be 1:2, but $\overline{AX}$ does have to be congruent to $\overline{DY}$ and $\overline{XB}$ does have to be congruent to $\overline{YC}$ for the relationship to be true.

Problem Set B, *continued*

7 Find the area of the kite shown. 78

8 Find the area of a rhombus with a perimeter of 40 and one angle of 60°. $50\sqrt{3}$

9 a Find the areas of region I, region II, and region III. 8; 28; 18

  b Find the area of $\triangle OBD$. 30

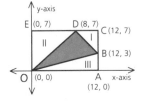

Problem Set C

10 Given a rhombus with diagonals 18 and 24, find the height. $\frac{72}{5}$

11 The formula for the area of a kite applies to *any* quadrilateral whose diagonals are perpendicular.
  Prove that the area of any quadrilateral with perpendicular diagonals equals half the product of the diagonals. (Hint: Use *w, x, y,* and *z* as marked to show that $A = \frac{1}{2}[w + x][y + z]$.)

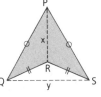

12 Observe the figure at the right. It resembles a kite, but it is not convex (it is "dented in"). Does the kite formula still hold? (That is, can it be shown that $A = \frac{1}{2}xy$?) Yes

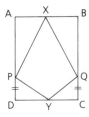

13 In rectangle ABCD, X and Y are midpoints of $\overline{AB}$ and $\overline{CD}$, and $\overline{PD} \cong \overline{QC}$.

  a Compare the area of quadrilateral XQYP with the area of ABCD. $\frac{1}{2}$

  b Prove your conjecture.

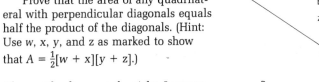

11.5 AREAS OF REGULAR POLYGONS

Objectives

After studying this section, you will be able to
- Find the areas of equilateral triangles
- Find the areas of other regular polygons

Part One: Introduction

The Area of an Equilateral Triangle

Equilateral triangles are encountered so frequently that a special formula for their areas will be useful.

Remember that the altitude of an equilateral triangle divides it into two 30°-60°-90° right triangles.

Thus, if $WY = s$, then $ZY = \frac{s}{2}$ and $WZ = \frac{s}{2}\sqrt{3}$.

Therefore, $A_{WXY} = \frac{1}{2}bh$

$$= \frac{1}{2}s\left(\frac{s}{2}\sqrt{3}\right) = \frac{s^2}{4}\sqrt{3}$$

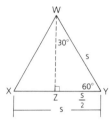

Theorem 106 *The area of an equilateral triangle equals the product of one-fourth the square of a side and the square root of 3.*

$$A_{eq.\,\triangle} = \frac{s^2}{4}\sqrt{3}$$

where s is the length of a side.

The Area of a Regular Polygon

Recall that in a regular polygon all interior angles are congruent and all sides are congruent.

In regular polygon PENTA,
- O is the center
- $\overline{OA}$ is a *radius*
- $\overline{OM}$ is an *apothem*

Vocabulary
apothem
radius

Class Planning

Time Schedule
Basic: 2 days
Average: 1 day
Advanced: 1 day

Resource References
Teacher's Resource Book
   Class Opener 11.5A, 11.5B

Class Opener

1 Find the area of an equilateral triangle with sides of length 8.

$\frac{1}{2} \cdot 4\sqrt{3} \cdot 8 = 16\sqrt{3}$

2 Find the area of a regular hexagon with sides of length 8.

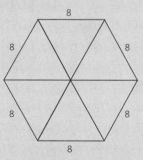

The hexagon is made up of six of the triangles from problem **1**. $6 \cdot 16\sqrt{3} = 96\sqrt{3}$

3 Find the ratio of the area of the hexagon to the area of the triangle.   6:1

Lesson Notes

- Students may need to be reminded that regular triangles and hexagons (and duodecagons) can be decomposed into 30°-60°-90° triangles.

Lesson Notes, continued

- Students who use the 30°-60°-90° triangle proficiently may choose to forgo the formula in Theorem 106 and, in effect, generate it when they need it.

- For each triangle, $A_\triangle = \frac{1}{2}a \cdot s$, where a is the apothem and s is the base. For a regular n-gon, the triangles are congruent and $n \cdot s = $ perimeter. Thus, $A_{\text{reg. poly.}} = n\left(\frac{1}{2}a \cdot s\right) = \frac{1}{2}a \cdot p$.

Communicating Mathematics

Have students write a lesson plan that explains how to derive the formula for the area of a regular polygon. Choose one or two students to present their lesson to the class and be responsible for answering questions.

Definition   A *radius* of a regular polygon is a segment joining the center to any vertex.

Definition   An *apothem* of a regular polygon is a segment joining the center to the midpoint of any side.

Here are some important observations about apothems and radii:
- All apothems of a regular polygon are congruent.
- Only regular polygons have apothems.
- An apothem is a radius of a circle inscribed in the polygon.
- An apothem is the perpendicular bisector of a side.
- A radius of a regular polygon is a radius of a circle circumscribed about the polygon.
- A radius of a regular polygon bisects an angle of the polygon.

If all of the radii of a regular polygon are drawn, the polygon is divided into congruent isosceles triangles. (What is an altitude of each triangle?) If you write an expression for the sum of the areas of those isosceles triangles, you can derive the following formula.

Theorem 107   *The area of a regular polygon equals one-half the product of the apothem and the perimeter.*
$$A_{\text{reg. poly.}} = \frac{1}{2}ap$$
where a is the length of an apothem and p is the perimeter.

Part Two: Sample Problems

Problem 1   *A regular polygon has a perimeter of 40 and an apothem of 5. Find the polygon's area.*

Solution $A_{\text{reg. poly.}} = \frac{1}{2}ap$
$$= \frac{1}{2}(5)(40) = 100$$

Problem 2   *An equilateral triangle has a side 10 cm long. Find the triangle's area.*

Solution $A_{\text{eq. }\triangle} = \frac{s^2}{4}\sqrt{3}$
$$= \frac{10^2}{4}\sqrt{3}$$
$$= 25\sqrt{3} \text{ sq cm}$$

Problem 3 A circle with a radius of 6 is inscribed in an equilateral triangle. Find the area of the triangle.

Solution Notice that $\overline{OP}$ is an apothem 6 units long and that AOP is a 30°-60°-90° triangle. Thus, OA = 12, AP = $6\sqrt{3}$, and the perimeter of △ABC is $36\sqrt{3}$. An equilateral triangle is a regular polygon, so

$A = \frac{1}{2}ap$
$= \frac{1}{2}(6)(36\sqrt{3}) = 108\sqrt{3}$

Problem 4 Find the area of a regular hexagon with sides 18 units long.

Solution AF = 18, so AP = 9.

Observe that OPA is a 30°-60°-90° triangle, so that apothem OP = $9\sqrt{3}$.

Perimeter = 6(18) = 108

$A = \frac{1}{2}ap$
$= \frac{1}{2}(9\sqrt{3})(108) = 486\sqrt{3}$

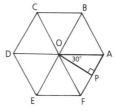

Part Three: Problem Sets

Problem Set A

1 The perimeter of a regular polygon is 24 and the apothem is 3. Find the polygon's area. 36

2 Find the areas of equilateral triangles with the following sides.
  a 6 $9\sqrt{3}$
  b 7 $\frac{49\sqrt{3}}{4}$
  c 8 $16\sqrt{3}$
  d $2\sqrt{3}$ $3\sqrt{3}$

3 Find the areas of equilateral triangles with the following apothems.
  a 6 $108\sqrt{3}$
  b 4 $48\sqrt{3}$
  c 3 $27\sqrt{3}$
  d $2\sqrt{3}$ $36\sqrt{3}$

4 Find, to the nearest tenth, the area of a regular hexagon whose
  a Side is 6 ≈93.5
  b Side is 8 ≈166.3
  c Apothem is 6 ≈124.7
  d Apothem is 8 ≈221.7

5 The radius of a regular hexagon is 12.
  Find: **a** The length of one side 12
    b The apothem $6\sqrt{3}$
    c The area $216\sqrt{3}$

Assignment Guide

Basic

Day 1   2, 3, 8, 10
Day 2   4a,c, 5–7, 9, 11

Average

2b,d, 3b,d, 4b,d, 5, 6, 15, 17, 18, 26

Advanced

2b,d, 3b,d, 10, 11, 15–20, 26

Cooperative Learning

Have students form small groups and compare the formula for the area of a regular polygon to the formula for the area of a circle. In what way can they describe a circle in order to derive the formula $A = \pi r^2$ from Theorem 107? (Hint: What happens to the radius and the apothem of a regular polygon as the number of sides increases?)
As the number of sides of a regular polygon increases, the radius and the apothem of the polygon become closer in value. $A = \frac{1}{2}(a)(p)$. For a circle, the perimeter is $2\pi r$. Thus, $A = \frac{1}{2}(r)(2\pi r) = \pi r^2$.

16

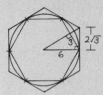

Since the 30°-60°-90° △ are similar, $\frac{A_1}{A_2} = \frac{3^2}{(2\sqrt{3})^2} = \frac{9}{12} = \frac{3}{4}$.

18 If the figures are numbered consecutively 1–6, the possibilities are
1,2
1,3 2,3
1,4 2,4 3,4
1,5 2,5 3,5 4,5
1,6 2,6 3,6 4,6 5,6
∴ Prob. $= \frac{6}{15} = \frac{2}{5}$

19a

$A = \frac{1}{2}ap$
$= \frac{1}{2}(15)(60\sqrt{3})$
$= 450\sqrt{3}$

b

$A = \frac{1}{2}\sqrt{3}$
$32\sqrt{3} = \frac{1}{2}(x\sqrt{3})(12x)$
$x = \frac{4}{\sqrt{3}}$
$x\sqrt{3} = 4$
∴ Span $= 2 \cdot 4 = 8$

c

$x\sqrt{3} = \frac{s}{2}$
$x = \frac{s\sqrt{3}}{6}$
$A = \frac{1}{2}ap$
$= \frac{1}{2}\left(\frac{s}{2}\right)\left(\frac{12s\sqrt{3}}{6}\right)$
$= \frac{s^2\sqrt{3}}{2}$

Problem Set A, *continued*

6 Find the area of a square whose
  a Apothem is 5 100   **c** Side is 7 49   **e** Radius is 6 72
  b Apothem is 12 576   **d** Diagonal is 10 50   **f** Perimeter is 12 9

7 Find the apothem of a square whose area is 36 sq mm. 3 mm

8 Find the side of an equilateral triangle whose area is $9\sqrt{3}$ sq km. 6 km

9 Find the area of a square if the radius of its inscribed circle is 9. 324

10 Find the area of an equilateral triangle if the radius of its inscribed circle is 3. $27\sqrt{3}$

11 Find the area of a regular hexagon if the radius of its inscribed circle is 12. $288\sqrt{3}$

Problem Set B

12 Find the area of
  a An equilateral triangle whose side is 9 $\frac{81\sqrt{3}}{4}$
  b A square whose apothem is $7\frac{1}{2}$ 225
  c A regular hexagon whose side is 7 $\frac{147\sqrt{3}}{2}$

13 Find the length of one side and of the apothem of
  a A square whose area is 121 11; 5.5
  b An equilateral triangle whose area is $36\sqrt{3}$ sq m 12 m; $2\sqrt{3}$ m
  c A regular hexagon whose perimeter is 24 cm 4 cm; $2\sqrt{3}$ cm

14 Find the perimeter of a regular polygon whose area is 64 and whose apothem is 4. 32

15 A circle of radius 12 is circumscribed about each regular polygon below. Find the area of each polygon.
  a $108\sqrt{3}$ **b** 288 **c** $216\sqrt{3}$

16 A circle is inscribed in one regular hexagon and circumscribed about another. If the circle has a radius of 6, find the ratio of the area of the smaller hexagon to the area of the larger hexagon. $\frac{3}{4}$

17 Find the area of the shaded region in each polygon. (Assume regular polygons.)

a
$36 - 9\sqrt{3}$

b
$27\sqrt{3}$

c
$36\sqrt{3}$

18 Suppose you are given a scalene triangle, an equilateral triangle, a kite, a square, a regular octagon, and a regular hexagon. If you choose two of the six figures at random, what is the probability that both have apothems? $\frac{2}{5}$

Problem Set C

19 a The span s of a regular hexagon is 30. Find the hexagon's area. $450\sqrt{3}$

b Find the span of a regular hexagon with an area of $32\sqrt{3}$. 8

c Find a formula for the area of a regular hexagon with a given span s. $A = \frac{s^2\sqrt{3}}{2}$

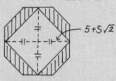

20 a Find the apothem of the regular octagon. $5 + 5\sqrt{2}$

b Find the area of the octagon. $200 + 200\sqrt{2}$

21 A square is formed by joining the midpoints of alternate sides of a regular octagon. A side of the octagon is 10.

a Find the area of the square. $150 + 100\sqrt{2}$

b Find the area of the shaded region. $50 + 100\sqrt{2}$

22 Given a set of four concentric regular hexagons, each with a radius 1 unit longer than that of the next smaller hexagon, find the total area of the shaded regions. $12\sqrt{3}$

23 A square is inscribed in an equilateral triangle as shown. Find the area of the shaded region. $1764\sqrt{3} - 3024$

Problem-Set Notes and Additional Answers, continued

20a

$10 = x\sqrt{2}$
$5\sqrt{2} = x$
$\therefore$ Apothem $= 5 + 5\sqrt{2}$

b $A = \frac{1}{2}(5 + 5\sqrt{2})(80)$
$= 200 + 200\sqrt{2}$

21a

From problem **20a,** the apothem of the octagon is $5 + 5\sqrt{2}$, so the diagonal of the square is $10 + 10\sqrt{2}$.
$A_{SQ} = \frac{1}{2}d_1d_2$
$= \frac{1}{2}(10 + 10\sqrt{2})^2$
$= 150 + 100\sqrt{2}$

b From problem **20b,**
$A_{oct} = 200 + 200\sqrt{2}$.
$\therefore A_{shaded} = (200 + 200\sqrt{2}) - (150 + 100\sqrt{2}) = 50 + 100\sqrt{2}$

22 $A_{shaded} = \frac{1}{2}A_{hex} = \frac{1}{2}(\frac{1}{2}ap)$
$A_{shaded} = \frac{1}{4}(2\sqrt{3})(24) = 12\sqrt{3}$

23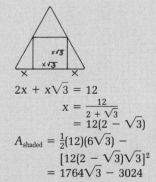

$2x + x\sqrt{3} = 12$
$x = \frac{12}{2 + \sqrt{3}}$
$= 12(2 - \sqrt{3})$
$A_{shaded} = \frac{1}{2}(12)(6\sqrt{3}) - [12(2 - \sqrt{3})\sqrt{3}]^2$
$= 1764\sqrt{3} - 3024$

Problem-Set Notes and Additional Answers, continued

24a

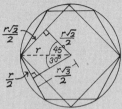

Side sq = $r\sqrt{2}$
Side hex = r
Ratio = $\dfrac{r\sqrt{2}}{r} = \dfrac{\sqrt{2}}{1}$

b $\dfrac{A_{sq}}{A_{hex}} = \dfrac{(r\sqrt{2})^2}{\frac{1}{2}\left(\frac{r\sqrt{3}}{2}\right)(6r)} = \dfrac{2r^2}{\frac{3r^2\sqrt{3}}{2}} =$

$\dfrac{4\sqrt{3}}{9}$

- The "proof" of the Pythagorean Theorem in problem **25** is credited to James A. Garfield in 1876. For more than 250 proofs of this theorem, see *The Pythagorean Proposition* by Elisha Scott Loomis (Reston, Va.: National Council of Teachers of Mathematics, 1968).

26 $A_{rect} = 6 \cdot 8 = 48$
$A_I = \frac{1}{2}(8)(2) = 8$
$A_{II} = \frac{1}{2}(4)(6) = 12$
$A_{III} = \frac{1}{2}(2)(6) = 6$
$\therefore A_{\triangle ABC} = 48 - (8 + 12 + 6)$
$= 22$

Problem Set C, *continued*

24 A square and a regular hexagon are inscribed in the same circle.
  a Find the ratio of a side of the square to a side of the hexagon. $\sqrt{2}:1$
  b Find the ratio of the area of the square to the area of the hexagon. $4\sqrt{3}:9$

25 a Express the area of ABCD as the sum of the areas of the three triangles. $\frac{1}{2}ab + \frac{1}{2}ab + \frac{1}{2}c^2$

  b Express the area of ABCD as the area of a trapezoid with bases $\overline{AB}$ and $\overline{CD}$. $\frac{1}{2}(a+b)(a+b)$

  c Equate your answers to parts **a** and **b** and simplify. Are you surprised? So was President James A. Garfield, who is said to have discovered this proof. $a^2 + b^2 = c^2$

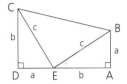

26 Find the area of $\triangle ABC$. 22

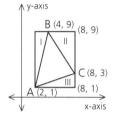

MATHEMATICAL EXCURSION

TILING AND AREA
Mathematics sheds light on chemistry problem

Finding the approximate area of a region by covering it with unit squares is one example of tiling. Tiling using squares is easy to imagine for anyone who has ever seen a checkerboard. Square tiling is an example of *periodic tiling* because the pattern repeats predictably throughout the region.

In the 1970's, Roger Penrose, a mathematical physicist at Oxford University in England, discovered tilings that could never be periodic. One such tiling consisted of kites and darts. Another consisted of fat diamonds and thin diamonds.

Penrose described specific rules governing which sides could come into contact with each other. These shapes, and portions of tilings using them, are shown here.

Penrose's tilings not only represented a mathematical breakthrough, they also have helped scientists better understand how molecules in certain complex crystal patterns "know" how to arrange themselves in such highly complicated ways.

If there were no restrictions regarding which sides could come into contact, how might the kites and darts be tiled periodically?

Kite   Dart   Thin diamond   Fat diamond

11.6 AREAS OF CIRCLES, SECTORS, AND SEGMENTS

Objectives

After studying this section, you will be able to
- Find the areas of circles
- Find the areas of sectors
- Find the areas of segments

Part One: Introduction

The Area of a Circle

You may already know the formula for the area of a circle.

Postulate   *The area of a circle is equal to the product of π and the square of the radius.*

$$A_{\odot} = \pi r^2$$

where r is the radius.

The Area of a Sector

The region bounded by a circle may be divided into **sectors**.

Definition   A **sector** of a circle is a region bounded by two radii and an arc of the circle.

Sector HOP

Just as the length of an arc is a fractional part of the circumference of a circle, the area of a sector is a fractional part of the area of the circle.

Theorem 108   *The area of a sector of a circle is equal to the area of the circle times the fractional part of the circle determined by the sector's arc.*

$$A_{\text{sector HOP}} = \left(\frac{m\widehat{HP}}{360}\right)\pi r^2$$

where r is the radius and $\widehat{HP}$ is measured in degrees.

Class Planning

Time Schedule
All levels: 2 days

Resource References
Teacher's Resource Book
   Class Opener 11.6A, 11.6B
   Additional Practice Worksheet 22
Evaluation
   Tests and Quizzes
     Quiz 5, Series 1
     Quizzes 4–6, Series 2
     Quiz 1, Series 3

Class Opener

An equilateral triangle has sides of length 6. A second triangle is similar to it, and has sides of length 18. Find the ratio of the area of the smaller triangle to the area of the larger triangle.

$A_{\text{small}} = \frac{6^2\sqrt{3}}{4} = 9\sqrt{3}$

$A_{\text{large}} = \frac{18^2\sqrt{3}}{4} = 81\sqrt{3}$

Ratio $= \frac{9\sqrt{3}}{81\sqrt{3}} = \frac{1}{9}$

This problem reviews concepts from Section **11.5** and previews Section **11.7**.

Lesson Notes

- Some students may prefer to set up the formula as a proportion.

$$\frac{A_{\text{sector}}}{\pi r^2} = \frac{\text{measure of arc}}{360}$$

Vocabulary
annulus
sector
segment

The Area of a Segment

Another way of dividing the interior of a circle produces a *segment*.

Definition	A *segment* of a circle is a region bounded by a chord of the circle and its corresponding arc.

By studying the diagram above, you may be able to see what to do to find the area of a segment. Sample problem 4 will illustrate the procedure in detail.

Part Two: Sample Problems

Problem 1 Find the area of a circle whose diameter is 10.

Solution The radius of the circle is 5 (half the diameter).

$$A_{\odot} = \pi r^2$$
$$= \pi(5^2) = 25\pi \text{ sq units}$$

Problem 2 Find the circumference of a circle whose area is 49π sq units.

Solution First find the radius, then use it to calculate the circumference.

$$A_{\odot} = \pi r^2 \qquad\qquad C = 2\pi r$$
$$49\pi = \pi r^2 \qquad\qquad\quad = 2\pi(7) = 14\pi$$
$$7 = r$$

Problem 3 Find the area of a sector with a radius of 12 and a 45° arc.

Solution $A_{\text{sector}} = \left(\dfrac{\text{m arc}}{360}\right)\pi r^2$

$$= \tfrac{45}{360}\pi(12^2) = 18\pi \text{ sq units}$$

Problem 4 The measure of the arc of the segment ($\overparen{AB}$) is 90. The radius of the circle is 10. Find the area of the segment.

Solution Draw radii to the endpoints of AB, forming sector AOB.

Area of segment = area of sector AOB − area of △AOB

$$= \left(\dfrac{m\overparen{AB}}{360}\right)\pi r^2 - \tfrac{1}{2}bh$$

$$= \tfrac{90}{360}\pi(10^2) - \tfrac{1}{2}(10)(10)$$

$$= 25\pi - 50$$

Part Three: Problem Sets

Problem Set A

1 Find the areas and circumferences of circles with the following radii.

    a 1 $\pi, 2\pi$ **b** 8 $64\pi, 16\pi$ **c** 15 $225\pi, 30\pi$

2 Find the radii of circles with the following areas.

    a 16π 4   **b** 169π 13

3 Find the circumference of a circle whose area is 100π sq cm. 20π cm

4 Find the area of a circle whose circumference is 18π dm. 81π sq dm

5 Find the area of each shaded sector.

a

π

b

32π

c

2π

d

27π

e

75π

f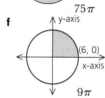

9π

6 The diagram shows a rectangular lawn and the circular regions watered by two sprinklers. Each circular region is 3 m in radius. Find, to the nearest square meter,

    a The total area that is watered ≈ 57 sq m

    b The area of the whole lawn 72 sq m

    c The area of lawn not watered (shaded) ≈ 15 sq m

7 Find the total area of the region shown. $150 + 25\pi$

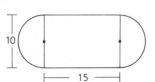

Problem Set B

8 Find, to the nearest tenth, the radii of circles with the following areas.

    a 24π ≈ 4.9 **b** 36 ≈ 3.4

Assignment Guide

Basic

Day 1	1–4, 6, 7, 12a
Day 2	5, 11, 14a,c,e,f, 15

Average

Day 1	1–4, 6, 7, 12, 15
Day 2	9–11, 14, 16, 17

Advanced

Day 1	3–5, 9–13, 15
Day 2	14, 16–18, 21, 23

Problem Set B, *continued*

9 Find the area of each sector.

a

24π

b

12π

c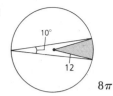

8π

10 If the area of a circle is 60π and the area of a sector of the circle is 24π, what is the measure of the sector's arc? 144

11 Find the area of each segment.

a

$16\pi - 32$

b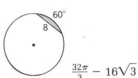

$\frac{32\pi}{3} - 16\sqrt{3}$

c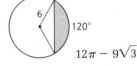

$12\pi - 9\sqrt{3}$

12 a Find the area of the shaded figure if the inner radius is 3 and the outer radius is 5. (Such a figure is called an **annulus**.) 16π

b If the inner circle has a radius r and the outer circle has a radius R, derive the formula for the area of any annulus. $(R^2 - r^2)\pi$

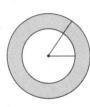

Cooperative Learning

Refer to problem **13**. Students should work in small groups to compare their answers to problem **13** and their observations. Do the observations hold true for all values of x? Can this be proven algebraically?
Yes; area of large circle $= x^2\pi$, and area of small circles $=$
$2\left(\frac{x}{2}\right)^2\pi$.
$x^2\pi - \frac{2x^2}{4}\pi = \frac{1}{2}x^2\pi$.

13 a What is the area of the shaded region if x = 6? If x = 10? If x = 7? 18π; 50π; $24\frac{1}{2}\pi$

b What observation can you make about the shaded region's area?
It is half the area of the large circle.

14 Find the area of the shaded part of each figure. (Assume regular polygons.)

a

$100 - 25\pi$

c

$100 - 25\pi$

e

$50\pi - 100$

b

$25\sqrt{3} - \frac{25\pi}{3}$

d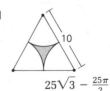

$25\sqrt{3} - \frac{25\pi}{2}$

f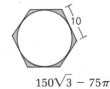

$150\sqrt{3} - 75\pi$

**Problem-Set Notes
and Additional Answers**

16c Prob. $= \frac{A_{\text{bull's-eye}}}{A_{\text{target}}} = \frac{1}{25}$

17 The pieces fit together into eight complete squares, with an area of 32 sq cm.

T 540

15 Find the area of the shaded region. 12π

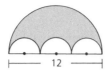

16 On the target, the radius of the bull's-eye is 5 cm, and each band is 5 cm wide.
  a Find the total shaded area to the nearest square centimeter. ≈ 1178 sq cm
  b Find the area of the unshaded bands to the nearest square centimeter. ≈ 785 sq cm
  c What is the probability that if you hit the target, you will get a bull's-eye? (Assume that no skill is involved.) $\frac{1}{25}$

17 In the square grid, each square is 2 cm wide. Find the area of the region bounded by the circular arcs. 32 sq cm

Problem Set C

18 Find the area of each shaded region.
  a $48\sqrt{3} - 16\pi$

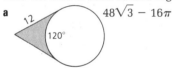

  b $18\sqrt{3} + 24\pi$

Diameter = 12

19 A rotor of a Wankel automotive engine has the geometric shape shown. The center of each arc is the opposite vertex of the equilateral triangle.
  a Find the figure's area. $50\pi - 50\sqrt{3}$
  b Find the figure's perimeter. 10π

20 Find the area of each shaded region.
  a $60\pi + 36\sqrt{3}$ **b** 27π **c** $18\pi + 36\sqrt{3}$

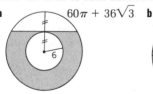
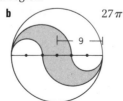
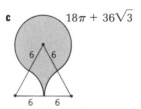

Problem Set Notes and Additional Answers, continued

18a

$A_{\text{sector}} = \frac{120}{360}\pi \cdot (4\sqrt{3})^2 = 16\pi$
$A_{30°\text{-}60°\text{-}90°\triangle} = \frac{1}{2}(12)(4\sqrt{3}) = 24\sqrt{3}$
$A_{\text{shaded}} = 48\sqrt{3} - 16\pi$

b

Shaded region = 2 sectors + 2 equilateral $\triangle$.
$\therefore A = 2\left(\frac{120}{360}\pi 36\right) + 2\left(\frac{36}{4}\sqrt{3}\right)$
$\quad = 24\pi + 18\sqrt{3}$

19a $A_{\text{segment}} = \frac{1}{6}(\pi)(10)^2 - \frac{1}{2}(10)(5\sqrt{3}) = \frac{50\pi}{3} - 25\sqrt{3}$
$A = 3\left(\frac{50\pi}{3} - 25\sqrt{3}\right) + 25\sqrt{3}$
$\quad = 50\pi - 50\sqrt{3}$

b $P = 3\left(\frac{1}{6} \cdot \pi \cdot 20\right) = 10\pi$

20a

$A_{\text{shaded}} = A_{\text{large} \odot} - A_{\text{small} \odot} - A_{\text{segment}}$
$\quad = 144\pi - 36\pi - (48\pi - 36\sqrt{3})$
$\quad = 60\pi + 36\sqrt{3}$

b $A = 2\left(\frac{1}{2} \cdot 36\pi - \frac{1}{2} \cdot 9\pi\right) = 27\pi$

c

The shaded area is formed by subtracting two sectors from the $\triangle$ and then adding five sectors in the $\odot$.
$A = \frac{144\sqrt{3}}{4} - 2 \cdot \frac{1}{6} \cdot 36\pi + 5 \cdot \frac{1}{6} \cdot 36\pi = 36\sqrt{3} + 18\pi$

Problem-Set Notes and
Additional Answers, continued

21a The shaded area is formed by adding the $\triangle$ and two smaller semicircles and then subtracting the larger semicircle.
$A = \frac{1}{2}(10 \cdot 24) + \frac{25\pi}{2} + \frac{144\pi}{2} - \frac{169\pi}{2} = 120$
(which is also the area of the $\triangle$!).

22

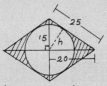

$\frac{1}{2} \cdot 15 \cdot 20 = \frac{1}{2} \cdot 25 \cdot h$
$\therefore h = 12$
$A_{shaded} = A_{rhom} - A_{\odot} = 600 - 144\pi$

23 Radius of $\odot$O = 4
$A_{\odot O} = 16\pi$
Radius of $\odot$P = 7
$A_{\odot P} = 49\pi$
$A_{shaded} = 33\pi$
$P = \frac{33\pi}{49\pi} = \frac{33}{49}$

Problem Set C, *continued*

21 Three arcs are drawn, centered at the midpoints of the sides of a triangle and meeting at the vertices, as shown.

a Find the total area of the shaded regions (which are called the lunes of Hippocrates). **120**

b Find the area of the triangle. **120**

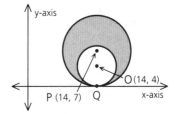

22 A circle is inscribed in a rhombus. Find the area of the shaded region if the diagonals of the rhombus are 30 and 40. **600 − 144π**

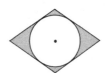

23 If $\odot$O and $\odot$P are tangent to the x-axis at Q and a point is selected at random in the interior of $\odot$P, what is the probability that the point is in the shaded region? $\frac{33}{49}$

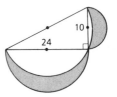

GEOMETRY IN VISUAL COMMUNICATION
William Field uses geometry to achieve simplicity and clarity

In a geometry textbook, lines and curves are the elements of more complex figures, such as angles, polygons, and circles. In the hands of a graphic designer, they are used to create images. The company logos shown on this page were

created by William Field, an award-winning graphic designer headquartered in Santa Fe, New Mexico. All bear Field's own trademarks:

cleanness of line and simplicity.

In an age when desktop publishing and computerized design have become commonplace, says Field, "I use a pen and a piece of paper." To achieve simplicity and clarity, Field uses a traditional grid system. He begins by dividing his page into, say, sixteen squares.

A Santa Fe native, Field earned a degree in anthropology from Harvard University. For ten years he served as the director of design for a camera company. Field has been presented with many of graphic design's most prestigious awards. Today he operates his own graphic design business in Santa Fe.

For each of the logos shown above, explain how geometry works in the image.

11.7 RATIOS OF AREAS

Objectives

After studying this section, you will be able to
- Find ratios of areas by calculating and comparing the areas
- Find ratios of areas by applying properties of similar figures

Part One: Introduction

Computing the Areas

One way of determining the ratio of the areas of two figures is to calculate the quotient of the two areas.

Example 1   Find the ratio of the area of the parallelogram to the area of the triangle.

$$\frac{A_{\square}}{A_{\triangle}} = \frac{b_1 h_1}{\frac{1}{2} b_2 h_2} = \frac{9 \cdot 10}{\frac{1}{2} \cdot 12 \cdot 8} = \frac{90}{48} = \frac{15}{8}, \text{ or } 15:8$$

Example 2   In the diagram, AB = 5 and BC = 2. Find the ratio of the area of $\triangle ABD$ to that of $\triangle CBD$.

Notice that the height of $\triangle ABD$ is the same as the height of $\triangle CBD$ and is labeled by the letter h.

$$\frac{A_{\triangle ABD}}{A_{\triangle CBD}} = \frac{\frac{1}{2} b_1 h}{\frac{1}{2} b_2 h} = \frac{b_1}{b_2} = \frac{5}{2}$$

Similar Figures

As you know, if two triangles are similar, the ratio of any pair of their corresponding altitudes, medians, or angle bisectors equals the ratio of their corresponding sides. Application of this concept leads to an interesting formula.

Section 11.7   Ratios of Areas   **543**

Class Planning

Time Schedule
All levels: 2 days

Resource References
Teacher's Resource Book
    Class Opener 11.7A, 11.7B
    Using Manipulatives 20
    Supposer Worksheet 13
Evaluation
    Tests and Quizzes
        Quizzes 6–9, Series 1
        Quiz 7, Series 2
        Quiz 2, Series 3

Class Opener

Given: $\odot O$, AB = 8, BC = 6
Find: Area of the shaded region
$\angle ABC = 90°$ because an $\angle$ inscribed in a semicircle is a rt. $\angle$. $\therefore$ AC = 10, and AO = 5.
$A_{shaded} = A_{semicircle} - A_{\triangle ABC}$
$\quad = \frac{1}{2}(5^2)\pi - \frac{1}{2} \cdot 6 \cdot 8$
$\quad = \frac{25\pi}{2} - 24$

Lesson Notes

- The two principles of this section may be difficult for some students.
- There are three main ideas in the problems in this section.
  1 Similar triangles (ratio of areas is the square of ratio of sides)
  2 Triangles with same base or same height (ratio of areas is equal to ratio of unequal parts)
  3 Figures transformable by "cut and paste" method (areas are equal)

T 543

11.7

The Geometric Supposer

Using *The Geometric preSupposer: Points and Lines* or *The Geometric Supposer: Triangles* or *Circles*, explore with students what happens to the area of a triangle when the altitude or the length of one side is constant.

Draw an obtuse triangle. Extend one of the sides to form the supplement of the obtuse angle. Make the extension unequal to the length of any of the sides of the triangle. (Do this by estimating. You do not need to measure the sides of the obtuse triangle in order to do this.) Draw a segment from the endpoint of this extension to a vertex of the obtuse triangle to form a second triangle. Ask students to estimate the ratio of the areas of these two triangles. Also ask them to estimate the ratio of the length of the extension to the length of the side of the obtuse triangle of which it is an extension. What is true of the altitudes of these two triangles? (They are congruent.)

Using the Measure option, find the two ratios students estimated. Ask students to make a generalization about the ratio of the areas of any two triangles with a common side. Have students repeat the procedure with several new pairs of triangles to see if this generalization seems valid.

Repeat the procedure described above (though not necessarily beginning with an obtuse triangle), using the common side as the base of each triangle. Compare the ratio of the areas of the two triangles with the ratio of the two altitudes. Again, have students make a generalization and describe any relationships they perceive. Repeat the procedure to test their conjectures.

T 544

Example 1 Given that $\triangle PQR \sim \triangle WXY$, find the ratio of their areas.

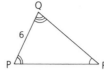

Notice that the ratio of the corresponding sides is $\frac{3}{2}$. The ratio of the areas is

$$\frac{A_{\triangle PQR}}{A_{\triangle WXY}} = \frac{\frac{1}{2}b_1 h_1}{\frac{1}{2}b_2 h_2} = \frac{b_1}{b_2} \cdot \frac{h_1}{h_2}$$

But $\dfrac{b_1}{b_2} = \dfrac{3}{2}$ and $\dfrac{h_1}{h_2} = \dfrac{3}{2}$, so $\dfrac{A_{\triangle PQR}}{A_{\triangle WXY}} = \dfrac{3}{2} \cdot \dfrac{3}{2} = \left(\dfrac{3}{2}\right)^2 = \dfrac{9}{4}$.

Notice that $\frac{9}{4}$ is the square of $\frac{3}{2}$.

The preceding example shows the key steps that can be used to prove a theorem about the areas of similar triangles. Because convex polygons can be divided into triangles, you may suspect that the areas of similar polygons have the same relationship. They do.

Theorem 109 *If two figures are similar, then the ratio of their areas equals the square of the ratio of corresponding segments. (Similar-Figures Theorem)*

$$\frac{A_1}{A_2} = \left(\frac{s_1}{s_2}\right)^2$$

where A_1 and A_2 are areas and s_1 and s_2 are measures of corresponding segments.

Corresponding segments can be any segments associated with the figures, such as sides, altitudes, medians, diagonals, or radii.

Example 2 Given the similar pentagons shown, find the ratio of their areas.

By the Similar-Figures Theorem,

$$\frac{A_I}{A_{II}} = \left(\frac{s_1}{s_2}\right)^2.$$

$$\frac{s_1}{s_2} = \frac{12}{9} = \frac{4}{3}$$

So $\dfrac{A_I}{A_{II}} = \left(\dfrac{4}{3}\right)^2 = \dfrac{16}{9}$, or $16:9$.

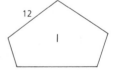

Part Two: Sample Problems

Problem 1   If $\triangle ABC \sim \triangle DEF$ (note the correspondences), find the ratio of the areas of the two triangles.

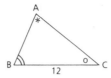

Solution   Use the Similar-Figures Theorem.

$$\frac{A_1}{A_2} = \left(\frac{s_1}{s_2}\right)^2$$

$$= \left(\frac{12}{8}\right)^2 = \left(\frac{3}{2}\right)^2 = \frac{9}{4}$$

Problem 2   If the ratio of the areas of two similar parallelograms is 49:121, find the ratio of their bases.

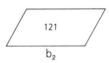

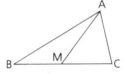

Solution   The Similar-Figures Theorem can be used.

$$\frac{A_1}{A_2} = \left(\frac{b_1}{b_2}\right)^2$$

$$\frac{49}{121} = \left(\frac{b_1}{b_2}\right)^2$$

$$\frac{7}{11} = \frac{b_1}{b_2}$$

Note that $\frac{7}{11}$ is the *square root* of $\frac{49}{121}$.

Problem 3 $\overline{AM}$ is a median of $\triangle ABC$. Find the ratio $A_{\triangle ABM} : A_{\triangle ACM}$.

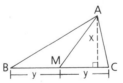

Solution   In this case, the Similar-Figures Theorem does not apply. Start by comparing the altitudes from A. They are the same! Call this common altitude x.
Now compare the bases. BM = MC, because $\overline{AM}$ is a median. Let y represent BM and MC.

$$\frac{A_{\triangle ABM}}{A_{\triangle ACM}} = \frac{\frac{1}{2}b_1 h_1}{\frac{1}{2}b_2 h_2} = \frac{xy}{xy} = 1, \text{ or } 1:1$$

Since the triangles have equal bases and equal heights, their areas are equal.

Communicating Mathematics

Refer to Sample Problem **3**. In $\triangle ABC$, have students draw median $\overline{CN}$. Have students verify that $A_{\triangle CBN} : A_{\triangle CAN} = 1:1$.

The result of sample problem 3 may be stated as a theorem.

Theorem 110   *A median of a triangle divides the triangle into two triangles with equal areas.*

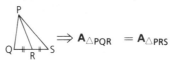 $\Rightarrow \mathbf{A}_{\triangle PQR} = \mathbf{A}_{\triangle PRS}$

Assignment Guide

Basic

| Day 1 | 1, 3–6, 9b |
| Day 2 | 2, 7, 8, 9a,c,d, 10a,b, 11 |

Average

| Day 1 | 1a,c, 2a,c, 3–8, 10–12 |
| Day 2 | 9, 13–17, 20 |

Advanced

| Day 1 | 1a,c, 2a,c, 3–8, 10–14 |
| Day 2 | 9, 15–18, 21, 22 |

Part Three: Problem Sets

Problem Set A

1 By computing the areas, find the ratio of the areas of each pair of figures shown.

a   1:1

c   15:8

b   1:2

d   5:6

2 By using the Similar-Figures Theorem, find the ratio of the areas of each pair of similar figures.

a   25:4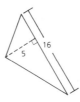

c   4:1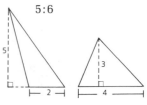

b 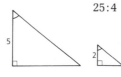   4:9

d 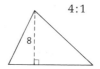   1:9

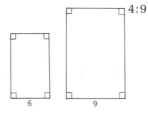

3 Given: $\overline{PM}$ is a median.

Find: **a** $A_{\triangle PQM} : A_{\triangle PRM}$ 1:1

b $A_{\triangle PQM} : A_{\triangle PQR}$ 1:2

c QR:MR 2:1

4 A pair of corresponding sides of two similar triangles are 4 and 9. Find the ratio of the triangles' areas. 16:81

5 If the ratio of the areas of two similar polygons is 9:16, find the ratio of a pair of corresponding altitudes. 3:4

6 Gladys Gardenia has a square garden, 3 m on a side. She wishes to make it exactly twice as large. Gladys decides to double the length and double the width. Does she succeed?
No; the new garden is 4 times as big.

7 Find the ratio of the areas of the regular hexagons. 1:16

8 Find the ratio of the areas of the triangles. 49:16

Problem Set B

9 For each pair of figures, find the ratio of area I to area II.

a

64:225

c

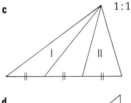

1:1

b

1:2

d

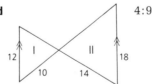

4:9

Cooperative Learning

Have students work in small groups and compare their answers to problem **6.** Then have students answer the following questions.

1 How long would Gladys have to make the sides of her garden in order to have a square garden that is twice as large?
$3\sqrt{2}$

2 If Gladys has a circular garden with a radius of 4, how long would the radius need to be in order to have a garden twice as large?
$4\sqrt{2}$

3 Gladys has a rectangular garden that is 8 feet by 2 feet. She wants to maintain the proportions of the sides, but double the size of the garden. How long should each side be?
$8\sqrt{2}$ and $2\sqrt{2}$

4 How are the answers to all these problems similar, and how do they relate to Theorem 109?
According to Theorem 109, if the ratio of the areas is 1:2, then the ratio of the sides would be $1:\sqrt{2}$.

Problem Set B, *continued*

10 Find the ratio of the area of the shaded triangle to that of the whole triangle.

a

4:81

c
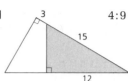
1:3

b 2:9
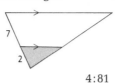

d 3

4:9
15
12

11 Find the ratio of the areas of the two triangles. 5:4

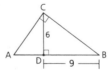
5   13
10
8

12 The ratio of the areas of two similar pentagons is 8:18.
   a Find the ratio of their corresponding sides. 2:3
   b Find the ratio of their perimeters. 2:3

13 The ratio of corresponding medians of two similar triangles is 5:2. Find the area of the larger triangle if the smaller triangle has an area of 40. 250

14 One triangle has sides 13, 13, and 10. A second triangle has sides 12, 20, and 16. Find the ratio of their areas. 5:8

15 Find the ratio of the areas of two circles if their radii are 4 and 9. 16:81

16 Find the ratio of the areas of two equilateral triangles with sides 6 and 8. 9:16

17 Find $A_{\triangle ACD}:A_{\triangle BCD}$. 4:9

C
6
A
D   9   B

**Problem-Set Notes
and Additional Answers**

18a,b Same bases and same heights

c Subtract △WPX from part **b.**

d △WPX ∼ △ZPY, so the ratio of the areas is $\left(\frac{WX}{ZY}\right)^2 = \left(\frac{3}{4}\right)^2 = \frac{9}{16}$.

e From part **d,** $\frac{WP}{PY} = \frac{3}{4}$. Since △WPX and △XPY have equal heights, the ratio of the areas is also $\frac{3}{4}$.

19a–d

Since all 8 small △ have equal areas, results in parts **a–d** follow immediately.

e Since $\overline{DE}$ is the median to the hypotenuse of △HEA, DE = 5. P = 2 · 10 + 2 · 5 = 30.

Problem Set C

18 Given trapezoid WXYZ, find the ratio of the areas of each pair of triangles.

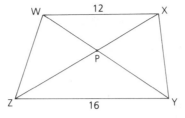

   a △WYZ and △XYZ 1:1

   b △WXZ and △WXY 1:1

   c △WPZ and △XPY 1:1

   d △WPX and △ZPY 9:16

   e △WPX and △XPY 3:4

19 Given △ABC is a right △.
   E and F are midpoints.
   D, H, and G divide $\overline{AC}$ into
   four ≅ segments.

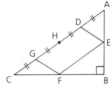

  Find: **a** The ratio of the areas of △ABC and △EBF 4:1

      b The ratio of the areas of △ABC and △GFC 8:1

      c The ratio of the areas of △ADE and △GFC 1:1

      d The ratio of the areas of ▱DEFG and △ABC 1:2

      e The perimeter of ▱DEFG if AC = 20 30

20 If the midpoints of the sides of a quadrilateral are joined in order, another quadrilateral is formed. Find the ratio of the area of the larger quadrilateral to that of the smaller quadrilateral. 2:1

21 Given: Trapezoid ABCD
   Find: The ratio of areas I and II 5:6

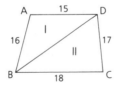

22 PQRS is a parallelogram.

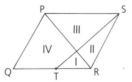

   a If T is a midpoint and the area of PQRS is 60, find the areas of regions I, II, III, and IV. 5; 10; 20; 25

   b If T divides $\overline{QR}$ such that $\frac{QT}{TR} = \frac{x}{y}$, find the ratio of the area of region I to that of ▱PQRS. $\frac{y^2}{2x^2 + 6xy + 4y^2}$

Section 11.7 Ratios of Areas **549**

Problem-Set Notes and Additional Answers, continued

20

$\overline{BH}$ is the base of △ABH and △XBH. The △ have ≅ altitudes, since $\overline{BH}$ is the midline of △ACG. Thus, the △ have equal areas. Similarly, other △ with equal areas are △GHF and △XHF, △EFD and △XFD, and △CBD and △XBD. So $A_{unshaded} = A_{shaded}$.

∴ $\frac{A_{large\ quad}}{A_{small\ quad}} = \frac{2}{1}$

21 Since I and II have the same height, $\frac{A_I}{A_{II}} = \frac{b_I}{b_{II}} = \frac{15}{18} = \frac{5}{6}$.

22a I + II = 15
   I + IV = 30
   II + III = 30
   IV + III = 45
   and $\frac{I}{III} = \left(\frac{1}{2}\right)^2 = \frac{1}{4}$ so
   4(I) = III
   Solving simultaneously yields I = 5, II = 10, III = 20, IV = 25.

b

$A_I = \frac{1}{2}wy$

$A_{▱PQRS} = (x + y)(w + z)$

$\frac{A_I}{A_{▱PQRS}} = \frac{\frac{1}{2}wy}{(x + y)(w + z)}$

However, III ~ I, so $\frac{w}{z} = \frac{y}{x + y}$, so $\frac{w}{w + z} = \frac{y}{x + 2y}$.

By substitution, $\frac{A_I}{A_{▱PQRS}} = \frac{\frac{1}{2}y \cdot y}{(x + y)(x + 2y)} = \frac{\frac{1}{2}y^2}{x^2 + 3xy + 2y^2} = \frac{y^2}{2x^2 + 6xy + 4y^2}$.

T 549

Class Planning

Time Schedule
Basic: 1 day (optional)
Average: 1 day
Advanced: 1 day

Resource References
Teacher's Resource Book
   Class Opener 11.8A, 11.8B
Evaluation
   Tests and Quizzes
      Quiz 10, Series 1
      Quiz 3, Series 3

Class Opener

Given: A(−3, −7), B(2, −3),
      C(12, 5), D(5, 7)
Find: Ratio of area of △ABD to
      area of △CBD

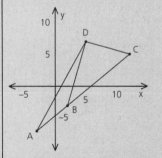

A, B, and C are collinear. The
heights of the △ are =, so
$A_{\triangle ABD}{:}A_{\triangle CBD} = AB{:}BC = 1{:}2$.

11.8 HERO'S AND BRAHMAGUPTA'S FORMULAS

Objective

After studying this section, you will be able to
- Find the areas of figures by using Hero's formula and Brahmagupta's formula

Part One: Introduction

A useful formula for finding the area of a triangle was developed nearly 2000 years ago by the mathematician Hero of Alexandria.

Theorem 111 $A_{\triangle} = \sqrt{s(s - a)(s - b)(s - c)}$,
where a, b, and c are the lengths
of the sides of the triangle and
$s = semiperimeter = \dfrac{a + b + c}{2}$.
(Hero's formula)

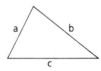

In about A.D. 628, a Hindu mathematician, Brahmagupta, recorded a formula for the area of an inscribed quadrilateral. This formula applies only to quadrilaterals that can be inscribed in circles (known as *cyclic quadrilaterals*).

Theorem 112 $A_{cyclic\ quad} = \sqrt{(s - a)(s - b)(s - c)(s - d)}$,
where a, b, c, and d are the sides of the quadri-
lateral and $s = semiperimeter = \dfrac{a + b + c + d}{2}$.
(Brahmagupta's formula)

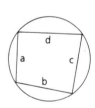

Part Two: Sample Problems

Problem 1   *Find the area of a triangle with sides 3, 6, and 7.*

Solution   First find the semiperimeter.

$$s = \frac{a + b + c}{2} = \frac{3 + 6 + 7}{2} = 8$$

Then use Hero's formula.

$$A_{\triangle} = \sqrt{s(s - a)(s - b)(s - c)}$$
$$= \sqrt{8(8 - 3)(8 - 6)(8 - 7)}$$
$$= \sqrt{8(5)(2)(1)} = \sqrt{16(5)} = 4\sqrt{5}$$

550   Chapter 11 Area

Vocabulary
Brahmagupta's formula
cyclic quadrilateral
Hero's formula

Problem 2   *Find the area of the inscribed quadrilateral with sides 2, 7, 6, and 9.*

Solution   First find the semiperimeter.

$$s = \frac{a + b + c + d}{2} = \frac{2 + 7 + 6 + 9}{2} = 12$$

Then use Brahmagupta's formula.

$$\begin{aligned} A_{\text{cyclic quad}} &= \sqrt{(s - a)(s - b)(s - c)(s - d)} \\ &= \sqrt{(12 - 2)(12 - 7)(12 - 6)(12 - 9)} \\ &= \sqrt{10(5)(6)(3)} = \sqrt{900} = 30 \end{aligned}$$

Part Three: Problem Sets

Problem Set A

1 Use Hero's formula to find the areas of triangles with sides of the following lengths.

a 3, 4, and 5 6
b 3, 3, and 4 $2\sqrt{5}$

c 5, 6, and 9 $10\sqrt{2}$
d 3, 7 and 8 $6\sqrt{3}$

e 8, 15, and 17 60
f 13, 14, and 15 84

2 Use Hero's formula to find the area of an equilateral triangle with a side 8 units long. $16\sqrt{3}$

3 Use Brahmagupta's formula to find the areas of inscribed quadrilaterals with sides of the following lengths.

a 5, 7, 4, and 10 36
b 2, 4, 5, and 9 $4\sqrt{15}$

c 3, 5, 9, and 5 24
d 1, 5, 9, and 11 $16\sqrt{3}$

Problem Set B

4 a Use Hero's formula to find the area of a (2, 5, 7) triangle. 0

b Use Hero's formula to find the area of a (4, 6, 12) triangle. $\sqrt{-385}$ (impossible)

c What explanation can you give for the results in parts **a** and **b**? **a** This is a collapsed $\triangle$, or line segment; **b** The two shorter sides have a sum that is less than the third side, so no such triangle exists.

5 Find the area of the figure to the nearest hundredth. ≈ 32.50

6 Find the area of each triangle.

a

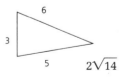

$2\sqrt{14}$

b

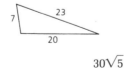

$30\sqrt{5}$

Lesson Notes

■

Deriving Hero's formula makes an excellent problem for the advanced student.

Use the Pythagorean Theorem to establish $h^2 = c^2 - x^2 = b^2 - (a - x)^2$, and thus, $x = \frac{a^2 + c^2 - b^2}{2a}$.

Substitute that value for x into $h^2 = c^2 - x^2 = (c - x)(c + x)$, and show $h^2 = \frac{(b - a + c)(b + a - c)(a + c - b)(a + c + b)}{4a^2}$.

Use $2s = a + b + c$, and show $4a^2 h^2 = 16s(s - a)(s - b)(s - c)$. Finally, find an expression for $\frac{1}{2}ah$, the area of the triangle.

Assignment Guide

Basic (optional)

1a,d,e, 3c,d, 5, 6, 10

Average

1d,e, 3c,d, 4, 7–10, 12

Advanced

4–10, 12, 13

Cooperative Learning

Refer to problem **1**. Parts **a** and **e** lend themselves to checking the answers using another method. Have students work in small groups and discuss what other method can be used and why. In parts **a** and **e**, the $\triangle$ are rt. $\triangle$. Use $A = \frac{1}{2}bh$.

Problem-Set Notes and Additional Answers

7

$$A_I = \frac{1}{2} \cdot 9 \cdot 12 = 54$$
$$A_{II} = \sqrt{18(18-10)(18-11)(18-15)}$$
$$= 12\sqrt{21} + 54$$

12

$$A_I = \sqrt{\left(\frac{25}{2} - 5\right)^3\left(\frac{25}{2} - 10\right)}$$
$$= \frac{75\sqrt{3}}{4}$$
$$A_{II} = \sqrt{12(12 - 6)(12 - 8)(12 - 10)}$$
$$= 24$$
$$A_{pent} = \frac{75\sqrt{3}}{4} + 24$$

13a E, the midpt. of $\overline{CD}$, is (6, 8). $\triangle CEO$ is a 30°-60°-90° rt. $\triangle$ with CE = 3, EO = $3\sqrt{3}$. So, O = (6, 8 − $3\sqrt{3}$).
  b $\triangle COD$ is equilateral and CO = 6. So, circumference = $12\pi \approx 37.7$.

Problem Set B, *continued*

7 Verify that the area of the quadrilateral shown is $12\sqrt{21} + 54$.

8 Find the measures of the three altitudes of the triangle at the right. (Hint: Use Hero's formula to find the area, and then use $A = \frac{1}{2}bh$ to find each altitude.) $\frac{168}{13}$; 12; $\frac{56}{5}$

9 Find the area of the quadrilateral shown. $2\sqrt{6}$

10 Find the area of the triangle. 76.5

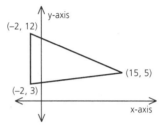

Problem Set C

11 a As $\overline{PQ}$ gets smaller and smaller, what happens to quadrilateral PQRS? It approaches a $\triangle$.

  b What happens to Brahmagupta's formula if P and Q become the same point? It becomes Hero's formula.

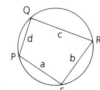

12 Find the area of the pentagon to the nearest tenth. ≈ 56.5

13 Given: $\odot O$, with C = (3, 8) and D = (9, 8), $\overset{\frown}{mCD} = 60$

  Find: **a** The coordinates of O (6, 8 − $3\sqrt{3}$)

   b The circumference of $\odot O$ to the nearest tenth ≈ 37.7

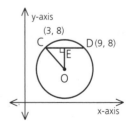

CHAPTER SUMMARY

CONCEPTS AND PROCEDURES

After studying this chapter, you should be able to

- Understand the concept of area (11.1)
- Find the areas of rectangles and squares (11.1)
- Use the basic properties of area (11.1)
- Find the areas of parallelograms (11.2)
- Find the areas of triangles (11.2)
- Find the areas of trapezoids (11.3)
- Use the measure of a trapezoid's median to find its area (11.3)
- Find the areas of kites (11.4)
- Find the areas of equilateral triangles (11.5)
- Find the areas of other regular polygons (11.5)
- Find the areas of circles (11.6)
- Find the areas of sectors (11.6)
- Find the areas of segments (11.6)
- Find ratios of areas by calculating and comparing the areas (11.7)
- Find ratios of areas by applying properties of similar figures (11.7)
- Find the areas of figures by using Hero's formula and Brahmagupta's formula (11.8)

VOCABULARY

annulus (11.6)

apothem (11.5)

area (11.1)

Brahmagupta's formula (11.8)

cyclic quadrilateral (11.8)

Hero's formula (11.8)

linear unit (11.1)

median (11.3)

radius (11.5)

sector (11.6)

segment (11.6)

square unit (11.1)

Chapter 11 Review

Class Planning

Time Schedule
All levels: 1 day

Resource References
Evaluation
    Tests and Quizzes
        Test 11, Series 1, 2, 3

Assignment Guide

Basic
1–10, 13, 14b, 15, 16, 22, 27

Average
16–29, 31

Advanced
16, 18–20, 22, 27–29, 31, 34, 38, 40, 43, 44

To integrate constructions, assign

Basic   4 and 6, page 680

Average   16 and 17, page 676
        15, page 681

Advanced   23 and 24, page 677
        17, 21, and 23, page 681
        29, page 685

REVIEW PROBLEMS

Problem Set A

1 Find the areas of the following polygons.
  a A rectangle with base 12 and height 7 84
  b A triangle with base 12 and height 7 42
  c A parallelogram with base 15 and height 5 75
  d A trapezoid with bases 3 and 10 and height 8 52
  e A kite with diagonals 5 and 8 20
  f A trapezoid with median 4 and height 2 8

2 Find the areas of rhombuses with the following dimensions.
  a A base of 9 and a height of 7 63
  b Diagonals of 6 and 11 33

3 Find the area of each shaded region.

  a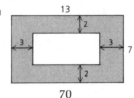
  70

  b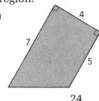
  24

  c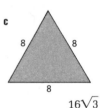
  $16\sqrt{3}$

4 A rectangular driveway is to be paved. The driveway is 20 m long and 4 m wide. The cost will be $15 per square meter. What is the total cost of paving the driveway. $1200.00

5 Find the area of a parallelogram with sides 12 and 8 and included angle 60°. $48\sqrt{3}$

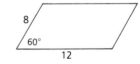

6 Find the area of an isosceles trapezoid with sides 8, 20, 40, and 20. 288

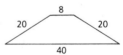

7 Find the area of the triangle shown at the right. 18

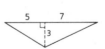

8 John has two sticks, 90 cm and 50 cm long, to use in making a paper kite. What will the cost of the kite be if the sticks and glue are gifts and the paper costs 3 cents per square decimeter? $67\frac{1}{2}$ cents

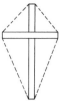

9 The apothem of a regular polygon is 7, and the polygon's perimeter is 56. Find the polygon's area. 196

10 Find the area of a circle if its circumference is 16π. 64π

11 Find the area of a square whose semiperimeter is 18 m. 81 sq m

12 Find, to the nearest tenth, the area of a semicircle whose diameter is 14 mm. ≈ 77.0 sq mm

13 Find the area of each shaded region.

a 9π

b 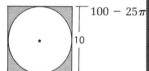 16π

c $100 - 25\pi$

14 Find the area of each sector.

a 24π

b 16π

c 4π

15 Find the ratio of the areas of each pair of figures.

a 5:8 :

b 9:16

16 Find the coordinates of B so that $\triangle ABD$ will have the same area as $\triangle ACD$. $(-2, 0)$

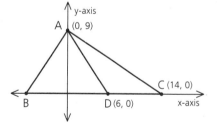

32a II ~ (I + II)

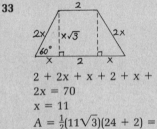

$$\frac{A_{II}}{A_{(I + II)}} = \left(\frac{6}{9}\right)^2 = \left(\frac{2}{3}\right)^2 = \frac{4}{9}$$

$$\therefore \frac{A_I}{A_{(I + II)}} = \frac{9}{9} - \frac{4}{9} = \frac{5}{9}$$

$$\frac{A_I}{A_{II}} = \frac{\frac{5}{9}}{\frac{4}{9}} = \frac{5}{4}$$

b I ~ (I + II)

$$\frac{A_I}{A_{(I + II)}} = \left(\frac{10}{16}\right)^2 = \left(\frac{5}{8}\right)^2 = \frac{25}{64}$$

$$\frac{A_{II}}{A_{(I + II)}} = \frac{64}{64} - \frac{25}{64} = \frac{39}{64}$$

$$\frac{A_I}{A_{II}} = \frac{\frac{25}{64}}{\frac{39}{64}} = \frac{25}{39}$$

33

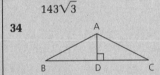

2 + 2x + x + 2 + x + 2x = 70

x = 11

$A = \frac{1}{2}(11\sqrt{3})(24 + 2) = 143\sqrt{3}$

34

Given: ∠BAC is supp. to ∠QPR.
△ABC and △PQR are isos.
$\overline{BC}$ and $\overline{QR}$ are bases.
$\overline{AB} \cong \overline{QR}$
Prove: $A_{\triangle ABC} = A_{\triangle PQR}$
We can let m∠BAC = 2x, which will make m∠QPR = 180 − 2x. Thus, m∠DAC = x and m∠SPR = 90 − x. They are thus comp. ∠s. Since △CDA is a rt. △, we know that ∠C is comp. to ∠DAC and thus ∠C ≅ ∠SPR. ∴ △ADC ≅ △RSP by AAS and thus have equal areas. Similarly, △BDA ≅ △PSQ and thus have equal areas. ∴ $A_{\triangle ABC} = A_{\triangle PQR}$.

Problem Set B

17 Find the area of a triangle with sides 41, 41, and 18. 360

18 Find the area of a parallelogram with sides 6 and 7 and included angle 45°. $21\sqrt{2}$

19 Find the area of a rhombus whose perimeter is 52 and longer diagonal is 24. 120

20 Find the area of an equilateral triangle with perimeter 21. $\frac{49}{4}\sqrt{3}$

21 Find the area and the perimeter of an isosceles trapezoid with lower base 18, upper base 4, and upper base angle 120°. $A = 77\sqrt{3}$; P = 50

22 Find, to the nearest tenth,

  a The circumference of the circle ≈40.2

  b The area of the circle ≈128.8

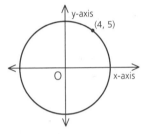

23 a The diagonal of a square is 26. Find the square's area. 338

  b Find the diagonal of a square whose area is 18. 6

24 Find the area of a regular hexagon whose span is 36. $648\sqrt{3}$

25 Find the area of the shaded region in each figure.

  a $\frac{168}{5}$ **b** $27\sqrt{3} - 9\pi$ **c** $6 - \pi$

  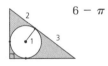

26 For each figure, find the ratio of the area of the whole figure to that of the shaded region.

  a   **b**   **c**

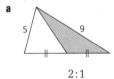 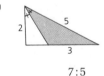

  2:1   7:5   4:1

27 Find the area of each shaded segment.

a $9\pi - 18$

b $6\pi - 9\sqrt{3}$

35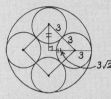

$$A = A_{\text{large}\odot} - 4 \cdot A_{\text{small}\odot}$$
$$= \pi(3 + 3\sqrt{2})^2 - 4\pi \cdot 3^2$$
$$= \pi(9 + 18\sqrt{2} + 18) - 36\pi$$
$$= 18\pi\sqrt{2} - 9\pi$$

28 For each figure, find the ratio of the area of region I to that of region II.

a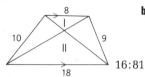
$16:81$

b
$4:9$

c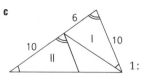
$1:1$

36a

$$A_\triangle = \frac{100\sqrt{3}}{4} = 25\sqrt{3}$$
$$A_{\text{segment}} = \frac{1}{6} \cdot 100\pi - 25\sqrt{3}$$
$$= \frac{50\pi}{3} - 25\sqrt{3}$$
$$A_{\text{shaded}} = 25\sqrt{3} - 3\left(\frac{50\pi}{3} - 25\sqrt{3}\right)$$
$$= 100\sqrt{3} - 50\pi$$

29 Which has a greater area, a circle with a circumference of 100 or a square with a perimeter of 100? Circle

30 BRAT is a trapezoid, with M the midpoint of one of the legs. Show that the area of $\triangle$CAT is equal to the area of BRAT.

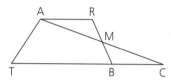

b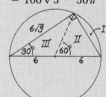

$$A_{\text{I}} = \frac{1}{6} \cdot 36\pi - \frac{36\sqrt{3}}{4} = 6\pi - 9\sqrt{3}$$
$$A_{(\text{II} + \text{III})} = \frac{1}{2} \cdot 6 \cdot 6\sqrt{3} = 18\sqrt{3}$$
$$A_{(\text{I} + \text{II} + \text{III})} = 6\pi + 9\sqrt{3}$$

31 TRAP is an isosceles trapezoid.
  a Find the coordinates of R. (12, 13)
  b Find the area of TRAP. 156

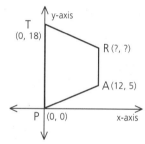

37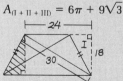

Since the area of the shaded region $= A_{\text{I}}$, $A_{\text{trap}} = A_{\text{rect}} = 24 \cdot 18 = 432$.

Problem Set C

32 In each figure, find the ratio of the area of region I to that of region II.

a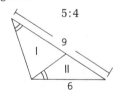
$5:4$

b
$25:39$

33 Given an isosceles trapezoid with a smaller base of 2, a perimeter of 70, and acute base angles of 60°, find the trapezoid's area. $143\sqrt{3}$

38a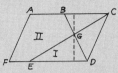

Since $\triangle BGC \sim \triangle DGE$, the altitudes are proportional to the bases. Thus, $\frac{A_I}{A_\square} = \frac{\frac{25hx}{2}}{48hx} = \frac{25}{96}$.

b $A_{II} = A_\square - A_{\triangle ECD} - A_{\triangle BGC}$
$= 48hx - 20hx - \frac{9hx}{2}$
$= \frac{47hx}{2}$

Prob. $= \frac{A_{II}}{A_\square} = \frac{\frac{47hx}{2}}{48hx} = \frac{47}{96}$

39

$A_{shaded} = A_\triangle - A_{sector}$
$= \frac{1}{2}(12 \cdot 12\sqrt{3}) - \frac{1}{6}\pi \cdot 12^2$
$= 72\sqrt{3} - 24\pi$

40

$A = \frac{3}{4}\pi \cdot 12^2 + \frac{1}{4}\pi 4^2 + \frac{1}{4}\pi \cdot 2^2$
$= 113\pi$

41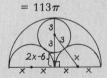

$(x + 3)^2 = x^2 + (2x - 3)^2$
$x = \frac{9}{2}$

$A = \frac{1}{2}(81\pi) - 2 \cdot \frac{1}{2}\left(\frac{81\pi}{4}\right) - 9\pi = \frac{45\pi}{4} = 11\frac{1}{4}\pi$

Review Problem Set C, *continued*

34 The legs of one isosceles triangle are congruent to those of another isosceles triangle, and the triangles' vertex angles are supplementary. Prove that the triangles' areas are equal. (Write a paragraph proof.)

35 Five circles are tangent as shown. If each small circle has a radius of 3, find the shaded area. $18\pi\sqrt{2} - 9\pi$

36 Find the shaded areas.

a 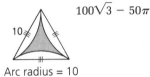 $100\sqrt{3} - 50\pi$

Arc radius = 10

b 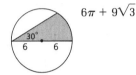 $6\pi + 9\sqrt{3}$

37 Find the area of a trapezoid whose diagonals are each 30 and whose height is 18. 432

38 Given: AB:BC = 1:1 and FE:ED = 1:5

a Find the ratio of the area of region I to the area of $\square$ACDF. $\frac{25}{96}$

b What is the probability that a gnat landing in $\square$ACDF would land in region II? $\frac{47}{96}$

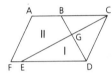

39 $\overline{AT}$ is tangent to $\odot P$ at T, and AB = 12. Find the shaded area. $72\sqrt{3} - 24\pi$

40 Archibold left his horse, Gremilda, tied to the corner of a barn by a 12-m rope. The barn measures 8 m by 10 m. Find the total grazing area for Gremilda. 113π sq m

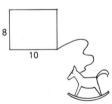

41 Find the area of the shaded region. $11\frac{1}{4}\pi$

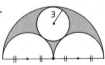

42 All 12 sides of the cross are congruent, each having a length of 4. All the angles are right angles. Find the area of the shaded region. $40\pi - 80$

43 Find the area of a trapezoid with sides 12, 17, 40, and 25 if the bases are the sides measuring 12 and 40. 390

44 Buster Bee lives on a honeycomb. What percentage of the honeycomb is made of wax? 36%

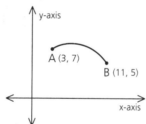

45 If m$\widehat{AB}$ = 90, find, to the nearest tenth, the area of the circle containing $\widehat{AB}$. ≈106.8

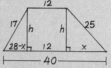

A (3, 7)

B (11, 5)

y-axis

x-axis

Problem-Set Notes and Additional Answers, continued

42

$A_{cross} = 5 \cdot 16 = 80$
$AB^2 = 4^2 + 12^2 = 16 + 144 = 160$
$AB = 4\sqrt{10}$
$\therefore r = 2\sqrt{10}$
$A_{\odot} = \pi \cdot (2\sqrt{10})^2 = 40\pi$
$A_{shaded} = 40\pi - 80$

43

$x^2 + h^2 = 25^2$
$(28 - x)^2 + h^2 = 17^2$
$x = 20;\ h = 15$
$A = \frac{1}{2}(15)(12 + 40) = 390$

44 $A_{large\ hex} = \frac{1}{2}\left(\frac{5\sqrt{3}}{2}\right)(30) = \frac{75\sqrt{3}}{2}$
$A_{small\ hex} = \frac{1}{2}(2\sqrt{3})(24) = 24\sqrt{3}$
$A_{wax} = \frac{75\sqrt{3}}{2} - 24\sqrt{3} = \frac{27\sqrt{3}}{2}$
$\therefore \% = \frac{\frac{27\sqrt{3}}{2}}{\frac{75\sqrt{3}}{2}} = \frac{27}{75} = 36\%$

45
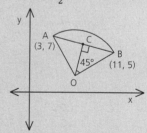

$AB = \sqrt{(11 - 3)^2 + (7 - 5)^2}$
$= \sqrt{64 + 4}$
$= \sqrt{68}$
$= 2\sqrt{17}$
$BC = \sqrt{17}$
$OB = \sqrt{17} \cdot \sqrt{2} = \sqrt{34}$
$A = 34\pi \approx 106.8$

Chapter 12 Schedule

Basic

Problem Sets:	7 days
Review:	1 day
Test:	1 day

Average

Problem Sets:	7 days
Review:	1 day
Test:	1 day

Advanced

Problem Sets:	6 days
Review:	1 day
Test:	1 day

MANY OF THE ideas in this chapter are extensions, into three dimensions, of concepts from Chapters **9–11**.

For some basic classes, this may be the last chapter covered. The Cumulative Review at the end of the chapter can be helpful in reviewing the course.

This chapter has three sections on surface area and three on volume.

Surface area is computed for

1 prisms,
2 pyramids, and
3 circular solids.

Volume is computed for

1 solids with parallel and congruent bases,
2 pointed solids, and
3 spheres.

SURFACE AREA AND VOLUME

The abstract concepts of surface area and volume become tangible in this solidscape by Pete Turner.

12.1 SURFACE AREAS OF PRISMS

Objective

After studying this section, you will be able to
- Find the surface areas of prisms

Part One: Introduction

Solids with flat faces are called ***polyhedra*** (meaning "many faces"). The faces are polygons, and the lines where they intersect are called edges.

One familiar type of polyhedron is the ***prism.*** Here are three examples:

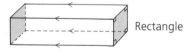

Triangle

Rectangle

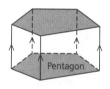

Pentagon

Triangular Prism   **Rectangular Prism**   **Pentagonal Prism**

Every prism has two congruent parallel faces (shaded in the examples) and a set of parallel edges that connect corresponding vertices of the two parallel faces.

The two parallel and congruent faces are called ***bases.*** The parallel edges joining the vertices of the bases are called ***lateral edges.*** The faces of the prism that are not bases are called ***lateral faces.*** The lateral faces of all prisms are parallelograms. Therefore, we name prisms by their bases—a prism with hexagonal bases, for example, is called a hexagonal prism.

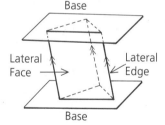

Base

Lateral
Face

Lateral
Edge

Base

Vocabulary

base
lateral edge
lateral face
lateral surface area

polyhedron
prism
total surface area

Class Planning

Time Schedule
All levels: 1 day

Resource References
Teacher's Resource Book
    Class Opener 12.1A, 12.1B
Transparency 19

Class Opener

A room is 32 feet by 20 feet, with a 10-foot-high ceiling. How many square feet of surface must be painted to cover the walls, the ceiling, and the floor?   2320 sq ft
This problem sets the stage for a discussion of surface area. This may also be an appropriate time to discuss volume.

Lesson Notes

- The terminology for solids and their surfaces is used throughout this chapter.

Definition   The *lateral surface area* of a prism is the sum of the areas of the lateral faces.

Definition   The *total surface area* of a prism is the sum of the prism's lateral area and the areas of the two bases.

If the lateral edges are perpendicular to the bases, then the lateral faces will be rectangles. (Why?) In such a case, we put the word *right* in front of the name of the prism. In this book, the word *box* will often be used to refer to a right prism.

Right Triangular Prism   **Right Pentagonal Prism**

Note   The base of a right triangular prism is not necessarily a right triangle.

Part Two: Sample Problem

Problem   Given: The right triangular prism shown

Find: **a** Its lateral area (L.A.)

b Its total area (T.A.)

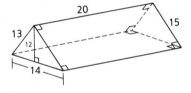

Solution   The right triangular prism can be divided into two triangles (the parallel bases) and three rectangles (the lateral faces).

a

L.A. = | A = 13 · 20 = 260 | 13 + | A = 14 · 20 = 280 | 14 + | A = 15 · 20 = 300 | 15

                20   20   20

Thus, L.A. = 260 + 280 + 300 = 840.

b T.A. = L.A.   +   [triangle: 12, 14, Base]   +   [triangle: 12, 14, Base]

Since the area of each base is $\frac{1}{2}(12)(14)$, or 84,

T.A. = 840 + 84 + 84 = 1008.

Part Three: Problem Sets

Problem Set A

1 Find the total surface area of a right rectangular prism with the given dimensions.

a $\ell = 15$ cm, $w = 5$ cm, $h = 10$ cm 550 sq cm
b $\ell = 12$ mm, $w = 7$ mm, $h = 3$ mm 282 sq mm
c $\ell = 18$ in., $w = 9$ in., $h = 9$ in. 810 sq in.

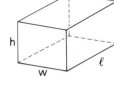

2 Find the lateral area of a right triangular prism with the given dimensions.

a $\ell = 10$, $a = 3$, $b = 5$, $c = 7$ 150
b $\ell = 14$, $a = 2$, $b = 3$, $c = 4$ 126

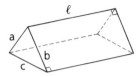

3 A right triangular prism has bases that are isosceles triangles. What is

a The prism's lateral area? 550
b The area of one base? 120
c The prism's total area? 790

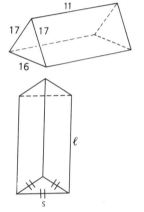

4 Find the total surface area of a right equilateral triangular prism with the given dimensions.

a $s = 6$, $\ell = 5$ $90 + 18\sqrt{3}$
b $s = 12$, $\ell = 10$ $360 + 72\sqrt{3}$

5 A cube is a rectangular prism in which each face is a square. What is the total surface area of a cube in which each edge has a measure of

a 5? 150
b 7? 294

Problem Set B

6 Find the total area of the pieces of cardboard needed to construct each open box shown.

a Open Top 236
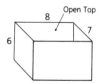

b Open Top $144 + 66\sqrt{3}$
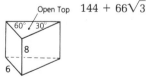

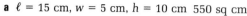

12.1

Assignment Guide

Basic
1b,c, 2b, 3, 4b, 5, 7b,c
Average
1c, 2b, 3, 4b, 5b, 6, 7c,d
Advanced
6, 7, 9–11

Communicating Mathematics

Refer to problem **5.** After students work parts **a** and **b**, have them rework the problem for a cube whose dimensions are twice as great. Have them write a generalization for what they discover.
The total surface area is four times as great.

Cooperative Learning

Students can work in small groups to solve problem **11** on page 564. A physical solid cube or a clay cube can be cut as a model.

Extend the problem by asking students to consider that the same 6-inch cube is cut into 64 smaller cubes. Would the total surface area of the unpainted surface change? If so, by how much?
The total surface area of the unpainted surface for four cuts per side would be 648 square inches. This is 216 more than in the original problem.

T 563

9 L.A. = 2(11 · 13) +
        2(11 · 15)
     = 616
Find the area $\mathcal{B}$ of each
base.

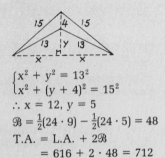

$\begin{cases} x^2 + y^2 = 13^2 \\ x^2 + (y + 4)^2 = 15^2 \end{cases}$
∴ x = 12, y = 5
$\mathcal{B} = \frac{1}{2}(24 \cdot 9) - \frac{1}{2}(24 \cdot 5) = 48$
T.A. = L.A. + 2$\mathcal{B}$
     = 616 + 2 · 48 = 712

10 L.A. = P · h = 17 · 10 = 170

11a

8 like I, 12 like II, 6 like III,
1 inside

b Prob. = $\frac{8 + 12}{27} = \frac{20}{27}$

c Each unpainted side has
area 4.

Cube Type	Blank Faces	Cubes	Total Blank Faces
I	3	8	24
II	4	12	48
III	5	6	30
Inside	6	1	6
Total			108

Total area = 108 · 4 = 432

Problem Set B, *continued*

7 Find the lateral area and the total area of each prism.

a Right Square Prism L.A. = 480;
                              T.A. = 552

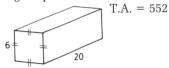

c Right Isosceles Triangular Prism

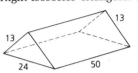

L.A. = 2500;
T.A. = 2620

b Right Triangular Prism L.A. = 120;
                                 T.A. = 132

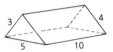

d Regular Hexagonal Prism

L.A. = 360;
T.A. = 360 + 108$\sqrt{3}$

8 Find the total area of the right prism
shown. 226

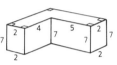

Problem Set C

9 Find the lateral area and the total area of
the right prism shown. L.A. = 616; T.A. = 712

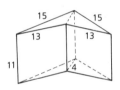

10 The perimeter of the scalene base of a pentagonal right prism is
17, and a lateral edge of the prism measures 10. Find the prism's
lateral area. 170

11 A 6-inch cube is painted on the outside and cut into 27 smaller
cubes.

a How many of the small cubes have six
faces painted? Five faces painted? Four
faces painted? Three faces painted?
Two faces painted? One face painted?
No face painted? 0; 0; 0; 8; 12; 6; 1

b If one of the small cubes is selected at
random, what is the probability that it
has at least two painted faces? $\frac{20}{27}$

c What is the total area of the unpainted
surfaces? 432 sq in.

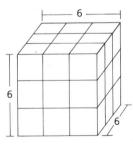

564 | Chapter 12   Surface Area and Volume

12.2 SURFACE AREAS OF PYRAMIDS

Objective

After studying this section, you will be able to
- Find the surface areas of pyramids

Part One: Introduction

Triangular Pyramid   Rectangular Pyramid   Pentagonal Pyramid

A *pyramid* has only one base. Its lateral edges are not parallel but meet at a single point called the vertex. The base may be any type of polygon, but the lateral faces will always be triangles. The diagrams above show three types of pyramids. Notice that each pyramid is named by its base.

A *regular pyramid* has a regular polygon as its base and also has congruent lateral edges. Thus, the lateral faces of a regular pyramid are congruent isosceles triangles.

Face

Recall from Section 9.8 that the altitude of a regular pyramid is a perpendicular segment from the vertex to the base. (The foot of the altitude is the center of the base.) Also recall that a regular pyramid's slant height is the height of a lateral face.

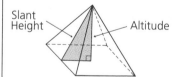

Slant Height   Altitude

Altitude

Lateral Edge

The altitude and a slant height determine a right triangle.

The altitude and a lateral edge determine a right triangle.

Section 12.2 Surface Areas of Pyramids **565**

Class Planning

Time Schedule
All levels: 1 day

Resource References
Teacher's Resource Book
  Class Opener 12.2A, 12.2B
Transparency 20

Class Opener

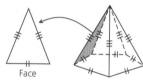

8 cm

A pyramid has a square base and equilateral triangles as faces. If the edges are each 8 cm long, find the total surface area of the pyramid.
The total surface area is
$64 + 64\sqrt{3}$ sq cm.

Communicating Mathematics

Have students write a paragraph that explains the difference between the altitude and the slant height of a regular pyramid.

Vocabulary
pyramid
regular pyramid

Part Two: Sample Problems

Problem 1   Given: The regular pyramid shown at the right

Find: **a** Its lateral area (L.A.)

   b Its total area (T.A.)

Solution   **a** The lateral area is the sum of the areas of four congruent isosceles triangles.

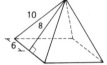

The Pythagorean Theorem shows the slant height to be 8. The area of each lateral face is $\frac{1}{2}(12)(8)$, or 48, so L.A. $= 4(48) = 192$.

b The total area is equal to the lateral area plus the area of the base. The area of the square base is 12^2, or 144, so T.A. $= 192 + 144 = 336$.

Problem 2   The base of rectangular pyramid ABCDE is 10 by 18. The altitude is 12. The lateral edges are congruent.

a Why is ABCDE not a regular pyramid?

b Find the pyramid's total surface area.

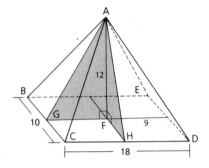

Solution   **a** The base is not regular, so ABCDE is not regular.

b AH and AG are the heights of the lateral faces. Applying the Pythagorean Theorem to $\triangle$AFH and $\triangle$AHG, we find that AH = 13 and AG = 15. There are five faces:

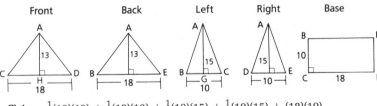

T.A. $= \frac{1}{2}(18)(13) + \frac{1}{2}(18)(13) + \frac{1}{2}(10)(15) + \frac{1}{2}(10)(15) + (18)(10)$

$= \quad 117 \quad + \quad 117 \quad + \quad 75 \quad + \quad 75 \quad + \quad 180$

$= 564$

Part Three: Problem Sets

Problem Set A

1 The pyramid shown is regular and has a square base.
  a Find the area of each lateral face. 60
  b Find the pyramid's lateral area. 240
  c Find the pyramid's total area. 340

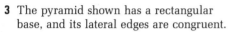

2 The pyramid shown is regular and has a triangular base. What is
  a The area of each lateral face? 120
  b The area of the base? $64\sqrt{3}$
  c The total area? $360 + 64\sqrt{3}$

3 The pyramid shown has a rectangular base, and its lateral edges are congruent.
  a Why is this pyramid not regular? The base is not regular.
  b What is its lateral area? 936
  c What is its total area? 1356

4 The diagram shows a solid that is a combination of a prism and a regular pyramid.
  a Is ABCD a face of the solid? No
  b How many faces does this solid have? 9
  c Find the total area. 740

5 PRXYZ is a regular pyramid. The midpoints of its lateral edges are joined to form a square, ABCD. PR = 10 and RX = 12.
  a Find the lateral area of PRXYZ. 192
  b Find the lateral area of pyramid PABCD. 48
  c What is the area of square ABCD? 36
  d What is the area of square RXYZ? 144
  e Find the ratio of the area of ABCD to the area of RXYZ. 1:4
  f What is the area of trapezoid ABXR? 36

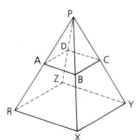

Problem Set B

6 A regular pyramid has a slant height of 8. The area of its square base is 25. Find its total area. **105**

7 A regular pyramid has a slant height of 12 and a lateral edge of 15. What is

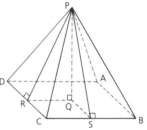

  a The perimeter of the base? **72**

  b The pyramid's lateral area? **432**

  c The area of the base? **324**

  d The pyramid's total area? **756**

8 PABCD is a regular square pyramid.

  a If each side of the base has a length of 14 and the altitude (PQ) is 24, find the pyramid's lateral area and total area. L.A. = 700

  b If each slant height is 17 and the alti- T.A. = 896
  tude is 15, find the pyramid's lateral area and total area. L.A. = 544; T.A. = 800

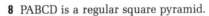

9 Suppose that the pyramid in problem 8 were not regular but had a rectangular base and congruent lateral edges.

  a Given that PQ = 8, CD = 12, and BC = 30, find PR (the slant height of face PCD), PS (the slant height of face PBC), and the lateral area and the total area of the pyramid. **17; 10; L.A. = 504; T.A. = 864**

  b If each lateral edge were 25 and the base were 24 by 30, what would the altitude (PQ) of the pyramid be? **16**

Problem Set C

10 Each lateral edge of a regular square pyramid is 3, and the height of the pyramid is 1. What is

  a The measure of a diagonal of the base? $4\sqrt{2}$

  b The pyramid's slant height? $\sqrt{5}$

  c The area of the base? **16**

  d The pyramid's lateral area? $8\sqrt{5}$

11 A regular *tetrahedron* ("four faces") is a pyramid with four equilateral triangular faces. If a regular tetrahedron has an edge of 6, what is

  a Its total surface area? $36\sqrt{3}$

  b Its height? $2\sqrt{6}$

**Problem-Set Notes
and Additional Answers**

10a–c

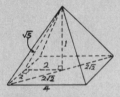

Use the Pythagorean Theorem.

d L.A. $= 4\left(\frac{1}{2} \cdot 4 \cdot \sqrt{5}\right) = 8\sqrt{5}$

11a

T.A. $= 4\left(\frac{6^2}{4}\sqrt{3}\right)$
$= 36\sqrt{3}$

b

AC $= 2\sqrt{3}$
AB $= 6$
∴ BC $= 2\sqrt{6}$ by the Pythagorean Theorem.

12 A regular *octahedron* is a solid with eight faces, each of which is an equilateral triangle. If each edge of the regular octahedron shown is 6 mm long, what is

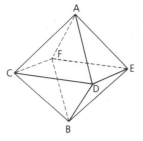

  a The solid's total surface area? $72\sqrt{3}$ sq mm

  b The distance from C to E? $6\sqrt{2}$ mm

  c The distance from A to B? $6\sqrt{2}$ mm

  d The shape of quadrilateral ACBE? Square

13 A regular *hexahedron* is a solid that does not have triangular faces. What is the common name for a regular hexahedron? Cube

Problem-Set Notes and Additional Answers, continued

12a   T.A. $= 8\left(\frac{6^2}{4}\sqrt{3}\right)$
         $= 72\sqrt{3}$

b–d Since ACBE, ADBF, and CDEF are squares with edge 6, diagonals CE and AB $= 6\sqrt{2}$.

CAREER PROFILE

PACKAGING IDEAS
Jolene Randby engineers design

Each year American workers produce more than $1 trillion worth of manufactured products. Nearly all must be packaged in some kind of box, carton, bag, or can. Besides basic size considerations, economic, environmental, health, and safety factors also affect the design of a package. The task of harmonizing all these factors falls to packaging engineers, employed by all major manufacturers.

"My job is to write the specifications and packaging standards and to design the packages for all of the industrial tapes that we produce," explains Jolene Randby, a packaging engineer with the 3M Corporation in St. Paul, Minnesota. "Our tapes are manufactured in rolls varying from 60 yards to 1000 yards in length. When a new container is needed, I use basic geometric formulas to find the roll dimensions and then calculate the size of the primary container and the shipping container." Formulas for volume and surface area of rectangular prisms and cylinders are commonly used by packaging

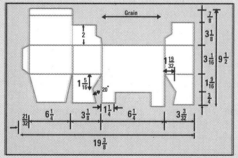

engineers. Randby also calculates the size of the pallet, the small platform on which shipping containers are stacked for storage and transportation.

Jolene Randby attended high school in her hometown of St. Paul. At the University of Wisconsin at Stout, where she earned her bachelor's degree, she majored in industrial technology, with a concentration in packaging engineering.

High on the list of benefits of her job is the necessary travel. "We have twelve converting plants where our packages are assembled," she explains. "I visit them all to oversee production." She also works closely with the marketing and purchasing departments at 3M.

Tape measuring 2 in. in width is manufactured in 250-foot rolls, each measuring $3\frac{3}{4}$ in. in diameter. A shipping container contains four stacks of tape arranged in a square array, with six rolls in each stack. How many shipping containers can fit in one layer on a pallet measuring 42 in. by 48 in.? The walls of each container are $\frac{1}{4}$ in. thick. 30 containers

SURFACE AREAS OF CIRCULAR SOLIDS

Class Planning

Time Schedule
All levels: 1 day

Resource References
Teacher's Resource Book
   Class Opener 12.3A, 12.3B
Transparency 21

Class Opener

8 in.

12 in.

Gandalf took a piece of paper 8 inches by 12 inches and made a cylindrical "can" out of it. He then put a top and bottom on the can. Compute the surface area of the can.

c = 8 = 2πr

12

Gandalf

Circumference of top/bottom is 8 in. Therefore, $c = 8 = 2\pi r$; $r = \frac{4}{\pi}$. So $A = \pi\left(\frac{4}{\pi}\right)^2 \cdot 2 + 8 \cdot 12$

$= \frac{16}{\pi} \cdot 2 + 96$

$= \frac{32}{\pi} + 96$

The surface area of Gandalf's cylinder is $\frac{32}{\pi} + 96$ sq in.

Frodo took a piece of paper 8 inches by 12 inches and did the same thing as Gandalf. Compute the surface area of Frodo's cylindrical can.

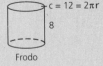

c = 12 = 2πr

8

Frodo

Objective

After studying this section, you will be able to
■ Find the surface areas of circular solids

Part One: Introduction

Consider the following three solids that are based on the circle.

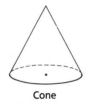
Cylinder

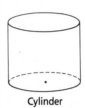
Cone

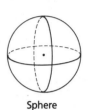
Sphere

A *cylinder* resembles a prism in having two congruent parallel bases. The bases of a cylinder, however, are circles. In this text, *cylinder* will mean a right circular cylinder— that is, one in which the line containing the centers of the bases is perpendicular to each base.

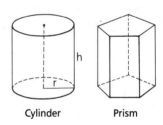
Cylinder   Prism

The lateral area of a cylinder can be visualized by thinking of a cylinder as a can, and the lateral area as the label on the can. If we cut the label and spread it out, we see that it is a rectangle. The height of the rectangle is the same as the height of the can. The base of the rectangle is the circumference of the can.

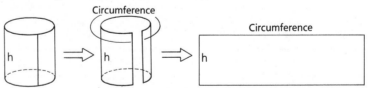
Circumference

Circumference

570   Chapter 12   Surface Area and Volume

Vocabulary

cone
cylinder
sphere

Theorem 113   *The lateral area of a cylinder is equal to the product of the height and the circumference of the base.*

$$L.A._{cyl} = Ch = 2\pi rh$$

where C is the circumference of the base, h is the height of the cylinder, and r is the radius of the base.

Definition   The total area of a cylinder is the sum of the cylinder's lateral area and the areas of the two bases.

$$T.A._{cyl} = L.A. + 2A_{base}$$

A ***cone*** resembles a pyramid, but its base is a circle. In a pyramid the slant height and the lateral edge are different; in a cone they are the same.

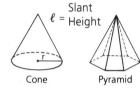

Slant
$\ell = $ Height

Cone   Pyramid

In this book, the word *cone* will mean a right circular cone— one in which the altitude passes through the center of the circular base.

Theorem 114   *The lateral area of a cone is equal to one-half the product of the slant height and the circumference of the base.*

$$L.A._{cone} = \tfrac{1}{2}C\ell = \pi r\ell$$

where C is the circumference of the base, ℓ is the slant height, and r is the radius of the base.

Definition   The total area of a cone is the sum of the lateral area and the area of the base.

$$T.A._{cone} = L.A. + A_{base}$$

A ***sphere*** is a special figure with a special surface-area formula. (A sphere has no lateral edges and no lateral area.) The proof of the formula requires the concept of limits and will not be given here.

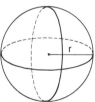

Postulate $T.A._{sphere} = 4\pi r^2$

where r is the sphere's radius.

Class Opener, continued
Circumference of top/bottom is 12 in. Therefore, $c = 12 = 2\pi r$; $r = \frac{6}{\pi}$. So $A = \pi \cdot \left(\frac{6}{\pi}\right)^2 \cdot 2 + 8 \cdot 12$

$= \frac{36}{\pi} \cdot 2 + 96$

$= \frac{72}{\pi} + 96$

The surface area of Frodo's cylinder is $\frac{72}{\pi} + 96$ sq in.

How much greater was the surface area of Frodo's cylindrical can than Gandalf's?

$\frac{40}{\pi}$ sq in., or ≈ 12.7 sq in.

Communicating Mathematics

Ask students to prepare a lesson that explains how to find the surface area of a cylinder. Choose one or two students to present their lesson. The student should be responsible for answering questions from the class.

Part Two: Sample Problem

Problem   *Find the total area of each figure.*

a   **b**   **c**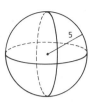

Solution

a $\text{T.A.}_{\text{cyl}} = \text{L.A.} + 2A_{\text{base}}$
$= 2\pi rh + 2\pi r^2$
$= 2\pi(5)(6) + 2\pi(5^2)$
$= 110\pi$

b $\text{T.A.}_{\text{cone}} = \text{L.A.} + A_{\text{base}}$
$= \pi r\ell + \pi r^2$
$= \pi(5)(6) + \pi(5^2)$
$= 55\pi$

c $\text{T.A.}_{\text{sphere}} = 4\pi r^2$
$= 4\pi(5^2)$
$= 100\pi$

Assignment Guide

Basic
1–5, 8, 10, 11

Average
2–6, 8–11

Advanced
6–14

Part Three: Problem Sets

Problem Set A

1 What is the total area of a sphere having
  a A radius of 7? 196π **c** A diameter of 6? 36π
  b A radius of 3? 36π **d** A diameter of 5? 25π

2 Find the lateral area and the total area of each solid.

a   **b**   **c** 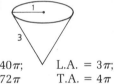   **d**

  L.A. = 24π;   L.A. = 140π;   L.A. = 40π;   L.A. = 3π;
  T.A. = 33π T.A. = 238π T.A. = 72π T.A. = 4π

3 Find the radius of a sphere whose surface area is 144π. 6

4 Find the total area of each solid. (Hint: Be sure that you include only outside surfaces and that you do not miss any.)

a 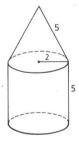   **b**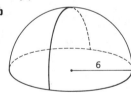

  34π 108π
This is a *hemisphere* ("half sphere"). The T.A. includes the area of the circular base.

5 ABCD is a parallelogram, with A = (3, 6), B = (13, 6), C = (7, −2), and D = (−3, −2).

   a Find the slopes of the diagonals, $\overline{AC}$ and $\overline{BD}$. −2; $\frac{1}{2}$

   b Use your answers to part **a** to identify □ABCD by its most specific name. Rhombus

Problem Set B

6 Find the total (including the rectangular face) surface area of a half cylinder with a radius of 5 and a height of 2. $35\pi + 20$

7 Find the total area of each solid.

   a 90π **b** 66π **c** 480π

  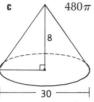

8 The total height of the tower shown is 10 m. If one liter of paint will cover an area of 10 sq m, how many 1-L cans of paint are needed to paint the entire tower? (Hint: First find the total area to be painted, using 3.14 for π.) 24 cans

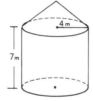

9 What size label (length and width) will just fit on a can 8 cm in diameter and 14 cm high? 8π cm by 14 cm

10 Find the total area of the solid. 93π

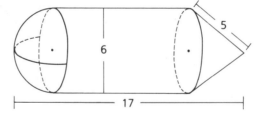

11 KITE is a kite.

   a Find the area of KITE. 64

   b Find the area of the rectangle formed when consecutive midpoints of the sides of KITE are connected. 32

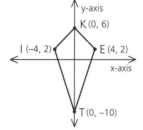

Section 12.3 Surface Areas of Circular Solids   **573**

Cooperative Learning

Have students work in small groups to solve problem **14** on page 574. In addition to finding the total surface area, ask the groups to draw three-dimensional pictures of the solids generated or to find an example of a similar solid in real life. You may also wish to have students draw the surfaces of the solids as a two-dimensional picture. Surface areas are in text. Drawings may vary.

**Problem-Set Notes
and Additional Answers**

12   T.A. = 4 · 80 + 60π +
      2(64 − 9π)
      = 448 + 42π

13

Since 4 is half of 8, we can
apply the converse of the
Midline Theorem to
conclude the slant height
of the top cone must be 5.
∴ T.A. = 16π + 64π + (π ·
8 · 10 − π · 4 · 5) = 140π

14a

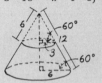

T.A. = π · 6 · 12 − π · 3 ·
6 + 9π + 36π = 99π

b

T.A. = 6π · 5 + 4π · 5 +
2(9π − 4π) = 60π

c

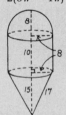

T.A. = ½(4π · 8²) + π · 16 ·
10 + π · 8 · 17 = 424π

Problem Set C

12 Find the total surface area of the solid
shown, including the surface inside the
hole.   448 + 42π

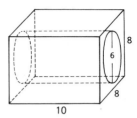

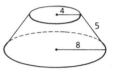

13 The solid at the right is called a *frustum*
of a cone. Find its total area if the radii
of the top and bottom bases are 4 and 8
respectively, and the slant height is 5.   140π

14 A *surface of rotation* is generated by revolving a shape about a
fixed line, called the *axis of rotation*. For example, revolving a
half circle about the line containing its endpoints produces a
sphere.

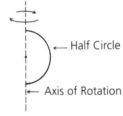

— Half Circle

— Axis of Rotation

Sphere

Identify the surface of rotation generated in each diagram below
and compute the total area of each of these surfaces.

a

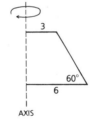

Frustum of cone;
99π

b

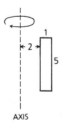

Cylindrical shell;
60π

c

Solid like that in
problem 10; 424π

12.4 VOLUMES OF PRISMS AND CYLINDERS

Objectives

After studying this section, you will be able to
- Find the volumes of right rectangular prisms
- Find the volumes of other prisms
- Find the volumes of cylinders
- Use the area of a prism's or a cylinder's cross section to find the solid's volume

Part One: Introduction

Volume of a Right Rectangular Prism

The measure of the space enclosed by a solid is called the solid's **volume**. In a way, volume is to solids what area is to plane figures.

Definition The **volume** of a solid is the number of cubic units of space contained by the solid.

A cubic unit is the volume of a cube with edges one unit long. A cube is a right rectangular prism with congruent edges, so all its faces are squares. In Section 12.1 we used the word *box* for a right prism. Thus, a right rectangular prism can also be called a rectangular box.

One linear unit

One Cubic Unit

Each layer has 4 rows of 6 cubic units each

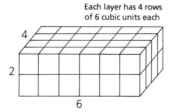

Rectangular Box

The rectangular box above contains 48 cubic units. The formula that follows is not only a way of counting cubic units rapidly but also works with fractional dimensions.

Class Planning

Time Schedule
Basic: 2 days
Average: 2 days
Advanced: 1 day

Resource References
Teacher's Resource Book
   Class Opener 12.4A, 12.4B
   Additional Practice Worksheet 23

Class Opener

Determine the total surface area of the hemisphere on top of the cylinder on top of the cone.
The key is that the radius is 12, because it is the radius of the hemisphere.

$T.A._{hemisphere} = \frac{1}{2} \cdot 4\pi \cdot 12^2$

$T.A._{cyl} = 2 \cdot 12 \cdot 12 \cdot \pi$

$T.A._{cone} = 12 \cdot 12\sqrt{2} \cdot \pi$

$T.A. = \frac{1}{2} \cdot 4\pi \cdot 12^2 + 2 \cdot 12 \cdot 12 \cdot \pi + 12 \cdot 12\sqrt{2} \cdot \pi$

$= (576 + 144\sqrt{2})\pi$

≈ 2449

Vocabulary
cross section
volume

Postulate *The volume of a right rectangular prism is equal to the product of its length, its width, and its height.*

$$V_{rect.\ box} = \ell wh$$

where ℓ is the length, w is the width, and h is the height.

Another way to think of the volume of a rectangular prism is to imagine the prism to be a stack of congruent rectangular sheets of paper. The area of each sheet is $\ell \cdot w$, and the height of the stack is h. Since the base of the prism is one of the congruent sheets, there is a second formula for the volume of a rectangular box.

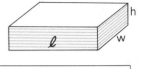

$$V = \ell wh$$
$$= (\ell w)h$$
$$= (\text{area of sheet}) \cdot h$$

Theorem 115 *The volume of a right rectangular prism is equal to the product of the height and the area of the base.*

$$V_{rect.\ box} = \mathcal{B}h$$

where $\mathcal{B}$ is the area of the base and h is the height.

Volumes of Other Prisms

The formula in Theorem 115 can be used to compute the volume of any prism, since any prism can be viewed as a stack of sheets the same shape and size as the base.

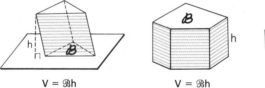

$V = \mathcal{B}h$ $V = \mathcal{B}h$ $V = \mathcal{B}h$

Theorem 116 *The volume of any prism is equal to the product of the height and the area of the base.*

$$V_{prism} = \mathcal{B}h$$

where $\mathcal{B}$ is the area of the base and h is the height.

Notice that the height of a *right* prism is equivalent to the measure of a lateral edge.

576   Chapter 12   Surface Area and Volume

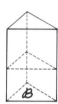

Volume of a Cylinder

The stacking property applies to a cylinder as well as to a prism, so the formula for a prism's volume can also be used to find a cylinder's. Furthermore, since the base of a cylinder is a circle, there is a second, more popular formula.

Theorem 117   *The volume of a cylinder is equal to the product of the height and the area of the base.*

$$V_{cyl} = \mathcal{B}h = \pi r^2 h$$

where $\mathcal{B}$ is the area of the base, h is the height, and r is the radius of the base.

Cross Section of a Prism or a Cylinder

When we visualize a prism or a cylinder as a stack of sheets, all the sheets are congruent, so the area of any one of them can be substituted for $\mathcal{B}$. Each of the sheets between the bases is an example of a ***cross section***.

Definition   A ***cross section*** is the intersection of a solid with a plane.

In this book, unless otherwise noted, all references to cross sections will be to cross sections *parallel to the base*. We can now combine Theorems 116 and 117, using the symbol $\mathcal{C}$ to represent the area of a cross section parallel to the base.

Theorem 118   *The volume of a prism or a cylinder is equal to the product of the figure's cross-sectional area and its height.*

$$V_{prism \ or \ cyl} = \mathcal{C}h$$

where $\mathcal{C}$ is the area of a cross section and h is the height.

Part Two: Sample Problems

Problem 1   Find the volume of the rectangular prism.

Solution $V = \ell wh$ or $V = \mathcal{B}h$
                 $= 20(5)(15)$ $= (5 \cdot 20)(15)$ (Using a 5 × 20 face as base)
                 $= 1500$ $= 1500$

Problem 2   Find the volume of the triangular prism.

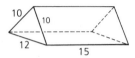

Solution   Notice that the base of the prism is the triangle at the right.

$A_\triangle = \frac{1}{2}(12)(8) = 48$
$V = \mathcal{B}h$
   $= 48(15)$
   $= 720$

Problem 3   Find the volume of a cylinder with a radius of 3 and a height of 12.

Solution $V = \pi r^2 h$
               $= \pi(3^2)(12)$
               $= \pi(9)(12)$
               $= 108\pi$

Problem 4   Find the volume of the right prism shown. (Take the left face as a representative cross section.)

Solution   Use either of two methods.

a Divide and Conquer:

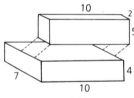

$V_{\text{top box}} = 2(5)(10) = 100$
$V_{\text{bottom box}} = 7(10)(4) = 280$
$V_{\text{solid}} = 280 + 100 = 380$

b Cross Section Times Height:

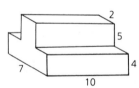

$\mathcal{C} = 10 + 28 = 38$
$V = \mathcal{C}h$
   $= 38(10)$
   $= 380$

Problem 5　A box (rectangular prism) is sitting in a corner of a room as shown.

a Find the volume of the prism.

b If the coordinates of point A in a three-dimensional coordinate system are (4, 5, 6), what are the coordinates of B?

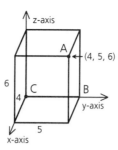

Solution　**a** $V = \mathcal{B}h$
$= (4 \cdot 5)6$
$= 120$

b Point C = (0, 0, 0) is the corner. To get from C to B, you would travel 0 units in the x direction, 5 units in the y direction, and 0 units in the z direction. So the coordinates of B are (0, 5, 0).

Part Three: Problem Sets

Problem Set A

1 Find the volume of each solid.

a 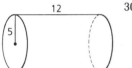　　300π

b 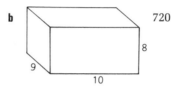　720

2 Find the volume of cement needed to form the concrete pedestal shown. (Leave your answer in π form.) (6 + 5π) cu m

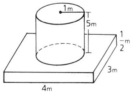

3 The area of the shaded face of the right pentagonal prism is 51. Find the prism's volume. 357

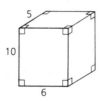

4 Find the volume and the total surface area of the rectangular box shown.
V = 300; T.A. = 280

Cooperative Learning

Have students work in small groups on this extension of Sample Problem **5**.
If three different colored dice were thrown, one representing the x-coordinate, one the y-coordinate, and one the z-coordinate, what is the probability that the resulting set of coordinates would be

1 On an edge of the prism? $\frac{13}{216}$

2 On a face of the prism? $\frac{47}{216}$

3 Inside the prism? $\frac{5}{18}$

4 Outside the prism? $\frac{4}{9}$

Assignment Guide

Basic	
Day 1	1–6
Day 2	7–10, 12, 16
Average	
Day 1	1–8
Day 2	9–18
Advanced	

7, 8, 10–13, 15, 16, 19, 20, 22

Problem Set A, *continued*

5 a Find the volume of a cube with an edge of 7. 343
  b Find the volume of a cube with an edge of e. e^3
  c Find the edge of a cube with a volume of 125. 5

6 Find the length of a lateral edge of a right prism with a volume of 286 and a base area of 13. 22

Problem Set B

7 Traci's queen-size waterbed is 7 ft long, 5 ft wide, and 8 in. thick.

  a Find the bed's volume to the nearest cubic foot. ≈23 cu ft

  b If 1 cu ft of water weighs 62.4 lb, what is the weight of the water in Traci's bed to the nearest pound? ≈1456 lb

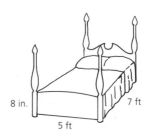

8 in.   7 ft
5 ft

8 If point A's coordinates in the three-dimensional coordinate system are (2, 5, 8), what are the coordinates of B? (2, 0, 8)

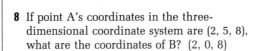

9 Find the volume and the total area of the prism. V = 600; T.A. = 620

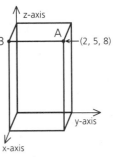

10 Find the volume and the total area of each right cylindrical solid shown.

  a 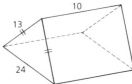 V = 243π   T.A. = 148.5π

  b 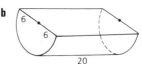 V = 360π   T.A. = 156π + 240

11 When Hilda computed the volume and the surface area of a cube, both answers had the same numerical value. Find the length of one side of the cube. 6

12 Find the volume and the surface area of the regular hexagonal right prism. $V = 540\sqrt{3}$; T.A. $= 360 + 108\sqrt{3}$

13 Find the volume of a cube in which a face diagonal is 10. $250\sqrt{2}$

14 A rectangular cake pan has a base 10 cm by 12 cm and a height of 8 cm. If 810 cu cm of batter is poured into the pan, how far up the side will the batter come? 6.75 cm

15 A rectangular container is to be formed by folding the cardboard along the dotted lines. Find the volume of this container. 189

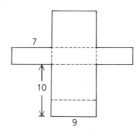

16 The cylindrical glass is full of water, which is poured into the rectangular pan. Will the pan overflow? Yes, by ≈ 3.6 cu cm

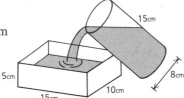

17 Jim's lunch box is in the shape of a half cylinder on a rectangular box. To the nearest whole unit, what is

a The total volume it contains? ≈ 621 cu in.

b The total area of the sheet metal needed to manufacture it? 439 sq in.

18 A cistern is to be built of cement. The walls and bottom will be 1 ft thick. The outer height will be 20 ft. The inner diameter will be 10 ft. To the nearest cubic foot, how much cement will be needed for the job? ≈ 770 cu ft

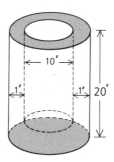

19 $V = \mathcal{B}h$

$V = \left(\frac{40}{360}\right)\pi(20)^2 \cdot 15$

$V = \frac{2000}{3}\pi$

T.A. $= 2\left[\frac{40}{360}\pi(20)^2 + 20 \cdot 15\right] + \left(\frac{40}{360}\right)(2\pi \cdot 20) \cdot 15$

T.A. $= 600 + \frac{1400}{9}\pi$

20a $V = 4^3 - \pi(1)^2 \cdot 4$

$V \approx 51.44$

 b $10(.89)(51.44)$

 c $6 \cdot 4^2 - 2\pi(1)^2 + 2\pi(1) \cdot 4$

 d $\frac{114.84}{96} = 1.20$

21 $V = \mathcal{B}h$

$V = \left[\frac{1}{2} \cdot \pi \cdot 6^2 - 2 \cdot \frac{1}{2}(\pi \cdot 3^2)\right]10$

$V = 90\pi$

22

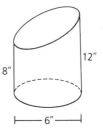

$V = \mathcal{B}h + \frac{1}{2}\mathcal{B}h$

$V = \pi \cdot 3^2 \cdot 8 + \frac{1}{2}\pi \cdot 3^2 \cdot 4$

$V = 90\pi$

Problem Set C

19 A wedge of cheese is cut from a cylindrical block. Find the volume and the total surface area of this wedge.
$V = \frac{2000}{3}\pi$, T.A. $= 600 + \frac{1400}{9}\pi$

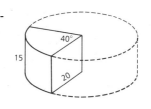

20 An ice-cube manufacturer makes ice cubes with holes in them. Each cube is 4 cm on a side and the hole is 2 cm in diameter.

a To the nearest tenth, what is the volume of ice in each cube? ≈ 51.4 cu cm

b To the nearest tenth, what will be the volume of the water left when ten cubes melt? (Water's volume decreases by 11% when it changes from a solid to a liquid.) ≈ 457.8 cu cm

c To the nearest tenth, what is the total surface area (including the inside of the hole) of a single cube? ≈ 114.8 sq cm

d The manufacturer claims that these cubes cool a drink twice as fast as regular cubes of the same size. Verify whether this claim is true by a comparison of surface areas. (Hint: The ratio of areas is equal to the ratio of cooling speeds.)
The claim is untrue (the ratio of cooling speeds is about 1.2 : 1).

21 Find the volume of the solid at the right. (A representative cross section is shown.) 90π

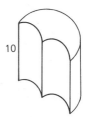

22 A cylinder is cut on a slant as shown. Find the solid's volume. 90π cu in.

12.5 VOLUMES OF PYRAMIDS AND CONES

Objectives

After studying this section, you will be able to
- Find the volumes of pyramids
- Find the volumes of cones
- Solve problems involving cross sections of pyramids and cones

Part One: Introduction

Volume of a Pyramid

The volume of a pyramid is related to the volume of a prism having the same base and height. At first glance, many people would guess that the volume of the pyramid is half that of the prism.

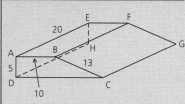

Such a guess would be wrong, though. The volume of a pyramid is actually *one third* of the volume of a prism with the same base and height.

Theorem 119   ***The volume of a pyramid is equal to one third of the product of the height and the area of the base.***

$$V_{pyr} = \tfrac{1}{3}\mathcal{B}h$$

where $\mathcal{B}$ is the area of the base and h is the height.

The proof of this formula is complex and will not be shown here.

Volume of a Cone

Because a cone is a close relative of a pyramid, although its base is circular rather than polygonal, the formula for its volume is similar to the formula for a pyramid's volume.

Vocabulary

frustum

Class Planning

Time Schedule
All levels: 1 day

Resource References
Teacher's Resource Book
   Class Opener 12.5A, 12.5B
   Additional Practice Worksheet 24

Class Opener

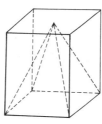

Given: ABCD is a trapezoid.
     AB = 10, BC = 13, AD = 5, EFGH ≅ ABCD,
     $\overleftrightarrow{AE} \parallel \overleftrightarrow{BF} \parallel \overleftrightarrow{GC} \parallel \overleftrightarrow{DH}$,
     $\overleftrightarrow{AE} \perp \overleftrightarrow{AB}$, $\overleftrightarrow{AE} \perp \overleftrightarrow{AD}$
Find: The volume of the solid

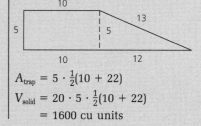

$A_{trap} = 5 \cdot \tfrac{1}{2}(10 + 22)$
$V_{solid} = 20 \cdot 5 \cdot \tfrac{1}{2}(10 + 22)$
     $= 1600$ cu units

Cooperative Learning

Have students work in small groups to demonstrate the validity of Theorem 119. Several ways of demonstrating are listed.
1   Fill cones and cylinders, or pyramids and prisms, with water or sand and compare the amounts.
2   Use water displacement.
3   Use model clay.
Each group should be able to demonstrate the volume formula to the class.

Lesson Notes

- Students should realize that the volume formulas for pyramids and cones are the same ($V = \frac{1}{3}\mathcal{B}h$).
- In general, for similar solids,
$$\frac{V_1}{V_2} = \left(\frac{s_1}{s_2}\right)^3, \frac{A_1}{A_2} = \left(\frac{s_1}{s_2}\right)^2,$$
where s, A, and V stand for side, area, and volume.

Theorem 120 *The volume of a cone is equal to one third of the product of the height and the area of the base.*
$$V_{cone} = \tfrac{1}{3}\mathcal{B}h = \tfrac{1}{3}\pi r^2 h$$
where $\mathcal{B}$ is the area of the base, h is the height, and r is the radius of the base.

Cross Section of a Pyramid or a Cone

Unlike a cross section of a prism or a cylinder, a cross section of a pyramid or a cone is not congruent to the figure's base. Observe that the cross section parallel to the base is *similar* to the base in each solid shown below.

Cross Section of a Cone

Cross Section of a Pyramid

The Similar-Figures Theorem (p. 544) suggests that in these solids the area of a cross section is related to the square of its distance from the vertex.

Theorem 121 *In a pyramid or a cone, the ratio of the area of a cross section to the area of the base equals the square of the ratio of the figures' respective distances from the vertex.*
$$\frac{\mathcal{C}}{\mathcal{B}} = \left(\frac{k}{h}\right)^2$$
where $\mathcal{C}$ is the area of the cross section, $\mathcal{B}$ is the area of the base, k is the distance from the vertex to the cross section, and h is the height of the pyramid or cone.

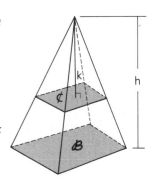

A proof of Theorem 121 is asked for in problem 15.

Part Two: Sample Problems

Problem 1 If the height of a pyramid is 21 and the pyramid's base is an equilateral triangle with sides measuring 8, what is the pyramid's volume?

Solution $V_{pyr} = \tfrac{1}{3}\mathcal{B}h$
Since the base is an equilateral triangle, $\mathcal{B} = \frac{s^2}{4}\sqrt{3} = 16\sqrt{3}$.
So $V = \tfrac{1}{3}\left(16\sqrt{3}\right)(21) = 112\sqrt{3}$.

Problem 2   Find the volume of a cone with a
base radius of 6 and a slant height
of 10.

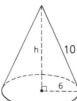

Solution   Using the right triangle shown, we
find that the height of the cone is 8.

$$V_{cone} = \frac{1}{3}\pi r^2 h$$
$$= \frac{1}{3}\pi(6^2)(8) = 96\pi$$

Problem 3   A pyramid has a base area of 24 sq
cm and a height of 12 cm. A cross
section is cut 3 cm from the base.

a Find the volume of the upper pyramid (the solid above the cross section).

b Find the volume of the *frustum*
(the solid below the cross section).

Solution   **a** Since the cross section is 3 cm from the base, its distance, k, from
the peak is 9 cm.

$$\frac{\mathscr{C}}{\mathscr{B}} = \left(\frac{k}{h}\right)^2 \qquad\qquad V_{upper\ pyramid} = \frac{1}{3}\mathscr{C}k$$
$$\frac{\mathscr{C}}{24} = \left(\frac{9}{12}\right)^2 \qquad\qquad\qquad = \frac{1}{3}\cdot\frac{27}{2}\cdot 9$$
$$\mathscr{C} = \frac{27}{2} \qquad\qquad\qquad\qquad = 40.5 \text{ cu cm}$$

b To find the volume of the frustum, we subtract the volume of the
upper pyramid from the volume of the whole pyramid.

$$V_{frustum} = V_{whole\ pyramid} - V_{upper\ pyramid}$$
$$= \frac{1}{3}(24)(12) - 40.5$$
$$= 96 - 40.5$$
$$= 55.5 \text{ cu cm}$$

Part Three: Problem Sets

Problem Set A

1 Find the volume of a pyramid whose
base is an equilateral triangle with sides
measuring 14 and whose height is 30. $490\sqrt{3}$

2 Find, to two decimal places, the volume
of a cone with a slant height of 13 and a
base radius of 5. ≈ 314.16

Assignment Guide

Basic
1–8
Average
1, 5, 6, 8–10, 12–14
Advanced
5, 8–10, 12–14, 17, 19, 20

Refer to problem **3**. Have students work parts **a** and **b**, then rework the problem with dimensions twice as great as the given dimensions. Have students compare the answers with the original answers, and write a generalization about what they find.

In problem **3**, the area is four times as great as the original area, and the volume is eight times as great. In general, $\frac{s_1}{s_2} = \frac{2}{1}$ for each problem, so $A = \left(\frac{2}{1}\right)^2 = 4$ and $V = \left(\frac{2}{1}\right)^3 = 8$.

Problem Set A, *continued*

3 The pyramid shown has a square base and a height of 12.

  a Find its volume. 400

  b Find its total area. 360

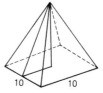

4 The volume of a pyramid is 42. If its base has an area of 14, what is the pyramid's height? 9

5 Given: The right circular cone shown

  Find: **a** Its volume 1080π

      b Its lateral area 369π

      c Its total area 450π

6 A tower has a total height of 24 m. The height of the wall is 20 m. The base is a rectangle with an area of 25 sq m. Find the total volume of the tower to the nearest cubic meter. ≈ 533 cu m

7 A well has a cylindrical wall 50 m deep and a diameter of 6 m. The tapered bottom forms a cone with a slant height of 5 m. Find, to the nearest cubic foot, the volume of water the well could hold. ≈ 1451 cu ft

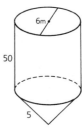

Problem Set B

8 Find, to the nearest tenth, the volume of a cone with a 60° vertex angle and a slant height of 12. ≈ 391.8

9 A pyramid has a square base with a diagonal of 10. Each lateral edge measures 13. Find the volume of the pyramid. 200

10 PABCD is a regular square pyramid.

  a Find the coordinates of C. (0, 10, 0)

  b Find the volume of the pyramid. 400

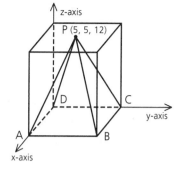

11 Find the volume remaining if the smaller cone is removed from the larger. 72π

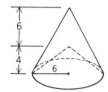

12 A rocket has the dimensions shown. If 60% of the space in the rocket is needed for fuel, what is the volume, to the nearest whole unit, of the portion of the rocket that is available for nonfuel items? ≈4423

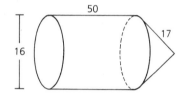

13 A gazebo (garden house) has a pentagonal base with an area of 60 sq m. The total height to the peak is 16 m. The height of the pyramidal roof is 6 m. Find the gazebo's total volume. 720 cu m

14 Use the diagram at the right to find

  a x 15

  b The radii of the circles 9 and 15

  c The volume of the smaller cone ≈1018

  d The volume of the larger cone ≈4712

  e The volume of the frustum ≈3695

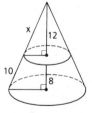

15 Set up and complete a proof of Theorem 121. (Hint: First prove that the ratio of corresponding segments of a cross section and a base equals the ratio of h to k.)

**Problem-Set Notes
and Additional Answers**

■ See *Solution Manual* for answer to problem **15**.

17a

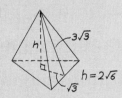

$V = \frac{1}{3}\mathcal{B}h$

$V = \frac{1}{3}\left(\frac{6^2}{4}\sqrt{3}\right) \cdot 2\sqrt{6}$

$V = 18\sqrt{2}$

b $h = \frac{s\sqrt{6}}{3}$

$V = \frac{1}{3}\mathcal{B}h$

$V = \frac{1}{3}\left(\frac{s^2}{4}\sqrt{3}\right)\left(\frac{s\sqrt{6}}{3}\right)$

$V = \frac{s^3\sqrt{2}}{12}$

T 587

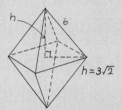

18

Consider the octahedron as
two square pyramids.

$V = 2 \cdot \frac{1}{3}\mathcal{B}h$

$V = 2 \cdot \frac{1}{3} \cdot 6^2 \cdot 3\sqrt{2}$

$V = 72\sqrt{2}$

19

$\mathcal{B} = \frac{1}{2}(6)(4) = 12$

$V = \frac{1}{3}\mathcal{B}h$

$V = \frac{1}{3}(12)(12)$

$V = 48$

20

$\frac{x}{x+5} = \frac{6}{9}$

$x = 10$

By Pythagorean Theorem,
$h_{\text{small cone}} = 8$, $h'_{\text{large cone}} = 12$.

V_{frustum}
$= V_{\text{large cone}} - V_{\text{small cone}}$
$= \frac{1}{3}\mathcal{B}h - \frac{1}{3}\mathcal{B}h'$
$= \frac{1}{3} \cdot \pi \cdot 9^2 \cdot 12 - \frac{1}{3} \cdot \pi \cdot 6^2 \cdot 8$
$= 228\pi$

Problem Set B, *continued*

16 Find the volume of a cube whose total surface area is 150 sq in. 125 cu in.

Problem Set C

17 A regular tetrahedron is shown. (Each of the four faces is an equilateral triangle.) Find the tetrahedron's total volume if each edge measures

a 6 $18\sqrt{2}$ **b** s $\frac{s^3\sqrt{2}}{12}$

18 A regular octahedron (eight equilateral faces) has an edge of 6. Find the octahedron's volume. $72\sqrt{2}$

19 Find the volume of the pyramid shown. 48

20 Find the volume of the frustum shown. 228π

12.6 | VOLUMES OF SPHERES

Objective

After studying this section, you will be able to
- Find the volumes of spheres

Part One: Introduction

The following theorem can be proved with the help of *Cavalieri's principle* (discussed in Problem Set D of this section), but we shall present it without proof.

Theorem 122 *The volume of a sphere is equal to four thirds of the product of π and the cube of the radius.*

$$V_{sphere} = \tfrac{4}{3}\pi r^3$$

where r is the radius of the sphere.

Some of the problems in this section require you to find both the volumes and the surface areas of spheres. Recall from Section 12.3 that the total surface area of a sphere is equal to $4\pi r^2$.

Part Two: Sample Problem

Problem   Find the volume of a hemisphere with a radius of 6.

Solution   First we find the volume of a sphere with a radius of 6.

$$V_{sphere} = \tfrac{4}{3}\pi r^3$$
$$= \tfrac{4}{3}\pi(6)^3 = 288\pi$$

The hemisphere's volume is half that of the sphere. Thus,
$V_{hemisphere} = 144\pi$, or ≈ 452.39.

Part Three: Problem Sets

Problem Set A

1 Find the volume of a sphere with

    a A radius of 3 36π **b** A diameter of 18 972π **c** A radius of 5 $\frac{500}{3}\pi$

12.6

Class Planning

Time Schedule
All levels: 1 day

Resource References
Teacher's Resource Book
   Class Opener 12.6A, 12.6B
Evaluation
   Tests and Quizzes
     Quizzes 1–2, Series 1
     Quizzes 1–3, Series 2
     Quizzes 1–2, Series 3

Class Opener

The cone has diameter 16 and slant height 10. Find the volume.

$V = \tfrac{1}{3}\pi r^2 h$
$= \tfrac{1}{3} \cdot \pi \cdot 8^2 \cdot 6$
$= 128\pi$ cu units

Assignment Guide

Basic		
1–5, 7		
Average		
2–7, 9–12		
Advanced		
6, 7, 10–17		

Vocabulary
Cavalieri's principle

Problem Set A, *continued*

2 Find the volume and the surface area of a sphere with a radius of 6. V = 288π, T.A. = 144π

3 Find the volume of the grain silo to the nearest cubic meter. ≈481 cu m

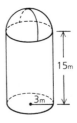

15m

3m

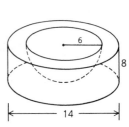

6

8

14

4 A plastic bowl is in the shape of a cylinder with a hemisphere cut out. The dimensions are shown.

  a What is the volume of the cylinder? 392π

  b What is the volume of the hemisphere? 144π

  c What is the volume of plastic used to make the bowl? 248π

5 What volume of gas, to the nearest cubic foot, is needed to inflate a spherical balloon to a diameter of 10 ft? ≈523 cu ft

Problem Set B

6 A rubber ball is formed by a rubber shell filled with air. The shell's outer diameter is 48 mm, and its inner diameter is 42 mm. Find, to the nearest cubic centimeter, the volume of rubber used to make the ball. 19 cu cm

7 Given: A cone and a hemisphere as marked

  Find: **a** The total volume of the solid 240π

      b The total surface area of the solid 132π

8

6

8 A hemispherical dome has a height of 30 m.

  a Find, to the nearest cubic meter, the total volume enclosed. ≈56,549 cu m

  b Find, to the nearest square meter, the area of ground covered by the dome (the shaded area). ≈2827 sq m

  c How much more paint is needed to paint the dome than to paint the floor? Twice as much

  d Find, to the nearest meter, the radius of a dome that covers double the area of ground covered by this one. ≈42 m

Chapter 12 Surface Area and Volume

Cooperative Learning

Refer to problem **10** on page 591. Have students work in small groups to find the volume and surface area of spheres with a different ratio of radii, such as 2:5, 3:4, 1:3, or 1:2.

After they compare their answers, ask groups to discover the general rule for the ratio of the volume and surface areas of two spheres with a given ratio of radii.

For spheres with radii in a ratio of x:y, the ratio of their surface areas will be $x^2:y^2$ and the ratio of their volumes will be $x^3:y^3$.

9 A cold capsule is 11 mm long and 3 mm in diameter. Find, to the nearest cubic millimeter, the volume of medicine it contains. ≈71 cu mm

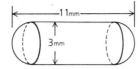

10 The radii of two spheres are in a ratio of 2:5.
  a Find the ratio of their volumes. 8:125
  b Find the ratio of their surface areas. 4:25

11 A minisubmarine has the dimensions shown.
  a What is the sub's total volume? 138π cu ft
  b Knowing the sub's surface area is important in determining how much pressure it will withstand. What is the sub's total surface area? 105π sq ft

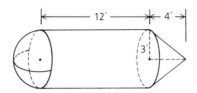

Problem Set C

12 An ice-cream cone is 9 cm deep and 4 cm across the top. A single scoop of ice cream, 4 cm in diameter, is placed on top. If the ice cream melts into the cone, will it overflow? (Assume that the ice cream's volume does not change as it melts.) Justify your answer.
No; the cone's volume is 12π cu cm, and the ice cream's is only $10\frac{2}{3}\pi$ cu cm.

13 The volume of a cube is 1000 cu m.
  a To the nearest cubic meter, what is the volume of the largest sphere that can be inscribed inside the cube? ≈524 cu m
  b To the nearest cubic meter, what is the volume of the smallest sphere that can be circumscribed about the cube? ≈2721 cu m

14 Find the ratio of the volume of a sphere to the volume of the smallest right cylinder that can contain it. 2:3

15 In the diagram, ABGH is a rectangle and $\overline{AB} \cong \overline{BC} \cong \overline{CD}$. To the nearest whole number, what percentage of the area of ⊙O is the area of ABGH? ≈11%

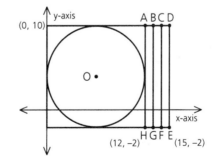

Section 12.6   Volumes of Spheres   **591**

Problem-Set Notes and Additional Answers

12 $V_{ice\ cream} = \frac{4}{3}\pi(2)^3$
  $= 10\frac{2}{3}\pi$
  $V_{cone} = \frac{1}{3}\mathcal{B}h$
  $= \frac{1}{3}(\pi \cdot 2^2) \cdot 9$
  $= 12\pi$

13a Radius of sphere will be half the side of cube, 5.
  $V = \frac{4}{3}\pi(5)^3$
  $V = \frac{500}{3}\pi$

  b Radius of sphere will be half the diagonal of the cube, $5\sqrt{3}$.
  $V = \frac{4}{3}\pi(5\sqrt{3})^3$
  $V = 500\pi\sqrt{3}$

14

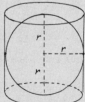

  $V_{sphere} = \frac{4}{3}\pi r^3$
  $V_{cyl} = \mathcal{B}h = \pi r^2 \cdot 2r = 2\pi r^3$
  $\frac{V_{sphere}}{V_{cyl}} = \frac{\frac{4}{3}\pi r^3}{2\pi r^3} = \frac{2}{3}$

15 $A_{\odot} = \pi r^2$
  $= \pi \cdot 6^2$
  $= 36\pi$
  $A_{rect\ ABGH} = 1 \cdot 12 = 12$
  $\therefore \left(\frac{12}{36\pi} \cdot 100\right)\% = \frac{100}{3\pi}\%$
  $\approx 11\%$

16

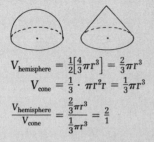

$$V_{hemisphere} = \frac{1}{2}\left[\frac{4}{3}\pi r^3\right] = \frac{2}{3}\pi r^3$$
$$V_{cone} = \frac{1}{3} \cdot \pi r^2 r = \frac{1}{3}\pi r^3$$

$$\frac{V_{hemisphere}}{V_{cone}} = \frac{\frac{2}{3}\pi r^3}{\frac{1}{3}\pi r^3} = \frac{2}{1}$$

17 $V_{shell} = \pi \cdot (2r)^2 \cdot h - \pi r^2 h$
$$= 3\pi r^2 h$$
$$V_{cyl} = \pi(r\sqrt{3})^2 h$$
$$= 3\pi r^2 h$$

18a $A_{annulus} = \pi r^2 - \pi d^2$
$$A_{\odot} = \pi(\sqrt{r^2 - d^2})^2$$
$$= \pi r^2 - \pi d^2$$
   b   By Cavalieri's principle,
$$V_{hemisphere} = V_{cyl} - V_{cone}.$$
$$V_{hemisphere} = \pi r^2 \cdot r - \frac{1}{3}\pi r^2 \cdot r$$
$$= \frac{2}{3}\pi r^3.$$
   Thus, $V_{sphere} = \frac{4}{3}\pi r^3$.

Communicating Mathematics

Problem Set C, *continued*

16 Compare the volumes of a hemisphere and a cone with congruent bases and equal heights. $V_{hemisphere} = 2V_{cone}$

Problem Set D

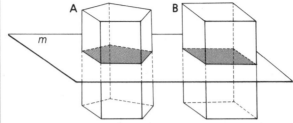

Plane m is a cross-sectional plane through solids A and B.

Cavalieri's Principle
If two solids A and B can be placed with their bases coplanar and if the area of every cross section of A is equal to the area of the coplanar cross section of B, then A and B have equal volumes.

17 Show that the volume of cylindrical shell
A is equal to the volume of cylinder B by
using Cavalieri's principle.

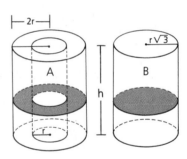

18 a Compare the cross-sectional areas of the solids shown below.
   b Use Cavalieri's principle to derive the formula for the volume
   of a sphere. (Hint: Use the diagrams to find a formula for the
   volume of a hemisphere.)

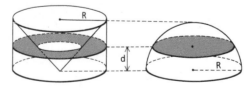

CHAPTER SUMMARY

CONCEPTS AND PROCEDURES

After studying this chapter, you should be able to
- Find the surface areas of prisms (12.1)
- Find the surface areas of pyramids (12.2)
- Find the surface areas of circular solids (12.3)
- Find the volumes of right rectangular prisms (12.4)
- Find the volumes of other prisms (12.4)
- Find the volumes of cylinders (12.4)
- Use the area of a prism's or a cylinder's cross section to find the solid's volume (12.4)
- Find the volumes of pyramids (12.5)
- Find the volumes of cones (12.5)
- Solve problems involving cross sections of pyramids and cones (12.5)
- Find the volumes of spheres (12.6)

VOCABULARY

base (12.1)
Cavalieri's principle (12.6)
cone (12.3)
cross section (12.4)
cylinder (12.3)
frustum (12.5)
lateral edge (12.1)
lateral face (12.1)

lateral surface area (12.1)
polyhedron (12.1)
prism (12.1)
pyramid (12.2)
regular pyramid (12.2)
sphere (12.3)
total surface area (12.1)
volume (12.4)

12   REVIEW PROBLEMS

Problem Set A

1 Find the lateral area and the total area of the regular pyramid and the cylinder.

a 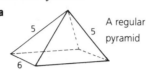 A regular pyramid

L.A. = 48; T.A. = 84

b 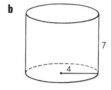   L.A. = 56π, T.A. = 88π

2 Find the volume of

  a A cube with a side of 8   512

  b A rectangular box that measures 3 by $4\frac{1}{2}$ by 8   108

  c A cylinder with a radius of 7 and a height of 2 98π

  d A pyramid with a height of 5 and a base area of 12   20

  e A prism with a height of 5 and a base area of 12   60

  f A sphere with a radius of 2 $\frac{32}{3}\pi$

3 Find the volume and the total surface area of each solid.

a 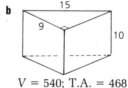   **b** 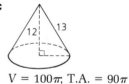   **c**

V = 360π, T.A. = 192π V = 540; T.A. = 468   V = 100π, T.A. = 90π

4 Find the volume of the solid that is formed when folds are made along the dotted lines.   90

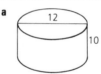

5 Find the height of

  a A box with a volume of 100, a length of 15, and a width of $1\frac{1}{3}$ 5

  b A cube with a volume of 216   6

6 Find the volume of a cylindrical glass if its height is 15 cm and a 17-cm straw just fits inside it as shown. 240π

7 A concrete staircase is to be built. Each step is 15 cm high, 25 cm deep, and 1 m wide. The top platform is square. What volume of concrete is needed? 562,500 cu cm

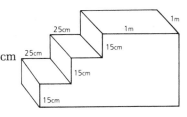

8 Given: A regular square pyramid with a slant height of 10 and a base measuring 16 by 16

Find: **a** The pyramid's lateral area 320

     b The pyramid's total area 576

     c The pyramid's volume 512

9 Find the volume of a sphere whose surface area is 36π. 36π

Problem Set B

10 Find the lateral area and the total area of each solid.

a The base is equilateral.

L.A. = 180
T.A. = 180 + 25√3

b

L.A. = 60π
T.A. = 96π

c

L.A. = 195π
T.A. = $279\frac{1}{2}\pi$

11 Find the total surface area of each solid. (Don't forget the flat faces.)

a 75π

b $45\pi + 60$

Review Problem Set B, *continued*

12 Find the volume of a cylinder formed from the pattern at the right. The area of each circle is 16π. The rectangle has an area of 24π. 48π

13 A pyramid has a height of 5. Its base is a rhombus with diagonals measuring 7 and 6. Find the volume of the pyramid. 35

14 A cross section of a hatbox is a regular hexagon with a side 12 cm long. The height of the box is 20 cm. Find the box's surface area and volume. $\left(1440 + 432\sqrt{3}\right)$ sq cm; $4320\sqrt{3}$ cu cm

15 Find the volume of the wedge. 5040

16 Find the total volume of the castle, including the towers.

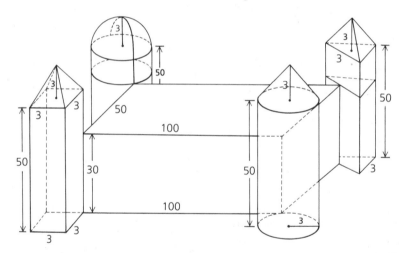

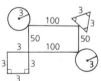

$150,459 + 927\pi + \frac{459\sqrt{3}}{4}$

Problem-Set Notes and Additional Answers

19 $A = \frac{1}{4}(12\pi \cdot 10) +$
   $2(6 \cdot 10) + 2\left(\frac{\pi}{4} \cdot 36\right)$
   $= 120 + 48\pi$

17 A hole with a diameter of 2 in. is drilled through a block as shown. Find the volume of the resulting solid to the nearest cubic inch. 215 cu in.

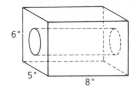

18 Find the volume of the prism shown at the right. (Hint: If you solve this problem, you will be a Hero.) $70\sqrt{2}$

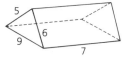

Problem Set C

19 A cylinder is cut into four equal parts. Find the total area of the part shown. $120 + 48\pi$

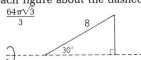

20 A right cylindrical log was cut parallel to the axis. Find the volume and the total surface area of the piece shown.
$V = 500\pi - 750\sqrt{3}$; T.A. $= 300 + \frac{400\pi}{3} - 50\sqrt{3}$

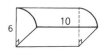

21 A frustum of a cone is shown. Find the volume of this solid. $333\pi\sqrt{3}$

22 Find the volume of the surface of rotation generated by rotating each figure about the dashed line.

a $\frac{64\pi\sqrt{3}}{3}$

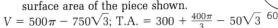

b 35π

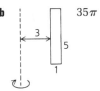

Review Problems **597**

Problem-Set Notes and Additional Answers, continued

20

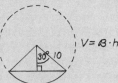

$$\mathscr{B} = \left(\frac{60}{360}\right)\pi(10)^2 - \frac{10^2}{4}\sqrt{3}$$
$$\mathscr{B} = \frac{50}{3}\pi - 25\sqrt{3}$$
$$V = \left(\frac{50}{3}\pi - 25\sqrt{3}\right)30$$
$$V = 500\pi - 750\sqrt{3}$$
$$\text{T.A.} = 2\mathscr{B} + 10\cdot 30 + 30\left(\frac{60}{360}\right)2\pi\cdot 10$$
$$\text{T.A.} = 300 + \frac{400}{3}\pi - 50\sqrt{3}$$

21

$$\frac{x}{x+6} = \frac{9}{12}$$
$$x = 18$$

By Pythagorean Theorem,
$h_{\text{small cone}} = 9\sqrt{3}$ and
$h_{\text{large cone}} = 12\sqrt{3}$.
$$V_{\text{frustum}} = V_{\text{large cone}} - V_{\text{small cone}}.$$
$$V_{\text{frustum}} = \frac{1}{3}\mathscr{B}h - \frac{1}{3}\mathscr{B}h'$$
$$= \frac{1}{3}\pi\cdot 12^2\cdot 12\sqrt{3} - \frac{1}{3}\pi\cdot 9^2\cdot 9\sqrt{3}$$
$$= 333\pi\sqrt{3}$$

22a

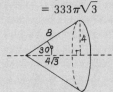

$$V = \frac{1}{3}\mathscr{B}h$$
$$= \frac{1}{3}\pi\cdot 4^2\cdot 4\sqrt{3} = \frac{64\pi\sqrt{3}}{3}$$

b

$$V = \mathscr{B}h$$
$$V = [\pi\cdot 4^2 - \pi\cdot 3^2]\cdot 5$$
$$V = 35\pi$$

T 597

CUMULATIVE REVIEW
CHAPTERS 1–12

**Cumulative Review
Chapters 1–12**

Time Schedule
Basic: 2 days
Average: 1 day
Advanced: 1 day

Resource References
Evaluation
   Tests and Quizzes
      Cumulative Review Test
      Chapters 1–12
      Series 1, 2, 3

Assignment Guide

Basic

Day 1   1–11
Day 2   12–18, 20–23, 24a, 26,
         27, 31

Average

7, 9, 12, 17, 23, 24a, 25–31, 35

Advanced

24–26, 29–31, 34, 37–39, 41

For classes that are approaching
the end of the term, these 42
problems provide a comprehen-
sive review of the first 12 chapters. Additional end-of-term
review problems can be selected
from the Cumulative Review for
Chapters **1–16** (pp. 706–711).

**Problem-Set Notes
and Additional Answers**

■ See *Solution Manual* for answer
   to problem **3.**

Problem Set A

1 The measure of one of the acute angles of a right triangle is nine
times the measure of the other acute angle. Find the measure of
the larger acute angle. 81

2 The perimeter of △ABC is 28. If AB = 2x + 3, BC = 4x − 5, and
CA = 8x − 19, is △ABC scalene, isosceles, or equilateral? Isosceles

3 Given: $\overline{BD} \perp \overline{AD}$, $\overline{BD} \perp \overline{BC}$, $\overline{AB} \cong \overline{CD}$
   Prove: ABCD is a ▱.

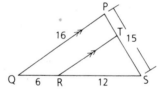

4 Given: $\overline{PQ} \parallel \overline{TR}$
   Find: **a** PT 5
         b TR $\frac{32}{3}$

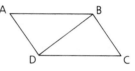

5 Find the value of x in each figure.

a

c

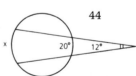

b $3\frac{1}{5}$

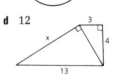

d 12

6 Two similar triangles have areas of 9 and 25.
   a What is the ratio of a pair of corresponding sides? $\frac{3}{5}$
   b What is the ratio of the triangles' perimeters? $\frac{3}{5}$

7 Given: Diagram as marked.

Find: **a** WY $\frac{50}{3}$

      b YZ $\frac{32}{3}$

      c XZ 8

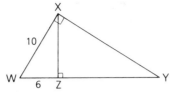

8 Find the areas of the trapezoid, the triangle, and the circle.

a 168   15   **b** $25\sqrt{3}$ **c**   24 169π

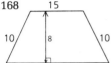 

9 Given: SPQR is an isosceles trapezoid.

      $\angle S = (x + 40)^\circ$,

      $\angle Q = (2x - 7)^\circ$

Find: $\angle R$. 89°

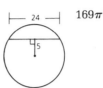

10 The numbers 3.14 and $3\frac{1}{7}$ are frequently used as approximations of π. Use your calculator to determine which of these approximations is the more accurate. $3\frac{1}{7}$

11 Find p, q, r, and s.

a 60   **b**   83   **c** 50   **d**   78

  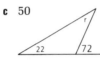

12 a Find the fourth proportional in a proportion whose first three terms are 5, 3, and 30. 18

     b Find the mean proportionals between 8 and 18. ±12

13 Given: ⊙O with tangent $\overline{CD}$,

      CD = 15,

      BC = 9

Find: **a** AC **25**

      b The diameter of ⊙O 16

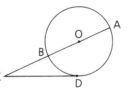

14 Given: $\overline{AR}$ is tangent to ⊙P.

      $\overline{RS}$ is a diameter of ⊙Q.

Prove: $\triangle PAR \sim \triangle SBR$

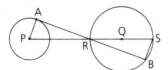

Problem-Set Notes and Additional Answers, continued

■ See *Solution Manual* for answer to problem **14.**

Cumulative Review Problems   **599**

Cumulative Review Problem Set A, *continued*

15 In △ABC, D and E are the midpoints of $\overline{AB}$ and $\overline{AC}$, DE = 4x, and BC = 2x + 48. Find BC. 64

16 Find the area of each polygon.

a 80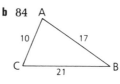

b 84

c $28\sqrt{5}$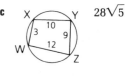

17 Find the number of sides of an equiangular polygon if each interior angle is 170°. 36

18 The perimeter of an isosceles triangle is 36. One side is 10. What are the possible lengths of the base? 16 or 10

19 Each polygon shown is regular.

   a Find the measure of ∠1. 60

   b Find the measure of ∠2. 135

   c Find the measure of ∠3. 120

   d Find the measure of ∠4. 45

   e Will a regular pentagon fit at ∠5? No

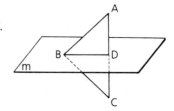

Problem-Set Notes and Additional Answers, continued

■ See Solution Manual for answers to problems **21** and **22**.

20 Given: Parallelogram as marked

   Find: x 65

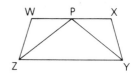

21 Given: WXYZ is an isosceles trapezoid, with $\overline{WZ} \cong \overline{XY}$.
      △PZY is isosceles.
   Prove: P is the midpoint of $\overline{WX}$.

22 Given: $\overleftrightarrow{AC} \perp m$, $\overline{BC} \cong \overline{BA}$
   Prove: D is the midpoint of $\overline{AC}$.

23 Find the length of a 45° arc of a circle whose radius is 8. 2π

Problem Set B

24 What is the angle formed by the hands of a clock at

    a 11:30? 165°   **b** 2:05? $32\frac{1}{2}°$ **c** 3:24? 42°

25 Given: Rectangle RECT in ⊙R,
             RT = 5, TQ = 2

    Find: ET 7

26 If M is the midpoint of $\overline{AB}$, what is the
area of the shaded region? 20π

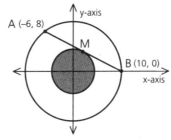

27 A woman walks 20 m west, 100 m south, another 8 m west, and
then 4 m north. How far is she from her starting point? 100 m

28 Given: ⊙O, CB = 9,
            ∠C = 30°, $\overline{BC} \cong \overline{BD}$;
            $\overleftrightarrow{CD}$ is tangent to ⊙O.

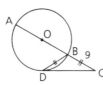

    Find: **a** m$\overset{\frown}{AD}$ 120   **b** CD $9\sqrt{3}$ **c** The radius of ⊙O 9

29 Given: The radius of ⊙O is 0.7.
            The radius of ⊙P is 1.1.
            $\overleftrightarrow{AB}$ is a common internal tangent.
            AB = 2.4

    Find: **a** OP 3

         b The distance between the circles 1.2

30 Given: Diagram as marked, with $\overleftrightarrow{PA}$ and
            $\overleftrightarrow{PD}$ tangent to ⊙O

    Find: **a** $\overset{\frown}{AD}$ 98°

         b m∠P 82

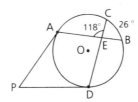

Cumulative Review Problem Set B, *continued*

31 Given: Triangle as marked

Find: m∠WYA 35

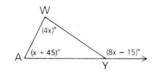

32 The water in a drainpipe is 18 cm deep. The width of the surface of the water is 48 cm. Find the radius of the pipe. 25 cm

33 Given: Diagram as marked, with $\overleftrightarrow{RQ}$ tangent to the circle

Find: x 3

34 Given: ABCD is a ▱.

Find: **a** AE 8

   b x:y 9:1

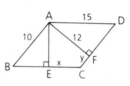

35 Given: m$\widehat{AB}$:m$\widehat{CD}$ = 5:2, ∠P = 24°

Find: $\widehat{CD}$ 32°

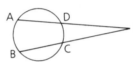

Problem Set C

36 A sled dog traveled 6 mi east, then 6 mi northeast, then another 6 mi east. How far was the dog from her starting point? $6\sqrt{5 + 2\sqrt{2}}$ mi

37 Given: $\overline{BC}$ is the base of isosceles △ABC. $\overline{DE}$ is the base of isosceles △AED. ∠BAE = 40°

Find: m∠DEC 20

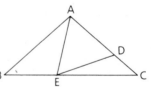

38 In this set of three semicircles, B can be any point between A and C. Prove that the shaded area is equal to π times the product of the radii of the unshaded semicircles.

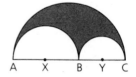

Problem-Set Notes and Additional Answers, *continued*

36

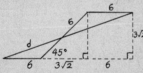

$d^2 = (12 + 3\sqrt{2})^2 + (3\sqrt{2})^2$

$d = 6\sqrt{5 + 2\sqrt{2}}$ mi

37

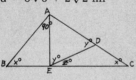

$z + y = x + 40$

$z = x + 40 - y$

∴ $y - x = x + 40 - y$

$2y - 2x = 40$

$y - x = 20$

$z = y - x$

∴ $z = 20$

38

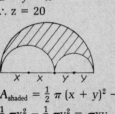

$A_{shaded} = \frac{1}{2}\pi(x + y)^2 -$

$\frac{1}{2}\pi x^2 - \frac{1}{2}\pi y^2 = \pi xy$

39 Two sides of one triangle are congruent to two sides of a second triangle, and the included angles are supplementary. The area of one triangle is 41. Can the area of the second triangle be found? **Yes**

40 Given: In quadrilateral QUAD, $\overline{QU} \cong \overline{AD}$, ∠A is supp. to ∠Q, and $\overline{QD} \neq \overline{AU}$.

Prove: QUAD is an isosceles trapezoid.

41 The lengths of the sides of a hexagon are in an arithmetic progression. The hexagon's perimeter is 30, and its longest side measures 7. Find the length of the next longest side. $6\frac{1}{5}$

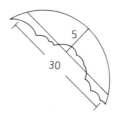

42 Clarence bragged that he ate most of a pizza, but he could not remember the pizza's diameter. On the remaining piece, however, he made the measurements shown. The distance from the midpoint of the arc to the midpoint of the corresponding chord was 5 cm. The chord measured 30 cm. Find the diameter of the pizza Clarence ate. **50 cm**

HISTORICAL SNAPSHOT

THE SHAPE OF THE UNIVERSE

Kepler and the orbits of the planets

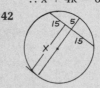

The ancient Greeks discovered that of all the possible polyhedra, only five consist of faces that are congruent regular polygons. In the *Timaeus*, a dialogue on the creation of the universe, the philosopher Plato used these regular polyhedra to explain the mathematical structure of the cosmos. He associated the cube and the regular icosahedron, octahedron, and tetrahedron with the four elements that were thought to make up all substances—earth, water, air, and fire. The fifth—the regular dodecahedron—he thought to represent the form of the whole universe. Perhaps this was because it is nearly a sphere, to the Greeks the most perfect of all solids.

In the late 1500's, the astronomer Johannes Kepler began to think about the distances between the planets. He recalled the *Timaeus* and the regular polyhedra. Only six planets were known in Kepler's time, so it occurred to him that their orbits might correspond to a series of circles alternately circumscribed about and inscribed in the five regular solids (see illustration).

Kepler was unable to devise an accurate model of the solar system based on regular polyhedra. But he later discovered one of the fundamental laws of the solar system: that the square of the time it takes a planet to complete an orbit is directly proportional to the cube of the planet's distance from the sun.

Problem-Set Notes and Additional Answers, continued

39 The areas are equal. Put the triangles together, and you have a median.

40 Since ∠Q is supp. to ∠A, then ∠D must be supp. to ∠U, and thus, QUAD may be inscribed in a ⊙. The figure may then be shown to be a trap. by getting $\overline{AU} \parallel \overline{QD}$.

41 $\begin{cases} x + (x + k) + (x + 2k) + \\ (x + 3k) + (x + 4k) + \\ (x + 5k) = 30 \end{cases}$

$x + 5k = 7$

$x = 3,\ k = \frac{4}{5}$

∴ $x + 4k = 6\frac{1}{5}$

42

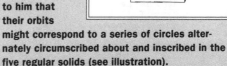

$15 \cdot 15 = 5x$

$45 = x$

∴ Diameter = 50

COORDINATE GEOMETRY EXTENDED

Chapter 13 Schedule

Basic
Problem Sets:	10 days
Review:	2 days
Test:	1 day

Average
Problem Sets:	10 days
Review:	2 days
Test:	1 day

Advanced
Problem Sets:	10 days
Review:	2 days
Test:	1 day

*C*OORDINATE GEOMETRY INTE-
GRATES aspects of algebra and geometry. Therefore, a review of
first-year algebra topics is built
into this chapter. For most students, this review is useful and
even necessary. This chapter also
prepares students for analytic
proofs, which can be concise and
powerful.

Some courses will finish the year
at the end of Chapter **12** or Chapter **13.** The Cumulative Review at
the end of Chapter **12** can be
helpful in reviewing the course.

These massive sections of a coordinate grid
show a Yale professor's playful conception.

13.1 GRAPHING EQUATIONS

Objective

After studying this section, you will be able to
- Draw lines and circles that represent the solutions of equations

Part One: Introduction

Throughout your work with this book, you have seen problems involving coordinate geometry. You have dealt extensively with the formulas used to determine midpoints, slopes, and distances on the coordinate plane. In this section, you will review what you learned about graphing equations in your algebra studies and prepare yourself for the topics covered later in the chapter.

Part Two: Sample Problems

Problem 1 Draw a graph of the equation $y = 2x - 1$.

Solution To make a graph, we frequently construct a ***table of values***. We choose values for either x or y and then substitute each value in the equation to find the other member of each ordered pair. For example, if $x = -1$, then $y = 2(-1) - 1 = -3$.

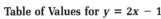

Table of Values for $y = 2x - 1$

x	−1	0	1	2	3
y	−3	−1	1	3	5

We then plot each ordered pair and draw a line through these points. This line represents the solution set of the equation.

Section 13.1 Graphing Equations | **605**

Vocabulary

intercept
table of values

Class Planning

Time Schedule
All levels: 1 day

Resource References
Teacher's Resource Book
  Class Opener 13.1A, 13.1B

Class Opener

The points A(−2, −3), B(1, 3), and C(2, y) are collinear.

1 Find y by graphing. Plot the points and draw a line through the points. By counting, when x = 2, y = 5.

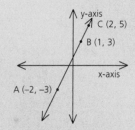

2 Find y by using the slope formula. Since the points are collinear, slope of $\overline{AB}$ = slope of $\overline{BC}$.

$$\frac{3 - (-3)}{1 - (-2)} = \frac{y - 3}{2 - 1}$$
$$\frac{6}{3} = y - 3$$
$$6 = 3(y - 3)$$
$$6 = 3y - 9$$
$$15 = 3y$$
$$5 = y$$

Lesson Notes

- The text assumes that students can plot and label points in the coordinate plane.
- While the use of graph paper may help ensure accurate plotting, the use of unlined paper prevents overdependence on counting boxes and can promote careful measurements and complete labeling.

Problem 2 *Draw a graph of the equation x = 4.*

Solution Notice that no y appears in the equation x = 4. Thus, y can have any real value, but x is always equal to 4. The graph is the vertical line shown.

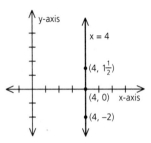

Problem 3 Karen reviewed the distance formula and then claimed that the circle at the right was the graph of the equation $(x - 1)^2 + (y - 2)^2 = 25$.

   a Confirm that (6, 2) and (4, 6) are on the circle.

   b Draw the radius and the tangent that intersect at (4, 6) and find the slope of the tangent.

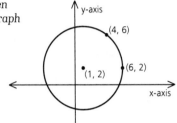

Solution **a** Test (6, 2).

$$(6 - 1)^2 + (2 - 2)^2 \overset{?}{=} 25$$
$$25 + 0 \overset{?}{=} 25$$
$$25 \overset{\checkmark}{=} 25$$

Test (4, 6).

$$(4 - 1)^2 + (6 - 2)^2 \overset{?}{=} 25$$
$$9 + 16 \overset{?}{=} 25$$
$$25 \overset{\checkmark}{=} 25$$

b Find the slope of the radius.

$$\text{Slope} = \frac{6 - 2}{4 - 1}$$

$$= \frac{4}{3}$$

Since the radius is perpendicular to the tangent, the slope of the tangent is $-\frac{3}{4}$.

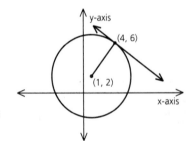

Problem 4 *Find the **intercepts** of the graph of the equation y = 4x − 2.*

Solution

x-intercept:
(Substitute 0 for y.)

$$y = 4x - 2$$
$$0 = 4x - 2$$
$$2 = 4x$$
$$0.5 = x$$

y-intercept:
(Substitute 0 for x.)

$$y = 4x - 2$$
$$= 4(0) - 2$$
$$= 0 - 2$$
$$= -2$$

Thus, the x-intercept is 0.5, and the y-intercept is − 2.

Note These results mean that the graph of the equation passes through the points (0.5, 0) and (0, − 2).

606 Chapter 13 Coordinate Geometry Extended

Part Three: Problem Sets

Problem Set A

1 Make a table of values for each of the following equations and graph the two equations on the same set of axes.

$$y = x + 3 \qquad y = x - 1$$

2 Make a table of values for each of the following equations and graph the two equations on the same set of axes.

$$y = 2x - 5 \qquad y = 2x - 7$$

3 Graph $y - 1 = 2x$.

4 Graph $y - 1 = 2(x + 1)$.

5 Verify that the three points shown lie on the circle whose equation is $x^2 + y^2 = 9$.

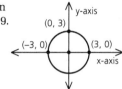

6 Verify that the three points shown lie on the circle whose equation is $x^2 + y^2 = 16$.

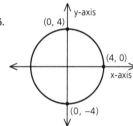

7 Find the x- and y-intercepts of the graph of $y = 2x - 6$.
x-intercept: 3; y-intercept: -6

8 Is (5, 4) on the graph of $y = 2x - 3$? No

9 Is $(-4, 6)$ on the V-shaped graph of $y = |x - 2|$? Yes

10 Consider the equations of three lines:

$$y = x + 4 \qquad y = 3x \qquad y = 2x - 2$$

If two of the three lines are selected at random, what is the probability that both contain the point (2, 6)? $\frac{1}{3}$

Assignment Guide

Basic
1, 3–5, 7–10, 15, 18–20
Average
1, 3–5, 7–10, 12, 14, 15, 18–20
Advanced
3–5, 7, 10, 12, 14, 15, 21–24

Problem-Set Notes and Additional Answers

■ See *Solutions Manual* for answers to problems **1–6**.

Have the students work in groups to solve this extension of problem **11**.

1 Consider the circle whose equation is $x^2 + y^2 = 100$. Find the coordinates of the point(s) where the graph of the given equation intersects the circle.
  a $y = 8$ $(-6, 8)$ and $(6, 8)$
  b $y = 6$ $(-8, 6)$ and $(8, 6)$
  c $x = 10$ $(10, 0)$
  d $y = x$
  $(-5\sqrt{2}, -5\sqrt{2})$, $(5\sqrt{2}, 5\sqrt{2})$
2 What is the significance of the line in part **c** of problem **1** with respect to the circle? $x = 10$ is a tangent to the circle.

Problem Set A, *continued*

11 Is (6, 8) on the graph of $x^2 + y^2 = 100$? Yes

Problem Set B

12 Write an equation that represents the circle shown. $x^2 + y^2 = 36$

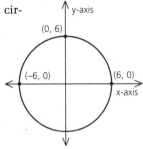

13 Find the area of a triangle with vertices $(-2, 0)$, $(4, 0)$, and $(2, 3)$. 9

14 The vertices of a right triangle are $(0, 0)$, $(3, 0)$, and $(3, 4)$.
  a Find the lengths of the three sides. 3, 4, 5
  b Find the length of the altitude to the hypotenuse. $\frac{12}{5}$
  c Find the length of the median to the hypotenuse. $2\frac{1}{2}$

15 Consider the isosceles trapezoid shown.
  a Find the coordinates of vertex B. (2, 5)
  b Find the lengths of the bases. 14 and 4
  c Find the length of the median. 9
  d Find the trapezoid's area. 63

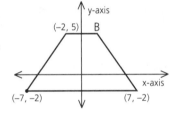

16 A parallelogram has vertices $(-5, -1)$, $(4, -1)$, and $(7, 6)$. Find the fourth vertex if two sides are parallel to the x-axis. $(-2, 6)$ or $(16, 6)$

17 Find, to the nearest tenth, the area of the shaded region. ≈21.5

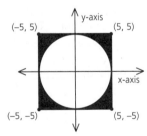

18 If $y = mx + b$, $(x, y) = (2, 4)$, and $m = -3$, find b. 10

19 If a line containing point (x_1, y_1) and having slope m can represent the equation $y - y_1 = m(x - x_1)$, find an equation that corresponds to the line containing point (5, 2) and having a slope of 6. $y - 2 = 6(x - 5)$

20 In the diagram, the point (2, 3) is the center of the circle. What is the slope of the tangent to the circle at (7, 8)? -1

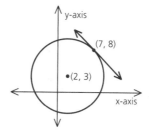

Problem Set C

21 $\overleftrightarrow{PT}$ is tangent to circle O at P.

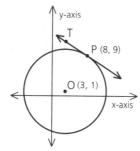

  a Find the slope of $\overleftrightarrow{PT}$. $-\frac{5}{8}$
  b Verify that $y - 9 = -\frac{5}{8}(x - 8)$ is an equation that represents $\overleftrightarrow{PT}$.
  c Verify that $y = -\frac{5}{8}x + 14$ is an equation that represents $\overleftrightarrow{PT}$.

22 Given that A = (10, 1) and B = (2, 9), reflect B across the y-axis to its image B'. If $\overline{AB'}$ intersects the y-axis at C, verify that the slope of $\overline{AC}$ is $-\frac{2}{3}$.

23 Find, to the nearest tenth, the area of the shaded region. ≈ 2.6

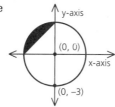

24 △OBD is *encased* in rectangle OACE as shown.

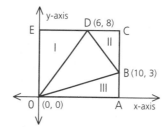

  a Find the areas of regions I, II, and III. 24; 10; 15
  b Find the area of △OBD. 31

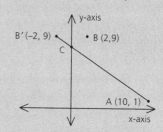

Section 13.1 Graphing Equations **609**

EQUATIONS OF LINES

Class Planning

Time Schedule
All levels: 2 days

Resource References
Teacher's Resource Book
  Class Opener 13.2A, 13.2B
  Additional Practice Worksheet 25
Evaluation
  Tests and Quizzes
  Quiz 1, Series 1–3

Class Opener

In problems **1–4,** write an equation of the line that contains $(-4, -3)$ and meets the given condition.

1. The line is perpendicular to the y-axis.
   $y = -3$
2. The line is parallel to the y-axis.
   $x = -4$
3. The line is perpendicular to the graph of $y = 4x - 3$.
   Possible answer: $y = -\frac{1}{4}x - 4$
4. The line is parallel to the graph of $y = 4x - 3$.
   Possible answer: $y = 4x + 13$

Lesson Notes

- Many students find this the most difficult section in the chapter. However, the ability to write the equation for a line is basic for most applications of algebra to geometry.

Objectives

After studying this section, you will be able to
- Write equations that correspond to nonvertical lines
- Write equations that correspond to horizontal lines
- Write equations that correspond to vertical lines
- Identify various forms of linear equations

Part One: Introduction

Equations of Nonvertical Lines

Consider a line with a y-intercept of b and a slope of m. You know that $(0, b)$ is one point on the line. Let (x, y) represent any other point on the line and substitute the two sets of coordinates in the slope formula.

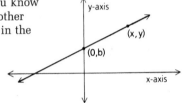

$$\frac{y - b}{x - 0} = m$$
$$y - b = mx + 0$$
$$y = mx + b$$

Theorem 123   *The y-form, or slope-intercept form, of the equation of a nonvertical line is*

$$y = mx + b$$

where b is the y-intercept of the line and m is the slope of the line.

Example   *Use the y-form to write an equation of the line containing $(-1, 4)$ and $(1, 8)$.*

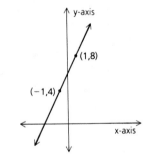

First we find the slope:
$m = \frac{8 - 4}{1 - (-1)} = \frac{4}{2} = 2$.
Since the line has a slope, we use the y-form, $y = mx + b$.
We now substitute 2 for m and $(1, 8)$ for (x, y).

$$8 = 2(1) + b$$
$$6 = b$$

Vocabulary

general linear form   slope-intercept form
intercept form   two-point form
point-slope form   y-form

Therefore, the equation is $y = 2x + 6$. You will get the same equation if you use $(-1, 4)$ instead of $(1, 8)$ for (x, y). Try it and see!

Equations of Horizontal Lines

Since a horizontal line is nonvertical, the y-form can be used to develop a formula for the equation of any horizontal line.

$\overleftrightarrow{AB}$ is a horizontal line. Every point has the same y-coordinate (ordinate). The y-intercept is 4, so $b = 4$. The slope of a horizontal line is zero, so $m = 0$.

$y = mx + b$
$y = 0 \cdot x + 4$
The equation of $\overleftrightarrow{AB}$ is $y = 4$.

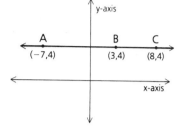

In general, all the points on a horizontal line have the same y-coordinate, b, but their x-coordinates are the set of real numbers. Since the slope m is zero, x does not appear in the equation of the line.

Theorem 124   *The formula for an equation of a horizontal line is*

$$y = b$$

where b is the y-coordinate of every point on the line.

The trick is to recognize a horizontal line, which may be disguised in a problem.

Example   *Find an equation that corresponds to*

  a *The line containing $(2, 5)$ and $(24, 5)$*
   $y = 5$

  b *The x-axis*
   $y = 0$

  c *The line $7\frac{1}{2}$ units below the x-axis*
   $y = -7\frac{1}{2}$

  d *The line perpendicular to the y-axis and passing through $(11, \sqrt{3})$*
   $y = \sqrt{3}$

Equations of Vertical Lines

A vertical line has no slope. Therefore, the previous formulas cannot apply. However, every point on a vertical line has the same x-coordinate (abscissa), while its y-coordinate may be any real number.

Lesson Notes, continued

■ The equations for horizontal and vertical lines are treated as special cases of the general equation for a line.

Theorem 125   ***The formula for the equation of a vertical line is***

$$x = a$$

where a is the x-coordinate of every point on the line.

The trick is to recognize when a line is vertical.

Example   *Find an equation that corresponds to*

a The line containing $(-1, 6)$ and $(-1, 7)$

$$x = -1$$

b The line having an x-intercept of 8 and passing through $\left(8, 5\sqrt{2}\right)$

$$x = 8$$

c The line that contains $(-10, 4)$ and is perpendicular to the graph of $y = 7$

$$x = -10$$

WE NEED TO BE
RECOGNIZED!

Forms of the Equations of Lines

In this book, we shall emphasize the y-form, but you may find other forms listed in the following table helpful.

Equations of Lines		
Form	**Formula**	**Used for**
Slope-intercept (*y-form*)	$y = mx + b$ (m = slope; b = y-intercept)	Nonvertical lines only
Point-slope	$y - y_1 = m(x - x_1)$ [m = slope; (x_1, y_1) = known point]	Nonvertical lines only
Two-point	$\dfrac{y - y_1}{x - x_1} = \dfrac{y_2 - y_1}{x_2 - x_1}$ [(x_1, y_1) and (x_2, y_2) are known points.]	Nonvertical lines only
General linear	$ax + by + c = 0$ (a, b, and c are real numbers.)	Any line
Intercept	$\dfrac{x}{a} + \dfrac{y}{b} = 1$ (a = x-intercept; b = y-intercept)	Lines not passing through the origin (nonzero intercepts)

Lesson Notes, continued

■ This section uses the $y = mx + b$ form of the equation of a line. Any of the forms listed on page 612 would serve as well. Students may wonder why *m* is used for slope. Here are three possible reasons:

1   *m* is the first letter in the French word *monter*, which means "to mount," "to climb," and "to slope up."

2   *m* is the first letter in the word *modulus*, the constant ratio that is the coefficient of the variable of a linear equation.

3   "It just happened," says Howard W. Eves in the book *Mathematical Circles Revisited* (Boston: Prindle, Weber, and Schmidt, 1971, pp. 141, 267).

■ Since a vertical line has no slope, students cannot simply substitute a value for *m* in a formula. Students who learn to watch for pairs of points with equal abscissas can quickly apply the $x = a$ formula.

■ It is not recommended that each of these alternate forms for a linear equation be used in a geometry course. They are presented here to illustrate the variety of forms that students may see in other mathematics courses.

612   Chapter 13   Coordinate Geometry Extended

Part Two: Sample Problems

Problem 1   *Write an equation of the line contain-*
ing (7, −3) and (4, 1).

Solution   First find the slope.

$$m = \frac{1-(-3)}{4-7} = \frac{4}{-3} = -\frac{4}{3}$$

Then substitute values in the y-form
formula, using either (7, −3) or (4, 1)
for (x, y).

$$y = mx + b$$
$$1 = -\frac{4}{3}(4) + b$$
$$\frac{3}{3} = -\frac{16}{3} + b$$
$$\frac{19}{3} = b$$

Thus, $y = -\frac{4}{3}x + \frac{19}{3}$ is an equation of
the line.

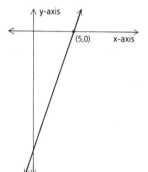

Problem 2   *Find an equation of the line with a slope of 3 and an x-intercept of 5.*

Solution   If the line has an x-intercept of 5, it must contain the point (5, 0).
Therefore, (5, 0) can be substituted for (x, y) and the given slope for
m in the y-form formula.

$$y = mx + b$$
$$0 = 3(5) + b$$
$$0 = 15 + b$$
$$-15 = b$$

An equation of the line is $y = 3x - 15$.

Problem 3   **a** *Find an equation of the line passing through (2, 5) and (17, 5).*
   b *Find an equation of the line that is parallel to the y-axis and*
   contains $\left(-\sqrt{6}, 1\right)$.

Solution   **a** The line is horizontal, so it corresponds to the equation $y = 5$.
   b The line is vertical, so it corresponds to the equation $x = -\sqrt{6}$.

Problem 4   In △ABC, A = (−3, 2), B = (9, 4), and C = (5, 12).

a *Find an equation of the median to* $\overline{AB}$.

b *Find an equation of the perpendicular bisector of* $\overline{AB}$.

c *Find an equation of the altitude to* $\overline{AB}$.

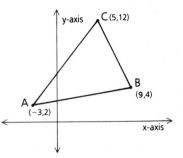

Solution

a By using the midpoint formula, we can find that the midpoint of $\overline{AB}$ is (3, 3). Let (3, 3) = (x_1, y_1) and (5, 12) = (x_2, y_2) in the two-point formula.

$$\frac{y - y_1}{x - x_1} = \frac{y_2 - y_1}{x_2 - x_1}$$

$$\frac{y - 3}{x - 3} = \frac{12 - 3}{5 - 3} = \frac{9}{2}$$

$$2y - 6 = 9x - 27$$
$$2y = 9x - 21$$

$$y = \frac{9}{2}x - \frac{21}{2}$$

Note   Actually, $y = \frac{9}{2}x - \frac{21}{2}$ is the equation of the *line containing* the median. The median itself is a segment. Unless otherwise stated, when we refer to the equation of a segment or ray, we mean the equation of the containing line.

b Slope of $\overleftrightarrow{AB} = \frac{4 - 2}{9 - (-3)} = \frac{1}{6}$

Since the slopes of two perpendicular lines are opposite reciprocals (except in the case of a horizontal and a vertical line), the slope of the perpendicular bisector is −6. Let the midpoint (3, 3) = (x_1, y_1) in the point-slope formula.

$$y - y_1 = m(x - x_1)$$
$$y - 3 = -6(x - 3)$$

Note   Since a line has infinitely many points, the point-slope formula does not produce a unique equation.

c The altitude contains C = (5, 12) and has a slope of −6 (it is perpendicular to $\overline{AB}$.) We can use the y-form formula or the point-slope formula.

y-form:

$$y = mx + b$$
$$12 = -6(5) + b$$
$$42 = b$$
$$y = -6x + 42$$

Point-slope form:

$$y - y_1 = m(x - x_1)$$
$$y - 12 = -6(x - 5)$$

The two equations are equivalent, so either is acceptable.

Part Three: Problem Sets

Problem Set A

1 Find the slope and the y-intercept of the graph of each equation.

a $y = 3x + 7$ 3; 7

d $y = 13 - 6x$ -6; 13

b $y = 4x$ 4; 0

e $y = -5x - 6$ -5; -6

c $y = \frac{1}{2}x - \sqrt{3}$ $\frac{1}{2}$; $-\sqrt{3}$

f $y = 7$ 0; 7

2 Rewrite each equation in y-form and find the slope and the y-intercept of its graph.

a $y - 3x = 1$ $y = 3x + 1$; 3; 1

c $2x + 3y = 6$ $y = \frac{-2}{3}x + 2$; $\frac{-2}{3}$; 2

b $y + 5x = 2$ $y = -5x + 2$; -5; 2

d $7 - (6 - 2x) = 4y$ $y = \frac{1}{2}x + \frac{1}{4}$; $\frac{1}{2}$, $\frac{1}{4}$

3 Write an equation of a line 6 units below, and parallel to, the x-axis. $y = -6$

4 Write an equation of a line that is perpendicular to the x-axis and passes through (8, 1). $x = 8$

5 Which two of the following three lines are parallel? **a** and **c**

a $y = 5x - 1$

b $y = 7x + 2$

c $y = 2 + 5x$

6 Write a y-form equation of each line.

a y-intercept of 2; slope = 4 $y = 4x + 2$

b $m = 5$; passes through (0, -2) $y = 5x - 2$

c Parallel to graph of $y = 10x - 6$; y-intercept of 1 $y = 10x + 1$

d Perpendicular to graph of $2y = x + 16$; passes through (0, -5) $y = -2x - 5$

e y-intercept of 2; perpendicular to line containing (-4, 6) and (1, 11) $y = -x + 2$

7 Use the graph to find

a The slope of $\overleftrightarrow{AB}$ 1

b An equation of $\overleftrightarrow{AB}$ $y = x + 5$

c The slope of $\overleftrightarrow{CD}$ $-\frac{1}{2}$

d An equation of $\overleftrightarrow{CD}$ $y = -\frac{1}{2}x - \frac{5}{2}$

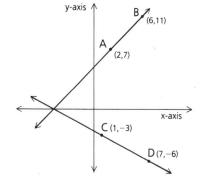

Assignment Guide

Basic

Day 1   1–6

Day 2   7–9, 13–16

Average

Day 1   2–7

Day 2   8–10, 13–16, 21, 23

Advanced

Day 1   7–16

Day 2   19–24, 27

Cooperative Learning

Have students work in small groups to find the equation for the line in problem **8a** (page 616) in each of the five forms shown on page 612.

point-slope: $y - 1 = 3(x - 2)$ or $y - 4 = 3(x - 3)$

slope-intercept: $y = 3x - 5$

two-point: $\frac{y - 1}{x - 2} = \frac{3}{1}$ or $\frac{y - 4}{x - 3} = \frac{3}{1}$

general linear: $3x - y - 5 = 0$

intercept: $\frac{x}{\frac{5}{3}} + \frac{y}{-5} = 1$

Problem-Set Notes and Additional Answers

■ In problem **8f**, the equation for a vertical line cannot be written in point-slope form.

11 and 12 $ax + by + c = 0$

$by = -ax - c$

$y = \frac{-a}{b}x - \frac{c}{b}$

$\therefore m = \frac{-a}{b}$ and the y-intercept is $\frac{-c}{b}$

■ Problems **13–17** can lead to a good discussion about the slopes of parallel and perpendicular lines and how to select the points that are substituted into the formulas.

20b (0, 0), (1, 1), (2, 2), and (3, 3) are winners: $\frac{4}{16} = \frac{1}{4}$.

c

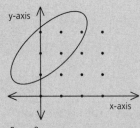

d $\frac{6}{16} = \frac{3}{8}$

21 (12, −3) doesn't work in the equation $y = -\frac{3}{4}x + 5$.

22

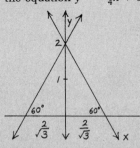

Problem Set A, *continued*

8 Write (if possible, in point-slope form) an equation of the line

a Containing (2, 1) and (3, 4) $y - 1 = 3(x - 2)$

b Containing (−6, 3) and (2, −1) $y - 3 = -\frac{1}{2}(x + 6)$

c Containing (1, 5) and (−3, 5) $y - 5 = 0$

d With an x-intercept of 2 and a slope of 7 $y = 7(x - 2)$

e That has an x-intercept of 3 and passes through (1, 8) $y = -4(x - 3)$

f That passes through (−3, 6) and (−3, 10) $x = -3$

g That passes through (8, 7) and is perpendicular to the graph of $3y = -2x + 24$ $y - 7 = \frac{3}{2}(x - 8)$

Problem Set B

9 The line that represents the equation $y = 8x - 1$ contains the point (k, 5). Find k. $\frac{3}{4}$

10 Line $\overleftrightarrow{CD}$ is perpendicular to the graph of $2x + 3y = 8$. If $C = (1, 4)$, find the equation of $\overleftrightarrow{CD}$. $y = \frac{3}{2}x + \frac{5}{2}$

11 Show that $-\frac{a}{b}$ is the slope of the graph of $ax + by + c = 0$.

12 Show that $-\frac{c}{b}$ is the y-intercept of the graph of $ax + by + c = 0$.

In problems 13–17, use △ABC in the diagram.

13 Write, in point-slope form, an equation of a line through C parallel to $\overleftrightarrow{AB}$.
$y - 12 = \frac{1}{7}(x - 4)$

14 Write an equation of the perpendicular bisector of $\overline{AB}$. $y = -7x + 65$

15 Write an equation of the altitude from C to $\overline{AB}$. $y = -7x + 40$

16 Write an equation of the median from C to $\overline{AB}$. $y = -2x + 20$

17 Find the slope of the line passing through the midpoints of $\overline{AC}$ and $\overline{BC}$. $\frac{1}{7}$

18 A line passes through a point 3 units to the left of and 2 units above the origin. Write an equation of the line if it is parallel to

a The x-axis $y = 2$ **b** The y-axis $x = -3$

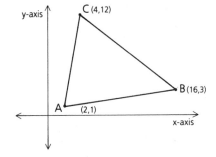

616 | Chapter 13   Coordinate Geometry Extended

19 If P = (−2, 5) and R = (0, 9), write, in point-slope form, an equation of the perpendicular bisector of $\overline{PR}$. $y − 7 = −\frac{1}{2}(x + 1)$

Problem Set C

20 Two numbers x and y (not necessarily different) are chosen at random from the set {0, 1, 2, 3}. The possible pairs (x, y) are illustrated by dots. Copy the graph.

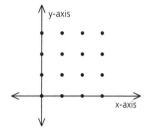

a How many pairs of numbers are there? 16

b What is the probability that x = y? $\frac{1}{4}$

c Show the points on the graph for which x < y.

d What is the probability that x < y? $\frac{3}{8}$

21 Does the point (12, −3) lie on the line whose slope is $−\frac{3}{4}$ and whose y-intercept is 5? Support your answer. No

22 A line has a y-intercept of 2 and forms a 60° angle with the x-axis. Find equations of the two possible lines. $y = x\sqrt{3} + 2$; $y = −x\sqrt{3} + 2$

23 Find an equation of the line whose intercepts are twice those of the graph of 2x + 5y = 10. $y = −\frac{2}{5}x + 4$

24 In △ABC, A = (0, 0), B = (4, 0), and C = (2, 6). Show that the medians of △ABC all intersect at (2, 2).

Note It can be shown that the medians of any triangle are concurrent at a point called the *centroid* of the triangle.

25 Find the center of the circle containing D = (−3, 5), E = (3, 3), and F = (11, 19). (3, 13)

Note The center of this circle is called the *circumcenter* of △DEF.

26 Find the reflection of the point (−9, 7) over the reference line y = x. (7, −9)

27 Find an equation of the reflection of the graph of $y = \frac{3}{4}x − 1$ over

a The x-axis
$y = −\frac{3}{4}x + 1$

b The y-axis
$y = −\frac{3}{4}x − 1$

c The line y = x
$y = \frac{4}{3}x + \frac{4}{3}$

Problem-Set Notes and
Additional Answers, continued

23   x-int. #1 = 5
y-int. #1 = 2
x-int. #2 = 10
y-int. #2 = 4
Slope #2 = $−\frac{2}{5}$
∴ Equation #2: $y = −\frac{2}{5}x + 4$

24

Equations of medians:
x = 2
y = x
y = −x + 4
These lines intersect at (2, 2).

25   ⊥ bisectors of chords of a ⊙ pass through the center.
⊥ bis. of $\overline{DE}$: y = 3x + 4
⊥ bis. of $\overline{EF}$: $y = −\frac{1}{2}x + \frac{29}{2}$
∴ Intersection = center = (3, 13)

26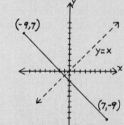

27a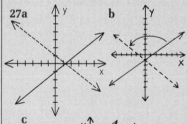

c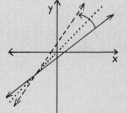

T 617

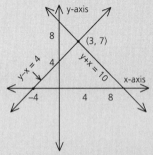
Objective

After studying this section, you will be able to
- Use two methods to solve systems of equations

Part One: Introduction

When two linear equations are graphed on the same coordinate
plane, the resulting lines may be

Parallel (a ∥ b)
Intersecting (c intersects d.)
Identical (e and f coincide.)

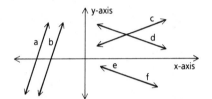

Each pair of lines may be represented by a ***system of equations***.

Systems of Equations		
System	**Graph**	**Intersection**
$\begin{cases} y = x + 8 \\ y = x + 13 \end{cases}$	Parallel lines	Empty set
$\begin{cases} y = x + 3 \\ y = -2x + 21 \end{cases}$	Intersecting lines	One point
$\begin{cases} y = \frac{1}{2}x - 10 \\ 2x - 4y = 40 \end{cases}$	Identical lines	All points on the line

    Most of the problems in this section require solving a system of
two linear equations. The sample problems illustrate two methods of
solving such systems:

1 Addition or subtraction
2 Substitution

618 | Chapter 13 Coordinate Geometry Extended

Part Two: Sample Problems

Problem 1 Find the intersection of the two lines corresponding to x = 4 and y = 2x + 8.

Solution *Substitution method:* Since x = 4 is the first equation, we can substitute 4 for x in the second equation.

y = 2x + 8
y = 2(4) + 8
y = 8 + 8
y = 16

Thus, the intersection is (4, 16).

Problem 2 Find the intersection of the lines corresponding to the following system.

$$\begin{bmatrix} 8x - 3y = 7 \\ 10x + 4y = 1 \end{bmatrix}$$

Solution *Addition-subtraction method:*

32x − 12y = 28	Multiply both sides of first equation by 4.
30x + 12y = 3	Multiply both sides of second equation by 3.
62x + 0 = 31	Add the equations.

$$x = \frac{1}{2}$$

Now substitute $\frac{1}{2}$ for x in the first or the second equation.

8x − 3y = 7
$8\left(\frac{1}{2}\right) - 3y = 7$
y = −1

The lines intersect at $\left(\frac{1}{2}, -1\right)$.

Problem 3 Find the intersection of the lines corresponding to the following system.

$$\begin{bmatrix} y = 3x + 1 \\ 6x - 2y = -2 \end{bmatrix}$$

Solution *Substitution method:*

Substitute 3x + 1 for y in the second equation.

6x − 2y = −2
6x − 2(3x + 1) = −2
6x − 6x − 2 = −2
−2 = −2

Since the statement −2 = −2 is always true, the intersection is the entire graph of the first equation, which is therefore identical to the graph of the second equation. The solution set is {(x, y):y = 3x + 1}.

Lesson Notes

■ Students who have not previously studied systems of equations may need additional help to understand the three sample problems.

Communicating Mathematics

After presenting the sample problems, ask students to describe, orally or in writing, how to solve each of the following systems by eliminating the x term.

1 $\begin{cases} 3x - 2y = -4 \\ 5x + 3y = -1 \end{cases}$
   Answers will vary. Multiply 3x − 2y = −4 by 5 and multiply 5x + 3y = −1 by −3. Then add the resulting equations. If 5x + 3y = −1 is multiplied by 3, then subtract the resulting equations.

2 $\begin{cases} 3x - 2y = 8 \\ 3x - 5y = -10 \end{cases}$
   Answers will vary. Multiply either equation, but not both, by −1. Then add the resulting equations. Students should understand that multiplying the second equation by −1 and adding is equivalent to subtracting the second equation from the first equation.

Assignment Guide

Basic

1–4

Average

1a,c, 2–6, 10

Advanced

5–12, 14, 15

Problem-Set Notes
and Additional Answers

8 $x^2 + y^2 = 25$
$\dfrac{x^2 - y^2 = \quad 7}{2x^2 \quad = 32}$
$x^2 = 16$
$x = \quad 4 \text{ or} \qquad x = -4$
$16 + y^2 = 25 \mid 16 - y^2 = 7$
$y = \pm 3 \mid \qquad y = \pm 3$

13 Solve the equations simultaneously by elimination or by substitution.

14 Equation of $\overline{BC}$: $y = -3x + 4$
Equation of altitude:
$y = \frac{1}{3}x - \frac{8}{3}$
Point of intersection: $(2, -2)$
∴ Length of altitude =
$\sqrt{(5 - 2)^2 + (-1 + 2)^2} =$
$\sqrt{10}$

Part Three: Problem Sets

Problem Set A

1 Determine the point of intersection of the graphs of each system.

a $\begin{bmatrix} x + y = 10 \\ x - y = 2 \end{bmatrix}$
(6, 4)

b $\begin{bmatrix} y = 5 \\ x + y = 7 \end{bmatrix}$
(2, 5)

c $\begin{bmatrix} y = 2x - 1 \\ y = 4x + 5 \end{bmatrix}$
(−3, −7)

d $\begin{bmatrix} x + 2y = 7 \\ 4x - y = 10 \end{bmatrix}$
(3, 2)

2 Determine the intersection of the graphs of each system.

a $\begin{bmatrix} x = 4 \\ x^2 + y^2 = 25 \end{bmatrix}$
(4, 3) and (4, −3)

b $\begin{bmatrix} y = 3 \\ |y - 2| = x \end{bmatrix}$
(1, 3)

3 Where do the lines intersect?

a $\begin{bmatrix} x + 2y = 12 \\ x\text{-axis} \end{bmatrix}$
(12, 0)

b $\begin{bmatrix} y = 3x - 7 \\ 9x - 3y = 21 \end{bmatrix}$
$\{(x, y): y = 3x - 7\}$

4 Find the points each pair of lines has in common.

a $\begin{bmatrix} 2x + y = 10 \\ 8x + 4y = 17 \end{bmatrix}$
{ }

b $\begin{bmatrix} y = 4x + 1 \\ \text{The line to the right of the y-axis,} \\ \text{parallel to it, and 4 units from it} \end{bmatrix}$
(4, 17)

Problem Set B

5 Where does $\overleftrightarrow{DE}$ intersect $\overleftrightarrow{FH}$?
(−2, 3)

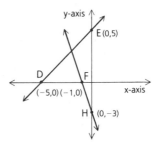

6 Find the intersection of the graphs of
$x = a$ and $3x + 2y = 12$. $(a, 6 - 1.5a)$

7 Show that the graphs of the following three equations are concurrent (intersect at a single point). What are the coordinates of the point of intersection? $(4, -2)$
$\begin{bmatrix} 2x + 3y = 2 \\ y = 2x - 10 \\ 3x - y = 14 \end{bmatrix}$

8 The graph of $x^2 + y^2 = 25$ is a circle. (Circular graphs will be studied later in this chapter.) The graph of $x^2 - y^2 = 7$ is a hyperbola. (Hyperbolas are normally studied in a later math course.) Use one of the methods of solving a system of equations to find the intersection of the circle and the hyperbola.
$(4, -3)$, $(4, 3)$, $(-4, 3)$, and $(-4, -3)$

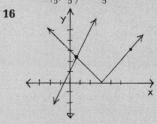

9 Find, in point-slope form, an equation of the line containing (2, 1) and the point of intersection of the graphs of $3x - y = 3$ and $x + 2y = 15$. $y - 1 = 5(x - 2)$

10 Find an equation of the line that is parallel to the graph of $2x + 3y = 5$ and contains the point of intersection of the graphs of $y = 4x + 8$ and $y = x + 5$. $y - 4 = -\frac{2}{3}(x + 1)$

11 Find the point of intersection of the graphs of $y - 3 = \frac{1}{2}(x - 1)$ and $y + 1 = -\frac{3}{2}(x - 1)$. $(-1, 2)$

12 Consider the line corresponding to $y = 2x + 1$. Line 2 contains (5, 3) and is parallel to the given line. Line 3 contains (5, 16) and has the same y-intercept as the given line. Find the intersection of lines 2 and 3. $(-8, -23)$

Problem Set C

13 If the equations $ax + by = c$ and $dx + ey = f$ represent two intersecting lines, what are the coordinates of their point of intersection? $\left(\frac{ce - bf}{ae - bd}, \frac{af - cd}{ae - bd}\right)$

14 In $\triangle ABC$, $A = (5, -1)$, $B = (1, 1)$, and $C = (5, -11)$. Find the length of the altitude from A to $\overline{BC}$. $\sqrt{10}$

15 Find the distance between the parallel lines corresponding to $y = 2x + 3$ and $y = 2x + 7$. (Hint: Start by choosing a convenient point on one of the lines.) $\frac{4\sqrt{5}}{5}$

16 Find the intersection of the V-shaped graph of $y = |x - 3|$ and the graph of $y = 2x + 1$. $\left(\frac{2}{3}, \frac{7}{3}\right)$

17 Find the area of the triangle whose sides lie on the graphs of $3x + y + 1 = 0$, $x + 4y - 7 = 0$, and $-5x + 2y + 13 = 0$. 11

18 Find the reflection of the point $(-6, 5)$ over the graph of $2y - x = 6$. $(-2, -3)$

Problem-Set Notes and
Additional Answers, continued

15 The equation of the $\perp$ from (0, 7) to $y = 2x + 3$ is $y = -\frac{1}{2}x + 7$. The intersection of $y = -\frac{1}{2}x + 7$ and $y = 2x + 3$ is $\left(\frac{8}{5}, \frac{31}{5}\right)$. The distance between (0, 7) and $\left(\frac{8}{5}, \frac{31}{5}\right)$ is $\frac{4\sqrt{5}}{5}$.

16

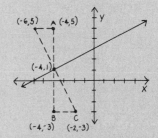

The intersection is the intersection of $y = -x + 3$ and $y = 2x + 1$, which is $\left(\frac{2}{3}, \frac{7}{3}\right)$.

17 Vertices are (3, 1), (−1, 2), and (1, −4). Foot of altitude from (3, 1) is $\left(\frac{-3}{10}, \frac{-1}{10}\right)$. $\therefore b = 2\sqrt{10}$ and $h = \frac{11\sqrt{10}}{10}$. So $A = \frac{1}{2}(2\sqrt{10})\left(\frac{11\sqrt{10}}{10}\right) = 11$.

18 The equation of the $\perp$ from (−6, 5) to $2y - x = 6$ is $y = -2x - 7$. The lines intersect at (−4, 1). The coordinates of A, B, and C can be found: (−4, 5), (−4, −3), and (−2, −3), respectively.

GRAPHING INEQUALITIES

Class Planning

Time Schedule
All levels: 1 day

Resource References
Teacher's Resource Book
   Class Opener 13.4A, 13.4B
Evaluation
   Tests and Quizzes
      Quiz 2, Series 1–3

Class Opener

1   Graph x + y = 10. Then shade the region to the right of the line.

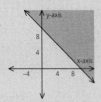

2   Graph x − y = −4 on a separate set of axes. Then shade the region to the right of the line.

3   Now graph the equations from problems **1** and **2** on the same set of axes. Using the answers to problems **1** and **2**, shade the region that is common to both.

4   What does the graph from problem **3** represent?
   The graph of this system of inequalities:

$$\begin{cases} x + y \geq 10 \\ x - y \geq -4 \end{cases}$$

Objective

After studying this section, you will be able to
■ Graph inequalities

Part One: Introduction

Inequalities and systems of inequalities can be graphed by means of the following procedure.

Two-Part Procedure for Graphing Inequalities
1 Pretend that the inequality is an equation. Graph this equation as a **boundary line.**
2 In the inequality, test the coordinates of points in the various regions separated by the boundary line. Shade the region(s) whose points satisfy the inequality.

In the final graph, the boundary line is dashed if it is not included in the graph of the inequality.
   Study the following sample problems closely.

Part Two: Sample Problems

Problem 1   *Graph y > 2x + 8.*

Solution   *Boundary line: Pretend that y = 2x + 8.*

x	y
0	8
1	10
−1	6
−2	4

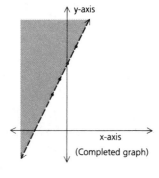

(Completed graph)

Vocabulary
boundary line

The boundary line is dashed, since there is no equal sign in the original inequality.

Test of Regions: In the inequality, test a convenient point not on the boundary line—for instance, (0, 0).

$y > 2x + 8$	Since $0 > 8$ is false, do not shade the
Is $0 > 2(0) + 8$?	region to the right of the line.
Is $0 > 8$?	

Now test a point in the other region, such as $(-10, 10)$.

Is $10 > 2(-10) + 8$?	Since $10 > -12$ is true, do shade the
$10 > -12$	region containing $(-10, 10)$.

Problem 2   *Graph $y \geq |x - 2|$.*

Solution   *Boundary line: $y = |x - 2|$.*

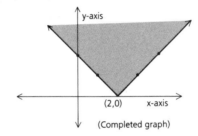

x	y
0	2
1	1
2	0
3	1
4	2

(2,0)   x-axis

(Completed graph)

The boundary line (a V) is solid because there is an equal sign in the original inequality. Two regions are formed.

Test of Regions: Test (0, 0) in the inequality.

$y > |x - 2|$	
Is $0 > |0 - 2|$?	$0 > 2$ is false.
Is $0 > |-2|$?	Do not shade the region containing (0, 0).
Is $0 > 2$?	Further tests confirm that the other region should be shaded.

Problem 3   *Determine the solution set of the system by graphing.*
$$\begin{cases} y \leq \frac{2}{5}x + 4 \\ y \geq -\frac{1}{2}x + 4 \\ 2x + y \leq 16 \end{cases}$$

Solution   We follow the two-part procedure three times.

Lesson Notes

■ This section is optional, but it is recommended for average and advanced classes.

Cooperative Learning

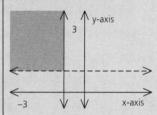

Have students work in small groups to write the system that is shown on this graph.
$$\begin{cases} y > 1 \\ x \leq -1 \end{cases}$$
Then have each group draw a system of inequalities on an overhead transparency to present to the class. Have students continue to work in groups to write the system represented by the graphs.

Communicating Mathematics

Have students write a paragraph that explains the similarities and differences between solving a system of equations and solving a system of inequalities.

Assignment Guide

Basic
1–3, 5a,c
Average
1, 2, 5
Advanced
5a,b, 6–9

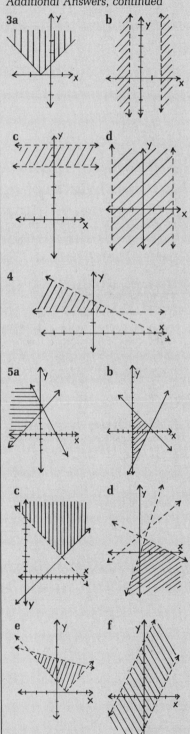

Boundary line:

$$y = \frac{2}{5}x + 4$$

x	y
0	4
5	6
10	8
15	10

After testing regions, we shade below the boundary line.

Boundary line:

$$y = -\frac{1}{2}x + 4$$

x	y
0	4
2	3
4	2
6	1

After testing regions, we shade above the boundary line.

Boundary line:

$$2x + y = 16$$

x	y
0	16
1	14
2	12
3	10

After testing regions, we shade below the boundary line.

The solution consists of the union of the triangle and its interior, as shown in the final graph.

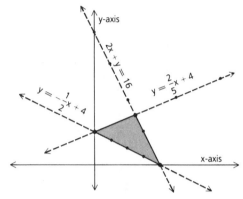

Part Three: Problem Sets

Problem Set A

1 Graph each inequality.

a $2x - 3y < 6$

b $y < \frac{1}{2}x - 1$

c $5x + 2y \geq 10$

d $x < -2$

e $y \geq 2x + 3$

2 Write the inequality represented by each graph.

a $y < -3$

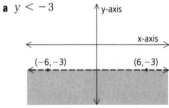

b $y \geq \frac{1}{2}x + 2$

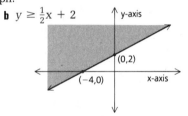

624 Chapter 13 Coordinate Geometry Extended

3 Graph each of the following.

a $y \geq |x + 1|$ **c** $\{(x, y):5 < y < 7\}$

b $\{(x, y):x > 2 \text{ or } x < -1\}$ **d** $\{(x, y):|x| < 3\}$

Problem Set B

4 Determine the intersection of the solution sets of the two inequalities $y > 2$ and $x + 2y < 6$ by graphing.

5 Graph the solution of each system of inequalities.

a $\begin{bmatrix} y \geq x + 4 \\ y \leq -2x + 6 \end{bmatrix}$ **c** $\begin{bmatrix} x + y > 12 \\ x - y \leq 4 \end{bmatrix}$ **e** $\begin{bmatrix} y > |x - 1| \\ x + 3y < 12 \end{bmatrix}$

b $\begin{bmatrix} x + y \leq 4 \\ 2x - y \leq 6 \\ x \geq 0 \end{bmatrix}$ **d** $\begin{bmatrix} 4y - 3x < 6 \\ y < 3x \\ 2x < 6 - 3y \end{bmatrix}$ **f** $\begin{bmatrix} y < 2x + 5 \\ 2x - y < 3 \end{bmatrix}$

6 Determine the *union* of the solution sets of the inequalities $x + y > 4$ and $y < 2x - 6$.

Problem Set C

7 Graph the solution set of each system of inequalities.

a $\begin{bmatrix} y < x^2 + 8 \\ y > -x + 12 \end{bmatrix}$ **b** $\begin{bmatrix} x^2 + y^2 \leq 25 \\ y \geq |x| \end{bmatrix}$ **c** $\begin{bmatrix} xy < 12 \\ x^2 + y^2 < 16 \end{bmatrix}$

8 Graph each inequality.

a $|x + y| \leq 4$ **b** $|x| + |y| \leq 4$

9 The graph of $y = x^2$ for the values of $0 \leq x \leq 3$ is shown.

a Find the coordinates of A, B, and C. (1, 1); (2, 4); (3, 9)

b We can estimate the area of the region between the graph of $y = x^2$ and the x-axis (when $0 \leq x \leq 3$) by adding the areas of $\triangle AOD$, trapezoid ABED, and trapezoid BCFE. Find this sum. $9\frac{1}{2}$

Note If you study calculus, you will learn that the actual area of this region is 9.

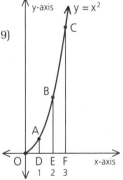

Problem-Set Notes and Additional Answers, continued

6
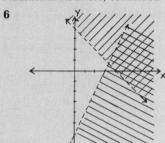

7a
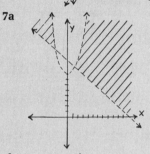

b
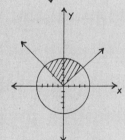

c
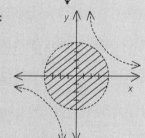

8a   **b**
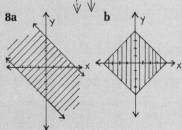

9b   Area of $\triangle AOD$ =   Area of BCFE =
       $\frac{1}{2}(1)(1) = \frac{1}{2}$ $\frac{1}{2}(1)(4 + 9) = 6\frac{1}{2}$
       Area of ABED =   Total area = $9\frac{1}{2}$
       $\frac{1}{2}(1)(4 + 1) = 2\frac{1}{2}$

9a   Substitute the x values in $y = x^2$.
       A: $x = 1$, $y = 1$; B: $x = 2$, $y = 4$; C: $x = 3$, $y = 9$

13.5 THREE-DIMENSIONAL GRAPHING AND REFLECTIONS

Class Planning

Time Schedule
All levels: 1 day

Resource References
Teacher's Resource Book
   Class Opener 13.5A, 13.5B
Transparency 22

Class Opener

Given: △AOB ≅ △DCB;
       $\overline{AB}$ is determined by the
       equation $y = -2x + 8$.
Find: The coordinates of D and
       the equation for $\overline{BD}$

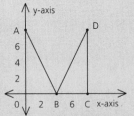

Given $y = -2x + 8$, when $x = 0$,
$y = 8$. Thus, the coordinates of
A are (0, 8). When $y = 0$, $x = 4$.
Thus, the coordinates of B are
(4, 0). Given that △AOB ≅
△DCB, $\overline{OB} ≅ \overline{CB}$. Since OB = 4,
CB = 4. Thus, OC = 8, and the
coordinates of C are (8, 0). Since
$\overline{OA} ≅ \overline{CD}$, it follows that the
coordinates of D are (8, 8).
The slope of $\overline{BD}$ is $\frac{8-0}{8-4} = \frac{8}{4} = 2$.
So $y - 8 = 2(x - 8)$ is the
equation for $\overline{BD}$.
This problem previews reflections and reviews writing equations of lines and solving
equations.

Lesson Notes

■ If the teacher walks through
   the reading of this section with
   illustrations on the board or
   overhead projector, student
   comprehension will be en-
   hanced.

Objectives

After studying this section, you will be able to
■ Graph in three dimensions
■ Apply the properties of reflections

Part One: Introduction

Three-Dimensional Graphing

As an extension to graphing in a coordinate plane, here is a brief
introduction to three-dimensional graphing.

In a three-dimensional coordinate system, there are three axes that are mutually
perpendicular. The point P = (3, 4, 12) may
be graphed in "3-D" by drawing the axis
system shown. A rectangular box should be
drawn as an aid in locating and visualizing
the point. The sides of the box are drawn
parallel to the axes. The x-axis is drawn at
an angle but should be visualized as being
perpendicular to the plane of the paper.

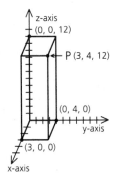

The distance between two points in
space can be found with the ***3-D distance
formula***, which is a logical extension of the
two-dimensional distance formula.

Theorem 126 *If $P = (x_1, y_1, z_1)$ and $Q = (x_2, y_2, z_2)$ are any two
points, then the distance between them can be
found with the formula*

$$PQ = \sqrt{(x_2 - x_1)^2 + (y_2 - y_1)^2 + (z_2 - z_1)^2}$$

Reflections

Since the beginning of this book, you have been solving problems
involving rotations, reflections, and translations of points. Now you
will work with a few applications of reflections that you might find
useful.

Vocabulary
3-D distance formula

Many of you have probably played mini-ature golf. In the diagram to the right, you see a type of hole you may have encountered. Obviously, it won't work to aim directly for the hole. You must aim to hit the barrier so that the ball will bounce off at the proper angle.

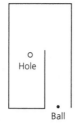

To find the point to shoot for, you can reflect point E (the pre-image) over the barrier to point E' (the image). If you aim at point E', the ball should strike the barrier at N and bounce directly to E. If your aim is good, you will have a hole in ONE.

Why does this reflection work? The answer depends on a law of physics. By the principles of reflection, $\overline{E'S} \cong \overline{ES}$ and $\angle ESN \cong \angle E'SN$. Since $\overline{SN} \cong \overline{SN}$, $\triangle SNE' \cong \triangle SNE$ by SAS. Hence, $\angle 2 \cong \angle 3$ by CPCTC. Since $\angle 1 \cong \angle 3$ (vertical angles), $\angle 1 \cong \angle 2$. A law of physics states that the angle of incidence ($\angle 1$) is equal in measure to the angle of reflection ($\angle 2$). Thus, your ball should bounce directly into the hole.

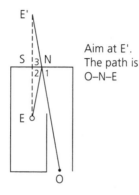

Aim at E'. The path is O–N–E

We can also use this diagram to prove an interesting physical fact: The path of the ball from O to N to E is the shortest possible path from O to E.

Pick any other point on the barrier, such as P, and consider the hypothetical path from O to P to E. By CPCTC, $\overline{PE'} \cong \overline{PE}$ and $\overline{NE'} \cong \overline{NE}$. So the path from O to P to E has the same length as the path from O to P to E'. In the same way, the distance from O to N to E is the same as that from O to N to E'. But $\overline{ONE'}$ is a straight line segment, so its length must be less than OP + PE' by the triangle-inequality principle.

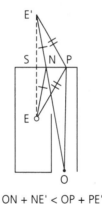

ON + NE' < OP + PE'

You can also solve situations involving several reflections over several barriers, such

Cooperative Learning

Using a cardboard box to help visualize the three-dimensional coordinate system, have students work in small groups to answer these questions.

1 Describe the graph of each equation.
  a x = 5   A plane parallel to the y- and z-axes, passing through (5, 0, 0)
  b y = 2   A plane parallel to the x- and z-axes, passing through (0, 2, 0)
  c z = 9   A plane parallel to the x- and y-axes, passing through (0, 0, 9)

2 Consider the coordinates (x, y, z). Describe the graph of each set of points.
  a Let x = 3 and y = 4, but let z vary.   A line parallel to the z-axis, passing through (3, 4, 0)
  b Let x = 6 and z = 1, but let y vary.   A line parallel to the y-axis, passing through (6, 0, 1)
  c Let y = 2 and z = 5, but let x vary.   A line parallel to the x-axis, passing through (0, 2, 5)

3 Describe the graph of the equation 2x + 3y + z = 6. (Hint: Find some points that make the equation true and visualize or model it.)
  A plane that passes through (0, 0, 6), (3, 0, 0), and (0, 2, 0)

4 The graph shown on page 626 (or the cardboard box you are using as a model) represents only a portion of the three-dimensional coordinate system. What parts are missing?   The model on page 626 shows only $\frac{1}{8}$ of the three-dimensional coordinate system. Each of the seven missing parts (octants) has negative values for x, y, and/or z.

as the complicated hole shown in Figure 1 at the right. First determine which barriers you wish the ball to strike. Then reflect the target point over these barriers, one by one, as shown in Figure 2. (Reflect E over the lower barrier to F; then reflect F over the line containing the left-hand barrier to G; then reflect G over the line containing the upper barrier to H.) If you putt the ball in the direction of point H, it should follow a path from A to R to N to I to E. You may not want to go to this much trouble when actually playing, but it is fun to know the principle involved.

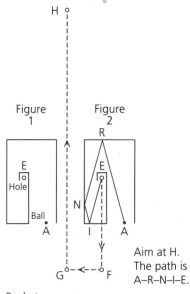

Aim at H.
The path is
A–R–N–I–E.

The reflection principle can be extended to the game of pool, or pocket billiards. With the cue ball at P, use the principles of reflection to knock the ball at T into the pocket.

One way is to reflect T to T' as shown. If you aim the cue ball at T', it should travel from P to R to T. There are some complications, however. You must strike T in such a way that

- T goes into the pocket
- The cue ball does not go into a pocket (If it does, you have "scratched" and lost your turn.)

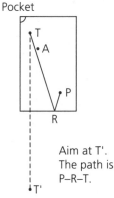

Aim at T'.
The path is
P–R–T.

Reflections are also useful in the game of three-cushion billiards.

The concepts of rotations and reflections are important in such mathematical studies as trigonometry and calculus. In addition, many professionals—structural engineers and architects, for example—use these concepts extensively in their work.

Part Two: Sample Problems

Problem 1   *Graph the point* A = (2, 5, 7) *on a 3-D graph.*

Solution   Use a rectangular box as shown to aid you in locating and visualizing point A at (2, 5, 7).

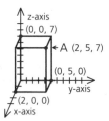

Problem 2 Find the distance from A = (2, 5, 7) to B = (3, −2, 4).

Solution Use the 3-D distance formula.

$$AB = \sqrt{(x_2 - x_1)^2 + (y_2 - y_1)^2 + (z_2 - z_1)^2}$$
$$= \sqrt{(3 - 2)^2 + (-2 - 5)^2 + (4 - 7)^2}$$
$$= \sqrt{59}$$
$$\approx 7.68$$

Problem 3 Show where the ball should strike the barrier for you to have the best chance of making a hole in one.

Solution Reflect the hole at E over the barrier to E′. Then aim the ball at the imaginary E′ so that it strikes the barrier at N.

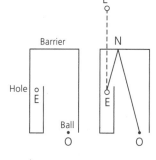

Problem 4 On the miniature-golf hole shown, the ball is at J and the hole is at Y. Jan wants the ball to strike the barrier at y = 8, the barrier at x = 0, and the barrier at y = −1 before it goes into the hole at (4, 2).

  a Show the reflections from the pre-image Y to the image at Y‴, where the ball should be aimed.

  b Find the coordinates of Y‴.

  c Find the coordinates of A, the point at which the ball should strike the first barrier.

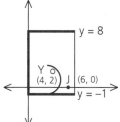

Solution **a** Reflect Y over the line y = −1 to Y′ = (4, −4). Then reflect Y′ over the y-axis to Y″ = (−4, −4). Finally, reflect Y″ over the line y = 8 to Y‴ = (−4, 20).

  b See part **a**.

  c Since J, A, and Y‴ are collinear, use slopes. Let A = (x, 8).
Slope of $\overleftrightarrow{JY'''}$ = slope of $\overleftrightarrow{JA}$

$$\frac{20 - 0}{-4 - 6} = \frac{8 - 0}{x - 6}$$
$$20(x - 6) = -10(8)$$
$$20x - 120 = -80$$
$$20x = 40$$
$$x = 2$$

So the coordinates of A are (2, 8).

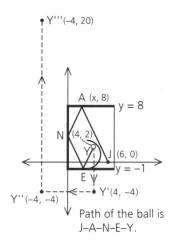

Have students discuss and identify the three-dimensional figure that is formed when each of the following two-dimensional figures is rotated 360° around the y-axis.

1 Right △   Cone

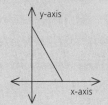

2 Rectangle   Cylinder

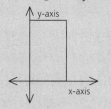

3 Semicircle   Sphere

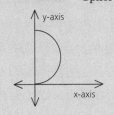

Section 13.5   Three-Dimensional Graphing and Reflections   **629**

Part Three: Problem Sets

Problem Set A

1 Graph the point A = (2, 4, 6) on a 3-D graph. Use a rectangular box as an aid in locating and visualizing point A.

2 Find the distance from P = (3, 4, 12) to the origin. 13

3 Find, to the nearest tenth, the distance from P = (3, 4, 12) to D = (−1, −2, 9). ≈7.8

4 Find, to the nearest tenth, the perimeter of a triangle with vertices at (0, 0, 6), (0, 8, 0), and (15, 0, 0). ≈43.2

5 On a 3-D graph, draw the rectangular solid whose base has vertices at D = (0, 0, 0), A = (4, 0, 0), B = (4, 5, 0), and C = (0, 5, 0) and whose height is 7.

  a Find the area of the base. 20

  b Find the volume of the solid. 140

  c Find the diagonal of the solid. $3\sqrt{10}$

  d Is (4, 5, 7) a vertex of the solid? Yes

  e If the solid were rotated 90° downward about $\overline{AB}$, what would the new coordinates of the vertex be? (11, 5, 0) or (−3, 5, 0)

6 Two famous geometry teachers, Mr. Ripple and Mr. Wood, were playing the miniature-golf hole shown at the right. Mr. Wood shot first, hitting the barrier at W but missing the hole. After a moment of reflection, Mr. Ripple hit his ball to strike the barrier at R, and the ball bounced straight into the hole.

    "How'd you do that?" asked Mr. Wood. "You just need the right image," Mr. Ripple replied. Draw a diagram to show Mr. Wood what Mr. Ripple meant.

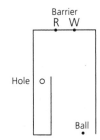

7 Reflect the hole over the barrier and give the coordinates of the image of the hole. (3, 10)

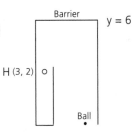

8 Are you more likely to make a hole in one by aiming at A, at B, at C, or at D? Show the reflections on a diagram to justify your answer. C

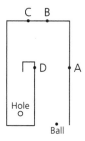

Problem Set B

9 Consider the points A = (2, 3, −5), B = (8, 9, 1), and C = (3, 17, 1).

   a Find the midpoint of $\overline{AB}$. (5, 6, −2)

   b Find, to the nearest tenth, the length of the median from C to $\overline{AB}$. ≈11.6

10 Suppose that P = (3, 5) and that point P is reflected over the graph of x = 1 to P′. Find in point-slope form, the equation of $\overleftrightarrow{JP'}$, if J = (5, 6). $y - 6 = \frac{1}{6}(x - 5)$

11 Point Q′ = (3, 7) is the image of a point after reflection over the y-axis. Find the pre-image. (−3, 7)

12 Verify that if the path from S to H to O is the shortest distance from S to the y-axis to O, then H = (0, 6).

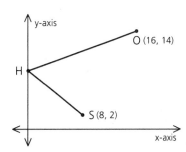

13 In a circle whose center is at P, the image of A = (4, 6) over P is (−2, −2). Find the image of B = (−3, 5) over P. (5, −1)

Problem Set C

14 The base of a triangular pyramid has vertices at (6, 0, 0), (0, 6, 0), and (0, 0, 6). If the peak of the pyramid is at (0, 0, 0), find the volume of the pyramid. 36

Problem-Set Notes and Additional Answers, continued

■ See *Solution Manual* for answer to problem **12**.

14

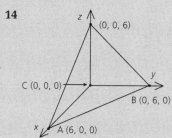

Use △ABC as the base.
$A_{\triangle ABC} = \frac{1}{2}(6)(6) = 18$
$V = \frac{1}{3}\mathscr{B}h = \frac{1}{3}(18)(6) = 36$

Problem-Set Notes and
Additional Answers, continued

15

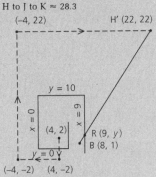

a $A'B' = \sqrt{(1-0)^2 + (-1+4)^2}$
  $= \sqrt{10}$
  Area $= (\sqrt{10})^2 = 10$

b Slope of $A'C' = \frac{-5+1}{3-1} = -2$
  $y = mx + b$
  $-1 = -2(1) + b$
  $b = 1$, so $y = -2x + 1$

16

Slope of $\overleftrightarrow{HK'}$ = slope of $\overleftrightarrow{HJ}$
$\frac{17-2}{-14-10} = \frac{y-2}{2-10}$
$8y - 16 = 40$
$y = 7$
$JH = \sqrt{(7-2)^2 + (2-10)^2} = \sqrt{89}$
$JK = \sqrt{(18-2)^2 + (17-7)^2} = \sqrt{356}$
H to J to K ≈ 28.3

17

Reflect the hole four times
as shown to H'(22, 22).
Slope of $\overleftrightarrow{BH'}$ = slope of $\overleftrightarrow{BR}$
$\frac{22-1}{22-8} = \frac{y-1}{9-8}$
$\frac{3}{2} = \frac{y-1}{1}$

Problem Set C, *continued*

15 A square with vertices at A = (1, 1), B = (0, 4), C = (3, 5), and
D = (4, 2) is reflected over the x-axis to produce a new square
with vertices A′, B′, C′, and D′.

  a Find the area of square A′B′C′D′. 10
  b Find, in y-form, the equation of $\overleftrightarrow{A'C'}$. $y = -2x + 1$

16 If H = (10, 2) and K = (18, 17) and if J is any point on the graph
of x = 2, find, to the nearest tenth, the minimum distance from
H to J to K. ≈ 28.3

17 On a miniature-golf course, the ball is at
(8, 1) and the hole is at (4, 2). A player
can make a hole in one by hitting the
ball along the path indicated by the ar-
rows. Find the coordinates of point R. $\left(9, 2\frac{1}{2}\right)$

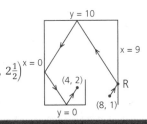

$y = 2\frac{1}{2}$
So R = $(9, 2\frac{1}{2})$.

IMAGE-PRODUCING WAVES
Sound in the coordinate plane

Sound travels in waves. The number of wave
cycles that pass a given point in one second is
the *frequency* of the wave. Human ears are ca-
pable of hearing frequencies from about 16 cycles per second to 20,000 cycles per second.
Sound with a frequency greater than 20,000 is
ultrasound. The reflection of ultrasound waves
directed into the body produces a moving image
that can be useful in diagnostic medicine.

Kathy Gurnee is an ultrasonographer at Love-
lace Medical Center in Albuquerque, New Mexico. She reads ultrasound echoes using a grid
that resembles three-dimensional coordinate
axes. Height and width can be read directly.
Depth and density can be gauged from the
shading of the image. The sonographer can
zoom in on objects with dimensions as small as
1 millimeter.

Gurnee finds her work extremely challenging.
"Unlike an X-ray," she says, "an ultrasound image is a continuously moving picture. This
makes it a much more powerful diagnostic
tool."

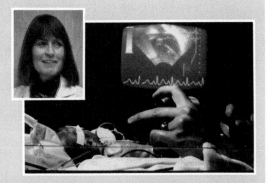

Originally from Yellow Springs, Ohio, Gurnee
moved to Albuquerque, where she earned a degree in education from the University of New
Mexico and then taught multiply-handicapped
children for several years. She became interested in ultrasound when she learned that the
technique had been used to diagnose the dis-
abilities of some of her students before their
births. After a year of further study at the University of New Mexico, she became a registered
diagnostic medical sonographer.

632 Chapter 13 Coordinate Geometry Extended

13.6 | CIRCLES

Objective

After studying this section, you will be able to
■ Write equations that correspond to circles

Part One: Introduction

The equation of a circle is based on the distance formula (Section 9.5) and the fact that all points on a circle are equidistant from the circle's center.

Theorem 127   ***The equation of a circle whose center is (h, k) and whose radius is r is***
$$(x - h)^2 + (y - k)^2 = r^2$$

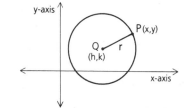

This *circle formula* may be used in several ways.

Example 1   *Find the equation of a circle whose center is (1, 5) and whose radius is 4.*

$$(x - 1)^2 + (y - 5)^2 = 16$$

Example 2   *Find the center and radius of the graph of $(x - 2)^2 + (y + 7)^2 = 64$.*

We rewrite the given equation in the same form as the circle equation.
$$(x - h)^2 + (y - k)^2 = r^2$$
$$(x - 2)^2 + [y - (-7)]^2 = 8^2$$
Hence, $h = 2$, $k = -7$, and $r = 8$. The center is $(2, -7)$ and the radius is 8.

The next example uses the circle formula in the same way as example 2, but the preparation is more complicated.

Section 13.6   Circles   **633**

Class Planning

Time Schedule
All levels: 2 days

Resource References
Teacher's Resource Book
    Class Opener 13.6A, 13.6B
    Additional Practice
    Worksheet 26
Evaluation
    Tests and Quizzes
    Quiz 3, Series 1–3

Class Opener

Find the missing term.
1   $x^2 + 4x + 4 = (x + \underline{})^2$ 2
2   $x^2 - 10x + 25 = (x - \underline{})^2$
    5
3   $x^2 + 10x + \underline{} = (x + 5)^2$
    25
4   $x^2 - \underline{} + 16 = (x - 4)^2$ 8
5   $x^2 - 20x + \underline{} = (x - 10)^2$
    100

Complete the square. Then write the trinomial as the square of a binomial.

6   $x^2 + 8x + \underline{}$ 16; $(x + 4)^2$
7   $x^2 - 14x + \underline{}$ 49; $(x - 7)^2$

Solve by completing the square.

8   $x^2 + 4x = 12$ $x = -6$ or $x = 2$
9   $x^2 + 10x = -16$ $x = -8$ or $x = -2$

Lesson Notes

■ Since the formula for a circle contains five variables and constants, students may need several examples before its meaning is clear.

Vocabulary

circle formula
completing the square
imaginary circle
point circle

Lesson Notes, continued

■ In example **3**, the algebraic steps of completing the square (actually, two squares) can also look confusing to the students, who may need additional explanation of the procedure.

Communicating Mathematics

Ask students to derive the equation of a circle from the distance formula. Refer to the diagram.

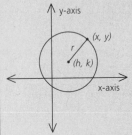

The distance r from the center (h, k) to (x, y), a point on the circle, is
$r = \sqrt{(x - h)^2 + (y - k)^2}$.
Squaring both sides yields
$r^2 = (x - h)^2 + (y - k)^2$,
the equation of a circle whose center is (h, k) and whose radius is r.

Example 3   Is $x^2 - 8x + y^2 - 10y = 8$ an equation of a circle?

We use the process of **completing the square**, which you have probably studied in algebra class, to rewrite the equation in the form of the circle equation.

$$x^2 - 8x + y^2 - 10y = 8$$
$$x^2 - 8x + \mathbf{16} + y^2 - 10y + \mathbf{25} = 8 + \mathbf{16} + \mathbf{25}$$

Key Number	Key Number

$$(x^2 - 8x + 16) + (y^2 - 10y + 25) = 49$$
$$(x - 4)^2 + (y - 5)^2 = 49$$

Yes, the solution set is a circle. The center is (4, 5) and the radius is 7. The two key numbers, 16 and 25, were introduced to complete the squares—to make the terms on the left-hand side of the equation form two perfect-square trinomials. Notice that 16 is the square of half of -8 and that 25 is the square of half of -10.

Part Two: Sample Problems

Problem 1   Find the equation of the circle with center $(0, -2)$ and a radius of 3.

Solution   Use the circle formula.
$$(x - h)^2 + (y - k)^2 = r^2$$
$$(x - 0)^2 + [y - (-2)]^2 = 3^2$$
$$x^2 + (y + 2)^2 = 9$$

Problem 2   Find, to the nearest tenth, the circumference of the circle represented by $3x^2 + 3y^2 + 6x - 18y = 15$.

Solution
$$3x^2 + 3y^2 + 6x - 18y = 15$$
$$x^2 + y^2 + 2x - 6y = 5 \qquad \text{Divide both sides by 3.}$$
$$x^2 + 2x + y^2 - 6y = 5 \qquad \text{Rearrange the terms.}$$
$$x^2 + 2x + 1 + y^2 - 6y + 9 = 5 + 1 + 9 \quad \text{Complete the squares.}$$
$$(x + 1)^2 + (y - 3)^2 = 15$$

The radius is $\sqrt{15}$ and the circumference $= 2\pi r = 2\pi\sqrt{15} \approx 24.3$.

Problem 3   **a** Describe the graph of $(x - 2)^2 + (y + 5)^2 = 0$.
   b Describe the graph of $x^2 + (y - 4)^2 = -25$.

Solution   **a** The form of this equation indicates a circle with its center at $(2, -5)$ and a radius of 0. This is sometimes called a **point circle**, a circle that has shrunk to a single point—in this case, the point $(2, -5)$.

   b The form of this equation indicates a circle with its center at $(0, 4)$ and a radius of $\sqrt{-25}$. However, $\sqrt{-25}$ is not a real number, so such a circle cannot be drawn on the coordinate plane. The equation is said to represent an **imaginary circle**.

Part Three: Problem Sets

Problem Set A

1 Write an equation of each circle.
a Center (0, 0); radius 4 $x^2 + y^2 = 16$
b Center (−2, 1); radius 5
   $(x + 2)^2 + (y − 1)^2 = 25$

   $x^2 + (y + 2)^2 = 12$
c Center (0, −2); radius $2\sqrt{3}$
d Center (−6, 0); radius $\frac{1}{2}$
   $(x + 6)^2 + y^2 = \frac{1}{4}$

2 Graph each equation.
a $x^2 + y^2 = 9$

b $(x − 1)^2 + (y + 2)^2 = 16$

3 Find the center, the radius, the diameter, the circumference, and the area of the circle represented by each equation.
a $x^2 + y^2 = 36$ (0, 0); 6; 12; 12π; 36π
b $(x + 5)^2 + y^2 = \frac{9}{4}$ (−5, 0); $\frac{3}{2}$; 3; 3π; $\frac{9}{4}\pi$

c $(x − 3)^2 + (y + 6)^2 = 100$ (3, −6); 10; 20; 20π; 100π
d $\frac{(x + 5)^2}{3} + \frac{(y − 2)^2}{3} = 27$
   (−5, 2); 9; 18; 18π; 81π

4 Write an equation of each circle. (Hint: Find the value of r and use the circle formula.)

a
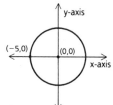
$x^2 + y^2 = 25$

b
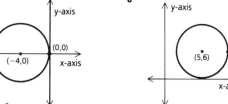
$(x + 4)^2 + y^2 = 16$

c
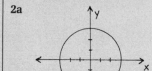
$(x − 5)^2 + (y − 6)^2 = 36$

5 Consider the equation $(x − 3)^2 + (y + 2)^2 = 17$.
a Is (4, 2) on the graph of the equation?   Yes
b Is (3, −2) on the graph of the equation?   No

6 a What type of "circle" is represented by $(x − 3)^2 + (y + 1)^2 = 0$?   Point circle
b What type of "circle" is represented by $(x + 5)^2 + y^2 = −100$?   Imaginary circle

7 The radius of circle P is 7. $\overline{AB}$ is the horizontal diameter and $\overline{CD}$ is the vertical diameter. Find the coordinates of A, B, C, and D.   (−6, 3); (8, 3); (1, 10); (1, −4)

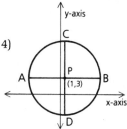

Assignment Guide

Basic

Day 1   1a,c, 2, 3a,c, 4a,c, 5, 8a,c
Day 2   3b,d, 4b, 7, 8b,d, 9a,c, 14

Average

Day 1   1a,c, 2, 3a,c, 4a,c, 5, 8a,c, 9a,c, 11a
Day 2   3b,d, 4b, 7, 8b,d, 11c, 14, 15

Advanced

Day 1   3a,b, 4–7, 8a,c, 9a,c, 11a,c, 13
Day 2   8b,d, 14–18

Problem-Set Notes and Additional Answers

2a

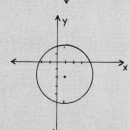

b

10

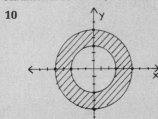

15 Center: (2, −3)

Slope of radius: $\frac{5}{-6}$

Slope of tangent: $\frac{6}{5}$

∴ Equation of tangent:

$y + 8 = \frac{6}{5}(x - 8)$

16 $x^2 + 4x + 4 + y^2 - \frac{5}{3}y + \frac{25}{36}$

$= \frac{2}{3} + 4 + \frac{25}{36}$

$(x + 2)^2 + \left(y - \frac{5}{6}\right)^2 = \frac{193}{36}$

17 The center of the circle is equidistant from (0, 0) and (8, 0). Thus, it lies on the ⊥ bis., x = 4. Since the center (4, y) is also equidistant from (0, 0) and (5, 3),

$\sqrt{(4 - 5)^2 + (y - 3)^2} = \sqrt{4^2 + y^2}$.

So y = −1 and radius = $\sqrt{17}$.

∴ A = 17π

Problem Set A, *continued*

8 Determine the equation of each circle.

a The center is the origin, and the circle passes through (0, −5). $x^2 + y^2 = 25$

b The endpoints of a diameter are (−2, 1) and (8, 25). $(x - 3)^2 + (y - 13)^2 = 169$

c The center is (−1, 7), and the circle passes through the origin. $(x + 1)^2 + (y - 7)^2 = 50$

d The center is (2, −3), and the circle passes through (3, 0). $(x - 2)^2 + (y + 3)^2 = 10$

9 For each given point, indicate whether the point is on, outside, or inside the circle with the given equation.

a (2, 5); $x^2 + y^2 = 29$ On

b (3, 0); $x^2 + y^2 = 100$ Inside

c Origin; $(x - 2)^2 + (y + 5)^2 = 16$ Outside

d (−2, 1); $x^2 + (y + 6)^2 = 23$ Outside

10 Graph the solution of the system.

$$\begin{cases} x^2 + y^2 \geq 9 \\ x^2 + y^2 \leq 25 \end{cases}$$

Problem Set B

11 Find the center and the radius of the circle represented by each equation.

a $x^2 + y^2 - 8y = 9$ (0, 4); 5

b $(x + 7)^2 + y^2 + 6y = 27$ (−7, −3); 6

c $x^2 + 10x + y^2 - 12y = -10$ (−5, 6); $\sqrt{51}$

d $x^2 + y^2 = 8x - 14y + 35$ (4, −7); 10

12 Find the solution set of each system.

a $\begin{cases} x^2 + y^2 = 25 \\ x = 3 \end{cases}$ {(3, 4), (3, −4)}

b $\begin{cases} x^2 + y^2 = 25 \\ x^2 - y^2 = 7 \end{cases}$

{(4, 3), (4, −3), (−4, 3), (−4, −3)}

c $\begin{cases} x^2 + y^2 = 34 \\ x + y = 8 \end{cases}$ {(5, 3), (3, 5)}

d $\begin{cases} |y| = 6 \\ x^2 + y^2 = 100 \end{cases}$

{(8, 6), (8, −6), (−8, 6), (−8, −6)}

13 Find the distance between the points of intersection of the graph of $x^2 + y^2 = 17$ and the graph of x + y = 3. $5\sqrt{2}$

14 Use the diagram of circle Q as marked to find

a An equation of the tangent to the circle at (6, 8) $y = -\frac{3}{4}x + \frac{25}{2}$

b The circumference of the circle 10π

c The distance from A to Q $\sqrt{53}$

d The distance from A to the circle (to the nearest tenth) ≈2.3

e The area of the shaded sector (to the nearest tenth) ≈13.1

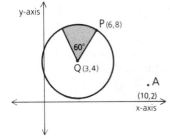

Chapter 13 Coordinate Geometry Extended

15 Consider the circle represented by $(x - 2)^2 + (y + 3)^2 = 61$. Write, in point-slope form, the equation of the tangent to the circle at point $(8, -8)$. $y + 8 = \frac{6}{5}(x - 8)$

Problem Set C

16 Find the center and the radius of the graph of $3x^2 + 12x + 3y^2 - 5y = 2$. $\left(-2, \frac{5}{6}\right); \frac{\sqrt{193}}{6}$

17 Find the area of the circle shown. 17π

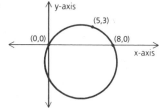

18 Find the equation of the path of a point that moves so that its distance from the point $(3, 0)$ is always twice its distance from the point $(-3, 0)$. $(x + 5)^2 + y^2 = 16$

19 A marble was placed at point $\left(2, 4\sqrt{3}\right)$ and rolled clockwise around the graph of $x^2 - 12x + y^2 = 28$ until it stopped at the intersection of the circle with the positive x-axis.

   a Find the distance the marble traveled. $\frac{16}{3}\pi$

   b Find, to the nearest hundredth, the distance that would have been saved if the marble had rolled in a straight line. ≈ 2.90

Problem Set D

20 The quadrilateral region bounded by the graphs of $y = mx + 3$, $x = 2$, $x = 5$, and $y = 1$ has an area of 2. Find the maximum value of m. $-\frac{8}{21}$

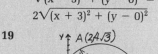
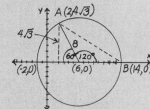
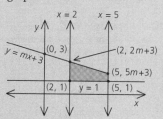

Class Planning

Time Schedule
All levels: 2 days

Resource References
Teacher's Resource Book
   Class Opener 13.7A, 13.7B

Class Opener

Find the area of the circle whose equation is
$x^2 + 8x + y^2 + 4y = 6$.
$x^2 + 8x + 16 + y^2 + 4y + 4 = 6 + 20$
$$r^2 = 26$$
$\therefore A = 26\pi$
Completing the square is an important mathematical process. This problem reviews completing the square.

Lesson Notes

- This section combines many of the concepts in the course. It can also give an opportunity to review earlier chapters in coordinate terms.

Assignment Guide

Basic	
Day 1	1–8
Day 2	9–12, 15–17
Average	
Day 1	1–12
Day 2	13–20, 24
Advanced	
Day 1	2–15
Day 2	16–20, 22–25, 27, 28

13.7

COORDINATE-GEOMETRY PRACTICE

Objective

After studying this section, you will be able to
- Apply the principles of coordinate geometry in a variety of situations

The problems that follow will give you a chance to use what you have learned about coordinate geometry throughout your study of this book. As you examine each problem, try to determine which of the formulas and properties you have learned provides the best means of solving it. You will find that coordinate-geometry skills will become more and more important as you continue your study of mathematics.

Problem Sets

Problem Set A

1 Write an equation of circle P. $x^2 + y^2 = 25$
  a Find the area of the circle 25π
  b Find the circle's circumference. 10π

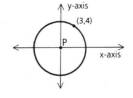

2 Find the area of the shaded sector. 3π

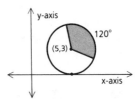

3 Find, to the nearest tenth, the area of the shaded region in each diagram.
  a ABCD is a square. ≈ 13.7

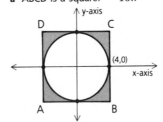

  b ≈ 30.5

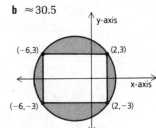

4 Find the area of the square with vertices at (1, 2), (6, 2), (6, 7), and (1, 7). **25**

5 Given: Diagram as marked;
  M is the midpoint of $\overline{EF}$.

Find: **a** OM **5**
  b EM **5**
  c FM **5**

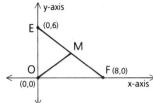

6 △ABC is equilateral. Find the coordinates of C. **(3, 3√3)**

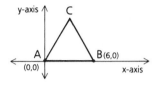

7 In rectangle ABCD, A = (2, 7) and C = (8, 15). Find BD. **10**

8 Find the area of the triangle with vertices at (0, 8) (0, 0), and (3, 0). **12**

9 Find the slope of the tangent of ⊙P at point H. **−0.5**

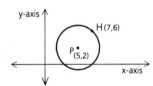

10 $\overline{JK}$ is a chord of ⊙Q, and $\overline{QM} \perp \overline{JK}$.
Find QM. **5**

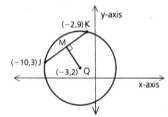

11 Given: $\overline{AC}$ and $\overline{DE}$ are chords,
  A = (8, 0), B = (18, 0),
  C = (24, 0),
  D = (22, −3)

Find: BE. **12**

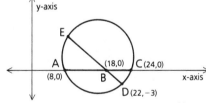

12 Given: Rectangle HJKM, with H = (4, 2),
  J = (14, 4), and M = (3, 7)

Find: **a** The coordinates of K **(13, 9)**
  b The area of the rectangle **52**

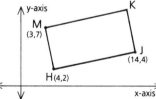

Communicating Mathematics

Have students write a paragraph that defines the term *coordinate geometry*.

Cooperative Learning

Any of the problems from Problem Set B, C, or D lend themselves to be worked cooperatively. Each of these problems requires a synthesis of the material covered in this and previous chapters.

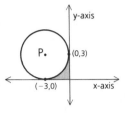

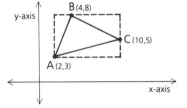

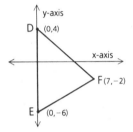

Problem-Set Notes
and Additional Answers

■ These problems contain three important terms for students who will be going on in mathematics: the encasement principle (problem **14**), lattice point (problem **27**), and the reflection principle (problem **28**).

21 Since $M = \left(\frac{x+4}{2}, \frac{y}{2}\right)$ lies on the circle, $\left(\frac{x+4}{2} - 0\right)^2 + \left(\frac{y}{2} - 0\right)^2 = 9$.
∴ $(x + 4)^2 + y^2 = 36$

22a Equation of $\overline{CD}$:
$y = \frac{1}{7}x + \frac{52}{7}$
Equation of $\overline{AB}$:
$y = -\frac{3}{5}x + \frac{78}{5}$
∴ $D = (11, 9)$

b $\frac{A_{\triangle ACD}}{A_{\triangle AOB}} = \left(\frac{AC}{AO}\right)^2 = \left(\frac{\sqrt{20}}{\sqrt{180}}\right)^2 = \frac{1}{9}$

23 Take the point $(0, 1)$ on $y = 2x - 1$. The equation of the ⊥ to $y = 2x + 7$ through $(0, -1)$ is
$y = -\frac{1}{2}x - 1$.
$\left.\begin{array}{l} y = -\frac{1}{2}x - 1 \\ y = 2x + 7 \end{array}\right\}$ int. at $\left(\frac{-16}{5}, \frac{3}{5}\right)$
$d = \sqrt{\left(-\frac{16}{5} - 0\right)^2 + \left(\frac{3}{5} + 1\right)^2}$
$= \frac{8\sqrt{5}}{5}$

24a $V = \frac{1}{3}\mathcal{B}h = \frac{1}{3} \cdot 8 \cdot 8\pi \cdot 6 = 128\pi$

b $V = \frac{1}{3}\mathcal{B}h = \frac{1}{3} \cdot 6^2 \cdot \pi \cdot 8 = 96\pi$

c The figure consists of two cones. Draw the altitude to the hypotenuse.
$V = \frac{1}{3}\mathcal{B}_1h_1 + \frac{1}{3}\mathcal{B}_2h_2$
$= \frac{1}{3} \cdot \frac{24}{5} \cdot \frac{24}{5} \cdot \frac{32}{5}\pi$
$+ \frac{1}{3} \cdot \frac{24}{5} \cdot \frac{24}{5} \cdot \frac{18}{5}\pi$
$= \frac{384\pi}{5}$
T.A. $= \pi r\ell_1 + \pi r\ell_2$
$= \pi \cdot \frac{24}{5} \cdot 8 + \pi \cdot \frac{24}{5} \cdot 6$
$= \frac{336\pi}{5}$

Problem Set B

13 ⊙P is tangent to the x-axis and the y-axis at the points shown.

a Find an equation of the circle.

b Find, to the nearest tenth, the area of the shaded region bounded by the circle and the axes.
a $(x + 3)^2 + (y - 3)^2 = 9$; **b** ≈ 1.9

14 In the figure as marked, what is the area of $\triangle ABC$? (Your solution should suggest a concept known as the *encasement principle*.) 18

15 In the figure as marked, what is the area of $\triangle DEF$? 35

16 The point $(13, 9)$ is on a circle centered at $(7, 1)$.

a Write an equation of the circle. $(x - 7)^2 + (y - 1)^2 = 100$

b What is the circle's area? 100π

c What is the circle's circumference? 20π

d Find the coordinates of the point on the circle directly opposite $(13, 9)$. $(1, -7)$

e Write, in point-slope form, an equation of the line tangent to the circle at $(13, 9)$. $y - 9 = -\frac{3}{4}(x - 13)$

f Find the distance between $(19, 6)$ and the center of the circle. 13

g Find the distance between $(19, 6)$ and the circle. 3

17 Find the area of the isosceles trapezoid with vertices at $(4, 8)$, $(2, 3)$, $(14, 3)$, and $(12, 8)$. 50

18 $\triangle ABC$ is an isosceles right triangle with base $\overline{AB}$. If $A = (-3, -2)$ and $B = (-3, 4)$, what are the two possibilities for the coordinates of C? $(-6, 1)$ and $(0, 1)$

19 In $\triangle DEF$, $D = (1, 2)$, $E = (7, 2)$, and $F = (1, 10)$. Find the length of the altitude from D to $\overline{EF}$. (Hint: First find the area of $\triangle DEF$.) $\frac{24}{5}$

20 Consider the circle represented by $(x - 4)^2 + (y + 2)^2 = 50$. Let P be the center of the circle and T be a point on chord $\overline{AB}$ such that $\overline{PT}$ is perpendicular to $\overline{AB}$. If A = (11, −1) and B = (5, −9), what is

a PT? 5
b m∠TPA? 45

Problem Set C

21 $\overline{OA}$ is a fixed line segment. M can lie anywhere on a circle with a radius of 3 and its center at O. B moves so that M is always the midpoint of $\overline{AB}$. Find an equation of the circle on which B lies.
$(x + 4)^2 + y^2 = 36$

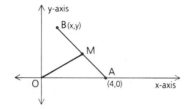

22 △AOB is placed on a coordinate system so that A = (6, 12), B = (21, 3), and O = (0, 0). A segment, $\overline{CD}$, is drawn parallel to $\overline{OB}$ so that C lies on $\overline{AO}$, D lies on $\overline{AB}$, and C = (4, 8).

a Find the coordinates of D. (11, 9)
b Find the ratio of the area of △ACD to the area of △AOB. 1:9

23 Find the distance between the lines represented by $y = 2x - 1$ and $y = 2x + 7$. $\frac{8\sqrt{5}}{5}$

24 In △AOB, A = (6, 0), B = (0, 8), and O = (0, 0).

a Find, to the nearest tenth, the volume of the solid formed by rotating the triangle about $\overline{OA}$. ≈402.1

b Find, to the nearest tenth, the volume of the solid formed by rotating the triangle about $\overline{OB}$. ≈301.6

c Find, to the nearest tenth, the volume and the total surface area of the solid formed by rotating the triangle about $\overline{AB}$. ≈241.3; ≈211.1

25 Given the circles represented by $(x + 9)^2 + (y - 4)^2 = 52$ and $(x - 12)^2 + (y - 3)^2 = 13$, find the length of a

a Common internal tangent $5\sqrt{13}$
b Common external tangent $\sqrt{429}$

26 Find the area of the quadrilateral with vertices at $(-3, 2)$, $(15, 6)$, $(7, 12)$, and $(-7, 8)$. 120

Problem Set D

27 A *lattice point* is a point whose coordinates are integers. How many lattice points are on the boundary and in the interior of the region bounded by the positive x-axis, the positive y-axis, the graph of $x^2 + y^2 = 25$, and the line passing through $(-3, 0)$ and $(0, 2)$? 20

28 A green billiard ball is located at (3, 1), and a gray billiard ball at (8, 9). Fats Tablechalk wants to strike the green ball so that it bounces off the y-axis and hits the gray ball. At what point on the y-axis should he aim? (Hint: Use the reflection principle.) $\left(0, \frac{35}{11}\right)$

Problem-Set Notes and Additional Answers, continued

25a

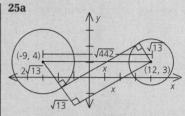

Use the common-tangent procedure. The distance from $(-9, 4)$ to $(12, 3)$ = $\sqrt{442}$. By the Pythagorean Theorem,
$$(3\sqrt{13})^2 + x^2 = (\sqrt{442})^2$$
$$x = 5\sqrt{13}$$

b

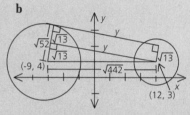

By the Pythagorean Theorem,
$$(\sqrt{13})^2 + y^2 = (\sqrt{442})^2$$
$$y = \sqrt{429}$$

26

Area of rectangle:
$22(10) = 220$
Area of △I: 12
Area of △II: 36
Area of △III: 24
Area of △IV: 28
Area of quadrilateral:
$220 - (12 + 36 + 24 + 28)$
$= 120$

27

28 Reflect point (8, 9) to $(-8, 9)$. The y-intercept, $\frac{35}{11}$, of the line joining $(-8, 9)$ to (3, 1) yields the answer $\left(0, \frac{35}{11}\right)$.

Problem Set D, *continued*

29 The points of $\overline{A_1B_1}$ are "mapped" onto a new coordinate system (with shorter units) in such a way that A_1B_1 is turned around, with A_1 becoming A_2 and B_1 becoming B_2.

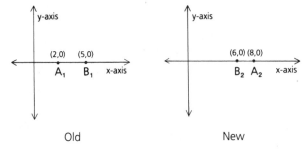

Old   New

a Find the coordinates of C_2, a point on the new coordinate system, if $C_1 = \left(3\frac{1}{2}, 0\right)$. **(7, 0)**

b Find the coordinates of D_2 if $D_1 = (4, 0)$. $\left(\frac{20}{3}, 0\right)$

c Find, in terms of x_1, the coordinates of E_2 if $E_1 = (x_1, 0)$. $\left(\frac{28 - 2x_1}{3}, 0\right)$

HISTORICAL SNAPSHOT

THE SERPENT AND THE PEACOCK
A problem from medieval India

From ancient to medieval times, the scholars of Egypt, Mesopotamia, India, and China developed a variety of useful algebraic techniques. Unlike the Greeks, they showed little interest in the more abstract aspects of geometry.

Nevertheless, these scholars *were* aware of the Pythagorean Theorem and devised clever problems involving its application. The following is found in the treatise *Lilavati* of the Indian mathematician Bhaskara (A.D. 1114–c. 1185).

A peacock perched atop a pillar sees a snake slithering toward its den, which is at the base of the pillar. The snake is three times as far from its den as the pillar is high. If the peacock swoops down on the snake in a straight line and if the peacock and the snake travel equal distances before they meet, how far is the snake from its den when the peacock pounces on it?

Can you find a linear relationship between the height of the pillar and the distance y by which the snake fails to reach its den, for any height *x*? $y = \frac{4}{3}x$

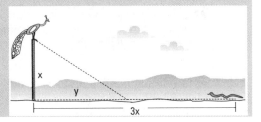

642 | Chapter 13   Coordinate Geometry Extended

13 CHAPTER SUMMARY

CONCEPTS AND PROCEDURES

After studying this chapter, you should be able to
- Draw lines and circles that represent the solutions of equations (13.1)
- Write equations that correspond to nonvertical lines (13.2)
- Write equations that correspond to horizontal lines (13.2)
- Write equations that correspond to vertical lines (13.2)
- Identify various forms of linear equations (13.2)
- Use two methods to solve systems of equations (13.3)
- Graph inequalities (13.4)
- Graph in three dimensions (13.5)
- Apply the properties of reflections (13.5)
- Write equations that correspond to circles (13.6)
- Apply the principles of coordinate geometry in a variety of situations (13.7)

VOCABULARY

boundary line (13.4)
circle formula (13.6)
completing the square (13.6)
general linear form (13.2)
imaginary circle (13.6)
intercept (13.1)
intercept form (13.2)
point circle (13.6)

point-slope form (13.2)
slope-intercept form (13.2)
system of equations (13.3)
table of values (13.1)
3-D distance formula (13.5)
two-point form (13.2)
y-form (13.2)

Chapter 13 Review
Class Planning

Time Schedule
All levels: 2 days

Resource References
Evaluation
  Tests and Quizzes
    Test 13 Series 1, 2, 3

Assignment Guide

Basic

Day 1	1–8, 9a,b,c,d
Day 2	9e,f,g, 10–12, 16–18

Average

Day 1	1–9
Day 2	10, 13, 14, 17, 18, 20, 23, 24

Advanced

Day 1	13, 14, 17, 18, 20, 23–26
Day 2	27–37

Lesson Notes

- Additional review problems can be found in the Cumulative Review for Chapters **1–15** on pages 706–711.

---

13 REVIEW PROBLEMS

Problem Set A

1 Is the point (7, 5) on the graph of $2x + 3y = 62$? No

2 In which quadrant are both coordinates negative? III

3 If H is reflected over the barrier at $y = 10$ to H′, find the slope of $\overleftrightarrow{BH'}$. -4

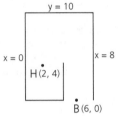

4 Find the coordinates of point D in each figure.

a ABCD is a rectangle. (10, 0)

b ABCD is an isosceles trapezoid. (2, 6)

c ABCD is a parallelogram. (−4, 4)

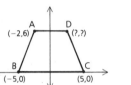 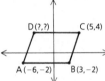

5 If P = (4, −2) and Q = (10, 6), what is

a PQ? 10   **b** The midpoint of $\overline{PQ}$? (7, 2)   **c** The slope of $\overleftrightarrow{PQ}$? $\frac{4}{3}$

6 Use the diagram of △ABC to find

a The slope of $\overline{AC}$ $\frac{4}{3}$

b The midpoint of $\overline{AC}$ (1, 6)

c The slope of the median from B $-\frac{3}{5}$

d The length of the median from B $\sqrt{34}$

e The slope of the altitude from B $-\frac{3}{4}$

f The slope of a line through A and parallel to $\overleftrightarrow{BC}$ $\frac{1}{8}$

g The slope of the perpendicular bisector of $\overline{AC}$ $-\frac{3}{4}$

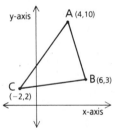

7 Find the area of each shaded region.

a 50

(4,5) (14,5)

(0,0) (10,0)

b 27

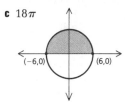

(2,9)

(−4,3) (5,3)

c 18π

(−6,0) (6,0)

8 Write an equation of each circle.

a Center at (2, −3); radius of 4 $(x - 2)^2 + (y + 3)^2 = 16$

b Center at origin; passes through (6, 8) $x^2 + y^2 = 100$

c Endpoints of a diameter are (0, 0) and (10, 0).
$(x - 5)^2 + y^2 = 25$

9 Write an equation of each line.

a Slope of 2; y-intercept of 1 $y = 2x + 1$

b Contains the points (2, 3) and (2, 7) $x = 2$

c Parallel to, and 5 units to the left of, the y-axis $x = -5$

d Contains the points (2, 4) and (6, 16) $y = 3x - 2$

e Slope of $\frac{1}{2}$; x-intercept of 4 $y = \frac{1}{2}x - 2$

f Parallel to the graph of $y = 3x + 1$, with the same y-intercept
as the graph of $y = 2x - 7$ $y = 3x - 7$

g x-intercept of 6; y-intercept of −3 $y = \frac{1}{2}x - 3$

10 Find the slope of the graph of each equation. Are the lines
perpendicular, parallel, or neither? The lines are ⊥.

a $x + 2y = 10$ $-\frac{1}{2}$

b $y = 2x + 3$ 2

11 Are the points (2, 4), (5, 13), and (26, 76) collinear? Yes

12 Find the slope of $\overleftrightarrow{AB}$. $-\frac{2}{3}$

Problem Set B

13 Find the x-intercept of the line joining (−2, 3) and (5, 7). $-\frac{29}{4}$

14 The points (2, 1), (4, 0), and $(-4, k^2)$ are collinear. What is the
value of k? ±2

15 $A = (-6, 1)$ and $B = (2, 3)$. If B is the midpoint of $\overline{AC}$, find C. (10, 5)

16 Find the coordinates of the point one-fourth of the way from
(−5, 0) to (7, 8). (−2, 2)

Review Problems | **645**

*Problem-Set Notes
and Additional Answers*

19a

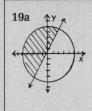

b

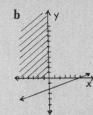

27 Any point equidistant from
the axes lies on the line
$y = x$ or the line $y = -x$.

$\left.\begin{array}{r} 2x - 5y = 10 \\ y = x \end{array}\right\}$ $\left(-\frac{10}{3}, \frac{-10}{3}\right)$ in III

$\left.\begin{array}{r} 2x - 5y = 10 \\ y = x \end{array}\right\}$ $\left(\frac{10}{7}, \frac{-10}{7}\right)$ in IV

T 645

Problem-Set Notes and
Additional Answers, continued

28

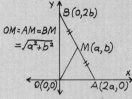

$OM = AM = BM$
$= \sqrt{a^2 + b^2}$

29 Equation of $\overline{AB}$:
$y = 3x - 17$
Equation of altitude:
$y = -\frac{1}{3}x + 3$
Point of intersection: (6, 1)
Distance between (6, 1)
and (12, −1): $2\sqrt{10}$

30

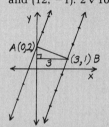

$\overline{AB}$ is $\perp$ to the $\parallel$ lines, so
$AB = \sqrt{10}$ = distance
between the lines.

31

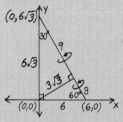

$V = \frac{1}{3}\pi(3\sqrt{3})^2 \cdot 9$
$+ \frac{1}{3}\pi(3\sqrt{3})^2 \cdot 3$
$= 108\pi$

32

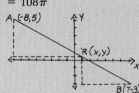

Students may need a hint
to work with the
coordinate separately.
$\frac{AR}{RB} = \frac{3}{2} = \frac{x + 8}{7 - x} = \frac{y - 5}{-3 - y}$
$\therefore x = 1, y = \frac{1}{5}$

Review Problem Set B, *continued*

17 In $\triangle ABC$, A = (2, 3), B = (12, 5), and C = (9, 8).
 a Find the length of the median from C to $\overline{AB}$. $2\sqrt{5}$
 b Write an equation of the median to $\overline{AB}$. $y = 2x - 10$
 c Write, in point-slope form, an equation of the perpendicular
 bisector of $\overline{AB}$. $y - 4 = -5(x - 7)$
 d Write, in point-slope form, an equation of the altitude from C
 to $\overline{AB}$. $y - 8 = -5(x - 9)$
 e Write, in point-slope form, an equation of the line containing
 C and parallel to $\overline{AB}$. $y - 8 = \frac{1}{5}(x - 9)$

18 a Write an equation of circle P.
 b What is the area of circle P? 40π
 c What are the coordinates of V? $(-3, 0)$
 d Write, in point-slope form, an equation
 of the tangent $\overleftrightarrow{RT}$. $y - 4 = -3(x - 9)$
 e Find PT. 20
 f Find, to the nearest tenth, the distance
 from T to the circle. ≈ 13.7
 g What is the area of $\triangle PRT$? 60

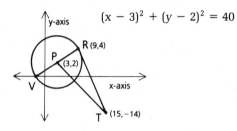

$(x - 3)^2 + (y - 2)^2 = 40$

19 Graph the solution of each of the following systems of inequalities.
 a $\begin{bmatrix} y \geq 2x + 1 \\ x^2 + y^2 \leq 25 \end{bmatrix}$
 b $\begin{bmatrix} y \geq 0 \\ x \leq 0 \\ x - 3y \leq 6 \end{bmatrix}$

20 If two of the five points (2, 1), (6, 4), (5, 17), (−2, −2), and (2, 10)
 are selected at random, what is the probability that
 a Both points lie in Quadrant I? $\frac{3}{5}$
 b The two points will be collinear with one of the other points? $\frac{3}{10}$

21 Find the intersection of the graphs of the equations in each system.
 a $\begin{bmatrix} y = 4x - 1 \\ y = 2x + 3 \end{bmatrix}$ **b** $\begin{bmatrix} x - 3y = 10 \\ 2x + y = 13 \end{bmatrix}$ **c** $\begin{bmatrix} y = 4 \\ (x - 1)^2 + (y - 5)^2 = 17 \end{bmatrix}$
 $\{(2, 7)\}$ $\{(7, -1)\}$ $\{(-3, 4), (5, 4)\}$

22 Find the center and the radius of the graph of
 $x^2 + 6x + y^2 - 4y = 12$. $(-3, 2); 5$

23 Find, to the nearest tenth, the distance between the centers of
 the circles represented by $(x - 5)^2 + (y - 2)^2 = 29$ and
 $x^2 + 8x + y^2 = 31$. ≈ 9.2

24 The bases of an isosceles trapezoid are parallel to the y-axis. If
 three vertices are (5, 2), (5, 12), and (−1, 10), find the trapezoid's
 area. 48

25 Describe the graph of $x^2 + 2x + y^2 - 6y = -10$. Point circle

26 Quadrilateral PQRS has vertices (3, 1), (15, −3), (9, 7), and (5, 7).

    a Find the quadrilateral's area by using the encasement principle. (See Section 13.7, problem 14.) 60

    b What is true about the diagonals? They are ⊥.

Problem Set C

27 Consider the graph of $2x - 5y = 10$. In which quadrant(s) is there a point that is on this line and equidistant from the x-axis and the y-axis? Find the point(s). III and IV; $\left(-\frac{10}{3}, -\frac{10}{3}\right)$ and $\left(\frac{10}{7}, -\frac{10}{7}\right)$

28 Use a triangle with vertices at (0, 0), (2a, 0), and (0, 2b) to show that the midpoint of the hypotenuse of any right triangle is equidistant from the vertices of the triangle.

29 If A = (0, −17), B = (4, −5), and C = (12, −1), what is the length of the altitude from C to $\overline{AB}$? $2\sqrt{10}$

30 Find the distance between the graphs of $y = 3x - 8$ and $y = 3x + 2$. $\sqrt{10}$

31 A triangle with vertices at (0, 0), (6, 0), and $\left(0, 6\sqrt{3}\right)$ is rotated around its longest side. Find, to the nearest tenth, the volume of the solid formed. ≈339.3

32 If A = (−8, 5) and B = (7, −3), where is the point R that divides $\overline{AB}$ so that AR:RB = 3:2? $\left(1, \frac{1}{5}\right)$

33 In ▱PQRS, M, N, and X are midpoints of $\overline{PQ}$, $\overline{PS}$, and $\overline{QR}$ respectively. Find the intersection of $\overline{MN}$ and $\overline{PX}$ if P = (−8, 1), Q = (0, 5), and S = (4, 1). $\left(-\frac{10}{3}, \frac{7}{3}\right)$

34 How many lattice points are in the intersection of this system?
$$\begin{cases} x > 0 \\ y > 0 \\ y < -|x - 4| + 10 \end{cases}$$ 66

35 A = (2, 10) and C = (8, 4). Find point B if it lies on the x-axis and AB + BC is a minimum. $\left(\frac{44}{7}, 0\right)$

36 Find the image of point (−5, 10) when it is reflected over

    a The x-axis   **b** The point (−3, 1)   **c** The graph of $y = 2x$

    (−5, −10)   (−1, −8)   (11, 2)

37 Find the intersection.
$$\begin{cases} x + y = 16 \\ y = |2x + 10| \end{cases}$$ {(2, 14), (−26, 42)}

Problem-Set Notes and Additional Answers, continued

33 R = (12, 5), M = (−4, 3)
N = (−2, 1), X = (6, 5)
Equation of $\overline{MN}$:
$y = -x - 1$
Equation of $\overline{PX}$:
$y = \frac{2}{7}x + \frac{23}{7}$
Point of intersection:
$\left(\frac{-10}{3}, \frac{7}{3}\right)$

34

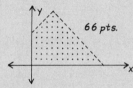

66 pts.

Lattice point is defined on page 641, problem **27.**

35

A(2,10)
C(8,4)
B(x,0)
C′

Reflect C to C′(8, −4).
Since AC′ = AB + BC, the equation of $\overline{AC'}$ is
$y = -\frac{7}{3}x + \frac{44}{3}$.
x-int. = $\frac{44}{7}$, so B = $\left(\frac{44}{7}, 0\right)$.

36b (−3, 1) is the midpoint of (−5, 10) and the desired point.

  c The slope of the ⊥ is $-\frac{1}{2}$, so it crosses $y = 2x$ at the midpoint (3, 6). Therefore, the other endpoint is (11, 2).

37 $y = |2x + 10|$ is equivalent to $y = 2x + 10$ if $x \geq -5$ or to $y = -2x - 10$ if $x < -5$.
Solving simultaneously,
$\left.\begin{array}{l} x + y = 16 \\ y = 2x + 10 \end{array}\right\}$ (2, 14)
$\left.\begin{array}{l} x + y = 16 \\ y = -2x - 10 \end{array}\right\}$ (−26, 42)

Chapter 14 Schedule

Basic (optional)

Problem Sets:	8 days
Review:	2 days
Test:	1 day

Average

Problem Sets:	12 days
Review:	2 days
Test:	1 day

Advanced

Problem Sets:	11 days
Review:	2 days
Test:	1 day

*M*OST BASIC COURSES will finish the term with Chapter **13**; Chapters **14–16** are considered optional. However, section schedules and assignment guides have been included for basic courses.

It is also anticipated that most average courses will finish the term at the end of Chapter **13** or Chapter **14**.

The six basic constructions (Section **14.4**) can be presented at any time during the course. They can all be justified after Chapter **4**.

Every home in each locus of summer homes in the small community of Brøndbyvester, Denmark, is equidistant from a parking lot.

14.1 LOCUS

Class Planning

Time Schedule
Basic: 2 days (optional)
Average: 2 days
Advanced: 2 days

Resource References
Teacher's Resource Book
   Class Opener 14.1A, 14.1B
   Using Manipulatives 21
Evaluation
   Tests and Quizzes
      Quiz 1, Series 1

Objective

After studying this section, you will be able to
- Use the four-step locus procedure to solve locus problems

Part One: Introduction

Mathematicians sometimes find it convenient to describe a figure as a ***locus***. (*Locus* is a Latin word meaning "place" or "position." Its plural form is *loci*.)

Definition	A ***locus*** is a set consisting of all the points, and only the points, that satisfy specific conditions.

In this book, all loci are to be considered sets of *coplanar* points unless specified otherwise.

Example 1 Find the locus of points that are 1 in. from a given point O.

Find one point, P_1, that is 1 in. from O. Then find a second such point, P_2. Continue finding such points until a pattern is formed—in this case, a circle.

Draw the circle and finish with a written description: "The locus of points 1 in. from a given point O is a circle having O as its center and a radius of 1 in."

Vocabulary
locus

Class Opener

Describe all points equidistant from the two given points, A and B.
From Chapter **4**, the two-dimensional solution is the perpendicular bisector of $\overline{AB}$.

Lesson Notes

- The term *locus* will be new to most students, but they can overcome their initial confusion by following the four-step locus procedure (p. 650). For many students, solving locus problems is fun.

Four-Step Procedure for Locus Problems

1 Find a single point that satisfies the given condition(s).
2 Find a second such point, and a third, and so on, until you can identify a pattern.
3 Look outside the pattern for points you may have overlooked. Look within the pattern to exclude points that do not meet the conditions.
4 Present the answer by drawing a diagram and writing a description of the locus.

Lesson Notes, continued

■ If points outside the angle are allowed, example **2** has another interpretation:

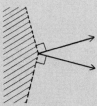

The distance between any point in the shaded region (including its boundary) and either side of the angle is the distance from that point to the vertex. Therefore, every point in the shaded region is equidistant from the sides of the angle.

Example 2   *What is the locus of all points equidistant from the sides of an angle?*

Step 1: Locate a point P_1 that is equidistant from the sides of the angle.

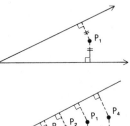

Step 2: Similarly, locate points P_2, P_3, P_4, ... The pattern appears to be the ray that bisects the angle.

Step 3: By sketching points outside the pattern, we can determine that the only points in the locus are those on the angle bisector.

Step 4: The locus of all points equidistant from the sides of an angle is the bisector of the angle.

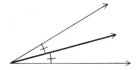

The following example draws special attention to the importance of step 3 of the four-step procedure.

Example 3   *What is the locus of points 3 cm from a given line?*

Step 1: Find a point P_1 that is 3 cm from a line ℓ.

Step 2: Find a second point, a third, and so on, until a recognizable pattern appears.

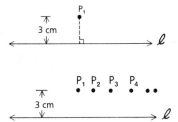

---

650   Chapter 14   Locus and Constructions

Step 3: The pattern appears to be a line parallel to ℓ and 3 cm above it. But by checking other points, we can see that points 3 cm below ℓ are also in the locus.

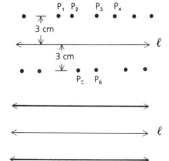

Step 4: The locus of points 3 cm from a given line is two lines parallel to the given line, 3 cm on either side of it.

Part Two: Sample Problems

Remember that loci are plane figures unless specified otherwise.

Problem 1   *What is the locus of points 2 in. from a given circle whose radius is 5 in.?*

Answer   The locus of points 2 in. from the given circle is two circles that are concentric with that circle and have radii of 3 in. and 7 in.

Problem 2   *Find the locus of points less than 3 cm from a given point A.*

Answer   The locus of points less than 3 cm from a given point A is the *interior* of a circle with its center at A and a radius of 3 cm. (The circle itself is not part of the locus. Do you see why?)

Problem 3   *What is the locus of the centers of all circles that have a fixed radius r and are tangent to a given line ℓ?*

Solution   Sketch a few circles tangent to the given line ℓ. Then consider the pattern of their centers only.

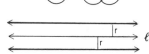

The locus of the centers of all circles that have a fixed radius r and are tangent to a given line ℓ is two parallel lines on opposite sides of ℓ, each at the distance r from ℓ.

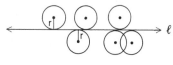

Lesson Notes, continued

■ Class time may be well used by having students consider these loci in space as well as in a plane.

Problem 4   Write the equation *of the locus of points equidistant from points* A = (1, 4) *and* B = (7, 8).

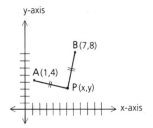

Solution   *Method One:* If P = (x, y) is any point on the locus, then PA = PB.

$$\sqrt{(x-1)^2 + (y-4)^2} = \sqrt{(x-7)^2 + (y-8)^2}$$
$$(x-1)^2 + (y-4)^2 = (x-7)^2 + (y-8)^2$$
$$x^2 - 2x + 1 + y^2 - 8y + 16 = x^2 - 14x + 49 + y^2 - 16y + 64$$
$$-2x - 8y + 17 = -14x - 16y + 113$$
$$y = -\frac{3}{2}x + 12$$

Method Two: We know that the locus of all points equidistant from A and B is the perpendicular bisector of $\overline{AB}$.

Midpoint of $\overline{AB}$ = (4, 6), and slope of $\overline{AB} = \frac{8-4}{7-1} = \frac{2}{3}$.

∴ Slope of perpendicular bisector $= -\frac{3}{2}$.

$$y = mx + b$$
$$6 = -\frac{3}{2}(4) + b$$
$$12 = b$$
$$y = -\frac{3}{2}x + 12$$

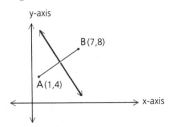

Part Three: Problem Sets

Problem Set A

Remember that loci are plane figures unless specified otherwise.

In problems 1–8, draw a sketch and write a description of each locus.

1 The locus of points that are 3 cm from a given line, $\overleftrightarrow{AB}$
The locus is two lines parallel to $\overleftrightarrow{AB}$ and 3 cm on each side of $\overleftrightarrow{AB}$.

2 The locus of the midpoints of the radii of a given circle
The locus is a circle with the same center as the given circle and with a radius half as long as the given circle's.

3 The locus of points equidistant from two given points
The locus is the perpendicular bisector of the segment joining the two given points.

652   Chapter 14   Locus and Constructions

Assignment Guide

Basic (optional)

Day 1   1–4
Day 2   5–9

Average

Day 1   1–10
Day 2   11–17, 20–23

Advanced

Day 1   11–21
Day 2   22–28, 30

4 The locus of points occupied by the center of a dime as it rolls around the edge of a quarter The locus is a circle with the same center as the quarter and with a radius equal to the sum of the radii of the two coins.

5 The locus of points that are 10 in. from a circle with a radius of 1 ft
The locus is two circles with the same center as the given circle and with radii of 2 in. and 22 in.

6 The locus of the centers of all circles tangent to both of two given parallel lines
The locus is a line parallel to the two parallel lines and halfway between them.

7 The locus of points equidistant from two given concentric circles (If the radii of the circles are 3 and 8, what is the size of the locus?) The locus is a circle concentric with the given circles and with a radius equal to the average of the radii of the given circles. If the radii of the given circles are 3 and 8, the radius of the locus is $5\frac{1}{2}$.

8 The locus of points less than or equal to 14 units from a fixed point P
The locus is the union of a circle and its interior, with the circle having point P as center and a radius of 14.

9 Write an equation for the locus of points that are 4 units from the origin.
$x^2 + y^2 = 16$

10 a Find the locus of points that are 5 units from both a and b.

b Find the locus of points that are 4 units from both a and b.
The locus is the empty set.

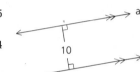

The locus is a line halfway between a and b.

Problem Set B

11 What is the locus of the midpoints of all chords that can be drawn from a given point of a given circle?
The locus does not include the given point but otherwise is a circle tangent internally to the given circle at the given point and passing through the center of the given circle.

12 What is the locus of the midpoints of all chords congruent to a given chord of a given circle?
The locus is a circle with the same center as the given circle and with a radius equal to the distance from the center to one of the chords.

13 Determine the locus of centers of all circles passing through two given points. Give an accurate, simple description of the locus.
The locus is the perpendicular bisector of the segment joining the two points.

14 Write an equation for the locus of points 6 units from $(-1, 3)$.
$(x + 1)^2 + (y - 3)^2 = 36$

Communicating Mathematics

Refer to problem **9**. Have students explain how to write an equation for the locus of points that are 4 units from $(-2, 6)$.
$(x + 2)^2 + (y - 6)^2 = 16$

Problem-Set Notes and Additional Answers

20a

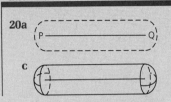

Cooperative Learning

Have students work in small groups to discuss what the loci of problems **1, 3, 8, 9,** and **18** would be if the loci need not be coplanar.

1 A cylinder of radius 3 with $\overleftrightarrow{AB}$ as the axis

3 A plane that is the perpendicular bisector of the segment

8 A solid sphere with radius 14 and center at P

9 $x^2 + y^2 + z^2 = 16$

18 Two intersecting circles—one determined by the points (0, 0, 5), (5, 5, 0), (0, 0, −5), and (−5, −5, 0), and the other determined by (0, 0, 5), (5, −5, 0), (0, 0, −5), and (−5, 5, 0).

24

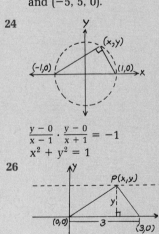

$$\frac{y-0}{x-1} \cdot \frac{y-0}{x+1} = -1$$
$$x^2 + y^2 = 1$$

26

$$A_\triangle = \tfrac{1}{2}bh$$
$$2 = \tfrac{1}{2} \cdot 3y$$
$$\tfrac{4}{3} = y$$

Problem Set B, *continued*

15 What is the locus of the midpoints of all segments drawn from one vertex of a triangle to the opposite side of the triangle? **The locus is a segment joining the midpoints of the sides containing the given vertex.**

16 What is the locus in *space* of points that are

  a 5 units from a given point? **The locus is a sphere with the given point as center and a radius of 5.**

  b 5 units from a given line? **The locus is a cylinder with a radius of 5.**

17 Write an equation for the locus of points equidistant from the lines whose equations are x = −2 and x = 7 **x = 2.5**

18 Find the locus of points that are 5 units from both the x-axis and the y-axis. **The locus is the four points (5, 5), (5, −5), (−5, 5), and (−5, −5).**

19 Given a circle Q with a radius of 9, find the locus of points 9 units from the circle Q. **The locus is the union of point Q and the circle with center Q and radius 18.**

20 a Sketch the locus of points 5 units from a segment, $\overline{PQ}$.

  b Find the area of the locus sketched in part **a** if PQ = 6. **$60 + 25\pi$**

  c Sketch the locus in *space* of points 5 units from segment $\overline{PQ}$.

  d Find the volume of the locus sketched in part **c** if PQ = 6. **$\frac{950\pi}{3}$**

21 Point P is 4 units above plane m. Find the locus of points that lie in plane m and are 5 units from P. **The locus is a circle with its center at the foot of the perpendicular from P to m and with a radius of 3.**

22 Write an equation for the locus of points equidistant from (3, 5) and (1, −9). **$y = -\tfrac{1}{7}x - \tfrac{12}{7}$**

23 Points T and V are fixed. Find the locus of P such that $\overline{PT} \perp \overline{PV}$. **The locus does not include T and V but otherwise is a circle with center at the midpoint of $\overline{TV}$ and diameter $\overline{TV}$.**

Problem Set C

24 Write an equation for the locus of points each of which is the vertex of the right angle of a right triangle whose hypotenuse is the segment joining (−1, 0) and (1, 0). Describe the set geometrically. **$x^2 + y^2 = 1$, x ≠ ±1; the locus does not include (1, 0) or (−1, 0) but otherwise is a circle with center at the origin and a radius of 1.**

25 a The locus of points equidistant from the vertices of a triangle is the point of intersection of the __?__ of the triangle. **Perpendicular bisectors of the sides**

  b The locus of points equidistant from the sides of a triangle is the point of intersection of the __?__ of the triangle. **Angle bisectors**

26 Write an equation for the locus of points (x, y) such that the area of the triangle with vertices (x, y), (0, 0), and (3, 0) is 2. **$y = \pm\tfrac{4}{3}$**

27 Given: P = (−3, 4)

  a Sketch the locus of points that are 2 or more units from P and at the same time are no more than 5 units from P.

  b Describe the locus algebraically. $\{(x, y): 4 \leq (x + 3)^2 + (y - 4)^2 \leq 25\}$

  c Find the area of the locus. 21π

28 A ladder 6 m long leans against a wall. Describe the locus of the midpoint of the ladder in all possible positions. Prove that your answer is correct.
The locus is a quarter circle with center at the foot of the wall and radius of 3 m.

29 Write an equation for the locus of points each of which is twice as far from (−2, 0) as it is from (1, 0). $(x - 2)^2 + y^2 = 4$

30 PQRS is a rectangle with $\overline{PQ}$ twice as long as $\overline{QR}$. T is the midpoint of $\overline{RS}$. $\overline{TQ}$ is drawn. Sketch the locus of the midpoints of segments that are parallel to $\overline{TQ}$ and end on the sides of the rectangle.

HISTORICAL SNAPSHOT

THE GEOMETRY OF MUSIC
Pythagoras and the harmonious blacksmiths

The philosopher Pythagoras (c. 540 B.C.) and his followers believed that the whole numbers were the key to the structure of the universe. In part, this belief was based on discoveries they made about mathematical relationships among the tones of the musical scale.

As the sixth-century writer Macrobius tells the story, one day Pythagoras was walking by a workshop where two blacksmiths were beating out a piece of hot iron. Noticing that the workers' hammers rang with different but harmonious sounds, Pythagoras went inside to investigate the reason. He determined that the tones produced by the hammers depended not on the force with which the hammers were wielded but only on their sizes and weights.

In later experiments, the Pythagoreans plucked stretched strings to produce musical tones. It was discovered that by treating a string as a line segment and dividing it in ratios corresponding to quotients of whole numbers, all the tones of the musical scale can be produced. For instance, if a string is bisected, each half sounds a tone one octave higher than that produced by the whole string. Similarly, when a string is divided in the ratio 2:3, each part sounds the musical interval known as a fifth. The ratio 3:4 produces the interval known as a fourth.

The Pythagoreans' research forms the basis for the construction of many modern musical instruments. It is important, however, as one of the earliest cases in which mathematics was used to explain natural phenomena.

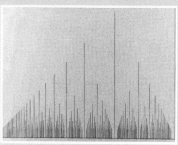

Problem-Set Notes and Additional Answers, continued

27a

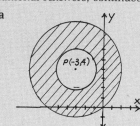

  c $A = \pi 5^2 - \pi 2^2$
        $= 21\pi$

28

$\sqrt{x^2 + y^2} = 3$
$x^2 + y^2 = 9 \; x \geq 0, \; y \geq 0$

29

$\sqrt{(x + 2)^2 + y^2} =$
$2\sqrt{(x - 1)^2 + y^2}$
$(x - 2)^2 + y^2 = 4$

30

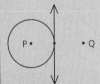
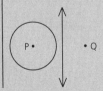
14.2 COMPOUND LOCUS

Objective
After studying this section, you will be able to
- Apply the compound-locus procedure

Part One: Introduction

Many locus problems involve combining two or more loci in one *compound locus*.

Example   *If points A and B are 5 units apart, what is the locus of points 3 units from A and 4 units from B?*

The locus of points 3 units from A is the circle shown in Figure 1.

The locus of points 4 units from B is the circle shown in Figure 2.

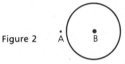

The locus of points that are both 3 units from A and 4 units from B is the two blue points in Figure 3.

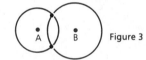

Notice that the compound locus illustrated in Figure 3 is the intersection of the loci in Figure 1 and Figure 2.

Compound-Locus Procedure
1 Solve each part of the compound locus problem separately.   **2** Find all possible intersections of the loci.

Part Two: Sample Problems

Problem 1   Find the locus of points that are a fixed distance from a given line and lie on a given circle.

Solution   Follow the compound-locus procedure.

Step 1: Find each locus individually.

The locus of points that are a fixed distance from a given line is two lines that are parallel to the given line.

The locus of points that lie on a given circle is simply the circle itself.

Step 2: Find all possible intersections of the loci. Thus, the locus of points that are a fixed distance from a given line and lie on a given circle is 4 points, 3 points, 2 points, 1 point, or the empty set.

4 points   3 points   2 points   1 point   ∅

Note To solve a compound-locus problem, keep any fixed distance or fixed figure the same in all drawings. Change given distances and the size and position of given figures to show all possible situations.

Problem 2   *Find the locus in space of points that are a fixed distance from a given plane and a given distance from a fixed point on the plane.*

Solution   Follow the compound-locus procedure.

Step 1: Find each locus individually.

The locus of points that are a fixed distance from a given plane is two planes that are parallel to the given plane.

The locus in space of points that are a given distance r from a fixed point P is a sphere with center P and radius r.

Step 2: Find all possible intersections of the two loci.

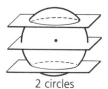 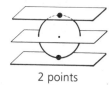 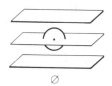

2 circles   2 points   ∅

Thus, the locus in space of points that are a fixed distance from a given plane and a given distance from a fixed point on the plane is two circles or two points or the empty set.

Lesson Notes, continued

- The fixed distance and the radius are not given; hence, they must be varied relative to each other.

Communicating Mathematics

Have students write a paragraph that explains how to solve a compound-locus problem. Encourage students to share their explanations with the class.

Part Three: Problem Sets

Problem Set A

1 Sketch all possible intersections for each compound locus. Then describe the compound locus.

   a The locus of points equidistant from two given points and lying on a given circle { }, 1 point, or 2 points

   b The locus of points that are a given distance from a point A and another given distance from a point B { }, 1 point, or 2 points

   c The locus of points on both the graph of $y = 5$ and the graph of $x^2 + y^2 = r^2$, where $r > 0$ { }, 1 point, or 2 points

   d The locus of points equidistant from two parallel lines and lying on a third line { }, 1 point, or a line

   e The locus of points equidistant from two intersecting lines and a fixed distance from their point of intersection 4 points

   f The locus of points equidistant from the sides of an angle and equidistant from two parallel lines { }, 1 point, or a ray

2 Find the locus of points that are 1 cm from a 4-cm-long segment and 2 cm from the midpoint of the segment. 4 points

3 How many points are equidistant from two given parallel lines and equidistant from two fixed points on one of those lines? 1 point

4 Given a regular hexagon, find the locus of points that are a given distance from its center and lie on the vertices of the hexagon. { } or 6 points

Problem Set B

5 a What is the locus of points that are less than or equal to a fixed distance from a given point and lie on a given line? { }, 1 point, or a segment

   b What is the locus of points that are less than a fixed distance from a given point and lie on a given line? { } or a segment

6 Find the locus of points equidistant from two concentric circles and on a diameter of the larger circle. 2 points

7 Find all the points on a given line that are a fixed distance from a given circle if the fixed distance is less than the circle's radius. { }, 1 point, 2 points, 3 points, or 4 points

8 Find the locus of points 10 units from the origin of a coordinate system and 6 units from the y-axis. 4 points

9 Transversal t intersects parallel lines m and n. Find the locus of points equidistant from m and n and 1 unit from t. 2 points

10 Given a regular pentagon, find the locus of points that are a given distance from its center and lie on it. { }, 5 points, or 10 points

Problem Set C

11 Given three points, A, B, and C, find the locus of points equidistant from all three points. { } or 1 point

12 a Find the locus *in space* of points that are 3 in. from a given plane and 5 in. from a fixed point on the plane. 2 circles

b Find the area of the figure(s) found in part **a**. 32π

13 Find all the points equidistant from two given points and at a given distance from a given circle.
{ }, 1 point, 2 points, 3 points, or 4 points

14 Find the locus *in space* of points that are equidistant from two given points and at a given distance from a given line.
{ }, a line, 2 lines, a circle, or an ellipse

15 Find the locus of points that lie on a given square and also lie on a given circle with its center in the interior of the square.
{ }, 1 point, 2 points, 3 points, . . . 8 points

16 Given ∠A and ∠B, find the locus of points that are equidistant from the sides of ∠A and the sides of ∠B.
{ }, 1 point, a ray, or a segment

17 Find the locus *in space* of a line segment revolving about its midpoint. A sphere and its interior

Problem-Set Notes and
Additional Answers, continued

■ See *Solution Manual* for diagrams for problems **10, 11, 13, 15–17.**

■ In problem **11**, if the three points are collinear, the locus is the empty set. If the three points are noncollinear, the locus is the intersection of the ⊥ bisectors of the segments joining the three points.

12a

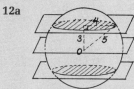

b $A = 2(\pi \cdot 4^2)$
$= 32\pi$

■ In problem **13**, the locus of points equidistant from two given points is the ⊥ bisector of the segment joining the two points. The locus of points at a given distance from a given circle is, in general, two circles. The set of possibilities of intersections of these two loci is the final locus.

■ In problem **14**, the locus of points in space equidistant from two given points is the plane that is the ⊥ bisector of the segment joining the two points. The locus of points in space at a given distance from a given line is a cylinder. The set of possibilities of intersections of these two loci is the final locus.

■ In problem **16**, each individual locus is a ray. The final locus is the set of possibilities of intersections of these two loci.

Class Planning

Time Schedule
Basic: Omit
Average: 2 days
Advanced: 1 day

Resource References
Teacher's Resource Book
  Class Opener 14.3A, 14.3B
  Using Manipulatives 22
  Supposer Worksheet 14
Transparency 23
Evaluation
  Tests and Quizzes
    Quizzes 1–2, Series 2
    Quiz 1, Series 3

Class Opener

Find the locus of all points equi-distant from three noncollinear points. (Hint: Suggest that students consider two points at a time.)
A point that is the intersection of the ⊥ bisectors of the segments formed by two of the three points at a time
This problem previews Theorem 128.

Lesson Notes

- This section, with its dependence on complex vocabulary, is not recommended for basic courses.
- The concurrence theorems are important for justifying descriptions of loci. Students will see them again in advanced geometry courses.
- Theorem 128 shows that a circle can be drawn through the three vertices of any triangle and locates the center of that circumscribed circle.

14.3 THE CONCURRENCE THEOREMS

Objective

After studying this section, you will be able to
- Identify the circumcenter, the incenter, the orthocenter, and the centroid of a triangle

Part One: Introduction

Lines that have exactly one point in common are said to be **concurrent**.

Definition **Concurrent lines** are lines that intersect in a single point.

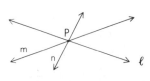

ℓ, m and n are
Concurrent at P.

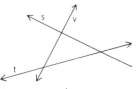

s, t, and v are
not Concurrent.

With this definition and an understanding of compound loci, we can investigate some theorems of advanced geometry.

Theorem 128 **The perpendicular bisectors of the sides of a triangle are concurrent at a point that is equidistant from the vertices of the triangle. (The point of concurrency of the perpendicular bisectors is called the circumcenter of the triangle.)**

Given: ℓ is the ⊥ bisector of $\overline{BC}$.
  m is the ⊥ bisector of $\overline{AC}$.
  n is the ⊥ bisector of $\overline{AB}$.

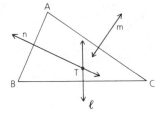

Vocabulary
center of gravity
centroid
circumcenter

concurrent lines
incenter
orthocenter

Prove: **a** ℓ, m, and n are concurrent at point T.

    b T is equidistant from A, B, and C.

Proof: Let T be the point of intersection of ℓ and n.
(How do we know that ℓ and n intersect?)
We must show that m passes through T.

Because T is on line ℓ, the perpendicular bisector of $\overline{BC}$, T is
equidistant from points B and C. (Any point on the perpendicular bisector of a line segment is equidistant from the endpoints of that segment.)

Similarly, T is equidistant from points A and B because it lies
on n, the perpendicular bisector of $\overline{AB}$.

By transitivity, T is equidistant from A and C. Since m is the
locus of *all* points equidistant from A and C, T must lie on m.

    The bisectors of the angles of a triangle are also concurrent. This
statement is formalized in the following theorem, which is presented
without proof.

Theorem 129   ***The bisectors of the angles of a triangle are concur-***
          rent at a point that is equidistant from the sides of
          the triangle. (The point of concurrency of the angle
          bisectors is called the incenter of the triangle.)

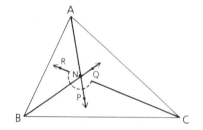

Given: $\overrightarrow{AP}$ bisects ∠BAC.
       $\overrightarrow{BQ}$ bisects ∠ABC.
       $\overrightarrow{CR}$ bisects ∠ACB.

Prove: **a** $\overrightarrow{AP}$, $\overrightarrow{BQ}$, and $\overrightarrow{CR}$ are concurrent at point N.

    b N is equidistant from $\overline{AB}$, $\overline{BC}$, and $\overline{AC}$.

The proof of Theorem 129 depends on the fact that the locus of all
points equidistant from the sides of an angle is the bisector of the
angle (example 2 in Section 14.1). The organization of the proof is
much like that of Theorem 128.

Theorem 130   ***The lines containing the altitudes of a triangle are***
          concurrent. (The point of concurrency of the lines
          containing the altitudes is called the orthocenter of
          the triangle.)

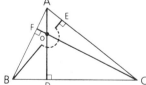

Given: $\overline{AD}$ is the altitude to $\overline{BC}$.
       $\overline{BE}$ is the altitude to $\overline{AC}$.
       $\overline{CF}$ is the altitude to $\overline{AB}$.

Prove: $\overleftrightarrow{AD}$, $\overleftrightarrow{BE}$, and $\overleftrightarrow{CF}$ are concurrent at O.

Lesson Notes, continued

  ■ Theorem 129 shows that a cir-
     cle can be inscribed in any tri-
     angle and locates the center of
     that inscribed circle.

The Geometric Supposer

Before beginning the lesson or
after introducing the definition
of concurrence, have students
use *The Geometric Supposer: Triangles* to draw a triangle and all
its angle bisectors. Have students
record or state their observations. They should note that the
angle bisectors all intersect at a
point, or are concurrent.

Have students erase the angle bisectors, draw all the altitudes of
the same triangle, and note their
observations. Now, have students
erase the altitudes and draw the
medians.

Ask students to make a generalization about all the angle bisectors, altitudes, and medians of a
triangle and then test their generalization by repeating the steps
above using different types of triangles.

Now ask students to predict
when any two of the three different points of concurrence will
coincide. When will all three coincide? Have students test their
predictions using the program.

Communicating Mathematics

Discuss the proof of Theorem
130. See page 665, problem **19.**

The proof of this theorem is asked for in Problem Set C, problem 19.

Note The orthocenter of a triangle is not always inside the triangle, as you can see in the following figures.

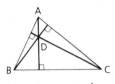

△ABC is an acute △.
D is the orthocenter.

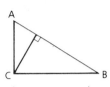

△ABC is a right △.
C is the orthocenter.

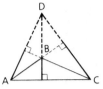

△ABC is an obtuse △.
D is the orthocenter.

Lesson Notes, continued
- The centroid is one of the tri-section points of each median. Its location has uses in problem solving, especially with engineering problems. Students dealt with the centroid in problem **24,** page 617.

- The three medians divide a triangle into six regions of equal area. Thus a triangular region will seesaw on each of its medians and will balance on its centroid.

Theorem 131 *The medians of a triangle are concurrent at a point that is two thirds of the way from any vertex of the triangle to the midpoint of the opposite side. (The point of concurrency of the medians of a triangle is called the **centroid** of the triangle.)*

Given: Medians $\overline{AM}$, $\overline{BN}$, and $\overline{CP}$
Prove: **a** $\overline{AM}$, $\overline{BN}$, and $\overline{CP}$ are concurrent at T.
  b $\dfrac{AT}{AM} = \dfrac{CT}{CP} = \dfrac{BT}{BN} = \dfrac{2}{3}$

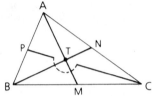

The proof of Theorem 131 is asked for in Problem Set D, problem 21. The centroid of a triangle is important in physics because it is the **center of gravity** of the triangle.

Part Two: Sample Problem

Problem In △PQR, medians $\overline{QT}$ and $\overline{PS}$ are concurrent at C.
PC = 4x − 6,
CS = x
Find: **a** x **b** PS

Solution **a** The medians of a triangle are concurrent at a point that is two thirds of the way from any vertex of the triangle to the midpoint of the opposite side. Thus,
PC = $\frac{2}{3}$(PS), or PC = 2(CS).

$$4x - 6 = 2x$$
$$x = 3$$
b PC = 4x − 6
$$= 4(3) − 6$$
$$= 12 − 6$$
$$= 6$$
Thus, PS = PC + CS = 6 + 3 = 9.

Part Three: Problem Sets

Problem Set A

1 Trace △ABC on a piece of paper. Use a ruler to locate the centroid of △ABC.

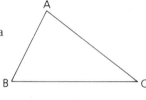

2 Find the orthocenter of right triangle PQR.

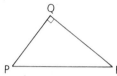

3 Given scalene △ DEF, explain how to find the locus of points equidistant from $\overline{DE}$, $\overline{EF}$, and $\overline{DF}$. Find the incenter.

4 Trace right △ RST on a piece of paper.

a Use a ruler to estimate the location of the circumcenter.

b Use your result in part **a** to guess the exact location of the circumcenter of any right triangle.
Midpoint of the hypotenuse

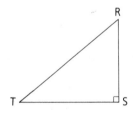

5 Every triangle has a circumcenter, an orthocenter, a centroid, and an incenter. Which of the four points will always lie in the interior of the triangle? Centroid and incenter

6 Given: △ABC, with medians $\overline{AM}$, $\overline{BN}$, and $\overline{CP}$.

a If AM = 9, find AT. 6
b If TN = 5, find BN. 15
c If TC = 8, find PT. 4
d If BN = $\sqrt{18}$, find TN. $\sqrt{2}$

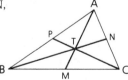

Assignment Guide

Basic
Omit
Average
Day 1   1–10
Day 2   12–15, 17
Advanced
3, 4, 6, 8–10, 12–18

Problem-Set Notes and Additional Answers

■ See *Solution Manual* for answers to problems **1–4a.**

Problem-Set Notes and Additional Answers, continued

■ In problem **15**, the equations of the ⊥ bisectors of the sides are $y = x - 4$, $y = -\frac{1}{4}x + 3$, and $y = -4x + 24$.

■ In problem **16**, △RST is a right △, so the orthocenter will be at the right angle, S.

18

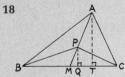

a △ATM ~ △PQM by AA ~.
$$\frac{AM}{PM} = \frac{AT}{PQ} = \frac{3}{1}$$

b $\dfrac{A_{\triangle PBC}}{A_{\triangle ABC}} = \dfrac{\frac{1}{2} \cdot BC \cdot PQ}{\frac{1}{2} \cdot BC \cdot AT} = \dfrac{1}{3}$

Problem Set A, *continued*

7 If a triangle is cut from cardboard and the circumcenter, the orthocenter, the centroid, and the incenter are located, upon which point could the triangle be balanced? Centroid

Problem Set B

8 In what kind of triangle is the orthocenter a vertex of the triangle? Right

9 In what kind of triangle is the orthocenter the same point as the circumcenter? Equilateral

10 In what kind of triangle does the centroid lie outside the triangle? None

11 Sketch three noncollinear points. Then sketch and describe the locus of points equidistant from all three points.
The locus is the circumcenter of the triangle formed by joining the three points.

12 Given: △RST with medians
$\overline{RM}$ and $\overline{TN}$ intersecting at P,
RP = 2y − x, TP = 2y,
PM = y − 2, PN = x + 2
Find: The longer of the two medians TN = 18

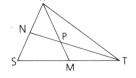

13 Given: △PLO with centroid V,
VT = 6, AT = 9, OT = 18
Find: **a** PA $9\sqrt{3}$
    b The area of △PLO $162\sqrt{3}$
    c The area of △POT $81\sqrt{3}$
    d m∠APT 30

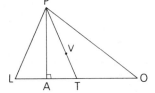

14 Given △PQR, with P = (0, 0), Q = (5, 12), and R = (10, 0), find the coordinates of its centroid. (5, 4)

Problem Set C

15 Given △ABC, with A = (1, 3), B = (7, −3), and C = (9, 5), find the circumcenter of the triangle. $\left(\frac{28}{5}, \frac{8}{5}\right)$

16 Given △RST, with R = (−3, 2), S = (4, 5), and T = (7, −2), find the coordinates of its orthocenter. (4, 5)

17 Recall that the coordinates of the midpoint of a side of a triangle are the averages of the coordinates of the endpoints. As an extension of this idea, it can be shown that the coordinates of the centroid of a triangle are the averages of the coordinates of the three vertices of the triangle.

Given: $\triangle ABC$, with A = (−2, 8), B = (−6, −2), and C = (12, 6)

Find: **a** The coordinates of the centroid of $\triangle ABC$ $\left(\frac{4}{3}, 4\right)$

b The coordinates of the centroid of the triangle formed by joining the midpoints of the sides of $\triangle ABC$ $\left(\frac{4}{3}, 4\right)$

18 Given: $\triangle ABC$, with median $\overline{AM}$ and centroid P

a Using $\overline{BC}$ as the base of each triangle, prove that the altitude of $\triangle PBC$ is one third of the altitude of $\triangle ABC$.

b Find the ratio of the area of $\triangle PBC$ to the area of $\triangle ABC$. $\frac{1}{3}$

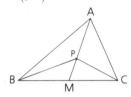

19 Given: $\triangle ABC$

Prove: The lines containing the altitudes of $\triangle ABC$ are concurrent (Theorem 130). (Hint: Through each vertex of the triangle, draw a line parallel to the opposite side, obtaining the diagram shown. Then apply Theorem 128.)

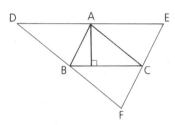

20 Sketch a triangle and its medians. As you know, the centroid of the triangle is one of the trisection points of each median. Now form another triangle by joining the *other* trisection points of the medians.

a Find the ratio of the area of this triangle to the area of the original triangle. $\frac{1}{4}$

b What is the relationship of this triangle to the triangle formed by joining the midpoints of the sides of the original triangle. Congruent

Problem Set D

21 Given: $\triangle ABC$

Prove: The medians of $\triangle ABC$ are concurrent at a point that is two thirds of the way from any vertex of $\triangle ABC$ to the midpoint of the opposite side (Theorem 131). (Hint: Use the coordinates shown in the diagram.)

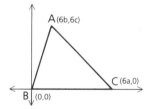

Problem-Set Notes and Additional Answers, continued

19 The altitudes of $\triangle ABC$ can be shown to be the $\perp$ bisectors of the sides of $\triangle DEF$.

20

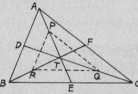

a $\overline{RQ}$ is the midline of $\triangle TBC$, so $\frac{A_{\triangle TRQ}}{A_{\triangle TBC}} = \frac{1}{4}$. The same is true for $\triangle TQP$ and $\triangle TPR$. Thus, $\frac{A_{\triangle PRQ}}{A_{\triangle ABC}} = \frac{1}{4}$.

b $\overline{RQ}$ is the midline for $\triangle TBC$, so $RQ = \frac{1}{2}BC$. $\overline{DF}$ is the midline for $\triangle ABC$, so $DF = \frac{1}{2}BC$. $\therefore \overline{RQ} \cong \overline{DF}$ In a similar manner, $\overline{PR} \cong \overline{EF}$ and $\overline{PQ} \cong \overline{DE}$. Thus, $\triangle PRQ \cong \triangle EFD$ by SSS.

21

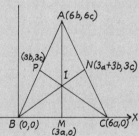

The equations of the medians are
$\overleftrightarrow{AM}$: $y = \frac{2c}{2b - a}x - \frac{6ac}{2b - a}$
$\overleftrightarrow{BN}$: $y = \frac{c}{a + b}x$
$\overleftrightarrow{CP}$: $y = \frac{c}{b - 2a}x - \frac{6ac}{b - 2a}$
Solve any two of the equations to find their intersection and then substitute into the third equation. The intersection is at $(2a + 2b, 2c)$.

BASIC CONSTRUCTIONS

Class Planning

Time Schedule
Basic: 3 days (optional)
Average: 2 days
Advanced: 2 days

Resource References
Teacher's Resource Book
    Class Opener 14.4A, 14.4B

Class Opener

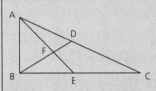

If $\overline{AB} \perp \overline{BC}$, $\overline{BD}$ and $\overline{AE}$ are medians, AB = 18, and BC = 24, find BF, FD, AF, and FE.
BF = 10, FD = 5, AF = $4\sqrt{13}$, FE = $2\sqrt{13}$

Lesson Notes

- Making classical constructions with straightedge and compass can be one of the most enjoyable activities of geometry. The six basic constructions provide a foundation for many more complex constructions.

Objectives

After studying this section, you will be able to
- Identify the tools and procedures used in constructions
- Interpret the shorthand notation used in describing constructions
- Perform six basic constructions

Part One: Introduction

Constructions

A *construction* is a drawing made with the help of only two simple tools. The procedures used for constructions are based on ones developed by ancient Greek geometricians. The two tools needed are

1 A compass, to construct circles or arcs of circles
2 A straightedge, to draw lines or rays (A straightedge differs from a ruler only by the absence of marks for measuring distances.)

These tools can produce accurate drawings when correctly used. (A sharp pencil and good paper on flat, firm cardboard are necessities.) Admittedly, modern drafting machines can produce more-accurate drawings in less time, but constructions are still worth studying for reasons such as the following:

- The tools are simple and portable.
- There is an orderly progression of steps. Nothing is accepted just because the result looks correct.
- Analyzing constructions strengthens understanding of theorems.
- The restrictions on equipment and the strictly defined rules make producing constructions a challenging game, one enjoyed by most people who learn it. The game has a practical bonus for some, because users of drafting machines must analyze problems, and their analyses are often the same as those used for constructions.

Vocabulary
construction

Shorthand Notation for Constructions

So that the step-by-step instructions will be clear and concise, the following notation for constructions will be used in this book:

1 ⊙ **(P, PB)** represents a circle with center P and radius of length PB.

2 **arc (P, PB)** represents an arc with center P and radius of length PB.

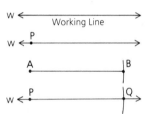

Six Basic Constructions

The following six constructions are the basis of all further work with constructions. Because a construction has meaning only in terms of how it is developed, we urge you to redraw these constructions, following the instructions step by step.

Construction 1: Segment Copy

Construction of a line segment congruent to a given segment.

Given: $\overline{AB}$ (In the setup on your paper, draw segment $\overline{AB}$ of any length you want.)

Construct: A segment $\overline{PQ}$ that is congruent to $\overline{AB}$

Procedure:
1 Draw a *working line*, w.
2 Let P be any point on w.
3 On the given segment, construct arc (A, AB).
4 Construct arc (P, AB) intersecting w at some point Q.
5 $\overline{PQ} \cong \overline{AB}$

Notice that in constructions lengths are not measured with rulers; they are matched by compass settings.

Your finished paper should look something like this:

Given: $\overline{AB}$

Construct: $\overline{PQ}$, $\cong$ to $\overline{AB}$ w ← P Q → $\therefore \overline{PQ} = \overline{AB}$

Note Do not erase any arc marks in any construction problems.

Construction 2: Angle Copy

Construction of an angle congruent to a given angle.

Given: ∠ABC (Make your setup big enough for easy use of the compass, yet not so big that everything won't fit on your paper.)

Construct: An angle PQR that is congruent to ∠ABC

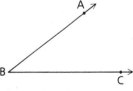

Lesson Notes, continued

- The constructions on the following pages are described using this notation.
- All geometric principles necessary to justify the six basic constructions were covered by the end of Chapter **4**.
- If students use colored pencils in their compasses, construction marks will show clearly.

Cooperative Learning

As you work through the constructions in this section, have students work in pairs or groups of three. As they watch demonstrations or read directions from the text, they should help each other with the physical constructions and compare drawings to determine why each one works.

Procedure:

1 On the setup, use any radius r to construct arc (B, r) intersecting ∠ABC at two points. Call them D and E.

2 Let Q be any point on a working line, w.

3 Construct arc (Q, r) to intersect w at some point R.

4 Construct arc (E, ED).

5 Construct arc (R, ED) intersecting arc (Q, r) at some point P.

6 Draw $\overrightarrow{QP}$.

7 ∠PQR ≅ ∠ABC

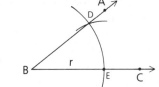

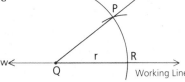

If we drew $\overline{DE}$ and $\overline{PR}$, we would form △BDE and △QPR. Do you see how SSS is the basis of this construction?

Construction 3: Angle Bisection

Construction of the bisector of a given angle.

Given: ∠ABC

Construct: $\overrightarrow{BP}$, the bisector of ∠ABC

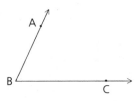

Procedure:

1 Use any radius r to construct arc (B, r) intersecting the sides of ∠ABC at two points, Q and T.

2 Use any radius s (which may or may not be equal to r) to construct arc (Q, s) and arc (T, s), intersecting each other at a point P.

3 Draw $\overrightarrow{BP}$.

4 $\overrightarrow{BP}$ bisects ∠ABC.

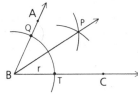

If we drew $\overline{QP}$ and $\overline{PT}$, each would be s units long. Can you see how SSS is the basis of this construction?

Construction 4: Perpendicular Bisector

Construction of the perpendicular bisector of a given line segment.

Given: $\overline{AB}$

Construct: $\overleftrightarrow{PQ}$, the perpendicular bisector of $\overline{AB}$

668 Chapter 14 Locus and Constructions

Procedure:

1. Use any radius *r* that is more than half the length of $\overline{AB}$ to construct arc (A, r).
2. Construct arc (B, r) intersecting arc (A, r) at P and Q.
3. Draw $\overleftrightarrow{PQ}$.
4. $\overleftrightarrow{PQ}$ is the perpendicular bisector of $\overline{AB}$.

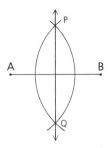

Construction 5: Erecting a Perpendicular

Construction of a line perpendicular to a given line at a given point on the line.

Given: $\overleftrightarrow{AB}$ and point P on the line

Construct: $\overleftrightarrow{PQ}$ perpendicular to $\overleftrightarrow{AB}$ at P

Procedure:

1. Use any radius *r* to construct arc (P, r) intersecting $\overleftrightarrow{AB}$ at V and T.
2. Use any radius *s* that is greater than *r* to construct arc (V, s) and arc (T, s), intersecting each other at a point Q.
3. Draw $\overleftrightarrow{PQ}$.
4. $\overleftrightarrow{PQ} \perp \overleftrightarrow{AB}$.

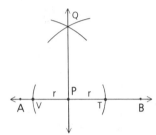

Construction 6: Dropping a Perpendicular

Construction of a line perpendicular to a given line from a given point not on the line.

Given: $\overleftrightarrow{AB}$ and point P not on $\overleftrightarrow{AB}$

Construct: $\overleftrightarrow{PQ}$ perpendicular to $\overleftrightarrow{AB}$

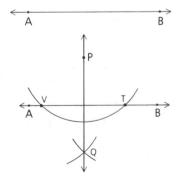

Procedure:

1. Use any radius *r* to construct arc (P, r) intersecting $\overleftrightarrow{AB}$ at V and T.
2. Use any radius *s* (which may or may not be equal to *r*) to construct arc (V, s) and arc (T, s), intersecting each other at a point Q.
3. Draw $\overleftrightarrow{PQ}$.
4. $\overleftrightarrow{PQ} \perp \overleftrightarrow{AB}$.

Part Two: Sample Problems

Problem 1   Given: ∠A and ∠B as shown

Construct: An angle whose measure is equal to (x + y)

Solution   1 On a working line w, use the angle-copy procedure to construct ∠VTS ≅ to ∠A.

2 With $\overleftrightarrow{TV}$ as a new working line, use the angle-copy procedure to construct ∠QTV ≅ to ∠B.

3 ∠QTS is the required angle.

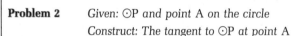

Problem 2   Given: ⊙P and point A on the circle

Construct: The tangent to ⊙P at point A

Solution   Make a freehand sketch of the required construction. Analyze the geometric relationships between the parts of the sketch to determine the required procedure.

For this problem, the sketch will look like the one at the right. Do you see what needs to be done?

1 Draw $\overrightarrow{PA}$.

2 Construct the ⊥ to $\overleftrightarrow{PA}$ at A. (See Construction 5.)

3 $\overleftrightarrow{TA}$ is the required tangent.

Problem 3   Given: ⊙O and point P outside the circle

Construct: A tangent to ⊙O from point P

Solution   At the point of tangency, a radius and the required tangent will form a right angle. (See Problem Set D, problem 19.)

1 Draw $\overline{OP}$.

2 Find the midpoint M of $\overline{OP}$ by the perpendicular-bisector procedure.

3 Construct ⊙ (M, MP).

4 Label A and B, the intersections of ⊙O and ⊙M.

5 Draw $\overline{PA}$.

6 $\overline{PA}$ is tangent to ⊙O.

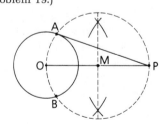

Part Three: Problem Sets

Problem Set A

1 Construct the locus of points equidistant from two fixed points A and B.

2 Draw two segments, $\overline{AB}$ and $\overline{CD}$, with AB > CD.
- **a** Construct a segment whose length is the sum of AB and CD.
- **b** Construct a segment whose length is the difference of AB and CD.
- **c** Locate the midpoint of $\overline{AB}$ by construction.
- **d** Construct an equilateral triangle whose sides are congruent to $\overline{CD}$.
- **e** Construct an isosceles triangle, making its base congruent to $\overline{CD}$ and each leg congruent to $\overline{AB}$.
- **f** Construct a square whose sides are congruent to $\overline{AB}$.
- **g** Construct a circle whose diameter is congruent to $\overline{CD}$.

3 Draw an acute angle ABC and an obtuse angle WXY.
- **a** Construct $\angle FGH$ congruent to $\angle WXY$.
- **b** Construct the complement of $\angle ABC$.
- **c** Construct the supplement of $\angle WXY$.
- **d** Construct an angle whose measure is the difference of $\angle WXY$ and $\angle ABC$.
- **e** Construct an angle whose measure is double that of $\angle ABC$.

4 Construct the following angles.
- **a** 90°
- **b** 45°
- **c** 60°
- **d** 75°

5 Draw an obtuse triangle. Construct the bisector of each angle.

Problem Set B

6 If a and b are the lengths of two segments and $a < b$, construct a segment whose length is equal to $\frac{1}{2}(b - a)$.

7 Given $\angle A$ and $\angle B$, construct an angle equal to $\frac{1}{2}(m\angle A + m\angle B)$.

8 Construct an angle with each given measure.
- **a** 135
- **b** $112\frac{1}{2}$
- **c** 165

9 Inscribe a square in a given circle. (Hint: Use the diagonals.)

10 Construct the three medians of a given $\triangle PQR$.

11 Construct the three altitudes of an acute $\triangle ABC$.

See page 670 for Assignment Guide.

Problem-Set Notes and Additional Answers

- For students inexperienced with a compass, two simple circle constructions are to inscribe a regular hexagon or an equilateral triangle in a given circle.
- See *Solution Manual* for answers to problems **1–8** and **11**.

9

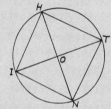

Draw any diameter $\overline{HN}$, then construct $\overline{IT} \perp \overline{HN}$ at center O.

10

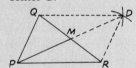

The $\perp$ bisector can locate a midpoint, but another method is to construct D so that PQDR is a $\square$.

- See *Solution Manual* for answers to problems **12** and **13**.

14 Connect the midpoints of two sides of the given triangle. The triangle formed will be the required triangle.

15

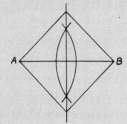

Construct the $\perp$ bisector of $\overline{AB}$ and mark off $\frac{1}{2}AB$ above and below the midpoint of $\overline{AB}$.

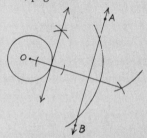
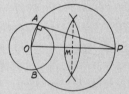
Communicating Mathematics

Problem **19** is a good exercise in
communicating mathematics.

Problem Set B, *continued*

12 Given circle P with point Q in the interior of the circle, construct
a chord of the circle having Q as its midpoint.

13 ∠A is the vertex angle of an isosceles
triangle. Find, by construction, one of
the base angles of the triangle. (Can you
do this without drawing a triangle?)

Problem Set C

14 Draw any triangle. Construct a second triangle similar to the first
such that the ratio of the perimeters is 1:2.

15 Construct a square whose diagonal is
equal to AB.

16 Explain how you would construct each angle.

a 32½° (if given an angle of 80°)   **b** 41¼°

17 Construct two parallel lines.

18 Construct a line, $\overleftrightarrow{CD}$, that is parallel to
$\overleftrightarrow{AB}$ and tangent to ⊙O.

Problem Set D

19 Write a paragraph proof to show that the construction of a tangent to a circle from an external point, as shown in sample
problem 3, is valid.

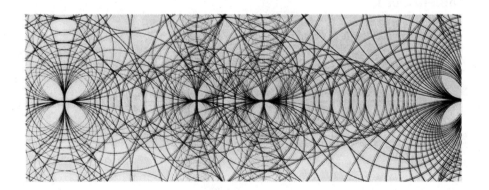

672   Chapter 14   Locus and Constructions

14.5 APPLICATIONS OF THE BASIC CONSTRUCTIONS

Objective

After studying this section, you will be able to
- Perform four other useful constructions

Part One: Introduction

The six basic constructions may be used to develop more-complicated constructions. Four of these are presented in this section. Once you have mastered all ten constructions, you will enjoy the challenge of future problem sets.

Construction 7: Parallels

Construction of a line parallel to a given line through a point not on the given line.

Given: $\overleftrightarrow{AB}$ with point P not on $\overleftrightarrow{AB}$

Construct: A line, $\overleftrightarrow{PQ}$, that is parallel to $\overleftrightarrow{AB}$

Procedure:
1. Draw any line t through P, intersecting $\overleftrightarrow{AB}$ at some point C.
2. Use the angle-copy procedure to construct $\angle QPR \cong$ to $\angle PCB$.
3. $\overleftrightarrow{PQ} \parallel \overleftrightarrow{AB}$ by corr. $\angle s \cong \Rightarrow \parallel$ lines.

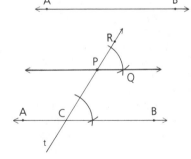

Class Planning

Time Schedule
Basic: 1 day (optional)
Average: 2 days
Advanced: 2 days

Resource References
Teacher's Resource Book
Class Opener 14.5A, 14.5B
Additional Practice Worksheet 27
Evaluation
Tests and Quizzes
Quiz 2, Series 3

Class Opener

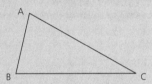

Construct a triangle similar to △ABC, with area four times as great.
For the area to be four times as great, the lengths of the sides would have to be twice as long. Construct $\overline{B'C'}$ twice the length of $\overline{BC}$ on a working line. Copy $\angle B$ at B' and $\angle C$ at C'.

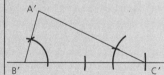

Construction 8: Segment Division

Division of a segment into a given number of congruent segments.

Given: $\overline{AB}$

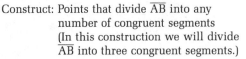

Construct: Points that divide $\overline{AB}$ into any number of congruent segments (In this construction we will divide $\overline{AB}$ into three congruent segments.)

If parallel lines cut off ≅ segments on some transversal, they cut off ≅ segments on any other. Work backwards, transversals first. $\overleftrightarrow{AB}$ is the "any other" transversal. On some other line through A, mark off three ≅ segments of any length. Call their sum AR. Then $\overleftrightarrow{RB}$ determines the direction of the parallel lines.

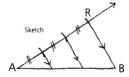

Procedure:
1. Draw any line ℓ through point A.
2. With a radius r, construct arc (A, r) intersecting line ℓ at a point P.
3. Construct arc (P, r) intersecting ℓ at Q.
4. Construct arc (Q, r) intersecting ℓ at R.
5. Draw $\overline{RB}$.
6. Using Construction 7, construct lines through P and Q parallel to $\overleftrightarrow{RB}$. Call the intersections of these lines with $\overline{AB}$ points S and T.
7. $\overline{AS} \cong \overline{ST} \cong \overline{TB}$. (Do you know the reason why?)

Construction 9: Mean Proportional

Construction of a segment whose length is the mean proportional between the lengths of two given segments.

Given: $\overline{AB}$ and $\overline{CD}$

Construct: $\overline{VR}$ such that $(VR)^2 = (AB)(CD)$

Mean proportional suggests an altitude on a hypotenuse. We can find h if we recall that an angle inscribed in a semicircle is a right angle.

Procedure:
1. On a working line w, use the segment-copy procedure to construct a segment of length AB + CD. (Make TV = AB and VZ = CD.)
2. Use the perpendicular-bisector procedure to find the midpoint M of $\overline{TZ}$.
3. Construct semicircle (M, MT).
4. At V, erect a perpendicular to $\overleftrightarrow{TZ}$. The perpendicular will intersect ⊙M at R, and ∠TRZ will be a right angle.
5. $h^2 = xy$, so $(VR)^2 = (AB)(CD)$.

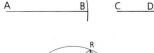

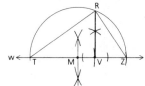

Construction 10: Fourth Proportional

Construction of a segment whose length is the fourth proportional to the lengths of three given segments.

Given: $\overline{AB}$ $\overline{CD}$ $\overline{EF}$ _____ c _____

Construct: $\overline{TV}$ such that $\dfrac{a}{b} = \dfrac{c}{TV}$

Procedure:

1 On a working line w, use the segment-copy procedure to construct $\overline{PS}$ of length $a + b$.
2 Draw any other line ℓ through P.
3 On ℓ, construct $\overline{PT} \cong$ to $\overline{EF}$ by the segment-copy procedure.
4 Draw $\overline{TR}$.
5 Through S, construct a line parallel to $\overleftrightarrow{RT}$, intersecting ℓ at V.
6 $\overline{TV}$ is the required segment, since $\frac{a}{b} = \frac{c}{TV}$.

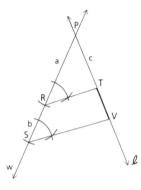

Communicating Mathematics

Have each student prepare to demonstrate and explain one of the constructions from this section. Choose students at random to make a presentation to the class. (You may want to limit students in basic courses to Construction 7.)

Part Two: Sample Problems

Problem 1   *Inscribe a circle in a given △ABC.*

Solution   The center of an inscribed circle is equidistant from the sides, so it is the point of concurrency of the angle bisectors.

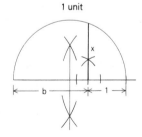

Sketch

1 Construct the angle bisectors of ∠A and ∠C.
2 Their intersection T is the incenter of △ABC.
3 Construct a perpendicular from T to $\overline{BC}$. Call the foot F.
4 Construct ⊙ (T, TF).
5 ⊙T is inscribed in △ABC.

Problem 2   Given: P _____ b _____ Q
Construct: *A segment whose length is $\sqrt{b}$.*

Solution   Since $x = \sqrt{b}$ is equivalent to $x^2 = b$ or $\frac{b}{x} = \frac{x}{1}$, use the mean-proportional procedure. To represent 1, choose any segment as a unit segment. Then b is the number of those units that are in the given segment $\overline{PQ}$.

Using $\overline{PQ}$ and the unit segment, construct the mean proportional, x, between b and 1.

Thus, $x^2 = b \cdot 1$ and $x = \sqrt{b}$

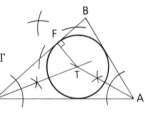

1 unit

Assignment Guide

Basic (optional)
1, 6, 8

Average

Day 1   1–8
Day 2   9, 11–16, 18

Advanced

Day 1   5–9, 11–15
Day 2   16, 18, 20–25

Problem-Set Notes and Additional Answers

■ See *Solution Manual* for answers to problems **1–9** and **11–17**.

10

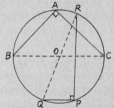

Draw any chord $\overline{AB}$ (not a diameter) and construct chord $\overline{AC} \perp \overline{AB}$. Then $\overline{BC}$ is a diameter. Repeat the steps to construct diameter $\overline{RQ}$. $\overline{RQ}$ intersects $\overline{BC}$ at the center of the circle.

■ See *Solution Manual* for answers to problems **18** and **19**.

20   Construct the fourth proportional to a, 1, and 1.

21   Construct the mean proportional to 3 and 1.

22

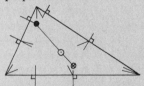

The points of concurrency for the medians (centroid ⊙), the ⊥ bisectors of the sides (circumcenter ⊗), and the altitudes (orthocenter •) are collinear. Also, the centroid is a trisection point of the segment between the other two points.

Part Three: Problem Sets

Problem Set A

1   Given △ABC, construct a line parallel to $\overleftrightarrow{AB}$ and passing through C.

2   Given △PQR, trisect $\overline{QR}$.

3   Given $\overline{AB}$, with point C between A and B, construct a segment whose length is the mean proportional between AC and BC.

4   Given acute ∠DEF, with H between E and F, find by construction a point J between E and D such that $\frac{EJ}{JD} = \frac{EH}{HF}$.

5   Construct an equilateral triangle and its inscribed circle.

6   Construct a parallelogram, given two sides and an angle.

7   Construct an isosceles right triangle and its circumscribed circle.

8   Construct a rectangle, given the base and a diagonal.

9   Construct the centroid of a given triangle.

10   Use an object with a circular surface to trace the outline of a circle. By construction, locate the center of the circle.

11   Given a point P anywhere on a line w, construct a circle of radius r that is tangent to w at P.

Problem Set B

12   Given a segment of length b (make it about 14 cm long), solve $5x = b$ for x by a geometric method.

13   Construct a rhombus, given its diagonals.

14   Construct an isosceles trapezoid, given the bases and the altitude.

15   Given three noncollinear points, construct a circle that passes through all three points.

16   Given ▱ABCD as shown, construct

   a   A rectangle with the same area as ▱ABCD

   b   A triangle with the same area as ▱ABCD

17   Where should a straight fence be located to divide a given triangular field into two fields whose areas are in the ratio 2:1?

18 Given a segment of length a, construct the geometric mean between $2a$ and $3a$.

19 Given $\triangle ABC$, find by construction a point M on $\overline{AC}$ that divides $\overline{AC}$ in a ratio equal to $\frac{AB}{BC}$.

Problem Set C

20 Given: a ————————   1 ————

Construct: A segment whose length is $\frac{1}{a}$

21 Given a unit segment, construct a segment whose length is $\sqrt{3}$.

22 Find the centroid, the circumcenter, and the orthocenter of a large scalene triangle. What seems to be true about these three points?

23 Construct a square equal in area to a given parallelogram.

24 Construct a square that has an area twice as great as the area of a given square.

25 Circumscribe a regular hexagon about a given circle.

Problem Set D

26 Suppose you wanted to construct a triangle equal in area to the given quadrilateral PQRS.

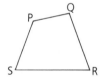

Procedure:
  1 Draw diagonal $\overline{PR}$.
  2 Construct a line parallel to $\overline{PR}$ through Q, intersecting $\overleftrightarrow{SR}$ at some point T.
  3 Draw $\overline{PT}$.
  4 Area ($\triangle PST$) = area (quad PQRS)

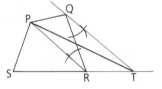

a Write a paragraph proof showing that this procedure is valid.

b Construct a triangle that is equal in area to a given pentagon.

Problem-Set Notes and Additional Answers, continued

23   Construct the mean proportional between the altitude and the base of the parallelogram. This will be the side of the square.

24   Use the diagonal of the original square as a side of the new square.

25

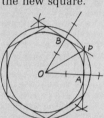

Using OA as a length, mark another point B on the circle such that $\overline{OA} \cong \overline{AB}$. Then construct tangents to $\odot O$ at A and B. These tangents will intersect each other at P. Now construct a new circle concentric with the original circle and having radius OP. Inscribe a hexagon in this new $\odot$, using PO as side length.

26a   Consider $\triangle QPR$ and $\triangle PRT$. They have the same base, $\overline{PR}$, and their heights are $\cong$, since parallel lines are everywhere equidistant. Thus, $A_{\triangle QPR} = A_{\triangle PRT}$ and $A_{\triangle PST} = A_{PQRS}$.

b

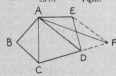

Draw diagonals $\overline{AC}$ and $\overline{AD}$. Construct $\overline{EF} \parallel \overline{AD}$ so that $A_{\triangle ACF} = A_{AEDC}$. Repeat the procedure for quadrilateral AFCB.

T 677

TRIANGLE CONSTRUCTIONS

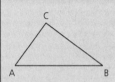
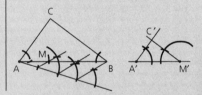
Objective

After studying this section, you will be able to
■ Construct triangles with given side lengths and angle measures

Part One: Introduction

In this section, you will construct triangles, given various combinations of parts and conditions. The following notation for parts and their associated measures will be helpful:

Side lengths: a, b, c
Angles: A, B, C
Altitudes: h_a, h_b, h_c
Medians: m_a, m_b, m_c
Angle bisectors: t_a, t_b, t_c

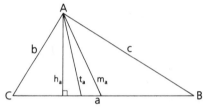

The side opposite vertex A is a units long.
The side opposite vertex B is b units long.
The side opposite vertex C is c units long.
The length of the altitude to the side opposite vertex A is h_a.
The length of the altitude to the side opposite vertex B is h_b.
Medians and angle bisectors have similar labeling.

The sample problems illustrate the importance of beginning with a sketch that shows the given parts and conditions.

Part Two: Sample Problems

Problem 1   Construct: △ABC, given {a, C, b}.

Given: **a** _____

      b _____

Construct: △ABC

Solution   This construction is based on SAS.

1 Sketch the construction.

2 Copy length a on a working line w. Label the endpoints of the segment C and B.

3 Using $\overrightarrow{CB}$ as one side, copy $\angle C$.

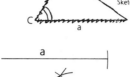

4 Copy length b on the other side of $\angle C$. Label the other endpoint of the segment A.

5 Draw $\overline{AB}$.

6 $\triangle ABC$ is the required triangle.

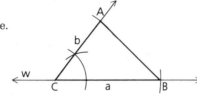

Problem 2   *Construct:* $\triangle ABC$, given $\{a, h_a, B\}$

Given: **a** ———————

   h_a ———————

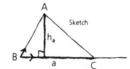

Construct: $\triangle ABC$

Solution   1 Sketch the required triangle. Use a compound locus to locate point A.

One locus containing A is $\overrightarrow{BA}$, the side of $\angle B$ not containing C. The other locus containing A is the set of all points h_a units from $\overleftrightarrow{BC}$. A is the intersection of the two loci.

2 Copy length a (side $\overline{BC}$) on a working line w. (See next page.)

3 At some point P on w, construct a $\perp$ to w.

4 Use the segment-copy procedure to copy length h_a on the $\perp$. Call the segment $\overline{PQ}$.

5 Construct a $\perp$ to $\overleftrightarrow{PQ}$ at Q.

6 Copy $\angle B$. The intersection of $\angle B$ and the parallel to w is A.

Cooperative Learning

As you work through the constructions in this section, have students work in pairs or groups of three. As they watch demonstrations or read directions from the text, they should help each other with the physical constructions and compare drawings to determine why each one works.

Assignment Guide

Basic (optional)

Day 1	1a, 2a, 5, 6
Day 2	1b,c, 3a, 4, 7

Average

Day 1	1–6
Day 2	7, 8, 12, 13, 14a,b,c, 15

Advanced

Day 1	1–9
Day 2	12, 13, 14a,b,c, 15–18

Communicating Mathematics

Refer to problem **1** on page 680. Students can discuss how to construct a triangle by ASA, SSS, and HL.
See *Solution Manual* for answer to problem **1**.

T 679

See page 679 for Assignment Guide.

Problem-Set Notes and Additional Answers

■ See *Solution Manual* for answers to problems **1–16**.

17 Construct the mean proportional between the base and height of the original triangle. Use this segment as the leg of the required right △.

18

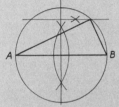

Construct the ⊥ bisector of $\overline{AB}$. Then construct a semicircle with $\overline{AB}$ as the diameter. Construct the locus of points that are of a distance equal to the altitude from $\overline{AB}$. The intersection of the line and the semicircle will be the vertex of the right ∠.

19a

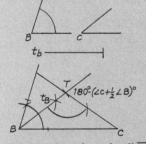

Bisect ∠B and mark off $\overline{BT}$ equal in length to t_b. At T, construct an angle equal to $180 - (\angle C + \frac{1}{2}\angle B)$.

b

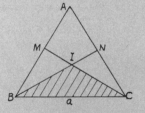

7 Draw $\overline{AC}$.

8 △ABC is the required triangle.

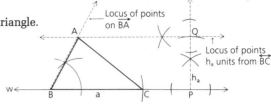

Part Three: Problem Sets

Problem Set A

1 In sample problem 1, a triangle was constructed by SAS. In a similar manner, construct a triangle by each of these methods.

   a ASA (Hint: Draw two different angles and a segment. Then construct a triangle in which the segment is the side included by the angles.)

   b SSS

   c HL

2 Construct an isosceles triangle, given

   a The vertex angle and a leg

   b The base and the altitude to the base

3 Construct an isosceles right triangle, given

   a A leg

   b The hypotenuse

4 Construct a triangle equal in area to a given square.

5 Construct a 30°−60°−90° triangle.

6 Given △ABC, construct a triangle whose area is twice as great as the area of △ABC.

7 Given a triangle, construct a triangle that is similar but not congruent to the given triangle.

Problem Set B

8 Construct a 30°−60°−90° triangle, given the hypotenuse.

9 Construct an isosceles triangle, given the length b of the base and the radius R of its circumscribed circle, where $b < 2R$.

10 Construct an isosceles triangle, given the length b of the base and the radius r of its inscribed circle, where $b > 2r$.

Construct △ABC by first constructing △IBC by using a, $\frac{2}{3}m_b$, and $\frac{2}{3}m_c$ as sides. Then extend $\overline{BI}$ and $\overline{CI}$ to points N and M, respectively. Draw $\overleftrightarrow{BM}$ and $\overleftrightarrow{CN}$ intersecting at A.

11 Construct an isosceles right triangle, given the median to the hypotenuse.

12 Construct a triangle by AAS. (Hint: Begin by constructing the third angle of the triangle.)

13 Construct an isosceles triangle, given the vertex angle and the altitude to the base.

14 For each given set, construct a triangle.
  a $\{a, c, m_c\}$ **d** $\{a, b, h_c\}$
  b $\{A, B, h_a\}$ **e** $\{A, B, h_c\}$
  c $\{h_b, t_b, a\}$, $h_b < t_b < a$ **f** $\{a, c, h_c\}$

15 Construct an isosceles triangle equal in area to a given triangle.

16 Construct an equilateral triangle, given the altitude.

Problem Set C

17 Construct an isosceles right triangle equal in area to a given triangle.

18 Construct a right triangle, given the hypotenuse and the altitude to the hypotenuse.

19 For each given set, construct a triangle.
  a $\{B, C, t_b\}$ **b** $\{a, m_b, m_c\}$ **c** $\{h_a, m_a, B\}$

20 Construct $\triangle ABC$, given a, b, and the point on the given length b where t_b intersects side $\overline{AC}$.

21 By construction, divide a given scalene triangle into a triangle and a trapezoid such that the ratio of the area of the triangle to the area of the trapezoid is 1:8.

Problem Set D

22 Construct a triangle, given the three medians.

23 Construct a regular hexagon. Then construct an equilateral triangle whose area is equal to that of the hexagon.

Problem Set E

24 Given an acute angle with a point P in the interior of the angle, construct a circle that is tangent to the sides of the angle and passes through P.

Problem-Set Notes and
Additional Answers, continued

19c

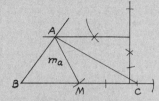

Construct $\angle B$ on the working line. Then construct the locus of points at a distance of h_a from the working line. They intersect at A. Using A as center and m_a as radius, find point M. Then find C such that $\overline{BM} \cong \overline{MC}$.

20 Use the Angle Bisector Theorem and construct the fourth proportional. Then use SSS.

21

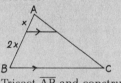

Trisect $\overline{AB}$ and construct a parallel to $\overline{BC}$.

22

23 Connect every other vertex of the hexagon. Construct a square whose side is equal to that of the equilateral triangle formed. The diagonal of the square is a side of the required equilateral triangle.

24

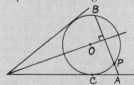

Point C can be found by using $AC^2 = AP \cdot AB$. (There is another circle based on the same principle.)

CHAPTER SUMMARY

CONCEPTS AND PROCEDURES

After studying this chapter, you should be able to
- Use the four-step locus procedure to solve locus problems (14.1)
- Apply the compound-locus procedure (14.2)
- Identify the circumcenter, the incenter, the orthocenter, and the centroid of a triangle (14.3)
- Identify the tools and procedures used in constructions (14.4)
- Interpret the shorthand notation used in describing constructions (14.4)
- Perform six basic constructions (14.4)
- Perform four other useful constructions (14.5)
- Construct triangles with given side lengths and angle measures (14.6)

VOCABULARY

center of gravity (14.3)
centroid (14.3)
circumcenter (14.3)
compound locus (14.2)
concurrent lines (14.3)

construction (14.4)
incenter (14.3)
locus (14.1)
orthocenter (14.3)

14 REVIEW PROBLEMS

Problem Set A

1 Given segment $\overline{AB}$, find the locus of points that are the vertices of isosceles triangles having $\overline{AB}$ as a base. The locus is the perpendicular bisector of $\overline{AB}$ (minus the midpoint of $\overline{AB}$).

2 Find the locus of the centers of all circles that pass through two fixed points. The locus is the perpendicular bisector of the segment joining the two points.

3 Find the locus of points 3 units from a given line and 5 units from a given point on the line. 4 points

4 What is the name of the surface *in space* every point of which is a fixed distance from a given line? Cylinder

5 What is the locus of points 2 in. from a circle with a radius of 2 in.? The locus is a point and a circle.

6 A circle of given radius rolls around the perimeter of a given equilateral triangle. Sketch the locus of its center.

7 Write the equation of the locus of points 5 units from the origin in the coordinate plane. $x_2 + y_2 = 25$

8 Given scalene $\triangle ABC$, construct each of the following.

  a The incenter   **b** The circumcenter   **c** The centroid   **d** The orthocenter

9 Given a 3-cm line segment, draw the locus of points 1 cm from the segment. (Each point of the locus must be 1 cm from the point of the segment nearest to it.)

10 Construct a parallelogram, given two sides and the angle they form.

11 Given a segment, construct an equilateral triangle with a perimeter equal to the segment's length.

12 What is the locus of points that are a fixed distance from a fixed point and equidistant from two given points. { }, 1 point, or two points

Chapter 14 Review
Class Planning

Time Schedule
All levels: 2 days

Resource References
Evaluation
  Tests and Quizzes
    Test 14, Series 1, 2, 3

Assignment Guide

Basic (optional)

Day 1	1–3
Day 2	5, 7, 10

Average

Day 1	1–12
Day 2	13, 15–24

Advanced

Day 1	13–24
Day 2	Page 681: 19–21
	Pages 684–685: 25, 27, 31

For end-of-course review, also see the Cumulative Review for Chapters **1–15,** pages 706–711.

Problem-Set Notes
and Additional Answers

■ See *Solution Manual* for answers to problems **6** and **8–11.**

Problem Set B

13 Write the equation of the locus of points for which the ordinate is 5 more than 3 times the abscissa. $y = 3x + 5$

14 What is the locus *in space* of points equidistant from all the points on a given circle? The locus is the line perpendicular to the plane of the circle at the circle's center.

15 Given segment $\overline{PQ}$, find the locus of points each of which is the intersection of the diagonals of a rectangle that has $\overline{PQ}$ as a base. The locus is the perpendicular bisector of $\overline{PQ}$ (minus the midpoint of $\overline{PQ}$).

16 Given a circle with center P and a radius of 10 cm, find the locus of the midpoints of all possible 12-cm chords in the circle. The locus is a circle with center P and a radius of 8 cm.

17 Find the locus of midpoints of all chords of a circle that have a fixed point of the circle as an endpoint. The locus is a circle that is tangent internally to the given circle at the fixed point and whose radius is half the original circle's but which excludes the fixed point.

18 A point outside a square 3 units on a side moves so that it is always 2 units from the point of the square nearest to it. Find the area enclosed by the locus of this moving point. $33 + 4\pi$

19 If the radius of a given circle is 10 cm, describe the locus of points 2 cm from the circle and equidistant from the endpoints of a given diameter of the circle. 4 points

20 Using coordinate-geometry methods, find the locus of points 5 units from the origin and 4 units from the y-axis. $\{(4, 3), (4, -3), (-4, 3), (-4, -3)\}$

21 Inscribe a regular octagon in a given circle.

22 Construct a parallelogram, given two sides and an altitude.

23 Explain how to construct an angle with each measure.
a 30   **b** $18\frac{3}{4}$

24 Given three points A, B, and C, describe the locus of points that are equidistant from A and B and also equidistant from B and C. { } or 1 point

Problem Set C

25 Find the locus of the intersections of the diagonals of all possible rhombuses having a fixed segment $\overline{PQ}$ as a side. The locus does not include P and Q but otherwise is a circle with $\overline{PQ}$ as diameter.

26 Prove that the angle bisectors of a kite are concurrent.

27 Given a chord of a circle, construct another chord parallel to the given chord and half its length.

Problem-Set Notes and Additional Answers, continued

■ See *Solution Manual* for answers to problems **21–23**.

26

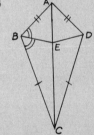

We know that $\overline{AC}$ bisects two ∠s of the kite. Bisect ∠ABC, with the bisector intersecting $\overline{AC}$ at E. Draw $\overline{ED}$. $\overline{ED}$ can be shown to be the bisector of ∠ADC.

27

$\overline{AB}$ is the required chord.

28 Given scalene △ ABC, construct in the exterior of the triangle a circle that is tangent to one side and to extensions of the other two sides.

29 Construct a square whose area is equal to the sum of the areas of two given squares.

30 Given two parallel lines and a point P between them, construct a circle that is tangent to both lines and passes through point P.

31 Inscribe a square in a given rhombus.

Problem Set D

32 Given two circles, construct a common external tangent.

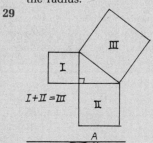

CAREER PROFILE

DARKNESS VISIBLE

Anne Dunn looks at the geometry of things unseen

Objects emit radiant energy in the form of electromagnetic waves. Normally we can see only about 3 percent of that energy, the energy we call visible light. Our eyes are not sensitive to X-rays, ultraviolet rays, and many other types of waves that, along with visible light, make up the electromagnetic spectrum. How, then, can we know what an object really looks like?

Optical engineers design instruments sensitive to different parts of the spectrum. Anne Dunn, a senior engineer with Nichols Research Corporation in Huntsville, Alabama, specializes in *infrared* optics. Infrared waves are longer than visible waves. An infrared-sensitive instrument can produce an image at night when there is no visible light.

"My main interest is in radiometry," says Dunn. "In radiometry we measure the strengths and characteristics of faint signals in the infra-red part of the spectrum." She explains that geometry is an important component of optics. "For example, the formula for magnification is derived from similar triangles. By using two sensors each measuring the rate and trajectory of a moving object, I can use triangulation to deduce the object's location."

Dunn, an Urbana, Illinois, native, majored in physics at Beloit College in Beloit, Wisconsin. She earned a master's degree and a doctorate in astrophysics at Rensselaer Polytechnic Institute in Troy, New York.

According to Dunn, a popular misconception about optics concerns magnification. "Often I look at essentially dimensionless point sources, which cannot be magnified. For seeing such an object, the light-gathering ability of a telescope is much more important than its magnification." The light-gathering power of a telescope is the ratio of the area of its objective lens to the area of the pupil of the eye of the observer. Find the light-gathering power of a telescope with a circular objective lens 10 inches in diameter if the observer's pupil has a diameter of $\frac{1}{5}$ inch.

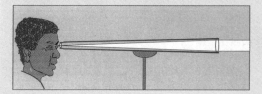

Problem-Set Notes and Additional Answers, continued

28 Bisect the two exterior angles and drop a perpendicular to determine the radius.

29

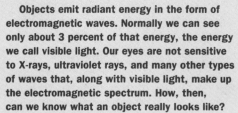

30

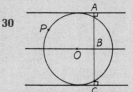

PO = AB and O is the center of the ⊙.

31 Bisect the right angles formed by the intersection of the diagonals. This gives the diagonals of the square required.

32

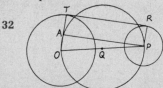

Construct ⊙Q, with Q at the midpoint of $\overline{OP}$ and radius $\overline{QO}$. Mark off $\overline{OA}$ such that OA = OT − PR. Extend $\overline{OA}$ to T. Make TR = AP. $\overline{TR}$ is the required common tangent.

Chapter 15 Schedule

Basic (optional)

Problem Sets:	4 days
Review:	4 days
Test:	1 day

Average (optional)

Problem Sets:	7 days
Review:	5 days
Test:	1 day

Advanced

Problem Sets:	5 days
Review:	3 days
Test:	1 day

SECTION SCHEDULES AND assignment guides for basic and average courses are considered optional for Chapter **15**, as most will have finished the term before reaching this chapter.

This chapter adds inequality to the geometric relations that the students have learned (i.e., congruence, similarity, equality) and uses this concept to explore dynamic properties of triangles and other polygons.

15 | INEQUALITIES

How these sprinklers in a painting by David Hockney represent inequalities may seem a mystery. But consider the consequences of changing the angle of the spray.

15.1 NUMBER PROPERTIES

Objective

After studying this section, you will be able to
- Use algebraic properties of inequality to solve inequalities

Part One: Introduction

The statement $a < b$ (a is less than b) is an inequality involving the numbers a and b. $a < b$ is equivalent to $b > a$ (b is greater than a).
Here is a review of some of the properties of inequality.

Postulate   *For any two real numbers x and y, exactly one of the following statements is true: $x < y$, $x = y$, or $x > y$. (**Law of Trichotomy**)*

Postulate   *If $a > b$ and $b > c$, then $a > c$. Similarly, if $x < y$ and $y < z$, then $x < z$. (**Transitive Property of Inequality**)*

If the lengths of $\overline{AB}$, $\overline{PQ}$, and $\overline{XY}$ are such that $AB < PQ$ and $PQ < XY$, then $AB < XY$.

A ———— B

P ———————— Q

X ——————————— Y

Postulate   *If $a > b$, then $a + x > b + x$. (**Addition Property of Inequality**)*

If $4 > -7$, then $4 + 9 > -7 + 9$
                          $13 > 2$

Section 15.1   Number Properties   **687**

Class Planning

Time Schedule
Basic: 2 days (optional)
Average: 2 days (optional)
Advanced: 1 day

Resource References
Teacher's Resource Book
   Class Opener 15.1A, 15.1B

Class Opener

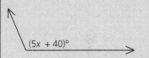

$(5x + 40)°$

Find all values of x for which the angle shown is obtuse.
$10 < x < 28$

Cooperative Learning

Give groups of students the following problems.

a $x + 7 < 5$	**e** $3x < 12$
b $x + 3 \geq 17$	**f** $-16x \geq 48$
c $x - 6 \leq 10$	**g** $\frac{x}{2} > 12$
d $x - 8 > 12$	**h** $\frac{x}{-4} \leq 9$

Ask students to work together to solve the problems and check their solutions. After students complete their work, each group should be able to write rules for solving inequalities. All members of each group should understand the procedure.

a $x < -2$	**e** $x < 4$
b $x \geq 14$	**f** $x \leq -3$
c $x \leq 16$	**g** $x > 24$
d $x > 20$	**h** $x \geq -36$

Postulate   *If $x < y$ and $a > 0$, then $a \cdot x < a \cdot y$. (Positive Multiplication Property of Inequality)*

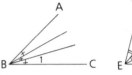

$m\angle 1 < m\angle 2$
$3 \cdot m\angle 1 < 3 \cdot m\angle 2$
$\angle ABC < \angle DEF$

When we say that one angle is greater than (or less than) another angle, we refer to their measures. Thus "$\angle ABC < \angle DEF$" means that $m\angle ABC < m\angle DEF$.

Postulate   *If $x < y$ and $a < 0$, then $a \cdot x > a \cdot y$. (Negative Multiplication Property of Inequality)*

Notice that the direction of the inequality sign reverses.

$-5x < 15$
$-\frac{1}{5}(-5x) > -\frac{1}{5}(15)$
$x > -3$

Part Two: Sample Problem

Problem   *A given angle is greater than twice its supplement. Find the possible measures of the given angle.*

Solution   Let x = the measure of the given angle and $180 - x$ = the measure of the supplement.

$$x > 2(180 - x)$$
$$x > 360 - 2x$$
$$x + 2x > 360 - 2x + 2x \quad \text{Addition Property of Inequality}$$
$$3x > 360$$
$$\tfrac{1}{3}(3x) > \tfrac{1}{3}(360) \qquad \text{Positive Multiplication Property of Inequality}$$
$$x > 120$$

Thus, the given angle is greater than 120°. Since it has a supplement, the given angle is also less than 180°. Therefore, $120 < x < 180$.

Part Three: Problem Sets

Problem Set A

1 Solve each inequality for x.

a $\frac{3}{5}x > 15$ $\{x \mid x > 25\}$

b $5x - 4 > 26$ $\{x \mid x > 6\}$

c $-4x \le 28$ $\{x \mid x \ge -7\}$

d $10 - x < 8x - (2x - 3)$ $\{x \mid x > 1\}$

2 a If $x + y < 30$ and $y = 12$, what is true about x? $x < 18$

   b If $x + y = 30$ and $y < 12$, what is true about x? $x > 18$

3 If x exceeds y by 5 and y exceeds z by 3, how is x related to z? x exceeds z by 8.

4 If x is twice y and y is three times z, how is x related to z? x is 6 times z.

5 If $\angle A = \angle 1 + \angle 2$, what is the relation between $\angle A$ and $\angle 2$? $\angle A > \angle 2$

6 a If X is between P and Q, how is PX related to PQ? $PX < PQ$

   b If X is the midpoint of $\overline{PQ}$, write the relation between PX and PQ as an inequality. $PX < PQ$

   c Using the situation in part **b**, write the relation between PX and PQ as an equality. $2(PX) = PQ$

Problem Set B

7 The complement of an angle is smaller than the angle. Find the restrictions on the measure of the original angle. $45 < x < 90$

8 If $\angle X < \angle Y$, what is the relation between their complements? Complement of $\angle X >$ complement of $\angle Y$

9 If $\frac{1}{x} > 5$, what two numbers is x between? $0 < x < \frac{1}{5}$

10 An angle is greater than twice its complement. Find the restrictions on the angle and on the complement.
The angle is between 60° and 90°; the complement is between 0° and 30°.

11 If $x \not< 3$ and $x \neq 3$, what can be concluded about x? $x > 3$

12 a What is the relation between an exterior angle of a triangle and the two remote interior angles? Equal

   b What, then, is the relation between an exterior angle and one of the remote interior angles? Exterior > interior

13 Given: $\angle ABC > \angle ACB$,
   $\overrightarrow{BD}$ bisects $\angle ABC$.
   $\overrightarrow{CD}$ bisects $\angle ACB$.
   Find and justify the relation between $\angle DBC$ and $\angle DCB$.
   $\angle DBC > \angle DCB$

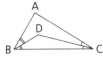

Problem Set C

14 Given: Real numbers a, b, and c, with $a > b$
   Prove: $c - a < c - b$ (Special Subtraction Property)

15 Solve $x^2 + x < 6$. $\{x : -3 < x < 2\}$

16 If $x > 3y + 7$ and $y > 6 - x$, find the restrictions on
   a x $x > \frac{25}{4}$ **b** y y is any real number.

14
$$a > b$$
$$a - c > b - c$$
$$(-1)(a - c) < (-1)(b - c)$$
$$c - a < c - b$$

15 Suppose $x^2 + x = 6$.
$$x^2 + x - 6 = 0$$
$$(x - 2)(x + 3) = 0$$
$$x = 2 \text{ or } x = -3$$

Test the three regions:
If $x < -3$, then $x^2 + x > 6$.
If $-3 < x < 2$, then
$x^2 + x < 6$.
If $x > 2$, then $x^2 + x > 6$.
Thus. $-3 < x < 2$.

16

$$\begin{cases} x > 3y + 7 \\ y > 6 - x \end{cases}$$
$$x - 7 > 3y$$
$$\frac{x - 7}{3} > y$$
$$\therefore \frac{x - 7}{3} > 6 - x$$
$$x > \frac{25}{4}$$
y is any real number.

17 $\begin{cases} x = 1.2y \\ y = 1.2z \end{cases}$
$$x = 1.2(1.2z)$$
$$x = 1.44z$$

18 $18 - 3x > 3$
$$x < 5$$
$\{2, 4\}$

Problem-Set Notes and Additional Answers, continued

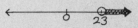
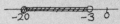

Problem Set C, *continued*

17 If x exceeds y by 20% and y exceeds z by 20%, by what percentage does x exceed z? 44

18 Solve $18 - 3x > 3$ over the positive even integers. {2, 4}

19 ∠A is greater than its complement, and the complement of ∠A is greater than ∠B.

  a Compare the complement of ∠A with the complement of ∠B. Comp. ∠B > comp. ∠A

  b Compare the complement of ∠B with ∠A. Comp. ∠B > ∠A

  c List ∠A, ∠B, and their complements in order of size, from largest to smallest. Comp. ∠B, ∠A, comp. ∠A, ∠B

Problem Set D

20 Solve $|2x - 7| > |x + 20| - 4$ for x. $\{x : x < -3 \text{ or } x > 23\}$

CAREER PROFILE

DEDUCTIONS FROM SEISMIC WAVES
Yvonna Pardus unravels the mysteries contained in rocks

The earth's crust is composed of a complex series of rock layers, or *strata*, one pile atop the next. Within the rock are spaces that are filled with petroleum, the end product of the decay of plants that were deposited there millions of years ago. To find the petroleum, some of which may be thousands of feet below the surface, petroleum geologists set off explosive charges. They then record the reflected seismic waves at receivers distributed over a certain area.

Geologist Yvonna Pardus explains: "Seismic waves act like light waves. They reflect off discontinuities between rock strata. By analyzing the angle of reflection, and other characteristics of the reflected wave, we can learn a great deal about the nature of the rock the wave has traveled through."

Pardus points out that the velocity of the wave depends on the density of the rock. Compiling a complete picture of what Pardus calls "the subsurface geometry of the earth" requires a series of seismic shots and reflections recorded on as many as forty-eight receivers.

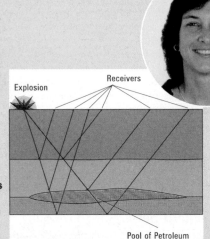

Yvonna Pardus attended Murray State University in Murray, Kentucky, where she earned a bachelor's degree in geology. A geologist needs a strong background in mathematics and physics, she says.

15.2 INEQUALITIES IN A TRIANGLE

Objectives

After studying this section, you will be able to
- Apply the Triangle Inequality Postulate
- Apply the Exterior-Angle-Inequality Theorem
- Use the Pythagorean Theorem test to classify a triangle as acute, right, or obtuse
- Recognize the relationships between the side lengths and the angle measures of a triangle

Part One: Introduction

The Triangle Inequality Postulate

The following postulate is a formal expression of an idea we have been using throughout this book.

Postulate   ***The sum of the measures of any two sides of a triangle is always greater than the measure of the third side.***

In other words, traveling from A to B along $\overline{AB}$ is shorter than going first to X along $\overline{AX}$ and then to B along $\overline{XB}$—that is, AX + XB > AB.

Exterior Angle Inequality

The following theorem, introduced in Section 5.2, can now be more easily proved.

Theorem 30   ***The measure of an exterior angle of a triangle is greater than the measure of either remote interior angle. (Exterior-Angle-Inequality Theorem)***

Given: △ABC, with exterior ∠1

Prove: ∠1 > ∠A and ∠1 > ∠B

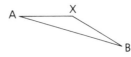

Proof: In Chapter 7 you learned that ∠1 = ∠A + ∠B. Clearly, ∠A + ∠B > ∠A, so ∠1 > ∠A by substitution. In a similar manner, ∠A + ∠B > ∠B, so ∠1 > ∠B.

Class Planning

Time Schedule
Basic: 2 days (optional)
Average: 2 days (optional)
Advanced: 2 days

Resource References
Teacher's Resource Book
   Class Opener 15.2A, 15.2B
   Supposer Worksheet 15
   Additional Practice Worksheet 29
Transparency 24
Evaluation
   Tests and Quizzes
     Quiz 1, Series 3

Class Opener

Given: △PQR ~ △FMT,
     PQ = 8, QR = 12,
     FM = 20, FT = 60
Find: PR and MT
By similarity, PR = 24 and MT = 30; therefore, △PQR and FMT are impossible by the Triangle Inequality Postulate.

Lesson Notes

- Note that the topics dealt with in this section have all been presented previously. Triangle inequality was previewed in Section **1.3** (page 19); the Exterior-Angle-Inequality Theorem was developed in Section **5.2** (page 216); the "Pythagorean" test for acute, right, and obtuse triangles was presented in Section **9.4** (p. 385); and Theorems 132 and 133 were introduced in Section **3.7** (page 149).

Classifying Triangles

As you discovered in Section 9.4, the converse of the Pythagorean Theorem can be used to prove that a triangle is a right triangle. You may recall that it also suggested the following way of finding whether a triangle is acute, right, or obtuse.

> ### The Pythagorean Theorem Test
>
> To classify a triangle as acute, right, or obtuse, compute a^2, b^2, and c^2, where c is the longest of the three sides a, b, and c.
>
> If $a^2 + b^2 = c^2$, then $\triangle ABC$ is right ($\angle C$ is right).
> If $a^2 + b^2 > c^2$, then $\triangle ABC$ is acute.
> If $a^2 + b^2 < c^2$, then $\triangle ABC$ is obtuse ($\angle C$ is obtuse).

Side and Angle Relationships

The following theorems, the inverses of Theorems 20 and 21, were presented in Chapter 3. Now we are in a position to prove them.

Theorem 132 *If two sides of a triangle are not congruent, then the angles opposite them are not congruent, and the larger angle is opposite the longer side. (If △, then △.)*

Given: $\triangle ABC$,
    $AC > AB$
Conclusion: $\angle B > \angle C$

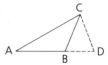

Proof: Since $AC > AB$, extend $\overline{AB}$ to a point D so that $AD = AC$. Draw $\overline{DC}$.
    $\angle ABC > \angle D$ by the Exterior-Angle-Inequality Theorem.
    $\angle D \cong \angle ACD$ (If △, then △.)
    $\angle ABC > \angle ACD$ by substitution.
    $\angle ACD > \angle ACB$ (See diagram.)
    $\therefore \angle ABC > \angle ACB$ by the Transitive Property of Inequality.

Theorem 133 *If two angles of a triangle are not congruent, then the sides opposite them are not congruent, and the longer side is opposite the larger angle. (If △, then △.)*

Given: $\triangle ABC$,
    $\angle B > \angle C$
Conclusion: $AB < AC$

Proof: According to the Law of Trichotomy there are exactly three possible conclusions: AB > AC, AB = AC, or AB < AC. We must test them.

Case 1: If AB > AC, then by Theorem 132, ∠C > ∠B, which contradicts the given information.
Thus, AB > AC cannot be the correct conclusion.
Case 2: If AB = AC, then ∠C ≅ ∠B (If △, then △.)
The given information is again contradicted.
Thus, AB = AC cannot be the correct solution.

All that is left is AB < AC, which must be true by the Law of Trichotomy.

A simple extension of Theorem 133 enables us to say that the longest side of any triangle is the side opposite the largest angle.

Part Two: Sample Problems

Problem 1   *Does a triangle with sides 2, 5, and 10 exist?*

Solution   The sum of any two sides must be greater than the third side, and 2 + 5 ≯ 10. Therefore, the answer is no.

Problem 2   *Find the restrictions on ∠A.*

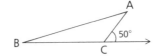

Solution   50 > m∠A because an exterior angle of a triangle exceeds either remote interior angle. An angle of a triangle must be greater than 0°, so m∠A > 0. Thus, 0 < m∠A < 50.

Problem 3   *In △ABC, ∠A = 40° and ∠B = 65°. List the sides in order of their lengths, starting with the smallest.*

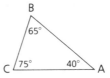

Solution   Draw a diagram listing all the angles. (∠C is easily found to be 75°.) The shortest side, $\overline{BC}$, is opposite the smallest angle, ∠A. The longest side, $\overline{BA}$, is opposite the largest angle, ∠C. Therefore, the correct order is $\overline{BC}$, $\overline{AC}$, $\overline{BA}$.

Communicating Mathematics

Refer to problem **9** on page 695. Suppose that ∠B is between 40° and 88°. Explain how to find the possible measures of vertex ∠A.
4 < m∠A < 100

Assignment Guide

Basic

Day 1   2a,b, 3, 5a
Day 2   2c,d, 5b, 19

Average

Day 1   1–6, 8
Day 2   9–11, 14, 18, 19

Advanced

Day 1   6–14, 25
Day 2   16–22

Problem-Set Notes and Additional Answers

■ See Solution Manual for answers to problems **4, 6,** and **15–18.**

20b Use trigonometry.
In $\triangle BDC$, $\sin 70° = \frac{BD}{BC}$.
In $\triangle ABC$, $\tan 20° = \frac{AC}{BC}$.
Since $\sin 70° > \tan 20°$,
$BD > AC$.

21

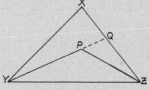

By the Triangle Inequality Postulate,
$$XY + XQ > YP + PQ$$
$$PQ + QZ > PZ$$
$$\overline{XY + XQ + QZ + PQ > YP + PQ + PZ}$$
$$\therefore XY + XZ > YP + PZ$$

22

By the Triangle Inequality Postulate,
(1) $z < x + y$
(2) $x + z > y$
    $z > y - x$
(3) $y + z > x$
    $z > x - y$
Thus, $|x - y| < z < x + y$

Problem 4   Given: $\triangle ABC$, with $AB < AC$;
$\overrightarrow{BD}$ bisects $\angle ABC$.
$\overrightarrow{CD}$ bisects $\angle ACB$.

Prove: $BD < DC$

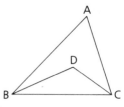

Proof

1 $AB < AC$	1 Given
2 $\angle ABC > \angle ACB$	2 If ⟁, then ⟁.
3 $\overrightarrow{BD}$ bisects $\angle ABC$.	3 Given
4 $\overrightarrow{CD}$ bisects $\angle ACB$.	4 Given
5 $\angle DBC > \angle DCB$	5 Positive Multiplication Property of Inequality (by $\frac{1}{2}$)
6 $BD < DC$	6 If ⟁, then ⟁.

Part Three: Problem Sets

Problem Set A

1 What are the restrictions on $\angle 1$?
$80 < m\angle 1 < 180$

2 Which of these sets can be the lengths of sides of a triangle? **b** and **d**

   a 3, 6, 9   **b** 4, 5, 8   **c** 2, 3, 8   **d** $\sqrt{2}, \sqrt{3}, \sqrt{6}$

3 In $\triangle PQR$, $\angle P = 67°$ and $\angle Q = 23°$.
   a Name the shortest and the longest side. $\overline{PR}$; $\overline{PQ}$
   b What name is given to side $\overline{PQ}$? Hypotenuse

4 Given: $AB > BC$, $BC > AC$
   Prove: B is the smallest angle in $\triangle ABC$.

5 Name the longest segment in each diagram.

   a $\overline{AB}$

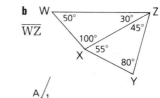

   b $\overline{WZ}$

6 Given: $\angle 1$ is an exterior angle of $\triangle ACD$.
       $\angle 2$ is an exterior angle of $\triangle ABC$.
   Prove: $\angle 1 > \angle 3$

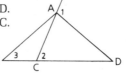

Problem Set B

7 A scalene triangle has a 60° angle. Is this angle opposite the longest, the shortest, or the other side? **The other**

8 The sides of a triangle are 14, 6, and x. Find the set of possible values of x. **{x:8 < x < 20}**

9 Vertex angle A of isosceles triangle ABC is between 40° and 88°. Find the possible values for ∠B. **46 < m∠B < 70**

10 a Name the longest segment in the figure below. **$\overline{AD}$**

b Name the shortest segment in the figure below. **$\overline{XY}$**

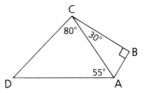

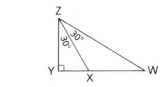

11 a List the angles in order of size, beginning with the smallest. **R, Q, P**

b At which vertex is the exterior angle the largest? **R**

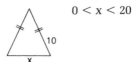

12 Find the restrictions on x.

a

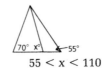

55 < x < 110

b

0 < x < 90

c **0 < x < 20**

d **x = 7.5√3**

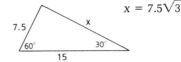

13 A stick 8 cm long is cut into three pieces of integral lengths to be assembled as a triangle. What is the length of the shortest piece? **2 cm**

14 For each set of numbers, tell whether the numbers represent the lengths of the sides of an acute triangle, a right triangle, an obtuse triangle, or no triangle.

a 12, 13, 14   **b** 11, 5, 18   **c** 9, 15, 18   **d** $\frac{1}{2}, 1\frac{1}{5}, 1\frac{3}{10}$
 Acute   None   Obtuse   Right

15 Prove that an altitude of an acute triangle is shorter than either side that is not the base.

Problem-Set Notes and Additional Answers, continued

23

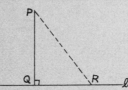

Let $\overline{PQ}$ be the ⊥ to line ℓ and $\overline{PR}$ be any other segment. Then △PQR is a right △ with hypotenuse $\overline{PR}$. Thus, PQ < PR.

24

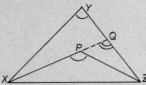

Consider △PQZ. ∠XPZ > ∠PQZ by the Exterior-Angle-Inequality Theorem. Consider △XQY. ∠XQZ > ∠Y by the Exterior-Angle-Inequality Theorem. ∴ ∠XPZ > ∠Y

Cooperative Learning

Problem **25**, page 696, is a good problem for students to work cooperatively. It may be helpful for the students to have a foldable cardboard model of the rectangular prism.

25   Abigail:

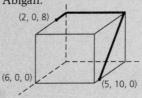

(2, 0, 8) to (0, 0, 8) to (0, 10, 8) to (5, 10, 0)

Ben:

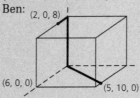

(2, 0, 8) to (0, 0, 8) to (0, 0, 0) to (5, 10, 0)

T 695

Carol:

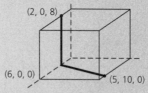

(2, 0, 8) to (2, 0, 0) to
(5, 10, 0)
Deanna unfolded part of
the room in her mind so
that the ceiling, one wall,
and the floor were coplanar.

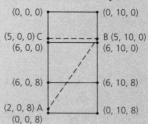

Deanna thought the path of
the spider should be $\overline{AB}$.
$(AC)^2 + (BC)^2 = (AB)^2$
$13^2 + 10^2 = (AB)^2$
$AB \approx 16.40$
Deanna's path was that
shown below, with
$P = (6, \approx 3.08, 8)$ and
$Q = (6, \approx 9.23, 0)$.

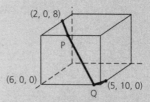

If students explore opening
the box in other ways, they
will find that there is an
even shorter path,
measuring only ≈ 16.28, as
shown below.

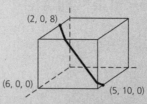

Problem Set B, *continued*

16 Prove that if ABCD is a quadrilateral, then AB + BC + CD > AD.

17 Given the diagram shown, prove or dis-
prove that $\angle 2 > \angle 1$.

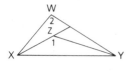

18 Given: $\overrightarrow{AC}$ bisects $\angle BAD$.
Prove: AD > CD

19 The pattern shown can be folded to form a
prism with a regular hexagonal base. Find,
to the nearest tenth of a unit, the prism's

  a Lateral surface area 714.0

  b Total surface area 968.6

  c Volume ≈ 2164.2

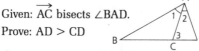

Problem Set C

20 $\angle ACB$ and $\angle CDB$ are right $\angle$s, and
$\angle B = 20°$.

  a List $\overline{AC}$, $\overline{CB}$, $\overline{AB}$, $\overline{AD}$, and $\overline{CD}$ in order
of size, starting with the smallest. $\overline{AD}$, $\overline{CD}$, $\overline{AC}$, $\overline{BC}$, $\overline{AB}$

  b Where would $\overline{DB}$ fit into this list? Either right before or
right after $\overline{AC}$ with current knowledge; actually AC < DB < BC.

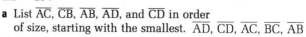

21 Given a point P in the interior of $\triangle XYZ$, prove that
PY + PZ < XY + XZ.

22 If two sides of a triangle have lengths x and y, what is the range
of possible values of the length of the third side? $|x - y| <$ third side $< x + y$

23 Prove: The shortest segment between a point and a line is the
segment perpendicular to the line.

24 Given a point P in the interior of $\triangle XYZ$, prove that $\angle XPZ > \angle Y$.

25 Deanna watched a spider crawl over the
interior surfaces of a room from point
(2, 0, 8) to point (5, 10, 0). The next day,
she asked three of her classmates if they
knew the length of the shortest path the
spider could have taken.
Abigail said, "$12 + \sqrt{89} \approx 21.43$."
Ben said, "$10 + \sqrt{125} \approx 21.18$."
Carol said, "$8 + \sqrt{109} \approx 18.44$."
Deanna responded, "Actually, it was
≈ 16.40." Explain the reasoning of each student.

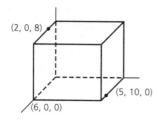

696 Chapter 15 Inequalities

THE HINGE THEOREMS

Objective

After studying this section, you will be able to
- Use the hinge theorems to determine the relative measures of sides and angles

Part One: Introduction

Thus far we have discussed inequalities involving the sides and the angles of a single triangle. We now turn our attention to two triangles. If the size of angle AHC is changed from that in Figure 1 to that in Figure 2, what happens to the length of a spring connecting A and C?

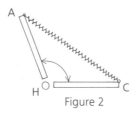

Figure 1   Figure 2

Theorem 134 **The Hinge Theorem:** *If two sides of one triangle are congruent to two sides of another triangle and the included angle in the first triangle is greater than the included angle in the second triangle, then the remaining side of the first triangle is greater than the remaining side of the second triangle. (SAS ≠)*

The following setup of Theorem 134 should help you see how the theorem can be applied.

Given: $\overline{AB} \cong \overline{XY}$,
$\overline{BC} \cong \overline{YZ}$,
$\angle B > \angle Y$
Conclusion: AC > XZ

The converse of the Hinge Theorem is also true.

Theorem 135 **The Converse Hinge Theorem:** *If two sides of one triangle are congruent to two sides of another triangle and the third side of the first triangle is greater than the third side of the second triangle, then the angle opposite the third side in the first triangle is greater than the angle opposite the third side in the second triangle. (SSS ≠)*

Given: $\overline{AB} \cong \overline{WX}$,
$\overline{BC} \cong \overline{XY}$,
AC > WY
Conclusion: $\angle B > \angle X$

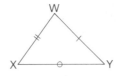

Section 15.3   The Hinge Theorems   **697**

Class Planning

Time Schedule
Basic: Omit
Average: 3 days (optional)
Advanced: 2 days

Resource References
Teacher's Resource Book
   Class Opener 15.3A, 15.3B
   Additional Practice Worksheet 30
Evaluation
   Tests and Quizzes
   Quiz 2, Series 3

Class Opener

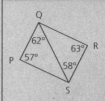

List the five segments in the figure in order, from shortest to longest.
In △PQS, QS < PQ < PS.
In △QSR, QR < SR < QS.
Therefore the order is $\overline{QR}$, $\overline{SR}$, $\overline{QS}$, $\overline{PQ}$, $\overline{PS}$.

Part Two: Sample Problems

Problem 1   Given: $\overline{BD}$ is a median.
               AD > CD

Which is greater, ∠1 or ∠2?

Solution   Since $\overline{BD}$ is a median, $\overline{AB} \cong \overline{BC}$.
Also, $\overline{BD} \cong \overline{BD}$ and AD > CD.
Thus, by the Converse Hinge Theorem, ∠1 > ∠2.

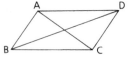

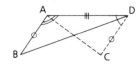

Problem 2   Given: △ABC is isosceles, with base $\overline{BC}$.
               C is the midpoint of $\overline{BD}$.

Prove: AD > AB

Proof

1 C is the midpoint of $\overline{BD}$.	1 Given
2 $\overline{BC} \cong \overline{CD}$	2 A midpoint divides a segment into two congruent segments.
3 △ABC is isosceles, with base $\overline{BC}$.	3 Given
4 $\overline{AB} \cong \overline{AC}$	4 The legs of an isosceles △ are ≅.
5 ∠1 > ∠2	5 Exterior-Angle-Inequality Theorem (∠1, △ABC)
6 AD > AC	6 Hinge Theorem (SAS≠)
7 AD > AB	7 Substitution (4 in 6)

Problem 3   Given: ABCD is a parallelogram.
               ∠BAD > ∠ADC

Which diagonal is longer, $\overline{AC}$ or $\overline{BD}$?

Solution   Consider the overlapping triangles as shown.

$\overline{AB} \cong \overline{DC}$ because the opposite sides of a parallelogram are ≅.
Also, $\overline{AD} \cong \overline{AD}$ and ∠BAD > ∠ADC.
So, BD > AC by the Hinge Theorem.

Part Three: Problem Sets

Problem Set A

1 Which is longer, $\overline{AC}$ or $\overline{DF}$? $\overline{AC}$

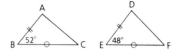

698   Chapter 15   Inequalities

2 Which is larger, ∠R or ∠Y? ∠R

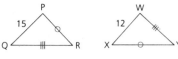

3 Given: $\overline{AB} \parallel \overline{CD}$, $\overline{AB} \cong \overline{AD}$,
∠DAC = 75°, ∠DCA = 45°
Which is longer, $\overline{BC}$ or $\overline{DC}$? $\overline{DC}$

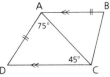

4 Compare AB in △ABC with XZ in △XYZ, where BC = 7,
AC = 9, ∠C = 75°, YZ = 7, XY = 9, and ∠Y = 80°. XZ > AB

5 Given: $\overline{WX} \cong \overline{WZ}$,
∠XWY > ∠ZWY
Prove: XY > ZY

6 Given: ⊙O,
AB < CD
Prove: ∠1 < ∠2

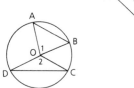

7 In △WXY, WX = 10, WY = 4, and XY = 7.
a Name the largest and the smallest angle. ∠Y; ∠X
b Is the triangle acute, right, or obtuse? Obtuse

Problem Set B

8 Given: ▱WXYZ, XZ > WY
Prove: **a** ∠XWZ > ∠WZY (Use a two-column proof.)
b ∠XWZ is obtuse. (Use a paragraph proof.)

9 Given: $\overline{PQ} \cong \overline{PR} \cong \overline{RS}$
Prove: QR < PS

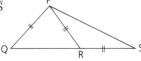

10 △WXY and △ABC are isosceles, with bases $\overline{WY}$ and $\overline{AB}$ respectively.
If ∠X and ∠B are each 50° and $\overline{WX} \cong \overline{BC}$, which triangle has
a The longer base? △ABC
b The longer altitude to the base? △WXY

11 Given: $\overline{AB}$ and $\overline{BC}$ are tangent to ⊙Q.
AD > DC
Conclusion: ∠ABD > ∠DBC

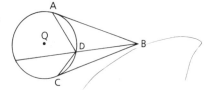

Communicating Mathematics

Refer to problem **3**. If ∠ACB = 65°, list the five segments in the figure in order, from longest to shortest. Then explain how you arrived at your answer.
In △ACD, DC > AC > AD.
In △ABC, AC > AB > BC.
Since AD = AB, the order is $\overline{DC}$, $\overline{AC}$, $\overline{AD}$ and $\overline{AB}$, $\overline{BC}$.

Problem-Set Notes and Additional Answers

■ See *Solution Manual* for answers to problems **5, 6, 8, 9,** and **11**.

19 A possible extension of this
problem is to ask, How
many of these triangles are
obtuse?
Since $700^2 + 800^2 \approx$
1063.014581^2, the 436
triangles with third-side
measures 1064–1499 are
obtuse.
Since $800^2 - 700^2 \approx$
387.2983346^2, the 287
triangles with third-side
measures 101–387 are
obtuse.
$436 + 287 = 723$

20 If AB > BC, then BM > BN,
so in △BMN, ∠BNM >
∠BMN. Since it follows that
∠OMN > ∠ONM by the
Special Subtraction Property
(Section **15.1**, problem **14**),
ON > OM.

21 △ABF ≅ △DEA by ASA.
$\overline{AB} \cong \overline{DE}$ by CPCTC.
$\overline{AB} \cong \overline{AD}$, so $\overline{DE} \cong \overline{AD}$ by
the Transitive Property.
Thus, △DEA is isosceles,
with leg $\overline{DE} \cong$ hypotenuse
$\overline{AD}$—an impossibility.

Problem Set B, *continued*

12 Given: ∠1 < ∠3,
$\overline{BA} \parallel \overline{CD}$,
AC > AD

Prove: BC > AD

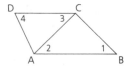

13 In △PQR, PQ = $1\frac{1}{2}$, QR = $2\frac{1}{2}$, and PR = 2. Is △PQR acute, right,
or obtuse? Right

14 In △ACE, AC < AE, and D is the midpoint of $\overline{CE}$. Point B is on
$\overline{AC}$, and point F is on $\overline{AE}$, with $\overline{CB} \cong \overline{FE}$. Prove that BD > FD.

15 $\overline{AD}$ is a median of △ABC, m∠ADC = 2x + 35, and
m∠ADB = 5x − 65.

  a Which side is longer, $\overline{AC}$ or $\overline{AB}$? $\overline{AC}$ **b** Which is larger, ∠B or ∠C? ∠B

16 Given: ∠C > ∠A, ∠D > ∠B

Prove: AB > CD

17 List ∠X, ∠Y, ∠XWY, ∠XWZ, and
∠XZW in order of size, starting with the
largest.
∠XZW, ∠X, ∠XWY, ∠Y, ∠XWZ

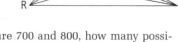

18 Given: QT > TR;
$\overrightarrow{QS}$ and $\overrightarrow{QT}$ trisect ∠PQR.
$\overrightarrow{RS}$ and $\overrightarrow{RT}$ trisect ∠PRQ.

Prove: PQ > PR

19 If two sides of a triangle measure 700 and 800, how many possible triangles exist such that all sides are integers? 1399

Problem Set C

20 Prove that if chords $\overline{AB}$ and $\overline{BC}$ are in
⊙O and AB > BC, then $\overline{AB}$ is closer to
the center than $\overline{BC}$. (Hint: Draw $\overline{MN}$.)

21 Given: ABCD is a square.
$\overline{AF} \perp \overline{DE}$, $\overline{AE} \cong \overline{BF}$

Which of the following is correct? **b**

  a DE < AF

  b The figure is overdetermined.

22 $\overline{WX}$ is a diameter of a circle with center P, and $\overline{YZ}$ is a diameter of a larger concentric circle. W, X, Y, and Z are noncollinear. Prove that ∠YWZ > ∠XYW.

23 Given: B, C, and D lie on plane m.
△BCD is isosceles, with base $\overline{CD}$.
∠ABD > ∠ABC

Conclusion: ∠ACD > ∠ADC

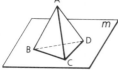

24 Given: $\widehat{AD} \cong \widehat{CD}$,
AE < EC

Prove: $\widehat{AB} < \widehat{BC}$

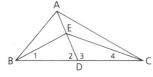

25 Given: $\overline{AD}$ is a median.
∠ABD > ∠ACD

Conclusion: ∠1 > ∠4

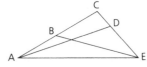

26 Given: AC > CE,
$\overline{AB} \cong \overline{DE}$

Prove: AD > BE

MATHEMATICAL EXCURSION

INEQUALITIES
A bicycle excursion

Triangle inequalities help explain why your bicycle feels and handles the way it does. Geometrically speaking, you can see from the diagram that a bicycle frame and rider form three triangles (one with an understood side) and a quadrilateral that is nearly a triangle. The properties of these triangles govern responsiveness, traction, riding position, and handling for a given bicycle. In general, racing bikes are more responsive, but handling is better on road bikes.

Some measurements that affect a bicycle's characteristics are shown on the diagram. *Chainstay length* affects uphill traction. *Wheelbase*, *fork rake*, and *head-tube angle* all determine how responsive the bicycle will be. A more responsive bike is more difficult to handle. *Top-tube length* determines how far the rider must lean over to reach the handlebars. Notice that the top tube and the rider's upper body and arms also form a triangle. Describe the effects on angles increasing and decreasing lengths of parts of the frame.

Problem-Set Notes and Additional Answers, continued

22

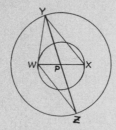

△WPZ ≅ △XPY by SAS, so $\overline{WZ} \cong \overline{YX}$. Now apply the Converse Hinge Theorem to △WYZ and △XYW.

23 Apply the Hinge Theorem to △ABD and △ABC to show that AD > AC. In △ADC, therefore, ∠ACD > ∠ADC (If ⧌, then ⧍).

24 ∠1 < ∠2 by the Converse Hinge Theorem applied to △ADE and △CDE. Thus, $\widehat{AB} < \widehat{BC}$.

25 Since ∠ABD > ∠ACD, AC > AB. Apply the Converse Hinge Theorem to △ADB and △ADC to show that ∠3 > ∠2. Then apply the Hinge Theorem to △EDB and △EDC to find that EC > EB. In △EBC, therefore, ∠1 > ∠4 (If ⧌, then ⧍).

26 In △ACE, ∠CEA > ∠CAE (If ⧌, then ⧍). Apply the Hinge Theorem to △BAE and △DEA to show that AD > BE.

15 | CHAPTER SUMMARY

CONCEPTS AND PROCEDURES

After studying this chapter, you should be able to
- Use algebraic properties of inequality to solve inequalities (15.1)
- Apply the Triangle Inequality Postulate (15.2)
- Apply the Exterior-Angle-Inequality Theorem (15.2)
- Use the Pythagorean Theorem test to classify a triangle as acute, right, or obtuse. (15.2)
- Recognize the relationships between the side lengths and the angle measures of a triangle (15.2)
- Use the hinge theorems to determine the relative measures of sides and angles (15.3)

Properties of Inequality
- For any two real numbers x and y, exactly one of the following statements is true: $x < y$, $x = y$, or $x > y$. (15.1)
- If $a > b$ and $b > c$, then $a > c$. Similarly, if $x < y$ and $y < z$, then $x < z$. (15.1)
- If $a > b$, then $a + x > b + x$. (15.1)
- If $x < y$ and $a > 0$, then $a \cdot x < a \cdot y$. (15.1)
- If $x < y$ and $a < 0$, then $a \cdot x > a \cdot y$. (15.1)
- If $a > b$, then $c - a < c - b$. (15.1)

Review Problems

Problem Set A

1 In △ABC, AB > AC > BC. List the angles in order, from smallest to largest. ∠A, ∠B, ∠C

2 In each case, decide which of the segments named is longest and state the reason for your decision.

a

$\overline{WZ}$ or $\overline{ZY}$?
$\overline{WZ}$; Hinge Thm.

b

$\overline{AB}$ or $\overline{BC}$?
$\overline{AB}$; Hypotenuse is longest side.

c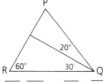

$\overline{PQ}$, $\overline{QR}$, or $\overline{PR}$? $\overline{QR}$: from size of ∠s in △PQR

3 In each case, tell which angle is largest and give the reason.

a

∠ABD or ∠CBD?
∠CBD; Converse Hinge Thm.

b

∠X, ∠Y, or ∠Z?
∠X; If △, then △.

c

∠1, ∠2, or ∠P?
∠1; Exterior-Angle-Ineq. Thm.

4 If x > 4 and x < y, what is the relationship between y and 4? y > 4

5 If x ≠ 6 and x ≮ 6, what can we conclude? x > 6

6 a Name all pairs of segments that we know to be congruent. $\overline{AE} \cong \overline{AB}$; $\overline{ED} \cong \overline{EC}$

b Which is shorter, $\overline{BE}$ or $\overline{EC}$? $\overline{BE}$

c What is the name of side $\overline{BC}$ in △BEC? Hypotenuse

d Which is longer, $\overline{AE}$ or $\overline{DE}$? $\overline{DE}$

e Which is the shortest segment in the figure? $\overline{BE}$

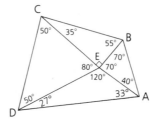

7 Which of these sets cannot represent the sides of a triangle? **a**

a 20, 40, 20   **b** 30, 40, 20   **c** 20, 20, 20   **d** 30, 40, 50

Chapter 15 Review
Class Planning

Time Schedule
Basic: 1 day
Average: 2 days
Advanced: 1 day

Resource References
Evaluation
  Tests and Quizzes
    Test 15, Series 1, 2, 3

Assignment Guide

Basic (optional)
1, 2b, 3b, 6a,b,c,d, 7

Average (optional)
Day 1 1–8
Day 2 9, 11, 13, 14, 16

Advanced
9–14, 16–20

18 (2, 5, 6), (3, 4, 6), (3, 5, 6), (4, 4, 6)

19

△ABC: AC > AB = BC;
AD > BD
△BCD: BC > CD > BD
△ACD: AC > CD > AD
Thus, the order is $\overline{AC}$, $\overline{AB}$ and $\overline{BC}$, $\overline{CD}$, $\overline{AD}$, $\overline{BD}$.

20

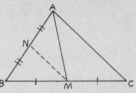

By the Triangle Inequality Postulate, in △AMN,
AM < AN + MN
∴ AM < BN + MN
$\overline{2(AM) < AN + BN + 2(MN)}$
Since 2(MN) = AC
and AN + BN = AB,
AM < $\frac{1}{2}$(AB + AC).

21

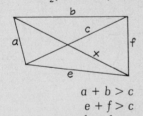

$a + b > c$
$e + f > c$
$b + f > x$
$a + e > x$
$\overline{2a + 2b + 2e + 2f > 2c + 2x}$
∴ $a + b + e + f > c + x$

22 In △WZX and △WZY, apply the Converse Hinge Theorem to show that ∠WZY > ∠WZX. Thus, ∠XZV > ∠YZV by the Special Subtraction Property (Section **15.1**, problem **14**).

Review Problem Set A, *continued*

8 Given: $\overline{PQ} \cong \overline{RS}$
Prove: PS > QR

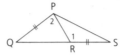

Problem Set B

9 Given △ABC as shown, list $\overline{AB}$, $\overline{AC}$, and $\overline{BC}$ in order of size, from longest to shortest.
$\overline{BC}$, $\overline{AC}$, $\overline{AB}$

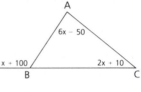

10 What are the restrictions on x?
60 < x < 150

11 △ABC is isosceles, with ∠C obtuse and AB = 6.
a Which side is longest? $\overline{AB}$
b The perimeter must be between what two numbers? $12 < P < 6 + 6\sqrt{2}$

12 In △ABC, AB > BC, m∠C = 4x − 4, and m∠A = x + 9. Find the minimum integral value of x. 5

13 A triangle has vertices P = (−1, −2), Q = (4, 1), and R = (6, −2).
a Find PQ, QR, and PR. $\sqrt{34}$; $\sqrt{13}$; 7
b Is △PQR acute, right, or obtuse? Obtuse
c List the angles in order of size, smallest first. ∠P, ∠R, ∠Q

14 Given: AB < AD,
$\overrightarrow{BE}$ bisects ∠ABC.
$\overrightarrow{DE}$ bisects ∠ADC.
Prove: ED > EB

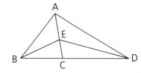

15 Given: $\overline{AB} \cong \overline{AD}$
Prove: AC < AD

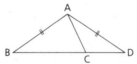

16 Given: PR < QS,
$\overline{PQ} \cong \overline{SR}$
Prove: ∠PQR < ∠SRQ

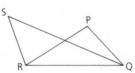

By the Hinge Theorem, in △VZX and △VZY, therefore, $\overline{XV} > \overline{YV}$.

17 Given: $\overline{BC} \cong \overline{EC}$,
$\angle A \cong \angle C$
Prove: AE < EC

Problem Set C

18 In an obtuse triangle, the side opposite the obtuse angle is 6. If all sides are integral, how many such triangles exist? 4

19 Given △ABC as shown, list the sides $\overline{AB}$, $\overline{AC}$, $\overline{AD}$, $\overline{BC}$, $\overline{BD}$, and $\overline{CD}$ in order, from longest to shortest.
$\overline{AC}$, $\overline{AB} \cong \overline{BC}$, $\overline{CD}$, $\overline{AD}$, $\overline{BD}$

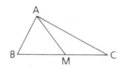

20 Prove that the measure of the median of a triangle is less than half of the sum of the measures of the two adjacent sides— that is, prove AM < $\frac{1}{2}$(AB + AC). (Hints: [1] Draw an appropriate midline, or [2] extend $\overline{AM}$ to point P so that $\overline{AM} \cong \overline{MP}$, then form □ABPC.)

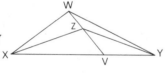

21 Prove that in any quadrilateral, the perimeter is greater than the sum of the diagonals.

22 Given: $\overline{ZX} \cong \overline{ZY}$,
WX < WY
Conclusion: XV > VY

23 The sides of triangle ABC are integers, with AB = 5 and AC = 13. If one of the possible values of BC is picked at random, what is the probability that the resulting triangle will be obtuse? $\frac{7}{9}$

24 Given: Quadrilateral PQRS,
PQ > PS,
$\angle Q \cong \angle S$
Prove: RS > RQ

25 P is any point inside quadrilateral WXYZ. Prove that the sum of the distances from P to the four vertices (PW + PX + PY + PZ) is greater than or equal to the sum of the diagonals. (Consider all three cases for the position of point P within the quadrilateral.)

Review Problems | **705**

Problem-Set Notes and Additional Answers, continued

23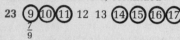
$\frac{7}{9}$

24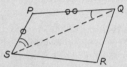

In △PSQ, $\angle$PSQ > $\angle$PQS. Since $\angle$PQR $\cong$ $\angle$PSR, $\angle$QSR < $\angle$SQR by the Special Subtraction Property (Section **15.1**, problem **14**). Thus, by "If △, then △," in △SQR, RS > RQ.

25 *Case I: P on neither diagonal*

In △PXZ, PX + PZ > XZ. In △PWY, PW + PY > WY. ∴ PX + PZ + PW + PY > XZ + WY

Case II: P on one diagonal

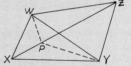

In △PWY, PW + PY > WY. Also, PX + PZ = XZ. ∴ PW + PY + PX + PZ > WY + XZ

Case III: P on both diagonals

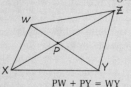

PW + PY = WY
PX + PZ = XZ
———————————
PW + PY + PX + PZ = WY + XZ
∴ PW + PY + PX + PZ ≥ WY + XZ

T 705

CUMULATIVE REVIEW
CHAPTERS 1–15

Problem Set A

1 Find the volume and the total surface area of a circular cone
  with a height of 4 and a base radius of 3. $V = 12\pi, A = 24\pi$

2 If you graphed the equation $2x + 3y = 12$,
  a What would the graph's x-intercept be? 6
  b What would the graph's slope be? $-\frac{2}{3}$
  c Would the point $(37, -21)$ lie on the graph? No

3 Given: Concentric circles, with $\overset{\frown}{CD} = 70°$
       and $\overset{\frown}{QR} = 54°$
  Find: $\overset{\frown}{AB}$ 16°

4 How far from the center of a circle with a diameter of 26 is a
  chord with a length of 24? 5

5 If the length of tangent $\overline{JM}$ is 10 and
  JP = 4, find PQ. 21

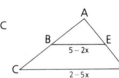

6 Solve for x. 3

7 In $\triangle ADC$, $\overline{BE} \parallel \overline{CD}$, AB = 8,
  BC = 4, AE = 6, and BE = 9.
  a Find DE. 3
  b Find CD. $13\frac{1}{2}$
  c Is $\triangle ABE$ a right triangle? No

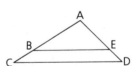

8 Given: ABCD is a trapezoid,
       with AD $\parallel$ BC.
  Prove: AE · BE = DE · EC

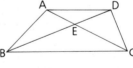

9 B and E are midpoints of $\overline{AC}$ and $\overline{AD}$
  respectively. Find CD. 42

10 If a base angle of an isosceles triangle is twice the vertex angle, then find the measure of the vertex angle. **36**

11 Draw a graph of △ABC with vertices A = (3, 8), B = (8, −4), and C = (−6, −4).

   a Find the lengths and the slopes of $\overline{AB}$, $\overline{BC}$, and $\overline{AC}$. **13, 14, 15; $-\frac{12}{5}, 0, \frac{4}{3}$**

   b Is △ABC acute, right, or obtuse? **Acute**

   c Find the equation of $\overleftrightarrow{AC}$ and its x- and y-intercepts. **$y = \frac{4}{3}x + 4$; −3; 4**

   d Find the equation of $\overleftrightarrow{BC}$. **$y = -4$**

   e Where does the altitude to $\overline{BC}$ intersect $\overline{BC}$? **(3, −4)**

   f What is the equation of the altitude to $\overline{BC}$? **x = 3**

   g Find the length of the altitude to $\overline{BC}$. **12**

   h Find the midpoint of $\overline{CB}$ and the slope of the median to $\overline{CB}$. **(1, −4); 6**

   i Find the area of △ABC. **84**

12 Two regular pentagons have areas 8 and 18. What is the ratio of their perimeters? **2:3**

13 Each interior angle of a regular polygon is 160°. Find the number of diagonals. **135**

14 Find the area of the sector formed by the hands of a clock at 2 o'clock if the diameter of the clock is 12 in. **6π sq in.**

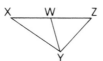

15 Find the area of an equilateral triangle whose height is 6. **$12\sqrt{3}$**

16 Write the converse, the inverse, and the contrapositive of the statement, "If a parallelogram is inscribed in a circle, then it is not a 'plain old parallelogram.'"

17 Given: ∠X ≇ ∠Z;
       W is the midpoint of $\overline{XZ}$.
   Prove: $\overline{WY}$ is not an altitude to $\overline{XZ}$.

Problem Set B

18 Find, to the nearest tenth,

   a The area of the shaded region (a half washer) **≈94.2**

   b The figure's perimeter (Hint: There are two semicircles and two segments.) **≈43.4**

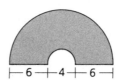

Problem-Set Notes and Additional Answers, continued

16 *Converse:* If a parallelogram is not a "plain old parallelogram," then it is inscribed in a circle. *Inverse:* If a parallelogram is not inscribed in a circle, then it is a "plain old parallelogram." *Contrapositive:* If a parallelogram is a "plain old parallelogram," then it is not inscribed in a circle.

■ *See Solution Manual for answer to problem* **17**.

Cumulative Review Problem Set B, *continued*

19 Given: $\overset{\frown}{BD} = \overset{\frown}{CE} = 80°$,
   $\angle CAB = 75°$

   Find: $\overset{\frown}{BE}$ 175°

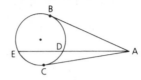

20 Given: $\overline{BD}$ is a diameter.
   $\overset{\frown}{AB} = 10°$, $\angle C = 40°$,
   $\angle GFC = 80°$

   Find: **a** $\overset{\frown}{CD}$ 110°
      b $\overset{\frown}{ED}$ 90°

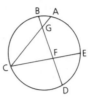

21 a Find RS.   11
   b Find $\overset{\frown}{QTS}$.   180°

22 Find x. $3\sqrt{5}$

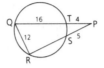

23 Find the area of ABCD.   36

24 ABCD is a square with a side of 12. The
   midpoints of the sides of the square are
   the centers of arcs tangent to the
   diagonals. Find the shaded area. $72 - 18\pi$

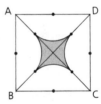

25 The vertices of △ABC are A = (5, 4), B = (11, 6), and C = (9, 10).
   a Find the length of the median to $\overline{AB}$. $\sqrt{26}$
   b Find the equation of the median to $\overline{AB}$. $y = 5x - 35$
   c Find the equation of the altitude to $\overline{AB}$. $y = -3x + 37$
   d Find the equation of the perpendicular bisector of $\overline{AB}$. $y = -3x + 29$

26 Given a kite with diagonals 6 and 14, find, to the nearest tenth,
   the length of the segment joining the midpoints of two opposite
   sides. ≈ 7.6

27 Roger is 2 m tall. He is standing atop a tower, and the total length of his shadow and the tower's shadow is 14 m. If he were standing on the ground, his shadow would be 1 m long. How high is the tower? 26 m

28 The diagonals of a rhombus are 8 and 12. Find its altitude. $\frac{24\sqrt{13}}{13}$

29 Quadrilateral PQRS is inscribed in ⊙O. The measures of $\overset{\frown}{PQ}$, $\overset{\frown}{QR}$, $\overset{\frown}{RS}$, and $\overset{\frown}{SP}$ are in the ratio $7:12:6:5$. Find the acute angle formed by the diagonals of the quadrilateral. 78°

30 Find the equation of the circle with center (2, 4) that passes through (1, 7). $(x-2)^2 + (y-4)^2 = 10$

31 Is △ARO acute, right, or obtuse? Acute

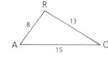

32 $\overline{CD}$ is the altitude to the hypotenuse of △ABC. The coordinates of points A, B, and D are given. Find the coordinates of point C. $(-2, 5)$

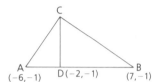

33 Find the ratio of the length of arc ARC to the length of diameter $\overline{AC}$. π:2

34 How far above the ground does the small ball touch the wall if the balls have radii of 4 cm and 9 cm? 21 cm

35 Given: Diagram as shown
Prove: ∠1 > ∠4

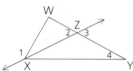

36 Given: $\overline{AD} \cong \overline{DC}$,
∠ADB < ∠BDC
Prove: ∠A > ∠C

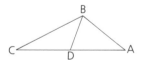

Cumulative Review Problems **709**

Problem-Set Notes and Additional Answers, continued

■ See *Solution Manual* for answer to problems **35** and **39**.

36

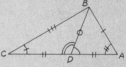

By the Hinge Theorem in △ADB and △CDB, AB < CB. Thus, ∠A > ∠C (If △, then △).

41

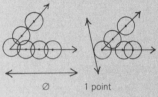

∅   1 point

2 points   3 points

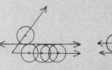

4 points   1 ray

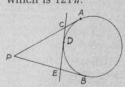

1 ray and 1 point   1 line

42 The difference between the areas is the area of ⊙R, which is 121π.

43

By the Two-Tangent Theorem, CD = CA and ED = EB. Thus, PC + CD = PA, and PE + ED = PB.
∴ PC + CD + ED + PE = PA + PB
PC + CE + PE = PA + PB

Cumulative Review Problem Set B, *continued*

44

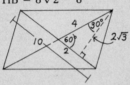

$HF = AF = 8\sqrt{2} - 8$
$FE = 8 - 4\sqrt{2}$
$BE = 4\sqrt{2}$
$HB = 8\sqrt{2} - 8$

45

$A = 2\left[\frac{1}{2}(10)(2\sqrt{3})\right] = 20\sqrt{3}$

46   8 have 3 faces painted
24 have 2 faces painted
24 have 1 face painted
 8 have 0 faces painted
64

a P(10 unpainted faces)
= P(2 painted faces)
= $\frac{24}{64} \cdot \frac{8}{63} + \frac{8}{64} \cdot \frac{24}{63} + \frac{24}{64} \cdot \frac{23}{63}$
= $\frac{13}{56}$

b P(at least 10 faces
unpainted)
= P(no more than 2 faces
painted)
= P(0 or 1 or 2)
= $\frac{8}{64} \cdot \frac{7}{63} + \frac{24}{64} \cdot \frac{8}{63}$
+ $\frac{8}{64} \cdot \frac{24}{63} + \frac{13}{56}$
= $\frac{43}{126}$

47

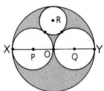

The locus is a cone with a
conical hole in it.

a $V = \frac{1}{3}\pi(225)(8) -$

b $\frac{1}{3}\pi(36)(8) = 504\pi$

---

37 Describe the locus of points a fixed distance from a given point and equidistant from the sides of an angle. { }, 1 point, or 2 points

38 Two pipes are used to fill a pool with water. Their diameters are 12 and 16. Peter Plumber is hired to replace the two pipes with a single pipe having the same capacity as the other two combined.

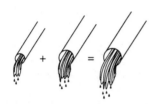

   a What is the diameter of the pipe that Peter must put in? 20

   b What general relationship exists between the diameters of the three pipes? $(d_1)^2 + (d_2)^2 = (d_3)^2$

39 a Prove that ABCD is a kite.

   b Prove analytically that the figure formed by joining consecutive midpoints of the sides of ABCD is a rectangle.

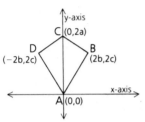

40 What can we conclude from the following statements?

   If *r* is red, then *b* is blue.
   If *q* is not green, then *y* is yellow.
   If *r* is not red, then *y* is not yellow.
   b is not blue. *q* is green.

Problem Set C

41 Describe the locus of points that are centers of congruent circles of a given radius if the circles are tangent to a given line and their centers lie on a given angle. { }, 1 point, 2 points, 3 points, 4 points, a ray, a ray and a point, or a line

42 Circles O, P, Q, and R are tangent as shown. If the radius of ⊙R is 11, find the difference between the areas of the shaded regions above and below $\overline{XY}$. 121π

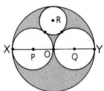

43 Prove: If two tangent segments are drawn to a circle from an external point, the triangle formed by these two tangents and any tangent to the minor arc included by them has a perimeter equal to the sum of the measures of the two original tangent segments.

44 In square ABCD, $\overline{HF} \perp \overline{AC}$.
If the perimeter of the square is 32 and
FC = BC, find the perimeter of quadri-
lateral HFEB. $16\sqrt{2} - 8$

45 The diagonals of a parallelogram have measures of 8 and 10 and
intersect at a 60° angle. Find the area of the parallelogram. $20\sqrt{3}$

46 Sixty-four $1 \times 1 \times 1$ cubes are stacked together to form a
$4 \times 4 \times 4$ cube. The large cube is painted and then broken up
into the original sixty-four cubes. If two of the small cubes are
selected at random, what is the probability that

a Exactly ten of the twelve faces will be unpainted? $\frac{13}{56}$

b At least ten of the twelve faces will be unpainted? $\frac{43}{126}$

47 a What is the locus in space of points
generated by obtuse △ABC if it is ro-
tated about the altitude from A to $\overleftrightarrow{BC}$?

b Find the volume of the locus. 504π

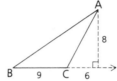

48 Prove that the shortest segment from an
exterior point P to the circle is the segment along the line from P to O.

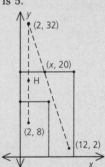

49 The point A = $(-3, 3)$ is rotated 90° clockwise about the origin
to A'. If C = $(-2, 8)$ and D = $(16, 4)$, how far is A' from the
midpoint of $\overline{CD}$? **5**

50 On a miniature-golf course, the hole is at
(2, 18) and the ball is at (12, 2), with
barriers as shown.

a You can make a hole in one by bounc-
ing the ball off the barrier y = 20 to
the barrier y = 13 and into the hole.
At what point must the ball strike the
barrier y = 20? **(6, 20)**

b Can you go directly to the barrier y = 20
and then directly into the hole? **No**

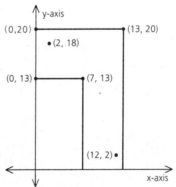

*Problem-Set Notes and
Additional Answers, continued*

48 Assume that PR < PQ.
Then PR + RO < PQ +
QO, but this is impossible
by the Triangle Inequality
Postulate.

49

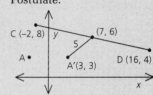

A(−3, 3) rotates to A'(3, 3).
The midpoint of $\overline{CD}$ is
(7, 6). Use the distance
formula to find that the
distance from (3, 3) to (7, 6)
is 5.

50a

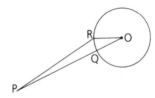

Reflect H across y = 13 to
(2, 8). Then reflect (2, 8)
across y = 20 to (2, 32). For
the points (12, 2), (x, 20)
and (2, 32),
$$\frac{32 - 2}{2 - 12} = \frac{20 - 2}{x - 12}$$
$$x = 6$$
Thus, the point on y = 20 is
(6, 20).

b No. By reflecting (2, 18)
across y = 20 to (2, 22), we
can see that such a shot
would have to strike y =
20 at (3, 20). But this shot
is impossible, since it
would strike the barrier
x = 7 at (7, 12) and be
deflected.

*T*HIS CHAPTER IS intended as enrichment. Therefore, the authors have not included a chapter schedule or suggestions for daily assignments.

The authors who have taught this material in advanced classes suggest the following procedure:

1 One or two students should be in charge of each topic.
2 For each topic, the student(s) who are in charge should prove the theorem or formula involved and give two or three examples of its use.
3 After completing the assignment, the student(s) in charge should be able to answer questions about the topic.
4 The student(s) in charge of a topic should submit two test questions to the teacher. The student(s) must also provide the solutions for these questions. The teacher should examine the questions and the solutions and compile a test, making modifications and additions as needed.

16 | ENRICHMENT TOPICS

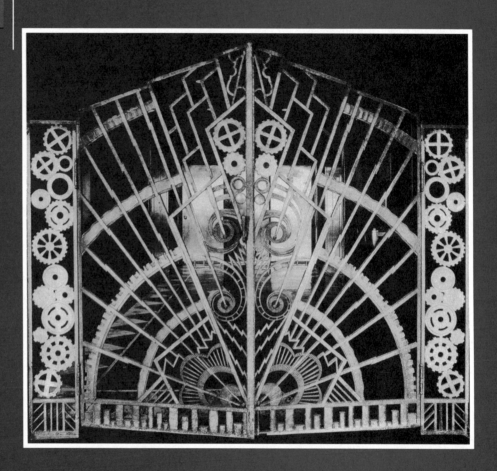

*O*pen the wrought-iron-and-bronze gates to discover enrichment topics in geometry.

16.1 THE POINT-LINE DISTANCE FORMULA

Objective

After studying this section, you will be able to
- Use a formula to determine the distance from a point to a line in the coordinate plane

Part One: Introduction

In this chapter, we shall present some advanced geometry topics that you may enjoy exploring. Unlike the problem sets in the other chapters of this book, those in this chapter are not divided into A, B, and C groups.

We shall begin by developing a formula that you will find useful in solving a variety of coordinate-geometry problems. As you know, it is easy to determine the distance from a given point to a horizontal or vertical line in the coordinate plane. The ***point-line distance formula,*** however, can be used to find the distance from a given point to *any* line in the plane.

Theorem 136 ***The distance d from any point P = (x₁, y₁) to a line whose equation is in the form ax + by + c = 0 can be found with the formula***

$$d = \frac{|ax_1 + by_1 + c|}{\sqrt{a^2 + b^2}}$$

Proof: Remember, the distance from a point to a line is the length of the perpendicular segment from the point to the line. In the diagram, $Q = (x_2, y_2)$ is the foot of the perpendicular from point P to the line represented by $ax + by + c = 0$. The slope of the given line is $-\frac{a}{b}$, so the slope of $\overleftrightarrow{PQ}$ is $\frac{b}{a}$. Thus, we can write the system

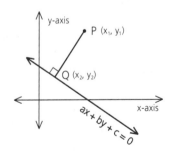

$$\begin{cases} y - y_1 = \frac{b}{a}(x - x_1) & \left(\text{Equation of } \overleftrightarrow{PQ}\right) \\ ax + by + c = 0 & \begin{array}{l}\text{(Equation of given} \\ \text{line)}\end{array} \end{cases}$$

Lesson Notes

- The point-line distance formula is an extension of earlier work with finding the distance from a point to a horizontal or vertical line in the coordinate plane.

Vocabulary

point-line distance formula

Now, by substituting x_2 and y_2 for x and y in the two equations and solving the system, we can express the coordinates of Q in terms of x_1 and y_1.

$$x_2 = \frac{b^2 x_1 - aby_1 - ac}{a^2 + b^2}$$

$$y_2 = \frac{-abx_1 + a^2 y_1 - bc}{a^2 + b^2}$$

Using these expressions in the distance formula to determine the distance from P to Q, we find that

$$d = \sqrt{\frac{a^2(ax_1 + by_1 + c)^2}{(a^2 + b^2)^2} + \frac{b^2(ax_1 + by_1 + c)^2}{(a^2 + b^2)^2}}$$

$$= \frac{|ax_1 + by_1 + c|}{\sqrt{a^2 + b^2}}$$

Part Two: Sample Problems

Problem 1   *Find the distance from the point $(2, -3)$ to the graph of $3x + 4y - 10 = 0$.*

Solution $$d = \frac{|ax_1 + by_1 + c|}{\sqrt{a^2 + b^2}}$$

$$= \frac{|3(2) + 4(-3) - 10|}{\sqrt{3^2 + 4^2}}$$

$$= \frac{16}{5}$$

Problem 2   *Find, to the nearest hundredth, the distance between the graphs of $y = 3x - 10$ and $y = 3x + 1$.*

Solution   Each of the two lines has a slope of 3, so the lines are parallel. We can choose a point on the first line—for example, $(0, -10)$—rewrite the second line's equation as $3x - y + 1 = 0$, and use the point-line distance formula.

$$d = \frac{|ax_1 + by_1 + c|}{\sqrt{a^2 + b^2}}$$

$$= \frac{|3(0) - (-10) + 1|}{\sqrt{3^2 + (-1)^2}}$$

$$= \frac{11}{\sqrt{10}}, \text{ or } \approx 3.48$$

714   Chapter 16   Enrichment Topics

Problem 3   Write equations of the two lines that are parallel to the graph of $5x - 12y = 17$ and tangent to the circle whose center is at $(4, -3)$ and whose radius is 5.

Solution   Since the lines are parallel to the graph of $5x - 12y = 17$, the slope of each is $\frac{5}{12}$. Their equations are therefore of the form $5x - 12y + c = 0$. We can substitute the coefficients of this equation, the coordinates of the circle's center, and the distance between the center and the required lines (the circle's radius) in the point-line distance formula.

$$d = \frac{|ax_1 + by_1 + c|}{\sqrt{a^2 + b^2}}$$

$$5 = \frac{|5(4) - 12(-3) + c|}{\sqrt{5^2 + 12^2}}$$

$$65 = |56 + c|$$

$$65 = 56 + c \text{ or } 65 = -56 - c$$
$$c = 9 \text{ or } c = -121$$

The lines can thus be represented by the equations
$5x - 12y + 9 = 0$ and $5x - 12y - 121 = 0$.

Part Three: Problem Set

1 Find the distance from the origin to the graph of
$4x - 3y + 15 = 0$. 3

2 Find the distance from the point $(4, 2)$ to the graph of
$3x + 4y - 10 = 0$. 2

3 Find the distance from the point $(2, 3)$ to the graph of
$7x - 24y + 2 = 0$. $\frac{56}{25}$

4 Find the distance from the point $(6, -4)$ to the graph of
$3x - 4y = 14$. 4

5 Find, to the nearest hundredth, the distance from the point
$(-2, 6)$ to the line having a slope of 2 and passing through the
point $(2, 1)$. ≈ 5.81

6 Find, to the nearest hundredth, the distance from the point
$(-2, 4)$ to the graph of $x \cos 30° + y \sin 30° - 8 = 0$. ≈ 7.73

7 Find, to four significant digits, the distance between the graphs
of $2x - 3y + 4 = 0$ and $2x - 3y + 15 = 0$. ≈ 3.051

8 Show that the graph of $12x + 5y = 12$ is tangent to the circle
having its center at $(6, 1)$ and passing through $(9, -3)$.

1 $d = \dfrac{|4(0) - 3(0) + 15|}{\sqrt{4^2 + (-3)^2}} = \dfrac{15}{5} = 3$

2 $d = \dfrac{|3(4) + 4(2) - 10|}{\sqrt{3^2 + 4^2}} = \dfrac{10}{5} = 2$

3 $d = \dfrac{|7(2) - 24(3) + 2|}{\sqrt{7^2 + (-24)^2}} = \dfrac{56}{25}$

4 Rewrite $3x - 4y = 14$ as
$3x - 4y - 14 = 0$.
$d = \dfrac{|3(6) - 4(-4) - 14|}{\sqrt{3^2 + (-4)^2}} = \dfrac{20}{5} = 4$

5 The equation of the line is
$y - 1 = 2(x - 2)$, or
$2x - y - 3 = 0$.
$d = \dfrac{|2(-2) - 1(6) - 3|}{\sqrt{2^2 + (-1)^2}}$
$= \dfrac{13}{\sqrt{5}} \approx 5.81$

6 $x \cos 30° + y \sin 30° - 8 = 0$
$x\left(\dfrac{\sqrt{3}}{2}\right) + y\left(\dfrac{1}{2}\right) - 8 = 0$
$\sqrt{3}x + y - 16 = 0$
$d = \dfrac{|(-2)\sqrt{3} + 4 - 16|}{\sqrt{(\sqrt{3})^2 + 1^2}}$
$= \dfrac{12 + 2\sqrt{3}}{2} = 6 + \sqrt{3} \approx 7.73$

7 Use $(-2, 0)$ on $2x - 3y + 4 = 0$.
$d = \dfrac{|2(-2) - 3(0) + 15|}{\sqrt{2^2 + (-3)^2}}$
$= \dfrac{11}{\sqrt{13}} \approx 3.051$

8 Radius $=$
$\sqrt{(6 - 9)^2 + [1 - (-3)]^2} = 5$
Rewrite $12x + 5y = 12$ as
$12x + 5y - 12 = 0$.
$5 = \dfrac{|12(6) + 5(1) - 12|}{\sqrt{12^2 + 5^2}}$
$\nleq 5$

9
$$2 = \frac{|3(5) - 4(1) + k|}{\sqrt{3^2 + (-4)^2}}$$
$$10 = |k + 11|$$
$$k + 11 = 10$$
$$k = -1 \text{ or}$$
$$-k - 11 = 10$$
$$k = -21$$

10 $d = \frac{|3(5) - 7(2) + 5(1) + 13|}{\sqrt{3^2 + (-7)^2 + 5^2}}$
$$= \frac{19}{\sqrt{83}} \approx 2.086$$

11 $\frac{|x - 2y + 5|}{\sqrt{5}} = \frac{|2x - y - 3|}{\sqrt{5}}$
$$|x - 2y + 5| = |2x - y - 3|$$
Therefore, the equations are
$$x - 2y + 5 = 2x - y - 3$$
and
$$-x + 2y - 5 = 2x - y - 3,$$
or
$$x + y - 8 = 0$$
and
$$3x - 3y + 2 = 0.$$

12
$$6 = \frac{|2(3) + 4b + 3|}{\sqrt{2^2 + b^2}}$$
$$6 = \frac{|6 + 4b + 3|}{\sqrt{2^2 + b^2}}$$
$$36(4 + b^2) = (9 + 4b)^2$$
$$20b^2 - 72b + 63 = 0$$
$$(2b - 3)(10b - 21) = 0$$
$$b = 1.5 \text{ or } b = 2.1$$

Problem Set, *continued*

9 If the graph of $3x - 4y + k = 0$ is tangent to the circle with a radius of 2 and a center at (5, 1), what is the value of k? -1 or -21

10 It can be shown that in three dimensions, the distance from a point (x_1, y_1, z_1) to the plane represented by the equation $ax + by + cz + d = 0$ can be found with the formula

$$d = \frac{|ax_1 + by_1 + cz_1 + d|}{\sqrt{a^2 + b^2 + c^2}}$$

Find, to four significant digits, the distance from the point (5, 2, 1) to the graph of $3x - 7y + 5z + 13 = 0$. ≈ 2.086

11 Write equations of the bisectors of the angles formed by the graphs of $x - 2y + 5 = 0$ and $2x - y - 3 = 0$.
$x + y - 8 = 0$ and $3x - 3y + 2 = 0$

12 Find the possible values of b if the point (3, 4) is six units from the graph of $2x + by + 3 = 0$. 1.5 or 2.1

CAREER PROFILE

PLANETARY PORTRAITS
Kim Poor paints the heavens

Kim Poor is a landscape painter, but the landscapes he paints in his Tucson, Arizona, studio are of places no one has ever visited. He creates scenes showing planets and moons as they would appear to a cosmic explorer.

To make his works as realistic as possible, Kim regularly travels to Kitt Peak Observatory and the University of Arizona, where he makes sketches based on his conversations with astronomers and his reading in the observatories' libraries. Back in his studio, he carefully plans a painting, using trigonometry to determine how large each object in the picture would appear from the point of view he has selected. Then he gets to work with his airbrush.

According to Kim, the most important shape for the space artist is the ellipse, because circular features on the planets appear elliptical when viewed at an oblique angle. He gives an example: "We see Saturn's rings as anything from a line segment to an ellipse corresponding to a 20° tilt. And if the rings are shown as 20° ellipses, the cloud bands on the planet must also be 20° ellipses or the painting will look wrong." Kim uses protractors, compasses, and a variety of drafting techniques to produce the most accurate representations he can.

Kim's work can be seen in planetariums, magazines, encyclopedias, and textbooks. Along with his fellow space artists, he plays an important role in the interpretation and communication of the latest discoveries in planetary and stellar astronomy.

16.2 TWO OTHER USEFUL FORMULAS

Objectives

After studying this section, you will be able to
- Use a formula to find the area of a triangle when only the coordinates of its vertices are known
- Use a formula to find the diameter of a triangle's circumscribed circle

Part One: Introduction

Area of a Triangle

In Chapter 13, you used the encasement principle to find the areas of triangles in the coordinate plane. (See, for example, Section 13.7, problem 14.) Now we can use the point-line distance formula to develop a general formula for the area of a triangle with given vertices.

Theorem 137   **The area A of a triangle with vertices at (x_1, y_1), (x_2, y_2), and (x_3, y_3) can be found with the formula**

$$A = \frac{1}{2}|x_1y_2 + x_2y_3 + x_3y_1 - x_1y_3 - x_2y_1 - x_3y_2|$$

Lesson Notes

- Theorem 137 provides a general formula for finding the area of a triangle with given vertices.

Proof: The point-slope form of the equation of $\overleftrightarrow{BC}$ in the diagram at the right is

$$y - y_2 = \frac{y_3 - y_2}{x_3 - x_2}(x - x_2)$$

which can be rewritten as

$$x(y_2 - y_3) + y(x_3 - x_2) + x_2y_3 - x_3y_2 = 0$$

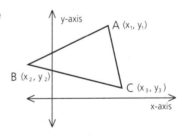

We can now find the distance from $A = (x_1, y_1)$ to $\overleftrightarrow{BC}$ (the altitude to base $\overline{BC}$) by using the point-line distance formula.

$$d = \frac{|(y_2 - y_3)x_1 + (x_3 - x_2)y_1 + x_2y_3 - x_3y_2|}{\sqrt{(y_2 - y_3)^2 + (x_3 - x_2)^2}}$$

Section 16.2   Two Other Useful Formulas   **717**

Vocabulary
circumcircle

Substituting this length and the length of base $\overline{BC}$ (determined by means of the distance formula) in the familiar formula for the area of a triangle, we find that

$$A = \frac{1}{2}bh$$

$$= \frac{1}{2}\sqrt{(y_3 - y_2)^2 + (x_3 - x_2)^2}\ \frac{|(y_2 - y_3)x_1 + (x_3 - x_2)y_1 + x_2y_3 - x_3y_2|}{\sqrt{(y_2 - y_3)^2 + (x_3 - x_2)^2}}$$

$$= \frac{1}{2}|x_1y_2 + x_2y_3 + x_3y_1 - x_1y_3 - x_2y_1 - x_3y_2|$$

Note If you are familiar with determinants, you may recognize that the formula in Theorem 137 can be written in the form

$$A = \frac{1}{2}\begin{Vmatrix} 1 & x_1 & y_1 \\ 1 & x_2 & y_2 \\ 1 & x_3 & y_3 \end{Vmatrix}$$

Diameter of a Circumscribed Circle

In solving certain problems, it is useful to be able to calculate the diameter of a circle circumscribed about a triangle (the triangle's *circumcircle*). The theorem that follows is an extension of the Law of Sines, which was presented in Section 9.10, problem 2.1. (Recall that when we describe the parts of a triangle, we use *a* to represent the length of the side *opposite* vertex A, and so forth.)

Theorem 138 *In any triangle ABC, with side lengths a, b, and c,*

$$\frac{a}{sin \angle A} = D \qquad \frac{b}{sin \angle B} = D \qquad \frac{c}{sin \angle C} = D$$

where D is the diameter of the triangle's circumcircle.

Proof: The diagrams below show the five possible cases for $\angle A$ in an inscribed triangle.

In the first two cases, we have drawn diameter $\overline{A'B}$, forming right triangle A'BC. (Remember, an angle inscribed in a semicircle is a right angle.) Since inscribed angles A and A' intercept the same arc, they are congruent, and therefore $sin \angle A = sin \angle A' = \frac{a}{D}$. Thus,

$$\frac{a}{sin \angle A} = \frac{a}{\frac{a}{D}} = D$$

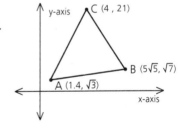

In the third case, we can obtain the same result without drawing an auxiliary line, since △ABC is already a right triangle.

In the fourth case, $\angle A = 90°$. From trigonometry, we know that $\sin 90° = 1$, so the formula follows immediately.

In the last case, where $\angle A$ is obtuse, we again draw diameter $\overline{A'B}$. Since opposite angles of an inscribed quadrilateral are supplementary, $\angle A$ is supplementary to $\angle A'$. It is a basic principle of trigonometry that the sine of any angle is equal to the sine of its supplement, so $\sin \angle A = \sin \angle A'$. Hence, we obtain the same result as in the first two cases.

Similar reasoning can be used to show that for any $\angle B$ and $\angle C$ in an inscribed triangle, $\frac{b}{\sin \angle B} = D$ and $\frac{c}{\sin \angle C} = D$.

Part Two: Sample Problems

Problem 1   *Find, to the nearest tenth, the diameter of the circle circumscribed about △ABC.*

Solution   According to the Law of Cosines (see Section 9.10, problem 20),

$$(BC)^2 = 5^2 + 6^2 - 2(5)(6)(\cos 30°)$$

$$= 25 + 36 - \frac{60\sqrt{3}}{2}$$

$$= 61 - 30\sqrt{3}$$

$$BC \approx 3.006$$

We now use the formula for the diameter of a circumcircle.

$$D = \frac{a}{\sin \angle A}$$

$$\approx \frac{3.006}{\sin 30°}$$

$$\approx \frac{3.006}{\frac{1}{2}} \approx 6.0$$

Problem 2   *Find the area of △ABC to the nearest tenth.*

Solution   We use the formula for the area of a triangle (Theorem 137).

$$A = \frac{1}{2}|x_1 y_2 + x_2 y_3 + x_3 y_1 - x_1 y_3 - x_2 y_1 - x_3 y_2|$$

$$= \frac{1}{2}|1.4\sqrt{7} + 5\sqrt{5}(21) + 4\sqrt{3} - 1.4(21) - 5\sqrt{5}\sqrt{3} - 4\sqrt{7}|$$

$$\approx 93.0$$

Problem-Set Notes and Additional Answers

1 By the Law of Cosines,
$(BC)^2 = 4^2 + 6^2 - 2(4)(6) \cos 60°$
$BC = 2\sqrt{7}$
By Theorem 138,
$D = \frac{a}{\sin \angle A} = \frac{2\sqrt{7}}{\sin 60°} \approx 6.110$

2 By Theorem 138, $D = \frac{a}{\sin \angle A}$,
so $D = \frac{10}{\sin 45°} = 10\sqrt{2}$.
$\therefore R = 5\sqrt{2}$ and $A = \pi R^2 = 50\pi$

3

By the Pythagorean Theorem, the altitude has a length of $\sqrt{16 - x^2}$.
$D = \frac{a}{\sin \angle A} = \frac{16}{\sqrt{16 - x^2}}$, so
$R = \frac{8}{\sqrt{16 - x^2}}$. Since $A = \pi R^2 = \frac{64\pi}{16 - x^2}$, $a = 64$, $b = 16$
and $c = 1$. Thus, $a + b + c = 64 + 16 + 1 = 81$.

4 By Theorem 137,
$A = \frac{1}{2}|-8(-2) + 12(25)$
$\quad + 3(-7) - (-8)(25)$
$\quad - 12(-7) - 3(-2)|$
$\quad = \frac{1}{2}|16 + 300 - 21 + 200$
$\quad + 84 + 6|$
$\quad = 292\frac{1}{2}$

5 By Theorem 137,
$A = \frac{1}{2}|-2(17) + 4(11) + 6x$
$\quad - (-2)(11) - 4(6) - 17x|$
$42 = \frac{1}{2}|-34 + 44 + 22 - 24$
$\quad - 11x|$
$\quad\quad 84 = |8 - 11x|$
$8 - 11x = 84$
$\quad x = -\frac{76}{11}$ or
$-8 + 11x = 84$
$\quad x = \frac{92}{11}$

Part Three: Problem Set

1 Find, to the nearest thousandth, the diameter of the circle circumscribed about $\triangle ABC$. ≈ 6.110

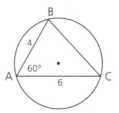

2 Find the area of the circle in the diagram at the right. 50π

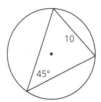

3 The area of the circumcircle of $\triangle CAT$ can be written as a simplified expression of the form

$$\frac{a\pi}{b - cx^2}$$

What is the value of $a + b + c$? 81

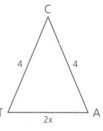

4 Find the area of $\triangle TRY$. $292\frac{1}{2}$

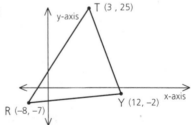

5 The coordinates of the vertices of a triangle are $(-2, 6)$, $(4, 17)$, and $(x, 11)$, and the triangle's area is 42. Find the possible values of x. $-\frac{76}{11}$ or $\frac{92}{11}$

16.3 STEWART'S THEOREM

Objective

After studying this section, you will be able to
- Recognize a relationship among the parts of a triangle with a segment drawn from a vertex to the opposite side

Part One: Introduction

The following theorem is usually called Stewart's Theorem, after the eighteenth-century Scottish mathematician Matthew Stewart, although forms of the theorem were known as long ago as the fourth century A.D.

Theorem 139 *In any triangle ABC, with side lengths a, b, and c,*

$$a^2n + b^2m = cd^2 + cmn$$

where d is the length of a segment from vertex C to the opposite side, dividing that side into segments with lengths m and n. (Stewart's Theorem)

Lesson Notes

- The work with triangles is extended with the introduction of Stewart's Theorem.

Given: Diagram as marked
Prove: $a^2n + b^2m = cd^2 + cmn$

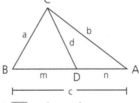

Proof: We draw $\overline{CE}$ perpendicular to $\overline{AB}$ and use the Pythagorean Theorem. In $\triangle BCE$, $a^2 = h^2 + (m - p)^2$, or $a^2 = h^2 + m^2 - 2mp + p^2$; and in $\triangle CED$, $d^2 = h^2 + p^2$. By subtraction, we find that $a^2 - d^2 = m^2 - 2mp$, or

$$a^2 = d^2 + m^2 - 2mp \qquad (1)$$

In $\triangle CEA$, $b^2 = h^2 + (p + n)^2$, or $b^2 = h^2 + p^2 + 2pn + n^2$. Since we know that $h^2 = d^2 - p^2$ (see the preceding paragraph), we can substitute $d^2 - p^2$ for h^2 to obtain the equation $b^2 = d^2 - p^2 + p^2 + 2pn + n^2$, or

$$b^2 = d^2 + 2pn + n^2 \qquad (2)$$

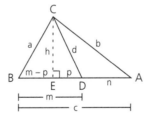

Section 16.3   Stewart's Theorem   **721**

Equation (1) can be rewritten as $a^2n = d^2n + m^2n - 2mnp$, and equation (2) can be rewritten as $b^2m = d^2m + 2mnp + mn^2$. Adding these equations, we find that

$$a^2n + b^2m = d^2n + d^2m + m^2n + mn^2$$
$$= d^2(n + m) + mn(m + n)$$
$$= cd^2 + cmn$$

Part Two: Sample Problems

Problem 1   *If the sides of a triangle have measures of 3, 5, and 6, what is the length of the bisector of the angle included by the sides measuring 3 and 5?*

Solution   By the Angle Bisector Theorem (see Section 8.5),

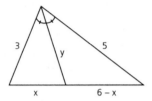

$$\frac{3}{5} = \frac{x}{6 - x}$$

$$5x = 18 - 3x$$

$$x = \frac{9}{4}$$

Therefore, $6 - x = \frac{15}{4}$. We now apply Stewart's Theorem.

$$3^2\left(\frac{15}{4}\right) + 5^2\left(\frac{9}{4}\right) = y^2(6) + 6\left(\frac{9}{4}\right)\left(\frac{15}{4}\right)$$

$$\frac{135}{4} + \frac{225}{4} = 6y^2 + \frac{405}{8}$$

$$y^2 = \frac{105}{16}$$

$$y = \frac{\sqrt{105}}{4}$$

Problem 2   *Prove that in a right triangle the sum of the squares of the segments from the vertex of the right angle to the trisection points of the hypotenuse is equal to five-ninths the square of the hypotenuse.*

Proof   According to Stewart's Theorem, in the diagram shown,

$$2a^2x + b^2x = d^2c + 2cx^2$$

and

$$a^2x + 2b^2x = ce^2 + 2cx^2$$

Adding these equations, we find that

$$3a^2x + 3b^2x = cd^2 + ce^2 + 4cx^2$$
$$3x(a^2 + b^2) = cd^2 + ce^2 + 4cx^2$$

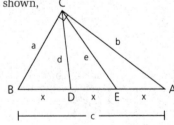

Problem-Set Notes and Additional Answers

1 Use Stewart's Theorem:
$$5^2x + 8^2(3) = 7^2(3 + x)$$
$$+ (3 + x)3x$$
$$25x + 192 = 147 + 49x + 9x$$
$$+ 3x^2$$
$$x^2 + 11x - 15 = 0$$
$$x = \frac{-11 \pm \sqrt{121 + 60}}{2}$$
$$= \frac{-11 \pm \sqrt{181}}{2}$$
Reject the negative answer.
$$\therefore x = \frac{-11 + \sqrt{181}}{2}$$

2 Use Stewart's Theorem:
$$3^2(3) + 4^2(1) = y^2(4)$$
$$+ 4(1)(3)$$
$$27 + 16 = 4y^2 + 12$$
$$y^2 = \frac{31}{4}$$
$$y = \frac{\sqrt{31}}{2}$$

By substituting c^2 for $a^2 + b^2$, we obtain

$3xc^2 = c(d^2 + e^2 + 4x^2)$

Now we substitute c for $3x$.

$c^3 = c(d^2 + e^2 + 4x^2)$
$c^2 = d^2 + e^2 + 4x^2$

Since $2x = \frac{2}{3}c$, we can substitute $\frac{4}{9}c^2$ for $4x^2$ to obtain

$c^2 = d^2 + e^2 + \frac{4}{9}c^2$

$d^2 + e^2 = \frac{5}{9}c^2$

Part Three: Problem Set

1 Find the value of x in the figure at the right. $\frac{-11 + \sqrt{181}}{2}$

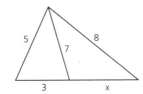

2 Find the value of y in the figure at the right. $\frac{\sqrt{31}}{2}$

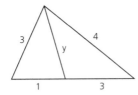

3 A parallelogram has sides with measures of 7 and 9, and the measure of its shorter diagonal is 8. Find the measure of the parallelogram's longer diagonal. **14**

4 Two sides of a triangle have measures of 9 and 18. If the bisector of the angle included by these sides has a measure of 8, what is the measure of the third side of the triangle? **21**

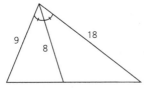

5 Find the measure of a side of a triangle if the other two sides and the bisector of the angle they include have measures of 3, 5, and 2 respectively. $\frac{8\sqrt{165}}{15}$

Problem-Set Notes and Additional Answers, continued

3 Use Stewart's Theorem, remembering that the diagonals of a parallelogram bisect each other.
$7^2x + 9^2x = 4^2(2x) + 2x(x)(x)$
$49x + 81x = 32x + 2x^3$
$x^3 = 49x$
$x^2 = 49$
$x = 7$
The measure of the longer diagonal is 14.

4

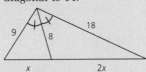

Since the angle bisector divides the opposite side into segments proportional to the sides of the angle, the figure can be labeled as shown above. By Stewart's Theorem,
$9^2(2x) + 18^2(x) = 8^2(3x) + 3x(x)(2x)$
$162x + 324x = 192x + 6x^3$
$x^3 = 49x$
$x^2 = 49$
$x = 7$
The third side (3x) therefore has a measure of 21.

5

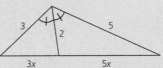

Since the angle bisector divides the opposite side into segments proportional to the sides of the angle, the figure can be drawn as shown above.
By Stewart's Theorem,
$3^2(5x) + 5^2(3x) = 2^2(8x) + 8x(3x)(5x)$
$45x + 75x = 32x + 120x^3$
$120x^3 = 88x$
$x^2 = \frac{11}{15}$
$x = \frac{\sqrt{165}}{15}$
The measure of the side (8x) is therefore $\frac{8\sqrt{165}}{15}$.

PTOLEMY'S THEOREM

The Geometric Supposer

You can use either the *Triangles* and *Quadrilaterals* programs or the *Circles* program in *The Geometry Supposer* series to preview the relationships expressed by Ptolemy's Theorem.
To illustrate Ptolemy's Theorem, have students draw a number of irregular quadrilaterals inscribed in circles and then measure and calculate the products of the lengths of the diagonals and the sums of the products of the lengths of opposite sides.

Objective

After studying this section, you will be able to
- Recognize a relationship involving the sides and the diagonals of a cyclic quadrilateral

Part One: Introduction

Ptolemy's Theorem is named for a famous Alexandrian mathematician, astronomer, and geographer (often referred to by the Latin form of his name, Claudius Ptolemaeus) who lived from about 85 to 165 A.D.

Theorem 140   *If a quadrilateral is inscribable in a circle, the product of the measures of its diagonals is equal to the sum of the products of the measures of the pairs of opposite sides. (Ptolemy's Theorem)*

Given: Quadrilateral ABCD inscribed in ⊙O
Prove: $(AC)(BD) = (AB)(CD) + (AD)(BC)$

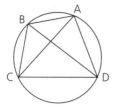

Proof: We extend $\overline{CD}$ to a point P so that ∠DAP ≅ ∠BAC. Since opposite angles of a cyclic quadrilateral are supplementary, ∠ABC is supplementary to ∠ADC; so ∠ABC ≅ ∠ADP because supplements of the same angle are congruent. Therefore, △BAC ~ △DAP (by AA), and

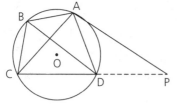

$$\frac{AB}{AD} = \frac{BC}{DP}$$

$$DP = \frac{(AD)(BC)}{AB} \qquad (1)$$

By the Addition Property, $\angle BAD \cong \angle CAP$; so $\angle ABD \cong \angle ACP$, since inscribed angles that intercept the same arc are congruent. Therefore, $\triangle ABD \sim \triangle ACP$ (by AA), and

$$\frac{AB}{AC} = \frac{BD}{CP}$$

$$CP = \frac{(AC)\,(BD)}{AB} \qquad\qquad (2)$$

We know that $CP = CD + DP$, and when we substitute the equivalent expressions from equations (1) and (2) for DP and CP in this equation, we obtain

$$\frac{(AC)\,(BD)}{AB} = CD + \frac{(AD)\,(BC)}{AB}$$

$$(AC)\,(BD) = (AB)\,(CD) + (AD)\,(BC)$$

Part Two: Sample Problems

Problem 1

Given: Inscribed quadrilateral ABCD,
AB = 3, BC = 5, CD = 4,

$$AD = 6,\ AC = \frac{2\sqrt{1729}}{13}$$

Find: BD

$AC = \frac{2\sqrt{1729}}{13}$

Solution

We use Ptolemy's Theorem.

$$(AC)\,(BD) = (AB)\,(CD) + (AD)\,(BC)$$

$$\frac{2\sqrt{1729}}{13}(BD) = 3(4) + 6(5)$$

$$BD = \frac{3\sqrt{1729}}{19}$$

Problem 2

A quadrilateral, PQRS, is inscribed in a circle, O. If PQ = 6, PS = 3, and diagonal $\overline{PR}$ has a measure of 10 and is a diameter of the circle, what is the measure of diagonal $\overline{SQ}$?

Solution

Since $\overline{PR}$ is a diameter, $\triangle PSR$ and $\triangle PQR$ are right triangles. Thus, by the Pythagorean Theorem, QR = 8 and RS = $\sqrt{91}$. By Ptolemy's Theorem,

$$(PR)\,(SQ) = (PQ)\,(SR) + (PS)\,(QR)$$

$$10(SQ) = 6\sqrt{91} + 3(8)$$

$$SQ = \frac{3\sqrt{91} + 12}{5}$$

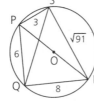

Problem-Set Notes and Additional Answers

One must be very careful in making up problems based on Ptolemy's Theorem. The opposite angles of a cyclic quadrilateral are supplementary, so one must use the Law of Cosines either to check that the opposite angles of a proposed quadrilateral are supplementary or to find appropriate lengths when constructing a quadrilateral.

1 By Ptolemy's Theorem,
$$(AC)(BD) = (AB)(CD) + (AD)(BC)$$
$$(AC)\left(\tfrac{18\sqrt{51}}{17}\right) = 2(6) + 8(3)$$
$$AC = \tfrac{2\sqrt{51}}{3}$$

2

In right $\triangle ABF$, $AF = 6$ and $BF = 8$, so $AB = 10$. In isosceles right $\triangle ABE$, $AE = BE = 5\sqrt{2}$. Since $\angle AFB$ and $\angle AEB$ are both right angles, $AEBF$ is a cyclic quadrilateral. (If opposite angles of a quadrilateral are supplementary, the quadrilateral is cyclic.) Now use Ptolemy's Theorem:
$$(AB)(EF) = (AF)(BE) + (AE)(BF)$$
$$10(EF) = 6(5\sqrt{2}) + 8(5\sqrt{2})$$
$$EF = 7\sqrt{2}$$

3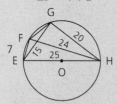

Since $\overline{EH}$ is a diameter, $\triangle EFH$ and $\triangle EGH$ are right triangles. By the Pythagorean Theorem, $FH = 24$ and $EG = 15$. Now use Ptolemy's Theorem:

$$(EF)(GH) + (FG)(EH) = (EG)(FH)$$
$$7(20) + (FG)(25) = 24(15)$$
$$FG = 8.8$$
$$\therefore P = 7 + 8.8 + 20 + 25 = 60.8$$

Problem 3 $\angle CFD$ is inscribed in the circumcircle of rectangle ABCD, with $\overline{CF}$ intersecting $\overline{DA}$ at E. If $DC = 6$, $DE = 6$, and $EA = 2$, find BF.

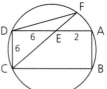

Solution   In right triangle CED, $CE = 6\sqrt{2}$. Because inscribed angles intercepting the same arc are congruent, $\angle CAD \cong \angle DFC$ and $\angle FDA \cong \angle FCA$. Thus, $\triangle DEF \sim \triangle CEA$ (by AA), and

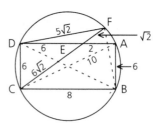

$$\frac{6}{6\sqrt{2}} = \frac{DF}{10}$$
$$DF = 5\sqrt{2}$$

In a similar way, it can be shown that $EF = \sqrt{2}$. We now apply Ptolemy's Theorem to quadrilateral BCDF.

$$(BD)(CF) = (DC)(BF) + (DF)(BC)$$
$$10(7\sqrt{2}) = 6(BF) + 5\sqrt{2}(8)$$
$$BF = 5\sqrt{2}$$

Part Three: Problem Set

1 Given: Cyclic quadrilateral ABCD,
$$AB = 2, \ BC = 3, \ CD = 6,$$
$$AD = 8, \ BD = \frac{18\sqrt{51}}{17}$$
Find: AC $\tfrac{2\sqrt{51}}{3}$

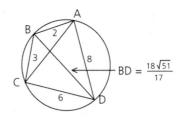

$BD = \dfrac{18\sqrt{51}}{17}$

2 In the diagram at the right, ABCD is a square, and E is the point of intersection of its diagonals. If a point, F, is located in the exterior of the square so that $\triangle ABF$ is a right triangle with hypotenuse $\overline{AB}$, $AF = 6$, and $BF = 8$, what is distance EF? (Hint: Apply Ptolemy's Theorem to quadrilateral AEBF.) $7\sqrt{2}$

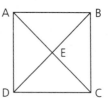

3 Given: ⊙O, with inscribed quadrilateral
EFGH, EF = 7, GH = 20, EH = 25

Find: The perimeter of EFGH 60.8

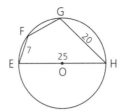

4 Diagonals $\overline{AC}$ and $\overline{BD}$ of quadrilateral
ABCD intersect at E. If AE = 2, BE = 5,
CE = 10, DE = 4, and BC = 7.5, what is
distance AB? (Hint: Look for similar triangles that you can use to find AD and
the ratio of AB to CD. Then apply
Ptolemy's Theorem.) $\frac{3\sqrt{19}}{2}$

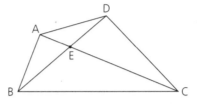

5 A triangle is inscribed in a circle with a radius of 5. The measures of two of the triangle's sides are 5 and 6. What are the
possible measures of the third side? (Hint: There are two possible
triangles.) $3\sqrt{3} - 4$ or $3\sqrt{3} + 4$

HISTORICAL SNAPSHOT

DYNAMIC GEOMETRY
Buckminster Fuller and the geodesic dome

One day while serving in the United States
Navy during World War I, young R. Buckminster
Fuller stood observing the bubbles that boiled
up in the wake of his ship. Noticing that they
were constantly changing in shape, always ap-
proximating spheres but never settling into a
stable spherical form, he concluded that "nature doesn't use pi." He decided that what gov-
erns the natural world is not the static figures
and relationships of traditional geometry but the
geometric interactions of forces that shape the
objects around us.

Fuller was to base a new approach to architecture and design on this insight. He liked to
call his approach "energetic-synergetic geometry." By studying the patterns of forces that
hold molecules together, he developed a system
of basic forms that could be used to produce
structures that combine maximum strength with
minimal materials. The most famous of these
structures is the geodesic dome, an unsupport-

ed framework of tensed triangular forms with a
remarkable property: the larger the dome, the
greater its total strength. Hence, there is no
limit to the possible size of a geodesic dome.
Fuller suggested that whole cities could be covered with domes to allow complete control of
their climates.

While no domes large enough to cover cities
have been constructed yet, the myriad ideas of
Buckminster Fuller, who died in 1983, continue
to exert influence in diverse fields from map-
making to environmental science.

*Problem-Set Notes and
Additional Answers, continued*

4

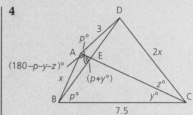

△AED ~ △BEC by SAS ~, so
$\frac{AE}{BE} = \frac{AD}{BC}$
$\frac{2}{5} = \frac{AD}{7.5}$
AD = 3
△AEB ~ △DEC by SAS ~, so
$\frac{AE}{DE} = \frac{AB}{DC}$
$\frac{2}{4} = \frac{AB}{DC}$
Thus, if AB = x then DC =
2x. To verify that quadrilateral ABCD is cyclic, use
properties of triangles to
determine the angle
relationships labeled on the
figure, showing that ∠BAD is
supplementary to ∠BCD.
According to Ptolemy's
Theorem, 9(12) = 3(7.5) +
x(2x), so x = $\frac{\sqrt{171}}{2}$ = $\frac{3\sqrt{19}}{2}$.

5 The following figure shows
one possibility.

By the Pythagorean Theorem,
BD = 5√3 and CD = 8. Now
use Ptolemy's Theorem with
ABCD:
(AC)(BD) = (AB)(CD) + (AD)(BC)
6(5√3) = 5(8) + 10(BC)
BC = 3√3 − 4
The following figure shows
the second possibility.

Again, use Ptolemy's
Theorem with ABDC:
(AD)(BC) = (AB)(CD) + (AC)(BD)
10(BC) = 5(8) + 6(5√3)
BC = 4 + 3√3

MASS POINTS

Objective

After studying this section, you will be able to
- Use the concept of mass points to solve problems

Lesson Notes

- The work with mass-point theory applies a physical principle to the solution of geometric problems.

Part One: Introduction

Some people claim that the theory of ***mass points*** was developed by students in New York as a way of simplifying the solutions of many mathematics problems. The theory is based on what might be called the balance principle (or the fulcrum principle or the teeter-totter principle).

In the diagram of a lever above, w_1 and w_2 are weights, and d_1 and d_2 are their respective distances from the fulcrum. For the lever to be in balance, the product $w_1 d_1$ must be equal to the product $w_2 d_2$. Mass-point theory is simply an application of this physical principle to geometric problems. Consider, for example, a segment divided in the ratio $2:3$.

We can assign "weights" of 3 and 2 to points A and C respectively to "balance" the segment ($3 \cdot 2x = 2 \cdot 3x$). We can then assign a weight of 5—the sum of the weights of the endpoints—to point B, the "fulcrum." The completed mass-point diagram of the segment will look like this:

As the sample problems and the problem set in this section illustrate, the mass-point procedure can be used to solve a variety of problems in two or more dimensions. If you wish to investigate the topic of mass points further, you can consult the following two sources:

> Hausner, Melvin. "The Center of Mass and Affine Geometry," *American Mathematical Monthly*, Vol. 69 (1962), pp. 724–737.
> Sitomer, Harry, and Steven R. Conrad. "Mass Points," *Eureka*, Vol. 2, No. 4 (April 1976), pp. 55–62.

Part Two: Sample Problems

Problem 1 Given: *Diagram as marked*
Find: $\frac{BF}{FE}$

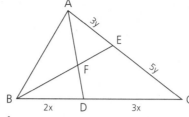

Solution We use the mass-point procedure, assigning a weight of 3 to point B and a weight of 2 to C, as in the diagram at the right. To find the weight at A, which we will symbolize w_A, we use the formula $w_1d_1 = w_2d_2$.

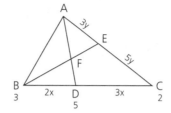

$$2(5y) = w_A(3y)$$

$$w_A = \frac{10}{3}$$

The weight at E, w_E, is thus $2 + \frac{10}{3}$, or $\frac{16}{3}$. Turning our attention to $\overline{BE}$, we find that since $w_B = 3$ and $w_E = \frac{16}{3}$,

$$\frac{BF}{FE} = \frac{\frac{16}{3}}{3} = \frac{16}{9}$$

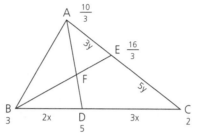

Problem 2 *Show that in a tetrahedron, the line segments joining the vertices to the centroids of the opposite faces are concurrent and divide each other in the ratio 3:1.*

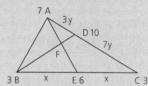

Solution We will assume that mass points can be applied to solid figures in the same way that they can be applied to plane figures. In this case, we assume that a tetrahedron has a unique "center of gravity" and that if we assign equal weights to the vertices, that point's weight will be the sum of the vertices' weights. If we assign a weight of 1 to each vertex, the centroid of each face will represent the mass-point sum of that face's vertices, so it will have a weight of 3. The sum of the weights at the four vertices of the tetrahedron will therefore lie on *each* of the segments connecting the vertices to the centroids of the opposite faces. Thus, these segments are concurrent at the summation point, and since the weights at each segment's endpoints are 1 and 3, the summation point divides each in the ratio 3:1.

Problem-Set Notes and Additional Answers

1

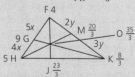

Assign ratios and weights as shown. Notice that for each labeled segment, $w_1d_1 = w_2d_2$. By the mass-point principle, the weights at B and D indicate that $\frac{BF}{FD} = \frac{10}{3}$.

2

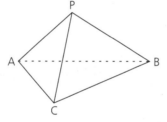

Assign ratios and weights as shown. By the mass-point principle, the weights at H and K indicate that

$\frac{HJ}{JK} = \frac{\frac{8}{3}}{5} = \frac{8}{15}$. Similarly, the weights at F and J indicate

that $\frac{FO}{FJ} = \frac{\frac{23}{3}}{\frac{35}{3}} = \frac{23}{35}$.

Problem 3 In the figure shown, $\frac{AB}{BC} = \frac{3}{4}$ and $\frac{CD}{DE} = \frac{2}{5}$. Find $\frac{CG}{GF}$.

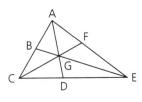

Solution We assign a weight of 4 to A and a weight of 3 to C so that $w_A(AB) = w_C(BC)$, as shown in the diagram at the right. We now find the weight at E.

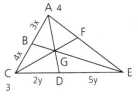

$$3(2y) = w_E(5y)$$

$$w_E = \frac{6}{5}$$

Thus, $w_F = 4 + \frac{6}{5}$, or $\frac{26}{5}$. Since $w_C = 3$ and $w_F = \frac{26}{5}$,

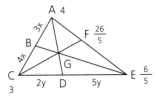

$$\frac{CG}{GF} = \frac{\frac{26}{5}}{3} = \frac{26}{15}$$

Part Three: Problem Set

1 Given: $\overline{AE}$ is a median of $\triangle ABC$.
   AD:DC = 3:7
   Find: BF:FD 10:3

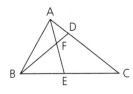

2 In the figure shown, $\frac{HG}{HF} = \frac{4}{9}$ and $\frac{FM}{MK} = \frac{2}{3}$.
   Find $\frac{HJ}{JK}$ and $\frac{FO}{FJ}$. $\frac{8}{15}$; $\frac{23}{35}$

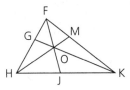

3 In a triangle ABC, $\overline{BD}$ is a median, F is a point on $\overline{AB}$, and $\overline{CF}$ intersects $\overline{BD}$ at E. If BE = 4(ED) and BF = 20, what is AF? 10

4 In a triangle ABC, $\angle A = 45°$, $\angle C = 60°$, and altitude $\overline{BH}$ intersects median $\overline{AM}$ at point P. If AP = 4, what is AM? $4 + \frac{2\sqrt{3}}{3}$

5 Given: Trapezoid ABCD ($\overleftrightarrow{AD} \parallel \overleftrightarrow{BC}$),
   $AD = \frac{1}{4}(BC)$, $\frac{DF}{FC} = \frac{2}{3}$
   Find: $\frac{GE}{GF}$ $\frac{4}{11}$

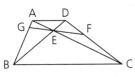

Problem-Set Notes and Additional Answers, continued

3

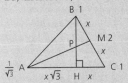

Draw the figure and assign ratios and weights as shown. Notice that for each labeled segment, $w_1d_1 = w_2d_2$. By the mass-point principle, the weights at A and B indicate that $\frac{BF}{AF} = \frac{1}{\frac{1}{2}} = \frac{2}{1}$. Since BF = 20, therefore, AF = 10.

4

Draw and label the figure as shown, according to the mass-point procedure. It can be seen that $\frac{AP}{PM} = \frac{2}{\frac{1}{\sqrt{3}}} = 2\sqrt{3}$.

Since AP = 4, $2\sqrt{3}y = 4$, so $y = \frac{2\sqrt{3}}{3}$. Thus, AM = $4 + \frac{2\sqrt{3}}{3}$.

5

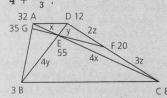

Assign weights so that E is the center of mass of trapezoid ABCD. One suitable set of weights is shown on the figure above. It can be seen that $\frac{GE}{GF} = \frac{20}{55} = \frac{4}{11}$.

T 730

16.6 INRADIUS AND CIRCUMRADIUS FORMULAS

Objective

After studying this section, you will be able to

- Use formulas to calculate the radii of a triangle's inscribed circle and a triangle's circumscribed circle

Part One: Introduction

In Section 16.2, we presented a formula that can be used to find the diameter of a triangle's circumcircle when one angle and the measure of the side opposite that angle's vertex are known. In this section, you will work with two other useful formulas—one for determining the radius of a triangle's inscribed circle (the triangle's *inradius*) and the other for determining the radius of a triangle's circumscribed circle (the triangle's *circumradius*).

Theorem 141 **The inradius r of a triangle can be found with the formula**

$$r = \frac{A}{s}$$

where A is the triangle's area and s is the triangle's semiperimeter.

Given: △ABC, with inscribed circle O and
an inradius (r) drawn to each side

Prove: $r = \frac{A}{s}$

Proof: We draw $\overline{IA}$, $\overline{IC}$, and $\overline{IB}$. The area of
△AIC is $\frac{1}{2}r(AC)$, the area of △AIB is $\frac{1}{2}r(AB)$,
and the area of △BIC is $\frac{1}{2}r(BC)$. Thus, in △ABC,

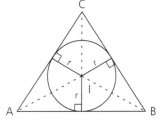

$$A = \frac{1}{2}r(AC) + \frac{1}{2}r(AB) + \frac{1}{2}r(BC)$$

$$= \frac{1}{2}r(AC + AB + BC)$$

$$= r\left[\frac{1}{2}(AC + AB + BC)\right]$$

$$= rs$$

Therefore, $r = \frac{A}{s}$.

The Geometric Supposer

You can use either the *Triangles* and *Quadrilaterals* programs or the *Circles* program in *The Geometric Supposer* series to preview the relationships expressed by the inradius and circumradius formulas.

To illustrate Theorem 141, have students draw a number of triangles with inscribed circles. For each triangle, have them measure the radius of the inscribed circle, then measure the triangle and divide its area by one half of its perimeter.

To illustrate Theorem 142 with *The Geometric Supposer*, have students draw a number of triangles with circumscribed circles. Have them measure the radius of each circle and the lengths of the sides of its inscribed triangle, then divide the product of the three sides by four times the area of the triangle.

Vocabulary

circumradius

inradius

Theorem 142 *The circumradius R of a triangle can be found with the formula*

$$R = \frac{abc}{4A}$$

where a, b, and c are the lengths of the sides of the triangle and A is the triangle's area.

Given: $\triangle ABC$, with circumcircle O
Prove: $R = \frac{abc}{4A}$

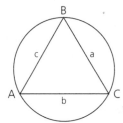

Proof: We draw diameter $\overline{BD}$ and chord $\overline{DC}$.
In $\triangle BDC$,

$$\sin \angle D = \frac{BC}{BD}$$

$$= \frac{a}{2R}$$

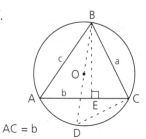

$$AC = b$$

Since $\angle A$ and $\angle D$ are inscribed angles intercepting the same arc, they are congruent, so $\sin \angle A = \frac{a}{2R}$. We now draw altitude $\overline{BE}$, with a length that we shall refer to as h_b. Therefore,

$$\sin \angle A = \frac{h_b}{c}$$

$$\frac{a}{2R} = \frac{h_b}{c}$$

$$R = \frac{ac}{2h_b}$$

$$= \frac{abc}{2h_b b} \quad \left(\text{Multiplying by } \frac{b}{b} \right)$$

$$= \frac{abc}{2(2A)} \quad \left(\text{Since } A = \frac{1}{2} bh_b \right)$$

$$= \frac{abc}{4A}$$

**Problem-Set Notes
and Additional Answers**

1   By Hero's formula,
$$A = \sqrt{7(4)(2)(1)} = 2\sqrt{14}$$
$$r = \frac{A}{s} = \frac{2\sqrt{14}}{7}$$
$$R = \frac{abc}{4A} = \frac{3(5)(6)}{8\sqrt{14}} = \frac{45\sqrt{14}}{56}$$

Part Two: Sample Problems

Problem 1   *Find the inradius and the circumradius of a (7, 8, 11) triangle.*

Solution   First, we use Hero's formula (see Section 11.8) to find the triangle's area.

$$A = \sqrt{13\,(6)\,(5)\,(2)} = 2\sqrt{195}$$

By the inradius formula,

$$r = \frac{A}{s} = \frac{2\sqrt{195}}{13}$$

By the circumradius formula,

$$R = \frac{abc}{4A} = \frac{7\,(8)\,(11)}{4\left(2\sqrt{195}\right)} = \frac{77\sqrt{195}}{195}$$

Problem 2   *Find the inradius and the circumradius of a (12, 35, 37) triangle.*

Solution   Be alert! By using the converse of the Pythagorean Theorem, we can establish that this is a right triangle. Therefore, $A = \frac{1}{2}(12)\,(35) = 210$.

By the inradius formula,

$$r = \frac{A}{s} = \frac{210}{42} = 5$$

By the circumradius formula,

$$R = \frac{abc}{4A} = \frac{12\,(35)\,(37)}{4\,(210)} = \frac{37}{2}$$

In this problem, is it a coincidence that the circumradius is half the hypotenuse? Is it a coincidence that the inradius of this right triangle is $\frac{a + b - c}{2}$?

Part Three: Problem Set

1 Find the inradius and the circumradius of a (3, 5, 6) triangle. $\frac{2\sqrt{14}}{7}$; $\frac{45\sqrt{14}}{56}$

2 Find the inradius and the circumradius of a (9, 40, 41) triangle. 4; $\frac{41}{2}$

3 Find the inradius and the circumradius of a (6, 8, 12) triangle. $\frac{\sqrt{455}}{13}$; $\frac{144\sqrt{455}}{455}$

4 Two of the sides of a triangle have measures of 10 and 12. If the triangle is inscribed in a circle with a diameter of 15, what is the altitude to the third side? (Hint: Substitute values in the circumradius formula.)   8

5 a Find the length of the third side of the triangle shown. $\frac{13\sqrt{10}}{10}$
   b Find the circumradius of the triangle to four significant digits. ≈ 4.744

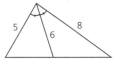

Problem-Set Notes and Additional Answers, continued

2   Since a (9, 40, 41) triangle is a right triangle, $A = \frac{1}{2}(9)(40) = 180$.

$$r = \frac{A}{s} = \frac{180}{45} = 4$$
$$R = \frac{abc}{4A} = \frac{9(40)(41)}{4(180)} = \frac{41}{2}$$

Note   It can be proved that in any right triangle the inradius is $\frac{a + b - c}{2}$ and the circumradius is half the hypotenuse. You may wish to have students derive these formulas.

3   By Hero's formula,
$$A = \sqrt{13(7)(5)(1)} = \sqrt{455}$$
$$r = \frac{A}{s} = \frac{\sqrt{455}}{13}$$
$$R = \frac{abc}{4A} = \frac{6(8)(12)}{4\sqrt{455}} = \frac{144\sqrt{455}}{455}$$

4   Let h represent the altitude to side c (the third side). Since $A = \frac{1}{2}ch$, $R = \frac{abc}{4A} = \frac{abc}{4\left(\frac{1}{2}ch\right)} = \frac{ab}{2h}$.

Thus, $\frac{15}{2} = \frac{10(12)}{2h}$
$$30h = 240$$
$$h = 8$$

5a   By the Angle Bisector Theorem and by Stewart's Theorem,
$$5^2(8x) + 8^2(5x) = 6^2(13x) + 13x(5x)(8x)$$
$$200x + 320x = 468x + 520x^3$$
$$x^2 = \frac{1}{10}$$
$$x = \frac{\sqrt{10}}{10}$$

Hence, $13x = \frac{13\sqrt{10}}{10}$.

b   First, use Hero's formula to find the area.
$$A = \sqrt{s(s - a)(s - b)(s - c)}$$
$$\approx 8.665989567$$

Now use the circumradius formula to complete the solution.
$$R = \frac{abc}{4A} \approx \frac{\frac{5(8)13\sqrt{10}}{10}}{4(8.665989567)}$$
$$\approx \frac{13\sqrt{10}}{8.665989567}$$
$$\approx 4.743787107$$
$$\approx 4.744$$

FORMULAS FOR YOU TO DEVELOP

Objective

After studying this section, you will be able to
- Find or prove five additional formulas

In this section, you are asked to establish the validity of five formulas. In each case, there is a problem or two for you to solve by applying the formula.

Three Triangle Formulas

I. Consider a right triangle with legs a and b and hypotenuse c. Find a formula for the perimeter P of the triangle in terms of its hypotenuse and its area A. $P = c + \sqrt{c^2 + 4A}$

Now use your formula to find the perimeter of a right triangle if the triangle's area is 40 and the altitude to its hypotenuse is 5. $16 + 4\sqrt{26}$

II. Find a formula relating a triangle's inradius, r, to its three altitudes, h_a, h_b, and h_c. $\frac{1}{h_a} + \frac{1}{h_b} + \frac{1}{h_c} = \frac{1}{r}$

Now use your formula to find the inradius of a triangle whose three altitudes are 4, 5, and 6. $\frac{60}{37}$

III. Prove that the area A of any triangle ABC, with side lengths a, b, and c, can be found with the formula

$$A = \frac{1}{2}ab(\sin \angle C)$$

Now use this formula to solve the following problems:

a Find the area of a regular dodecagon inscribed in a circle with a diameter of 20. 300

b Given: Diagram as marked

Find: The area of $\triangle ADE$ $\frac{576}{25}$

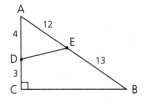

Ceva's Theorem

The following theorem is known as Ceva's Theorem, after the Italian mathematician Giovanni Ceva (c. 1647–1734).

Theorem 143 **If ABC is a triangle with D on $\overline{BC}$, E on $\overline{AC}$, and F on $\overline{AB}$, then the three segments $\overline{AD}$, $\overline{BE}$, and $\overline{CF}$ are concurrent if, and only if,**

$$\left(\frac{BD}{DC}\right)\left(\frac{CE}{EA}\right)\left(\frac{AF}{FB}\right) = 1$$

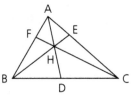

734 | Chapter 16 Enrichment Topics

Vocabulary

sensed magnitude

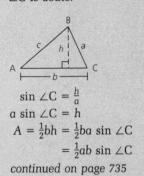

Copy the following proof of Ceva's Theorem and see if you can fill in the missing reasons.

Proof:

Part One ("Only if" part)

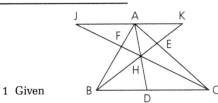

1. $\overline{AD}$, $\overline{BE}$, and $\overline{CF}$ are concurrent at H.
2. Draw through A a line parallel to $\overleftrightarrow{BC}$, and extend $\overline{CH}$ and $\overline{BH}$ to meet the line at J and K respectively.
3. $\triangle JAH \sim \triangle CDH$; $\triangle BDH \sim \triangle KAH$
4. $\frac{BD}{AK} = \frac{DC}{AJ}$, or $\frac{BD}{DC} = \frac{AK}{AJ}$
5. $\triangle KAE \sim \triangle BCE$, so $\frac{CE}{EA} = \frac{BC}{AK}$.
6. $\triangle JAF \sim \triangle CFB$, so $\frac{AF}{FB} = \frac{JA}{BC}$.
7. $\left(\frac{BD}{DC}\right)\left(\frac{CE}{EA}\right)\left(\frac{AF}{FB}\right) = \left(\frac{AK}{AJ}\right)\left(\frac{BC}{AK}\right)\left(\frac{JA}{BC}\right) = 1$

1. Given
2. Parallel Postulate; a seg. can be extended as far as desired.

3. AA
4. Corr. sides of $\sim$ $\triangle$ are proportional; algebra
5. AA; corr. sides of $\sim$ $\triangle$ are proportional.
6. Same as 5
7. Mult. Prop. of Eq.; Mult. Inverse Prop.

Part Two ("If" part)

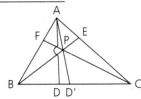

1. $\left(\frac{BD}{DC}\right)\left(\frac{CE}{EA}\right)\left(\frac{AF}{FB}\right) = 1$
2. Let $\overline{BE}$ and $\overline{FC}$ intersect at P.
3. Draw $\overline{AP}$ and extend it to intersect $\overline{BC}$ at D'.
4. $\left(\frac{BD'}{D'C}\right)\left(\frac{CE}{EA}\right)\left(\frac{AF}{FB}\right) = 1$
5. $\frac{BD}{DC} = \frac{BD'}{D'C}$
6. Point D is the same as point D'.
7. $\overline{AD}$, $\overline{BE}$, and $\overline{CF}$ are concurrent.

1. Given
2. Two coplanar, nonparallel lines intersect in a pt.
3. Two pts. determine a line; a seg. can be extended as far as desired.
4. Part One above
5. Div. Prop. of Eq.; Trans. Prop. of Eq.
6. A seg. is divided into a given ratio by a unique pt.
7. Def. of concurrent (since $\overline{AD}$ is $\overline{AD'}$)

Now use Ceva's Theorem to prove the medians of a triangle concurrent.

Theorem of Menelaus

The following theorem is known as the Theorem of Menelaus. (Menelaus was an Alexandrian mathematician of the first century A.D.) It is important to note that this theorem involves the concept of **sensed magnitudes**—that is, the measure of a segment in one direction is considered to be the opposite of its measure in the other direction. (For example, AB = − BA.)

Section 16.7   Formulas for You to Develop   **735**

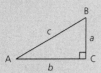
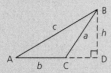

Given: △ABC with medians $\overline{CE}$, $\overline{BD}$, and $\overline{AF}$

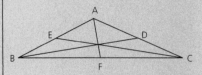

Prove: $\overline{CE}$, $\overline{BD}$, and $\overline{AF}$ are concurrent.

Proof:

1 △ABC with medians $\overline{CE}$, $\overline{BD}$, and $\overline{AF}$ (Given)

2 $\frac{AE}{EB} = 1$; $\frac{BF}{FC} = 1$; $\frac{CD}{DA} = 1$ (Defs. of *median* and *midpoint*)

3 $\left(\frac{AE}{EB}\right)\left(\frac{BF}{FC}\right)\left(\frac{CD}{DA}\right) = 1$ (Mult.)

4 $\overline{CE}$, $\overline{BD}$, and $\overline{AF}$ are concurrent. (Ceva's Theorem)

By the Theorem of Menelaus,
$\left(\frac{AP}{PB}\right)\left(\frac{BQ}{QC}\right)\left(\frac{CR}{RA}\right) = -1$
$\left(\frac{AP}{PB}\right)\left(\frac{7}{-2}\right)\left(\frac{1}{1}\right) = -1$
$7(AP) = 2(PB)$
$AP{:}BP = 2{:}7$

Note This problem can also be solved by using mass points.

Theorem 144 *If ABC is a triangle and F is on $\overline{AB}$, E is on $\overline{AC}$, and D is on an extension of $\overline{BC}$, then the three points D, E, and F are collinear if, and only if,*

$$\left(\frac{BD}{DC}\right)\left(\frac{CE}{EA}\right)\left(\frac{AF}{FB}\right) = -1$$

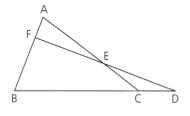

Once again, copy the proof and see if you can supply the reasons for the major steps.

Proof:

Part One ("Only if" part)

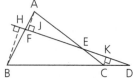

1 D, E, and F are collinear.
2 Draw $\overline{BH}$, $\overline{AJ}$, and $\overline{CK}$, each perpendicular to $\overleftrightarrow{FD}$.
3 $\overleftrightarrow{CK} \parallel \overleftrightarrow{AJ} \parallel \overleftrightarrow{BH}$
4 △CKE ~ △AJE; △BHF ~ △AJF; △DKC ~ △DHB
5 $\frac{BD}{DC} = \frac{BH}{KC}; \frac{CE}{EA} = \frac{CK}{AJ}; \frac{AF}{FB} = \frac{AJ}{BH}$
6 $\left(\frac{BD}{DC}\right)\left(\frac{CE}{EA}\right)\left(\frac{AF}{FB}\right) = \left(\frac{BH}{KC}\right)\left(\frac{CK}{AJ}\right)\left(\frac{AJ}{BH}\right) = -1$

1 Given
2 From a pt. outside a line, only one ⊥ can be drawn to the line.
3 In a plane, 2 lines ⊥ to a third are ‖.
4 AA
5 Corr. sides of ~ △ are proportional.
6 Mult. Prop. of Eq.; Mult. Inverse Prop.; algebra

Part Two ("If" part)

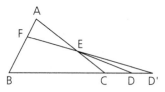

1 $\left(\frac{BD}{DC}\right)\left(\frac{CE}{EA}\right)\left(\frac{AF}{FB}\right) = -1$
2 Let $\overleftrightarrow{FE}$ intersect $\overleftrightarrow{BC}$ at D'.
3 $\left(\frac{BD'}{D'C}\right)\left(\frac{CE}{EA}\right)\left(\frac{AF}{FB}\right) = -1$
4 $\frac{BD'}{D'C} = \frac{BD}{DC}$
5 Point D is the same as point D'.
6 D, E, and F are collinear.

1 Given
2 Two coplanar, nonparallel lines intersect in a pt.
3 Part One above
4 Div. Prop. of Eq.; Trans. Prop. of Eq.
5 A seg. is divided into a given ratio by a unique pt.
6 Def. of *collinear* (since $\overleftrightarrow{FED}$ is $\overleftrightarrow{FED'}$)

In the diagram at the right, R is the midpoint of $\overline{AC}$, and $\overline{BC}$ is extended to point Q so that BC:CQ = 5:2. Use the Theorem of Menelaus to find AP:PB. 2:7

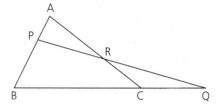

CHAPTER SUMMARY

CONCEPTS AND PROCEDURES

After studying this chapter, you should be able to

- Use a formula to determine the distance from a point to a line in the coordinate plane (16.1)
- Use a formula to find the area of a triangle when only the coordinates of its vertices are known (16.2)
- Use a formula to find the diameter of a triangle's circumscribed circle (16.2)
- Recognize a relationship among the parts of a triangle with a segment drawn from a vertex to the opposite side (16.3)
- Recognize a relationship involving the sides and the diagonals of a cyclic quadrilateral (16.4)
- Use the concept of mass points to solve problems (16.5)
- Use formulas to calculate the radii of a triangle's inscribed circle and a triangle's circumscribed circle (16.6)
- Find or prove five additional formulas (16.7)

VOCABULARY

circumcircle (16.2)

circumradius (16.6)

inradius (16.6)

mass point (16.5)

point-line distance formula (16.1)

sensed magnitude (16.7)

Problem-Set Notes and Additional Answers

1 By Stewart's Theorem,
$3x^2 + 4y^2 = 343 + 84$, and
$6x^2 + 2y^2 = 200 + 96$.

$$\begin{cases} 3x^2 + 4y^2 = 427 \\ 3x^2 + y^2 = 148 \end{cases}$$

The solutions of the system are $x = \frac{\sqrt{165}}{3}$ and $y = \sqrt{93}$.

2 Using the mass-point principle, label the figure as shown.

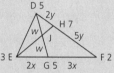

It can be seen that $\frac{DJ}{DG} = \frac{1}{2}$.

3 Using the mass-point principle, label the figure as shown.

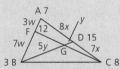

Since $3w = 12$, $w = 4$. Thus, $BF = 7(4) = 28$.

4 A special case of Stewart's
Theorem states that in a right
triangle the sum of the
squares of the segments from
the vertex opposite the
hypotenuse to the trisection
points of the hypotenuse is $\frac{5}{9}$
the square of the hypotenuse.

$7^2 + 8^2 = \frac{5}{9}c^2$

$c^2 = \frac{1017}{5}$

Therefore, $\frac{2}{3}c \approx 9.508$.

5 Use the inradius and
circumradius formulas.

$\frac{r}{R} = \frac{\frac{A}{s}}{\frac{abc}{4A}} = \frac{4A^2}{s(abc)}$

$\frac{4}{6\sqrt{2}} = \frac{4(8)(8)}{s(64\sqrt{2})}$

$4s\sqrt{2} = 4(6\sqrt{2})$

$s = 6$

6

$\angle$RAO is a right angle. In
$\triangle$RAO, $5^2 + x^2 = 12^2$, so
$x = \sqrt{119}$. In $\triangle$RHO, $2^2 +$
$y^2 = 12^2$, so $y = 2\sqrt{35}$. Now
use Ptolemy's Theorem.

$12(AH) = 5(2\sqrt{35}) + 2\sqrt{119}$

$AH = \frac{5\sqrt{35} + \sqrt{119}}{6}$

7

Use the Angle Bisector
Theorem to place 2x and 3x
as shown in the figure. Now
use Ptolemy's Theorem.

$6^2(2x) + 4^2(3x) = 4^2(5x)$
$\qquad + 5x(2x)(3x)$

$72x + 48x = 80x + 30x^3$

$40 = 30x^2$

$x^2 = \frac{4}{3}$

$x = \frac{2\sqrt{3}}{3}$

Therefore, $5x = \frac{10\sqrt{3}}{3}$.

---

1 Use the figures below to solve for x
and y. $x = \frac{\sqrt{165}}{3}$; $y = \sqrt{93}$

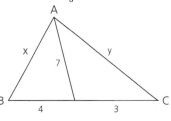

2 In the figure shown, EG:GF = 4:6 and
DH:HF = 2:5. Find DJ:DG.   1:2

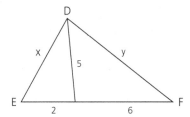

3 In the figure shown, AF = 12, AD:DC =
8:7, and $\overline{CF}$ intersects $\overline{BD}$ at G so that
BG = 5(GD). Find BF.   28

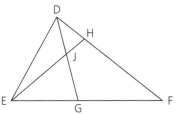

4 Given: $\angle$DAC = 90°,
$\qquad \overline{DE} \cong \overline{EF} \cong \overline{FC}$,
$\qquad$ AE = 7, AF = 8

Find: DF, to four significant digits ≈ 9.508

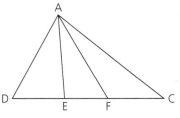

5 The ratio of a triangle's inradius to its circumradius is $4:6\sqrt{2}$. If
the triangle's area is 8 and the product of the measures of its
sides is $64\sqrt{2}$, what is its semiperimeter?   6

6 A quadrilateral, RHOA, is inscribed in a circle, D. If RO = 12, AR = 5, RH = 2, and the radius of ⊙D is 6, what is AH? $\frac{5\sqrt{35} + \sqrt{119}}{6}$

7 In the diagram at the right, the measures of two sides and an angle bisector of a triangle are shown. Find the measure of the third side of the triangle. $\frac{10\sqrt{3}}{3}$

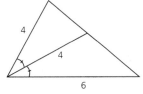

8 Find, to four significant digits, the distance from the point (−2, 5) to the graph of x − 3y = 7. ≈7.589

9 Find the distance in space from the point (3, 4, 2) to the plane represented by 3x − 4y + 12z = 20. $\frac{3}{13}$

10 Find the distance between the graphs of 3x − 4y + 10 = 0 and 6x − 8y + 15 = 0. $\frac{1}{2}$

11 Write equations of the two lines that are parallel to the graph of x − 4y = 7 and three units from the point (5, 1).
x − 4y − 1 + 3√17 = 0 and x − 4y − 1 − 3√17 = 0

12 Find the area of a triangle with vertices at (5, 1), (16, −4), and (3, 12). $55\frac{1}{2}$

13 What is the area of the circle in the diagram at the right? 48π

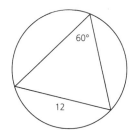

Problem-Set Notes and Additional Answers, continued

8 Rewrite x − 3y = 7 as x − 3y − 7 = 0. By the point-line distance formula,
$d = \frac{|1(-2) - 3(5) - 7|}{\sqrt{1^2 + (-3)^2}} \approx 7.589$

9 By the point-plane distance formula (see Section **16.1**, problem **10**),
$d = \frac{|3(3) - 4(4) + 12(2) - 20|}{\sqrt{3^2 + (-4)^2 + 12^2}} = \frac{3}{13}$

10 Since the lines are parallel, pick a point (−6, −2) on 3x − 4y + 10 = 0 and use the point-line distance formula.
$d = \frac{|6(-6) - 8(-2) + 15|}{\sqrt{6^2 + (-8)^2}} = \frac{1}{2}$

11 Since the lines are parallel to the graph of x − 4y = 7, they must have the same slope as that line. Hence, their equations are of the form x − 4y + c = 0. Use the point-line distance formula, substituting (5, 1) for (x₁, y₁) and 3 for d.
$$3 = \frac{|5 - 4(1) + c|}{\sqrt{17}}$$
and
$$3\sqrt{17} = |c + 1|$$
c + 1 = 3√17 or
c = −1 + 3√17
−c − 1 = 3√17
c = −1 − 3√17
So the lines are represented by x − 4y − 1 ± 3√17 = 0.

12 Use Theorem 137.
A = ½|5(−4) + 16(12) + 3(1) − 5(12) − 16(1) − 3(−4)|
= $55\frac{1}{2}$

13 By the formula for the diameter of a circumcircle,
$\frac{a}{\sin \angle A} = D$
$\frac{12}{\sin 60°} = D$
$\frac{12}{\frac{\sqrt{3}}{2}} = D$
D = 8√3
A = πR² = π(4√3)² = 48π

LIST OF POSTULATES AND THEOREMS

Postulates

Any segment or angle is congruent to itself. (Reflexive Property)	**112**
If there exists a correspondence between the vertices of two triangles such that three sides of one triangle are congruent to the corresponding sides of the other triangle, the two triangles are congruent. (SSS)	**116**
If there exists a correspondence between the vertices of two triangles such that two sides and the included angle of one triangle are congruent to the corresponding parts of the other triangle, the two triangles are congruent. (SAS)	**117**
If there exists a correspondence between the vertices of two triangles such that two angles and the included side of one triangle are congruent to the corresponding parts of the other triangle, the two triangles are congruent. (ASA)	**117**
Two points determine a line (or ray or segment).	**132**
If there exists a correspondence between the vertices of two right triangles such that the hypotenuse and a leg of one triangle are congruent to the corresponding parts of the other triangle, the two right triangles are congruent. (HL)	**156**
A line segment is the shortest path between two points.	**184**
Through a point not on a line there is exactly one parallel to the given line. (Parallel Postulate)	**224**
Three noncollinear points determine a plane.	**270**
If a line intersects a plane not containing it, then the intersection is exactly one point.	**271**
If two planes intersect, their intersection is exactly one line.	**271**
If there exists a correspondence between the vertices of two triangles such that the three angles of one triangle are congruent to the corresponding angles of the other triangle, then the triangles are similar. (AAA)	**339**
A tangent line is perpendicular to the radius drawn to the point of contact.	**459**
If a line is perpendicular to a radius at its outer endpoint, then it is tangent to the circle.	**459**
Circumference of a circle = $\pi \cdot$ diameter.	**499**
The area of a rectangle is equal to the product of the base and the height for that base.	**512**
Every closed region has an area.	**512**
If two closed figures are congruent, then their areas are equal.	**512**

If two closed regions intersect only along a common boundary, then the area of their union is equal to the sum of their individual areas. **512**

The area of a circle is equal to the product of π and the square of the radius. **537**

Total area of a sphere $= 4\pi r^2$, where r is the sphere's radius. **571**

The volume of a right rectangular prism is equal to the product of its length, its width, and its height. **576**

For any two real numbers x and y, exactly one of the following statements is true: $x < y$, $x = y$, or $x > y$. (Law of Trichotomy) **687**

If $a > b$ and $b > c$, then $a > c$. Similarly, if $x < y$ and $y < z$, then $x < z$. (Transitive Property of Inequality) **687**

If $a > b$, then $a + x > b + x$. (Addition Property of Inequality) **687**

If $x < y$ and $a > 0$, then $a \cdot x < a \cdot y$. (Positive Multiplication Property of Inequality) **688**

If $x < y$ and $a < 0$, then $a \cdot x > a \cdot y$. (Negative Multiplication Property of Inequality) **688**

The sum of the measures of any two sides of a triangle is always greater than the measure of the third side. **691**

Theorems

1 If two angles are right angles, then they are congruent. **24**

2 If two angles are straight angles, then they are congruent. **24**

3 If a conditional statement is true, then the contrapositive of the statement is also true. (If p, then $q \Leftrightarrow$ If $\sim q$, then $\sim p$.) **46**

4 If angles are supplementary to the same angle, then they are congruent. **76**

5 If angles are supplementary to congruent angles, then they are congruent. **77**

6 If angles are complementary to the same angle, then they are congruent. **77**

7 If angles are complementary to congruent angles, then they are congruent. **77**

8 If a segment is added to two congruent segments, the sums are congruent. (Addition Property) **82**

9 If an angle is added to two congruent angles, the sums are congruent. (Addition Property) **83**

10 If congruent segments are added to congruent segments, the sums are congruent. (Addition Property) **83**

List of Theorems, *continued*

11	If congruent angles are added to congruent angles, the sums are congruent. (Addition Property)	83
12	If a segment (or angle) is subtracted from congruent segments (or angles), the differences are congruent. (Subtraction Property)	84
13	If congruent segments (or angles) are subtracted from congruent segments (or angles), the differences are congruent. (Subtraction Property)	84
14	If segments (or angles) are congruent, their like multiples are congruent. (Multiplication Property)	89
15	If segments (or angles) are congruent, their like divisions are congruent. (Division Property)	90
16	If angles (or segments) are congruent to the same angle (or segment), they are congruent to each other. (Transitive Property)	95
17	If angles (or segments) are congruent to congruent angles (or segments), they are congruent to each other. (Transitive Property)	95
18	Vertical angles are congruent.	101
19	All radii of a circle are congruent.	126
20	If two sides of a triangle are congruent, the angles opposite the sides are congruent. (If $\triangle$, then $\triangle$.)	148
21	If two angles of a triangle are congruent, the sides opposite the angles are congruent. (If $\triangle$, then $\triangle$.)	149
22	If A = (x_1, y_1) and B = (x_2, y_2), then the midpoint M = (x_m, y_m) of $\overline{AB}$ can be found by using the midpoint formula: $$M = (x_m, y_m) = \left(\frac{x_1 + x_2}{2}, \frac{y_1 + y_2}{2}\right)$$	171
23	If two angles are both supplementary and congruent, then they are right angles.	180
24	If two points are each equidistant from the endpoints of a segment, then the two points determine the perpendicular bisector of that segment.	185
25	If a point is on the perpendicular bisector of a segment, then it is equidistant from the endpoints of that segment.	185
26	If two nonvertical lines are parallel, then their slopes are equal.	200
27	If the slopes of two nonvertical lines are equal, then the lines are parallel.	200
28	If two lines are perpendicular and neither is vertical, each line's slope is the opposite reciprocal of the other's.	200
29	If a line's slope is the opposite reciprocal of another line's slope, the two lines are perpendicular.	200
30	The measure of an exterior angle of a triangle is greater than the measure of either remote interior angle.	216

31	If two lines are cut by a transversal such that two alternate interior angles are congruent, the lines are parallel. (Alt. int. ∠s ≅ ⇒ ∥ lines)	217
32	If two lines are cut by a transversal such that two alternate exterior angles are congruent, the lines are parallel. (Alt. ext. ∠s ≅ ⇒ ∥ lines)	217
33	If two lines are cut by a transversal such that two corresponding angles are congruent, the lines are parallel. (Corr. ∠s ≅ ⇒ ∥ lines)	217
34	If two lines are cut by a transversal such that two interior angles on the same side of the transversal are supplementary, the lines are parallel.	218
35	If two lines are cut by a transversal such that two exterior angles on the same side of the transversal are supplementary, the lines are parallel.	218
36	If two coplanar lines are perpendicular to a third line, they are parallel.	218
37	If two parallel lines are cut by a transversal, each pair of alternate interior angles are congruent. (∥ lines ⇒ alt. int. ∠s ≅)	225
38	If two parallel lines are cut by a transversal, then any pair of the angles formed are either congruent or supplementary.	225
39	If two parallel lines are cut by a transversal, each pair of alternate exterior angles are congruent. (∥ lines ⇒ alt. ext. ∠s ≅)	226
40	If two parallel lines are cut by a transversal, each pair of corresponding angles are congruent. (∥ lines ⇒ corr. ∠s ≅)	226
41	If two parallel lines are cut by a transversal, each pair of interior angles on the same side of the transversal are supplementary.	226
42	If two parallel lines are cut by a transversal, each pair of exterior angles on the same side of the transversal are supplementary.	226
43	In a plane, if a line is perpendicular to one of two parallel lines, it is perpendicular to the other.	227
44	If two lines are parallel to a third line, they are parallel to each other. (Transitive Property of Parallel Lines)	227
45	A line and a point not on the line determine a plane.	271
46	Two intersecting lines determine a plane.	271
47	Two parallel lines determine a plane.	271
48	If a lie is perpendicular to two distinct lines that lie in a plane and that pass through its foot, then it is perpendicular to the plane.	277
49	If a plane intersects two parallel planes, the lines of intersection are parallel.	283
50	The sum of the measures of the three angles of a triangle is 180.	295
51	The measure of an exterior angle of a triangle is equal to the sum of the measures of the remote interior angles.	296

List of Theorems, *continued*

52 A segment joining the midpoints of two sides of a triangle is parallel to the third side, and its length is one-half the length of the third side. (Midline Theorem) — 296

53 If two angles of one triangle are congruent to two angles of a second triangle, then the third angles are congruent. (No-Choice Theorem) — 302

54 If there exists a correspondence between the vertices of two triangles such that two angles and a nonincluded side of one are congruent to the corresponding parts of the other, then the triangles are congruent. (AAS) — 302

55 The sum S_i of the measures of the angles of a polygon with n sides is given by the formula $S_i = (n - 2)180$. — 308

56 If one exterior angle is taken at each vertex, the sum S_e of the measures of the exterior angles of a polygon is given by the formula $S_e = 360$. — 308

57 The number d of diagonals that can be drawn in a polygon of n sides is given by the formula $d = \frac{n(n-3)}{2}$. — 308

58 The measure E of each exterior angle of an equiangular polygon of n sides is given by the formula $E = \frac{360}{n}$. — 315

59 In a proportion, the product of the means is equal to the product of the extremes. (Means-Extremes Products Theorem) — 327

60 If the product of a pair of nonzero numbers is equal to the product of another pair of nonzero numbers, then either pair of numbers may be made the extremes, and the other pair the means, of a proportion. (Means-Extremes Ratio Theorem) — 327

61 The ratio of the perimeters of two similar polygons equals the ratio of any pair of corresponding sides. — 334

62 If there exists a correspondence between the vertices of two triangles such that two angles of one triangle are congruent to the corresponding angles of the other, then the triangles are similar. (AA) — 339

63 If there exists a correspondence between the vertices of two triangles such that the ratios of the measures of corresponding sides are equal, then the triangles are similar. (SSS~) — 340

64 If there exists a correspondence between the vertices of two triangles such that the ratios of the measures of two pairs of corresponding sides are equal and the included angles are congruent, then the triangles are similar. (SAS~) — 340

65 If a line is parallel to one side of a triangle and intersects the other two sides, it divides those two sides proportionally. (Side-Splitter Theorem) — 351

66 If three or more parallel lines are intersected by two transversals, the parallel lines divide the transversals proportionally. — 351

67 If a ray bisects an angle of a triangle, it divides the opposite side into segments that are proportional to the adjacent sides. (Angle Bisector Theorem) — 352

68 If an altitude is drawn to the hypotenuse of a right triangle, then   **378**
   a. The two triangles formed are similar to the given right triangle and to each other
   b. The altitude to the hypotenuse is the mean proportional between the segments of the hypotenuse
   c. Either leg of the given right triangle is the mean proportional between the hypotenuse of the given right triangle and the segment of the hypotenuse adjacent to that leg (i.e., the projection of that leg on the hypotenuse)

69 The square of the measure of the hypotenuse of a right triangle is equal to the sum   **384**
   of the squares of the measures of the legs. (Pythagorean Theorem)

70 If the square of the measure of one side of a triangle equals the sum of the squares   **385**
   of the measures of the other two sides, then the angle opposite the longest side is a
   right angle.

71 If $P = (x_1, y_1)$ and $Q = (x_2, y_2)$ are any two points, then the distance between them   **393**
   can be found with the formula

$$PQ = \sqrt{(x_2 - x_1)^2 + (y_2 - y_1)^2} \text{ or } PQ = \sqrt{(\Delta x)^2 + (\Delta y)^2}$$

72 In a triangle whose angles have the measures 30, 60, and 90, the lengths of the sides   **405**
   opposite these angles can be represented by x, $x\sqrt{3}$, and $2x$ respectively. (30°–60°–90°–Triangle Theorem)

73 In a triangle whose angles have the measures 45, 45, and 90, the lengths of the sides   **406**
   opposite these angles can be represented by x, x, and $x\sqrt{2}$ respectively. (45°–45°–90°–Triangle Theorem)

74 If a radius is perpendicular to a chord, then it bisects the chord.   **441**

75 If a radius of a circle bisects a chord that is not a diameter, then it is perpendicular   **441**
   to that chord.

76 The perpendicular bisector of a chord passes through the center of the circle.   **441**

77 If two chords of a circle are equidistant from the center, then they are congruent.   **446**

78 If two chords of a circle are congruent, then they are equidistant from the center of   **446**
   the circle.

79 If two central angles of a circle (or of congruent circles) are congruent, then their   **453**
   intercepted arcs are congruent.

80 If two arcs of a circle (or of congruent circles) are congruent, then the correspond-   **453**
   ing central angles are congruent.

81 If two central angles of a circle (or of congruent circles) are congruent, then the   **453**
   corresponding chords are congruent.

82 If two chords of a circle (or of congruent circles) are congruent, then the corre-   **453**
   sponding central angles are congruent.

83 If two arcs of a circle (or of congruent circles) are congruent, then the correspond-   **453**
   ing chords are congruent.

List of Theorems, *continued*

84	If two chords of a circle (or of congruent circles) are congruent, then the corresponding arcs are congruent.	**453**
85	If two tangent segments are drawn to a circle from an external point, then those segments are congruent. (Two-Tangent Theorem)	**460**
86	The measure of an inscribed angle or a tangent-chord angle (vertex on a circle) is one-half the measure of its intercepted arc.	**469**
87	The measure of a chord-chord angle is one-half the sum of the measures of the arcs intercepted by the chord-chord angle and its vertical angle.	**470**
88	The measure of a secant-secant angle, a secant-tangent angle, or a tangent-tangent angle (vertex outside a circle) is one-half the difference of the measures of the intercepted arcs.	**471**
89	If two inscribed or tangent-chord angles intercept the same arc, then they are congruent.	**479**
90	If two inscribed or tangent-chord angles intercept congruent arcs, then they are congruent.	**479**
91	An angle inscribed in a semicircle is a right angle.	**480**
92	The sum of the measures of a tangent-tangent angle and its minor arc is 180.	**480**
93	If a quadrilateral is inscribed in a circle, its opposite angles are supplementary.	**487**
94	If a parallelogram is inscribed in a circle, it must be a rectangle.	**488**
95	If two chords of a circle intersect inside the circle, then the product of the measures of the segments of one chord is equal to the product of the measures of the segments of the other chord. (Chord-Chord Power Theorem)	**493**
96	If a tangent segment and a secant segment are drawn from an external point to a circle, then the square of the measure of the tangent segment is equal to the product of the measures of the entire secant segment and its external part. (Tangent-Secant Power Theorem)	**493**
97	If two secant segments are drawn from an external point to a circle, then the product of the measures of one secant segment and its external part is equal to the product of the measures of the other secant segment and its external part. (Secant-Secant Power Theorem)	**494**
98	The length of an arc is equal to the circumference of its circle times the fractional part of the circle determined by the arc.	**500**
99	The area of a square is equal to the square of a side.	**512**
100	The area of a parallelogram is equal to the product of the base and the height.	**516**
101	The area of a triangle is equal to one-half the product of a base and the height (or altitude) for that base.	**517**

102	The area of a trapezoid equals one-half the product of the height and the sum of the bases.	523
103	The measure of the median of a trapezoid equals the average of the measures of the bases.	524
104	The area of a trapezoid is the product of the median and the height.	524
105	The area of a kite equals half the product of its diagonals.	528
106	The area of an equilateral triangle equals the product of one-fourth the square of a side and the square root of 3.	531
107	The area of a regular polygon equals one-half the product of the apothem and the perimeter.	532
108	The area of a sector of a circle is equal to the area of the circle times the fractional part of the circle determined by the sector's arc.	537
109	If two figures are similar, then the ratio of their areas equals the square of the ratio of corresponding segments. (Similar-Figures Theorem)	544
110	A median of a triangle divides the triangle into two triangles with equal areas.	546
111	Area of a triangle $= \sqrt{s(s-a)(s-b)(s-c)}$, where a, b, and c are the lengths of the sides of the triangle and $s =$ semiperimeter $= \frac{a+b+c}{2}$. (Hero's formula)	550
112	Area of a cyclic quadrilateral $= \sqrt{(s-a)(s-b)(s-c)(s-d)}$, where a, b, c, and d are the sides of the quadrilateral and $s =$ semiperimeter $= \frac{a+b+c+d}{2}$. (Brahmagupta's formula)	550
113	The lateral area of a cylinder is equal to the product of the height and the circumference of the base.	571
114	The lateral area of a cone is equal to one-half the product of the slant height and the circumference of the base.	571
115	The volume of a right rectangular prism is equal to the product of the height and the area of the base.	576
116	The volume of any prism is equal to the product of the height and the area of the base.	576
117	The volume of a cylinder is equal to the product of the height and the area of the base.	577
118	The volume of a prism or a cylinder is equal to the product of the figure's cross-sectional area and its height.	577
119	The volume of a pyramid is equal to one third of the product of the height and the area of the base.	583
120	The volume of a cone is equal to one third of the product of the height and the area of the base.	584

List of Theorems, *continued*

121	In a pyramid or a cone, the ratio of the area of a cross section to the area of the base equals the square of the ratio of the figures' respective distances from the vertex.	**584**
122	The volume of a sphere is equal to four thirds of the product of π and the cube of the radius.	**589**
123	The y-form, or slope-intercept form, of the equation of a nonvertical line is $y = mx + b$, where b is the y-intercept of the line and m is the slope of the line.	**610**
124	The formula for an equation of a horizontal line is $y = b$, where b is the y-coordinate of every point on the line.	**611**
125	The formula for the equation of a vertical line is $x = a$, where a is the x-coordinate of every point on the line.	**612**
126	If $P = (x_1, y_1, z_1)$ and $Q = (x_2, y_2, z_2)$ are any two points, then the distance between them can be found with the formula $$PQ = \sqrt{(x_2 - x_1)^2 + (y_2 - y_1)^2 + (z_2 - z_1)^2}$$	**626**
127	The equation of a circle whose center is (h, k) and whose radius is r is $(x - h)^2 + (y - k)^2 = r^2$.	**633**
128	The perpendicular bisectors of the sides of a triangle are concurrent at a point that is equidistant from the vertices of the triangle. (The point of concurrency of the perpendicular bisectors is called the circumcenter of the triangle.)	**660**
129	The bisectors of the angles of a triangle are concurrent at a point that is equidistant from the sides of the triangle. (The point of concurrency of the angle bisectors is called the incenter of the triangle.)	**661**
130	The lines containing the altitudes of a triangle are concurrent. (The point of concurrency of the lines containing the altitudes is called the orthocenter of the triangle.)	**661**
131	The medians of a triangle are concurrent at a point that is two thirds of the way from any vertex of the triangle to the midpoint of the opposite side. (The point of concurrency of the medians of a triangle is called the centroid of the triangle.)	**662**
132	If two sides of a triangle are not congruent, then the angles opposite them are not congruent, and the larger angle is opposite the longer side. (If $\triangle$, then $\triangle\!\triangle$.)	**692**
133	If two angles of a triangle are not congruent, then the sides opposite them are not congruent, and the longer side is opposite the larger angle. (If $\triangle\!\triangle$, then $\triangle$.)	**692**
134	If two sides of one triangle are congruent to two sides of another triangle and the included angle in the first triangle is greater than the included angle in the second triangle, then the remaining side of the first triangle is greater than the remaining side of the second triangle. (SAS $\neq$)	**697**
135	If two sides of one triangle are congruent to two sides of another triangle and the third side of the first triangle is greater than the third side of the second triangle, then the angle opposite the third side in the first triangle is greater than the angle opposite the third side in the second triangle. (SSS $\neq$)	**697**

136   The distance d from any point $P = (x_1, y_1)$ to a line whose equation is in the form $ax + by + c = 0$ can be found with the formula   **713**

$$d = \frac{|ax_1 + by_1 + c|}{\sqrt{a^2 + b^2}}$$

137   The area A of a triangle with vertices at (x_1, y_1), (x_2, y_2), and (x_3, y_3) can be found with the formula   **717**

$$A = \frac{1}{2}|x_1y_2 + x_2y_3 + x_3y_1 - x_1y_3 - x_2y_1 - x_3y_2|$$

138   In any triangle ABC, with side lengths a, b, and c,   **718**

$$\frac{a}{\sin \angle A} = D \qquad \frac{b}{\sin \angle B} = D \qquad \frac{c}{\sin \angle D} = D$$

where D is the diameter of the triangle's circumcircle.

139   In any triangle ABC, with side lengths a, b, and c,   **721**

$$a^2n + b^2m = cd^2 + cmn$$

where d is the length of a segment from vertex C to the opposite side, dividing that side into segments with lengths m and n. (Stewart's Theorem)

140   If a quadrilateral is inscribable in a circle, the product of the measures of its diagonals is equal to the sum of the products of the measures of the pairs of opposite sides. (Ptolemy's Theorem)   **724**

141   The inradius r of a triangle can be found with the formula $r = \frac{A}{s}$, where A is the triangle's area and s is the triangle's semiperimeter.   **731**

142   The circumradius R of a triangle can be found with the formula $R = \frac{abc}{4A}$, where a, b, and c are the lengths of the sides of the triangle and A is the triangle's area.   **732**

143   If ABC is a triangle with D on $\overline{BC}$, E on $\overline{AC}$, and F on $\overline{AB}$, then the three segments $\overline{AD}$, $\overline{BE}$, and $\overline{CF}$ are concurrent if, and only if,   **734**

$$\left(\frac{BD}{DC}\right)\left(\frac{CE}{EA}\right)\left(\frac{AF}{FB}\right) = 1$$

144   If ABC is a triangle and F is on $\overline{AB}$, E is on $\overline{AC}$, and D is on an extension of $\overline{BC}$, then the three points D, E, and F are collinear if, and only if,   **736**

$$\left(\frac{BD}{DC}\right)\left(\frac{CE}{EA}\right)\left(\frac{AF}{FB}\right) = -1$$

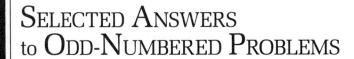

1.1 Getting Started

1 $\overleftrightarrow{AB}$, $\overleftrightarrow{BA}$, line ℓ **3** No **5 a** B **b** $\overleftrightarrow{AC}$ or ∠CEA **c** E
d { } **e** $\overrightarrow{EC}$ **f** ∠ABC **g** △BEC **9** J **11 a** $37\frac{1}{2}$ **b** 28

1.2 Measurement of Segments and Angles

1 a 61°40′ **b** 71°42′ **3** ∠1 and ∠2 **5 a** 87°10″
b 82°49′ **7 a** ∠5 **b** Same size **c** ∠4 **9 a** 90 **b** 45
c 100 **d** $142\frac{1}{2}$ **11** 51°53′50″ **13** 22 **15** $y = x + 17$
17 a 0 < m∠P < 90 **b** 20 < x < 50 **19** No
21 −10; 40 **23** $72\frac{3}{8}°$

1.3 Collinearity, Betweenness, and Assumptions

1 134 **3 a** B and D **b** No; yes **c** $\overline{AB}$ and $\overline{BC}$ **d** Yes
e Not necessarily **f** B **g** G **h** $\overline{AF}$ **i** $\overline{EB}$, $\overrightarrow{ED}$ **j** E and
B **5** 7 **7 a** e.g., 33° and 40° **b** e.g., 60° and 70° **c** e.g.,
45° and 45° **9** 135 **13 a** 15 **b** 3 **15** 80; 100; 80
17 1 : 05 : 27

1.4 Beginning Proofs

9 110° **13 b** Yes **c** It is the midpt. of $\overline{AA'}$.

1.5 Division of Segments and Angles

1 a $\overline{CO} \cong \overline{DO}$ **b** $\overline{WX} \cong \overline{WV}$ **3 a** $\overrightarrow{JG}$ **b** $\overrightarrow{OK}$ **5 a** 2 ; 9
b 14 **7** 43 **9** Yes **19** 16 **21 a** 2 **b** No **23** 60

1.6 Paragraph Proofs

5 Cannot be proved

1.7 Deductive Structure

1 Undefined terms; assumptions (postulates);
definitions; theorems and other conclusions **3 a** Yes
b No
5 a **i** If B, then A.
   ii Wet ⇒ ran
   iii If an angle is acute, then it is a 45° angle.
   iv If a point divides a segment into two congruent
      segments, it is the midpoint of the segment.
b i and ii are not necessarily true; iii is not true; iv, a
definition, is true. **7** Not true if it is a "fair" coin.
Probabilities don't "grow." **9** Not correct, since we do
not know about ∠C. **11** Not correct, since we were
reasoning from the reverse statement.

1.8 Statements of Logic

1 If a person is 18 years old, then (s)he may vote in
federal elections. **b** If two angles are opposite angles
of a parallelogram, then the two angles are congruent.
3 a 5 **a** $d \Rightarrow f$ **b** $s \Rightarrow {\sim}p$ **c** If bobcats begin to
browse, then horses head for home. **7** Original
Statement: If a polygon is a square, then it is a
quadrilateral with four congruent sides.
Converse: If a quadrilateral has four congruent sides,
then it is a square.
Inverse: If a polygon is not a square, then it is not a
quadrilateral with four congruent sides.
Contrapositive: If a quadrilateral does not have four
congruent sides, then it is not a square. **9** $p \Rightarrow {\sim}g$

1.9 Probability

1 $\frac{3}{5}$ **3** $\frac{1}{5}$ **5** $\frac{1}{3}$ **7** $\frac{2}{5}$ **9** $\frac{9}{25}$ **11** $\frac{5}{12}$ **13** 18 **15 a** $\frac{1}{5}$ **b** $\frac{4}{5}$

Review Problems

1 a $\overleftrightarrow{AR}$, $\overleftrightarrow{AD}$, $\overleftrightarrow{RA}$, $\overleftrightarrow{RD}$, $\overleftrightarrow{DA}$, $\overleftrightarrow{DR}$ **b** $\overrightarrow{BA}$, $\overrightarrow{BC}$ **c** $\overrightarrow{DF}$ **d** $\overrightarrow{CB}$
e 60°; 52°; 120° **f** No **g** No angle can be called ∠B,
since 3 angles have B as a vertex. **h** $\overleftrightarrow{AC}$ **i** $\overline{EF}$ **j** ∠1
k A **l** $\overline{FE}$ **3 a** 69°4′35″ **b** 50°59′43″ **5 a** $\overline{BC} \cong \overline{RT}$
b ∠A ≅ ∠S **7** 20 **9** No **17 a** −3 **b** −13 **19 a** $\frac{2}{5}$
b $\frac{1}{10}$ **21 a** 30° **b** 140° **c** $127\frac{1}{2}°$ **23** 14 **25** 15
29 48°50′44″ **31 a** ≈44.5° **b** ≈44°33′
33 7 < PR < 31 **35 a** 90 < m∠Q < 180
b 59 < x < 104 **37 a** 6 < w < 10 **41** ≈ 3 : 32 : 44
43 ≈ 2 : 38 : 11

2.1 Perpendicularity

1 a ∠s A, B, C, D **b** ∠s EHF, GHF, EFG
3 a 21°42′26″ **b** 45 **5** (4, 0) **11** 15; 30; 45 **15 a** $\frac{9}{10}$
b $\frac{9}{10}$ **c** $\frac{4}{5}$

2.2 Complementary and Supplementary Angles

1 ∠A and ∠C **3** $(90 − y)°$ **5** 30 and 60 **11** 125
13 a Each angle is a right angle. (If 2x = 180, then
x = 90.) **b** Each angle is a 45° angle. (If 2x = 90, then
x = 40.) **15** ≈94.84 **19** 27 **21** 30 **23** 12 **25** 70

2.4 Congruent Supplements and Complements

1 a 49 **b** 131 **c** 49 **d** 41 **e** 139 **f** 41 **g** 139
5 35 and 55 **15** 37 or 61 **17** 165° **19** 98 **21** 3:2

2.5 Addition and Subtraction Properties

1 a $\overline{AD} \cong \overline{AC}$ **b** $\angle JFG \cong \angle JHG$ **9** 12; 21 **15** Bisector
of $\angle ABC$ **17 a** Yes **b** $\angle ABC = 180°$; $\overrightarrow{BF} \perp \overleftrightarrow{AC}$
19 $152\frac{1}{2}$

2.6 Multiplication and Division Properties

7 a $x = 6$ **b** $y = 8$ **9** (7, 2) **15** $x = -5$; $4 < y < 6$ or
$-4 < y < -2$

2.7 Transitive and Substitution Properties

9 70 **13 a** $180 - x - y$ **b** $180 - y$ **c** $180 - x$
17 Can be proved false **19** $\frac{1}{3}$

2.8 Vertical Angles

1 a $\overrightarrow{FE}$, $\overrightarrow{FC}$; $\overrightarrow{FD}$, $\overrightarrow{FA}$; $\overrightarrow{BA}$, $\overrightarrow{BC}$ **b** $\angle EFA$ and $\angle CFD$;
$\angle EFD$ and $\angle CFA$ **3** 43 **7** No **13** They are right $\angle$s.
15 $132\frac{3}{4}$ or 140

Review Problems

7 6 cm **13** $22\frac{1}{2}$ **15** 4 **17** (2, −5), (6, −11) **19 a** 1
b $\frac{1}{2}$ **21** 43°43′ **29** 50 **31** 110 **33 a** $x - 2y = y$ or
$x = 3y$ **b** $y = 124 - x$ **c** 31 **35 a** 28 **b** Same
37 \$14

3.1 What Are Congruent Figures?

1 $\triangle TCR$; Reflexive Property **3** $\triangle YTW$; Vert. $\angle$s are $\cong$

3.2 Three Ways to Prove Triangles Congruent

1 a $\overline{GH} \cong \overline{KO}$, $\angle J \cong \angle M$ **b** $\overline{PS} \cong \overline{TR}$, $\angle PVS \cong \angle TVR$
c $\overline{BZ} \cong \overline{AX}$, $\angle BWZ \cong \angle AYX$ **21** $\angle 1 \cong \angle 2$

3.3 CPCTC and Circles

7 $A \approx 490.0$ sq cm; $C \approx 78.5$ cm **19 a** (18, 0) **b** 100π,
or ≈ 314

3.4 Beyond CPCTC

1 a Median **b** Altitude **c** Altitude **d** Both **13** At
any point (x, y) where $y = 11$ or $y = 1$

3.6 Types of Triangles

1 Scalene **3 a** Right **b** Obtuse **c** Right **d** Acute
e Right **f** Acute **11** $\overline{VY}$ **13** 4

3.7 Angle-Side Theorems

7 $\angle B$, $\angle A$, $\angle C$ **11** No **25** 22; 8; 60°

3.8 The HL Postulate

15 a (4, 0) **b** CPCTC **c** (10, 0) **d** 54

Review Problems

1 a S **b** A **c** N **d** N **e** N **7 a** 28; 60 **b** 114 **9** 8
15 $x = -5$ or $x = 11$ **17** 60

Cumulative Review Problems, Chapters 1–3

1 a C **b** $\overline{GJ}$ **c** $\angle FAB$ **d** D **e** ϕ **3** 46°42′09″ **5** 94
15 72; 60; 48 **17** −14; 28 **19** 35 m

4.1 Detours and Midpoints

5 (7, 2) **7** 6

4.3 A Right-Angle Theorem

5 (−1, 8) **7** ≈ 12.6 **15** 45°; 60°

4.4 The Equidistance Theorems

7 4 **9** (15, 3) **19 a** 6 units **b** (8, −2)

4.5 Introduction to Parallel Lines

1 a $\angle 3$ and $\angle 7$, $\angle 4$ and $\angle 8$ **b** $\angle 1$ and $\angle 5$, $\angle 2$ and $\angle 6$
c $\angle 1$ and $\angle 7$, $\angle 2$ and $\angle 4$, $\angle 3$ and $\angle 5$, $\angle 8$ and $\angle 6$
d $\angle 7$ and $\angle 8$, $\angle 3$ and $\angle 4$ **e** $\angle 1$ and $\angle 6$, $\angle 2$ and $\angle 5$
3 a (4, 8) **b** (9, 10) **c** Parallel **d** Congruent
e $\angle ANM$ and $\angle ACB$ **5 a** $\frac{5}{4}$ **b** $\frac{5}{4}$ **c** Parallel

4.6 Slope

1 a $\frac{8}{9}$ **b** $\frac{1}{7}$ **c** $\frac{11}{6}$ **d** 0 **e** No slope **f** $\frac{5c}{a}$ **3** −4
5 a $\frac{1}{4}$ **b** −4 **c** $-\frac{7}{2}$ **d** $\frac{1}{4}$ **9 a** $\frac{1}{2}$ **b** $\frac{1}{2}$ **c** Collinear
11 a $\angle ECD$ **b** $\angle CEB$ **13** $\overline{AC}$ **15** (7, 7)

Review Problems

9 a (9, 4) **b** $\frac{1}{2}$ **c** No; different slopes **d** −2
e 7 units **19** $m\angle Q = 90$

5.1 Indirect Proof

7 a $(2a, 2b)$ **b** Yes **9 a** Corr. **b** Alt. int.

5.2 Proving That Lines Are Parallel

1 a Corr. $\angle$s $\cong \Rightarrow \parallel$ lines **b** Alt. int. $\angle$s $\cong \Rightarrow \parallel$ lines
c Alt. ext. $\angle$s $\cong \Rightarrow \parallel$ lines **3** (1, 5), (1, 7), (2, 6), (2, 8),
(3, 7), (3, 5), (4, 8), (4, 6) **5** $\overleftrightarrow{BC} \parallel \overleftrightarrow{DE}$; corr. $\angle$s $\cong \Rightarrow$
$\parallel$ lines **7** ≈ 0.4 **9** $\overleftrightarrow{BE} \parallel \overleftrightarrow{DF}$; alt. ext. $\angle$s $\cong \Rightarrow \parallel$ lines
11 $0 < x < 110$ **15** No, because the slopes are not
equal. **17** $16 < x < 66$ **23** No, because if $x = 25$,
then $\angle QBD$ supp. $\angle RCD$ and $\overleftrightarrow{PQ} \parallel \overleftrightarrow{RS}$, a contradiction

5.3 Congruent Angles Associated with Parallel Lines

5 41 **7** No **11** $\angle 2 \cong \angle 5$ **13** $x = 19.72$;
$m\angle 1 = 42.2008$ **17** Yes; 60 **23** All but (4, 11)

5.4 Four-Sided Polygons

1 a–d **3** f **7 a** $\overline{ST}$ and $\overline{RV}$ **b** $\overline{SV}$ and $\overline{RT}$ **c** $\overline{RS}$ and $\overline{VT}$ **d** $\angle SRV$ and $\angle TVR$ **e** $\angle RST$ and $\angle VTS$ **f** $\angle STR \cong \angle VRT$, $\angle TSV \cong \angle RVS$ **9** A polygon consists entirely of segments. **11 a** 360 **b** 540 **19** 112 sq units **21 a** 0 **b** 2 **c** 5 **d** 9 **e** $n - 3$ **f** n **g** $\frac{n(n-3)}{2}$

5.5 Properties of Quadrilaterals

5 44 **7** 8; 5; 5 **11** 240 **13 a** m **b** p **c** q and r **19** 28 **21** $\overline{KI} \cong \overline{KE}$, $\overline{IT} \cong \overline{ET}$ **23 a** (19, 15)
b Slope $\overleftrightarrow{HM} = -\frac{3}{2}$; slope $\overleftrightarrow{RO} = \frac{2}{3}$ **25** 1 **27** ≈ 22.7
29 a $a = 180 - y + x$ **b** $y - x < 90$

5.6 Proving That a Quadrilateral Is a Parallelogram

7 a Yes **b** Yes **c** Yes **9 a** 9 **b** 17 **c** 56 **11 a** S **b** S **c** A **d** A **17** 145 **21 a** $\frac{2}{7}$ **b** No. You will win $\frac{24}{49}$ of the time and lose $\frac{25}{49}$ of the time.

5.7 Proving That Figures Are Special Quadrilaterals

1 Rectangle **3** 21 **5 a** $18x^2 - 45x$ **b** $22x - 10$ **c** $A = 128.5$; $P = 82.4$ **13 a** Parallelogram **b** Trapezoid **c** Isosceles trapezoid **d** Rectangle **e** Rectangle **f** Rhombus **g** Kite **h** Quadrilateral **17 a** rhombus **b** kite **c** Square **19** $A = x^2 + 6x$; $P = 52$ m **21** ≈ 29.61 **23** $-\frac{1}{4}$; 4 **25** $\frac{1}{2}$ **29 a** $324 - x^2$ **b** $0 < $ area < 144

Review Problems

1 a Parallelogram **b** Kite **c** Trapezoid **d** Square **e** Square **3** $0 < x < 25$ **7 a** Yes **b** No **c** Yes **13** $\frac{1}{3}$ **17** ≈ 78.540 sq units **19 a** S **b** A **c** S **d** S **25** 70 **27** $\frac{3}{5}$

6.1 Relating Lines to Planes

1 No; no **3** No; yes **11** Yes; no **13** Not necessarily **15** 7; ABC, ABP, BCP, CDP, DAP, ACP, BDP

6.2 Perpendicularity of a Line and a Plane

1 12

6.3 Basic Facts About Parallel Planes

1 a F **b** T **c** F **d** T **e** T **5** $\frac{1}{2}$; -2; right; the slopes are opposite reciprocals, so $\overline{EF} \perp \overline{FG}$.

Review Problems

1 a F **b** T **c** F **d** F **e** F **f** F **g** F **h** T **i** F **j** F **k** F **5** Not necessarily **7 a** No **b** No **c** Yes **9 a**

Cumulative Review Problems, Chapters 1–6

1 a Rhombus **b** Parallelogram **c** Right triangle **d** Square **3** 131 **11 a** 140 **b** 58 **15** 110

7.1 Triangle Application Theorems

1 70 **3** 40; 140 **5** 48; 60; 72 **7** 9 **9 a** A **b** A **c** N **d** A **e** N **11** 40; 50 **15** 110 **17 a** 40 **b** Rhombus **21** $25\frac{5}{7}$

7.2 Two Proof-Oriented Triangle Theorems

9 50 **17 a** Rectangle **b** Rectangle **c** Square **d** Rhombus **e** Parallelogram **f** Parallelogram **g** Rhombus

7.3 Formulas Involving Polygons

1 a 360 **b** 900 **c** 1080 **d** 1800 **e** 16,380 **3 a** 5 **b** 9 **c** 2 **d** 0 **5 a** 70 **b** 45 **c** 65 **7** 3 **11 a** Quadrilateral **b** Hexagon **13 a** Heptagon **b** Decagon **c** 22-gon **15** $18 < x < 36$ **17** 48 **19** $\frac{3}{4}$ **23** 40; 50; 130

7.4 Regular Polygons

1 a 120 **b** 90 **c** 45 **d** 24 **e** $15\frac{15}{23}$ **3 a** 6 **b** 9 **c** 10 **d** 180 **e** 48 **7** Equiangular decagon **11** Pentagon **13 a** A **b** S **c** A **d** S **e** S **f** N **17** $x = 30$; $y = 12$

Review Problems

7 45 **9** 50 **11** 20 **13 a** 5580 **b** 360 **15** 90 **19 a** 32.5 **b** 122.5 **c** 25 **21 a** A **b** S **c** S **d** N **23** No **29** 32

8.1 Ratio and Proportion

1 a 9 **b** 4 and 9; 3 and 12 **3 a** 4 **b** $\frac{54}{7}$ **c** $\frac{47}{3}$ **5 a** $-\frac{1}{2}$ **b** $-\frac{1}{2}$ **c** Yes **7** $\frac{5}{36}$ **9** Rectangle **11 a** ± 10 **b** $\pm\sqrt{15}$ **c** $\pm\sqrt{ab}$ **13** 8; 12; 20; 28 **15** $\frac{8}{3}$ **17** 150 **19** $\frac{c+d}{a+b}$ **21** $\frac{x-3}{x+4}$ **25** 75 **27** $\frac{1}{3}$

8.2 Similarity

1 b, c, and d **3** 90; 30; 7.5 **5** 6; 9 **7** $\frac{5}{3}$ **9** 5.6; 7.5 **11** 205 **13** $\frac{11}{34}$ **15** 11 ft by 14 ft **17** 180 cm by 240 cm **19** $10\sqrt{3}$; $7.5\sqrt{3}$

8.3 Methods of Proving Triangles Similar

7 Cannot be proved. **19 a** Yes, by SAS~ **b** Yes, by corr. $\angle$s $\cong \Rightarrow \parallel$ lines

8.4 Congruences and Proportions in Similar Triangles

7 4; 14 **9** 25 m **17** $6\frac{6}{7}$ **19** 12 **21 a** SAS~ **b** $\angle$BDC
c 18 **23** 6

8.5 Three Theorems Involving Proportions

1 a 8 **b** $\frac{35}{3}$ **3** 3; 9 **5** 1 **7** 9 **9** $\frac{5}{2}$; 10 **11** 51 **13** $\frac{27}{11}$; $\frac{72}{11}$

19 a 24 cm **b** $4x + 8$ **21** 10.5; 17.5 **27** $8\frac{3}{4}$ m

Review Problems

1 b and c; a and d **3** ± 10 **5** $\frac{4}{9}$ **7** $\frac{15}{2}$ **9** $\frac{2br}{3a}$ **11** 4; 15

13 $\frac{20}{3}$ **15** ≈ 879 ft **19** $(-13, 0)$ **25** 30 **27** $\frac{104}{5}$
29 8; 9; 12 **31** 225 **35** 32

9.1 Review of Radicals and Quadratic Equations

1 a 2 **b** $3\sqrt{3}$ **c** $6\sqrt{2}$ **d** $4\sqrt{2}$ **e** $7\sqrt{2}$ **f** $10\sqrt{2}$ **g** $2\sqrt{5}$
h $2\sqrt{6}$ **3 a** $\frac{\sqrt{2}}{2}$ **b** $\frac{\sqrt{5}}{5}$ **c** $2\sqrt{2}$ **d** $2\sqrt{3}$ **5 a** ± 5

b ± 12 **c** ± 13 **d** $\pm\frac{1}{2}$ **e** $\pm 2\sqrt{3}$ **f** $\pm 3\sqrt{2}$ **7 a** $\{6, -1\}$
b $\{2, -6\}$ **c** $\{5, 3\}$ **d** $\{6, -3\}$ **e** $\{12, -3\}$ **f** $\{9, -4\}$
9 a 20 **b** $2\sqrt{3}$ **c** ≈ 8.2 **11** $\{\frac{5}{2}, 5\}$ **13 a** $-h$ **b** $3 - x$
c pq **d** $-xy\sqrt{x}$

9.2 Introduction to Circles

1 $C = 19.6\pi$, or ≈ 61.58; $A = 96.04\pi$, or ≈ 301.72
3 a 180 **b** ≈ 12.6 **5** 12.5π, or ≈ 39.27 **7 a** 40 **b** 40

9 (3, 3) **11 a** $(-3, 14)$ **b** -1 **13** (7, 1); $(-3, 7)$ **15** $2\frac{2}{5}$

9.3 Altitude-on-Hypotenuse Theorems

1 a $\sqrt{21}$ **b** $\sqrt{77}$ **c** 4 **3 a** 6 **b** 8 **c** $4\sqrt{3}$ **d** 4 **e** 9
f Impossible **5 a** 9 **b** 54 **c** $15 + 6\sqrt{3}$ **9** 25; 25; 25
11 25π, or ≈ 78.54 **13** 60; 30 **17 a** $2\sqrt{7}$ **b** $16\frac{2}{3}$
c $4\sqrt{6}$ **d** $7\frac{11}{12}$ **19** 1.8 **21** $2\sqrt{5}$

9.4 Geometry's Most Elegant Theorem

1 a $\sqrt{41}$ **b** 8 **c** 12 **d** 5 **e** 10 **f** 2 **g** $3\sqrt{2}$ **3** 40 km
5 9 **7** 10 **9 a** (2, 3) **b** 8; 6 **c** 10 **d** Yes
11 a $\sqrt{x^2 + y^2}$ **b** $\sqrt{4 + x^2}$ **c** $5a$ **d** $12c$ **13** 10 km
15 60 **17** $\sqrt{5}$ **19** ≈ 5.56 m **21** $6\sqrt{5}$ **23 a** 1500 sq m
b ≈ 36 m **25 a** 8 **b** No **27** 7 ft **29** 50

31 a Parallelogram **b** 54 **33** $12\sqrt{2}$ **35** $\frac{5}{12}$

9.5 The Distance Formula

1 a 2 **b** 4 **c** 5 **d** 10 **e** $\sqrt{29}$ **f** $2\sqrt{5}$ **5** 225π **7** $\frac{9}{2}$; 10
11 a 13π **b** ≈ 132.7 **13 a** ≈ 42.4 **b** ≈ 0.8 **15** ≈ 24.9;
kite **21** 40 **27** $(4 + 2\sqrt{3}, 3 - 2\sqrt{3})$ or $(4 - 2\sqrt{3},$
$3 + 2\sqrt{3})$

9.6 Families of Right Triangles

1 a 25 **b** 36 **c** 21 **d** $\frac{5}{3}$ **e** 60 **3 a** 250 **b** 48 **c** 28
d 2.4 **e** 264 **5 a** 12 **b** $2\sqrt{7}$ **c** 10 **d** 0.5 **e** 34
f $5\sqrt{7}$ **g** 72 **h** 45 **i** $12\sqrt{7}$ **7** 50 dm **9 a** 24
b $300\sqrt{5}$ **11** 40; 60 **13** ≈ 28 km **15 a** 144 **b** $\frac{3}{8}$
c $\sqrt{7}$ **17 a** Isosceles trapezoid **b** 48 **c** $4\sqrt{10}$ **21** 140
cm **23 a** $\frac{2}{5}$ **b** $\frac{4}{15}$ **25** (22, 99, 101); (20, 48, 52); (20, 21,
29); (15, 20, 25); (12, 16, 20)

9.7 Special Right Triangles

1 a 7; $7\sqrt{3}$ **b** 20; $10\sqrt{3}$ **c** 10; 5 **d** 346; $173\sqrt{3}$ **e** 114;
$114\sqrt{3}$ **3 a** $2\sqrt{3}$ **b** $14\sqrt{3}$ **c** $13\sqrt{3}$ **5** $11\sqrt{2}$ **7** $3\sqrt{3}$ mm
9 a $5\sqrt{2}$ **b** 13 **13 a** (1, 1) **b** 1 **c** 1 **15** 38
17 a $3\sqrt{3}$ **b** 9 **c** $6\sqrt{3}$ **d** 1:2 **19** ≈ 57.9 **21 a** 48
b $6 + 6\sqrt{2}$ **23** $(0, 2\sqrt{5})$ **25 a** $2 + 2\sqrt{3}$ **b** $2\sqrt{6}$
27 $\frac{40(12 - 5\sqrt{3})}{23}$

9.8 The Pythagorean Theorem and Space Figures

1 a 5 **b** 13 **3** $5\sqrt{3}$; 10 **5 a** 14 **b** 7 **c** 25 **d** 56
e $14\sqrt{2}$ **7** $\overleftrightarrow{PB}$ and $\overleftrightarrow{PD}$ **9** 30 **11 a** $(-3, 0)$ **b** ≈ 7.1
c ≈ 4.7 **13 a** $5\sqrt{13}$ **b** $9\sqrt{11}$ **15** $2\sqrt{3}$ **17** 4
19 $d = \sqrt{a^2 + b^2 + c^2}$ **21** Impossible

9.9 Introduction to Trigonometry

1 a $\frac{8}{17}$ **b** $\frac{15}{17}$ **c** $\frac{8}{15}$ **d** $\frac{15}{17}$ **e** $\frac{8}{17}$ **f** $\frac{15}{8}$ **3 a** $\frac{\sqrt{2}}{2}$ **b** $\frac{\sqrt{2}}{2}$
c 1 **5** $4\frac{4}{5}$ **7 a** $2\sqrt{6}$ **b** $\frac{2\sqrt{6}}{7}$ **c** $\frac{5\sqrt{6}}{12}$ **9 a** $\frac{7}{25}$ **b** $\frac{8}{17}$ **c** $\frac{4}{5}$
11 a 45 **b** 30 **13 a** $\frac{2}{3}$ **b** $\frac{3\sqrt{13}}{13}$ **15** $\frac{4}{5}$ **21** $\frac{1}{6}$

9.10 Trigonometric Ratios

1 a ≈ 0.3584 **b** ≈ 1.2799 **c** ≈ 0.9962 **d** 1.0000
e ≈ 0.8660 **3 a** 45 **b** 30 **c** 60 **5** ≈ 15 **7** $\approx 24°$
9 $\approx 37°$, $\approx 53°$, 90° **11 a** $\approx 67°$ **b** $5\sqrt{11}$ **13** 10
15 ≈ 104 dm **17 a** ≈ 36.4 **b** ≈ 37.37 **c** $\approx 74°$ **d** $\approx 68°$
19 a $\frac{2}{3}$ **b** ≈ 34 **21 a** $\approx 52°$ **b** ≈ 11 **d** $\frac{a}{\sin \angle A} = \frac{b}{\sin \angle B}$

Review Problems

1 a 9 **b** 8 **c** 5 **d** $\sqrt{13}$ **3 a** 30 **b** $5\sqrt{3}$; 5 **c** 7 **d** 15
e $4\sqrt{5}$ **f** 9 **g** $5\sqrt{3}$; $10\sqrt{3}$ **h** $\frac{25}{2}$ **i** 26 **j** $4\sqrt{2}$; $4\sqrt{2}$
5 $3\sqrt{3}$ **7** 7 ft **9** 48 **11** $\sqrt{85}$ **13** 13 **15** ≈ 14.5 **17** 5
19 a 120 **b** 45 **c** 55 **21 a** 90 **b** 180 **c** 10π, or
≈ 31.42 **23** 75 km **25** Swim directly across and walk
27 7.5 **29 a** 12 **b** $10\sqrt{13}$ **33** 1:2 **35** 7 **37 a** $\approx 35°$
b 60°

Cumulative Review Problems, Chapters 1–9

1 $67\frac{1}{2}$ **3 a** 1260 **b** 30 **c** 14 **5** 14.4 m **7** 52 **9 a** $\pm\frac{7}{2}$
b $\frac{20}{3}$ **13 a** 55° **b** $5\frac{1}{2}$° **15** 15 **17** 27 **21 a** 7°
b $(2\sqrt{3}, -4)$ **25** 25 **27** $2\frac{2}{7}$ **29** No; opposite sides are
not of equal length **31** $\frac{1}{2}$

10.1 The Circle

5 8 mm **11** 8 m **17 a** 13 **b** 5 **c** 24 **23** 2 **25** 24

10.2 Congruent Chords

1 Same distance **7 a** ≈283.53 sq mm **b** ≈59.69 mm
11 a 8 **b** 5 **13** 10 **15 a** 8 **b** $24 + 6\sqrt{7}$

10.3 Arcs of a Circle

1 a 6 **b** 2 **c** 5 **d** 4 **e** 3 **f** 7 **g** 1 **3 a** 90 **b** 130
c 230 **d** 180 **e** 220 **9 a** $\frac{1}{45}$ **b** $\frac{2}{3}$ **c** $\frac{2}{5}$ **d** $\frac{7}{8}$ **11** 132
13 a 15π **b** 18π **19 a** $5\sqrt{2}$ **b** 135 **23 a** $\frac{n(n-1)}{2}$
b $n(n-1)$ **c** $\frac{360}{n}$

10.4 Secants and Tangents

1 17 cm **5 a** $Q = (16, 0)$; $S = (38, 0)$ **b** 3 **11 a** $2\sqrt{13}$,
or ≈7.21 **b** $-\frac{3}{2}$ **13 a** 12 cm **b** Yes **17 a** 9 **b** 17
19 a $\frac{2}{7}$ **b** $\frac{3}{7}$ **c** $\frac{2}{7}$ **23** 60 mm **25** 10 **27** $\frac{8\sqrt{10}}{3}$
29 $\frac{c - a + b}{2}$

10.5 Angles Related to a Circle

1 62 **3 a** 35 **b** 75 **5 a** 140 **b** 80 **c** 20 **d** $81\frac{1}{2}$ **e** 100
7 85°; 23° **9** 10° **11** 81° **13** 125° **15** 75 **17 a** $\frac{1}{2}$
b $\frac{5}{18}$ **c** $\frac{1}{6}$ **d** $\frac{11}{36}$ **19** 64; 120; 116 **21 a** 132 **b** 10 **c** 20
d 30 **23** 98° **25** 25° **27** $175\frac{1}{2}$ **29 a** 70 **b** 110
33 20° **35** 90

10.6 More Angle-Arc Theorems

3 10 mm **5** 97° **7** 137° **9** 42 **11 a** 6 cm
b $18 + 6\sqrt{3}$ cm **13** 15 **15** $10\sqrt{3}$ **17 a** 29° **b** 81°
23 4 **27** 47°

10.7 Inscribed and Circumscribed Polygons

1 113°; 76° **3 a** 70° **b** 110° **c** 85° **d** 80° **5** No
7 a Square **b** Isosceles **9 a** 65 **b** 85° **c** 85° **d** 95°
11 46 **15** 116 or 80 **17 a** Yes **b** No **c** Yes **19 a** 20
b $\frac{1}{2}$ **21 a** Midpoint of hypotenuse **b** Interior of $\triangle$
c Exterior of $\triangle$ **23 a** A **b** S **c** S **d** N **e** S **f** S
25 48 **27** If the equiangular polygon has an odd
number of sides, it will be equilateral.

10.8 The Power Theorems

1 a 15 **b** 8 **c** 25 **3 a** 20 **b** 21 **c** 48 **5** 3.5 **7** 6
9 1 or 6 **11** 5 **13** 2.5 **15** 900 mi **17** 26; 39 **19 a** 4
b $12 - 4\sqrt{5} < y < 8$

10.9 Circumference and Arc Length

1 a 21π mm; ≈65.97 mm **b** 12π mm; ≈37.70 mm
3 a 4π **b** 5π **c** $\frac{10\pi}{3}$ **d** 10π **5 a** $40 + 6\pi$ **b** $24 + 4\pi$
c $12 + 3\pi$ **d** $6 + 7\pi$ **7** 138 m **9 a** $4\pi\sqrt{3}$ **b** 6π
c 9π **11** ≈214 m **13** ≈96.5 m; ≈71.4 m **15** 10π in.
17 $(24\sqrt{3} + 22\pi)$ cm

Review Problems

1 a 94 **b** 94 **c** 43 **3 a** 16 **b** 8 **c** 4 **5** 125 **7 a** $\frac{3\pi}{2}$
b $12 + 1.5\pi$ **13** $4\frac{2}{3}$ **15 a** $\frac{2}{9}$ **b** $\frac{7}{12}$ **17** 45°, 105°, 135°, 75°
19 20 **23** 6; 18 **25** ≈258 cm **29** 8:5 **31** ≈16.68 ft

11.1 Understanding Area

1 a 207 **b** 78 **c** 80 **3 a** 120 **b** 84 **5 a** 144 **b** 50
c 7 **d** 36 **e** 81 **7 a** 10 **b** 7 **9** 256 sq cm
11 9 m by 15 m **13** (18, 8) **15** $1.3 < x < 2.3$ **17** $\frac{27}{65}$

11.2 Areas of Parallelograms and Triangles

1 a 198 sq mm **b** 102 sq cm **c** 35 **3 a** 35 **b** 144
5 48 **7** 14 **9 a** $18\sqrt{3}$ **b** 72 **c** $36\sqrt{3}$ **11 a** 84 **b** 128
13 12 **15** 85 **17** 33 **19 a** 168 **b** 180 **c** 81
21 a ≈72.7 **b** ≈120.2 **c** 120.0 **23** 12 **25 a** 30 **b** 12
27 a 6 **b** $A = \frac{s^2}{4}\sqrt{3}$ **29** 50 **31** 120

11.3 The Area of a Trapezoid

1 a 104 **b** 13 **3** 10.5; 73.5 **5** 11 **7 a** 396 **b** 78 **9** 24
11 89 **13** 72 **15** 16 sq units **17 a** $42\sqrt{3}$ **b** $27\sqrt{3}$

11.4 Areas of Kites and Related Figures

1 60 **3** 5 **5 a** 336 **b** 500 **7** 78 **9 a** 8; 28; 18 **b** 30

11.5 Areas of Regular Polygons

1 36 **3 a** $108\sqrt{3}$ **b** $48\sqrt{3}$ **c** $27\sqrt{3}$ **d** $36\sqrt{3}$ **5 a** 12
b $6\sqrt{3}$ **c** $216\sqrt{3}$ **7** 3 mm **9** 324 **11** $288\sqrt{3}$ **13 a** 11;
5.5 **b** 12 m; $2\sqrt{3}$ m **c** 4 cm; $2\sqrt{3}$ cm **15 a** $108\sqrt{3}$
b 288 **c** $216\sqrt{3}$ **17 a** $36 - 9\sqrt{3}$ **b** $27\sqrt{3}$ **c** $36\sqrt{3}$
19 a $450\sqrt{3}$ **b** 8 **c** $A = \frac{s^2\sqrt{3}}{2}$ **21 a** $150 + 100\sqrt{2}$
b $50 + 100\sqrt{2}$ **23** $1764\sqrt{3} - 3024$
25 a $\frac{1}{2}ab + \frac{1}{2}ab + \frac{1}{2}c^2$ **b** $\frac{1}{2}(a + b)(a + b)$
c $a^2 + b^2 = c^2$

11.6 Areas of Circles, Sectors, and Segments

1 a π, 2π **b** 64π, 16π **c** 225π, 30π **3** 20π cm **5 a** π
b 32π **c** 2π **d** 27π **e** 75π **f** 9π **7** $150 + 25\pi$
9 a 24π **b** 12π **c** 8π **11 a** $16\pi - 32$ **b** $\frac{32\pi}{3} - 16\sqrt{3}$
c $12\pi - 9\sqrt{3}$ **13 a** 18π, 50π, $24\frac{1}{2}\pi$ **15** 12π
17 32 sq cm **19 a** $50\pi - 50\sqrt{3}$ **b** 10π **21 a** 120
b 120 **23** $\frac{33}{49}$

11.7 Ratios of Areas

1 a 1:1 **b** 1:2 **c** 15:8 **d** 5:6 **3 a** 1:1 **b** 1:2 **c** 2:1
5 3:4 **7** 1:16 **9 a** 64:225 **b** 1:2 **c** 1:1 **d** 4:9
11 5:4 **13** 250 **15** 16:81 **17** 4:9 **19 a** 4:1 **b** 8:1
c 1:1 **d** 1:2 **e** 30 **21** 5:6

11.8 Hero's and Brahmagupta's Formulas

1 a 6 **b** $2\sqrt{5}$ **c** $10\sqrt{2}$ **d** $6\sqrt{3}$ **e** 60 **f** 84 **3 a** 36
b $4\sqrt{15}$ **c** 24 **d** $16\sqrt{3}$ **5** ≈ 32.50 **9** $2\sqrt{6}$ **11 a** It
approaches a $\triangle$. **b** It becomes Hero's formula.
13 a $(6, 8 - 3\sqrt{3})$ **b** ≈ 37.7

Review Problems

1 a 84 **b** 42 **c** 75 **d** 52 **e** 20 **f** 8 **3 a** 70 **b** 24
c $16\sqrt{3}$ **5** $48\sqrt{3}$ **7** 18 **9** 196 **11** 81 sq m **13 a** 9π
b 16π **c** $100 - 25\pi$ **15 a** 5:8 **b** 9:16 **17** 360
19 120 **21** $A = 77\sqrt{3}$; $P = 50$ **23 a** 338 **b** 6
25 a $\frac{168}{5}$ **b** $27\sqrt{3} - 9\pi$ **c** $6 - \pi$ **27 a** $9\pi - 18$
b $6\pi - 9\sqrt{3}$ **29** Circle **31 a** (12, 13) **b** 156
33 $143\sqrt{3}$ **35** $18\pi\sqrt{2} - 9\pi$ **37** 432 **39** $72\sqrt{3} - 24\pi$
41 $11\frac{1}{4}\pi$ **43** 390 **45** ≈ 106.8

12.1 Surface Areas of Prisms

1 a 550 sq cm **b** 282 sq mm **c** 810 sq in. **3 a** 550
b 120 **c** 790 **5 a** 150 **b** 294 **7 a** L.A. = 480;
T.A. = 552 **b** L.A. = 120; T.A. = 132 **c** L.A. = 2500;
T.A. = 2620 **d** L.A. = 360; T.A. = 360 + $108\sqrt{3}$
9 L.A. = 616; T.A. = 712 **11 a** 0; 0; 0; 8; 12; 6; 1
b $\frac{20}{27}$ **c** 432 sq in.

12.2 Surface Areas of Pyramids

1 a 60 **b** 240 **c** 340 **3 a** The base is not regular.
b 936 **c** 1356 **5 a** 192 **b** 48 **c** 36 **d** 144 **e** 1:4
f 36 **7 a** 72 **b** 432 **c** 324 **d** 756 **9 a** 17; 10;
L.A. = 504; T.A. = 864 **b** 16 **11 a** $36\sqrt{3}$ **b** $2\sqrt{6}$
13 Cube

12.3 Surface Areas of Circular Solids

1 a 196π **b** 36π **c** 36π **d** 25π **3** 6 **5 a** -2; $\frac{1}{2}$
b Rhombus **7 a** 90π **b** 66π **c** 480π **9** 8π cm by
14 cm **11 a** 64 **b** 32 **13** 140π

12.4 Volumes of Prisms and Cylinders

1 a 300π **b** 720 **3** 357 **5 a** 343 **b** e^3 **c** 5 **7 a** ≈ 23
cu ft **b** ≈ 1456 lb **9** V = 600; T.A. = 620 **11** 6
13 $250\sqrt{2}$ **15** 189 **17 a** ≈ 621 cu in. **b** 439 sq in.
19 $V = \frac{2000}{3}\pi$; T.A. = $600 + \frac{1400}{.9}\pi$ **21** 90π

12.5 Volumes of Pyramids and Cones

1 $490\sqrt{3}$ **3 a** 400 **b** 360 **5 a** 1080π **b** 369π **c** 450π
7 ≈ 1451 cu ft **9** 200 **11** 72π **13** 720 cu m
17 a $18\sqrt{2}$ **b** $\frac{s^3\sqrt{2}}{12}$ **19** 48

12.6 Volumes of Spheres

1 a 36π **b** 972π **c** $\frac{500}{3}\pi$ **3** ≈ 481 cu m **5** ≈ 523 cu ft
7 a 240π **b** 132π **9** ≈ 71 cu mm **11 a** 138π cu ft
b 105π sq ft **13 a** ≈ 524 cu m **b** ≈ 2721 cu m
15 $\approx 11\%$

Review Problems

1 a L.A. = 48; T.A. = 84 **b** L.A. = 56π; T.A. = 88π
3 a V = 360π; T.A. = 192π **b** V = 540; T.A. = 468
c V = 100π; T.A. = 90π **5 a** 5 **b** 6 **7** 562,500 cu cm
9 36π **11 a** 75π **b** $45\pi + 60$ **13** 35 **15** 5040
17 215 cu in. **19** $120 + 48\pi$ **21** $333\pi\sqrt{3}$

Cumulative Review Problems, Chapters 1–12

1 81 **5 a** 8 **b** $\frac{31}{5}$ **c** 44 **d** 12 **7 a** $\frac{50}{3}$ **b** $\frac{32}{3}$ **c** 8
9 89° **11 a** 60 **b** 83 **c** 50 **d** 78 **13 a** 25 **b** 16
15 64 **17** 36 **19 a** 60 **b** 135 **c** 120 **d** 45 **e** No
23 2π **25** 7 **27** 100 m **29 a** 3 **b** 1.2 **31** 35 **33** 3
35 32° **37** 20 **39** Yes **41** $6\frac{1}{5}$

13.1 Graphing Equations

7 x-intercept: 3; y-intercept: -6 **9** Yes **11** Yes **13** 9
15 a (2, 5) **b** 14 and 4 **c** 9 **d** 63 **17** ≈ 21.5
19 $y - 2 = 6(x - 5)$ **21 a** $-\frac{5}{8}$ **23** ≈ 2.6

13.2 Equations of Lines

1 a 3; 7 **b** 4; 0 **c** $\frac{1}{2}$; $-\sqrt{3}$ **d** -6; 13 **e** -5; -6
f 0; 7 **3** $y = -6$ **5** a and c **7 a** 1 **b** $y = x + 5$
c $-\frac{1}{2}$ **d** $y = -\frac{1}{2}x - \frac{5}{2}$ **9** $3\frac{3}{4}$ **13** $y - 12 = \frac{1}{7}(x - 4)$
15 $y = -7x + 40$ **17** $\frac{1}{7}$ **19** $y - 7 = -\frac{1}{2}(x + 1)$
21 No **23** $y = -\frac{2}{5}x + 4$ **25** (3, 13)
27 a $y = -\frac{3}{4}x + 1$ **b** $y = -\frac{3}{4}x - 1$ **c** $y = \frac{4}{3}x + \frac{4}{3}$

13.3 Systems of Equations

1 a (6, 4) **b** (2, 5) **c** (−3, −7) **d** (3, 2) **3 a** (12, 0)
b $\{(x, y)|y = 3x - 7\}$ **5** (−2, 3) **7** (4, −2)
9 $y - 1 = 5(x + 2)$ **11** (−1, 2) **13** $\left(\dfrac{ce - bf}{ae - bd}, \dfrac{af - cd}{ae - bd}\right)$
15 $\dfrac{4\sqrt{5}}{5}$ **17** 11

13.4 Graphing Inequalities

9 a (1, 1); (2, 4); (3, 9) **b** $9\frac{1}{2}$

13.5 Three-Dimensional Graphing and Reflections

3 ≈7.8 **5 a** 20 **b** 140 **c** $3\sqrt{10}$ **d** Yes **e** (11, 5, 0) or
(−3, 5, 0) **7** (3, 10) **9 a** (5, 6, −2) **b** ≈11.6
11 (−3, 7) **13** (5, −1) **15 a** 10 **b** $y = -2x + 1$
17 $\left(9, 2\frac{1}{2}\right)$

13.6 Circles

1 a $x^2 + y^2 = 16$ **b** $(x + 2)^2 + (y - 1)^2 = 25$
c $x^2 + (y + 2)^2 = 12$ **d** $(x + 6)^2 + y^2 = \frac{1}{4}$ **3 a** (0, 0);
6; 12; 12π; 36π **b** (−5, 0); $\frac{3}{2}$; 3; 3π; $\frac{9}{4}\pi$ **c** (3, −6); 10;
20; 20π; 100π **d** (−5, 2); 9; 18; 18π; 81π **5 a** Yes
b No **7** (−6, 3); (8, 3); (1, 10); (1, −4) **9 a** On
b Inside **c** Outside **d** Outside **11 a** (0, 4); 5
b (−7, −3); 6 **c** (−5, 6); $\sqrt{51}$ **d** (4, −7); 10 **13** $5\sqrt{2}$
15 $y + 8 = \frac{6}{5}(x - 8)$ **17** 17π **19 a** $\frac{16}{3}\pi$ **b** ≈2.90

13.7 Coordinate-Geometry Practice

1 $x^2 + y^2 = 25$ **a** 25π **b** 10π **3 a** ≈13.7 **b** ≈30.5
5 a 5 **b** 5 **c** 5 **7** 10 **9** −0.5 **11** 12
13 a $(x + 3)^2 + (y - 3)^2 = 9$ **b** ≈1.9 **15** 35 **17** 50
19 $\frac{24}{5}$ **21** $(x + 4)^2 + y^2 = 36$ **23** $\frac{8\sqrt{5}}{5}$ **25 a** $5\sqrt{13}$
b $\sqrt{429}$ **27** 20 **29 a** (7, 0) **b** $\left(\frac{20}{3}, 0\right)$ **c** $\left(\frac{28 - 2x_1}{3}, 0\right)$

Review Problems

1 No **3** −4 **5 a** 10 **b** (7, 2) **c** $\frac{4}{3}$ **7 a** 50 **b** 27
c 18π **9 a** $y = 2x + 1$ **b** $x = 2$ **c** $x = -5$
d $y = 3x - 2$ **e** $y = \frac{1}{2}x - 2$ **f** $y = 3x - 7$
g $y = \frac{1}{2}x - 3$ **11** Yes **13** $-\frac{29}{4}$ **15** (10, 5) **17 a** $2\sqrt{5}$
b $y = 2x - 10$ **c** $y - 4 = -5(x - 7)$
d $y - 8 = -5(x - 9)$ **e** $y - 8 = \frac{1}{5}(x - 9)$ **21 a** $\{(2, 7)\}$
b $\{(7, -1)\}$ **c** $\{(-3, 4), (5, 4)\}$ **23** ≈9.2 **25** Point circle
27 III and IV; $\left(-\frac{10}{3}, -\frac{10}{3}\right)$ and $\left(\frac{10}{7}, -\frac{10}{7}\right)$ **29** $2\sqrt{10}$
31 ≈339.3 **33** $\left(-\frac{10}{3}, \frac{7}{3}\right)$ **35** $\left(\frac{44}{7}, 0\right)$ **37** {(2, 14),
(−26, 42)}

14.1 Locus

1 The locus is two lines parallel to $\overleftrightarrow{AB}$ and 3 cm on
each side of $\overleftrightarrow{AB}$. **3** The locus is the perpendicular
bisector of the segment joining the two given points.
5 The locus is two circles with the same center as the
given circle and with radii of 2 in. and 22 in. **7** The
locus is a circle concentric with the given circles and
with a radius equal to the average of the radii of the
given circles. If the radii of the given circles are 3 and
8, the radius of the locus is $5\frac{1}{2}$. **9** $x^2 + y^2 = 16$
11 The locus does not include the given point but
otherwise is a circle tangent internally to the given
circle at the given point and passing through the
center of the given circle. **13** The locus is the
perpendicular bisector of the segment joining the two
points. **15** The locus is a segment joining the
midpoints of the sides containing the given vertex.
17 $x = 2.5$ **19** The locus is the union of point Q and
the circle with center Q and radius 18. **21** The locus
is a circle with its center at the foot of the
perpendicular from P to m and with a radius of 3.
23 The locus does not include T and V but otherwise
is a circle with center at the midpoint of $\overline{TV}$ and
diameter TV. **25 a** Perpendicular bisectors of the
sides **b** Angle bisectors **29** $(x - 2)^2 + y^2 = 4$

14.2 Compound Locus

1 a ϕ, 1 point, or 2 points **b** ϕ, 1 point, or 2 points
c ϕ, 1 point, or 2 points **d** ϕ, 1 point, or a line **e** 4
points **f** ϕ, 1 point, or a ray **3** 1 point **5 a** ϕ, 1 point,
or a segment **b** ϕ or a segment **7** ϕ, 1 point, 2 points,
3 points, or 4 points **9** 2 points **11** ϕ or 1 point **13** ϕ,
1 point, 2 points, 3 points, or 4 points **15** ϕ, 1 point, 2
points, 3 points, . . . , or 8 points **17** A sphere and its
interior

14.3 The Concurrence Theorems

3 Find the incenter. **5** Centroid and incenter
7 Centroid **9** Equilateral **11** The locus is the
circumcenter of the triangle formed by joining the
three points. **13 a** $9\sqrt{3}$ **b** $162\sqrt{3}$ **c** $81\sqrt{3}$ **d** 30
15 $\left(\frac{28}{5}, \frac{8}{5}\right)$ **17 a** $\left(\frac{4}{3}, 4\right)$ **b** $\left(\frac{4}{3}, 4\right)$

Review Problems

1 The locus is the perpendicular bisector of $\overline{AB}$
(minus the midpoint of $\overline{AB}$). **3** 4 points **5** The locus
is a point and a circle. **7** $x^2 + y^2 = 25$ **13** $y = 3x + 5$
15 The locus is the perpendicular bisector of $\overline{PQ}$
(minus the midpoint of $\overline{PQ}$). **17** The locus is a circle
that is tangent internally to the given circle at the fixed
point and whose radius is half the original circle's but
which excludes the fixed point. **19** 4 points **25** The
locus does not include P and Q but otherwise is a
circle with $\overline{PQ}$ as diameter.

15.1 Number Properties

1 a $\{x|x > 25\}$ **b** $\{x|x > 6\}$ **c** $\{x|x \geq -7\}$ **d** $\{x|x > 1\}$
3 x exceeds z by 8. **5** $\angle A > \angle 2$ **7** $45 < x < 90$
9 $0 < x < \frac{1}{5}$ **11** $x > 3$ **13** $\angle DBC > \angle DCB$
15 $\{x|-3 < x < 2\}$ **17** 44 **19 a** Comp. $\angle B >$ comp.
$\angle A$ **b** Comp. $\angle B > \angle A$ **c** Comp. $\angle B$, $\angle A$, comp. $\angle A$,
$\angle B$

15.2 Inequalities in a Triangle

1 $80 < m\angle 1 < 180$ **3 a** $\overline{PR}$; $\overline{PQ}$ **b** Hypotenuse
5 a $\overline{AB}$ **b** $\overline{WZ}$ **7** The other **9** $46 < m\angle B < 70$
11 a $\angle R$, $\angle Q$, $\angle P$ **b** R **13** 2 cm **19 a** 714.0
b ≈ 968.6 **c** ≈ 2164.2

15.3 The Hinge Theorems

1 $\overline{AC}$ **3** $\overline{DC}$ **7 a** $\angle Y$; $\angle X$ **b** Obtuse **13** Right
15 a $\overline{AC}$ **b** $\angle B$ **17** $\angle XZW$, $\angle X$, $\angle XWY$, $\angle Y$, $\angle XWZ$
19 1399 **21** b

Review Problems

1 $\angle A$, $\angle B$, $\angle C$ **3 a** $\angle CBD$; Converse Hinge Theorem
b $\angle X$; If $\triangle$, then $\triangle$ **c** $\angle 1$; Exterior-Angle-Inequality
Theorem **5** $x > 6$ **7 a** **9** $\overline{BC}$, $\overline{AC}$, $\overline{AB}$ **11 a** $\overline{AB}$
b $12 < P < 6 + 6\sqrt{2}$ **13 a** $\sqrt{34}$; $\sqrt{13}$; 7 **b** Obtuse
c $\angle P$, $\angle R$, $\angle Q$ **19** $\overline{AC}$, $AB \cong \overline{BC}$, $\overline{CD}$, $\overline{AD}$, $\overline{BD}$ **23** $\frac{7}{9}$

Cumulative Review Problems, Chapters 1–15

1 $V = 12\pi$, $A = 24\pi$ **3** 16° **5** 21 **7 a** 3 **b** $13\frac{1}{2}$ **c** No
9 42 **11 a** 13, 14, 15; $-\frac{12}{5}$, 0, $\frac{4}{3}$ **b** Acute
c $y = \frac{4}{3}x + 4$; -3; 4 **d** $y = -4$ **e** $(3, -4)$ **f** $x = 3$
g 12 **h** $(1, -4)$; 6 **i** 84 **13** 135 **15** $12\sqrt{3}$ **19** 175°
21 a 11 **b** 180° **23** 36 **25 a** $\sqrt{26}$ **b** $y = 5x - 35$
c $y = 3x + 37$ **d** $y = -3x + 29$ **27** 26 m **29** 78°
31 Acute **33** π:2 **37** ϕ, 1 point, or 2 points **41** ϕ, 1
point, 2 points, 3 points, 4 points, a ray, a ray and a
point, or a line **45** $20\sqrt{3}$ **47 a** A cone with a conical
hole in it **b** 504π **49** 5

16.1 The Point-Line Distance Formula

1 3 **3** $3\frac{56}{25}$ **5** ≈ 5.81 **7** ≈ 3.051 **9** -1 or -21
11 $x + y - 8 = 0$ and $3x - 3y + 2 = 0$

16.2 Two Other Useful Formulas

1 ≈ 6.110 **3** 81 **5** $-\frac{76}{11}$ or $\frac{92}{11}$

16.3 Stewart's Theorem

1 $\frac{-11 + \sqrt{181}}{2}$ **3** 14 **5** $\frac{8\sqrt{165}}{15}$

16.4 Ptolemy's Theorem

1 $\frac{2\sqrt{51}}{3}$ **3** 60.8 **5** $3\sqrt{3} - 4$ or $3\sqrt{3} + 4$

16.5 Mass Points

1 10:3 **3** 10 **5** $\frac{4}{11}$

16.6 Inradius and Circumradius Formulas

1 $\frac{2\sqrt{14}}{7}$, $\frac{45\sqrt{14}}{56}$ **3** $\frac{\sqrt{455}}{13}$, $\frac{144\sqrt{455}}{455}$ **5 a** $\frac{13\sqrt{10}}{10}$ **b** ≈ 4.744

Review Problems

1 $x = \frac{\sqrt{165}}{3}$; $y = \sqrt{93}$ **3** 28 **5** 6 **7** $\frac{10\sqrt{3}}{3}$ **9** $\frac{3}{13}$
11 $x - 4y - 1 + 3\sqrt{17} = 0$ and
$x - 4y - 1 - 3\sqrt{17} = 0$ **13** 48π

GLOSSARY

acute angle An angle whose measure is greater than 0 and less than 90. (p. 11)

acute triangle A triangle in which all three angles are acute. (p. 143)

alternate exterior angles A pair of angles in the exterior of a figure formed by two lines and a transversal, lying on alternate sides of the transversal and having different vertices. (p. 194)

alternate interior angles A pair of angles in the interior of a figure formed by two lines and a transversal, lying on alternate sides of the transversal and having different vertices. (p. 193)

altitude (of triangle) A perpendicular segment from a vertex of a triangle to the opposite side, extended if necessary. (p. 132)

angle A figure formed by two rays with a common endpoint. (p. 5)

angle of depression The angle between a downward line of sight and the horizontal. (p. 423)

angle of elevation The angle between an upward line of sight and the horizontal. (p. 423)

annulus A region bounded by two concentric circles. (p. 540)

apothem A segment joining the center of a regular polygon to the midpoint of one of the polygon's sides. (p. 532)

arc A figure consisting of two points on a circle and all points on the circle needed to connect them by a single path. (p. 450)

area The number of square units of space within the boundary of a closed region. (p. 511)

arithmetic mean The average of two numbers. The arithmetic mean of the numbers a and b, for example, is $\frac{1}{2}(a + b)$. (p. 328)

auxiliary line A line introduced into a diagram for the purpose of clarifying a proof. (p. 132)

base (of isosceles triangle) In a nonequilateral isosceles triangle, the side that is congruent to neither of the other sides. (p. 142)

base (of trapezoid) Either of the two parallel sides of a trapezoid. (p. 236)

base angle In an isosceles triangle or trapezoid, the angle formed by a base and an adjacent side. (pp. 142, 236)

bisect To divide a segment or an angle into two congruent parts. (pp. 28, 29)

center (of arc) The center of the circle of which an arc is a part. (p. 450)

center (of circle) See *circle*.

central angle An angle whose vertex is at the center of a circle. (p. 450)

centroid The point of concurrency of the medians of a triangle. (p. 662)

chord A segment joining two points on a circle. (p. 440)

chord-chord angle An angle formed by two chords that intersect at a point inside a circle but not at the circle's center. (p. 470)

circle The set of all points in a plane that are a given distance from a given point in the plane. (That point is called the circle's center.) (p. 439)

circumcenter A point associated with a polygon, corresponding to the center of the polygon's circumscribed circle. (The circumcenter of a triangle is the point of concurrency of the perpendicular bisectors of the triangle's sides.) (p. 486)

circumference The perimeter of a circle. (p. 370)

circumscribed polygon A polygon each of whose sides is tangent to a circle. (p. 486)

collinear Lying on the same line. (p. 18)

common tangent A line tangent to two circles (not necessarily at the same point)—called a *common internal tangent* if it lies between the circles or a *common external tangent* if it does not. (p. 461)

complementary angles Two angles whose sum is 90° (a right angle). (p. 66)

compound locus The intersection of two or more loci. (p. 656)

concentric circles Two or more coplanar circles with the same center. (p. 439)

conclusion The "then" clause in a conditional statement. (p. 40)

concurrent lines Lines that intersect in a single point. (p. 660)

conditional statement A statement in the form "If p, then q," where p and q are declarative statements. (p. 40)

congruent angles Angles that have the same measure. (p. 12)

congruent arcs Arcs that have the same measure and are parts of the same circle or congruent circles. (p. 452)

congruent segments Segments that have the same length. (p. 12)

congruent triangles Triangles in which all pairs of corresponding parts (angles and sides) are congruent. (p. 111)

construction A drawing made with only a compass and a straightedge. (p. 666)

contrapositive A statement associated with a conditional statement "If p, then q," having the form "If not q, then not p." (p. 44)

converse A statement associated with a conditional statement "If p, then q," having the form "If q, then p." (p. 40)

convex polygon A polygon in which each interior angle has a measure less than 180. (p. 235)

coplanar Lying in the same plane. (p. 192)

corresponding angles In a figure formed by two lines and a transversal, a pair of angles on the same side of the transversal, one in the interior and one in the exterior of the figure, having different vertices. (p. 194)

cross section The intersection of a solid with a plane. (p. 577)

cyclic quadrilateral A quadrilateral that can be inscribed in a circle. (p. 550)

diagonal (of polygon) A segment that joins two nonconsecutive (nonadjacent) vertices of a polygon. (p. 235)

diagonal (of rectangular solid) A segment whose endpoints are vertices not in the same face of a rectangular solid. (p. 413)

diameter A chord that passes through the center of a circle. (p. 440)

distance The length of the shortest path between two objects. (p. 184)

equiangular Having all angles congruent. (p. 143)

equilateral Having all sides congruent. (p. 142)

exterior angle An angle that is adjacent to and supplementary to an interior angle of a polygon. (p. 296)

extremes The first and fourth terms of a proportion. In the proportion $a:b = c:d$ $\left(\text{or } \frac{a}{b} = \frac{c}{d}\right)$, for example, a and d are the extremes. (p. 327)

face One of the polygonal surfaces making up a polyhedron. (p. 413)

foot (of line) The point of intersection of a line and a plane. (p. 270)

fourth proportional The fourth term of a proportion. In the proportion $a:b = c:x$ $\left(\text{or } \frac{a}{b} = \frac{c}{x}\right)$, for example, x is the fourth proportional. (p. 328)

frustum The portion of a pyramid or a cone that lies between the base and a cross section of the figure. (p. 585)

geometric mean Either of the two means of a proportion in which the means are equal. Also called a *mean proportional*. (p. 327)

hypotenuse The side opposite the right angle in a right triangle. (p. 143)

hypothesis The "if" clause in a conditional statement. (p. 40)

incenter A point associated with a polygon, corresponding to the center of the polygon's inscribed circle. (The incenter of a triangle is the point of concurrency of the triangle's angle bisectors.) (p. 487)

inscribed angle An angle whose vertex is on a circle and whose sides are determined by two chords. (p. 469)

inscribed polygon A polygon each of whose vertices lies on a circle. (p. 486)

inverse A statement associated with a conditional statement "If p, then q," having the form "If not p, then not q." (p. 44)

isosceles trapezoid A trapezoid in which the nonparallel sides are congruent. (p. 236)

isosceles triangle A triangle in which at least two sides are congruent. (p. 142)

interior angle An angle whose sides are determined by two consecutive sides of a polygon. (p. 308)

kite A quadrilateral in which two disjoint pairs of consecutive sides are congruent. (p. 236)

lateral surface area The sum of the areas of a solid's lateral faces. (p. 562)

leg (of isosceles trapezoid) One of the nonparallel, congruent sides of an isosceles trapezoid. (p. 236)

leg (of isosceles triangle) One of the two congruent sides of a nonequilateral isosceles triangle. (p. 142)

leg (of right triangle) One of the sides that form the right angle in a right triangle. (p. 143)

locus A set consisting of all the points, and only the points, that satisfy specific conditions. (p. 649)

major arc An arc whose points are on or outside a central angle. (p. 451)

mean proportional See geometric mean.

means The second and third terms of a proportion. In the proportion $a:b = c:d$ $\left(\text{or } \frac{a}{b} = \frac{c}{d}\right)$, for example, b and c are the means. (p. 327)

median (of trapezoid) A segment joining the midpoints of the nonparallel sides of a trapezoid. (p. 523)

median (of triangle) A segment from a vertex of a triangle to the midpoint of the opposite side. (p. 131)

midpoint A point that divides a segment or an arc into two congruent parts. (pp. 28, 453)

minor arc An arc whose points are on or between the sides of a central angle. (p. 451)

oblique lines Two intersecting lines that are not perpendicular. (p. 65)

obtuse angle An angle whose measure is greater than 90 and less than 180. (p. 11)

obtuse triangle A triangle in which one of the angles is obtuse. (p. 143)

opposite rays Two collinear rays that have a common endpoint and extend in opposite directions. (p. 100)

orthocenter The point of concurrency of the altitudes of a triangle. (p. 661)

parallel lines Coplanar lines that do not intersect. (p. 195)

parallelogram A quadrilateral in which both pairs of opposite sides are parallel. (p. 236)

perimeter The sum of the lengths of the sides of a polygon. (p. 8)

perpendicular Intersecting at right angles. (p. 61)

perpendicular bisector A line that bisects and is perpendicular to a segment. (p. 185)

plane A surface such that if any two points on the surface are connected by a line, all points of the line are also on the surface. (p. 192)

postulate An unproved assumption. (p. 39)

prism A solid figure that has two congruent parallel faces whose corresponding vertices are joined by parallel edges. (p. 561)

proportion An equation stating that two or more ratios are equal. (p. 326)

protractor An instrument, marked in degrees, used to measure angles. (p. 9)

pyramid A solid figure that has a polygonal base and lateral edges that meet in a single point. (p. 565)

quadrilateral A four-sided polygon. (p. 236)

radius (of circle) A segment joining the center of a circle to a point on the circle. Also, the length of such a segment. (p. 439)

radius (of regular polygon) A segment joining the center of a regular polygon to one of the polygon's vertices. (p. 532)

ratio A quotient of two numbers. (p. 325)

ray A straight set of points that begins at an endpoint and extends infinitely in one direction. (p. 4)

rectangle A parallelogram in which at least one angle is a right angle. (p. 236)

rectangular solid A prism with six rectangular faces. (p. 413)

regular polygon A polygon that is both equilateral and equiangular. (p. 314)

rhombus A parallelogram in which at least two consecutive sides are congruent. (p. 236)

right angle An angle whose measure is 90. (p. 11)

right triangle A triangle in which one of the angles is a right angle. (p. 143)

scalene triangle A triangle in which no two sides are congruent. (p. 142)

secant A line that intersects a circle at exactly two points. (p. 459)

secant-secant angle An angle whose vertex is outside a circle and whose sides are determined by two secants. (p. 471)

secant segment The part of a secant that joins a point outside the circle to the farther point of intersection of the secant and the circle. (p. 460)

secant-tangent angle An angle whose vertex is outside a circle and whose sides are determined by a secant and a tangent. (p. 471)

sector A region bounded by two radii and an arc of a circle. (p. 537)

segment (*of circle*) A region bounded by a chord of a circle and its corresponding arc. (p. 538)

semicircle An arc whose endpoints are the endpoints of a diameter. (p. 451)

similar polygons Polygons in which the ratios of the measures of corresponding sides are equal and corresponding angles are congruent. (p. 333)

skew lines Two lines that are not coplanar. (p. 283)

slant height A perpendicular segment from the vertex of a pyramid to a side of the pyramid's base. (p. 413)

square A parallelogram that is both a rhombus and a rectangle. (p. 236)

straight angle An angle whose measure is 180. (p. 11)

supplementary angles Two angles whose sum is 180° (a straight angle). (p. 67)

tangent A line that intersects a circle at exactly one point. (p. 459)

tangent-chord angle An angle whose vertex is on a circle and whose sides are determined by a tangent and a chord that intersect at the tangent's point of contact. (p. 469)

tangent circles Circles that intersect at exactly one point. (p. 460)

tangent segment The part of a tangent line between the point of contact and a point outside the circle. (p. 460)

tangent-tangent angle An angle whose vertex is outside a circle and whose sides are determined by two tangents. (p. 471)

theorem A mathematical statement that can be proved. (p. 23)

transversal A line that intersects two coplanar lines in two distinct points. (p. 192)

trapezoid A quadrilateral with exactly one pair of parallel sides. (p. 236)

triangle A three-sided polygon. (p. 5)

trisect To divide a segment or an angle into three congruent parts. (pp. 29, 30)

vertex (*of angle*) The common endpoint of the two rays that form an angle. (p. 5)

vertex (*of polygon*) The common endpoint of two sides of a polygon. (p. 6)

vertex angle The angle opposite the base of a nonequilateral isosceles triangle. (p. 142)

vertical angles A pair of angles such that the rays forming the sides of one and the rays forming the sides of the other are opposite rays. (p. 100)

volume The number of cubic units of space contained by a solid figure. (p. 575)

INDEX

K

Kite, 189, 190, 236, 256
  area of, 528
  properties of, 242

L

Lateral area
  of cone, 571
  of cylinder, 570–71
  of prism, 562
  of pyramid, 566
Lateral edge, 561, 562, 565
Lattice point, 641
Law of Cosines, 427, 719
Law of Sines, 427, 718
Law of Trichotomy, 687
Legs
  of isosceles trapezoid, 236,
      242
  of isosceles triangle, 142
  of right triangle, 143, 156,
      384, 418–19
Line of centers, 461
Lines, 3
  auxiliary, 132
  concurrence of, 660
  coplanar, 192, 269–70
  determination of, 132
  distances from points to,
      174, 713–14
  equations of, 610–12
  feet of, 270, 276–77, 565
  horizontal
      equations of, 611
      slope of, 199
  intersecting, 61, 62, 100, 192,
      271, 618
  intersection of, with planes,
      269–270, 271, 276–77
  number, 3–4, 62
  oblique, 65
  parallel, 195, 200, 217–18,
      224–27, 271, 283,
      351–52
  parallel to plane, 282, 283
  perpendicular, 61–62, 180,
      185, 200, 218, 227
  perpendicular to plane,
      276–77, 283
  secant, 459
  skew, 195, 274, 282–83
  slopes of, 198–200
  tangent, 459, 461
  vertical
      equations of, 611–12
      slope of, 199
  working, 667
Line segments. See Segments.
Locus, 649–51
  compound, 656
Logic, 41, 44–46. See also Proof.
Lunes of Hippocrates, 542

M

Major arc, 451
Masères, rule of, 403
Mass points, 728
Mathematical Excursions, 17,
      81, 130, 318, 383, 492,
      536, 701
Mean
  arithmetic, 328, 383
  geometric, 327–28, 383. See
      also Mean
      proportional.
  harmonic, 383
Mean proportional, 327, 377–78
  construction of, 674
Means (of proportion), 327
Means-Extremes Products
      Theorem, 327, 345
Means-Extremes Ratio
      Theorem, 327
Measures
  of angles, 9–10, 11–12
  of arcs, 371, 451–52
  of exterior angles of polygon,
      216, 296, 308, 315, 691
  of interior angles of polygon,
      216, 295, 296, 308, 691
  of segments, 9
Median(s)
  of trapezoid, 523–24
  of triangle, 31, 131, 132, 546,
      662
Midline Theorem, 296–97
Midpoint
  of arc, 453
  of segment, 28–29, 170–71
  of side of polygon, 131,
      296–97, 523
Midpoint formula, 170–71
Minor arc, 451
Minute (angle measure), 11–12
Missing diagram, 176–77
Multiplication property
  of inequality, 688
  for segments and angles, 89

N

Negation, 44
n-gon, 307
No-Choice Theorem, 302
Nonagon, 307
Nonconvex polygon, 235, 312
Number line, 3–4, 62

O

Oblique lines, 65
Obtuse angle, 11
Obtuse triangle, 143
  test for, 385, 692
Octagon, 307
Octahedron, 569
Opposite rays, 100

P

Opposite reciprocals, 200, 257,
      614
Ordered pair, 62, 605
Ordinate, 611
Orthocenter, 661–62
Overlapping triangles, 138

Paragraph proof, 36–37
Parallelism
  of line and plane, 282, 283
  of lines, 195, 200, 217–18,
      224–27, 271, 283,
      351–52
  of planes, 282–83
Parallel Postulate, 224, 295
Parallelogram, 203, 219, 236,
      249, 487–88
  area of, 516–17
  properties of, 241
Pentadecagon, 307
Pentagon, 307
Pentagonal prism, 561, 562
Perimeter(s), 8
  circumference as, 370
  ratios of, 334
  of right triangle, 734
Perpendicular bisector, 185,
      440, 660
  construction of, 668–69
Perpendicularity. See also
      Perpendicular bisector.
  of line and plane, 276–77,
      283
  of lines, 61–62, 180, 200,
      218, 227
Pi, 126, 499
Planes, 192, 269
  determination of, 270–71
  intersection of, 271, 283
  intersection of, with lines,
      269–70, 271, 276–77
  parallel, 282–83
Plato, rule of, 403
Point circle, 634
Point-line distance formula,
      713–14
Point of tangency, 459
Points, 3
  betweenness of, 18
  bisection. See Midpoint.
  collinear, 18, 19, 735–36
  coordinates of, 62
  coplanar, 192, 269–70, 649
  distances between, 184,
      392–93, 626
  distances to lines, 179,
      713–14
  equidistant, 184–85
  lattice, 641
  loci of, 649–51, 656
  mass, 728
  trisection, 29

Point-slope form of equation, 612
Polygons, 234–35. *See also* Quadrilaterals; Triangles.
circumscribed, 486–87, 731
convex, 235
diagonals of, 235, 308
equiangular, 150, 314–15
equilateral, 140, 150, 314
exterior angles of, 308, 315
inscribed, 486–88, 550, 718–19, 724–25, 731–32
interior angles of, 308
names of, 235, 307
regular, 314
areas of, 531–32
as bases of regular pyramids, 565
similar, 333–34
Polyhedra, 561. *See also names of specific solids.*
Postulate, 39, 41, 72
Power theorems, 493–94
Principle of reduced triangle, 399
Prism, 561. *See also Rectangular solid.*
cross section of, 577
lateral area of, 562
total area of, 562
volume of, 576, 577
Probability, 49
tree diagrams and, 412
Progression, arithmetic, 322
Proof, 23–24, 39–41, 46, 61, 72
analytic, 393
coordinate, 393
detour, 169–70
indirect, 211–12
paragraph, 36–37
two-column, 23–24
working backwards in, 346
Properties
of equality
addition, 82–83
division, 90
multiplication, 89
reflexive, 112–13
substitution, 95–96
subtraction, 83–84
transitive, 95
of inequality, 687–88
of special quadrilaterals, 241–42
Proportional
fourth, 328
construction of, 675
mean, 327, 377–78
construction of, 674
Proportions, 326–28. *See also Ratios.*
shadow problems and, 346–47
in similar triangles, 345

theorems involving, 351–52, 377–78
Protractor, 9
Ptolemy's Theorem, 724–25
Pyramid, 413, 565
cross section of, 584
lateral area of, 566
total area of, 566
volume of, 583
Pythagoras, 385
rule of, 403
Pythagorean Theorem, 204, 384–85
converse of, 385, 692
distance formula and, 392–93
space figures and, 413–14
Pythagorean triples, 398–99
rules for generating, 403

Q

Quadratic equations, 367–68
Quadrilaterals, 236
cyclic, 550, 724–25
inscribed, 487–88, 550, 724–25
kite, 189, 190, 236, 256
area of, 528
properties of, 242
parallelogram, 203, 219, 236, 249, 487–88
area of, 516–17
properties of, 241
rectangle, 15, 236, 255, 488, 562
area of, 63, 512
properties of, 241
rhombus, 182, 189, 236, 256
area of, 528
properties of, 242
square, 236, 256
area of, 512, 528
properties of, 242
trapezoid, 219, 236, 256
area of, 523–24
isosceles, 236, 242, 256

R

Radicals, 367–68
Radius
of circle, 125, 126, 439, 441
of regular polygon, 531–32
Ratios, 31, 325–26
of areas, 543–44, 584
of parts of triangles, 405–6, 418–19
proportions and, 326
in similar figures, 333–34, 340, 345, 543–44, 584
trigonometric, 418–19, 423–24
Rays, 4, 5
opposite, 100

Reciprocals, opposite, 200, 257, 614
Rectangle, 15, 236, 255, 488, 562
area of, 63, 512
properties of, 241
Rectangular solid, 413, 575
Reduced triangle, 399
Reduction, 332
Reflection, 8, 112, 626–28
Reflexive Property, 112–13
Regular polygon, 314
area of, 531–32
as base of regular pyramid, 565
Regular pyramid, 413, 565
Rhombus, 182, 189, 236, 256
area of, 528
properties of, 242
Right angle, 11, 61, 66, 180, 480
Right triangles, 143, 156, 377–78
families of, 398, 406
45°-45°-45°, 406
perimeter of, 734
Pythagorean Theorem and, 204, 384–85
test for, 385, 692
30°-60°-90°, 405, 418
Rotation, 16, 64, 204
axis of, 574
surface of, 574, 597

S

SAS postulate, 116–17
Scalene triangle, 142
Secant, 459–60
Secant-secant angle, 470–72
Secant-Secant Power Theorem, 494
Secant-tangent angle, 470–72
Second (angle measure), 11–12
Sector, 371, 537
Segment (of circle), 538
Segments, 4, 6
addition property for, 82–83
bisectors of, 28–29
perpendicular, 185, 441, 660, 668–69
congruent, 12–13, 82–84, 89–90, 95, 112, 352, 667, 674
coplanar, 192, 269
copying, 667
distances and, 184, 392
division of, 674
external, 331
division property for, 90
measures of, 9
midpoints of, 28–29, 170–71
multiplication property for, 89
secant, 460, 493–94

substitution property for, 95–96
subtraction property for, 84
tangent, 460, 461, 493
transitive property for, 95
trisectors of, 29
Semicircle, 451, 479–80
Sensed magnitude, 735
Shadow problems, 346–47
Side-Splitter Theorem, 351
Similar-Figures Theorem, 545
Similarity, 332–34
  of polygons, 333–34
  ratios of areas and, 543–44, 584
  of triangles, 333, 339–40, 345–46, 377–78
Sine, 418–19, 423–24, 718–19, 734
Sines, Law of, 427, 718
Skew lines, 195, 274, 282–83
Slant height
  of cone, 571
  of regular pyramid, 413, 565
Slide, 113, 207
Slope, 198–200, 610–11
Slope-intercept form of equation, 610, 612
Sphere, 571
  surface area of, 571
  volume of, 589
Square, 236, 256
  area of, 512, 528
  properties of, 242
SSS postulate, 115–16
Stewart's Theorem, 721–22
Straight angle, 11, 67, 180
Straightedge, 666
Substitution method of solving systems, 618–19
Substitution Property, 95–96
Subtraction property
  for segments and angles, 83–84
  special, 689
Supplementary angles, 67, 180, 218, 225–26, 487
Surface area
  of cone, 571
  of cylinder, 571
  of prism, 562
  of pyramid, 566
  of sphere, 571
Surface of rotation, 574, 597
System
  of equations, 618–19
  of inequalities, 623–24

T

Table of values, 605
Tangent (trigonometric ratio), 418–19, 423–24

Tangent-chord angle, 468–69, 472, 479
Tangent circles, 460–61
Tangents, 459–60, 493
  common, 461–62
Tangent-Secant Power Theorem, 493
Tangent-tangent angle, 470–72, 480
Tetrahedron, 568
Theorem, 23–24, 39, 40–41, 72
Theorem of Menelaus, 735–36
30°-60°-90° triangle, 405, 418
Three-dimensional distance formula, 626
Three-dimensional graphs, 626
Tick marks, 12
Transitive property
  of inequality, 687
  of parallel lines, 227
  for segments and angles, 95
  of similar triangles, 344
Transversal, 192–94, 217–18, 224–27, 351
Trapezoid, 219, 236, 256
  area of, 523–24
  isosceles, 236, 256
    properties of, 242
Tree diagram, 412
Triangles, 5–6
  acute, 143, 385, 692
  altitudes of, 131–32, 377–78, 661–62, 734
  angle bisectors in, 352, 661
  areas of, 517, 717–18, 734
    encasement principle and, 609, 640, 717
    equilateral, 531
    Hero's formula for, 550
  centroids of, 617, 662, 729
  circumcenters of, 617, 660
  circumcircles of, 718–19, 732
  circumradii of, 731–32
  classification of, 142–43, 385, 692
  congruent, 111–13, 125
    by AAS, 302
    by ASA, 117
    by HL, 156
    by SAS, 116–17
    by SSS, 115–16
  construction of, 678–80
  equiangular, 143, 150
  equilateral, 142–43, 150, 531
  exterior angles of, 216, 296, 691
  incenters of, 661
  inequalities in, 19, 216, 691–93, 697
  inradii of, 731, 734
  isosceles, 142, 148–49, 532, 565

medians of, 31, 131, 132, 546, 662
midline of, 296–97
obtuse, 143, 385, 692
orthocenters of, 661–62
overlapping, 138
reduced, 399
right, 143, 156, 204, 377–78, 384–85, 398, 405–6, 418, 692, 734
scalene, 142
similar, 333, 345–46, 377–78
  by AA, 339
  by AAA, 339
  by SAS~, 340
  by SSS~, 340
  transitive property of, 344
sum of angle measures, 295
vertices of, 6
Trichotomy, Law of, 687
Trigonometric ratios, 418–19, 423
  table of, 424
Trisectors
  of angle, 30
  of segment, 29
Two-column proof, 23–24
Two-point form of equation, 612
Two-Tangent Theorem, 460

U

Undefined term, 39
Union, 6
Units of measure
  for angles, 9–10, 11–12
  for areas, 511
  for segments, 9
  for volumes, 575

V

Venn diagram, 45
Vertex
  of angle, 5
  of pyramid, 413, 565
  of triangle, 6
Vertex angle, 142
Vertical angles, 100–101
Vertical line
  equation of, 611–12
  slope of, 199
Volume(s), 575
  Cavalieri's principle and, 592
  of cone, 583–84
  of cylinder, 577
  "divide and conquer" method of calculating, 578
  of frustum, 585, 587, 588
  of prism, 576, 577

of pyramid, 583
of sphere, 589
units of, 575

W

Walk-around problems, 463
Working line, 667

X

x-axis, 62
x-coordinate, 611, 612
x-intercept, 606, 612

Y

y-axis, 62
y-coordinate, 611
y-form of equation, 610, 612
y-intercept, 606, 610, 612

Symbols Used in Geometry

Symbol	Meaning	Symbol	Meaning
$\overline{AB}$	segment $\overline{AB}$	$=$	is equal to
$\overleftrightarrow{AB}$	line AB	$\neq$	is not equal to
$\overrightarrow{AB}$	ray AB	$>$	is greater than
$\overarc{AB}$	arc AB	$<$	is less than
AB	length of $\overline{AB}$	$\not>$	is not greater than
$\angle$s	angles	$\geq$	is greater than or equal to
$\angle$A	angle A	$\leq$	is less than or equal to
m$\angle$A	measure of $\angle$ A	$\approx$	is approximately equal to
$\triangle$	triangle	$\Longleftrightarrow$	is equivalent to
$\triangle$	triangles	$\Longrightarrow$	implies
$\odot$	circle	$\sim p$	not p or p is false
$\odot$	circles	$\therefore$	therefore
$\cancel{C}$	cross-sectional area	$\{ \ \}$	set
$\square$	parallelogram	$\varnothing$	null set
$\cong$	congruent	$\cup$	union
$\not\cong$	not congruent	$\cap$	intersection
⊢⊣	congruent segments	$\sqrt{x}$	square root of x
$\angle\angle$	congruent angles	$|a|$	absolute value of a
$\perp$	perpendicular	$\triangle x$	change in x
$\not\perp$	not perpendicular	π	pi
$\llcorner$	right angle	$\circ$	degrees
$\parallel$	parallel	$'$	minutes
$\not\parallel$	not parallel	$''$	seconds
⤶⤶	parallel lines	$\frac{a}{b}$	$a \div b$, $a{:}b$; ratio of a to b
$\sim$	similar